ENCYCLOPEDIA OF SPECIAL FUNCTIONS: THE ASKEY–BATEMAN PROJECT

Volume 2: Multivariable Special Functions

This is the second of three volumes that form the *Encyclopedia of Special Functions*, an extensive update of the Bateman Manuscript Project.

Volume 2 covers multivariable special functions. When the Bateman project appeared, study of multivariable special functions was in an early stage, but revolutionary developments began to be made in the 1980s and have continued ever since. World-renowned experts survey these over the course of 12 chapters, each containing an extensive bibliography. The reader encounters different perspectives on a wide range of topics, from Dunkl theory, to Macdonald theory, to the various deep generalizations of classical hypergeometric functions to the several variables case, including the elliptic level. Particular attention is paid to the close relation of the subject with Lie theory, geometry, mathematical physics and combinatorics.

TOM H. KOORNWINDER is Professor Emeritus at the University of Amsterdam. He is an expert in special functions, orthogonal polynomials and Lie theory. He introduced the five-parameter extension of the BC-type Macdonald polynomials, which are nowadays called Koornwinder polynomials. He was co-author of the chapter on orthogonal polynomials in the *Digital Library of Mathematical Functions*, and is involved in its revision.

JASPER V. STOKMAN is Professor of Mathematics at the University of Amsterdam. He is an expert in special functions, Lie theory and integrable systems. He introduced BC-type extensions of several families of classical orthogonal polynomials, and nonpolynomial generalizations of Koornwinder polynomials. He linked multivariable special functions to harmonic analysis on quantum groups and Hecke algebras, and to statistical mechanics and analytic number theory.

ENCYCLOPEDIA OF SPECIAL FUNCTIONS: THE ASKEY–BATEMAN PROJECT

Volume 2: Multivariable Special Functions

Edited by

TOM H. KOORNWINDER
Universiteit van Amsterdam

JASPER V. STOKMAN
Universiteit van Amsterdam

CAMBRIDGE
UNIVERSITY PRESS

CAMBRIDGE
UNIVERSITY PRESS

University Printing House, Cambridge CB2 8BS, United Kingdom

One Liberty Plaza, 20th Floor, New York, NY 10006, USA

477 Williamstown Road, Port Melbourne, VIC 3207, Australia

314–321, 3rd Floor, Plot 3, Splendor Forum, Jasola District Centre, New Delhi – 110025, India

79 Anson Road, #06–04/06, Singapore 079906

Cambridge University Press is part of the University of Cambridge.

It furthers the University's mission by disseminating knowledge in the pursuit of education, learning, and research at the highest international levels of excellence.

www.cambridge.org
Information on this title: www.cambridge.org/9781107003736
DOI: 10.1017/9780511777165

First published 2021

Printed in the United Kingdom by TJ International Ltd, Padstow Cornwall

A catalogue record for this publication is available from the British Library.

ISBN – 3 Volume Set 978-1-108-88244-6 Hardback
ISBN – Volume 1 978-0-521-19742-7 Hardback
ISBN – Volume 2 978-1-107-00373-6 Hardback
ISBN – Volume 3 978-0-521-19039-8 Hardback

Contents

Contributors

Charles F. Dunkl *Department of Mathematics, University of Virginia, P.O. Box 400137, Charlottesville, VA 22904-4137, USA*

Jim Haglund *Department of Mathematics, University of Pennsylvania, Philadelphia, PA 19104, USA*

Gert J. Heckman *IMAPP, Radboud Universiteit, P.O. Box 9010, 6500 GL Nijmegen, Netherlands*

Tom H. Koornwinder *Korteweg-de Vries Institute for Mathematics, University of Amsterdam, P.O. Box 94248, 1090 GE Amsterdam, Netherlands*

Keiji Matsumoto *Department of Mathematics, Faculty of Science, Hokkaido University, Sapporo 060-0810, Japan*

Eric M. Opdam *Korteweg-de Vries Institute for Mathematics, University of Amsterdam, P.O. Box 94248, 1090 GE Amsterdam, Netherlands*

Hjalmar Rosengren *Department of Mathematical Sciences, Chalmers University of Technology and University of Gothenburg, SE-412 96 Göteborg, Sweden*

Michael J. Schlosser *Fakultät für Mathematik, Universität Wien, Oskar-Morgenstern-Platz 1, A-1090 Vienna, Austria*

Jasper V. Stokman *Korteweg-de Vries Institute for Mathematics, University of Amsterdam, P.O. Box 94248, 1090 GE Amsterdam, Netherlands*

Nobuki Takayama *Department of Mathematics, Kobe University, Rokko, Kobe 657-8501, Japan*

Vitaly Tarasov *Department of Mathematical Sciences, Indiana University–Purdue University Indianapolis, Indianapolis, IN 46202-3216, USA, and St. Petersburg Branch of Steklov Mathematical Institute, Fontanka 27, St. Petersburg, 191023, Russia*

Joris Van der Jeugt *Department of Applied Mathematics, Computer Science and Statistics, Faculty of Sciences, Ghent University, Krijgslaan 281-S9, 9000 Ghent, Belgium*

Alexander Varchenko *Department of Mathematics, University of North Carolina, Chapel Hill, NC 27599-3250, USA*

S. Ole Warnaar *School of Mathematics and Physics, The University of Queensland, Brisbane, QLD 4072, Australia*

Yuan Xu *Department of Mathematics, University of Oregon, Eugene, OR 97403-1222, USA*

Preface

This is the second volume of the *Encyclopedia of Special Functions* of the Askey–Bateman project. It is devoted to multivariable special functions.

As was explained in the preface to volume 1, the *Encyclopedia of Special Functions* aims to realize a vision that the late Richard Askey had in the 1970s: to update the Bateman project, in particular the three volumes of *Higher Transcendental Functions*, according to present knowledge and state of the art. As for multivariable special functions, the Bateman project contained material on Appell hypergeometric functions (part of Chapter V) and orthogonal polynomials in several variables (Chapter XII). These two most classical parts of multivariable special functions are treated in the present volume in Chapters 3 and 2, respectively.

In the past 65 years, since the Bateman project appeared, multivariable orthogonal polynomials and special functions have seen several revolutionary developments which partially interacted with each other and which also were fed by new insights into one-variable theory (notably basic and elliptic hypergeometric functions, and Askey–Wilson polynomials). One development was the successive introduction of zonal polynomials, hypergeometric functions of matrix argument, Jack polynomials, Hall–Littlewood polynomials, Heckman–Opdam polynomials (Chapter 8), Macdonald polynomials and Koornwinder's extension of Macdonald's BC case (Chapter 9), and Rains' elliptic generalization of the Koornwinder polynomials (Chapter 6). Dunkl's simultaneous introduction of the Dunkl operator (Chapter 7) and, a little later, Cherednik's double affine Hecke algebras (Chapter 9) gave important boosts to these theories. Macdonald theory was also in fruitful interaction with algebraic combinatorics (Chapter 10). Analysis on semisimple Lie groups (Chapter 8) and quantum groups was also an important inspiration.

A second line of development was the quest for multivariable analogues of hypergeometric functions (which should be deep enough that many one-variable formulas generalize). The Appell hypergeometric functions turned out to be special cases of the A-hypergeometric functions (Chapter 4), introduced by Gel'fand and coworkers. Work by, among others, Biedenharn and coworkers resulted in many classes of multivariable hypergeometric series with expansion coefficients patterned by root systems; see Chapter 5 for the classical and basic cases, and Chapter 6 for the elliptic case. A very different kind of hypergeometric function associated with root systems, generalizing the theory of spherical functions on noncompact Riemannian symmetric spaces, was developed by Heckman and Opdam (Chapter 8). Yet another source of multivariable hypergeometric functions, also closely connected with conformal field theory, comes from solving Knizhnik–Zamolodchikov-type equations (Chapter 11).

Wigner and Racah coefficients in the representation theory of SU(n) and their application to quantum mechanics were the historical context from which the theory on multivariable hypergeometric series described in Chapter 5 arose. The case of SU(2) is described in Chapter 12. Here the $9j$-coefficients give rise to still mysterious orthogonal polynomials in two variables.

A more detailed survey of the chapters and their interconnections is given in the introductory Chapter 1.

We hope that the volume will help the reader to oversee the global landscape of multivariable special functions and their applications, and will serve as a useful guide to the extensive literature. We are very grateful to the authors of the chapters for their contributions to this volume. The final editing of the individual chapters and the creation of the index to the volume was done by the first editor.

Tom H. Koornwinder and Jasper V. Stokman

1

General Overview of Multivariable Special Functions

Tom H. Koornwinder and Jasper V. Stokman

1.1 Introduction

The theory of one-variable (ordinary) hypergeometric and basic hypergeometric series goes back to work of Euler, Gauss and Jacobi. The theory of elliptic hypergeometric series is of a much more recent vintage (Frenkel and Turaev, 1997). The three theories deal with the study of series $\sum_{k\geq 0} c_k$ with $f(k) := c_{k+1}/c_k$ a rational function in k (*hypergeometric* theory), a rational function in q^k (*basic hypergeometric* theory) or a doubly periodic meromorphic function in k (*elliptic hypergeometric* theory; see Gasper and Rahman, 2004, Ch. 11 for an overview).

Examples of elementary functions admitting hypergeometric and basic hypergeometric series representations are

$$(1-z)^{-\alpha} = \sum_{k=0}^{\infty} \frac{(\alpha)_k}{k!} z^k, \qquad \frac{(az;q)_\infty}{(z;q)_\infty} = \sum_{k=0}^{\infty} \frac{(a;q)_k}{(q;q)_k} z^k \tag{1.1.1}$$

for $|z| < 1$ and $\alpha, a \in \mathbb{C}$, with $(\alpha)_k := \alpha(\alpha+1)\cdots(\alpha+k-1)$ for $k \in \mathbb{Z}_{\geq 0}$ the *shifted factorial* (or *Pochhammer symbol*), $(a;q)_k := (1-a)(1-qa)\cdots(1-q^{k-1}a)$ for $k \in \mathbb{Z}_{\geq 0} \cup \{\infty\}$ the *q-shifted factorial*. Here, and throughout the entire chapter, we assume for convenience that $0 < q < 1$. Note that the series in the second identity, with $a = q^\alpha$, tends to the series in the first identity as $q \uparrow 1$, at least formally, and that the identities (1.1.1) reduce to polynomial identities when $\alpha \in \mathbb{Z}_{\leq 0}$. Also note that the series in (1.1.1) are indeed hypergeometric and basic hypergeometric series, respectively, since $f(k) = \frac{k+\alpha}{k+1} z$ and $f(k) = \frac{1-q^k a}{1-q^{k+1}} z$ for the first and second series in (1.1.1). These are the well-known Newton (generalized) *binomial theorem* and its q-analogue (Gasper and Rahman, 2004, §1.3). They form, apart from the (q-)exponential series, the simplest nontrivial examples of an impressive scheme of hypergeometric and basic hypergeometric summation identities (Gasper and Rahman, 2004), with the members in the scheme related by limit transitions.

The summands of elliptic, basic and classical hypergeometric series are expressible in terms of products and quotients of elliptic, basic and classical shifted factorials. The basic and classical ones are the (q-)shifted factorials as defined in the previous paragraph. The *elliptic* (or *theta*) *shifted factorial* is given by $(z;q,p)_k := \prod_{i=0}^{k-1} \theta(zq^i;p)$ for $k \in \mathbb{Z}_{\geq 0}$ and $0 < p < 1$, with $\theta(z;p) := \prod_{i=0}^{\infty}(1-p^i z)(1-p^{i+1}/z)$ the *modified theta function*. These shifted factorials can be expressed as $\Gamma(q^k z)/\Gamma(z)$ (or, in the classical case, $\Gamma(z+k)/\Gamma(z)$) with $\Gamma(z)$ an appropriate analogue of the classical Gamma function. For the elliptic hypergeometric case this is Ruijsenaars' (1997) *elliptic Gamma function*

$$\prod_{i,j=0}^{\infty} \frac{1 - z^{-1} p^{i+1} q^{j+1}}{1 - z p^i q^j}, \qquad 0 < p, q < 1,$$

for the basic hypergeometric case the (modified) *q-Gamma function* $(z;q)_\infty^{-1}$ and for the classical hypergeometric case the classical *Gamma function*.

There is no "simple" elliptic analogue of (1.1.1). In fact, the first elliptic hypergeometric summation formula that was found (Frenkel and Turaev, 1997) generalizes the *top-level* terminating (basic) hypergeometric summation identity! This is a general pattern for the elliptic hypergeometric theory: the top levels of the (basic) hypergeometric theory admit elliptic versions, and there is little room for degenerations without falling outside the realm of elliptic hypergeometric series. Possibly this is one of the reasons for the late discovery of elliptic hypergeometric series.

Parallel to the theory of hypergeometric series there is a theory of hypergeometric integrals; see §1.2.3 and, in later chapters, §5.3 and §6.2. Such integrals can often be identified with hypergeometric series. But, certainly in the elliptic case, there are many instances where the hypergeometric integral is convergent while a possible corresponding hypergeometric series diverges (Rosengren, 2017, §2.10). Hypergeometric integrals naturally appear as coordinates of vector-valued solutions of Knizhnik–Zamolodchikov (*KZ*) and Knizhnik–Zamolodchikov–Bernard (*KZB*) equations and their q-analogues; see Chapter 11. The elliptic case, corresponding to solutions of *qKZB* equations, appeared in Felder et al. (1997, §7) (yet formally) and soon afterwards rigorously in Felder et al. (1999, §6).

This volume deals with multivariable generalizations of ordinary, basic and elliptic hypergeometric series and integrals. This includes various multivariable extensions of classical (bi)orthogonal polynomials and functions, which form an important subclass of hypergeometric series within the one-variable theory.

Various multivariable theories have emerged, each with its own characteristic features depending on the particular motivation for, and context behind, its multivariable extension. For instance, there are important multivariable theories motivated by special function theory itself (see Chapters 2–6), by representation theory and Lie theory (see Chapters 7–9 and 12), by combinatorics (see Chapter 10) and by theoretical physics (see Chapters 8–9 and 11–12).

In the remainder of this introductory chapter we give a short discussion of each of the types of multivariable special functions treated in this volume, and we highlight their interrelations and differences. In §1.2 we first discuss the multivariable series which may be seen as extensions of the three types of hypergeometric series. The different classes of multivariable extensions of classical (bi)orthogonal functions will be discussed in §1.3.

We hope that this short impression of the various classes of multivariable special functions and their interrelations helps the reader to oversee the chapters in this volume and how they are related.

1.2 Multivariable Classical, Basic and Elliptic Hypergeometric Series

1.2.1 Appell and Lauricella Hypergeometric Series

Gauss' hypergeometric series is given by

$$ {}_2F_1\left(\begin{matrix}a,b\\c\end{matrix};x\right) := \sum_{k=0}^{\infty}\frac{(a)_k(b)_k}{(c)_k k!}x^k, \tag{1.2.1}$$

which absolutely converges for $|x| < 1$. One of the oldest generalizations of the Gauss hypergeometric series to several variables was given by Appell, who introduced the four *Appell hypergeometric series* in two variables (see Appell and Kampé de Fériet, 1926; Erdélyi, 1953, §5.7), denoted by F_1, F_2, F_3, F_4. For instance,

$$F_2(a,b_1,b_2,c_1,c_2;x,y) := \sum_{m,n=0}^{\infty}\frac{(a)_{m+n}(b_1)_m(b_2)_n}{(c_1)_m(c_2)_n\, m!\, n!}x^m y^n, \quad (x,y)\in\mathbb{C}^2,\ |x|+|y|<1. \tag{1.2.2}$$

The series (1.2.2) are double series $\sum_{m,n=0}^{\infty} c_{m,n}/(m!n!)$ with $c_{m+1,n}/c_{m,n}$ and $c_{m,n+1}/c_{m,n}$ of the form $p_1(m,n)/r_1(m,n)$ and $p_2(m,n)/r_2(m,n)$ for suitable relative prime polynomials p_i and r_i in two variables ($i = 1,2$). This extends the property characterizing hypergeometric series in one variable, and such series are therefore also called *hypergeometric*. The highest degree of the four polynomials p_1, r_1, p_2, r_2 is called the *order* of the hypergeometric series in two variables. The Appell hypergeometric series have order two. Horn classified all hypergeometric series of order two; see the list of 34 series in Erdélyi (1953, §5.7.1). Lauricella defined n-variable analogues F_A, F_B, F_C, F_D of F_2, F_3, F_4, F_1, respectively. The Appell and Lauricella hypergeometric series are discussed in Chapter 3.

Many properties and formulas for Gauss hypergeometric series generalize to Appell and Lauricella hypergeometric series, but, not surprisingly, one has to deal with interesting complications concerning, for instance, the integral representations, systems of partial differential equations and monodromy; see Chapter 3. Furthermore, solutions of the system of partial differential equations for these series form a much richer collection than in the case of the Gauss hypergeometric series, where all local solutions at regular singularities are expressed in terms of series of the same type. For instance, for F_2 six different types of series occur as local solutions, including some that are hypergeometric series of order higher than two, or even not hypergeometric series at all; see Olsson (1977). Gel'fand's A-hypergeometric functions (see Chapter 4) offer a fruitful point of view for the study of Appell and Lauricella hypergeometric series. This can also give inspiration for a study of q-analogues; see Noumi (1992), where also a connection is made with quantum groups. Gasper and Rahman (2004, Chapter 10) give an account of q-series in two or more variables.

Appell and Lauricella hypergeometric series have several interrelations with other special functions in several variables. The first example (which may have been the motivating example for Appell) is the biorthogonal polynomials on the simplex and ball; see Chapter 2. Further examples deal with Heckman–Opdam hypergeometric functions (see Chapter 8). In the case of root system A, these functions can be identified for certain degenerate parameter values with a special Lauricella F_D (or, in two variables, with Appell F_1); see Shimeno and Tamaoka (2015). A special case of Heckman–Opdam hypergeometric functions for root system BC_2 can be written as the sum of two Appell F_4 functions (see Beerends, 1992, Theorems 3.3 and 2.3). In the polynomial case one of the F_4 terms vanishes, so that special BC_2 Jacobi polynomials can be written as a terminating F_4; see Koornwinder and Sprinkhuizen-Kuyper (1978, (7.15)). Another special case of the Heckman–Opdam functions for BC_n, now for general n, can be expressed as n-variable analogues of *Kampé de Fériet hypergeometric series*, certain hypergeometric series in two variables of order three (Beerends, 1992, (5.1) and Theorem 5.4).

1.2.2 A-Hypergeometric Functions

The A-hypergeometric (or *GKZ* hypergeometric) functions were introduced by Gel'fand et al. (1989), but there have been analogous approaches before. In particular, Miller Jr. (1973) described a new approach to the *hypergeometric differential equation*

$$z(1-z)f''(z) + (c-(a+b+1)z)f'(z) - abf(z) = 0, \tag{1.2.3}$$

of which the Gauss hypergeometric series (1.2.1) is a solution. He observed that if the parameters a, b, c in (1.2.3) are replaced by $s\partial_s$, $u\partial_u$, $t\partial_t$, then the resulting system of PDEs

$$QF = 0, \quad s\partial_s F = aF, \quad u\partial_u F = bF, \quad t\partial_t F = cF \tag{1.2.4}$$

with

$$Q := z(1-z)\partial_{zz} + t\partial_{tz} - z(s\partial_{sz} + u\partial_{uz} + \partial_z) - su\partial_{su}$$

has a solution

$$F(s,u,t,z) = s^a u^b t^c\, {}_2F_1\left(\begin{matrix} a, b \\ c \end{matrix}; z\right). \tag{1.2.5}$$

Miller defines the *dynamical symmetry algebra* $\mathcal{G}$ of Q as the set of all first-order PDEs L such that $QLf = 0$ whenever $Qf = 0$. It is a Lie algebra with a basis of operators acting on solutions of the form (1.2.5) (so-called *contiguity relations*). Then $\mathcal{G}$ is seen to be isomorphic to sl(4). Miller Jr. (1973) pointed out that a similar approach works for generalized hypergeometric series ${}_{r+1}F_r$ and for Appell and Lauricella hypergeometric series. This was elaborated on by him in several papers in 1972, 1973.

Kalnins et al. (1980) transformed systems like (1.2.4), in the case of Appell's and Horn's hypergeometric series in two variables, into so-called *canonical systems*. These systems coincide with special cases of the later-introduced A-hypergeometric systems (Gel'fand et al., 1989). M. Saito (1996, 2001) recognized the relevance of Kalnins et al. (1980) for the *GKZ* theory. He also worked with a symmetry algebra for operators Q which no longer requires that the operators in the algebra are first order.

A change of variables turns system (1.2.4) of PDEs into the following canonical (or A-hypergeometric) form:

$$(\partial_{xy} - \partial_{zw})f = 0, \quad (x\partial_x - y\partial_y)f = (1-c)f, \quad (x\partial_x + z\partial_z)f = -af, \quad (x\partial_x + w\partial_w)f = -bf, \tag{1.2.6}$$

with corresponding solution

$$f(x,y,z,w) = y^{c-1}z^{-a}w^{-b}\,{}_2F_1\left(\begin{matrix}a,b\\c\end{matrix};\frac{xy}{zw}\right); \tag{1.2.7}$$

see for instance Stienstra (2007, §2.6, §3.2.1). Note that the change of variables has transformed the second-order partial differential operator Q to $\partial_{xy} - \partial_{zw}$, which is essentially the four-dimensional Laplace operator. This makes it manifest that the dynamical symmetry algebra of Q is $\mathfrak{sl}(4)$; see also Dereziński and Majewski (2016).

The general A-hypergeometric system in n variables $x = (x_1, \ldots, x_n)$ depends on a $d \times n$ matrix $A = (a_{ij}) = (a_1 \ldots a_n)$ with integer column vectors $a_j \in \mathbb{Z}^d$ (from which the A in A-hypergeometric) such that the $\mathbb{Z}$-span of the a_j equals $\mathbb{Z}^d$. The A-hypergeometric system, depending on parameters $\beta_1, \ldots, \beta_d$, is given by

$$\Big(\prod_{u_i>0}\partial_{x_i}^{u_i}\Big)f = \Big(\prod_{u_i<0}\partial_{x_i}^{-u_i}\Big)f \quad (u \in L\backslash\{0\}), \qquad \Big(\sum_{j=1}^{n} a_{ij}x_j\partial_{x_j}\Big)f = \beta_i f \quad (i=1,\ldots,d) \tag{1.2.8}$$

with $L := \{u \in \mathbb{Z}^n \mid Au = 0\}$. It can be seen to have system (1.2.6) as the special case

$$A = \begin{pmatrix}1 & -1 & 0 & 0\\ 1 & 0 & 1 & 0\\ 1 & 0 & 0 & 1\end{pmatrix}, \qquad \beta = (1-c, -a, -b)^t. \tag{1.2.9}$$

For $v \in \mathbb{C}^n$ such that $Av = \beta$, we have a formal solution of the (v-independent) differential equations (1.2.8) given by the series

$$\sum_{u\in L}\prod_{j=1}^{n}\frac{x_j^{v_j+u_j}}{\Gamma(v_j+u_j+1)}, \tag{1.2.10}$$

called *A-hypergeometric series in Gamma function form*. With the choice (1.2.9) of A, β and with $v := (0, c-1, -a, -b)^t$, $u := k(1,1,-1,-1)^t$ $(k \in \mathbb{Z})$ the series (1.2.10) becomes

$$\sum_{k=-\infty}^{\infty}\frac{x_1^k x_2^{c-1+k}x_3^{-a-k}x_4^{-b-k}}{\Gamma(k+1)\Gamma(c+k)\Gamma(-a-k+1)\Gamma(-b-k+1)}$$
$$= \frac{x_2^{c-1}x_3^{-a}x_4^{-b}}{\Gamma(c)\Gamma(1-a)\Gamma(1-b)}\sum_{k=0}^{\infty}\frac{(a)_k(b)_k}{(c)_k\,k!}\left(\frac{x_1x_2}{x_3x_4}\right)^k,$$

which is (1.2.7) apart from the Gamma factors in the denominator in front of the summation.

Choices for A and v in (1.2.10) can be made such that the resulting series involves ${}_{r+1}F_r(z)$ or an Appell or Lauricella hypergeometric series. For instance, for Appell's F_2 one can take

$$A = \begin{pmatrix}1&0&0&0&0&1&1\\0&1&0&0&0&1&0\\0&0&1&0&0&0&1\\0&0&0&1&0&-1&0\\0&0&0&0&1&0&-1\end{pmatrix}, \qquad \begin{aligned}\beta &= (-a,-b_1,-b_2,c_1-1,c_2-1)^t,\\ u &= m(-1,-1,0,1,0,1,0)+n(-1,0,-1,0,1,0,1)^t,\\ v &= (-a,-b_1,-b_2,c_1-1,c_2-1,0,0)^t.\end{aligned}$$

Then the first part of system (1.2.8) is generated by $\partial_1\partial_2 f = \partial_4\partial_6 f$, $\partial_1\partial_3 f = \partial_5\partial_7 f$, and (1.2.10) becomes

$$\sum_{m,n=-\infty}^{\infty} \frac{x_1^{-a-m-n} x_2^{-b_1-m} x_3^{-b_2-n} x_4^{c_1+m-1} x_5^{c_2+n-1} x_6^m x_7^n}{\Gamma(m+1)\Gamma(n+1)\Gamma(c_1+m)\Gamma(c_2+n)\Gamma(1-a-m-n)\Gamma(1-b_1-m)\Gamma(1-b_2-n)}$$
$$= \frac{x_1^{-a} x_2^{-b_1} x_3^{-b_2} x_4^{c_1-1} x_5^{c_2-1}}{\Gamma(c_1)\Gamma(c_2)\Gamma(1-a)\Gamma(1-b_1)\Gamma(1-b_2)} F_2\left(a, b_1, b_2, c_1, c_2; \frac{x_4 x_6}{x_1 x_2}, \frac{x_5 x_7}{x_1 x_3}\right).$$

The *GKZ* theory, of which Chapter 4 gives a survey, not only unifies the study of many classes of multivariable special functions, but also exploits methods from algebra, geometry, D-module theory and combinatorics, far beyond the methods used in classical approaches.

1.2.3 Classical, Basic and Elliptic Hypergeometric Series and Integrals Associated with Root Systems

Hypergeometric integrals of classical, basic and elliptic type are integrals with integrand expressed in terms of products and quotients of Gamma factors $\Gamma(ax)$ (in the classical case, $\Gamma(a+x)$), with $\Gamma(x)$ the Gamma function of the appropriate type. In the classical case integrands involving products of the form $(1-x)^a$ are also considered to be hypergeometric ($(1-x)^a$ is formally the $q \to 1$ limit of the quotient $(q^x;q)_\infty/(q^{a+x};q)_\infty$ of q-Gamma functions). The singular set of the integrand of a hypergeometric integral is a union of geometric (in the classical case, arithmetic) progressions. Hypergeometric series naturally arise as the sum of residues of the integrand over such pole progressions.

Multidimensional hypergeometric integrals typically arise in contexts involving representation theory of algebraic and Lie groups. For instance, in harmonic analysis on compact symmetric spaces, the zonal spherical functions give rise to a family of multivariable orthogonal polynomials with respect to a measure on a compact torus that is absolutely continuous with respect to the Haar measure. The associated weight function admits a natural factorization in terms of the root system underlying the symmetric space. Such multivariable integrals often admit generalizations beyond the representation-theoretic context. They provide the prototypical examples of *hypergeometric integrals associated with root systems.*

Let us focus now more closely on the structure of such integrals. Suppose R is an irreducible root system in $\mathbb{R}^n$, and fix a choice R^+ of positive roots. The co-weight lattice $P^\vee$ of R is the lattice in $\mathbb{R}^n$ dual to the $\mathbb{Z}$-span of R. For the classical root systems we take the usual realization of $R = R^+ \cup (-R^+)$ in $\mathbb{R}^n$ with respect to the standard orthonormal basis $\{e_i\}_{i=1}^n$ of $\mathbb{R}^n$. Concretely, $R^+ = \{e_i - e_j\}_{1\le i<j\le n}$ for type A_{n-1}, $R^+ = \{e_i \pm e_j\}_{1\le i<j\le n}$ for type D_n, $R^+ = \{e_i \pm e_j\}_{1\le i<j\le n} \cup \{2e_i\}_{i=1}^n$ for type C_n and $R^+ = \{e_i \pm e_j\}_{1\le i<j\le n} \cup \{e_i, 2e_i\}_{i=1}^n$ for type BC_n.

Let $k_\alpha \in \mathbb{C}$ be parameters that depend only on the Weyl group orbit of the root $\alpha \in R$ (equivalently, k_α depends only on the root length $\|\alpha\|$ of the root $\alpha \in R$). The prototypical example of a classical hypergeometric integral associated with R is

$$\int_{A_R} w_{\mathbf{k}}(\mathbf{x})\,\mathrm{d}\mathbf{x}, \qquad w_{\mathbf{k}}(\mathbf{x}) := \prod_{\alpha\in R} (1 - e^{2\pi i(\alpha,\mathbf{x})})^{k_\alpha} \qquad (1.2.11)$$

with $A_R \subset \mathbb{R}^n$ a fundamental domain for the translation action of $P^\vee$ on $\mathbb{R}^n$, $\boldsymbol{x} = (x_1, \ldots, x_n)$, $\mathrm{d}\boldsymbol{x} = \mathrm{d}x_1 \cdots \mathrm{d}x_n$ and $\boldsymbol{k} := \{k_\alpha\}_{\alpha \in R}$ the collection of the parameters k_α (here the k_α should satisfy appropriate conditions to ensure convergence of the integral). Remarkably the integral (1.2.11) admits an explicit evaluation as a product of Gamma functions. The resulting identity is known as the *Macdonald constant term identity* (see Theorem 8.4.2(i)). It gives the volume of the orthogonality measure of root system generalizations of the Jacobi polynomials, also known nowadays as *Heckman–Opdam polynomials*; see Chapter 8 for a detailed discussion.

Of particular interest is the special case that the root system R is of type BC_n. In that case the Macdonald constant term identity reduces after the change of variables $z_j = \sin^2(\pi x_j)$ to the well-known *Selberg integral* (Selberg, 1944)

$$\int_{[0,1]^n} \prod_{i=1}^{n} z_i^{\alpha-1}(1-z_i)^{\beta-1} \prod_{1\le i<j\le n} |z_i - z_j|^{2\gamma}\, \mathrm{d}z = \prod_{j=0}^{n-1} \frac{\Gamma(\alpha + j\gamma)\Gamma(\beta + j\gamma)\Gamma(1+(j+1)\gamma)}{\Gamma(\alpha+\beta+(n+j-1)\gamma)\Gamma(1+\gamma)}$$

with parameters $\alpha = k_{\epsilon_1} + k_{2\epsilon_1} + \frac{1}{2}, \beta = k_{2\epsilon_1} + \frac{1}{2}$ and $\gamma = k_{\epsilon_1-\epsilon_2}$, which in turn is a multidimensional generalization of the beta integral. There are many applications of the Selberg integral, for instance in the theory of integrable systems (Chapters 8 and 9), in conformal field theory (Chapter 11) and in random matrix theory; see the overview article by Forrester and Warnaar (2008).

For basic hypergeometric integrals associated with root systems a similar story applies. The roles of Lie groups and root systems are taken over by quantum groups and affine root systems, although this time the representation-theoretic context came later. The *affine root system* associated to an irreducible reduced root system R is denoted by $R^{(1)}$ and consists of the collection of affine linear functionals $a\colon \mathbb{R}^n \to \mathbb{R}$ of the form $a(\boldsymbol{x}) = (\alpha, \boldsymbol{x}) + m$ (with $\alpha \in R$ and $m \in \mathbb{Z}$). The role of $w_{\boldsymbol{k}}(\boldsymbol{x})$ is now taken over by

$$w_{\boldsymbol{k},q}(\boldsymbol{x}) = \prod_{a\in R^{(1)};\, a(0)\ge 0} \left(\frac{1-q^{a(\boldsymbol{x})}}{1-q^{k_a + a(\boldsymbol{x})}} \right) = \prod_{\alpha\in R} \frac{(q^{(\alpha,\boldsymbol{x})}; q_\alpha)_\infty}{(q^{k_\alpha+(\alpha,\boldsymbol{x})}; q_\alpha)_\infty},$$

where $k_a = k_\alpha$ if α is the gradient of $a \in R^{(1)}$. Macdonald (1982) conjectured an explicit evaluation for the *basic hypergeometric integral*

$$\int_{A_R} w_{\boldsymbol{k},q}(\boldsymbol{x}/\tau)\, \mathrm{d}\boldsymbol{x}, \qquad q = \exp(2\pi i \tau) \tag{1.2.12}$$

associated with R, which was proved in full generality by Cherednik (1995) using the theory of double affine Hecke algebras. The evaluation formula gives the volume of the orthogonality measure of the Macdonald polynomials; see Chapter 9. The integral (1.2.12) and its evaluation generalize to arbitrary (possibly nonreduced) irreducible affine root systems and with milder equivariance conditions on $\boldsymbol{k} = \{k_a\}_{a\in R^{(1)}}$. In the case of the nonreduced affine root system of type $\mathrm{C}^\vee\mathrm{C}_n$, this leads to the multivariable analogue of the Askey–Wilson integral (Gustafson, 1990) which depends, apart from q, on five additional parameters. It gives the volume of the orthogonality measure of the Koornwinder polynomials; see Chapter 9.

A very general elliptic analogue of the Selberg integral and of Gustafson's multivariable analogue of the Askey–Wilson integral was conjectured by van Diejen and Spiridonov (2001, Theorem 4.2) and proved by Rains (2010, Theorem 6.1):

$$\begin{aligned}\frac{1}{(2\pi i)^n}\int_{T^n}\prod_{1\le i<j\le n}\frac{\Gamma(tz_iz_j, tz_iz_j^{-1}, tz_i^{-1}z_j, tz_i^{-1}z_j^{-1})}{\Gamma(z_iz_j, z_iz_j^{-1}, z_i^{-1}z_j, z_i^{-1}z_j^{-1})}\prod_{k=1}^{n}\frac{\prod_{i=1}^{6}\Gamma(t_iz_k, t_iz_k^{-1})}{\Gamma(z_k^2, z_k^{-2})}\frac{dz_1}{z_1}\cdots\frac{dz_n}{z_n}\\ = \frac{2^n n!}{(p;p)_\infty^n(q;q)_\infty^n}\prod_{m=1}^{n}\left(\frac{\Gamma(t^m)}{\Gamma(t)}\prod_{1\le i<j\le 6}\Gamma(t^{m-1}t_it_j)\right),\end{aligned} \tag{1.2.13}$$

with T the positively oriented unit circle in the complex plane, $\Gamma(x_1,\ldots,x_r) := \Gamma(x_1)\cdots\Gamma(x_r)$ a product of elliptic Gamma functions $\Gamma(x_i)$, and parameters $t, t_i \in \mathbb{C}$ satisfying $|t|, |t_i| < 1$ and $t^{2n-2}t_1\cdots t_6 = pq$. The integral (1.2.13) is an example of an *elliptic hypergeometric integral* associated with the root system of type C_n. For $n = 1$ it reduces to Spiridonov's elliptic beta integral (Spiridonov and Zhedanov, 2000). It is a special case of a family of transformation formulas that relate elliptic hypergeometric integrals associated with type-C root systems of different ranks (van de Bult, 2009). The basic analogue of (1.2.13) is a multivariable analogue of the *Nassrallah–Rahman integral* (Nassrallah and Rahman, 1985),

$$\frac{1}{2\pi i}\int_T\frac{(z^2, z^{-2}, Az, Az^{-1};q)_\infty}{\prod_{j=1}^{5}(t_jz, t_jz^{-1};q)_\infty}\frac{dz}{z} = \frac{2\prod_{j=1}^{5}(At_j^{-1};q)_\infty}{(q;q)_\infty\prod_{1\le j<k\le 5}(t_jt_k;q)_\infty}, \qquad |t_j| < 1 \tag{1.2.14}$$

where $A := t_1t_2t_3t_4t_5$ and $(a_1,\ldots,a_r;q)_\infty := \prod_{i=1}^{r}(a_i;q)_\infty$. Just as (1.2.14) gives the Askey–Wilson integral for $t_5 = 0$, its multivariable analogue yields Gustafson's (1990) integral by the same substitution. The identity (1.2.13) and some of its degenerations give the volumes of (bi)orthogonality measures for important families of multivariable (bi)orthogonal functions; see §1.4.

The multivariable elliptic integrals appearing in Felder et al. (1997, 1999) as coordinates of vector-valued solutions of *qKZB* equations are associated with the root system of type A_n. Their semiclassical limits, which provide solutions of the *KZB* equation, as well as their degenerations to the basic and classical hypergeometric level, are discussed in Chapter 11.

A further rough division of hypergeometric integrals associated with root systems involves the notion of types. Multidimensional integrals are said to be *type-II basic* (resp. *elliptic*) *hypergeometric integrals* associated with the root system R if the integrand contains a factor of the form $\prod_{\alpha\in R}(\Gamma(q^{(\alpha,x)})/\Gamma(q^{k_\alpha+(\alpha,x)}))$ with $\Gamma(x)$ the basic (resp. elliptic) Gamma function. It is called *type I* if it contains a factor of the form $\prod_{\alpha\in R}\Gamma(q^{(\alpha,x)})^{-1}$. Similarly, a multidimensional integral is said to be a *type-II classical hypergeometric integral* associated with the root system R if the integrand contains a factor of the form $\Delta_k(\boldsymbol{x})$ or $\prod_{\alpha\in R}(\Gamma(\alpha,\boldsymbol{x})/\Gamma(k_\alpha + (\alpha,\boldsymbol{x})))$, with $\Gamma(x)$ the classical Gamma function (and a similar adjustment for type I). The examples of multidimensional integral evaluations highlighted so far are type II. In Chapters 5 and 6 many examples of type-I and type-II multidimensional integral evaluations and transformations are discussed. Note that there are also hypergeometric integrals of mixed type; see (6.2.3) for an example.

Next we turn our attention to multivariable hypergeometric series. For a given root system R, we can define a *Weyl-type denominator* by

$$\Delta(\boldsymbol{x}) := \prod_{\alpha\in R^+} h((\alpha,\boldsymbol{x})) \quad \text{with } h(z) = \begin{cases} z & \text{(classical hypergeometric type)},\\ 1-q^z & \text{(basic hypergeometric type)},\\ \theta(z;p) & \text{(elliptic hypergeometric type)}. \end{cases}$$

In the basic hypergeometric case $\Delta(\boldsymbol{x})$ is the Weyl denominator of the semisimple Lie algebra associated with R, while in the elliptic case $\Delta(\boldsymbol{x})$ is closely related to the Weyl denominator of the affine Lie algebra associated with $R^{(1)}$; see §6.1.2.

Multivariable classical, basic and elliptic hypergeometric series $\sum_{\boldsymbol{k}\in D} f(\boldsymbol{k})$ $(D\subseteq\mathbb{Z}^n)$ are said to be *associated with the classical root system R* if $f(\boldsymbol{k})$ contains the factor $\Delta(\boldsymbol{y}+\boldsymbol{k})$ for some fixed $\boldsymbol{y}\in\mathbb{C}^n$ in a nontrivial way. First examples of multivariable classical hypergeometric series identities appeared in the work of Holman et al. (1976) on $6j$-symbols for $\mathrm{SU}(n)$ (the associated root system is of type A). An important nontrivial example of a multivariable basic hypergeometric series identity is the fundamental theorem of Milne (1985):

$$\sum_{\boldsymbol{k}\in D_N} \Delta(\boldsymbol{y}+\boldsymbol{k})\prod_{\ell=1}^n q^{(\ell-1)k_\ell}\prod_{i,j=1}^n \frac{(q^{\beta_i+y_j-y_i};q)_{k_j}}{(q^{1+y_j-y_i};q)_{k_j}} = \frac{(q^{\beta_1+\cdots+\beta_n};q)_N}{(q;q)_N}\Delta(\boldsymbol{y})$$

with $D_N := \{\boldsymbol{k}\in\mathbb{Z}^n_{\geq 0} \mid k_1+\cdots+k_n = N\}$ and $\Delta(\boldsymbol{x}) = \prod_{1\leq i<j\leq n}(1-q^{x_i-x_j})$ the Weyl-type denominator for the root system R of type A_{n-1}. An elliptic generalization is the elliptic Jackson summation formula (6.3.1a) due to Rosengren (2004, Theorem 5.1).

For classical root systems, identities and transformations for multivariable hypergeometric series naturally arise from related multidimensional hypergeometric integral identities and transformations through residue calculus. In this process, the Weyl denominator $\Delta(\boldsymbol{k})$ arises from the integrands of the multidimensional hypergeometric integrals through the formula (6.1.3). The residue calculus typically involves iterated small contour deformations per coordinate, avoiding at each step the poles of the factors of the integrand that do not depend on a single coordinate x_j. This technique was developed in Stokman (2000), where it was applied to type-II basic hypergeometric integrals associated with Koornwinder polynomials. When applied to the elliptic Selberg integral (1.2.13) one obtains a type-C elliptic hypergeometric series identity (see (6.3.6)) that reduces for $n=1$ to the *Frenkel–Turaev elliptic summation formula* (Frenkel and Turaev, 1997):

$$\begin{aligned}\sum_{m=0}^N \frac{\theta(aq^{2m};p)}{\theta(a;p)}\frac{(a,b,c,d,e,q^{-N};q,p)_m}{(q,aq/b,aq/c,aq/d,aq/e,aq^{1+N};q,p)_m}q^m \\ = \frac{(aq,aq/bc,aq/bd,aq/cd;q,p)_N}{(aq/b,aq/c,aq/d,aq/bcd;q,p)_N}\end{aligned} \tag{1.2.15}$$

for $bcde = a^2q^{N+1}$, where $(x_1,\ldots,x_r;q,p)_m = \prod_{i=1}^r(x_i;q,p)_m$. In this way, many of the hypergeometric series identities and transformations associated with classical root systems as discussed in Chapter 5 (classical and basic hypergeometric) and in Chapter 6 (elliptic hypergeometric) can be viewed as discrete versions of multidimensional hypergeometric integrals.

1.3 Multivariable (Bi)Orthogonal Polynomials and Functions

1.3.1 One-Variable Cases

The class of one-variable (bi)orthogonal polynomials and functions splits up into various natural subclasses, each subclass having its distinct features that are vital for the construction and study of its multivariable generalization.

(a) General theory of orthogonal polynomials (Szegő, 1975)
(b) Classical orthogonal polynomials
(c) Classical biorthogonal rational functions
(d) Bessel functions (Olver et al., 2010, §10.22(v)) and Jacobi functions (Koornwinder, 1984)

By *classical orthogonal polynomials* we mean (more generally than in Chapter 2) the one-variable orthogonal polynomials belonging to the Askey or q-Askey scheme (Koekoek et al., 2010, Chs. 9, 14). They are characterized as the orthogonal polynomials that are joint eigenfunctions of a suitable type of second-order differential or (q-)difference operator. The corresponding classification results are called (generalized) Bochner theorems (Grünbaum and Haine, 1996; Ismail, 2003; Vinet and Zhedanov, 2008). Prominent members are the Jacobi polynomials (Szegő, 1975, Ch. IV) and their top-level q-analogues, the Askey–Wilson polynomials (Askey and Wilson, 1985). By *classical biorthogonal rational functions* we refer to the generalizations of classical orthogonal polynomials due to Rahman (1986, 1991) and Wilson (1991), and their elliptic analogues due to Spiridonov and Zhedanov (2000).

Classical orthogonal polynomials and biorthogonal rational functions are expressible as ordinary, basic and elliptic hypergeometric series. The various classes admit (bi)orthogonality relations with respect to explicit measures whose total masses are the outcome of important integral evaluation formulas. For example, for the classical Jacobi polynomials the integral evaluation is the beta integral

$$\int_{-1}^{1}(1-x)^{\alpha}(1+x)^{\beta}\,dx = \frac{2^{\alpha+\beta+1}\Gamma(\alpha+1)\Gamma(\beta+1)}{\Gamma(\alpha+\beta+2)}, \qquad \alpha,\beta > -1,$$

with $\Gamma(x)$ the classical Gamma function, for Rahman's (1986) biorthogonal basic hypergeometric rational functions it is the Nassrallah–Rahman integral (1.2.14) and for Spiridonov–Zhedanov's (2000) elliptic biorthogonal rational functions it is Spiridonov's (2001) elliptic beta integral (the $n = 1$ case of (1.2.13)).

1.3.2 Multivariable Generalizations

The subclasses (a)–(d) of one-variable (bi)orthogonal polynomials and functions generalize to the multivariable case as follows. Note that the one-variable subclass (b) generalizes to two subclasses (b1) and (b2).

(a) General theory of multivariable orthogonal polynomials with respect to orthogonality measures on $\mathbb{R}^d$ (Dunkl and Xu, 2014). This is discussed in Chapter 2.
(b1) Multivariable orthogonal polynomials expressible as (nonstraightforward) products of one-variable classical orthogonal polynomials and elementary polynomials; see

Tratnik (1991) and Gasper and Rahman (2005) for the continuous cases. Examples are in Chapter 2.

(b2) Root system generalizations of classical orthogonal polynomials. Prominent examples are the BC-type Heckman–Opdam polynomials (Heckman, 1991) – see Chapter 8, and Koornwinder polynomials (Koornwinder, 1992) – see Chapter 9, which provide multivariable generalizations of the Jacobi polynomials and their top-level q-analogues in the q-Askey scheme (Askey–Wilson polynomials), respectively.

(c) By the extension of the class of classical orthogonal polynomials to biorthogonal rational functions, one arrives at a class of one-variable special functions that generalizes to the elliptic hypergeometric level as well as to the multivariable level. In the one-variable setup these are the biorthogonal elliptic rational functions from Spiridonov and Zhedanov (2000). The multivariable generalization is due to Rains (2010). This is discussed in Chapter 6. The Macdonald–Koornwinder polynomials associated with classical root systems are limit cases of Rains' elliptic biorthogonal rational functions.

(d) Root system generalizations of Bessel functions and the associated Fourier transforms (Dunkl, 1992) are discussed in Chapter 7. Root system generalizations of Jacobi functions and the associated harmonic analysis (Opdam, 1995) are discussed in Chapter 8. These functions can be thought of as nonpolynomial generalizations of the Heckman–Opdam polynomials so that they may be called *Heckman–Opdam functions*. Their basic hypergeometric analogues (Cherednik, 1997; Stokman, 2003) are not discussed in this volume.

In the next section, for these five subclasses we briefly discuss how these multivariable extensions came about, and what techniques are used to study them.

1.4 Multivariable (Bi)Orthogonal Polynomials and Functions, Some Details

Class (a): General Theory of Multivariable Orthogonal Polynomials

Important topics are the development of multivariable versions of Gram–Schmidt orthogonalization and of three-term recurrence relations (Favard's theorem). Much of the general theory deals with the space $\mathcal{V}_n$ of *orthogonal polynomials of degree n in d variables* $x_1, \ldots, x_d$, defined as the space of all polynomials of total degree $\leq n$ which are orthogonal to all polynomials of total degree $\leq n-1$ with respect to a fixed inner product on the space of polynomials in $x_1, \ldots, x_d$. The subspace $\mathcal{V}_n$ does not have a canonical orthogonal basis. In particular, the *monomial basis* of $\mathcal{V}_n$, consisting of the polynomials in $\mathcal{V}_n$ of the form $x^\alpha + Q_\alpha(x)$ ($\alpha \in \mathbb{Z}^d_{\geq 0}$ multi-index of degree n) with $Q_\alpha(x)$ of total degree $\leq n-1$, is usually not an orthogonal basis of $\mathcal{V}_n$. Still, any basis of $\mathcal{V}_n$, in particular the monomial basis, admits another basis which is biorthogonal to the first basis. If the subspaces $\mathcal{V}_n$ are the eigenspaces of some second-order partial differential operator L then the orthogonal polynomials are said to be *classical*. For two-variable examples, see §2.3.4. Remarkably, while the factorizable multivariable orthogonal polynomials (class (b1)) are usually classical, the root system generalizations of the classical orthogonal polynomials (class (b2)) are usually not. In both cases, for $q = 1$, there is a second-order differential operator around, but in the second case its eigenspaces are usually not the full spaces $\mathcal{V}_n$.

Class (b1): Factorizable Multivariable Orthogonal Polynomials

For two-dimensional regions like the disk and the triangle with obvious weight functions generalizing the ultraspherical or Jacobi weight function, the monomial basis and the basis biorthogonal to it were explicitly given by Appell in the late nineteenth century. These bases were expressed in terms of Appell hypergeometric functions. Explicit orthogonal bases for these cases were given much later, although they have a very simple factorized form. See the survey by Koornwinder (1975, §3.4). These bases usually came up in close connection with various kinds of applications. Very noteworthy are the *disk polynomials* of Zernike and Brinkman (1935), motivated by optics and still having important applications there.

The factorized orthogonal polynomials were extended to higher dimensions d (ball and simplex). In addition to the second-order operator L mentioned under class (a), there are $d-1$ further partial differential operators generating, together with L, a commutative algebra of which the orthogonal polynomials are the joint eigenfunctions. These orthogonal polynomials also occur in connection with spherical harmonics (Koornwinder, 1975, §3.5; Dunkl, 1988; Kalnins et al., 1991); see also Chapter 2. Furthermore, these polynomials naturally arise as coupling coefficients for tensor product representations of $SL_2(\mathbb{R})$, while multivariable basic hypergeometric orthogonal polynomials arise as coupling coefficients for tensor products of the associated quantum group (Rosengren, 1999, 2001).

Class (b2): Root System Generalizations of Classical Orthogonal Polynomials.

Restrictions of zonal spherical functions on compact symmetric spaces to the torus provide multivariable root system generalizations of Jacobi polynomials depending on special discrete parameter values, which come from the root multiplicities of the symmetric space (Helgason, 2000, Ch. V, §4). Heckman–Opdam polynomials (Heckman, 1991) provide generalizations without restrictions on the parameters. The techniques from geometric group theory now fail. Initially the only alternative was laborious analytic work, but the early nineties brought the great insight that there is another Lie-type setting for these polynomials, namely representation theory of degenerate affine Hecke algebras (Opdam, 1995) in terms of Heckman–Dunkl differential-reflection operators (Heckman, 1991), which generalize the Dunkl operators (Dunkl, 1989) treated in Chapter 7 (see also Cherednik's (1991, pp. 429–430) slight variants of the Heckman–Dunkl operators). As mentioned already in §1.2.3, the total mass of the orthogonality measure of the Heckman–Opdam polynomials is evaluated by the Macdonald constant term identity. A significant difference with multivariable orthogonal polynomials in class (b1) is the fact that explicit hypergeometric series expressions of the Heckman–Opdam polynomials are not available. However, there is a binomial formula in the BC_n case (see Koornwinder, 2015, (9.1), (11.5)), as we will discuss for the elliptic case in §1.4, class (c).

All key properties, including orthogonality and norm formulas, need representation-theoretic tools involving the Dunkl-type operators and related operators, such as intertwining and shift operators. The Heckman–Opdam polynomials are joint eigenfunctions of a commutative algebra generated by d commuting partial differential operators, where d is the rank of the associated root system. It includes the explicit second-order differential operator L arising as the radial part of the Laplace–Beltrami operator with the root multiplicities taken continuously; see Chapter 8. An important application and source of further applications is the fact that they produce the eigenstates for an important class of integrable one-dimensional quantum many-body systems in theoretical physics called quantum Calogero–Moser systems; see Opdam (1988).

In the q-case the Macdonald (2000) and Koornwinder (1992) polynomials provide multivariable root system generalizations of the Askey–Wilson polynomials (the top level in the q-Askey scheme) and its subclasses of continuous q-Jacobi and continuous q-ultraspherical polynomials (only for root systems of type BC do all three classes generalize). Just as for $q = 1$, they relate for special parameter values to harmonic analysis on quantum compact symmetric spaces (Noumi, 1996; Letzter, 2003) – although in this case it was not the original motivation for introducing these polynomials. Many properties of the Heckman–Opdam polynomials as described above generalize to Macdonald and Koornwinder polynomials, with the role of differential operators now taken over by q-difference operators. The deeper study of these polynomials involves Cherednik's theory on (double) affine Hecke algebras (Cherednik, 2005; Sahi, 1999; Macdonald, 2003); see Chapter 9.

Prior to the Macdonald polynomials for arbitrary root systems, Macdonald (1988; 1995, Ch. VI) introduced a version of these polynomials which is related to the general linear group (they relate to the Macdonald polynomials for root systems of type A as, for $q = 1$, the Jack polynomials relate to the A-type Heckman–Opdam polynomials). These GL_d Macdonald polynomials are homogeneous symmetric polynomials in d (or countably many) variables depending on two parameters q and t, which generalize various important classes of symmetric functions such as Jack polynomials, Hall–Littlewood polynomials and Schur functions. Their important role in modern algebraic combinatorics is discussed in Chapter 10.

Class (c): Biorthogonal Rational Functions

From the q-Askey scheme of classical basic hypergeometric orthogonal polynomials and their biorthogonal rational extensions, only Rahman's (1986) biorthogonal rational functions have been generalized to the elliptic regime (Spiridonov and Zhedanov, 2000). Multivariable generalizations in the class (b2) were introduced by Rains (2010). The associated explicit evaluation formula of the total mass of the biorthogonality measure is the type-II elliptic hypergeometric integral associated to the root system of type C given by (1.2.13). The elliptic variant of the theory of interpolation functions, which goes back to work of Kostant and Sahi (1991, 1993), has been particularly useful in the development of Rains' multivariable elliptic biorthogonal rational functions. These are classes of multivariable polynomials defined by explicit vanishing properties, and serve as a kind of monomial-type basis within the theory. In particular, one can write down explicit series expansions of the elliptic biorthogonal rational functions in terms of interpolation functions. These so-called *binomial formulas* (earlier given for Koornwinder polynomials by Okounkov, 1998) are for the moment the closest to explicit elliptic or basic hypergeometric series expressions one can get. See also §6.4.

The elliptic generalization of the double affine Hecke algebra is currently in development (Rains, 2018) from the point of view of algebraic geometry.

Class (d): Root System Generalizations of Bessel and Jacobi Functions

The discussion of class (b2) largely applies here too. For special parameter values Jacobi functions relate to zonal spherical functions on *noncompact* Riemannian symmetric spaces and multivariable Bessel functions to spherical functions for Cartan motion groups. The nonsymmetric versions of multivariable Bessel functions are called *Dunkl kernels*, and it is there that the Hecke algebraic approach, mentioned before in the discussion of class (b2), first arose with Dunkl's invention of a commuting family of differential-difference operators serving as deformations of directional derivatives. These operators are nowadays known as *Dunkl operators*; see Chapter 7.

It remains puzzling that nonsymmetric Jacobi functions and nonsymmetric Jacobi polynomials associated with root systems do not seem to live, for special parameter values, on Riemannian symmetric spaces, as symmetric Jacobi functions and polynomials do. However, partial interpretations of nonsymmetric special functions are known in connection with representations of affine Lie algebras and p-adic groups; see Sanderson (2000), Ion (2003, 2006) and Brubaker et al. (2015).

References

Appell, P., and Kampé de Fériet, J. 1926. *Fonctions Hypergéométriques et Hypersphériques. Polynômes d'Hermite*. Gauthier-Villars.

Askey, R., and Wilson, J. 1985. *Some Basic Hypergeometric Orthogonal Polynomials that Generalize Jacobi Polynomials*. Mem. Amer. Math. Soc., vol. 54, no. 319.

Beerends, R. J. 1992. Some special values for the *BC* type hypergeometric function. Pages 27–49 of: *Hypergeometric Functions on Domains of Positivity, Jack Polynomials, and Applications*. Contemp. Math., vol. 138. Amer. Math. Soc.

Brubaker, B., Bump, D., and Licata, A. 2015. Whittaker functions and Demazure operators. *J. Number Theory*, **146**, 41–68.

van de Bult, F. J. 2009. *An elliptic hypergeometric beta integral transformation*. arXiv:0912.3812.

Cherednik, I. 1991. A unification of Knizhnik–Zamolodchikov and Dunkl operators via affine Hecke algebras. *Invent. Math.*, **106**, 411–431.

Cherednik, I. 1995. Double affine Hecke algebras and Macdonald's conjectures. *Ann. of Math. (2)*, **141**, 191–216.

Cherednik, I. 1997. Difference Macdonald–Mehta conjecture. *Int. Math. Res. Not.*, 449–467.

Cherednik, I. 2005. *Double Affine Hecke Algebras*. London Mathematical Society Lecture Note Series, vol. 319. Cambridge University Press.

Dereziński, J., and Majewski, P. 2016. From conformal group to symmetries of hypergeometric type equations. *SIGMA*, **12**, Paper 108, 69 pp.

van Diejen, J. F., and Spiridonov, V. P. 2001. Elliptic Selberg integrals. *Int. Math. Res. Not.*, 1083–1110.

Dunkl, C. F. 1988. Reflection groups and orthogonal polynomials on the sphere. *Math. Z.*, **197**, 33–60.

Dunkl, C. F. 1989. Differential-difference operators associated to reflection groups. *Trans. Amer. Math. Soc.*, **311**, 167–183.

Dunkl, C. F. 1992. Hankel transforms associated to finite reflection groups. Pages 123–138 of: *Hypergeometric Functions on Domains of Positivity, Jack Polynomials, and Applications*. Contemp. Math., vol. 138. Amer. Math. Soc.

Dunkl, C. F., and Xu, Y. 2014. *Orthogonal Polynomials of Several Variables*. Second edn. Encyclopedia of Mathematics and Its Applications, vol. 155. Cambridge University Press.

Erdélyi, A. et al. 1953. *Higher Transcendental Functions, Vol. I*. McGraw-Hill.

Felder, G., Tarasov, V., and Varchenko, A. 1997. Solutions of the elliptic qKZB equations and Bethe ansatz. I. Pages 45–75 of: *Topics in Singularity Theory*. Amer. Math. Soc. Transl. Ser. 2, vol. 180. Amer. Math. Soc. Also available as arXiv:q-alg/9606005 (1996).

Felder, G., Tarasov, V., and Varchenko, A. 1999. Monodromy of solutions of the elliptic quantum Knizhnik–Zamolodchikov–Bernard difference equations. *Internat. J. Math.*, **10**, 943–975.

Forrester, P. J., and Warnaar, S. O. 2008. The importance of the Selberg integral. *Bull. Amer. Math. Soc. (N.S.)*, **45**, 489–534.

Frenkel, I. B., and Turaev, V. G. 1997. Elliptic solutions of the Yang–Baxter equation and modular hypergeometric functions. Pages 171–204 of: *The Arnold–Gelfand Mathematical Seminars*. Birkhäuser.

Gasper, G., and Rahman, M. 2004. *Basic Hypergeometric Series*. Second edn. Encyclopedia of Mathematics and Its Applications, vol. 96. Cambridge University Press.

Gasper, G., and Rahman, M. 2005. Some systems of multivariable orthogonal Askey–Wilson polynomials. Pages 209–219 of: *Theory and Applications of Special Functions*. Dev. Math., vol. 13. Springer.

Gel'fand, I. M., Zelevinsky, A. V., and Kapranov, M. M. 1989. Hypergeometric functions and toral manifolds. *Funct. Anal. Appl.*, **23**, 94–106. Correction in *Funct. Anal. Appl.* **27** (1993), p. 295.

Grünbaum, F. A., and Haine, L. 1996. The q-version of a theorem of Bochner. *J. Comput. Appl. Math.*, **68**, 103–114.

Gustafson, R. A. 1990. A generalization of Selberg's beta integral. *Bull. Amer. Math. Soc. (N.S.)*, **22**, 97–105.

Heckman, G. J. 1991. An elementary approach to the hypergeometric shift operators of Opdam. *Invent. Math.*, **103**, 341–350.

Helgason, S. 2000. *Groups and Geometric Analysis*. Amer. Math. Soc. Corrected reprint of the 1984 original.

Holman, III, W. J., Biedenharn, L. C., and Louck, J. D. 1976. On hypergeometric series well-poised in SU(n). *SIAM J. Math. Anal.*, **7**, 529–541.

Ion, B. 2003. Nonsymmetric Macdonald polynomials and Demazure characters. *Duke Math. J.*, **116**, 299–318.

Ion, B. 2006. Nonsymmetric Macdonald polynomials and matrix coefficients for unramified principal series. *Adv. Math.*, **201**, 36–62.

Ismail, M. E. H. 2003. A generalization of a theorem of Bochner. *J. Comput. Appl. Math.*, **159**, 319–324.

Kalnins, E. G., Manocha, H. L., and Miller, Jr., W. 1980. The Lie theory of two-variable hypergeometric functions. *Stud. Appl. Math.*, **62**, 143–173.

Kalnins, E. G., Miller, Jr., W., and Tratnik, M. V. 1991. Families of orthogonal and biorthogonal polynomials on the N-sphere. *SIAM J. Math. Anal.*, **22**, 272–294.

Koekoek, R., Lesky, P. A., and Swarttouw, R. F. 2010. *Hypergeometric Orthogonal Polynomials and their q-Analogues*. Springer.

Koornwinder, T. 1975. Two-variable analogues of the classical orthogonal polynomials. Pages 435–495 of: *Theory and Application of Special Functions*. Academic Press.

Koornwinder, T. H. 1984. Jacobi functions and analysis on noncompact semisimple Lie groups. Pages 1–85 of: *Special Functions: Group Theoretical Aspects and Applications*. Reidel.

Koornwinder, T. H. 1992. Askey–Wilson polynomials for root systems of type BC. Pages 189–204 of: *Hypergeometric Functions on Domains of Positivity, Jack Polynomials, and Applications*. Contemp. Math., vol. 138. Amer. Math. Soc.

Koornwinder, T. H. 2015. Okounkov's BC-type interpolation Macdonald polynomials and their $q = 1$ limit. *Sém. Lothar. Combin.*, **72**, B72a, 27 pp. Corrections in arxiv:1408.5993v5.

Koornwinder, T., and Sprinkhuizen-Kuyper, I. 1978. Generalized power series expansions for a class of orthogonal polynomials in two variables. *SIAM J. Math. Anal.*, **9**, 457–483.

Kostant, B., and Sahi, S. 1991. The Capelli identity, tube domains, and the generalized Laplace transform. *Adv. Math.*, **87**, 71–92.

Kostant, B., and Sahi, S. 1993. Jordan algebras and Capelli identities. *Invent. Math.*, **112**, 657–664.

Letzter, G. 2003. Quantum symmetric pairs and their zonal spherical functions. *Transform. Groups*, **8**, 261–292.

Macdonald, I. G. 1982. Some conjectures for root systems. *SIAM J. Math. Anal.*, **13**, 988–1007.

Macdonald, I. G. 1988. A new class of symmetric functions. *Sém. Lothar. Combin.*, **20**, Paper B20a, 41 pp.

Macdonald, I. G. 1995. *Symmetric Functions and Hall Polynomials*. Second edn. Oxford University Press.

Macdonald, I. G. 2000. Orthogonal polynomials associated with root systems. *Sém. Lothar. Combin.*, **45**, B45a, 40 pp. Text of handwritten preprint, 1987.

Macdonald, I. G. 2003. *Affine Hecke Algebras and Orthogonal Polynomials*. Cambridge Tracts in Mathematics, vol. 157. Cambridge University Press.

Miller Jr., W. 1973. Lie theory and generalizations of the hypergeometric functions. *SIAM J. Appl. Math.*, **25**, 226–235.

Milne, S. C. 1985. An elementary proof of the Macdonald identities for $A_l^{(1)}$. *Adv. Math.*, **57**, 34–70.

Nassrallah, B., and Rahman, M. 1985. Projection formulas, a reproducing kernel and a generating function for q-Wilson polynomials. *SIAM J. Math. Anal.*, **16**, 186–197.

Noumi, M. 1992. Quantum Grassmannians and q-hypergeometric series. *CWI Quarterly*, **5**, 293–307.

Noumi, M. 1996. Macdonald's symmetric polynomials as zonal spherical functions on some quantum homogeneous spaces. *Adv. Math.*, **123**, 16–77.

Okounkov, A. 1998. BC-type interpolation Macdonald polynomials and binomial formula for Koornwinder polynomials. *Transform. Groups*, **3**, 181–207.
Olsson, P. O. M. 1977. On the integration of the differential equations of five-parametric double-hypergeometric functions of second order. *J. Math. Phys.*, **18**, 1285–1294.
Olver, F. W. J., Lozier, D. W., Boisvert, R. F., and Clark, C. W. (eds). 2010. *NIST Handbook of Mathematical Functions*. Cambridge University Press. Online available as *Digital Library of Mathematical Functions*, `https://dlmf.nist.gov`.
Opdam, E. M. 1988. Root systems and hypergeometric functions. IV. *Compos. Math.*, **67**, 191–209.
Opdam, E. M. 1995. Harmonic analysis for certain representations of graded Hecke algebras. *Acta Math.*, **175**, 75–121.
Rahman, M. 1986. An integral representation of a ${}_{10}\varphi_9$ and continuous bi-orthogonal ${}_{10}\varphi_9$ rational functions. *Canad. J. Math.*, **38**, 605–618.
Rahman, M. 1991. Biorthogonality of a system of rational functions with respect to a positive measure on $[-1, 1]$. *SIAM J. Math. Anal.*, **22**, 1430–1441.
Rains, E. M. 2010. Transformations of elliptic hypergeometric integrals. *Ann. of Math. (2)*, **171**, 169–243.
Rains, E. M. 2018. *Elliptic double affine Hecke algebras*. arXiv:1709.02989v2.
Rosengren, H. 1999. Multivariable orthogonal polynomials and coupling coefficients for discrete series representations. *SIAM J. Math. Anal.*, **30**, 232–272.
Rosengren, H. 2001. Multivariable q-Hahn polynomials as coupling coefficients for quantum algebra representations. *Int. J. Math. Math. Sci.*, **28**, 331–358.
Rosengren, H. 2004. Elliptic hypergeometric series on root systems. *Adv. Math.*, **181**, 417–447.
Rosengren, H. 2017. *Elliptic hypergeometric functions*. arXiv:1608.06161v3.
Ruijsenaars, S. N. M. 1997. First order analytic difference equations and integrable quantum systems. *J. Math. Phys.*, **38**, 1069–1146.
Sahi, S. 1999. Nonsymmetric Koornwinder polynomials and duality. *Ann. of Math. (2)*, **150**, 267–282.
Saito, M. 1996. Symmetry algebras of normal $\mathcal{A}$-hypergeometric systems. *Hokkaido Math. J.*, **25**, 591–619.
Saito, M. 2001. Isomorphism classes of A-hypergeometric systems. *Compos. Math.*, **128**, 323–338.
Sanderson, Y. B. 2000. On the connection between Macdonald polynomials and Demazure characters. *J. Algebraic Combin.*, **11**, 269–275.
Selberg, A. 1944. Bemerkinger om et multipelt integral (in Norwegian). *Norsk Mat. Tidsskr.*, **26**, 71–78.
Shimeno, N., and Tamaoka, Y. 2015. *The hypergeometric function for the root system of type A with a certain degenerate parameter*. arXiv:1801.05178.
Spiridonov, V. P. 2001. On the elliptic beta function. *Russian Math. Surveys*, **56**, 185–186.
Spiridonov, V., and Zhedanov, A. 2000. Spectral transformation chains and some new biorthogonal rational functions. *Comm. Math. Phys.*, **210**, 49–83.
Stienstra, J. 2007. GKZ hypergeometric structures. Pages 313–371 of: *Arithmetic and Geometry around Hypergeometric Functions*. Progr. Math., vol. 260. Birkhäuser. Also available as arXiv:math/0511351.
Stokman, J. V. 2000. On *BC* type basic hypergeometric orthogonal polynomials. *Trans. Amer. Math. Soc.*, **352**, 1527–1579.
Stokman, J. V. 2003. Difference Fourier transforms for nonreduced root systems. *Selecta Math. (N.S.)*, **9**, 409–494.

Szegő, G. 1975. *Orthogonal Polynomials*. Fourth edn. Colloquium Publications, vol. XXIII. Amer. Math. Soc.

Tratnik, M. V. 1991. Some multivariable orthogonal polynomials of the Askey tableau—continuous families. *J. Math. Phys.*, **32**, 2065–2073.

Vinet, L., and Zhedanov, A. 2008. Generalized Bochner theorem: characterization of the Askey–Wilson polynomials. *J. Comput. Appl. Math.*, **211**, 45–56.

Wilson, J. A. 1991. Orthogonal functions from Gram determinants. *SIAM J. Math. Anal.*, **22**, 1147–1155.

Zernike, F., and Brinkman, H. C. 1935. Hypersphärische Funktionen und die in sphärischen Bereichen orthogonalen Polynome. *Proc. Akad. Wetensch. Amsterdam*, **38**, 161–170.

2

Orthogonal Polynomials of Several Variables

Yuan Xu

2.1 Introduction

Polynomials of d variables are indexed by the set $\mathbb{N}_0^d$ of multi-indices, where $\mathbb{N}_0 := \{0, 1, 2, \ldots\}$. The standard multi-index notation will be used throughout this chapter. For $\alpha = (\alpha_1, \ldots, \alpha_d) \in \mathbb{N}_0^d$ and $x = (x_1, \ldots, x_d)$, a monomial x^α is defined by $x^\alpha = x_1^{\alpha_1} \cdots x_d^{\alpha_d}$. The number $|\alpha| = \alpha_1 + \cdots + \alpha_d$ is called the (total) degree of x^α. The space of homogeneous polynomials of degree n is denoted by

$$\mathcal{P}_n^d := \operatorname{span}\left\{x^\alpha \;\middle|\; |\alpha| = n,\ \alpha \in \mathbb{N}_0^d\right\}$$

and the space of polynomials of degree at most n is denoted by

$$\Pi_n^d := \operatorname{span}\left\{x^\alpha \;\middle|\; |\alpha| \le n,\ \alpha \in \mathbb{N}_0^d\right\}.$$

Evidently, Π_n^d is a direct sum of $\mathcal{P}_k^d$ for $k = 0, 1, \ldots, n$. Furthermore,

$$\dim \mathcal{P}_n^d = \binom{n+d-1}{n} \quad \text{and} \quad \dim \Pi_n^d = \binom{n+d}{n}. \tag{2.1.1}$$

Let $\langle \cdot, \cdot \rangle$ be an inner product defined on the space of polynomials of d variables. A priori it may be indefinite or even degenerate. Usually it will be given by

$$\langle f, g \rangle_\mu := \int_{\mathbb{R}^d} f(x) g(x)\, d\mu(x),$$

where the *orthogonality measure* μ is a positive Borel measure on $\mathbb{R}^d$ such that the integral is well defined on polynomials. This inner product will be nondegenerate, and hence positive definite if μ is supported on a set Ω that has nonempty interior. This will almost always be the case in this chapter, apart from some exceptional cases in §2.8.

A polynomial $P \in \Pi_n^d$ is said to be an *orthogonal polynomial* of degree n with respect to $\langle \cdot, \cdot \rangle$ if P is orthogonal to all polynomials of degree $< n$:

$$\langle P, Q \rangle = 0 \quad \forall\, Q \in \Pi^d \quad \text{with } \deg Q < \deg P. \tag{2.1.2}$$

However, two linearly independent orthogonal polynomials of degree n are not necessarily orthogonal to each other. Let $\mathcal{V}_n^d$ be the space of orthogonal polynomials of degree n, that is,

$$\mathcal{V}_n^d := \left\{P \in \Pi_n^d \;\middle|\; \langle P, Q \rangle = 0 \quad \forall\, Q \in \Pi_{n-1}^d\right\}. \tag{2.1.3}$$

If the inner product is nondegenerate then $\dim \mathcal{V}_n^d = \dim \mathcal{P}_n^d = \binom{n+d-1}{n} := r_n^d$, and Π_n^d is a direct sum of $\mathcal{V}_k^d$ for $k = 0, 1, \ldots, n$.

Given a nondegenerate inner product, we can assign to the set $\{x^\alpha \mid \alpha \in \mathbb{N}_0^d\}$ a linear order $\succ$ which is *graded* (i.e., $x^\alpha \succ x^\beta$ if $|\alpha| > |\beta|$), and apply the Gram–Schmidt orthogonalization process to generate a sequence of orthogonal polynomials. In contrast to $d = 1$, however, there is no obvious natural graded order among monomials when $d > 1$. There are instead many well-defined orders. One example is given by the *graded lexicographic order*:

$$x^\alpha \succ x^\beta \text{ if } |\alpha| > |\beta| \text{ or if } |\alpha| = |\beta| \text{ and the first nonzero entry in the difference } \alpha - \beta \text{ is positive.}$$

In general, different orderings will lead to different orthogonal systems. Consequently, orthogonal polynomials of several variables are not unique. Moreover, any system of orthogonal polynomials obtained by an ordering of the monomials is necessarily unsymmetric in the variables $x_1, \ldots, x_d$. These were recognized as the essential difficulties in the study of orthogonal polynomials of several variables in Erdélyi (1953, Ch. XII), which contains a rather comprehensive account of the results up to 1950.

A sequence of polynomials $\{P_\alpha\} \in \mathcal{V}_n^d$ is called *orthogonal* if $\langle P_\alpha, P_\beta \rangle = 0$ whenever $\alpha \neq \beta$, and *orthonormal* if moreover $\langle P_\alpha, P_\alpha \rangle = 1$ for all α. The space $\mathcal{V}_n^d$ can have many different bases and a basis does not have to be orthogonal. One way to extend the theory of orthogonal polynomials of one variable to several variables is to state the results in terms of $\mathcal{V}_0^d, \mathcal{V}_1^d, \ldots, \mathcal{V}_n^d, \ldots$, rather than in terms of a particular basis in each $\mathcal{V}_n^d$.

This point of view will be prominent in our next section, which contains a brief account of the general properties of the orthogonal polynomials of several variables, mostly developed in the last two decades. In the later sections of this chapter, we will discuss in more detail specific systems of orthogonal polynomials in two and more variables that correspond to, or are generalizations of, the classical orthogonal polynomials of one variable. Most of these systems are of separated type, by which we mean that a basis of orthogonal polynomials can be expressed as products in some separation of variables in terms of classical orthogonal polynomials of one variable.

There are many points of contact with other chapters of this volume. Some of the orthogonal polynomials will be given in terms of Appell and Lauricella hypergeometric functions, which are the subject of Chapter 3. Orthogonal polynomials for weight functions invariant under a reflection group are addressed in Chapter 7. Orthogonal polynomials associated with root systems are treated in Chapter 8. q-Analogues of such orthogonal polynomials are discussed in Chapter 9.

2.2 General Properties of Orthogonal Polynomials of Several Variables

The general properties of orthogonal polynomials of several variables were studied as early as 1936 by Jackson. Most earlier studies dealt with two variables; see the references in Erdélyi (1953, Ch. XII) and Suetin (1999). The presentation below follows the book by Dunkl and Xu (2014).

2.2.1 Moments and Orthogonal Polynomials

Associated with each multisequence $s\colon \mathbb{N}_0^d \mapsto \mathbb{R}$, $s = (s_\alpha)_{\alpha\in\mathbb{N}_0^d}$, we can define a linear functional $\mathcal{L}$, called a *moment functional*, by

$$\mathcal{L}(x^\alpha) = s_\alpha, \quad \alpha \in \mathbb{N}_0^d.$$

A polynomial $P \in \Pi_n^d$ is called orthogonal with respect to $\mathcal{L}$ if it is orthogonal with respect to the bilinear form $\langle f, g\rangle = \mathcal{L}(fg)$, which however is not necessarily a positive definite or nondegenerate inner product.

For each $n \in \mathbb{N}_0$ let $\mathbf{x}^n$ denote the column vector

$$\mathbf{x}^n := (x^\alpha)_{|\alpha|=n} = (x^{\alpha_j})_{j=1}^{r_n^d},$$

where $\alpha_1, \alpha_2, \dots, \alpha_{r_n^d}$, $r_n^d = \dim \mathcal{P}_n^d$ is the arrangement of the elements in $\{\alpha \in \mathbb{N}_0^d \mid |\alpha| = n\}$ according to the lexicographical order. For $k, j \in \mathbb{N}_0$, define a vector of moments $\mathbf{s}_k$ and a matrix of moments $\mathbf{s}_{k,j}$ by

$$\mathbf{s}_k := \mathcal{L}(\mathbf{x}^k) \quad \text{and} \quad \mathbf{s}_{k,j} = \mathbf{s}_{\{k\}+\{j\}} := \mathcal{L}(\mathbf{x}^k(\mathbf{x}^j)^{\mathrm{tr}}). \tag{2.2.1}$$

By definition, $\mathbf{s}_{\{k\}+\{j\}}$ is a matrix of size $r_k^d \times r_j^d$, and its elements are $s_{\alpha+\beta}$ for $|\alpha| = k$ and $|\beta| = j$. Finally, for each $n \in \mathbb{N}_0$, we define a *moment matrix* by using $\mathbf{s}_{\{k\}+\{j\}}$ as its building blocks,

$$M_{n,d} := (\mathbf{s}_{\{k\}+\{j\}})_{k,j=0}^n \quad \text{and} \quad \Delta_{n,d} := \det M_{n,d}. \tag{2.2.2}$$

The elements of $M_{n,d}$ are $s_{\alpha+\beta}$ for $|\alpha| \le n$ and $|\beta| \le n$.

Theorem 2.2.1 (Dunkl and Xu, 2014, Theorem 3.2.6) *Let $\mathcal{L}$ be a moment functional. The corresponding inner product is nondegenerate if and only if $\Delta_{n,d} \neq 0$ for all $n \in \mathbb{N}_0$.*

From now on in this section we assume the above nondegeneracy condition. Then orthogonal bases of $\mathcal{V}_n^d$ exist. A special, usually not orthogonal, basis can be expressed in terms of moments $\mathcal{L}$ as follows. For $\alpha \in \mathbb{N}_0^d$ we denote by $\mathbf{s}_{\alpha,k}$ the column vector $\mathbf{s}_{\alpha,k} := \mathcal{L}(x^\alpha \mathbf{x}^k)$; in particular, $\mathbf{s}_{\alpha,0} = s_\alpha$.

Theorem 2.2.2 (Dunkl and Xu, 2014, Theorem 3.2.12) *For $|\alpha| = n$, the polynomials*

$$P_\alpha^n(x) := \frac{1}{\Delta_{n-1,d}} \det \left[\begin{array}{c|c} & \mathbf{s}_{\alpha,0} \\ & \mathbf{s}_{\alpha,1} \\ M_{n-1,d} & \vdots \\ & \mathbf{s}_{\alpha,n-1} \\ \hline 1, \mathbf{x}^{\mathrm{tr}} \cdots (\mathbf{x}^{n-1})^{\mathrm{tr}} & x^\alpha \end{array}\right] \tag{2.2.3}$$

form a basis for the space $\mathcal{V}_n^d$ of orthogonal polynomials of degree n with respect to $\mathcal{L}$.

The polynomial P_α^n can also be characterized as the unique polynomial in $\mathcal{V}_n^d$ of the form

$$P_\alpha^n(x) = x^\alpha + Q_{n-1}(x), \quad Q_{n-1} \in \Pi_{n-1}^d.$$

It is evident that a polynomial thus characterized exists. Because of the leading term x^α the basis $\{P_\alpha^n\}_{|\alpha|=n}$ is called a *monic* or *monomial* basis of orthogonal polynomials.

If $\mathcal{L}(p^2) > 0$ for all nonzero polynomials p, then the moment functional $\mathcal{L}$ is called *positive definite*. In that case the corresponding inner product $\langle f, g\rangle = \mathcal{L}(fg)$ is also positive definite and orthonormal bases of polynomials with respect to $\mathcal{L}$ will exist.

Theorem 2.2.3 (Dunkl and Xu, 2014, Lemma 3.2.8) *If $\mathcal{L}$ is positive definite, then $\Delta_{n,d} > 0$ for all $n \in \mathbb{N}_0$.*

For a sequence of polynomials $\{P_\alpha \mid |\alpha| = n\}$, we denote by $\mathbb{P}_n$ the polynomial (column) vector

$$\mathbb{P}_n := (P_\alpha^n)_{|\alpha|=n} = (P^n_{\alpha^{(1)}}, \ldots, P^n_{\alpha^{(r_n)}})^{\mathrm{tr}}, \tag{2.2.4}$$

where $\alpha^{(1)}, \ldots, \alpha^{(r_n)}$ is the arrangement of elements in $\{\alpha \in \mathbb{N}_0^d : |\alpha| = n\}$ according to a fixed monomial order. We also regard $\mathbb{P}_n$ as a set of polynomials $\{P_\alpha^n \mid |\alpha| = n\}$. Many properties of orthogonal polynomials of several variables can be expressed more compactly in terms of $\mathbb{P}_n$. For example, orthonormality of $\{P_\alpha \mid \alpha \in \mathbb{N}_0^d\}$ with respect to $\langle\cdot,\cdot\rangle$ can be written as

$$\langle \mathbb{P}_n, \mathbb{P}_m\rangle = \delta_{m,n} I_{r_n^d},$$

where I_k denotes the identity matrix of size $k \times k$.

The rows of $M_{n,d}$ are indexed by $\{\alpha : |\alpha| \le n\}$. The row indexed by α is

$$(\mathbf{s}_{\alpha,0}^{\mathrm{tr}}, \mathbf{s}_{\alpha,1}^{\mathrm{tr}}, \ldots, \mathbf{s}_{\alpha,n}^{\mathrm{tr}}) = \mathcal{L}\left(x^\alpha, x^\alpha \mathbf{x}^{\mathrm{tr}}, \ldots, x^\alpha (\mathbf{x}^n)^{\mathrm{tr}}\right).$$

For $|\alpha| = n$, let $\widetilde{M}_\alpha(x)$ be the matrix obtained from $M_{n,d}$ by replacing the above row of index α by $\left(1, \mathbf{x}^{\mathrm{tr}}, \ldots, (\mathbf{x}^n)^{\mathrm{tr}}\right)$. Define

$$\widetilde{P}_\alpha(x) := \frac{1}{\Delta_{n,d}} \det \widetilde{M}_\alpha(x), \quad |\alpha| = n, \ \alpha \in \mathbb{N}_0^d.$$

Let $N_{n,d}$ denote the principal submatrix of the inverse matrix $M_{n,d}^{-1}$ of size $r_n^d \times r_n^d$ at the lower-right corner. Then $N_{n,d}$ is positive definite.

Theorem 2.2.4 (Dunkl and Xu, 2014, Theorem 3.2.13, (3.2.14)) *Let $\mathcal{L}$ be a positive definite moment functional. Then*

$$\mathbb{P}_n(x) := (N_{n,d})^{-\frac{1}{2}} \widetilde{\mathbb{P}}_n(x) = G_n \mathbf{x}^n + \cdots \tag{2.2.5}$$

consists of orthonormal polynomials. Furthermore, the matrix G_n is positive definite and

$$G_n = (N_{n,d})^{1/2}, \quad \det G_n = (\Delta_{n-1,d}/\Delta_{n,d})^{1/2}.$$

As mentioned before, the space of orthogonal polynomials $\mathcal{V}_n^d$ of degree n has many different bases. The orthonormal bases, however, are unique up to orthogonal transformations.

Theorem 2.2.5 (Dunkl and Xu, 2014, Theorem 3.2.14) *Let $\mathcal{L}$ be positive definite and let $\{Q_\alpha^n\}$ be a sequence of orthonormal polynomials forming a basis of $\mathcal{V}_n^d$. Then there is an orthogonal matrix O_n such that $\mathbb{Q}_n = O_n \mathbb{P}_n$, where $\mathbb{P}_n$ are the orthonormal polynomials defined in* (2.2.5).

Let $\mathcal{M} = \mathcal{M}(\mathbb{R}^d)$ denote the set of nonnegative Borel measures on $\mathbb{R}^d$ such that $\int_{\mathbb{R}^d} |x^\alpha| \, d\mu(x) < \infty$ for all $\alpha \in \mathbb{N}_0^d$. Each $\mu \in \mathcal{M}$ defines a positive linear functional

$$\mathcal{L} f = \int_{\mathbb{R}^d} f(x) \, d\mu(x), \qquad f \in \Pi^d. \tag{2.2.6}$$

A polynomial p is orthogonal with respect to the *orthogonality measure* μ if it is orthogonal with respect to $\mathcal{L}$ defined in (2.2.6). Thus, all theorems in this subsection apply to the orthogonal polynomials with respect to $d\mu$ for $\mu \in \mathcal{M}$.

On the other hand, not every positive definite linear functional admits an integral representation (2.2.6). The moment problem asks when a linear functional, defined via its moments, admits an integral representation (2.2.6) for a $\mu \in \mathcal{M}$ and, if so, when the measure will be determinate. Here a measure μ is called *determinate* if no other measure in $\mathcal{M}$ has all its moments equal to those of μ. It is known that $\mathcal{L}$ admits an integral representation if, and only if, $\mathcal{L}$ is *positive* in the sense that $\mathcal{L} p \geq 0$ for every nonnegative polynomial p. Evidently, a positive linear functional is necessarily positive semidefinite (i.e, $\mathcal{L} p^2 \geq 0$ for every polynomial p), which also holds sufficiently for $d = 1$. For $d > 1$, however, a positive definite linear functional may not be positive: there exist nonnegative polynomials that cannot be written as the sum of squared polynomials. For moment problems of several variables, including various sufficient conditions on a measure being determinate, see Berg (1987), Fuglede (1983), and Schmüdgen (1990) and references therein.

2.2.2 Three-Term Relations

Every system of orthogonal polynomials of one variable satisfies a three-term relation, which can also be used to compute orthogonal polynomials recursively. For orthogonal polynomials of several variables, an analogue of the three-term relation is stated in terms of $\mathcal{V}_n^d$, or rather, in terms of $\mathbb{P}_n$.

Theorem 2.2.6 (Dunkl and Xu, 2014, Theorem 3.3.1) *Let $\mathbb{P}_n$ denote a basis of $\mathcal{V}_n^d$, $n = 0, 1, \ldots$ and let $\mathbb{P}_{-1}(x) := 0$. Then there exist unique matrices $A_{n,i} \colon r_n^d \times r_{n+1}^d$, $B_{n,i} \colon r_n^d \times r_n^d$, and $C_{n,i} \colon r_n^d \times r_{n-1}^d$ for $n = 0, 1, 2, \ldots$, such that*

$$x_i \mathbb{P}_n(x) = A_{n,i} \mathbb{P}_{n+1}(x) + B_{n,i} \mathbb{P}_n(x) + C_{n,i} \mathbb{P}_{n-1}(x), \qquad 1 \leq i \leq d. \tag{2.2.7}$$

In fact, let $H_n := \mathcal{L}(\mathbb{P}_n \mathbb{P}_n^{\mathrm{tr}})$; then the orthogonality shows that

$$A_{n,i} H_{n+1} = \mathcal{L}(x_i \mathbb{P}_n \mathbb{P}_{n+1}^{\mathrm{tr}}), \quad B_{n,i} H_n = \mathcal{L}(x_i \mathbb{P}_n \mathbb{P}_n^{\mathrm{tr}}), \quad A_{n,i} H_{n+1} = H_n C_{n+1,i}^{\mathrm{tr}}.$$

The coefficient matrices $A_{n,i}$ and $C_{n,i}$ in (2.2.7) have full rank. Indeed,

$$\operatorname{rank} A_{n,i} = \operatorname{rank} C_{n+1,i} = r_n^d, \tag{2.2.8}$$

$$\operatorname{rank} A_n = \operatorname{rank} C_{n+1}^{\mathrm{tr}} = r_{n+1}^d, \tag{2.2.9}$$

where $A_n = (A_{n,1}^{\mathrm{tr}}, \ldots, A_{n,d}^{\mathrm{tr}})^{\mathrm{tr}}$ and $C_{n+1}^{\mathrm{tr}} = (C_{n+1,1}, \ldots, C_{n+1,d})^{\mathrm{tr}}$ are matrices of size $d r_n^d \times r_{n+1}^d$. In the case of orthonormal polynomials, H_n is the identity matrix and the three-term relation takes a simpler form.

Theorem 2.2.7 (Dunkl and Xu, 2014, Theorem 3.3.2) *If $\mathbb{P}_n$ is an orthonormal basis of $\mathcal{V}_n^d$, $n = 0, 1, \ldots$, then*

$$x_i \mathbb{P}_n(x) = A_{n,i}\mathbb{P}_{n+1}(x) + B_{n,i}\mathbb{P}_n(x) + A_{n-1,i}^{\mathsf{tr}}\mathbb{P}_{n-1}(x), \quad 1 \le i \le d, \tag{2.2.10}$$

where $A_{n,i} = \mathcal{L}(x_i \mathbb{P}_n \mathbb{P}_{n+1}^{\mathsf{tr}})$, $B_{n,i} = \mathcal{L}(x_i \mathbb{P}_n \mathbb{P}_n^{\mathsf{tr}})$, and $B_{n,i}$ is symmetric.

As an analogue of the classical Favard theorem for orthogonal polynomials of one variable, the three-term relation and the rank conditions characterize orthogonality.

Theorem 2.2.8 (Dunkl and Xu, 2014, Theorem 3.3.8) *Let $\{\mathbb{P}_n\}_{n=0}^{\infty} = \{P_\alpha^n \mid |\alpha| = n, n \in \mathbb{N}_0\}$, $\mathbb{P}_0 = 1$ be an arbitrary sequence in Π^d such that $\{P_\alpha^m \mid |\alpha| = m \le n\}$ spans Π_n^d for each n. Then the following statements are equivalent:*

1. *There exists a positive definite linear functional $\mathcal{L}$ that makes $\{\mathbb{P}_n\}_{n=0}^{\infty}$ an orthonormal basis for Π^d.*
2. *For $n \ge 0$, $1 \le i \le d$, there exist matrices $A_{n,i}$ and $B_{n,i}$ such that*
   (i) $x_i \mathbb{P}_n(x) = A_{n,i}\mathbb{P}_{n+1}(x) + B_{n,i}\mathbb{P}_n(x) + A_{n-1,i}^{\mathsf{tr}}\mathbb{P}_{n-1}(x)$, $1 \le i \le d$;
   (ii) $\operatorname{rank} A_{n,i} = r_n^d$, $1 \le i \le d$ *and* $\operatorname{rank} A_n = r_{n+1}^d$.

Unlike in one variable, the characterization does not conclude regarding the existence of an orthogonality measure.

The coefficient matrices of the three-term relation for orthonormal polynomials satisfy a set of *commutativity conditions* (Dunkl and Xu, 2014, Theorem 3.4.1): for $1 \le i, j \le d$ and $k \ge 0$,

$$\begin{aligned} A_{k,i}A_{k+1,j} &= A_{k,j}A_{k+1,i}, \\ A_{k,i}B_{k+1,j} + B_{k,i}A_{k,j} &= B_{k,j}A_{k,i} + A_{k,j}B_{k+1,i}, \\ A_{k-1,i}^{\mathsf{tr}}A_{k-1,j} + B_{k,i}B_{k,j} + A_{k,i}A_{k,j}^{\mathsf{tr}} &= A_{k-1,j}^{\mathsf{tr}}A_{k-1,i} + B_{k,i}B_{k,j} + A_{k,j}A_{k,i}^{\mathsf{tr}}, \end{aligned} \tag{2.2.11}$$

where $A_{-1,i} := 0$, which is derived from computing, say, $\mathcal{L}(x_i x_j \mathbb{P}_k \mathbb{P}_{k-1})$, by applying the three-term relation in two different ways. These coefficient matrices also define a family of tridiagonal matrices J_i, $1 \le i \le d$:

$$J_i = \begin{bmatrix} B_{0,i} & A_{0,i} & & \bigcirc \\ A_{0,i}^{\mathsf{tr}} & B_{1,i} & A_{1,i} & \\ & A_{1,i}^{\mathsf{tr}} & B_{2,i} & \ddots \\ \bigcirc & & \ddots & \ddots \end{bmatrix}, \quad 1 \le i \le d, \tag{2.2.12}$$

called *block Jacobi matrices*. The entries of the J_i are matrices that have increasing sizes going down the main diagonal. The commutativity relations (2.2.11) are equivalent to the formal commutativity of J_i (Dunkl and Xu, 2014, Lemma 3.4.4), that is,

$$J_i J_j = J_j J_i, \quad 1 \le i \ne j \le d.$$

These block Jacobi matrices can be viewed as the realization of the multiplication operators $X_1, \ldots, X_d$ defined on the space of polynomials by

$$(X_i f)(x) = x_i f(x), \quad 1 \le i \le d.$$

The operators can be extended to a family of commuting self-adjoint operators on an L^2 space. This connection to operator theory allows the use of the spectral theory of commuting self-adjoint operators, and helps to answer the question of when the inner product, with respect to which the polynomials are orthogonal, is defined by a measure. It gives, for example, the following theorem, which strengthens Theorem 2.2.8:

Theorem 2.2.9 (Dunkl and Xu, 2014, Theorem 3.4.7) *Let $\{\mathbb{P}_n\}_{n=0}^{\infty} = \{P_\alpha^n \mid |\alpha| = n, n \in \mathbb{N}_0\}$, $\mathbb{P}_0 = 1$ be an arbitrary sequence in Π^d such that $\{P_\alpha^m \mid |\alpha| = m \le n\}$ spans Π_n^d for each n. Then the following statements are equivalent:*

1. *There exists a determinate measure $\mu \in \mathcal{M}$ with compact support in $\mathbb{R}^d$ such that $\{\mathbb{P}_n\}_{n=0}^{\infty}$ is orthonormal with respect to $d\mu$.*
2. *Statement 2 in Theorem 2.2.8 holds, together with*

$$\sup_{k\ge 0} \|A_{k,i}\|_2 < \infty \quad \text{and} \quad \sup_{k \ge 0} \|B_{k,i}\|_2 < \infty, \quad 1 \le i \le d. \tag{2.2.13}$$

If the measure $\mu \in \mathcal{M}$ is given by $d\mu(x) = W(x)\,dx$, W being a nonnegative measurable function, we call W a *weight function*. Let $\Omega \subset \mathbb{R}^d$ be the support set of W. A function W is called *centrally symmetric* if

$$x \in \Omega \Rightarrow -x \in \Omega \quad \text{and} \quad W(x) = W(-x) \text{ a.e.}$$

For example, the product weight function $\prod_{i=1}^d (1 - x_i)^{a_i}(1 + x_i)^{b_i}$ on the cube $[-1, 1]^d$ is centrally symmetric if and only if $a_i = b_i$. Furthermore, a linear functional $\mathcal{L}$ is called *centrally symmetric* if

$$\mathcal{L}(x^\alpha) = 0, \quad \alpha \in \mathbb{N}^d, \quad |\alpha| \text{ is an odd integer.}$$

The two notions are equivalent when $\mathcal{L}$ is given by $\mathcal{L}f = \int fW\,dx$.

Theorem 2.2.10 (Dunkl and Xu, 2014, Theorem 3.3.10) *Let $\mathcal{L}$ be a positive definite linear functional. Then $\mathcal{L}$ is centrally symmetric if and only if $B_{n,i} = 0$ for all $n \in \mathbb{N}_0$ and $1 \le i \le d$, where $B_{n,i}$ are given in* (2.2.10). *Furthermore, if $\mathcal{L}$ is centrally symmetric, then an orthogonal polynomial of degree n is the sum of monomials of even degrees if n is even, and the sum of monomials of odd degrees if n is odd.*

In one variable, the three-term relation can be used as a recurrence formula for computing orthogonal polynomials of one variable. For several variables, let $D_n^{\mathrm{tr}} = (D_{n,1}^{\mathrm{tr}}, \ldots, D_{n,d}^{\mathrm{tr}})$, where $D_{n,i}^{\mathrm{tr}}: r_{n+1}^d \times r_n^d$ is a matrix that satisfies

$$D_n^{\mathrm{tr}} A_n = \sum_{i=1}^d D_{n,i}^{\mathrm{tr}} A_{n,i} = I_{r_{n+1}^d}.$$

Such a matrix is not unique. The three-term relation (2.2.10) implies

$$\mathbb{P}_{n+1} = \sum_{i=1}^{d} x_i D_{n,i}^{\mathrm{tr}} \mathbb{P}_n + E_n \mathbb{P}_n + F_n \mathbb{P}_{n-1}, \tag{2.2.14}$$

where $E_n := -\sum_{i=1}^{d} D_{n,i}^{\mathrm{tr}} B_{n,i}$ and $F_n := -\sum_{i=1}^{d} D_{n,i}^{\mathrm{tr}} A_{n-1,i}^{\mathrm{tr}}$.

Given two sequences of matrices $A_{n,i}$ and $B_{n,i}$, (2.2.14) can be used as a recursive relation to generate a sequence of polynomials. These polynomials are orthogonal if the matrices satisfy certain relations:

Theorem 2.2.11 (Dunkl and Xu, 2014, Theorem 3.5.1) *Let $\{\mathbb{P}_n\}_{n=0}^{\infty}$ be defined by* (2.2.14). *Then there is a positive definite linear functional $\mathcal{L}$ that makes $\{\mathbb{P}_n\}_{n=0}^{\infty}$ an orthonormal basis for Π^d if and only if $B_{k,i}$ are symmetric, $A_{k,i}$ satisfy the rank conditions* (2.2.8) *and* (2.2.9), *and together they satisfy the commutativity conditions* (2.2.11).

Further results and references The idea of studying orthogonal polynomials of several variables in terms of $\mathcal{V}_0^d, \mathcal{V}_1^d, \ldots$ goes back to Krall and Sheffer (1967). The vector notion of the three-term relation and Favard's theorem were initiated by M. Kowalski (1982b; 1982a). The versions in this section were developed by Xu (1993, 1994b) and subsequent papers. Another earlier work is Gekhtman and Kalyuzhny (1994). See Dunkl and Xu (2014, Chapter 3) for references and further results.

For further study of three-term relations, see Cichoń et al. (2005). For an approach based on matrix factorization, see Ariznabarreta and Mañas (2016). Three-term relations are used for evaluating orthogonal polynomials of several variables in Barrio et al. (2010).

2.2.3 Zeros of Orthogonal Polynomials of Several Variables

An orthogonal polynomial of degree n in one variable has n distinct real zeros and the zeros are nodes of a Gaussian quadrature formula. For a polynomial of several variables, its set of zeros is an algebraic variety, an intrinsically difficult object. The correct notion for orthogonal polynomials of several variables is the common zeros of a family of polynomials, such as $\mathbb{P}_n$.

Let $\mathbb{P}_n$ be an orthonormal basis of $\mathcal{V}_n^d$. A point $x \in \mathbb{R}^d$ is a *zero* of $\mathbb{P}_n$ if it is a zero for every element in $\mathbb{P}_n$ (or all elements in $\mathcal{V}_n^d$), and it is a *simple zero* if at least one partial derivative of $\mathbb{P}_n$ is not zero at x. Let $A_{n,i}$ and $B_{n,i}$ be matrices in (2.2.10). For each $n \in \mathbb{N}_0$, define the *truncated block Jacobi matrices*

$$J_{n,i} := \begin{bmatrix} B_{0,i} & A_{0,i} & & & \bigcirc \\ A_{0,i}^{\mathrm{tr}} & B_{1,i} & A_{1,i} & & \\ & \ddots & \ddots & \ddots & \\ & & A_{n-3,i}^{\mathrm{tr}} & B_{n-2,i} & A_{n-2,i} \\ \bigcirc & & & A_{n-2,i}^{\mathrm{tr}} & B_{n-1,i} \end{bmatrix}, \qquad 1 \le i \le d.$$

These are symmetric matrices of order $N = \dim \Pi^d_{n-1}$. An element $\lambda \in \mathbb{R}^d$ is called a *joint eigenvalue* of $J_{n,1}, \ldots, J_{n,d}$ if there is a $\xi \neq 0$, $\xi \in \mathbb{R}^N$ such that $J_{n,i}\xi = \lambda_i \xi$ for $i = 1, \ldots, d$; the vector ξ is called a *joint eigenvector*.

Theorem 2.2.12 (Dunkl and Xu, 2014, Theorem 3.7.2) *A point $\lambda \in \mathbb{R}^d$ is a common zero of $\mathbb{P}_n$ if and only if it is a joint eigenvalue of $J_{n,1}, \ldots, J_{n,d}$; moreover, $(\mathbb{P}_0^{\mathrm{tr}}(\lambda), \ldots, \mathbb{P}_{n-1}^{\mathrm{tr}}(\lambda))^{\mathrm{tr}}$ is a joint eigenvector of λ.*

Many properties of zeros of $\mathbb{P}_n$ can be derived from this characterization.

Theorem 2.2.13 (Dunkl and Xu, 2014, Theorems 3.7.1, 3.7.5; Corollaries 3.7.3, 3.7.4) *All zeros of $\mathbb{P}_n$ are real, distinct, and simple. Further, $\mathbb{P}_n$ has at most $\dim \Pi^d_{n-1}$ distinct zeros, and $\mathbb{P}_n$ has $\dim \Pi^d_{n-1}$ zeros if and only if*

$$A_{n-1,i}A^{\mathrm{tr}}_{n-1,j} = A_{n-1,j}A^{\mathrm{tr}}_{n-1,i}, \quad 1 \leq i, j \leq d. \tag{2.2.15}$$

Theorem 2.2.14 (Dunkl and Xu, 2014, Corollary 3.7.7; Xu, 1994a, Theorem 3.1.1) *If L is centrally symmetric and $d \geq 2$, then $\mathbb{P}_n$ has fewer than $\dim \Pi^d_{n-1}$ common zeros.*
If, moreover, $d = 2$, then $\mathbb{P}_n$ has no zero if n is even and has one zero ($x = 0$) if n is odd.

As in the case of one variable, zeros of orthogonal polynomials are closely related to cubature formulas, which are finite sums that approximate integrals. A *cubature formula* $\mathcal{I}_n(f)$ is said to have degree $2n - 1$ if

$$\int_{\mathbb{R}^d} f(x)\, d\mu(x) = \sum_{k=1}^{N} \lambda_k f(x_k) =: \mathcal{I}_n(f) \quad \forall\, f \in \Pi^d_{2n-1}, \tag{2.2.16}$$

where $\lambda_k \in \mathbb{R}$ and $x_k \in \mathbb{R}^d$, and there is a polynomial $f^* \in \Pi^d_{2n}$ for which the equality does not hold. The number of nodes N of (2.2.16) satisfies a lower bound

$$N \geq \dim \Pi^d_{n-1}. \tag{2.2.17}$$

A cubature formula of degree $2n - 1$ with $\dim \Pi^d_{n-1}$ nodes is called *Gaussian*.

Theorem 2.2.15 (Dunkl and Xu, 2014, Theorem 3.8.4) *Let $\mu \in \mathcal{M}$ and $\mathbb{P}_n$ be an orthogonal basis of $\mathcal{V}^d_n$ with respect to $d\mu$. Then the integral $\int f(x)\, d\mu(x)$ admits a Gaussian cubature formula of degree $2n - 1$ if and only if $\mathbb{P}_n$ has $\dim \Pi^d_{n-1}$ common zeros.*

When combined with Theorem 2.2.14, this shows the following:

Corollary 2.2.16 *If μ is centrally symmetric, then no Gaussian cubature formulas exist.*

On the other hand, there are two families of weight functions, discussed in §2.9.1, for which Gaussian cubature formulas do exist.

The nonexistence of Gaussian cubature means that the inequality (2.2.17) is not sharp. There is, in fact, an improved lower bound, which we only give here for $d = 2$.

Theorem 2.2.17 (Xu, 1994a, (1.2.10) and Ch. 5) *Let $\mu \in \mathcal{M}(\mathbb{R}^2)$. Then a cubature formula of degree $2n - 1$ for $d\mu$ exists only if*

$$N \geq \dim \Pi^2_{n-1} + \tfrac{1}{2} \operatorname{rank}\left(A_{n-1,1}A^{\mathrm{tr}}_{n-1,2} - A_{n-1,2}A^{\mathrm{tr}}_{n-1,1}\right). \tag{2.2.18}$$

In particular, if μ is centrally symmetric, then (2.2.18) *specializes to*

$$N \geq \dim \Pi_{n-1}^2 + \left\lfloor \frac{n}{2} \right\rfloor. \tag{2.2.19}$$

The bounds (2.2.18), (2.2.19) were first obtained by Möller (1973, 1976). For general d see Möller (1979) and Xu (1994a, Chapter 5). The bound presented there for μ centrally symmetric is, for $d > 2$, sharper than the analogue of (2.2.18).

The condition under which the lower bound in (2.2.19) is attained is determined as follows.

Theorem 2.2.18 (Xu, 1994a, Theorems 4.1.4 and 5.3.1) *Let μ be centrally symmetric. A cubature formula of degree $2n-1$ attains the lower bound* (2.2.19) *if and only if its nodes are common zeros of $n+1-\lfloor \frac{n}{2} \rfloor$ orthogonal polynomials in $\mathcal{V}_n^2$.*

Further results and references The lower bound (2.2.17) is classical; see Stroud (1971) and Mysovskikh (1981). Theorem 2.2.15 was first proved in Mysovskikh (1976). The first example of a cubature formula that attains (2.2.19) was a degree 5 formula on the square constructed by Radon (1948). At the moment, the only weight functions for which (2.2.19) is attained for all n are given by the weight functions $W(x,y) = |x-y|^{2\alpha+1}|x+y|^{2\beta+1}(1-x^2)^{\pm 1/2}(1-y^2)^{\pm 1/2}$, where $\alpha, \beta \geq -1/2$, of which the case $\alpha = \beta = 1/2$ is classical (Morrow and Patterson, 1978; Xu, 1994a) and the general case is far more recent (Xu, 2012).

2.2.4 Reproducing Kernel and Fourier Orthogonal Expansion

Let $\mu \in \mathcal{M}$ and assume that the space of polynomials is dense in $L^2(d\mu)$. Define the projection operator $\operatorname{proj}_n : L^2(d\mu) \to \mathcal{V}_n^d$ by

$$(\operatorname{proj}_n f)(x) := \int_{\mathbb{R}^d} f(y) P_n(x,y)\, d\mu(y), \tag{2.2.20}$$

where $P_n(\cdot,\cdot)$ is the *reproducing kernel* of $\mathcal{V}_n^d$ satisfying

$$\int_{\mathbb{R}^d} f(y) P_n(x,y)\, d\mu(y) = \begin{cases} f(x), & f \in \mathcal{V}_n^d, \\ 0, & f \in \mathcal{V}_m^d,\ m \neq n. \end{cases} \tag{2.2.21}$$

Let $\{P_\alpha^n\}_{|\alpha|=n}$ be an orthonormal basis of $\mathcal{V}_n^d$. Then $P_n(\cdot,\cdot)$ can be expressed as

$$P_n(x,y) = \sum_{|\alpha|=n} P_\alpha^n(x) P_\alpha^n(y) = \mathbb{P}_n^{\mathrm{tr}}(x)\mathbb{P}_n(y). \tag{2.2.22}$$

The projection operator is independent of a particular basis, and so is the reproducing kernel, as also seen by Theorem 2.2.5. The standard Hilbert space argument shows that $f \in L^2(d\mu)$ has a Fourier orthogonal expansion

$$f = \sum_{n=0}^{\infty} \operatorname{proj}_n f \quad \forall f \in L^2(d\mu). \tag{2.2.23}$$

In terms of the orthonormal basis $\{P^n_\alpha\}$, the orthogonal expansion reads

$$f = \sum_{n=0}^{\infty} \sum_{|\alpha|=n} a^n_\alpha(f) P^n_\alpha \quad \text{with } a^n_\alpha(f) := \int_{\mathbb{R}^d} f(x) P^n_\alpha(x)\, d\mu(x). \tag{2.2.24}$$

For studying Fourier expansions, it is often important to have a closed formula for the kernel $P_n(\cdot,\cdot)$. Such formulas are often available for classical orthogonal polynomials in several variables.

The nth *partial sum* $S_n f$ of the Fourier orthogonal expansion of $f \in L^2(d\mu)$ is defined by

$$(S_n f)(x) := \sum_{k=0}^{n} (\operatorname{proj}_k f)(x) = \int_{\mathbb{R}^d} K_n(x,y) f(y)\, d\mu(y), \tag{2.2.25}$$

where the kernel $K_n(\cdot,\cdot)$ is defined by

$$K_n(x,y) := \sum_{k=0}^{n} \sum_{|\alpha|=k} P^k_\alpha(x) P^k_\alpha(y) = \sum_{k=0}^{n} P_k(x,y). \tag{2.2.26}$$

The kernel $K_n(\cdot,\cdot)$ is the reproducing kernel of Π^d_n in $L^2(d\mu)$. It satisfies a *Christoffel–Darboux formula*, deduced from the three-term relation.

Theorem 2.2.19 (Xu, 1993, Theorem 3; Dunkl and Xu, 2014, Theorem 3.6.3) *Let $\mathcal{L}$ be a positive definite linear functional, and let $\{\mathbb{P}_k\}_{k=0}^\infty$ be a sequence of orthonormal polynomials with respect to $\mathcal{L}$. Then, for any integer $n \ge 0$, $x, y \in \mathbb{R}^d$,*

$$K_n(x,y) = \frac{[A_{n,i}\mathbb{P}_{n+1}(x)]^{\mathrm{tr}}\mathbb{P}_n(y) - \mathbb{P}_n^{\mathrm{tr}}[A_{n,i}\mathbb{P}_{n+1}(y)]}{x_i - y_i}, \quad 1 \le i \le d, \tag{2.2.27}$$

where $x = (x_1, \ldots, x_d)$ and $y = (y_1, \ldots, y_d)$.

The right-hand side of (2.2.27) can also be stated in terms of orthogonal, instead of orthonormal, polynomials. A related function is the *Christoffel function* defined by

$$\Lambda_n(x) := [K_n(x,x)]^{-1}. \tag{2.2.28}$$

Theorem 2.2.20 (Dunkl and Xu, 2014, Theorem 3.6.6) *Let $\mu \in \mathcal{M}$. Then for any $x \in \mathbb{R}^d$,*

$$\Lambda_n(x) = \min_{P(x)=1,\ P \in \Pi^d_n} \int_{\mathbb{R}^d} P^2(y)\, d\mu(y).$$

2.3 Orthogonal Polynomials of Two Variables

Almost all that can be stated about the general properties of orthogonal polynomials of two variables also holds for orthogonal polynomials of more than two variables. This section contains results on various special systems of orthogonal polynomials of two variables and their properties.

A basis of $\mathcal{V}_n^2$ in two variables is often indexed by $\alpha = (k, n-k)$, or by a single index k, as $\{P_k^n\}_{k=0}^n$. Many examples below will be given in terms of classical orthogonal polynomials of one variable, which are listed, together with their associated weight function, orthogonality interval, and parameter constraints, in Table 2.1. The normalization given in the last column (here usually the value attained at an endpoint of the orthogonality interval) makes the definition precise. The notation $(a)_n := a(a+1)\cdots(a+n-1)$, shifted factorial, or Pochhammer symbol, will be used throughout the rest of the chapter.

Table 2.1 *Classical orthogonal polynomials of one variable*

Name	Notation	Weight	Interval	Constraint	Normalization
Hermite	$H_n(t)$	e^{-t^2}	$(-\infty,\infty)$		$H_n(t) = 2^n t^n + \cdots$
Laguerre	$L_n^\alpha(t)$	$t^\alpha e^{-t}$	$[0,\infty)$	$\alpha > -1$	$L_n^\alpha(0) = \frac{(\alpha+1)_n}{n!}$
Chebyshev 1st	$T_n(t)$	$(1-t^2)^{-1/2}$	$[-1,1]$		$T_n(1) = 1$
Chebyshev 2nd	$U_n(t)$	$(1-t^2)^{1/2}$	$[-1,1]$		$U_n(1) = n+1$
Gegenbauer	$C_n^\lambda(t)$	$(1-t^2)^{\lambda-1/2}$	$[-1,1]$	$\lambda > -1/2$	$C_n^\lambda(1) = \frac{(2\lambda)_n}{n!}$
Jacobi	$P_n^{(\alpha,\beta)}(t)$	$(1-t)^\alpha(1+t)^\beta$	$[-1,1]$	$\alpha,\beta > -1$	$P_n^{(\alpha,\beta)}(1) = \frac{(\alpha+1)_n}{n!}$

The Gegenbauer case $\lambda = 0$ can be obtained for $n > 0$ by the renormalization

$$\lim_{\lambda\to 0} \lambda^{-1} C_n^\lambda(x) = 2n^{-1} T_n(x), \qquad n > 0. \tag{2.3.1}$$

2.3.1 Product Orthogonal Polynomials

For the weight function $W(x,y) = w_1(x)w_2(y)$, where w_1 and w_2 are two weight functions of one variable, an orthogonal basis of $\mathcal{V}_n^d$ is given by

$$P_k^n(x,y) := p_k(x)q_{n-k}(y), \qquad 0 \le k \le n,$$

where $\{p_k\}$ and $\{q_k\}$ are sequences of orthogonal polynomials with respect to w_1 and w_2, respectively. If $\{p_k\}$ and $\{q_k\}$ are orthonormal, then so is $\{P_k^n\}$.

1. *Product Hermite polynomials.* For weight function $W(x,y) = e^{-x^2-y^2}$, a possible orthogonal basis is given by $P_k^n(x,y) = H_k(x)H_{n-k}(y)$, $0 \le k \le n$. This satisfies the differential equation

$$\tfrac{1}{2}(v_{xx} + v_{yy}) - (xv_x + yv_y) = -nv. \tag{2.3.2}$$

2. *Product Laguerre polynomials.* For weight function $W(x,y) = x^\alpha y^\beta e^{-x-y}$, a possible orthogonal basis is given by $P_k^n(x,y) = L_k^\alpha(x)L_{n-k}^\alpha(y)$, $0 \le k \le n$. This satisfies the differential equation

$$xv_{xx} + yv_{yy} + (1+\alpha-x)v_x + (1+\beta-y)v_y = -nv. \tag{2.3.3}$$

3. *Product Hermite–Laguerre polynomials.* For weight function $W(x, y) = y^\alpha e^{-x^2-y}$, a possible orthogonal basis is given by $P_k^n(x, y) = H_k(x)L_{n-k}^\alpha(y)$, $0 \le k \le n$. This satisfies the differential equation

$$\tfrac{1}{2}v_{xx} + yv_{yy} - xv_x + (1 + \alpha - y)v_y = -nv. \tag{2.3.4}$$

There are other bases and further results for these product weight functions. These three cases are the only product-type orthogonal polynomials that are eigenfunctions of a second-order differential operator with eigenvalues depending only on n. See §2.3.4 for further results.

2.3.2 Orthogonal Polynomial on the Unit Disk

On the unit disk $B^2 := \{(x, y) \in \mathbb{R}^2 \mid x^2 + y^2 \le 1\}$ consider the weight function

$$W_\mu(x, y) := \frac{2\mu + 1}{2\pi}(1 - x^2 - y^2)^{\mu-\frac{1}{2}}, \quad \mu > -1/2, \tag{2.3.5}$$

normalized such that its integral over B^2 is 1. There are several distinct bases of $\mathcal{V}_n^2$ that can be given explicitly.

1. *First orthonormal basis* This is the basis $\{P_k^n\}_{k=0}^n$ of $\mathcal{V}_n^2$ defined by

$$P_k^n(x, y) := (h_{k,n})^{-1} C_{n-k}^{k+\mu+\frac{1}{2}}(x)(1 - x^2)^{\frac{k}{2}} C_k^\mu\left(\frac{y}{\sqrt{1 - x^2}}\right), \tag{2.3.6}$$

$$[h_{k,n}]^2 := \frac{(2k + 2\mu + 1)_{n-k}(2\mu)_k(\mu)_k(\mu + \frac{1}{2})}{(n - k)!\,k!\,(\mu + \frac{1}{2})_k(n + \mu + \frac{1}{2})}, \tag{2.3.7}$$

where the case $\mu = 0$ can be obtained as a limit for $\mu \to 0$ after dividing, for $k > 0$, C_k^μ and $h_{k,n}$ by μ and by using (2.3.1).

2. *Second orthonormal basis* In polar coordinates $(x, y) = (r\cos\theta, r\sin\theta)$, let

$$\begin{aligned} P_{j,1}(x, y) &:= [h_{j,n}]^{-1} P_j^{(\mu-\frac{1}{2},n-2j)}(2r^2 - 1)r^{n-2j}\cos((n - 2j)\theta), \quad 1 \le j \le n/2, \\ P_{j,2}(x, y) &:= [h_{j,n}]^{-1} P_j^{(\mu-\frac{1}{2},n-2j)}(2r^2 - 1)r^{n-2j}\sin((n - 2j)\theta), \quad 1 \le j < n/2, \end{aligned} \tag{2.3.8}$$

$$[h_{j,n}]^2 := \frac{(\mu + \frac{1}{2})_j(n - j)!\,(n - j + \mu + \frac{1}{2})}{j!\,(\mu + \frac{3}{2})_{n-j}(n + \mu + \frac{1}{2})} \times \begin{cases} 2, & n \ne 2j, \\ 1, & n = 2j. \end{cases} \tag{2.3.9}$$

Then $\{P_{j,1}\}_{j=0}^{\lfloor n/2\rfloor} \cup \{P_{j,2}\}_{j=0}^{\lfloor (n-1)/2\rfloor}$ is an orthonormal basis of $\mathcal{V}_n^2$.

3. *Appell's biorthogonal polynomials* These are two families $\{U_k^n\}_{k=0}^n$ and $\{V_k^n\}_{k=0}^n$ of bases of $\mathcal{V}_n^2$ that satisfy

$$\int_{B^2} U_k^n(x, y)V_j^n(x, y)W_\mu(x, y)\,dx\,dy = h_k^n\delta_{k,j}, \quad 0 \le k, j \le n.$$

The first basis is defined via the *Rodrigues-type formula*

$$U_k^n(x, y) := \frac{(-1)^n(2\mu)_n}{2^n(\mu + \frac{1}{2})_n n!}(1 - x^2 - y^2)^{-\mu+\frac{1}{2}} \frac{\partial^n}{\partial x^k\,\partial y^{n-k}}((1 - x^2 - y^2)^{n+\mu-\frac{1}{2}}).$$

The second basis is monic, up to constant factors: $V_k^n(x,y) = \text{const.}\, x^k y^{n-k} + $ polynomial of degree $< n$. For an explicit expression of $V_k^n(x,y)$ and further properties of these two bases see §2.5.1, where they are given in the setting of the d-dimensional ball.

4. *An orthonormal basis of ridge polynomials for $\mu = 1/2$* Let

$$P_k^n(x,y) = \pi^{-1/2} U_n\left(x \cos \tfrac{k\pi}{n+1} + y \sin \tfrac{k\pi}{n+1}\right), \quad 0 \le k \le n. \tag{2.3.10}$$

Then $\{P_k^n\}_{k=0}^n$ is an orthonormal basis of $\mathcal{V}_n^2$ for the weight function $W_{1/2}(x,y) = \frac{1}{\pi}$ on B^2.

Differential equation All orthogonal polynomials of degree n for W_μ are eigenfunctions of a second-order differential operator. For $n \ge 0$,

$$(1-x^2)v_{xx} - 2xyv_{xy} + (1-y^2)v_{yy} - (2\mu+2)(xv_x + yv_y) = -n(n+2\mu+1)v, \quad v \in \mathcal{V}_n^2. \tag{2.3.11}$$

Further results and references For further properties, such as a closed formula for the reproducing kernel and convergence of orthogonal expansions, see §2.5.1, where the disk will be a special case ($d = 2$) of the d-dimensional ball. If the complex plane $\mathbb{C}$ is identified with $\mathbb{R}^2$, then the basis (2.3.8) can be written in variables $z = x + iy$ and $\bar{z} = x - iy$; see §2.3.5.

The first orthonormal basis goes back as far as Hermite and was studied in Appell and Kampé de Fériet (1926). Biorthogonal polynomials were studied in detail in Appell and Kampé de Fériet (1926); see also Erdélyi (1953). The basis of ridge polynomials in (2.3.10) was first discovered by Logan and Shepp (1975) (see also Xu, 2000) and it plays an important role in computer tomography (Logan and Shepp, 1975; Marr, 1974; Xu, 2006b). For further studies on the orthogonal polynomials on the disk, see Waldron (2008) and Wünsche (2005).

2.3.3 Orthogonal Polynomials on the Triangle

On the triangle $T^2 := \{(x,y) \mid 0 \le x, y, x+y \le 1\}$ consider an analogue of the Jacobi weight function

$$W_{\alpha,\beta,\gamma}(x,y) := \frac{\Gamma(\alpha+\beta+\gamma+\frac{3}{2})}{\Gamma(\alpha+\frac{1}{2})\Gamma(\beta+\frac{1}{2})\Gamma(\gamma+\frac{1}{2})} x^{\alpha-\frac{1}{2}} y^{\beta-\frac{1}{2}} (1-x-y)^{\gamma-\frac{1}{2}}, \quad \alpha,\beta,\gamma > -1/2, \tag{2.3.12}$$

normalized such that its integral over T^2 equals 1. Several distinct bases of $\mathcal{V}_n^2$ can be given explicitly.

1. *An orthonormal basis* $\{P_k^n\}_{k=0}^n$ of $\mathcal{V}_n^2$ with

$$P_k^n(x,y) := [h_{k,n}]^{-1} P_{n-k}^{(2k+\beta+\gamma,\alpha-\frac{1}{2})}(2x-1)(1-x)^k P_k^{(\gamma-\frac{1}{2},\beta-\frac{1}{2})}\left(\frac{2y}{1-x} - 1\right), \tag{2.3.13}$$

$$[h_{k,n}]^2 := \frac{(\alpha+\frac{1}{2})_{n-k}(\beta+\frac{1}{2})_k(\gamma+\frac{1}{2})_k(\beta+\gamma+1)_{n+k}}{(n-k)!\,k!\,(\beta+\gamma+1)_k(\alpha+\beta+\gamma+\frac{3}{2})_{n+k}} \frac{(n+k+\alpha+\beta+\gamma+\frac{1}{2})(k+\beta+\gamma)}{(2n+\alpha+\beta+\gamma+\frac{1}{2})(2k+\beta+\gamma)}.$$

Parametrizing the triangle differently leads to two more orthonormal bases. Indeed, denote the P_k^n in (2.3.13) by $P_{k,n}^{\alpha,\beta,\gamma}$ and define

$$Q_k^n(x,y) := P_{k,n}^{\gamma,\beta,\alpha}(1-x-y, y) \quad \text{and} \quad R_k^n(x,y) := P_{k,n}^{\alpha,\gamma,\beta}(x, 1-x-y).$$

Then $\{Q_k^n\}_{k=0}^n$ and $\{R_k^n\}_{k=0}^n$ are also orthonormal bases.

2. *Biorthogonal polynomials including Appell polynomials* A basis $\{U_k^n\}_{k=0}^n$ of $\mathcal{V}_n^2$ due to Appell is defined via the Rodrigues-type formula

$$U_k^n(x,y) := x^{-\alpha+\frac{1}{2}} y^{-\beta+\frac{1}{2}} (1-x-y)^{-\gamma+\frac{1}{2}} \frac{\partial^n}{\partial x^k \, \partial y^{n-k}} \left(x^{k+\alpha-\frac{1}{2}} y^{n-k+\beta-\frac{1}{2}} (1-x-y)^{n+\gamma-\frac{1}{2}} \right). \tag{2.3.14}$$

Biorthogonal to this basis is a basis $\{V_k^n\}_{k=0}^n$ of $\mathcal{V}_n^2$:

$$\int_{B^2} U_k^n(x,y) V_j^n(x,y) W_{\alpha,\beta,\gamma}(x,y)\, dx\, dy = h_k^n \delta_{k,j}, \qquad 0 \le k, j \le n.$$

This is a monic basis, up to constant factors: $V_k^n(x,y) = \text{const.}\, x^k y^{n-k}$ + polynomial of degree $< n$. For an explicit expression of $V_k^n(x,y)$ and further properties of these two bases see §2.5.2, where they are given in the setting of the d-dimensional simplex.

Differential equation Orthogonal polynomials of degree n with respect to $W_{\alpha,\beta,\gamma}$ are eigenfunctions of a second-order differential operator. For $n \ge 0$,

$$\begin{aligned} x(1-x)v_{xx} - 2xyv_{xy} + y(1-y)v_{yy} - \left((\alpha+\beta+\gamma+\tfrac{3}{2})x - (\alpha+\tfrac{1}{2})\right) v_x \\ - \left((\alpha+\beta+\gamma+\tfrac{3}{2})y - (\beta+\tfrac{1}{2})\right) v_y = -n\left(n+\alpha+\beta+\gamma+\tfrac{1}{2}\right) v, \quad v \in \mathcal{V}_n^2. \end{aligned} \tag{2.3.15}$$

Further results and references For further properties, such as biorthogonal polynomials, a closed formula for the reproducing kernel, and convergence of orthogonal expansions, see §2.5.2, where the triangle will be the special case $d = 2$ of the d-dimensional simplex.

The orthonormal polynomials in (2.3.13) were first introduced by Proriol (1957) (see also Koornwinder, 1975); the case $\alpha = \beta = \gamma = 0$ became known as Dubiner's polynomials much later in the finite element community (see Dubiner, 1991). Appell polynomials were studied in detail by Appell and Kampé de Fériet (1926). See §2.5.2 for further references on orthogonal polynomials on the simplex.

2.3.4 *Differential Equations and Orthogonal Polynomials of Two Variables*

A linear second-order partial differential operator

$$L := A(x,y)\partial_1^2 + 2B(x,y)\partial_1\partial_2 + C(x,y)\partial_2^2 + D(x,y)\partial_1 + E(x,y)\partial_2, \tag{2.3.16}$$

where $\partial_1 := \frac{\partial}{\partial x}$ and $\partial_2 := \frac{\partial}{\partial y}$, is called *admissible* if for each nonnegative integer n there exists a number λ_n such that the equation

$$Lu = \lambda_n u$$

has $n+1$ linearly independent solutions of polynomials of degree n and has no nonzero solutions of polynomials of degree less than n. For L in (2.3.16) to be admissible, its coefficients must be of the form

$$A(x,y) = Ax^2 + a_1x + b_1y + c_1, \quad B(x,y) = Axy + a_2x + b_2y + c_2,$$
$$C(x,y) = Ay^2 + a_3x + b_3y + c_3, \quad D(x,y) = Bx + d_1, \quad E(x,y) = By + d_2,$$

and, furthermore, for each $n = 0, 1, 2, \ldots$,

$$nA + B \neq 0 \quad \text{and} \quad \lambda_n = -n((n-1)A + B).$$

A classification of the admissible equations that have orthogonal polynomials as eigenfunctions was given by Krall and Sheffer (1967). Up to affine transformations, there are only nine equations. Five of them admit classical orthogonal polynomials. These are

1. product Hermite polynomials; see (2.3.2);
2. product Laguerre polynomials; see (2.3.3);
3. product Hermite and Laguerre polynomials; see (2.3.4);
4. orthogonal polynomials on the disk; see (2.3.11);
5. orthogonal polynomials on the triangle; see (2.3.15).

The other four admissible differential equations are

6. $3yv_{xx} + 2v_{yy} - xv_x - yv_y = \lambda u$;
7. $(x^2 + y + 1)v_{xx} + (2xy + 2x)v_{xy} + (y^2 + 2y + 1)v_{yy} + g(xv_x + yv_y) = \lambda u$;
8. $x^2v_{xx} + 2xyv_{xy} + (y^2 - y)v_{yy} + g[(x-1)v_x + (y-\alpha)v_y] = \lambda u$;
9. $(x+\alpha)v_{xx} + 2(y+1)v_{yy} + xv_x + yv_y = \lambda u$.

The solutions for the last four equations are weak orthogonal polynomials in the sense that the polynomials are orthogonal with respect to a linear functional that is not positive definite.

Another classification in Suetin (1999), based on Engélis (1974), listed fifteen cases: some of them are equivalent under affine transformations in Krall and Sheffer (1967) but are treated separately because of other considerations. The orthogonality of cases 6 and 7 is determined in Krall and Sheffer (1967), while cases 8 and 9 are determined in Berens et al. (1995b) and Kwon et al. (2001). For further results, including solutions of the last four cases and further discussion of the impact of affine transformations, see Littlejohn (1988), Lyskova (1991), Kwon et al. (2001), and references therein. Classical orthogonal polynomials in two variables were studied in the context of hypergroups by Connett and Schwartz (1995).

By the definition of the admissibility, all orthogonal polynomials of degree n are eigenfunctions of an admissible differential operator for the same eigenvalue. In other words, the eigenfunction space for each eigenvalue is $\mathcal{V}_n^2$. This requirement excludes, for example, the product Jacobi polynomial $P_k^{(\alpha,\beta)}(x)P_{n-k}^{(\gamma,\delta)}(y)$, which satisfies a second-order equation of the form $Lu = \lambda_{k,n}u$, where $\lambda_{k,n}$ depends on both k and n. The product Jacobi polynomials, and other classical orthogonal polynomials of two variables, satisfy a second-order matrix differential equation (see Fernández et al., 2005 and the references therein) and they also satisfy a matrix form of a Rodrigues-type formula (Álvarez de Morales et al., 2009).

2.3.5 Orthogonal Polynomials of Complex Variables

Orthogonal polynomials of two real variables can be given as polynomials of complex variables z and $\bar{z}$ by identifying $\mathbb{R}^2$ with the complex plane $\mathbb{C}$ and setting $z = x + iy$. For a real weight function W on $\Omega \in \mathbb{R}^2$, we consider polynomials in z and $\bar{z}$ that are orthogonal with respect to the inner product

$$\langle f, g\rangle_W^{\mathbb{C}} := \int_\Omega f(z,\bar{z})\overline{g(z,\bar{z})}w(z)\,dx\,dy, \tag{2.3.17}$$

where $w(z) = W(x,y)$. Let $\mathcal{V}_n^2(W,\mathbb{C})$ denote the space of orthogonal polynomials in z and $\bar{z}$ with respect to the inner product (2.3.17). In this subsection, we denote by $P_{k,n}(x,y)$ real orthogonal polynomials with respect to W and denote by $Q_{k,n}(z,\bar{z})$ orthogonal polynomials in $\mathcal{V}_n^2(W,\mathbb{C})$.

Proposition 2.3.1 *The space $\mathcal{V}_n^2(W,\mathbb{C})$ has a basis $Q_{k,n}$ that satisfies*

$$Q_{k,n}(z,\bar{z}) = \overline{Q_{n-k,n}(z,\bar{z})}, \quad 0 \le k \le n. \tag{2.3.18}$$

Furthermore, this basis is related to the basis of $\mathcal{V}_n^2(W)$ by

$$\begin{aligned} P_{k,n}(x,y) &= \tfrac{1}{2^{1/2}}\left(Q_{k,n}(z,\bar{z}) + Q_{n-k,n}(z,\bar{z})\right), \quad 0 \le k \le \tfrac{1}{2}n, \\ P_{k,n}(x,y) &= \tfrac{1}{2^{1/2}i}\left(Q_{k,n}(z,\bar{z}) - Q_{k-k,n}(z,\bar{z})\right), \quad \tfrac{1}{2}n < k \le n. \end{aligned} \tag{2.3.19}$$

Writing orthogonal polynomials in terms of complex variables often leads to more symmetric formulas. Below are two examples.

1. *Complex Hermite polynomials* For $j,k \in \mathbb{N}_0$, define

$$H_{k,j}(z,\bar{z}) := z^k\bar{z}^j\,{}_2F_0\left(\begin{matrix}-k,-j \\ -\end{matrix};\frac{1}{z\bar{z}}\right).$$

Then $H_{k,j} \in \mathcal{V}_{k+j}^2(w_H,\mathbb{C})$, where $w_H(z) = \pi^{-1}e^{-x^2-y^2} := \pi^{-1}e^{-|z|^2}$ with $z = x+iy \in \mathbb{C}$. These polynomials satisfy

(i) $H_{k,j}(z,\bar{z}) = \overline{H_{j,k}(z,\bar{z})}$;
(ii) $H_{k,j}(z,\bar{z}) = (-1)^j j!\,z^{k-j}L_j^{k-j}(|z|^2)$, $k \ge j$ (L_j^α Laguerre polynomial, for $k \le j$ use (i));
(iii) $\frac{\partial}{\partial z}H_{k,j} = \bar{z}H_{k,j} - H_{k,j+1}$, $\frac{\partial}{\partial \bar{z}}H_{k,j} = zH_{k,j} - H_{k+1,j}$;
(iv) $zH_{k,j} = H_{k+1,j} + jH_{k,j-1}$, $\bar{z}H_{k,j} = H_{k,j+1} + kH_{k-1,j}$;
(v) $\int_{\mathbb{C}} H_{k,j}(z,\bar{z})\overline{H_{m,l}(z,\bar{z})}w_H(z)\,dx\,dy = j!\,k!\,\delta_{k,m}\delta_{j,l}$.

2. *Disk polynomials* For $j,k \in \mathbb{N}_0$, define

$$P_{k,j}^\lambda(z,\bar{z}) := \frac{(\lambda+1)_{k+j}}{(\lambda+1)_k(\lambda+1)_j}z^k\bar{z}^j\,{}_2F_1\left(\begin{matrix}-k,-j \\ -\lambda-k-j\end{matrix};\frac{1}{z\bar{z}}\right),$$

normalized by $P_{k,j}^\lambda(1,1) = 1$. Then $P_{\kappa,j}^\lambda \in \mathcal{V}_{k+1}^2(w_\lambda,\mathbb{C})$, where $w_\lambda(z) := \frac{\lambda+1}{\pi}(1-|z|^2)^\lambda$, $\lambda > -1$. These polynomials satisfy

(i) $P^{\lambda}_{k,j}(z,\bar{z}) = \overline{P^{\lambda}_{j,k}(z,\bar{z})}$;
(ii) $P^{\lambda}_{k,j}(z,\bar{z}) = \frac{j!}{(\lambda+1)_j} P_j^{(\lambda,k-j)}(2|z|^2-1)z^{k-j}$, $k \ge j$ ($P_j^{(\alpha,\beta)}$ Jacobi polynomial, for $k \le j$ use (i));
(iii) $|P^{\lambda}_{k,j}(z,\bar{z})| \le 1$ for $|z| \le 1$ and $\lambda \ge 0$;
(iv) $zP^{\lambda}_{k,j} = \frac{\lambda+k+1}{\lambda+k+j+1} P^{\lambda}_{k+1,j} + \frac{j}{\lambda+k+j+1} P^{\lambda}_{k,j-1}$ and a similar relation holds for $\bar{z}P^{\lambda}_{k,j}$ upon using (i);
(v) $\int_D P^{\lambda}_{k,j}(z,\bar{z})\overline{P^{\lambda}_{m,l}(z,\bar{z})}w_{\lambda}(z)\,dx\,dy = \frac{\lambda+1}{\lambda+k+j+1}\frac{k!\,j!}{(\lambda+1)_k(\lambda+1)_j}\delta_{k,m}\delta_{j,l}$.

The complex Hermite polynomials were introduced in Itô (1952). They have been widely studied by many authors; see Ghanmi (2008), Intissar and Intissar (2006), and Ismail (2016) for some recent studies and the references therein. Disk polynomials were introduced by Zernike (1934) for $\alpha = \beta = 1/2$, and in a subsequent paper by Zernike and Brinkman (1935) in general. They are also called *Zernike polynomials* and they have applications in optics. They were used by Folland (1975) to expand the Poisson–Szegő kernel for the ball in $\mathbb{C}^d$. A Banach algebra related to disk polynomials was studied by Kanjin (1985). For further properties of disk polynomials, including the fact that for $\lambda = d-2$, $d = 2, 3, \ldots$, they are spherical functions for $\mathrm{U}(d)/\mathrm{U}(d-1)$, see Ikeda (1967), Koornwinder (1975), Vilenkin and Klimyk (1993) and Wünsche (2005).

The structure of complex orthogonal polynomials of two variables and its connection and contrast to its real counterpart was studied in Xu (2015a).

2.3.6 Jacobi Polynomials Associated with Root Systems and Related Orthogonal Polynomials

There are two families of Jacobi polynomials of two variables associated with a root system and a related family of orthogonal polynomials.

Jacobi Polynomials Associated with Root System BC_2 Consider the weight function

$$W_{\alpha,\beta,\gamma}(x,y) := (1-x+y)^{\alpha}(1+x+y)^{\beta}(x^2-4y)^{\gamma}, \tag{2.3.20}$$

where $\alpha,\beta,\gamma > -1$, $\alpha+\gamma > -\frac{3}{2}$, and $\beta+\gamma > -\frac{3}{2}$, defined on the domain

$$\Omega := \{(x,y) \mid |x| < y+1, x^2 > 4y\}, \tag{2.3.21}$$

which is depicted in the left-hand panel of Figure 2.1. After a change of variables $x = u+v$, $y = uv$ the domain and weight function become

$$\begin{aligned} \Omega^* &= \{(u,v) \mid -1 < u < v < 1\}, \\ W^*_{\alpha,\beta,\gamma}(u,v) &= (1-u)^{\alpha}(1+u)^{\beta}(1-v)^{\alpha}(1+v)^{\beta}(v-u)^{2\gamma+1}. \end{aligned} \tag{2.3.22}$$

Let $\mathcal{N} := \{(n,k) \mid 0 \le k \le n\}$. In $\mathcal{N}$ define $(j,m) \prec (k,n)$ if $m < n$ or $m = n$ and $j \le k$ (graded lexicographic order). Then an orthogonal polynomial P_k^n that satisfies

$$P_k^n(x,y) = x^{n-k}y^k + \sum_{(j,m)\prec(k,n)} a_{j,m}x^{m-j}y^j \tag{2.3.23}$$

and is orthogonal to all $x^{m-j}y^j$ for $(j,m) \prec (k,n)$ is uniquely determined.

In the case $\gamma = \pm\frac{1}{2}$, a basis of orthogonal polynomials can be given explicitly. In fact, such a basis can be given in the more general case where in (2.3.22) we twice replace the Jacobi weight function by an arbitrary weight function w. Then

$$W_{\pm\frac{1}{2}}(x,y) = w(u)w(v)(x^2-4y)^{\pm\frac{1}{2}} \quad \text{with } x = u+v,\ y = uv, \tag{2.3.24}$$

defined on the domain $\{(x,y) \mid x^2 > 4y,\ u,v \in \operatorname{supp} w\}$ for any weight function w on $\mathbb{R}$. Let $\{p_n\}$ denote an orthonormal basis with respect to w. Then an orthonormal basis of polynomials for $W_{\pm\frac{1}{2}}$ is given by

$$P_k^{n,(-\frac{1}{2})}(x,y) = \begin{cases} p_n(u)p_k(v) + p_n(u)p_k(v), & k < n, \\ 2^{1/2}p_n(u)p_n(v), & k = n, \end{cases} \tag{2.3.25}$$

$$P_k^{n,(\frac{1}{2})}(x,y) = \frac{p_{n+1}(u)p_k(v) - p_{n+1}(v)p_k(u)}{u-v}, \quad 0 \le k \le n, \tag{2.3.26}$$

where in both cases (u,v) is related to (x,y) by $x = u+v$, $y = uv$.

A Related Family of Orthogonal Polynomials

Consider the family of weight functions defined by

$$\mathcal{W}_{\alpha,\beta,\gamma}(x,y) := |x-y|^{2\alpha+1}|x+y|^{2\beta+1}(1-x^2)^{\gamma}(1-y^2)^{\gamma}, \quad (x,y) \in [-1,1]^2, \tag{2.3.27}$$

where $\alpha,\beta,\gamma > -1$, $\alpha+\gamma+\frac{3}{2} > 0$, and $\beta+\gamma+\frac{3}{2} > 0$. These weight functions are related to those in (2.3.20) by

$$\mathcal{W}_{\alpha,\beta,\gamma}(x,y) = 4^{\gamma}\left|x^2-y^2\right|W_{\alpha,\beta,\gamma}(2xy, x^2+y^2-1).$$

Let $\{P_{k,n}^{\alpha,\beta,\gamma}\}_{k=0}^n$ denote an orthogonal basis of $\mathcal{V}_n^2$ for $W_{\alpha,\beta,\gamma}$. Then an orthogonal basis of $\mathcal{V}_{2n}^2$ for $\mathcal{W}_{\alpha,\beta,\gamma}$ is given by

$$\begin{aligned} &P_{k,n}^{\alpha,\beta,\gamma}(2xy, x^2+y^2-1), && 0 \le k \le n, \\ &(x^2-y^2)P_{k,n-1}^{\alpha+1,\beta+1,\gamma}(2xy, x^2+y^2-1), && 0 \le k \le n-1, \end{aligned} \tag{2.3.28}$$

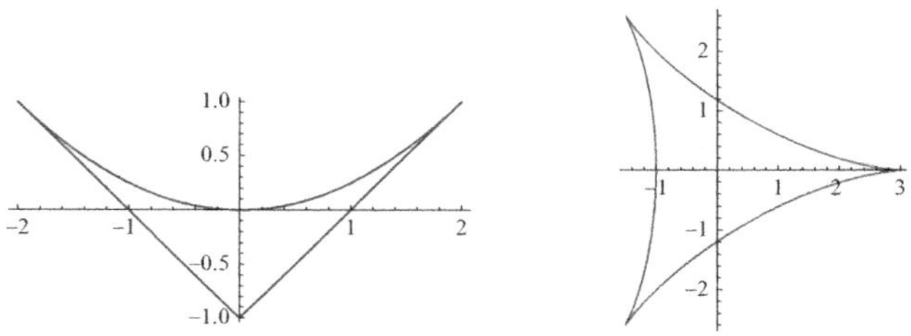

Figure 2.1 Regions for Koornwinder orthogonal polynomials. Left: BC_2. Right: A_2.

and an orthogonal basis of $\mathcal{V}_{2n+1}^2$ for $\mathcal{W}_{\alpha,\beta,\gamma}$ is given by

$$\begin{aligned} &(x+y)P_{k,n}^{\alpha,\beta+1,\gamma}(2xy, x^2+y^2-1), && 0 \le k \le n, \\ &(x-y)P_{k,n}^{\alpha+1,\beta,\gamma}(2xy, x^2+y^2-1), && 0 \le k \le n. \end{aligned} \tag{2.3.29}$$

In particular, when $\gamma = \pm\frac{1}{2}$, the basis can be given in terms of the Jacobi polynomials of one variable by using (2.3.25) and (2.3.26).

Jacobi Polynomials Associated with Root System A_2
These polynomials are orthogonal with respect to the weight function

$$W_\alpha(x,y) := \left(-3(x^2+y^2+1)^2 + 8(x^3-3xy^2) + 4\right)^\alpha \tag{2.3.30}$$

on the region bounded by $-3(x^2+y^2+1)^2 + 8(x^3-3xy^2) + 4 = 0$, which is called *Steiner's hypocycloid* and can be described as the curve

$$x + iy = (2e^{i\theta} + e^{-2i\theta})/3, \quad 0 \le \theta \le 2\pi.$$

This three-cusped region is depicted in the right-hand panel of Figure 2.1. Apart from $\alpha = \pm\frac{1}{2}$, orthogonal polynomials with respect to W_α are not explicitly known. In the case of $\alpha = \pm\frac{1}{2}$, a basis of orthogonal polynomials can be given in homogeneous coordinates as follows. Let

$$\mathbb{R}_H^3 := \{\mathbf{t} = (t_1,t_2,t_3) \in \mathbb{R}^3 \mid t_1 + t_2 + t_3 = 0\}.$$

For $\mathbf{t} \in \mathbb{R}_H^3$ and $\mathbf{k} = (k_1,k_2,k_3) \in \mathbb{R}_H^3 \cap \mathbb{Z}^3$, define $\phi_{\mathbf{k}}(\mathbf{t}) = e^{\frac{2\pi i}{3}\mathbf{k}\cdot\mathbf{t}}$ and

$$\mathsf{C}_{\mathbf{k}}^{\pm}(\mathbf{t}) := \tfrac{1}{6}\Big(\phi_{k_1,k_2,k_3}(\mathbf{t}) + \phi_{k_2,k_3,k_1}(\mathbf{t}) + \phi_{k_3,k_1,k_2}(\mathbf{t}) \pm \phi_{-k_1,-k_3,-k_2}(\mathbf{t}) \pm \phi_{-k_2,-k_1,-k_3}(\mathbf{t}) \pm \phi_{-k_3,-k_2,-k_1}(\mathbf{t})\Big),$$

which are analogues of cosine and sine functions. The region bounded by Steiner's hypocycloid is the image of the triangle $\Delta = \{(t_1,t_2) \in \mathbb{R}^2 \mid t_1,t_2 \ge 0,\ t_1+t_2 \le 1\}$ under the change of variables $(t_1,t_2) \mapsto (x,y)$, defined by

$$z := x + iy = \mathsf{C}_{0,1,-1}^{+}(\mathbf{t}) = \tfrac{1}{3}\big(\phi_{0,1,-1}(\mathbf{t}) + \phi_{1,-1,0}(\mathbf{t}) + \phi_{-1,0,1}(\mathbf{t})\big). \tag{2.3.31}$$

Under the change of variables (2.3.31), define

$$\begin{aligned} T_k^n(z,\bar z) &:= \mathsf{C}_{k,n-k,-n}^{+}(\mathbf{t}), && 0 \le k \le n, \\ U_k^n(z,\bar z) &:= \frac{\mathsf{C}_{k+1,n-k+1,-n-2}^{-}(\mathbf{t})}{\mathsf{C}_{1,1,-2}^{-}(\mathbf{t})}, && 0 \le k \le m. \end{aligned}$$

Then $\{T_k^n\}_{k=0}^n$ and $\{U_k^n\}_{k=0}^n$ are bases of $\mathcal{V}_n^2$ with respect to $W_{-\frac{1}{2}}$ and $W_{\frac{1}{2}}$, respectively. Both families of polynomials satisfy the relation $P_{n-k}^n(z,\bar z) = \overline{P_k^n(z,\bar z)}$, so that real-valued orthogonal bases can be derived from their real and imaginary parts. These polynomials are analogues of Chebyshev polynomials of the first and the second kind. They satisfy the following three-term relations:

$$P_k^{m+1}(z,\bar z) = 3zP_k^m(z,\bar z) - P_{k+1}^m(z,\bar z) - P_{k-1}^{m-1}(z,\bar z) \tag{2.3.32}$$

for $0 \le k \le m$ and $m \ge 1$, where

$$
\begin{aligned}
T^m_{-1}(z,\bar{z}) &:= T^{m+1}_1(z,\bar{z}), \qquad & T^m_{m+1}(z,\bar{z}) &:= T^{m+1}_m(z,\bar{z}),\\
U^m_{-1}(z,\bar{z}) &:= 0, & U^{m-1}_m(z,\bar{z}) &:= 0,
\end{aligned}
$$

and, moreover,

$$
\begin{aligned}
T^0_0(z,\bar{z}) &= 1, \qquad & T^1_0(z,\bar{z}) &= z, \qquad & T^1_1(z,\bar{z}) &= \bar{z},\\
U^0_0(z,\bar{z}) &= 1, & U^1_0(z,\bar{z}) &= 3z, & U^1_1(z,\bar{z}) &= 3\bar{z}.
\end{aligned}
$$

Further Results and References The Jacobi polynomials associated with BC_2 and A_2 were initially studied by Koornwinder (1974a,b) (see also Koornwinder, 1975), where it was shown that the orthogonal polynomials for $W_{\alpha,\beta,\gamma}$ in (2.3.20) are eigenfunctions of two commuting differential operators of second and fourth order, whereas the orthogonal polynomials associated with A_2 are eigenfunctions of two commuting differential operators of second and third order. The two families are rank-two cases of the Jacobi polynomials associated with root systems for general rank, the study of which was initiated by Heckman and Opdam; see Chapter 7. The Jacobi polynomials associated with A_2 can be identified with *Jack polynomials* of two variables.

The special case of $P^n_0(x,y)$ of the first family when $\gamma = 1/2$ was studied also by Eier and Lidl (1974). For further results on the first family, see Koornwinder and Sprinkhuizen-Kuyper (1978) and Sprinkhuizen-Kuyper (1976), including an explicit formula for P^n_k in (2.3.23) given in terms of power series and Rodrigues-type formulas, and Xu (2012) where an explicit formula for the reproducing kernel in the case of W in (2.3.24) with $\gamma = \pm 1/2$ was given in terms of the reproducing kernels of the orthogonal polynomials of one variable. For further results on the second family, see Shishkin (1997), Suetin (1999), and Li et al. (2008); the latter includes a connection with translation tiling and convergence of orthogonal expansions for $\mu = \pm\frac{1}{2}$.

Orthogonal polynomials with respect to $W_{\pm\frac{1}{2}}$ in (2.3.24) and the Jacobi polynomials associated with A_2 when $\gamma = \frac{1}{2}$ are remarkable for having the maximum number of common zeros, i.e., $\mathbb{P}_n = \{P^n_k \mid 0 \le k \le n\}$ has $\dim \Pi^2_{n-1}$ distinct real zeros (Li et al., 2008; Schmid and Xu, 1994). By Theorem 2.2.15, Gaussian cubature formulas exist for these weight functions. For their generalizations to higher dimension, see §§2.9.1, 2.9.2.

The orthogonal polynomials with respect to $\mathcal{W}$ in (2.3.27) were studied by Xu (2012). The reproducing kernel of Π^2_n in $L^2(\mathcal{W})$ can also be expressed in terms of the reproducing kernels of Π^2_n in $L^2(W)$ for W in (2.3.20). In particular, in the case of $\gamma = \pm\frac{1}{2}$, the kernel can be expressed in terms of the reproducing kernels of the Jacobi polynomials. When $\gamma = \pm\frac{1}{2}$, these weight functions admit minimal cubature rules that attain the lower bound (2.2.19).

2.3.7 Methods of Constructing Orthogonal Polynomials of Two Variables from one Variable

Let w_1 and w_2 be weight functions defined on the intervals (a, b) and (c, d), respectively. Let ρ be a positive function on (a, b). For the weight function

$$W(x,y) := w_1(x)w_2(\rho^{-1}(x)y), \quad (x,y) \in \Omega, \tag{2.3.33}$$

where the domain Ω is defined by

$$\Omega := \{(x,y) \in \mathbb{R}^2 \mid a < x < b,\ c\rho(x) < y < d\rho(x)\}, \tag{2.3.34}$$

a basis of orthogonal polynomials of two variables can be given in terms of orthogonal polynomials of one variable whenever either one of the following additional assumptions is satisfied:

Case 1. ρ is a polynomial of degree 1.
Case 2. $\rho = \sqrt{q}$ with q a nonnegative polynomial of degree at most 2, and further assume that $c = -d > 0$ and w_2 is an even function on $(-c, c)$.

For each $k \in \mathbb{N}_0$ let $\{p_{n,k}\}_{n=0}^{\infty}$ denote the system of orthonormal polynomials with respect to the weight function $(\rho(x))^{2k+1}w_1(x)$ on (a, b). And let $\{q_n\}$ be the system of orthonormal polynomials with respect to $w_2(x)$ on (c, d). Define polynomials of two variables by

$$P_k^n(x,y) := p_{n-k,k}(x)(\rho(x))^k q_k\left(\frac{y}{\rho(x)}\right), \quad 0 \le k \le n. \tag{2.3.35}$$

In Case 2 we see that P_k^n are polynomials of degree n because q_k has the same parity as k by evenness of w_2. Then $\{P_k^n\}_{k=0}^n$ is an orthonormal basis of $\mathcal{V}_n^2$ with respect to W on Ω.

Examples of orthogonal polynomials constructed by this method include product orthogonal polynomials, for which $\rho(x) = 1$, and also the following cases:

Jacobi Polynomials on the Disk Let $w_1(x) = w_2(x) := (1-x^2)^{\mu-1/2}$ on $[-1, 1]$ and $\rho(x) := (1-x^2)^{1/2}$. Then the weight function (2.3.33) and the basis (2.3.35) coincide up to constant factors with (2.3.5) and (2.3.6), respectively.

Jacobi Polynomials on the Triangle Let $w_1(x) := x^{\alpha-\frac{1}{2}}(1-x)^{\beta+\gamma-1}$ and $w_2(x) := x^{\beta-\frac{1}{2}}(1-x)^{\gamma-\frac{1}{2}}$, both defined on the interval $(0, 1)$, and let $\rho(x) := 1 - x$. Then the weight function (2.3.33) and the basis (2.3.35) coincide up to constant factors with (2.3.12) and (2.3.13), respectively.

Orthogonal Polynomials on the Parabolic Domain Let $w_1(x) := x^a(1-x)^b$ on $[0, 1]$, $w_2(x) := (1-x^2)^a$ on $[-1, 1]$, and $\rho(x) := \sqrt{x}$. Then the weight function (2.3.33) becomes

$$W_{a,b}(x,y) := (1-x)^b(x-y^2)^b, \quad y^2 < x < 1. \tag{2.3.36}$$

The domain $\{(x, y) \mid y^2 < x < 1\}$ is bounded by a straight line and a parabola. The orthogonal polynomials P_k^n in (2.3.35) are

$$P_k^n(x,y) = P_{n-k}^{(a,b+k+1/2)}(2x-1)x^{k/2}P_k^{(b,b)}(yx^{-1/2}), \quad 0 \le k \le n. \tag{2.3.37}$$

Further Results and References This method of generating orthogonal polynomials of two variables first appeared in Larcher (1959) and was used by Agahanov (1956) in certain special cases. It was presented systematically by Koornwinder (1975), where the two cases for ρ were stated. For further examples of explicit bases constructed in various domains, such as $\{(x,y) \mid x^2 + y^2 \le 1,\ -a \le y \le b\}$ $(0 < a, b < 1)$, see Suetin (1999).

The sequence of polynomials in (2.3.37) satisfies a product formula (Koornwinder and Schwartz, 1997) that generates a convolution structure for $L^2(W_{\alpha,\beta})$, which was used to study the convergence of orthogonal expansions (zu Castell et al., 2009).

2.3.8 Other Orthogonal Polynomials of Two Variables

This subsection contains miscellaneous results on orthogonal polynomials of two variables.

1. *Orthogonal polynomials for a radial weight function.* A weight function W is called *radial* if it is of the form $W(x,y) = w(r)$, where $r = \sqrt{x^2+y^2}$. For such a weight function, an orthonormal basis can be given in polar coordinates $(x,y) = (r\cos\theta, r\sin\theta)$. Let $p_m^{(k)}$ denote the orthogonal polynomial of degree m with respect to the weight function $r^{k+1}w(r)$ on $[0,\infty)$. Define

$$\begin{aligned} P_{j,1}(x,y) &= p_{2j}^{(2n-4j+1)}(r) r^{(n-2j)} \cos((n-2j)\theta), \qquad 0 \le j \le n/2, \\ P_{j,2}(x,y) &= p_{2j}^{(2n-4j+1)}(r) r^{(n-2j)} \sin((n-2j)\theta), \qquad 0 \le j < n/2. \end{aligned} \tag{2.3.38}$$

Then $\{P_{j,1}\}_{j=0}^{\lfloor n/2\rfloor} \cup \{P_{j,2}\}_{j=0}^{\lfloor (n-1)/2\rfloor}$ is an orthogonal basis of $\mathcal{V}_n^2$ with respect to W. For $W(x,y) = (1-r^2)^{\mu-1/2}$ this is the basis given in (2.3.8). Another example is the following:

Product Hermite weight function $W(x,y) = e^{-x^2-y^2}$. The basis (2.3.38) is given by

$$\begin{aligned} P_{j,1}(x,y) &= L_j^{n-2j}(r^2) r^{(n-2j)} \cos((n-2j)\theta), \qquad 0 \le j \le n/2, \\ P_{j,2}(x,y) &= L_{2j}^{n-2j}(r^2) a r^{(n-2j)} \sin((n-2j)\theta), \qquad 0 \le j < n/2 \end{aligned} \tag{2.3.39}$$

in terms of Laguerre polynomials.

2. *Bernstein–Szegő weight function of two variables.* Let $m \in \mathbb{N}$. For any $i = 0, \ldots, m$ let $h_i(y)$ be polynomials in y with real coefficients of degree at most $\frac{m}{2} - \left|\frac{m}{2} - i\right|$, with $h_0(y) = 1$, such that for all $y \in [-1,1]$,

$$h(z,y) = \sum_{i=0}^{m} h_i(y) z^i, \quad z \in \mathbb{C}, \tag{2.3.40}$$

is nonzero whenever $|z| \le 1$. Consider the two variable weight function

$$W(x,y) := \frac{4}{\pi^2} \frac{\sqrt{1-x^2}\,\sqrt{1-y^2}}{|h(z,y)|^2}, \quad x = \tfrac{1}{2}(z+z^{-1}). \tag{2.3.41}$$

For $\lceil \frac{m-2}{2} \rceil \le k \le n$, define the polynomials

$$P_k^n(x,y) = U_{n-k}(y) \sum_{i=0}^{m} h_i(y) U_{k-i}(x), \tag{2.3.42}$$

where it is understood that $U_n(x) = -U_{-n-2}(x)$ if $n < 0$. Then P_k^n is an orthogonal polynomial of degree n with respect to W. In particular, if $m \le 2$, then $\{P_k^n\}_{k=0}^n$ is an orthogonal basis of $\mathcal{V}_n^2$.

Example $h(z, y) = 1 - 2ayz + a^2z^2$, where $|a| < 1$ and $a \in \mathbb{R}$. Then

$$W(x, y) = \frac{4}{\pi^2} \frac{(1 - a^2)\sqrt{1 - x^2}\sqrt{1 - y^2}}{4a^2(x^2 + y^2) - 4a(1 + a^2)xy + (1 - a^2)^2}. \tag{2.3.43}$$

The orthogonal polynomials in (2.3.42) are given, up to a constant, by $P_0^n(x, y) = U_n(y)$ and

$$P_k^n(x, y) = \big(U_k(x) - 2ayU_{k-1}(x) + a^2U_{k-2}(x)\big)\, U_{n-k}(y), \qquad 1 \le k \le n.$$

For further examples and other properties of such polynomials, see Delgado et al. (2009).

3. *Orthogonal polynomials on the regular hexagon.* Orthogonal polynomials with respect to the constant weight function on the regular hexagon were studied by Dunkl (1987), who gave an algorithm for generating an orthogonal basis. No closed form of such a basis is known.

2.4 Spherical Harmonics

Here and later we will use the notation $\|x\| := (x_1^2 + \cdots + x_d^2)^{1/2}$ and $\langle x, y\rangle := x_1y_1 + \cdots + x_dy_d$ $(x, y \in \mathbb{R}^d)$. Spherical harmonics are an essential tool for Fourier analysis on the unit sphere $\mathbb{S}^{d-1} := \{x \in \mathbb{R}^d \mid \|x\| = 1\}$ in $\mathbb{R}^d$ $(d \ge 2)$. They are also building blocks for families of orthogonal polynomials with respect to radial weight functions on $\mathbb{R}^d$.

2.4.1 Ordinary Spherical Harmonics

Let $\Delta := \frac{\partial^2}{\partial x_1^2} + \cdots + \frac{\partial^2}{\partial x_d^2}$ be the *Laplace operator* on $\mathbb{R}^d$. A polynomial Y on $\mathbb{R}^d$ is called *harmonic* if $\Delta Y = 0$. For $n = 0, 1, 2, \ldots$, let $\mathcal{H}_n^d$ denote the linear space of homogeneous harmonic polynomials of degree n in d variables, i.e.,

$$\mathcal{H}_n^d := \{P \in \mathcal{P}_n^d \mid \Delta P = 0\}.$$

By definition, a *spherical harmonic* is the restriction of a homogeneous harmonic polynomial to the unit sphere. If $Y \in \mathcal{H}_n^d$, then $Y(x) = \|x\|^n Y(x')$ where $x' = x/\|x\| \in \mathbb{S}^{d-1}$. We shall also use $\mathcal{H}_n^d$ to denote the space of spherical harmonics of degree n. For $n \in \mathbb{N}_0$,

$$\dim \mathcal{H}_n^d = \dim \mathcal{P}_n^d - \dim \mathcal{P}_{n-2}^d = \binom{n+d-1}{d-1} - \binom{n+d-3}{d-1} = \frac{(2n+d-2)(n+d-3)!}{n!(d-2)!}. \tag{2.4.1}$$

Spherical harmonics of different degrees are orthogonal in $L^2(\mathbb{S}^{d-1}, d\sigma)$, where $d\sigma$ denotes the normalized spherical measure on $\mathbb{S}^{d-1}$. There is a unique decomposition

$$\mathcal{P}_n^d = \bigoplus_{j=0}^{\lfloor \frac{n}{2} \rfloor} \|x\|^{2j} \mathcal{H}_{n-2j}^d : \qquad P(x) = \sum_{j=0}^{\lfloor \frac{n}{2} \rfloor} \|x\|^{2j} Y_{n-2j}^d(x').$$

Orthonormal basis In terms of *spherical polar coordinates*

$$\begin{cases} x_1 = r \sin\theta_{d-1} \cdots \sin\theta_2 \sin\theta_1, \\ x_2 = r \sin\theta_{d-1} \cdots \sin\theta_2 \cos\theta_1, \\ \quad \cdots \\ x_{d-1} = r \sin\theta_{d-1} \cos\theta_{d-2}, \\ x_d = r \cos\theta_{d-1}, \end{cases} \tag{2.4.2}$$

where $r \geq 0$, $0 \leq \theta_1 < 2\pi$, $0 \leq \theta_i \leq \pi$ for $i = 2, \dots, d$, the normalized measure $d\sigma$ on $\mathbb{S}^{d-1}$ is given by

$$d\sigma := \omega_d^{-1} \prod_{j=1}^{d-2} (\sin\theta_{d-j})^{d-j-1} d\theta_1\, d\theta_2 \cdots d\theta_{d-1}, \tag{2.4.3}$$

where $\omega_d := \frac{2\pi^{d/2}}{\Gamma(d/2)}$ is the surface area of $\mathbb{S}^{d-1}$. For $d = 2$, $\dim \mathcal{H}_n^2 = 2$. Then an orthogonal basis for $\mathcal{H}_n^2$ is given in polar coordinates by

$$Y_n^{(1)}(x) := r^n \cos(n\theta), \qquad Y_n^{(2)}(x) := r^n \sin(n\theta).$$

For $d > 2$ and $\alpha \in \mathbb{N}_0^d$, define

$$Y_\alpha(x) := [h_\alpha]^{-1} r^{|\alpha|} g_\alpha(\theta_1) \prod_{j=1}^{d-2} (\sin\theta_{d-j})^{|\alpha^{j+1}|} C_{\alpha_j}^{\lambda_j}(\cos\theta_{d-j}), \tag{2.4.4}$$

where $g_\alpha(\theta_1) := \cos(\alpha_{d-1}\theta_1)$ (if $\alpha_d = 0$) and $:= \sin((\alpha_{d-1} + 1)\theta_1)$ (if $\alpha_d = 1$), where $|\alpha^j| := \alpha_j + \cdots + \alpha_d$, $\lambda_j := |\alpha^{j+1}| + (d - j - 1)/2$, and where

$$[h_\alpha]^2 := b_\alpha \prod_{j=1}^{d-2} \frac{\alpha_j! (\frac{1}{2}(d - j + 1))_{|\alpha^{j+1}|} (\alpha_j + \lambda_j)}{(2\lambda_j)_{\alpha_j} (\frac{1}{2}(d - j))_{|\alpha^{j+1}|} \lambda_j}$$

with $b_\alpha = 2$ if $\alpha_{d-1} + \alpha_d > 0$, else $= 1$. Then $\{Y_\alpha\}_{|\alpha|=n, \alpha_d=0,1}$ is an orthonormal basis of $\mathcal{H}_n^d$.

Projection Operator The operator $\mathrm{proj}_n \colon L^2(\mathbb{S}^{d-1}, d\sigma) \to \mathcal{H}_n^d$ satisfies

$$\begin{aligned} \mathrm{proj}_n P(x) &= \sum_{j=0}^{\lfloor \frac{n}{2} \rfloor} \frac{1}{4^j j! (-n + 2 - \frac{1}{2}d)_j} \|x\|^{2j} \Delta^h P(x) \\ &= \frac{(-1)^n}{2^n (\frac{1}{2}d - 1)_n} \|x\|^{2\|\alpha\| + d - 2} P\Big(\frac{\partial}{\partial x}\Big) \{\|x\|^{-d+2}\}. \end{aligned} \tag{2.4.5}$$

In particular, the projection of x^α is, up to a constant, *Maxwell's representation*, defined by

$$H_\alpha(x) := \|x\|^{2|\alpha|+d-2} \frac{\partial^\alpha}{\partial x^\alpha} \{\|x\|^{-d+2}\}, \quad \alpha \in \mathbb{N}_0^d. \tag{2.4.6}$$

The set $\{H_\alpha\}_{|\alpha|=n, \alpha_d=0,1}$ is a basis of $\mathcal{H}_n^d$. Furthermore, H_α satisfy a recursive relation

$$H_{\alpha+e_i}(x) = -(2|\alpha| + d - 2)x_i H_\alpha(x) + \|x\|^2 \frac{\partial}{\partial x_i} H_\alpha(x). \tag{2.4.7}$$

Let $P_n(\cdot, \cdot)$ denote the reproducing kernel of $\mathcal{H}_n^d$. Then the projection operator can be written as an integral operator

$$\operatorname{proj}_n f(x) = \frac{1}{\omega_d} \int_{\mathbb{S}^{d-1}} f(y) P_n(x, y)\, d\sigma(y).$$

Reproducing Kernel and Zonal Spherical Harmonics In terms of an orthonormal basis $\{Y_j\}_{j=1}^{\dim \mathcal{H}_n^d}$ of $\mathcal{H}_n^d$, by definition the reproducing kernel can be written as

$$P_n(x, y) = \sum_{1 \le j \le \dim \mathcal{H}_n^d} Y_j(x) Y_j(y). \tag{2.4.8}$$

The kernel is invariant under the action of the orthogonal group $O(d)$ and it depends only on the distance between x and y on the sphere. Moreover,

$$P_n(x, y) = \frac{n + \lambda}{\lambda} C_n^\lambda(\langle x, y \rangle), \quad x, y \in \mathbb{S}^{d-1}, \ \lambda = \tfrac{1}{2}d - 1. \tag{2.4.9}$$

For $y \in \mathbb{S}^{d-1}$ fixed, both sides of (2.4.9) as a function of $x \in \mathbb{S}^{d-1}$ are *zonal spherical harmonics*, i.e., spherical harmonics which are invariant under an orthogonal transformation leaving y fixed. The corresponding homogeneous polynomial $\|x\|^n C_n^{(d-2)/2}(\langle x, y \rangle / \|x\|)$ is then called a *zonal harmonic polynomial* in $x \in \mathbb{R}^d$. The combination of (2.4.8) and (2.4.9) is known as the *addition formula of spherical harmonics*. The reproducing property of the kernel leads to the *Funk–Hecke formula*

$$\int_{\mathbb{S}^{d-1}} f(\langle x, y \rangle) Y(x)\, d\sigma(x) = \lambda_n Y(y), \quad Y \in \mathcal{H}_n^{d-1}, \ y \in \mathbb{S}^{d-1}, \tag{2.4.10}$$

for all functions f for which the left-hand side is finite, where

$$\lambda_n = \frac{1}{\omega_{d-1}} \int_{-1}^1 f(t) \frac{C_n^\lambda(t)}{C_n^\lambda(1)} (1 - t^2)^{\lambda - \frac{1}{2}}\, dt, \quad \lambda = \tfrac{1}{2}d - 1.$$

The *Poisson summation kernel* satisfies, for $x, y \in \mathbb{S}^{d-1}$,

$$\sum_{n=0}^{\infty} P_n(x, y) r^n = \frac{1 - r^2}{(1 - \langle x, y \rangle r + r^2)^{\frac{1}{2}d}} = \frac{1 - r^2}{\|rx - y\|^d}. \tag{2.4.11}$$

Laplace–Beltrami Operator This is the operator Δ_0 defined by

$$(\Delta_0 f)(x) := (\Delta F)(x), \quad x \in \mathbb{S}^{d-1}, \tag{2.4.12}$$

where $F(y) := f(y/\|y\|)$ is the extension of f to $\mathbb{R}^d \backslash \{0\}$ which is homogeneous of degree 0. In terms of the spherical polar coordinates $x = rx'$, $r > 0$, and $x' \in \mathbb{S}^{d-1}$, the usual Laplace operator Δ is decomposed as

$$\Delta = \frac{d^2}{dr^2} + \frac{d-1}{r}\frac{d}{dr} + \frac{1}{r^2}\Delta_0. \tag{2.4.13}$$

The spherical harmonics are eigenfunctions of Δ_0:

$$\Delta_0 Y = -n(n+d-2)Y, \quad Y \in \mathcal{H}_n^d. \tag{2.4.14}$$

In terms of the spherical coordinates (2.4.2), Δ_0 is given by

$$\begin{aligned}\Delta_0 =& \frac{1}{(\sin\theta_{d-1})^{d-2}}\frac{\partial}{\partial\theta_{d-1}}\left((\sin\theta_{d-1})^{d-2}\frac{\partial}{\partial\theta_{d-1}}\right) \\ &+ \sum_{j=1}^{d-2}\frac{1}{(\sin\theta_{d-1})^2\cdots(\sin\theta_{j+1})^2\sin^{j-1}\theta_j}\frac{\partial}{\partial\theta_j}\left((\sin\theta_j)^{j-1}\frac{\partial}{\partial\theta_j}\right).\end{aligned} \tag{2.4.15}$$

Furthermore, it satisfies a decomposition

$$\Delta_0 = \sum_{1\le i<j\le d}\left(x_i\frac{\partial}{\partial x_j} - x_j\frac{\partial}{\partial x_i}\right)^2 \tag{2.4.16}$$

for $x \in \mathbb{S}^{d-1}$. The operator $A_{i,j} = x_i\frac{\partial}{\partial x_j} - x_j\frac{\partial}{\partial x_i}$ is the derivative with respect to the angle (Euler angle) in the polar coordinates of the (x_i, x_j)-plane, and it is also the infinitesimal operator of the regular representation $f \mapsto f(Q^{-1}x)$ of the rotation group $SO(d)$.

Further Results and References A number of books contain chapters or sections on spherical harmonics, treating the subject from various points of view. For earlier development, especially on $\mathbb{S}^2$, see Hobson (1931). A well-circulated early introductory is Müller (1966). The connection to Fourier analysis in Euclidean space is treated by Stein and Weiss (1971); see also Müller (1998). For the connection to the Radon transform, see Helgason (2000). For applications in integral geometry, see Groemer (1996). For the point of view of group representations, see Vilenkin (1968). The fact that the zonal polynomial is of the form $p_n(\langle x, y\rangle)$ can be used as a starting point to study properties of Gegenbauer polynomials; see Müller (1966) and Vilenkin (1968), as well as Andrews et al. (1999). Spherical harmonics are used as building blocks for orthogonal families on radial symmetric measures, see Dunkl and Xu (2014), Xu (2005a), and the next section.

2.4.2 *h-Harmonics for Product Weight Functions on the Sphere*

A far-reaching extension of spherical harmonics is Dunkl's h-harmonics associated with reflection groups; see Chapter 7. We consider the case $\mathbb{Z}_2^d$, since explicit formulas are available mostly in this case. Let

$$w_\kappa(x) := c_\kappa \prod_{i=1}^{d} |x_i|^{2\kappa_i}, \quad \text{with } c_\kappa := \frac{\pi^{\frac{1}{2}d}}{\Gamma(\frac{1}{2}d)} \frac{\Gamma(|\kappa| + \frac{1}{2}d)}{\Gamma(\kappa_1 + \frac{1}{2}) \cdots \Gamma(\kappa_d + \frac{1}{2})}, |\kappa| := \kappa_1 + \cdots + \kappa_d, \quad (2.4.17)$$

normalized such that $\int_{\mathbb{S}^{d-1}} w_\kappa(x)\, d\sigma = 1$. This weight function is invariant under the group $\mathbb{Z}_2^d$, for which the results for ordinary spherical harmonics can be extended in explicit formulas.

Definition The *h-harmonics* are homogeneous polynomials that satisfy $\Delta_h Y = 0$, where

$$\Delta_h := \mathcal{D}_1^2 + \cdots + \mathcal{D}_d^2 \quad (2.4.18)$$

is the *Dunkl Laplacian* and $\mathcal{D}_j$, $1 \le j \le d$ are the *Dunkl operators* associated with $\mathbb{Z}_2^d$,

$$\mathcal{D}_j f(x) := \frac{\partial}{\partial x_j} f(x) + \kappa_j \frac{f(x) - f(x_1, \ldots, -x_j, \ldots, x_d)}{x_j}. \quad (2.4.19)$$

The *spherical h-harmonics* are the restriction of h-harmonics to the sphere. Let $\mathcal{H}_n^d(w_\kappa)$ denote the space of h-harmonics of degree n. Then

$$\Delta_{h,0} Y = -n(n + 2|\kappa| + d - 2)Y, \quad Y \in \mathcal{H}_n^d(w_\kappa), \quad (2.4.20)$$

where $\Delta_{h,0}$ is the spherical h-Laplacian operator, and $\dim \mathcal{H}_n^d(w_\kappa) = \dim \mathcal{H}_n^d$.

Orthonormal Basis A basis of $\mathcal{H}_n^d(w_\kappa)$ can be given in spherical coordinates (2.4.2) and in terms of *generalized Gegenbauer polynomials* which are defined by

$$\begin{aligned} C_{2n}^{(\lambda,\mu)}(x) &:= \frac{(\lambda + \mu)_n}{(\mu + \frac{1}{2})_n} P_n^{(\lambda - 1/2, \mu - 1/2)}(2x^2 - 1), \\ C_{2n+1}^{(\lambda,\mu)}(x) &:= \frac{(\lambda + \mu)_{n+1}}{(\mu + \frac{1}{2})_{n+1}} x P_n^{(\lambda - 1/2, \mu + 1/2)}(2x^2 - 1). \end{aligned} \quad (2.4.21)$$

The polynomials $C_n^{(\lambda,\mu)}$ are orthogonal with respect to the weight function

$$w_{\lambda,\mu}(x) = |x|^{2\lambda}(1 - x^2)^{\mu - 1/2}, \quad x \in [-1, 1].$$

Let $h_n^{(\lambda,\mu)} := c \int_{-1}^{1} (C_n^{(\lambda,\mu)}(t))^2 w_{\lambda,\mu}(t)\, dt$, normalized such that $h_0^{(\lambda,\mu)} = 1$. Then

$$\begin{aligned} h_{2n}^{(\lambda,\mu)} &= \frac{(\lambda + \frac{1}{2})_n (\lambda + \mu)_n}{n!\, (\mu + \frac{1}{2})_n} \frac{\lambda + \mu}{\lambda + \mu + 2n}, \\ h_{2n+1}^{(\lambda,\mu)} &= \frac{(\lambda + \frac{1}{2})_n (\lambda + \mu)_{n+1}}{n!\, (\mu + \frac{1}{2})_{n+1}} \frac{\lambda + \mu}{\lambda + \mu + 2n + 1}. \end{aligned} \quad (2.4.22)$$

For $d \ge 2$ and $\alpha \in \mathbb{N}_0^d$, define

$$Y_\alpha(x) := [h_\alpha]^{-1} r^{|\alpha|} g_\alpha(\theta_1) \prod_{j=1}^{d-2} (\sin\theta_{d-j})^{|\alpha^{j+1}|} C_{\alpha_j}^{(\lambda_j,\kappa_j)}(\cos\theta_{d-j}), \tag{2.4.23}$$

where $g_\alpha(\theta) := C_{\alpha_{d-1}}^{(\kappa_d,\kappa_{d-1})}(\cos\theta)$ (if $\alpha_d = 0$) and $:= \sin\theta\, C_{\alpha_{d-1}-1}^{(\kappa_d+1,\kappa_{d-1})}(\cos\theta)$ if ($\alpha_d = 1$), and where $|\alpha^j| := \alpha_j + \cdots + \alpha_d$, $|\kappa^j| := \kappa_j + \cdots + \kappa_d$, $\lambda_j := |\alpha^{j+1}| + |\kappa^{j+1}| + \frac{1}{2}(d-j-1)$, and

$$[h_\alpha^n]^2 := \frac{a_\alpha}{(|\kappa| + \frac{1}{2}d)_n} \prod_{j=1}^{d-1} h_{\alpha_i}^{(\lambda_i,\kappa_i)}(\kappa_i + \lambda_i)_{\alpha_i}, \quad a_\alpha := \begin{cases} 1 & \text{if } \alpha_d = 0, \\ \kappa_d + \frac{1}{2} & \text{if } \alpha_d = 1. \end{cases}$$

Then $\{Y_\alpha\}_{|\alpha|=n,\alpha_d=0,1}$ is an orthonormal basis of $\mathcal{H}_n^d(w_\kappa)$.

Reproducing Kernel The reproducing kernel of $\mathcal{H}_n^d(w_\kappa)$ is given by

$$P_n(w_\kappa; x, y) = \sum_{1 \le j \le \dim \mathcal{H}_n^d(w_\kappa)} Y_j(x) Y_j(y),$$

where $\{Y_j\}_{j=1}^{\dim \mathcal{H}_n^d(w_\kappa)}$ is an orthonormal basis of $\mathcal{H}_n^d(w_\kappa)$. It satisfies a closed formula

$$\begin{aligned} P_n(w_\kappa; x, y) = \frac{n + |\kappa| + \frac{1}{2}d - 1}{|\kappa| + \frac{1}{2}d - 1} \int_{[-1,1]^d} C_n^{|\kappa|+\frac{1}{2}d-1}(x_1 y_1 t_1 + \cdots + x_d y_d t_d) \\ \times \left(\prod_{i=1}^d c_{\kappa_i}(1 + t_i)(1 - t_i^2)^{\kappa_i - 1} \right) dt. \end{aligned} \tag{2.4.24}$$

Further Results and References The h-harmonics associated with a finite reflection group were first studied by Dunkl (1988). Next (Dunkl, 1989), he defined his Dunkl operators. For an overview of the extensive theory of h-harmonics, see Dunkl and Xu (2014) and Chapter 7. The case $\mathbb{Z}_2^d$ was studied in detail by Xu (1997b), which paper contains (2.4.23) and (2.4.24), as well as a closed formula for an analogue of the Poisson integral. A Funk–Hecke-type formula was given by Xu (2000). For a monic h-harmonic basis and a biorthogonal basis, see Xu (2005b). For a connection to products of Heine–Stieltjes polynomials, see Volkmer (1999).

2.5 Classical Orthogonal Polynomials of Several Variables

General properties of orthogonal polynomials of several variables were given in §2.2. This section contains results for specific weight functions.

2.5.1 Classical Orthogonal Polynomials on the Unit Ball

On the unit ball $B^d := \{x \in \mathbb{R}^d \mid \|x\| \le 1\}$, consider the weight function

$$W_\mu(x, y) := \frac{\Gamma(\mu + \frac{1}{2}(d+1))}{\pi^{d/2}\Gamma(\mu + \frac{1}{2})}(1 - \|x\|^2)^{\mu - \frac{1}{2}}, \quad \mu > -\tfrac{1}{2}, \tag{2.5.1}$$

normalized such that its integral over B^d is 1. For $d = 2$, see §2.3.2.

Differential Operator Orthogonal polynomials of degree n with respect to W_μ are eigenfunctions of a second-order differential operator:

$$\left(\Delta - \sum_{j=1}^{d} \frac{\partial}{\partial x_j} x_j \left((2\mu - 1) + \sum_{i=1}^{d} x_i \frac{\partial}{\partial x_i}\right)\right) P = -(n + d)(n + 2\mu - 1)P, \quad P \in \mathcal{V}_n^d, \tag{2.5.2}$$

where Δ is the Laplace operator.

First Orthonormal Basis Associated with $x = (x_1, \ldots, x_d) \in \mathbb{R}^d$, define $\mathbf{x}_j := (x_1, \ldots, x_j)$ for $1 \le j \le d$ and $\mathbf{x}_0 := 0$. For $\alpha \in \mathbb{N}_0^d$ and $1 \le j \le d$, let $\alpha^j := (\alpha_j, \ldots, \alpha_d)$ and $\alpha^{d+1} := 0$. An orthonormal basis $\{P_\alpha\}_{|\alpha|=n}$ of $\mathcal{V}_n^d$ is given by

$$P_\alpha(x) := [h_\alpha]^{-1} \prod_{j=1}^{d} (1 - \|\mathbf{x}_{j-1}\|^2)^{\frac{1}{2}\alpha_j} C_{\alpha_j}^{\lambda_j}\left(\frac{x_j}{(1 - \|\mathbf{x}_{j-1}\|^2)^{1/2}}\right), \tag{2.5.3}$$

where $\lambda_j := \mu + |\alpha^{j+1}| + \frac{1}{2}(d - j)$ and h_α is given by

$$[h_\alpha]^2 := \frac{(\mu + \frac{1}{2}d)_{|\alpha|}}{(\mu + \frac{1}{2}(d + 1))_{|\alpha|}} \prod_{j=1}^{d} \frac{(\mu + \frac{1}{2}(d - j))_{|\alpha^j|}(2\mu + 2|\alpha^{j+1}| + d - j)_{\alpha_j}}{(\mu + \frac{1}{2}(d - j + 1))_{|\alpha^j|}\alpha_j!}, \tag{2.5.4}$$

and where the case $\mu = 0$ can be obtained as a limit for $\mu \to 0$ by using (2.3.1), in the same way as for (2.3.6), (2.3.7).

Second Orthonormal Basis Let $r_k^d := \dim \mathcal{H}_k^d$. For $0 \le j \le n/2$ let $\{Y_{\ell,n-2j} \mid 1 \le \ell \le r_{n-2j}^d\}$ be an orthonormal basis of $\mathcal{H}_{n-2j}^d$, the space of spherical harmonics of degree $n - 2j$, with respect to the normalized surface measure. For $0 \le j \le n/2$, define

$$P_{\ell,j}(x) := [h_{j,n}]^{-1} P_j^{(\mu - \frac{1}{2}, n-2j+\frac{1}{2}d-1)}(2\|x\|^2 - 1) Y_{\ell,n-2j}(x), \tag{2.5.5}$$

where

$$[h_{j,n}]^2 := \frac{(\mu + \frac{1}{2})_j (\frac{1}{2}d)_{n-j}}{j!\,(\mu + \frac{1}{2}(d + 1))_{n-j}} \frac{(n - j + \mu + \frac{1}{2}(d - 1))}{(n + \mu + \frac{1}{2}(d - 1))}.$$

Then $\{P_{\ell,j} \mid 1 \le \ell \le r_{n-2j}^d,\ 0 \le j \le n/2\}$ is an orthonormal basis of $\mathcal{V}_n^d$.

Appell's Biorthogonal Polynomials These are two families of polynomials yielding bases $\{U_\alpha\}_{|\alpha|=n}$ and $\{V_\alpha\}_{|\alpha|=n}$ of $\mathcal{V}_n^d$, where the second basis is the monic basis, up to constant factors, and the first basis is biorthogonal to it.

(i) The family $\{U_\alpha\}$ is defined by the generating function

$$\left((1 - \langle b, x\rangle)^2 + \|b\|^2(1 - \|x\|^2)\right)^{-\mu} = \sum_{\alpha \in \mathbb{N}_0^d} b^\alpha U_\alpha(x), \quad b \in \mathbb{R}^d,\ \|b\| < 1. \tag{2.5.6}$$

It satisfies the Rodrigues-type formula

$$U_\alpha(x) = \frac{(-1)^{|\alpha|}(2\mu)_{|\alpha|}}{2^{|\alpha|}(\mu+\frac{1}{2})_{|\alpha|}\alpha!}(1-\|x\|^2)^{-\mu+\frac{1}{2}}\frac{\partial^{|\alpha|}}{\partial x^\alpha}(1-\|x\|^2)^{|\alpha|+\mu-\frac{1}{2}}, \tag{2.5.7}$$

where $\frac{\partial^{|\alpha|}}{\partial x^\alpha} := \frac{\partial^{|\alpha|}}{\partial x_1^{\alpha_1}\cdots\partial x_d^{\alpha_d}}$. Furthermore, it can be explicitly given as

$$\begin{aligned} U_\alpha(x) &= \frac{(2\mu)_{|\alpha|}}{\alpha!}\sum_{\beta\le\alpha}\frac{(-1)^{|\beta|}(-\alpha)_{2\beta}}{2^{2|\beta|}\beta!(\mu+\frac{1}{2})_{|\beta|}}x^{\alpha-2\beta}(1-\|x\|^2)^{|\beta|} \\ &= \frac{(2\mu)_{|\alpha|}x^\alpha}{\alpha!}F_B\big(-\tfrac{1}{2}\alpha, -\tfrac{1}{2}(\alpha-\mathbf{1}); \mu+\tfrac{1}{2}; x_1^{-2}(1-\|x\|^2),\ldots,x_d^{-2}(1-\|x\|^2)\big), \end{aligned} \tag{2.5.8}$$

where $(-\alpha)_{2\beta} := (-\alpha_1)_{2\beta_1}\cdots(-\alpha_d)_{2\beta_d}$, $\mathbf{1} := (1,\ldots,1)$, $\beta\le\alpha$ means $\beta_i\le\alpha_i$ $(1\le i\le d)$, and F_B is Lauricella's hypergeometric series of type B (see Exton, 1976 and Chapter 3).

(ii) The family $\{V_\alpha\}$ is defined by the generating function

$$(1-2\langle b,x\rangle+\|b\|^2)^{-\mu-(d-1)/2} = \sum_{\alpha\in\mathbb{N}_0^d} b^\alpha V_\alpha(x), \quad b\in\mathbb{R}^d,\ \|b\|<1. \tag{2.5.9}$$

The generating function implies that

$$\|b\|^n C_n^{\mu+\frac{1}{2}(d-1)}\left(\frac{\langle b,x\rangle}{\|b\|}\right) = \sum_{|\alpha|=n} b^\alpha V_\alpha(x). \tag{2.5.10}$$

The polynomial V_α can be written explicitly as

$$\begin{aligned} V_\alpha(x) &= 2^{|\alpha|}x^\alpha\sum_{\gamma<\alpha}\frac{(\mu+\frac{1}{2}(d-1))_{|\alpha|-|\gamma|}(-\alpha+\gamma)_\gamma}{(\alpha-\gamma)!\gamma!}2^{-2|\gamma|}x^{-2\gamma} \\ &= \frac{2^{|\alpha|}(\mu+\frac{1}{2}(d-1))_{|\alpha|}}{\alpha!}x^\alpha F_B\big(-\tfrac{1}{2}\alpha, -\tfrac{1}{2}(\alpha-\mathbf{1}); -|\alpha|-\mu-\tfrac{1}{2}(d-3); x_1^{-2},\ldots,x_d^{-2}\big). \end{aligned} \tag{2.5.11}$$

Neither $\{U_\alpha\}_{|\alpha|=n}$ nor $\{V_\alpha\}_{|\alpha|=n}$ is an orthogonal basis of $\mathcal{V}_n^d$, but the two bases are biorthogonal to each other:

$$\int_{B^d} V_\alpha(x)U_\beta(x)W_\mu(x)\,dx = \frac{\mu+\frac{1}{2}(d-1)}{|\alpha|+\mu+\frac{1}{2}(d-1)}\frac{(2\mu)_{|\alpha|}}{\alpha!}\delta_{\alpha,\beta}. \tag{2.5.12}$$

The Monic Basis For each α define

$$R_\alpha(x) := \frac{\alpha!}{2^{|\alpha|}(\mu+\frac{1}{2}(d-1))_{|\alpha|}}V_\alpha(x). \tag{2.5.13}$$

Then the polynomials R_α $(|\alpha|=n)$ form a monic basis of $\mathcal{V}_n^d$, i.e., $R_\alpha(x) = x^\alpha - Q_\alpha(x)$ with $Q_\alpha\in\Pi_{n-1}^d$. The $L^2(W_\mu, B^d)$ norm, $\|\cdot\|_{2,\mu}$, of R_α is the error of the best approximation of x^α, which satisfies a closed formula

$$\min_{P\in\Pi_{n-1}^d}\|x^\alpha-P(x)\|_{2,\mu}^2 = \|R_\alpha\|_{2,\mu}^2 = \frac{\lambda\alpha!}{2^{n-1}(\lambda)_n}\int_0^1\left(\prod_{i=1}^d P_{\alpha_i}(t)\right)t^{n+2\lambda-1}\,dt, \tag{2.5.14}$$

where $\lambda = \mu + (d-1)/2 > 0$, $n = |\alpha|$, and P_{α_i} is the Legendre polynomial.

Reproducing Kernel The kernel $P_n(\cdot,\cdot)$ of $\mathcal{V}_n^d$ with respect to W_μ, defined in (2.2.22), satisfies a compact formula. For $\mu > 0$, $x, y \in B^d$,

$$P_n(x,y) = c_\mu \frac{2n+2\mu+d-1}{2\mu+d-1} \int_{-1}^{1} C_n^{\mu+\frac{1}{2}(d-1)}\left(\langle x,y\rangle + t\sqrt{1-\|x\|^2}\sqrt{1-\|y\|^2}\right)(1-t^2)^{\mu-1}\,dt, \tag{2.5.15}$$

where $[c_\mu]^{-1} := \int_{-1}^{1}(1-t^2)^{\mu-1}dt$. For $\mu = 0$, $x, y \in B^d$ this degenerates to

$$P_n(x,y) = \frac{n+\frac{1}{2}(d-1)}{d-1} \sum_{\epsilon=0,1} C_n^{\frac{1}{2}(d-1)}\left(\langle x,y\rangle + (-1)^\epsilon\sqrt{1-\|x\|^2}\sqrt{1-\|y\|^2}\right). \tag{2.5.16}$$

These formulas are essential for obtaining sharp results for convergence of orthogonal expansions. When $\mu = \frac{1}{2}(m-1)$, (2.5.15) can also be written as

$$P_n(x,y) = \frac{2n+m+d-2}{m+d-2} \int_{\mathbb{S}^{m-1}} C_n^{\frac{1}{2}(m+d)-1}\left(\langle x,y\rangle + \sqrt{1-\|x\|^2}\sqrt{1-\|y\|^2}\langle \xi, e_1\rangle\right) d\sigma_m(\xi), \tag{2.5.17}$$

where $d\sigma_m$ is the normalized surface measure on $\mathbb{S}^{m-1}$.

For the constant weight $W_{1/2}(x) = (d+1)/\pi$, there is another formula for the reproducing kernel,

$$P_n(x,y) = (2nd^{-1}+1) \int_{\mathbb{S}^{d-1}} C_n^{d/2}(\langle x,\xi\rangle) C_n^{d/2}(\langle \xi,y\rangle)\, d\sigma_d(\xi). \tag{2.5.18}$$

Further Results and References For orthogonal bases on the ball, see Appell and Kampé de Fériet (1926), Dunkl and Xu (2014), and Erdélyi (1953). There are further results on biorthogonal bases: see Appell and Kampé de Fériet (1926) and Erdélyi (1953). For the monic basis, see Xu (2005b). Orthogonal bases consisting of ridge polynomials were discussed by Xu (2000), together with a Funk–Hecke-type formula for orthogonal polynomials. The compact formulas (2.5.15) and (2.5.16) for the reproducing kernels were proved by Xu (1999) and used to study expansion problems, whereas the compact formula (2.5.17) was proved in (Xu, 2001a, Theorem 2.6) (there take $H_2(\eta) = 1$). Formula (2.5.18) was proved by Petrushev (1999) in the context of approximation by ridge functions, and by Xu (2007) in connection with Radon transforms. Atkinson et al. (2014) use three-term relations in order to develop an efficient numerical algorithm for the evaluation of orthogonal polynomials in (2.5.3) with $d = 2, 3$. For convergence and summability of orthogonal expansions, see §2.7.

2.5.2 Classical Orthogonal Polynomials on the Simplex

For $x \in \mathbb{R}^d$, let $|x| := x_1 + \cdots + x_d$. Let $T^d := \{x \in \mathbb{R}^d \mid x_1, \ldots, x_d, 1-|x| \geq 0\}$ be the simplex in $\mathbb{R}^d$. The classical weight function on T^d is defined by

$$W_\kappa(x) := \frac{\Gamma(|\kappa| + \frac{1}{2}(d+1))}{\prod_{i=1}^{d+1}\Gamma(\kappa_i+\frac{1}{2})} x_1^{\kappa_1-\frac{1}{2}} \cdots x_d^{\kappa_d-\frac{1}{2}} (1-|x|)^{\kappa_{d+1}-\frac{1}{2}}, \quad \kappa_1, \ldots, \kappa_{d+1} > -\tfrac{1}{2}. \tag{2.5.19}$$

Differential Operator Orthogonal polynomials of degree n with respect to W_κ are eigenfunctions of a second-order differential operator,

$$\left(\sum_{i=1}^{d} x_i(1-x_i)\frac{\partial^2}{\partial x_i^2} - 2\sum_{1\le i<j\le d} x_i x_j \frac{\partial^2}{\partial x_i \partial x_j} + \sum_{i=1}^{d}\left((\kappa_i+\tfrac{1}{2}) - (|\kappa|+\tfrac{1}{2}(d+1))x_i\right)\frac{\partial}{\partial x_i}\right)P$$
$$= -n\left(n+|\kappa|+\tfrac{1}{2}(d-1)\right)P, \quad P\in\mathcal{V}_n^d, \tag{2.5.20}$$

where $|\kappa| = \kappa_1 + \cdots + \kappa_{d+1}$.

An Orthonormal Basis To state this basis we use the notation $\mathbf{x}_j$ and α^j as in the first orthonormal basis on B^d of §2.5.1. We also put $\kappa^j := (\kappa_j, \ldots, \kappa_{d+1})$ $(j = 0, 1, \ldots, d+1)$ if $\kappa = (\kappa_1, \ldots, \kappa_{d+1})$. Then an orthonormal basis $\{P_\alpha\}_{|\alpha|=n}$ of $\mathcal{V}_n^d$ is given by

$$P_\alpha(x) := [h_\alpha]^{-1}\prod_{j=1}^{d}(1-|\mathbf{x}_{j-1}|)^{\alpha_j} P_{\alpha_j}^{(a_j,\kappa_j-\frac{1}{2})}\left(\frac{2x_j}{1-|\mathbf{x}_{j-1}|}-1\right), \tag{2.5.21}$$

where $a_j := 2|\alpha^{j+1}| + |\kappa^{j+1}| + \frac{1}{2}(d-j-1)$ and h_α is given by

$$[h_\alpha]^2 := \prod_{j=1}^{d}\frac{(\kappa_j+\frac{1}{2})_{\alpha_j}(|\kappa^{j+1}|+\frac{1}{2}(d-j+1))_{|\alpha^j|+|\alpha^{j+1}|}}{\alpha_j!(|\kappa^j|+\frac{1}{2}(d-j+2))_{|\alpha^j|+|\alpha^{j+1}|}}\,\frac{2(a_j+\kappa_j+\alpha_j)+1}{2(a_j+\kappa_j+2\alpha_j)+1}.$$

Appell's Biorthogonal Polynomials These are two families of polynomials yielding bases $\{U_\alpha\}_{|\alpha|=n}$ and $\{V_\alpha\}_{|\alpha|=n}$ of $\mathcal{V}_n^d$, where the second basis is the monic basis and the first basis is biorthogonal to it.

(i) The family $\{U_\alpha\}$ is defined by the Rodrigues-type formula

$$U_\alpha(x) := x_1^{-\kappa_1+\frac{1}{2}}\cdots x_d^{-\kappa_d+\frac{1}{2}}(1-|x|)^{-\kappa_{d+1}+\frac{1}{2}}\frac{\partial^{|\alpha|}}{\partial x^\alpha}\left(x_1^{\alpha_1+\kappa_1-\frac{1}{2}}\cdots x_d^{\alpha_d+\kappa_d-\frac{1}{2}}(1-|x|)^{|\alpha|+\kappa_{d+1}-\frac{1}{2}}\right). \tag{2.5.22}$$

(ii) The family $\{V_\alpha\}$ is explicitly defined by

$$V_\alpha(x) := \sum_{\beta\le\alpha}(-1)^{n+|\beta|}\left(\prod_{i=1}^{d}\binom{\alpha_i}{\beta_i}\frac{(\kappa_i+\frac{1}{2})_{\alpha_i}}{(\kappa_i+\frac{1}{2})_{\beta_i}}\right)\frac{(|\kappa|+\frac{1}{2}(d-1))_{n+|\beta|}}{(|\kappa|+\frac{1}{2}(d-1))_{n+|\alpha|}}x^\beta$$
$$= \frac{(-1)^n(\kappa+\frac{1}{2})_\alpha}{(n+|\kappa|+\frac{1}{2}(d-1))_{|\alpha|}}F_A\left(n+|\kappa|+\tfrac{1}{2}(d-1), -\alpha; \kappa+\tfrac{1}{2}, x\right) \tag{2.5.23}$$

where F_A denotes Lauricella's hypergeometric series of type A (see Exton, 1976 and Chapter 3).

Neither $\{U_\alpha\}_{|\alpha|=n}$ nor $\{V_\alpha\}_{|\alpha|=n}$ is an orthogonal basis of $\mathcal{V}_n^d$, but the two bases are biorthogonal to each other:

$$\int_{T^d} V_\beta(x)U_\alpha(x)W_\kappa(x)\,dx = \frac{(\kappa+\frac{1}{2})_\alpha(\kappa_{d+1}+\frac{1}{2})_{|\alpha|}}{(|\kappa|+\frac{1}{2}(d+1))_{2|\alpha|}}\alpha!\,\delta_{\alpha,\beta}. \tag{2.5.24}$$

Monic Orthogonal Basis By (2.5.23) and (2.5.24) the polynomials V_α ($|\alpha| = n$) form a monic basis: $V_\alpha(x) = x^\alpha - Q_\alpha(x)$, $Q_\alpha \in \Pi_{n-1}^d$. Such polynomials can be defined more generally in view of the observation that the simplex T^d is associated with the permutation group of $X := (x_1, \ldots, x_d, x_{d+1})$, where $x_{d+1} = 1 - |x|$. For $x \in T^d$ and $\alpha \in \mathbb{N}_0^{d+1}$, define $X^\alpha := x_1^{\alpha_1} \cdots x_d^{\alpha_d}(1-|x|)^{\alpha_{d+1}}$. For $|\alpha| = n$ let $R_\alpha(x) = X^\alpha - Q_\alpha(x)$ be the element of $\mathcal{V}_n^d$ such that Q_α is a polynomial of degree at most $|\alpha| - 1$. When $\alpha_{d+1} = 0$, $R_\alpha(x)$ agrees with $V_\alpha(x)$ in (2.5.23) up to a constant factor. The $L^2(W_\kappa, T^d)$ norm $\|\cdot\|_{2,\kappa}$ of R_α is the error of the best approximation of X^α, which satisfies a closed formula:

$$\min_{P \in \Pi_{n-1}^d} \|X^\alpha - P(x)\|_{2,\kappa}^2 = \|V_\alpha\|_{2,\kappa}^2 = \frac{(|\kappa| + \frac{1}{2}(d-1))(\kappa + \frac{1}{2})_\alpha}{|\kappa| + \frac{1}{2}(d-1)_{2|\alpha|}} \int_0^1 \left(\prod_{i=1}^{d+1} P_{\alpha_i}^{(0,\kappa_i - \frac{1}{2})}(2t-1) \right) t^{|\alpha| + |\kappa| + \frac{d-3}{2}} \, dt. \tag{2.5.25}$$

If $\alpha_{d+1} = 0$, (2.5.25) gives the error of best approximation to x^α.

Reproducing Kernel The kernel $P_n(\cdot, \cdot)$ of $\mathcal{V}_n^d$ with respect to W_κ, defined in (2.2.22), satisfies a compact formula. For $\kappa_i > 0$, $1 \le i \le d$, $x, y \in T^d$,

$$\begin{aligned} P_n(x,y) = c_\kappa & \frac{2(2n + |\kappa|) + d - 1}{2|\kappa| + d - 1} \frac{(|\kappa| + \frac{1}{2}(d-1))_n}{(\frac{1}{2})_n} \\ & \times \int_{[-1,1]^{d+1}} P_n^{(|\kappa| + (d-2)/2, -1/2)}\left(2z(x,y,t)^2 - 1\right) \left(\prod_{i=1}^{d+1} (1 - t_i^2)^{\kappa_i - 1} \right) dt, \end{aligned} \tag{2.5.26}$$

where $z(x,y,t) := \sqrt{x_1 y_1}\, t_1 + \cdots + \sqrt{x_d y_d}\, t_d + \sqrt{1-|x|}\sqrt{1-|y|}\, t_{d+1}$ and

$$[c_\kappa]^{-1} := \int_{[-1,1]^{d+1}} \left(\prod_{i=1}^{d+1} (1 - t_i^2)^{\kappa_i - 1} \right) dt.$$

If some $\kappa_i = 0$, then the formula holds under the limit relation

$$\lim_{\lambda \to 0} c_\lambda \int_{-1}^1 g(t)(1-t)^{\lambda - 1} \, dt = \tfrac{1}{2}(g(1) + g(-1)).$$

Further Results and References For $d = 2$ the polynomials U_α on T^d were defined by Appell and Kampé de Fériet (1926) and the polynomials V_α were studied by Fackerell and Littler (1974). For $d > 2$ they appeared in Grundmann and Möller (1978) when $W_\kappa(x) = 1$, and in Dunkl and Xu (2014) in general. The monic basis of polynomials R_α was studied by Xu (2005b). Formula (2.5.26) of the reproducing kernel appeared in Xu (1998b). A product formula for orthogonal polynomials on the simplex was established by Koornwinder and Schwartz (1997). The polynomials in (2.5.21) serve as generating functions for the Hahn polynomials in several variables (Karlin and McGregor, 1975; Xu, 2015b). For convergence and summability of orthogonal expansions see §2.7.

2.5.3 Hermite Polynomials of Several Variables

These are orthogonal polynomials with respect to the product weight

$$W_H(x) := \pi^{-d/2} e^{-\|x\|^2}, \quad x \in \mathbb{R}^d. \tag{2.5.27}$$

Many properties are inherited from Hermite polynomials of one variable.

Differential Operator Orthogonal polynomials of degree n with respect to W_H are eigenfunctions of a second-order differential operator:

$$\left(\Delta - 2\sum_{i=1}^{d} x_i \frac{\partial}{\partial x_i}\right) P = -2nP, \quad P \in \mathcal{V}_n^d. \tag{2.5.28}$$

Product Orthogonal Basis For $\alpha \in \mathbb{N}_0^d$, define

$$H_\alpha(x) := H_{\alpha_1}(x_1) \cdots H_{\alpha_d}(x_d). \tag{2.5.29}$$

Then $\{[c_\alpha]^{-1} H_\alpha\}_{|\alpha|=n}$ is an orthonormal basis of $\mathcal{V}_n^d$, where $[c_\alpha]^2 := 2^{|\alpha|}\alpha!$. As products of Hermite polynomials of one variable, they inherit a generating function and Rodrigues-type formula. Furthermore, they satisfy

$$\sum_{|\alpha|=n} \frac{H_\alpha(y)}{\alpha!} x^\alpha = \frac{1}{n!\pi^{\frac{1}{2}}} \int_{-\infty}^{\infty} H_n\Big(\langle x, y\rangle + s\sqrt{1 - \|x\|^2}\Big) e^{-s^2}\, ds, \quad x \in \mathbb{B}^d,\ y \in \mathbb{R}^d.$$

In particular,

$$\sum_{|\alpha|=n} \frac{H_\alpha(y)}{\alpha!} x^\alpha = \frac{1}{n!} H_n(\langle x, y\rangle)), \quad \|x\| = 1,\ y \in \mathbb{R}^d. \tag{2.5.30}$$

Second Orthonormal Basis For $0 \le j \le n/2$ and $r_k^d := \dim \mathcal{H}_k^d$, let $\{Y_{\ell,n-2j}\}_{\ell=1}^{r_{n-2j}^d}$ be an orthonormal basis of $\mathcal{H}_{n-2j}^d$ as in (2.5.5). Define

$$P_{\ell,j}^n(x) := [c_{j,n}]^{-1} L_j^{n-2j+\frac{1}{2}(d-2)}(\|x\|^2) Y_{\ell,n-2j}(x), \quad [c_{j,n}]^2 := \frac{(\frac{1}{2}d)_{n-j}}{j!}. \tag{2.5.31}$$

Then $\{P_{\ell,j}^n \mid 1 \le \ell \le r_{n-2j}^d,\ 0 \le j \le \frac{1}{2}n\}$ is an orthonormal basis of $\mathcal{V}_n^d$.

Mehler Formula The reproducing kernel $P_n(\cdot,\cdot)$ of $\mathcal{V}_n^d$, as defined in (2.2.22), satisfies, for $0 < z < 1$ and $x, y \in \mathbb{R}^d$,

$$\sum_{n=0}^{\infty} P_n(x, y) z^n = \frac{1}{(1 - z^2)^{d/2}} \exp\left(-\frac{z^2(\|x\|^2 + \|y\|^2) - 2z\langle x, y\rangle}{1 - z^2}\right). \tag{2.5.32}$$

Further Results and References The study of Hermite polynomials of several variables was started by Hermite and followed by many other authors; see Appell and Kampé de Fériet (1926) and Erdélyi (1953) for references. Analogues of Hermite polynomials can be defined more generally for the weight function

$$W(x) = (\det A)^{\frac{1}{2}} \pi^{-d/2} \exp\left(-x^{\mathrm{tr}} A x\right), \tag{2.5.33}$$

where A is a positive definite matrix. Two families of biorthogonal polynomials can be defined for W in (2.5.33), which coincide when A is an identity matrix. These were studied by Appell and Kampé de Fériet (1926); see also Erdélyi (1953). Since A is positive definite, it can be written as $A = B^{\mathrm{tr}} B$. Thus, orthogonal polynomials for W in (2.5.33) can be derived from Hermite polynomials for W_H by a change of variables.

2.5.4 Laguerre and Generalized Hermite Polynomials

Laguerre Polynomials of Several Variables Put $|x| := x_1 + \cdots + x_d$ and $\mathbb{R}^d_+ := \{x \in \mathbb{R}^d \mid x_1, \ldots, x_d \geq 0\}$. *Laguerre polynomials of several variables* are orthogonal polynomials with respect to the weight function

$$W_L(x) := \frac{1}{\prod_{k=1}^d \Gamma(\kappa_i + \frac{1}{2})} x^\kappa e^{-|x|}, \quad x \in \mathbb{R}^d_+, \tag{2.5.34}$$

which can be written as the product of weight functions in one variable. Many properties are inherited from Laguerre polynomials of one variable.

The polynomials in $\mathcal{V}_n$ are eigenfunctions of a differential operator:

$$\left(\sum_{i=1}^d x_i \frac{\partial^2}{\partial x_i^2} + \sum_{i=1}^d (\kappa_i + 1 - x_i) \frac{\partial}{\partial x_i} \right) P = -nP, \quad P \in \mathcal{V}^d_n. \tag{2.5.35}$$

An orthonormal basis of $\mathcal{V}^d_n$ is given by $\{L^\kappa_\alpha\}_{|\alpha|=n}$, where

$$L^\kappa_\alpha(x) := \binom{\alpha + \kappa}{\alpha}^{-1/2} L^{\kappa_1}_{\alpha_1}(x_1) \cdots L^{\kappa_d}_{\alpha_d}(x_d). \tag{2.5.36}$$

The reproducing kernel $P_n(\cdot, \cdot)$ of $\mathcal{V}^d_n$, defined in (2.2.22), satisfies, for $0 < z < 1$ and $x, y \in \mathbb{R}^d_+$,

$$\sum_{n=0}^\infty P_n(x, y) z^n = (1 - z)^{-1} \prod_{i=1}^d \exp\left(-\frac{z(x_i + y_i)}{1 - z} \right) (x_i y_i z)^{-\frac{1}{2}\kappa_i} I_{\kappa_i} \left(\frac{2\sqrt{x_i y_i z}}{1 - z} \right), \tag{2.5.37}$$

where I_κ denotes the modified Bessel function of order κ.

Generalized Hermite Polynomials of Several Variables These are orthogonal polynomials with respect to the product weight function

$$W_\kappa(x) := \prod_{i=1}^d |x_i|^{2\kappa_i} e^{-x_i^2}. \tag{2.5.38}$$

Let Δ_h be the Dunkl Laplacian (2.4.18). The orthogonal polynomials in $\mathcal{V}_n^d$ are eigenfunctions of a differential-difference operator:

$$\big(\Delta_h - 2\langle x, \nabla\rangle\big)P = -2nP, \quad P \in \mathcal{V}_n^d. \tag{2.5.39}$$

For $\kappa > 0$ let the *generalized Hermite polynomials* H_n^κ in one variable be defined by

$$\begin{aligned} H_{2n}^\kappa(x) &:= (-1)^n 2^{2n} n!\, L_n^{\kappa-\frac{1}{2}}(x^2), \\ H_{2n+1}^\kappa(x) &:= (-1)^n 2^{2n+1} n!\, x L_n^{\kappa+\frac{1}{2}}(x^2). \end{aligned} \tag{2.5.40}$$

Let $h_n^\kappa = \Gamma(\kappa+\frac{1}{2})^{-1} \int_{\mathbb{R}} [H_n^\kappa(x)]^2 x^\kappa e^{-x^2}\, dx$, normalized such that $h_0^\kappa = 1$. Then

$$h_{2n}^\kappa = 2^{4n} n! (\mu+\tfrac{1}{2})_n \quad \text{and} \quad h_{2n+1}^\kappa = 2^{4n+2} n! (\mu+\tfrac{1}{2})_{n+1}. \tag{2.5.41}$$

An orthonormal basis of $\mathcal{V}_n^d$ for W_κ is given by $\{P_\alpha\}_{|\alpha|=n}$, where

$$P_\alpha(x) := (h_{\alpha_1}^{\kappa_1} \cdots h_{\alpha_d}^{\kappa_d})^{-\frac{1}{2}} H_{\alpha_1}^{\kappa_1}(x_1) \cdots H_{\alpha_d}^{\kappa_d}(x_d). \tag{2.5.42}$$

Another orthonormal basis, analogous to (2.5.31), can be given in polar coordinates by Laguerre polynomials and h-harmonics.

Further Results and References As products of Laguerre polynomials of one variable, the polynomials $L_\alpha^\kappa(x)$ in (2.5.36) have a generating function and a Rodrigues-type formula, as well as a product formula that induces a convolution structure. The generalized Hermite polynomials can be defined for weight functions under other reflection groups. For those and further properties of these functions, including a Mehler-type formula, see Dunkl and Xu (2014), Rösler (1998), and Xu (2001b).

2.5.5 Jacobi Polynomials of Several Variables

These polynomials are orthogonal with respect to the weight function

$$W_{a,b}(x) := \prod_{i=1}^{d} (1-x_i)^{a_i}(1+x_i)^{b_i}, \quad x \in [-1,1]^d. \tag{2.5.43}$$

An orthogonal basis is formed by the *product Jacobi polynomials*,

$$P_\alpha(x) := P_{\alpha_1}^{(a_1,b_1)}(x_1) \cdots P_{\alpha_d}^{(a_d,b_d)}(x_d), \quad \alpha \in \mathbb{N}_0^d.$$

Most results for these orthogonal polynomials follow from properties of Jacobi polynomials of one variable. The reproducing kernel $P_n(\cdot,\cdot)$ of $\mathcal{V}_n^d$ satisfies

$$\sum_{n=0}^{\infty} P_n(x,\mathbf{1}) r^n = \prod_{i=1}^{d} \frac{(1-r)(1+r)^{a_i-b_i+1}}{(1-2rx_i+r^2)^{a_i+3/2}}\, {}_2F_1\left(\begin{matrix} \frac{1}{2}(b_i-a_i), \frac{1}{2}(b_i-a_i-1) \\ b_i+1 \end{matrix}; \frac{2r(1+x_i)}{(1+r)^2}\right), \tag{2.5.44}$$

where $0 < r < 1$ and $\mathbf{1} = (1,\ldots,1)$.

In the case of the product Chebyshev weight function, i.e., $a_i = b_i = -\frac{1}{2}$, $i = 1, \dots, d$, the reproducing kernel $P_n(\cdot,\cdot)$ satisfies a closed formula (Dunkl and Xu, 2014, Theorem 9.6.3). This is given in terms of a *divided difference* $[x_1, \dots, x_d]f$ defined by

$$[x_0]f := f(x_0) \quad \text{and} \quad [x_0, \dots, x_n]f := \frac{[x_0, \dots, x_{n-1}]f - [x_1, \dots, x_n]f}{x_0 - x_n},$$

which is a symmetric function in x. For $W(x) := \prod_{i=1}^{d}(1 - x_i^2)^{-\frac{1}{2}}$,

$$P_n(x, \mathbf{1}) = [x_1, \dots, x_d]G_n, \tag{2.5.45}$$

where

$$G_n(t) := 2(-1)^{[\frac{1}{2}(d+1)]}(1 - t^2)^{\frac{1}{2}(d-1)} \begin{cases} T_n(t) & \text{for } d \text{ even}, \\ U_{n-1}(t) & \text{for } d \text{ odd}. \end{cases}$$

The product Jacobi polynomials inherit a product formula from the one-variable case (Dunkl and Xu, 2014, Lemma 9.6.1), which allows one to define a convolution structure for orthogonal expansions.

2.6 Relation Between Orthogonal Polynomials on Classical Domains

By *classical domains* we mean the sphere, ball, simplex, $\mathbb{R}^d$, and $\mathbb{R}^d_+$. Orthogonal polynomials on these domains are closely related.

2.6.1 Orthogonal Polynomials on the Sphere and on the Ball

A nonnegative weight function H defined on $\mathbb{R}^{d+1}$ is called *S-symmetric* if $H(x', x_{d+1}) = H(x', -x_{d+1}) = H(-x', x_{d+1})$, where $x' \in \mathbb{R}^d$, and if the restriction W_H of H on the sphere $\mathbb{S}^d$ is a nontrivial weight function. Let $\mathcal{H}_n^{d+1}(H)$ denote the space of homogeneous polynomials of degree n that are orthogonal in $L^2(\mathbb{S}^d, H)$ to polynomials of lower degrees. Then, just as in the case (2.4.1) of $H = 1$,

$$\dim \mathcal{H}_n^{d+1}(H) = \dim \mathcal{P}_n^{d+1} - \dim \mathcal{P}_{n-2}^{d+1} = \binom{n+d}{d} - \binom{n+d-2}{d}. \tag{2.6.1}$$

Associated with an S-symmetric weight function H, define

$$W_H(x) := H\big(x, (1 - \|x\|^2)^{\frac{1}{2}}\big), \qquad x \in \mathbb{B}^d,$$

which is a centrally symmetric weight function on the ball. Let $\{P_\alpha\}_{|\alpha|=n, \alpha\in\mathbb{N}_0^d}$ be an orthogonal basis for $\mathcal{V}_n^d$ with respect to $(1 - \|x\|^2)^{-\frac{1}{2}} W_H(x)$, and let $\{Q_\beta\}_{|\beta|=n-1, \beta\in\mathbb{N}_0^d}$ be an orthogonal basis for $\mathcal{V}_{n-1}^d$ with respect to the weight function $(1 - \|x\|^2)^{\frac{1}{2}} W_H(x)$. For $y \in \mathbb{R}^{d+1}$ define

$$y = r(x, x_{d+1}), \qquad x \in \mathbb{B}^d, \ (x, x_{d+1}) \in \mathbb{S}^d, \ r \ge 0.$$

For $\alpha, \beta \in \mathbb{N}_0^d$, $|\alpha| = n$, and $|\beta| = n - 1$, define

$$Y_\alpha^{(1)}(y) = r^n P_\alpha(x) \quad \text{and} \quad Y_\beta^{(2)}(y) = r^n x_{d+1} Q_\beta(x).$$

Theorem 2.6.1 *The functions $Y_\alpha^{(1)}$ and $Y_\beta^{(2)}$ are homogeneous polynomials of degree n in the variable y. Furthermore, $\{Y_\alpha^{(1)}\}_{|\alpha|=n} \cup \{Y_\beta^{(2)}\}_{|\beta|=n-1}$ is an orthogonal basis for $\mathcal{H}_n^{d+1}(H)$.*

Let $P_n^H(\cdot,\cdot)$ denote the reproducing kernel of $\mathcal{H}_n^d(H)$ and let $P_n(\cdot,\cdot)$ denote the reproducing kernel of $\mathcal{V}_n^d$ with respect to $(1-\|x\|^2)^{-\frac{1}{2}} W_H(x)$. Then

$$P_n(x,y) = \tfrac{1}{2}\left(P_n^H((x,x_{d+1}),(y,y_{d+1})) + P_n^H((x,x_{d+1}),(y,-y_{d+1}))\right), \tag{2.6.2}$$

where $x_{d+1} = (1-\|x\|^2)^{\frac{1}{2}}$ and $y_{d+1} = (1-\|y\|^2)^{\frac{1}{2}}$. The relation is based on

$$\int_{\mathbb{S}^d} f(y)\,d\omega(y) = \int_{\mathbb{B}^d} \left(f(x,(1-\|x\|^2)^{\frac{1}{2}}) + f(x,-(1-\|x\|^2)^{\frac{1}{2}})\right)(1-\|x\|^2)^{-\frac{1}{2}}\,dx, \tag{2.6.3}$$

where $d\omega$ is the Lebesgue measure (not normalized) on $\mathbb{S}^d$. A further relation between orthogonal polynomials on $\mathbb{B}^d$ and those on $\mathbb{S}^{d+m-1}$ follows from

$$\int_{\mathbb{S}^{d+m-1}} f(y)\,d\omega(y) = \int_{\mathbb{B}^d} (1-\|x\|^2)^{\frac{1}{2}(m-1)} \left(\int_{\mathbb{S}^m} f(x,(1-\|x\|^2)^{\frac{1}{2}}\xi)\,d\omega_m(\xi)\right) dx.$$

As a consequence of these relations, properties of the orthogonal polynomials with respect to the weight function

$$W_\kappa(x) = \prod_{i=1}^d |x_i|^{2\kappa_i}(1-\|x\|^2)^{\kappa_{d+1}-\frac{1}{2}}, \quad k_i \ge 0 \tag{2.6.4}$$

on $\mathbb{B}^d$ can be derived from h-harmonics with respect to $w_\kappa(x) = \prod_{i=1}^{d+1}|x_i|^{2\kappa_i}$ on the sphere $\mathbb{S}^{d+1}$. In particular, following (2.4.23) and using generalized Gegenbauer polynomials (2.4.21), define

$$P_\alpha(x) := [h_\alpha]^{-1} \prod_{j=1}^d (1-\|\mathbf{x}_{j-1}\|^2)^{\frac{1}{2}\alpha_j} C_{\alpha_j}^{(a_j,\kappa_j)}\left((1-\|\mathbf{x}_{j-1}\|^2)^{-\frac{1}{2}}x_j\right), \quad \alpha \in \mathbb{N}_0^d, \tag{2.6.5}$$

where $|\alpha^j| := \alpha_j + \cdots + \alpha_d$, $|\kappa^j| := \kappa_j + \cdots + \kappa_{d+1}$, $a_j := |\alpha^{j+1}| + |\kappa^{j+1}| + \frac{d-j}{2}$, and

$$[h_\alpha^n]^2 := \frac{1}{(|\kappa| + \frac{1}{2}(d+1))_n} \prod_{j=1}^d h_{\alpha_i}^{(\alpha_i,\kappa_i)}(\kappa_i + a_i)_{\alpha_i}.$$

Then $\{P_\alpha\}_{|\alpha|=n}$ is an orthonormal basis of $\mathcal{V}_n^d$. A second orthonormal basis can be given in spherical-polar coordinates, analogous to (2.5.5) but using h-harmonics. The orthogonal polynomials in $\mathcal{V}_n^d$ are eigenfunctions of a second-order differential-difference operator,

$$\left(\Delta_h - \langle x,\nabla\rangle^2 - (2|\kappa|+d-1)\langle x,\nabla\rangle\right)P = -n(n+2|\kappa|+d-1)P, \tag{2.6.6}$$

where Δ_h is the Dunkl Laplacian given in §2.4.2. The reproducing kernel $P_n(\cdot,\cdot)$ of $\mathcal{V}_n^d$ for W_κ satisfies a closed formula:

$$\begin{aligned} P_n(x,y) = \frac{2n+2|\kappa|+d-1}{2|\kappa|+d-1} \int_{[-1,1]^{d+1}} & C_n^{|\kappa|+\frac{1}{2}(d-1)}(x_1y_1t_1 + \cdots + x_{d+1}y_{d+1}t_{d+1}) \\ & \times \left(\prod_{i=1}^d (1+t_i)\right)\left(\prod_{i=1}^{d+1} c_{\kappa_i}(1-t_i^2)^{\kappa_i-1}\right)dt, \end{aligned} \tag{2.6.7}$$

where $x_{d+1} = \sqrt{1-\|x\|^2}$ and $y_{d+1} = \sqrt{1-\|y\|^2}$.

Further Results and References The relation between orthogonal polynomials on the sphere and on the ball was explored by Xu (1998a, 2001a). The orthonormal basis (2.6.5) and the closed form of formula (2.6.7) were given by Xu (2001a). The monic orthogonal basis for W_κ was studied by Xu (2005b).

2.6.2 Orthogonal Polynomials on the Ball and on the Simplex

Let W be a weight function defined on the simplex T^d. Define

$$W_T(x) := W(x)/\sqrt{x_1\cdots x_d} \quad \text{and} \quad W_B(x) := W(x_1^2,\ldots,x_d^2)$$

on T^d and on the ball $\mathbb{B}^d$, respectively. There is a close relation between orthogonal polynomials for W_T and those for $\mathbb{B}^d$. Let $\mathcal{V}_n^d(W)$ denote the space of orthogonal polynomials of degree n with respect to W. Furthermore, let $\mathcal{V}_n(W_B,\mathbb{Z}_2^d)$ denote the subspace of $\mathcal{V}_n^d(W_B)$ which contains polynomials that are invariant under $\mathbb{Z}_2^d$, that is, even in each variable. Then the mapping

$$\psi : T^d \to \mathbb{B}^d : (x_1,\ldots,x_d) \mapsto (x_1^2,\ldots,x_d^2)$$

induces a one-to-one correspondence between $R \in \mathcal{V}_n^d(W_T)$ and $R\circ\psi \in \mathcal{V}_n(W_B,\mathbb{Z}_2^d)$. This is based on the relation

$$\int_{B^d} f(x_1^2,\ldots,x_d^2)\,dx = \int_{T^d} f(u_1,\ldots,u_d)\frac{du}{\sqrt{u_1\cdots u_d}}. \tag{2.6.8}$$

Let $P_n(W;\cdot,\cdot)$ denote the reproducing kernel of $\mathcal{V}_n^d(W)$. Then the correspondence also extends to the reproducing kernel:

$$P_n(W_T;x,y) = 2^{-d}\sum_{\varepsilon\in\mathbb{Z}_2^d} P_n(W_B;x^{1/2},\varepsilon y^{1/2}), \tag{2.6.9}$$

where $x^{1/2} := (x_1^{1/2},\ldots,x_d^{1/2})$. In particular, the identity (2.5.26) can be deduced from (2.6.7) by this relation.

Essentially all properties of orthogonal polynomials for W_T can be deduced from the corresponding results for W_B. In particular, all results in §2.5.2 can be deduced from the corresponding results with respect to W_κ in (2.6.4). In combination with §2.6.1 there is also a correspondence between orthogonal polynomials on the simplex and those on the sphere.

Further Results and References The relation between orthogonal polynomials on the ball and on the simplex was studied by Xu (1998b). The details on orthogonal systems for the classical weight functions were worked out by Xu (2001b). The connection extends to other aspects of analysis, including orthogonal expansion, approximation, and numerical integration.

2.6.3 Limit Relations

Two limit relations between orthogonal polynomials on two different domains are worth mentioning.

Limit of Orthogonal Polynomials on the Ball (Dunkl and Xu, 2014, Theorem 8.3.5)
Let $P_\alpha(W_\kappa)$ be the orthogonal polynomials (2.6.5) on the ball for W_κ in (2.6.4). Then

$$\lim_{\kappa_{d+1}\to 0} P_\alpha(W_\kappa, x/\kappa_{d+1}) = c_\alpha^{-1} H_{\alpha_1}^{\kappa_1}(x_1)\cdots H_{\alpha_d}^{\kappa_d}(x_d),$$

where the right-hand side is the generalized Hermite polynomial (2.5.42) on $\mathbb{R}^d$, a product of generalized Hermite polynomials in one variable.

Limit of Orthogonal Polynomials on the Simplex (Dunkl and Xu, 2014, Theorem 8.4.5)
Let $P_\alpha(W_\kappa)$ be the orthogonal polynomials in (2.5.21) on the simplex for W_κ in (2.5.19). Then

$$\lim_{\kappa_{d+1}\to 0} P_\alpha(W_\kappa, x/\kappa_{d+1}) = L_{\alpha_1}^{\kappa_1}(x_1)\cdots L_{\alpha_d}^{\kappa_d}(x_d) S_n$$

where the right-hand side is a Laguerre polynomial (2.5.36) on $\mathbb{R}_+^d$, a product of Laguerre polynomials of one variable.

2.7 Orthogonal Expansions and Summability

As long as polynomials are dense in $L^2(d\mu)$, the standard Hilbert space theory shows that the partial sum $S_n f$ in (2.2.25) converges to f in the $L^2(d\mu)$ norm. The convergence does not hold in general for $S_n f$ in other norms. The summability of the orthogonal expansions is often studied via the Cesàro mean. For $\delta > 0$, the *Cesàro* (C, δ) *mean* of the orthogonal expansion (2.2.23) is defined by

$$S_n^\delta f(x) := \binom{n+\delta}{n}^{-1} \sum_{k=0}^{n} \binom{n-k+\delta-1}{n-k} S_k f(x). \tag{2.7.1}$$

There are many results for orthogonal expansions for classical-type orthogonal polynomials. Below is a list of highlights with references.

Orthogonal Expansions on $\mathbb{S}^{d-1}$ The results are stated in terms of w_κ defined in (2.4.17) on the sphere. The case $\kappa = 0$ gives the result for ordinary spherical harmonics expansions. Let $\|\cdot\|_{\kappa,p}$ denote the $L^p(w_\kappa)$ norm; for $p = \infty$, the norm is the uniform norm of $C(\mathbb{S}^{d-1})$. Let

$$\sigma_\kappa := \tfrac{d-2}{2} + |\kappa| - \min_{1\le j\le d} \kappa_j, \quad \delta_\kappa(p) := \max\left((2\sigma_\kappa + 1)\left|\tfrac{1}{p} - \tfrac{1}{2}\right| - \tfrac{1}{2}, 0\right).$$

1. For $p = 1$ or ∞, the norm of the partial sum operator and the projection operator satisfy $\|S_n\|_{\kappa,p} \sim \|\operatorname{proj}_n\|_{\kappa,p} \sim n^{\sigma_\kappa}$.
2. The (C, δ) mean $S_n^\delta f \ge 0$ for all $f \ge 0$ if and only if $\delta \ge 2|\kappa| + d - 1$.

3. If $f \in C(\mathbb{S}^{d-1})$ then $S_n^\delta f$ converges to f pointwise in $\{x \in \mathbb{S}^{d-1} \mid x_1 \cdots x_d \neq 0\}$ if $\delta > \frac{d-1}{2}$; if $f \in L^1(w_\kappa)$ then $S_n^\delta f$ converges almost everywhere to f on $\mathbb{S}^{d-1}$ if $\delta > \sigma_\kappa$.
4. For $p = 1$ or ∞, $S_n^\delta f$ converges to f in $L^p(w_\kappa)$ if and only if $\delta > \sigma_\kappa$.
5. If $f \in L^p(w_\kappa)$, $1 \le p \le \infty$, $\left|\frac{1}{p} - \frac{1}{2}\right| \ge \frac{1}{2\sigma_\kappa+2}$, and $\delta > \delta_\kappa(p)$, then $S_n^\delta f$ converges to f in $L^p(w_\kappa^2)$.

For expansions of ordinary spherical harmonics, these were mostly proved by Bonami and Clerc (1973) and Sogge (1986); see Erdélyi (1953) for earlier results. For h-harmonics, these results were established by Dai and Xu (2009a,b), Li and Xu (2003), and Xu (1997a). A comprehensive account that includes many more results is given in the monograph of Dai and Xu (2013).

Orthogonal Expansions on $\mathbb{B}^d$ and T^d For orthogonal expansions with respect to W_κ in (2.6.4) on the ball $\mathbb{B}^d$ and W_κ in (2.5.19) on the simplex T^d, analogues of the above results hold. In fact, if we replace σ_κ by

$$\sigma_\kappa := \frac{d-1}{2} + |\kappa| - \min_{1 \le j \le d+1} \kappa_j,$$

then all five properties hold with obvious modification. This is no accident: the three cases are intimately connected. Some of the results on one domain can be deduced from the corresponding results on one of the other two domains. The study of summability on the ball started in a paper by Xu (1999). The connection between the three domains was first applied in the study of orthogonal expansions by Xu (2001a,b), and next in full power by Dai and Xu (2009a,b) and Li and Xu (2003). See Erdélyi (1953, Ch. 12) for earlier results on orthogonal expansions and Dai and Xu (2013) for further results.

Orthogonal Expansions on $[-1,1]^d$ On $[-1,1]^d$ we consider orthogonal expansions in the product Jacobi polynomials for $W_{a,b}$ in (2.5.43).

1. Let $a_j > -1$, $b_j > -1$, and $a_j + b_j \ge -1$ for $1 \le j \le d$. For f in $L^p(W_{a,b}; [-1,1]^d)$, $1 \le p < \infty$, or in $C([-1,1]^d)$, the (C,δ) mean $S_n^\delta f$ converges to f in norm as $n \to \infty$ if
$$\delta > \delta_0 := \sum_{j=1}^d \max(a_j, b_j) + \tfrac{1}{2}d + \max\left(0, -\sum_{j=1}^d \min(a_j, b_j) - \tfrac{1}{2}(d+2)\right).$$
2. Let $a_j \ge -1/2$, $b_j \ge -1/2$, and $a_j + b_j \ge -1$ for $1 \le j \le d$. Then $S_n^\delta f \ge 0$ whenever $f \ge 0$ if and only if $\delta \ge \sum_{i=1}^d (a_i + b_i) + 3d - 1$.

These were established by Li and Xu (2000). In comparison with the results on the sphere, ball, and simplex, far less is known for orthogonal expansions on $[-1,1]^d$. The difficulty lies in the lack of a closed form of the reproducing kernel.

Product Hermite and Laguerre Expansions Hermite expansions on $\mathbb{R}^d$ for W_H in (2.5.27) and Laguerre expansions on $\mathbb{R}^d_+$ for W_κ in (2.5.34) have been extensively studied. We mention just two results.

1. The Riesz mean S_R^δ of the product Hermite expansions converges in norm for $f \in L^p(\mathbb{R}^d)$ $(1 \le p < \infty)$ if $\delta > \frac{1}{2}(d-1)$. For every $f \in L^p(\mathbb{R}^d)$, the mean $S_R^\delta f$ converges to f almost everywhere if $\delta > \frac{1}{2}(d - \frac{1}{3})$.
2. Let $\kappa_j \ge 0$, $1 \le j \le d$. The (C,δ) mean $S_n^\delta f$ converges to $f \in L^p(W_\kappa)$ in norm if $\delta \ge \sum_{i=1}^d (a_i + b_i) + 3d - 1$. For $p = 1$ or ∞, the condition on δ is also necessary.

For extensive study on these expansions, see Thangavelu (1993) and the references therein.

2.8 Discrete Orthogonal Polynomials of Several Variables

Let V be a finite or countable set of points in $\mathbb{R}^d$ and let the *weight* W be a positive function on V. *Discrete orthogonal polynomials* are orthogonal polynomials with respect to the discrete inner product

$$\langle f, g\rangle := \sum_{x\in V} f(x)g(x)W(x). \tag{2.8.1}$$

In the case of infinite V assume that W decays fast enough such that the sum in (2.8.1) converges absolutely for all polynomials f, g.

It should be noted that $\langle\cdot,\cdot\rangle$ in (2.8.1) is an inner product only on $\Pi^d/I(V)$, where $I(V)$ is the polynomial ideal of polynomials which vanish on V. To be more specific, fix a monomial order and let $\Lambda(V)$ consist of all $\alpha \in \mathbb{N}_0^d$ such that $c\mathbf{x}^\alpha$ is not a leading monomial of any polynomial in $I(V)$. Then the inner product (2.8.1) is well defined on $\Pi_V := \operatorname{span}\{x^\beta\}_{\beta\in\Lambda(V)}$ and there exists a sequence of orthonormal polynomials $\{P_\alpha\}_{\alpha\in\Lambda(V)}$ where P_α is a polynomial in Π_V with leading monomial $c\mathbf{x}^\alpha$. See Xu (2004) for details and further discussions.

Notice, in particular, that the Gram–Schmidt method of generating an orthonormal basis can only be performed within Π_V. Besides the above precaution, general properties of discrete orthogonal polynomials are analogous to the continuous case, modulo $I(V)$ if necessary.

2.8.1 Classical Discrete Orthogonal Polynomials

These polynomials are expressed in terms of the classical discrete orthogonal polynomials of one variable, which are listed in Table 2.2. Here the Krawtchouk and Hahn polynomials have a finite domain $[0, N] := \{0, 1, \ldots, N\}$, and accordingly $n = 0, 1, \ldots, N$. The polynomials in the table can be expressed by hypergeometric functions, and their normalization means that the coefficient in front of the hypergeometric function is the constant 1.

There are several ways to extend classical discrete orthogonal polynomials to several variables. One way is to consider those families for which all orthogonal polynomials of degree

Table 2.2 *Classical discrete orthogonal polynomials of one variable*

Name	Notation	Weight	Domain	Constraint	Normalization
Charlier	$C_n(x; s)$	$s^x/x!$	$\mathbb{N}_0$	$s > 0$	$C_n(0; s) = 1$
Meixner	$M_n(x; \beta, c)$	$\beta^x (c)_x/x!$	$\mathbb{N}_0$	$\beta > 0, 0 < c < 1$	$M_n(0; \beta, c) = 1$
Krawtchouk	$K_n(x; p, N)$	$p^x(1-p)^{N-x}\binom{N}{x}$	$[0, N]$	$0 < p < 1$	$K_n(0; p, N) = 1$
Hahn	$Q_n(x; \alpha, \beta, N)$	$\binom{\alpha+x}{x}\binom{\beta+N-x}{N-x}$	$[0, N]$	$\alpha, \beta > -1$ or $< -N$	$Q_n(0; \alpha, \beta, N) = 1$

exactly n are eigenfunctions of a difference operator of specific form

$$\left(\sum_{1 \le i,j \le d} A_{i,j}(x) \Delta_i \nabla_j + \sum_{i=1}^{d} B_i(x) \Delta_i \right) \psi(x) = \lambda_n \psi(x), \tag{2.8.2}$$

where Δ_i and ∇_i denote the forward and backward difference operators

$$\Delta_i f(x) := f(x + e_i) - f(x) \quad \text{and} \quad \nabla_i f(x) := f(x) - f(x - e_i) \quad (e_1, \ldots, e_d \text{ standard basis}).$$

It follows readily that the $A_{i,j}(x)$ are necessarily quadratic polynomials and the $B_i(x)$ are necessarily linear polynomials in x. Some of these families are tensor products of classical polynomials of one variable, which will be discussed in the next subsection. Below are several families that do not come from products.

For a vector $x = (x_1, x_2, \ldots, x_d) \in \mathbb{R}^d$ we will denote

$$x^j := (x_j, x_{j+1}, \ldots, x_d) \quad \text{and} \quad X_j := (x_1, x_2, \ldots, x_j), \tag{2.8.3}$$

with the convention that $x^{d+1} := 0$ and $X_0 := 0$. Set $|x| := x_1 + \cdots + x_d$.

Meixner polynomials (Iliev and Xu, 2007, §6.1.2)
Let $0 < c_i < 1$ $(1 \le i \le d)$ such that $|c| < 1$ and let $s > 0$. For $\nu \in \mathbb{N}_0^d$ define the polynomials

$$M_\nu(x; s, c) = \prod_{j=1}^{d} (\delta_j)_{\nu_j} M_{\nu_j}\left(x_j; \delta_j, \frac{c_j}{1 - |c^{j+1}|}\right), \quad \text{where } \delta_j := s + |\nu^{j+1}| + |X_{j-1}|. \tag{2.8.4}$$

They satisfy the orthogonality relation

$$\sum_{x \in \mathbb{N}_0^d} M_\nu(x; s, c) M_\mu(x; s, c) (s)_{|x|} \prod_{i=1}^{d} \frac{c_i^{x_i}}{x_i!} = \frac{(s)_{|\nu|}}{(1 - |c|)^s} \left(\prod_{j=1}^{d} \nu_j! \left(\frac{1 - |c^{j+1}|}{c_j} \right)^{\nu_j} \right) \delta_{\nu,\mu}. \tag{2.8.5}$$

Furthermore, $\psi_\nu := M_\nu(\cdot; s, c)$ is an eigenfunction of a difference operator:

$$D\psi_\nu = -|\nu| \psi_\nu, \quad D := \sum_{1 \le i,j \le d} \left(\delta_{i,j} + \frac{c_i}{1 - |c|} \right) x_j \Delta_i \nabla_j + \sum_{i=1}^{d} \left(-x_i + \frac{c_i}{1 - |c|} s \right) \Delta_i. \tag{2.8.6}$$

Krawtchouk Polynomials (Iliev and Xu, 2007, §6.1.1)
Let $0 < p_i < 1$ $(1 \leq i \leq d)$ such that $|p| < 1$ and let $N \in \mathbb{N}$. For $\nu \in \mathbb{N}_0^d$, $|\nu| \leq N$, define the polynomials

$$\begin{aligned} K_\nu(x; p, N) := \frac{(-1)^{|\nu|}}{(-N)_{|\nu|}} \prod_{j=1}^{d} \frac{p_j^{\nu_j}}{(1-|P_j|)^{\nu_j}} (-N + |X_{j-1}| + |\nu^{j+1}|)_{\nu_j} \\ \times K_{\nu_j}\left(x_j; \frac{p_j}{1-|P_{j-1}|}, N - |X_{j-1}| - |\nu^{j+1}|\right). \end{aligned} \tag{2.8.7}$$

They satisfy the orthogonality relation

$$\sum_{|x|\leq N} K_\nu(x; p, N) K_\mu(x; p, N) \prod_{i=1}^{d+1} \frac{p_i^{x_i}}{x_i!} = \frac{(-1)^{|\nu|}}{(-N)_{|\nu|} N!} \left(\prod_{j=1}^{d} \frac{\nu_j!\, p_j^{\nu_j}}{(1-|P_j|)^{\nu_j - \nu_{j+1}}} \right) \delta_{\nu,\mu}, \tag{2.8.8}$$

where $x_{d+1} = N - |x|$, $p_{d+1} = 1 - |p|$, and $\nu_{d+1} = 0$. Furthermore, $\psi_\nu := K_\nu(\cdot; p, N)$ is an eigenfunction of a difference operator:

$$D\psi_\nu = -|\nu|\psi_\nu, \quad D := \sum_{1\leq i,j\leq d} (\delta_{i,j} - p_i) x_j \triangle_i \nabla_j + \sum_{i=1}^{d} (p_i N - x_i) \triangle_i. \tag{2.8.9}$$

Hahn Polynomials on the Parallelepiped (Iliev and Xu, 2007, §5.1.2)
Let $l_i \in \mathbb{N}$ $(1 \leq i \leq d)$, $\beta > -1$, and $r > 0$. For $\nu \in \mathbb{N}_0^d$, $\nu_i \leq l_i$, define the polynomials

$$\begin{aligned} \phi_\nu(x; \beta, r, l) := \prod_{i=1}^{d} (\alpha_{1,j} + 1)_{\nu_j} Q_{\nu_j}(x_j; \alpha_{1,j}, \alpha_{2,j}, l_j), \\ \alpha_{1,j} := \beta + |\nu^{j+1}| + |X_{j-1}| \quad \text{and} \quad \alpha_{2,j} := |L_{j-1}| - |X_{j-1}| + |\nu^{j+1}| + r - 1 \end{aligned} \tag{2.8.10}$$

(recall the notation (2.8.3)). They satisfy the orthogonality relation

$$\begin{aligned} \sum_{x\leq l} \phi_\nu(x; \beta, r, l) \phi_\mu(x; \beta, r, l) \left(\prod_{i=1}^{d} \frac{(-l_i)_{x_i}}{x_i!} \right) \frac{(\beta+1)_{|x|}}{(-|l| - r + 1)_{|x|}} \\ = \frac{(-1)^{|\nu|}(1+\beta)_{|\nu|}}{(r + |\nu|)_{|l|-|\nu|}} \left(\prod_{j=1}^{d} \frac{\nu_j!\,(\beta + r + 2|\nu^{j+1}| + \nu_j + |L_{j-1}|)_{l_j+1}}{(-l_j)_{\nu_j} (\beta + r + 2|\nu^j| + |L_{j-1}|)} \right) \delta_{\nu,\mu}. \end{aligned} \tag{2.8.11}$$

Furthermore, they satisfy the difference equation $D\phi_\nu = -|\nu|(|\nu| + \beta + r)\phi_\nu$ with

$$D = \sum_{1\leq i,j\leq d} x_j (l_i - x_i + r\delta_{i,j}) \triangle_i \nabla_j - \sum_{i=1}^{d} (x_i(r + \beta + 1) - l_i(1+\beta)) \triangle_i. \tag{2.8.12}$$

Furthermore, these relations also hold if $\beta < -|l|$ and $r < -|l| + 1$.

Hahn Polynomials on the Simplex (Iliev and Xu, 2007, §5.2.1)
Let $\sigma_i > -1$ $(1 \le i \le d+1)$ and $N \in \mathbb{N}$. For $\nu \in \mathbb{N}_0^d$ such that $|\nu| \le N$, define the polynomials

$$Q_\nu(x;\sigma,N) := \frac{(-1)^{|\nu|}}{(-N)_{|\nu|}} \prod_{j=1}^{d} \frac{(\sigma_j+1)_{\nu_j}}{(a_j+1)_{\nu_j}} (-N+|X_{j-1}|+|\nu^{j+1}|)_{\nu_j} Q_{\nu_j}(x_j;\sigma_j,a_j,N-|X_{j-1}|-|\nu^{j+1}|), \tag{2.8.13}$$

where $a_j := |\sigma^{j+1}| + 2|\nu^{j+1}| + d - j$. They satisfy the orthogonality relation

$$\begin{aligned} &\sum_{|x|\le N} Q_\nu(x;\sigma,N) Q_\mu(x;\sigma,N) \left(\prod_{i=1}^{d} \binom{x_i+\sigma_i}{x_i}\right) \binom{N-|x|+\sigma_{d+1}}{N-|x|} \\ &\quad = \frac{(-1)^{|\nu|}(|\sigma|+d+2|\nu|+1)_{N-|\nu|}}{(-N)_{|\nu|}N!} \left(\prod_{j=1}^{d} \frac{(\sigma_j+a_j+\nu_j+1)_{\nu_j}(\sigma_j+1)_{\nu_j}\nu_j!}{(a_j+1)_{\nu_j}}\right) \delta_{\nu,\mu}. \end{aligned} \tag{2.8.14}$$

Furthermore, they are eigenfunctions of a difference operator:

$$\begin{aligned} D\psi_\nu &= -|\nu|(|\nu|+|\sigma|+d)\psi_\nu, \\ D &:= \sum_{i=1}^{d} x_i(N-x_i+|\sigma|-\sigma_i+d)\Delta_i\nabla_i - \sum_{1\le i\ne j\le d} x_j(x_i+\sigma_i+1)\Delta_i\nabla_j \\ &\quad + \sum_{i=1}^{d} \big((N-x_i)(\sigma_i+1) - x_i(|\sigma|-\sigma_i+d)\big)\Delta_i. \end{aligned} \tag{2.8.15}$$

These relations also hold if $\sigma_i < -N$ for $i = 1, 2, \ldots, d+1$.

Hahn Polynomials on the Simplex-Parallelepiped (Iliev and Xu, 2007, §5.2.2)
Let S be a nonempty set in $\{1,2,\ldots,d\}$, let $l_i \in \mathbb{N}_0$ $(i \in S)$, and let $N \in \mathbb{N}$. Define $V_{N,S}^d := \{x \mid |x| \le N\} \cap \{x \mid x_i \le \ell_i \text{ for } i \in S\}$ and set $\sigma_i = -l_i - 1$ $(1 \le i \le d)$. For $\nu \in V_{N,S}^d$ the polynomials $Q_\nu(\cdot;\sigma,N)$ in (2.8.13) satisfy the orthogonality relation

$$\sum_{x\in V_{N,S}^d} Q_\nu(x;\sigma,N) Q_\mu(x;\sigma,N) \left(\prod_{i=1}^{d} \binom{x_i+\sigma_i}{x_i}\right) \binom{N-|x|+\sigma_{d+1}}{N-|x|} = A_\nu \delta_{\nu,\mu}, \tag{2.8.16}$$

where A_ν is the coefficient of $\delta_{\nu,\mu}$ in (2.8.14), and they satisfy the same difference equation as in (2.8.15).

Further Results and References The multivariate Krawtchouk polynomials were first studied by Milch (1968) and the Hahn polynomials on the simplex were first studied by Karlin and McGregor (1975). Both classes of polynomials are associated with a linear growth model of birth and death processes. Biorthogonal systems of Hahn polynomials were found by Tratnik (1989a). The Meixner and Krawtchouk polynomials were studied by Tratnik (1989b). They can be deduced as limits of biorthogonal or orthogonal Hahn polynomials (Tratnik, 1989b, 1991). For example, for Q_ν in (2.8.13) and K_ν in (2.8.7),

$$\lim_{t\to\infty} Q_\nu(x; p_1 t, \ldots, p_d t, (1-p_1-\cdots-p_d)t, N) = K_\nu(x; p, N)$$

follows from the one-variable case.

All polynomials in this section were also studied by Iliev and Xu (2007) in connection with difference equations. There is one more family of discrete orthogonal polynomials $\{R_\nu\}$ that resemble the Hahn polynomials studied in Iliev and Xu (2007). They satisfy the orthogonal relation

$$\sum_{x\in\mathbb{N}_0^d} R_\nu(x;\sigma,\beta,\gamma)R_\mu(x;\sigma,\beta,\gamma)\left(\prod_{i=1}^{d}\frac{(\sigma_i+1)_{x_i}}{x_i!}\right)\frac{(\beta+1)_{|x|}}{(\gamma+1)_{|x|}} = A_\nu\,\delta_{\nu,\mu}, \tag{2.8.17}$$

where $R_\nu(\cdot;\sigma,\beta,\gamma)$ is defined for $\nu\in\mathbb{N}_0^d$ such that $2|\nu| < \gamma - |\sigma| - \beta - d - 1$, and they are also eigenfunctions of a second-order difference operator. Furthermore, together with product-type polynomials to be discussed in the following subsection, the discrete orthogonal polynomials in this subsection yield all orthogonal polynomial eigenfunctions of a fairly general class of difference operators (2.8.2).

2.8.2 Product Orthogonal Polynomials

By taking products of classical discrete orthogonal polynomials in one variable one can generate many different products of orthogonal polynomials in several variables. Below is a list of such polynomials when $d = 2$, of the form $u(x_1,x_2) = p_k(x_1)q_{n-k}(x_2)$ $(0\le k\le n)$, that satisfy the difference equation (2.8.2) with eigenvalue $\lambda_n = n$.

Charlier–Charlier The polynomials $C_k(x_1;a_1)C_{n-k}(x_2;a_2)$ $(0\le k\le n)$ are orthogonal with respect to the weights $\frac{a_1^{x_1}}{x_1!}\frac{a_2^{x_2}}{x_2!}$ $(a_1,a_2>0)$ on $\mathbb{N}_0^2$. They satisfy

$$x_1\triangle_1\nabla_1 u + x_2\triangle_2\nabla_2 u + (a_1-x_1)\triangle_1 u + (a_2-x_2)\triangle_2 u = -nu. \tag{2.8.18}$$

Charlier–Meixner The polynomials $M_k(x_1;\beta,c)C_{n-k}(x_2;a)$ $(0\le k\le n)$ are orthogonal with respect to the weights $\frac{(\beta)_{x_1}}{x_1!}c^{x_1}\frac{(a)_{x_2}}{x_2!}$ $(a,\beta>0,\ 0<c<1)$ on $\mathbb{N}_0^2$. They satisfy

$$(c-1)^{-1}x_1\triangle_1\nabla_1 u - x_2\triangle_2\nabla_2 u + (c-1)^{-1}(c(x_1+\beta)-x_1)\triangle_1 u - (a-x_2)\triangle_2 u = nu. \tag{2.8.19}$$

Charlier–Krawtchouk The polynomials $K_k(x_1;p,N)C_{n-k}(x_2;a)$ $(0\le k\le n)$ are orthogonal with respect to the weights $\binom{N}{x_1}p^{x_1}(1-p)^{N-x_1}\frac{a^{x_2}}{x_2!}$ $(a>0,\ 0<p<1)$ on $[0,N]\times\mathbb{N}_0$. They satisfy

$$(1-p)x_1\triangle_1\nabla_1 u + x_2\triangle_2\nabla_2 u + (p(N-x_1)-(1-p)x_1)\triangle_1 u + (a-x_2)\triangle_2 u = -nu. \tag{2.8.20}$$

Meixner–Meixner The polynomials $M_k(x_1;\beta_1,c_1)M_{n-k}(x_2;\beta_2,c_2)$ $(0\le k\le n)$ are orthogonal with respect to the weights $\frac{(\beta_1)_{x_1}}{x_1!}c_1^{x_1}\frac{(\beta_2)_{x_2}}{x_2!}c_2^{x_2}$ $(\beta_1,\beta_2>0,\ c_1,c_2\in(0,1))$ on $\mathbb{N}_0^2$. They satisfy

$$\begin{aligned}(c_1-1)^{-1}&\triangle_1\nabla_1 u + (c_2-1)^{-1}x_2\triangle_2\nabla_2 u\\ &+ (c_1-1)^{-1}(c_1(x_1+\beta_1)-x_1)\triangle_1 u + (c_2-1)^{-1}(c_2(x_2+\beta_2)-x_2)\triangle_2 u = nu.\end{aligned} \tag{2.8.21}$$

Meixner–Krawtchouk The polynomials $M_k(x_1;\beta,c)K_{n-k}(x_2;p,N)$ $(0\le k\le n)$ are orthogonal with respect to the weights $\frac{(\beta)_{x_1}}{x_1!}c^{x_1}\binom{N}{x_2}p^{x_2}(1-p)^{N-x_2}$ $(\beta>0;\, c,p\in(0,1))$ on $\mathbb{N}_0\times[0,N]$. They satisfy

$$\begin{aligned}(c-1)^{-1}&x_1\triangle_1\nabla_1 u-(1-p)x_2\triangle_2\nabla_2 u\\&+(c-1)^{-1}(c(x_1+\beta)-x_1)\triangle_1 u-(p(N-x_2)-(1-p)x_2)\triangle_2 u=nu.\end{aligned}\tag{2.8.22}$$

Krawtchouk–Krawtchouk The polynomials $K_k(x_1;p_1,N_1)K_{n-k}(x_2;p_2,N_2)$ $(0\le k\le n)$ are orthogonal with respect to the weights $\binom{N_1}{x_1}p_1^{x_1}(1-p_1)^{N_1-x_1}\binom{N_2}{x_2}p_2^{x_2}(1-p_2)^{N_2-x_2}$ $(p_1,p_2\in(0,1))$ on $[0,N_1]\times[0,N_2]$. They satisfy

$$\begin{aligned}(1-p_1)&x_1\triangle_1\nabla_1 u+(1-p_2)x_2\triangle_2\nabla_2 u\\&+(p_1(N_1-x_1)-(1-p_1)x_1)\triangle_1 u+(p_2(N_2-x_2)-(1-p_2)x_2)\triangle_2 u=-nu.\end{aligned}\tag{2.8.23}$$

For $d>2$, there are many more product discrete orthogonal polynomials that satisfy the second-order difference equations. In fact, besides the product of classical one-variable polynomials, there are also products of classical polynomials of one variable and other lower-dimensional orthogonal polynomials. For example, the products of Meixner polynomials on the simplex with either Charlier, Meixner, or Krawtchouk polynomials are discrete orthogonal polynomials, of three variables, and they satisfy difference equations of the form $Du=\lambda_n u$, where n is the total degree of the orthogonal polynomials. For further discussions and details, see Iliev and Xu (2007), Xu (2004, 2005c).

There are also product orthogonal polynomials that are given by products which have a Hahn polynomial as one of their factors. Such polynomials, however, are eigenfunctions of a difference operator with eigenvalues not just depending on the total degree n but also (in the two-variable case) on k.

2.8.3 Further Results on Discrete Orthogonal Polynomials

Racah Polynomials. These are defined via ${}_4F_3$ functions and are orthogonal with respect to weights on $[0,N]$. They were extended to several variables by Tratnik (1991) for the weights

$$\begin{aligned}w(x)=w(x;c_1,\dots,c_{d+1},\gamma,N):=&\frac{N!\Gamma(|C_d|+N+1)}{\Gamma(c_{d+1}+N)\Gamma(|c|+N)}\frac{(c_1)_{x_1}(\gamma+1)_{x_1}}{x_1!(c_1-\gamma)_{x_1}}\\&\times\prod_{k=1}^{d}\frac{\Gamma(c_{k+1}+x_{k+1}-x_k)\Gamma(|C_{k+1}|+x_{k+1}+x_k)}{(x_{k+1}-x_k)!\Gamma(|C_k|+x_{k+1}+x_k+1)}\frac{|C_k|+2x_k}{|C_k|}\\&(x\in\mathbb{N}_0^d,\ 0\le x_1\le\cdots\le x_d\le N),\end{aligned}$$

where $c=(c_1,\dots,c_{d+1})$, C_k is defined as in (2.8.3), and $x_{d+1}=N$.

Tratnik (1991) defined multivariable dual Hahn polynomials as limit cases of Racah polynomials. The Hahn polynomials on the simplex are also contained as a limit case of the Racah family. The multivariable Racah polynomials are studied from the viewpoint of bispectrality by Geronimo and Iliev (2010).

Griffiths (1971) used a generating function to define polynomials in d variables orthogonal with respect to the multinomial distribution, which gives a family of Krawtchouk polynomials that satisfy several symmetric relations among their variables and parameters. These polynomials are related to character algebras and the Aomoto–Gel'fand hypergeometric function by Mizukawa and Tanaka (2004) (see also §4.5). The recurrence relations as well as the reductions which lead to the polynomials defined by Milch (1968) and Hoare and Rahman (2008) can be found in Iliev (2012a). Some of these properties are explored by Iliev and Terwilliger (2012), in which these polynomials are interpreted in terms of the Lie algebra $\mathrm{sl}_3(\mathbb{C})$. For applications of these bases of polynomials in probability, see Diaconis and Griffiths (2014).

Orthogonal polynomials for the negative multinomial distribution are Meixner polynomials. A general family of these polynomials was defined in terms of generating functions by Griffiths (1975) and their properties were studied by Iliev (2012b).

2.9 Other Orthogonal Polynomials of Several Variables

This section contains several families of orthogonal polynomials of several variables that are not classical type but can be constructed explicitly.

2.9.1 Orthogonal Polynomials from Symmetric Functions

A polynomial $f \in \Pi^d$ is called *symmetric* if f is invariant under any permutation of its variables. *Elementary symmetric polynomials* are given by

$$E_k(x_1, \ldots, x_d) := \sum_{1 \le i_1 < \cdots < i_k \le d} x_{i_1} \cdots x_{i_k}, \quad k = 1, 2, \ldots, d.$$

They generate the algebra of symmetric polynomials.

The mapping

$$x \mapsto u, \quad u_i := E_i(x_1, \ldots, x_d) \quad (i = 1, \ldots, d) \tag{2.9.1}$$

is a bijection from the region $S := \{x \in \mathbb{R}^d \mid x_1 < x_2 < \cdots < x_d\}$ onto its image Ω. The Jacobian of this mapping is $J(x) := \prod_{1 \le i < j \le d} (x_i - x_j)$. The square of $J(x)$ becomes a polynomial $\Delta(u)$ in u under the mapping (2.9.1),

$$\Delta(u) = \prod_{1 \le i < j \le d} (x_i - x_j)^2, \quad u \in \Omega.$$

Let $d\mu$ be a nonnegative measure on $\mathbb{R}$. Define the measure $d\nu$ on Ω as the image of the product measure $d\mu(x_1) \cdots d\mu(x_d)$ under the mapping (2.9.1). The orthogonal polynomials with respect to the measures $(\Delta(u))^{\pm 1/2} d\nu$ can be given in terms of orthogonal polynomials with respect to $d\mu$ on $\mathbb{R}$, as we will describe now.

Let $\{p_n\}$ be orthonormal polynomials for the measure $d\mu$ on $\mathbb{R}$. For $n \in \mathbb{N}_0$ and $\alpha \in \mathbb{N}_0^d$ such that $n = \alpha_d \ge \cdots \ge \alpha_1 \ge 0$, define

$$P_\alpha^{n, -\frac{1}{2}}(u) := \sum_\beta p_{\alpha_1}(x_{\beta_1}) \cdots p_{\alpha_d}(x_{\beta_d}), \quad u \in \Omega, \tag{2.9.2}$$

where the summation is performed over all permutations β of $\{1, 2, \ldots, d\}$. These are polynomials of degree n and satisfy

$$\int_\Omega P_\alpha^{n,-\frac{1}{2}}(u)P_\beta^{m,-\frac{1}{2}}(u)(\Delta(u))^{-\frac{1}{2}}\, dv(u) = m_1! \cdots m_{d'}!\,\delta_{n,m}\,\delta_{\alpha,\beta}, \tag{2.9.3}$$

where d' is the number of distinct elements in α and m_i is the number of occurrences of the ith distinct element in α.

For $n \in \mathbb{N}_0$ and $\alpha \in \mathbb{N}_0^d$ such that $n = \alpha_d \geq \cdots \geq \alpha_1 \geq 0$, define

$$P_\alpha^{n,\frac{1}{2}}(u) = \frac{J_\alpha^n(x)}{J(x)}, \quad \text{where } J_\alpha^n(x) := \det[p_{\alpha_i+d-i}(x_j)]_{i,j=1}^d. \tag{2.9.4}$$

These are indeed polynomials of degree n in u under (2.9.1) and satisfy

$$\int_\Omega P_\alpha^{n,\frac{1}{2}}(u)P_\beta^{m,\frac{1}{2}}(u)\,(\Delta(u))^{\frac{1}{2}}\, dv(u) = \delta_{n,m}\,\delta_{\alpha,\beta}. \tag{2.9.5}$$

Both these families of orthogonal polynomials satisfy the striking property that the polynomials in $\mathcal{V}_n^d$ have $\dim \Pi_{n-1}^d$ distinct real common zeros. In other words, the Gaussian cubature formula exists for $(\Delta(u))^{\pm 1/2}\, dv(u)$ by Theorem 2.2.15. For $d = 2$, these are Koornwinder's polynomials as discussed in §2.3.6. For $d > 2$ they were studied by Berens et al. (1995a).

2.9.2 Orthogonal Polynomials Associated with Root System A_d

Using homogeneous coordinates $\mathbf{t} = (t_1, \ldots, t_{d+1}) \in \mathbb{R}^{d+1}$ satisfying the relation $t_1 + \cdots + t_{d+1} = 0$, the space $\mathbb{R}^d$ can be identified with the hyperplane

$$\mathbb{R}_H^{d+1} := \{\mathbf{t} \in \mathbb{R}^{d+1} \mid t_1 + \cdots + t_{d+1} = 0\}$$

in $\mathbb{R}^{d+1}$. The reflection group $\mathcal{A}_d$ for the root system A_d is generated by the reflections σ_{ij}, under homogeneous coordinates, defined by

$$\mathbf{t}\sigma_{ij} := \mathbf{t} - 2\frac{\langle \mathbf{t}, \mathbf{e}_{i,j}\rangle}{\langle \mathbf{e}_{i,j}, \mathbf{e}_{i,j}\rangle}\mathbf{e}_{i,j} = \mathbf{t} - (t_i - t_j)\mathbf{e}_{i,j}, \quad \text{where } \mathbf{e}_{i,j} := e_i - e_j.$$

This group can be identified with the symmetric group of $d+1$ elements. Define the operators $\mathcal{P}^+$ and $\mathcal{P}^-$ by

$$\mathcal{P}^\pm f(\mathbf{t}) := \frac{1}{(d+1)!}\left(\sum_{\sigma\in\mathcal{A}^+} f(\mathbf{t}\sigma) \pm \sum_{\sigma\in\mathcal{A}^-} f(\mathbf{t}\sigma)\right),$$

where $\mathcal{A}^+$ (or $\mathcal{A}^-$) consists of an even (or odd) number of products of reflections σ_{ij}. They map f to $\mathcal{A}_d$-invariant ($\mathcal{P}^+$) or anti-invariant ($\mathcal{P}^-$) functions, respectively.

Let $\mathbb{H} := \{\mathbf{k} \in \mathbb{Z}^{d+1} \cap \mathbb{R}_H^{d+1} \mid k_1 \equiv \cdots \equiv k_{d+1} \mod (d+1)\}$. The functions

$$\phi_{\mathbf{k}}(\mathbf{t}) := e^{2\pi i(d+1)^{-1}\langle \mathbf{k},\mathbf{t}\rangle}, \quad \mathbf{k} \in \mathbb{H}$$

are periodic functions: $\phi_{\mathbf{k}}(\mathbf{t}) = \phi_{\mathbf{k}}(\mathbf{t}+\mathbf{j})$ ($\mathbf{j} \in \mathbb{Z}^{d+1} \cap \mathbb{R}_H^{d+1}$). Let

$$\Lambda := \{\mathbf{k} \in \mathbb{H} \mid k_1 \geq k_2 \geq \cdots \geq k_{d+1}\} \quad \text{and} \quad \Lambda^\circ := \{\mathbf{k} \in \mathbb{H} \mid k_1 > k_2 > \cdots > k_{d+1}\}.$$

Then the functions defined by

$$\mathsf{TC}_{\mathbf{k}}(\mathbf{t}) := \mathcal{P}^{+}\phi_{\mathbf{k}}(\mathbf{t}) \quad (\mathbf{k} \in \Lambda) \qquad \text{and} \qquad \mathsf{TS}_{\mathbf{k}}(\mathbf{t}) := i^{-1}\mathcal{P}^{-}\phi_{\mathbf{k}}(\mathbf{t}) \quad (\mathbf{k} \in \Lambda^{\circ})$$

are invariant and anti-invariant functions, respectively, and they are analogues of cosine and sine functions that are orthogonal on the simplex

$$\Delta := \{\mathbf{t} \in \mathbb{R}_H^{d+1} \mid 0 \le t_i - t_j \le 1 \; (1 \le i < j \le d+1)\}.$$

These functions become, under the change of variables $\mathbf{t} \mapsto z$, orthogonal polynomials, where $z_1, \ldots, z_d$ denote the first d elementary symmetric functions of $e^{2\pi i t_1}, \ldots, e^{2\pi i t_{d+1}}$. Indeed, for the index $\alpha \in \mathbb{N}_0^d$ associated to $\mathbf{k} \in \Lambda$ by

$$\alpha_i = \alpha_i(\mathbf{k}) := \frac{k_i - k_{i+1}}{d+1}, \qquad 1 \le i \le d,$$

we define under the change of variables $\mathbf{t} \mapsto z = (z_1, \ldots, z_d)$,

$$T_\alpha(z) := \mathsf{TC}_{\mathbf{k}}(\mathbf{t}) \quad \text{and} \quad U_\alpha(z) := \frac{\mathsf{TS}_{\mathbf{k}+\mathbf{v}^\circ}(\mathbf{t})}{\mathsf{TS}_{\mathbf{v}^\circ}(\mathbf{t})},$$

where $\mathbf{v}^\circ := (d+1)(\frac{1}{2}d, \frac{1}{2}d - 1, \ldots, -\frac{1}{2}d)$. Then T_α and U_α are polynomials in z of degree $|\alpha|$. These polynomials are analogues of Chebyshev polynomials of the first kind and the second kind, respectively. In particular, they are orthogonal on the domain Δ^*, the image of Δ under $\mathbf{t} \mapsto z$,

$$\Delta^* := \left\{x = x(\mathbf{t}) \in \mathbb{R}^d \;\middle|\; \mathbf{t} \in \mathbb{R}_H^{d+1}, \textstyle\prod_{1 \le i < j \le d+1} \sin(\pi(t_i - t_j)) \ge 0\right\}$$

with respect to the weight functions $W_{-1/2}$ and $W_{1/2}$, respectively, where

$$W_\alpha(z) := \prod_{1 \le \mu < \nu \le d+1} \left|\sin(\pi(t_\mu - t_\nu))\right|^{2\alpha}.$$

For $d = 2$, these are the second family of Koornwinder polynomials in §2.3.6.

These polynomials satisfy simple recurrence relations and the relation

$$\overline{P_\alpha(z)} = P_{\alpha_d, \alpha_{d-1}, \ldots, \alpha_1}(z), \qquad P_\alpha = T_\alpha \text{ or } U_\alpha, \; \alpha \in \mathbb{N}_0^d.$$

Together with the fact that $z_k = z_{d-k+1}$, one can derive a sequence of real orthogonal polynomials from either $\{T_\alpha\}$ or $\{U_\alpha\}$. The set of orthogonal polynomials $\{U_\alpha\}_{|\alpha|=n}$ of degree n has $\dim \Pi_{n-1}^d$ distinct real common zeros in Δ^*, so that the Gaussian cubature formula exists for $W_{1/2}$ on Δ^* by Theorem 2.2.15. Gaussian cubature, however, does not exist for $W_{-1/2}$.

Further Results and References These orthogonal polynomials were studied systematically by Beerends (1991), who extended earlier work by Koornwinder (1974b) for $d = 2$ and partial results in Dunn and Lidl (1982) and Eier and Lidl (1974, 1982). The presentation here follows Li and Xu (2010), which studied these polynomials from the viewpoint of tiling and discrete Fourier analysis and, in particular, studied their common zeros. Chebyshev polynomials of the second kind are closely related to *Schur functions*. In fact,

$$\mathsf{TS}_{\mathbf{k}+\mathbf{v}^\circ}(\mathbf{t}) = \det\left(z_j^{\lambda_k + \beta}\right)_{1 \le j,k \le d+1}, \qquad z_j = e^{2\pi i t_j},$$

where $\lambda := (k_1 - k_{d+1}, \ldots, k_d - k_{d+1}, 0)$ and $\beta = (d, d-1, \ldots, 1, 0)$. In terms of symmetric polynomials in $z_1, \ldots, z_d$, they are related to the BC_n-type orthogonal polynomials; see Chapter 8, Beerends and Opdam (1993), Vretare (1984), and the references therein.

2.9.3 Sobolev Orthogonal Polynomials

Despite extensive studies of Sobolev orthogonal polynomials in one variable, there have been few results in several variables until now, and what is known is mostly on the unit ball $\mathbb{B}^d$. Let $\mathcal{H}_n^d$ be the space of harmonic polynomials of degree n as in §2.4 and let $\mathcal{V}_n^d(W_\mu)$ denote the space of orthogonal polynomials on $\mathbb{B}^d$ with respect to $W_\mu(x) := (1 - \|x\|^2)^\mu$, which differs from (2.5.1) by a shift of $\frac{1}{2}$ in the index.

First Family on $\mathbb{B}^d$ Let $\Delta := \partial_1^2 + \cdots + \partial_d^2$ be the Laplace operator. Define the inner product on the unit ball $\mathbb{B}^d$ by

$$\langle f, g \rangle_\Delta := \frac{1}{4d^2 \operatorname{vol}(\mathbb{B}^d)} \int_{\mathbb{B}^d} \Delta\big((1 - \|x\|^2) f(x)\big) \Delta\big((1 - \|x\|^2) g(x)\big)\, dx, \tag{2.9.6}$$

which is normalized such that $\langle 1, 1 \rangle_\Delta = 1$. The space $\mathcal{V}_n^d$ of orthogonal polynomials of degree n for $\langle \cdot, \cdot \rangle_\Delta$ satisfies an orthogonal decomposition

$$\mathcal{V}_n^d = \mathcal{H}_n^d \oplus (1 - \|x\|^2) \mathcal{V}_{n-2}^d(W_2), \tag{2.9.7}$$

from which explicit orthonormal bases can be derived easily.

Second Family on $\mathbb{B}^d$ Let $\nabla := (\partial_1, \ldots, \partial_d)$. Define the inner product

$$\langle f, g \rangle_\nabla := \frac{\lambda}{\omega_d} \int_{\mathbb{B}^d} \nabla f(x) \cdot \nabla g(x)\, dx + \frac{1}{\omega_d} \int_{\mathbb{S}^{d-1}} f(x) g(x)\, d\sigma_d(x), \tag{2.9.8}$$

where $\lambda > 0$ such that $\langle 1, 1 \rangle_\nabla = 1$. The space of orthogonal polynomials of degree n for $\langle \cdot, \cdot \rangle_\nabla$ satisfies an orthogonal decomposition

$$\mathcal{V}_n^d = \mathcal{H}_n^d \oplus (1 - \|x\|^2) \mathcal{V}_{n-2}^d(W_1). \tag{2.9.9}$$

Moreover, the polynomials in $\mathcal{V}_n^d$ are eigenfunctions of a second-order differential operator that is exactly the limiting case of (2.5.2) with $\mu = -1/2$.

Third Family on $\mathbb{B}^d$ Define the inner product by

$$\langle f, g \rangle := \frac{\lambda}{\omega_d} \int_{\mathbb{B}^d} \Delta f(x) \Delta g(x)\, dx + \frac{1}{\omega_d} \int_{S^{d-1}} f(x) g(x)\, d\omega, \tag{2.9.10}$$

where $\lambda > 0$ such that $\langle 1, 1 \rangle = 1$. The space $\mathcal{V}_n^d$ of orthogonal polynomials of degree n satisfies an orthogonal decomposition

$$\mathcal{V}_n^d = \mathcal{H}_n^d \oplus (1 - \|x\|^2) \mathcal{H}_{n-2}^d \oplus (1 - \|x\|^2)^2 \mathcal{V}_{n-4}^d(W_2). \tag{2.9.11}$$

Further Results and References The first family was studied by Xu (2006a). The motivation for the inner product (2.9.6) came from a Galerkin method in the numerical solution of the Poisson equation on the disk. The second family was studied by Xu (2008). That reference also considered the inner product where the second integral in (2.9.8) is replaced by $f(0)g(0)$. The case where the second integral in (2.9.8) is replaced by an integral over the ball was studied by Pérez et al. (2013). The third family was studied by Piñar and Xu (2009), where the connection of orthogonal polynomials with the eigenfunctions of the differential operator was explored. Finally, Sobolev orthogonal polynomials with higher-order derivatives in the inner product are studied by Li and Xu (2014). They are used in connection with simultaneous approximation by polynomials on the unit ball. A first study of Sobolev orthogonal polynomials on the simplex was conducted by Aktaş and Xu (2013). A further reference is a paper by Lee and Littlejohn (2006), which, however, contains few concrete examples.

2.9.4 Orthogonal Polynomials with Additional Point Masses

Let $\langle p, q\rangle_\mu := \int_{\mathbb{R}^d} p(x)q(x)\,d\mu(x)$ be an inner product for which orthogonal polynomials exist. Let $\{\xi_1, \xi_2, \ldots, \xi_N\}$ be a set of distinct points in $\mathbb{R}^d$ and let Λ be a positive definite matrix of size $N \times N$. With the notation $\mathbf{p}(\xi) = \{p(\xi_1), p(\xi_2), \ldots, p(\xi_N)\}$, considered as a column vector, we define a new inner product

$$\langle p, q\rangle_\nu := \langle p, q\rangle_\mu + \mathbf{p}(\xi)^{\mathrm{tr}}\Lambda\mathbf{q}(\xi). \tag{2.9.12}$$

When $\Lambda = \mathrm{diag}\{\lambda_1, \ldots, \lambda_N\}$, the inner product $\langle\cdot,\cdot\rangle_\nu$ takes the form

$$\langle p, q\rangle_\nu = \langle p, q\rangle_\mu + \sum_{j=1}^{N} \lambda_j p(\xi_j)q(\xi_j). \tag{2.9.13}$$

The orthogonal polynomials with respect to $\langle\cdot,\cdot\rangle_\nu$ and their kernels can be expressed in terms of quantities associated with $\langle\cdot,\cdot\rangle_\mu$.

Let $\mathbb{P}_n$ denote a basis of orthogonal polynomials for $\mathcal{V}_n^d$ with respect to $\langle\cdot,\cdot\rangle_\mu$, as in (2.2.4), and let $P_n(\mu;\cdot,\cdot)$ and $K_n(\mu;\cdot,\cdot)$ denote the reproducing kernel of $\mathcal{V}_n^d$ and Π_n^d, respectively, with respect to $\langle\cdot,\cdot\rangle_\mu$, as defined in (2.2.22) and (2.2.26). Let $\mathsf{P}_n(\xi)$ be the matrix that has $\mathbb{P}_n(\xi_i)$ as columns,

$$\mathsf{P}_n(\xi) := (\mathbb{P}_n(\xi_1) \mid \mathbb{P}_n(\xi_2) \mid \ldots \mid \mathbb{P}_n(\xi_N)) \in \mathcal{M}_{r_n^d \times N},$$

let $\mathbf{K}_{n-1}$ be the matrix whose entries are $K_{n-1}(\mu;\xi_i,\xi_j)$,

$$\mathbf{K}_{n-1} := (K_{n-1}(\mu;\xi_i,\xi_j))_{i,j=1}^{N} \in \mathcal{M}_{N\times N},$$

and, finally, let $\mathbb{K}_{n-1}(\xi, x)$ be the column vector of functions

$$\mathbb{K}_{n-1}(\xi, x) = \{K_{n-1}(\mu;\xi_1,x), K_{n-1}(\mu;\xi_2,x), \ldots, K_{n-1}(\mu;\xi_N,x)\}.$$

Then the orthogonal polynomials $\mathbb{Q}_n$ associated with $\langle\cdot,\cdot\rangle_\nu$ are given by

$$\mathbb{Q}_n(x) = \mathbb{P}_n(x) - \mathsf{P}_n(\xi)(I_N + \Lambda\mathbf{K}_{n-1})^{-1}\Lambda\mathbb{K}_{n-1}(\xi, x), \quad n \geq 1, \tag{2.9.14}$$

and the reproducing kernel of Π_n^d associated with $\langle \cdot, \cdot \rangle_\nu$ is given by

$$K_n(\nu; x, y) = K_n(\mu; x, y) - \mathbb{K}_n^{\mathrm{tr}}(\xi, x)(I_N + \Lambda \mathbf{K}_n)^{-1} \Lambda \mathbb{K}_n(\xi, y). \tag{2.9.15}$$

These results were developed by Delgado et al. (2010), where the Jacobi weight on the simplex with mass points on its vertexes was studied as an example. The results can be modified to allow derivatives at the point masses; for example,

$$\langle p, q \rangle_\nu := \langle p, q \rangle_\mu + \sum_{j=0}^{N} \lambda_j p(\xi_j) q(\xi_j) + \sum_{j=0}^{N} \lambda'_j \nabla p(\xi_j) \cdot \nabla q(\xi_j).$$

2.9.5 Orthogonal Polynomials for Radial Weight Functions

Let w be a nonnegative function on the real line with support set $[a, b]$, where $0 \le a \le b \le \infty$. For a *radial weight function* $W(x) := w(\|x\|)$ the orthogonal polynomials can be constructed explicitly in polar coordinates. Indeed, let $p_{2n}^{(2n-4j+d-1)}$ denote the orthonormal polynomials with respect to the weight function $|t|^{2n-4j+d-1} w(t)$ and, for $0 \le j \le n/2$, let $\{Y_{n-2j,\beta}\}$ denote an orthonormal basis for $\mathcal{H}_{n-2j}^d$ of ordinary spherical harmonics. Then the polynomials

$$P_{\beta,j}(x) := p_{2j}^{(2n-4j+d-1)}(\|x\|) Y_{\beta,n-2j}(x) \tag{2.9.16}$$

form an orthonormal basis of $\mathcal{V}_n^d$ with $W(x) = w(\|x\|)$.

The classical examples of radial weight functions are W_μ in (2.5.1) on the unit ball $\mathbb{B}^d$ and the Hermite weight function W_H in (2.5.27). The orthogonal polynomials in (2.9.16) appeared in Xu (2005a) and they were used by Waldron (2009).

Acknowledgements I would like to thank Tom Koornwinder for his numerous comments and corrections.

References

Agahanov, C. A. 1956. A method of constructing orthogonal polynomials of two variables for a certain class of weight functions (in Russian). *Vestnik Leningrad Univ.*, **20**(19), 5–10.

Aktaş, R., and Xu, Y. 2013. Sobolev orthogonal polynomials on a simplex. *Int. Math. Res. Not.*, 3087–3131.

Álvarez de Morales, M., Fernández, L., E., Pérez T., and Piñar, M. A. 2009. A matrix Rodrigues formula for classical orthogonal polynomials in two variables. *J. Approx. Theory*, **157**, 32–52.

Andrews, G. E., Askey, R., and Roy, R. 1999. *Special Functions*. Encyclopedia of Mathematics and Its Applications, vol. 71. Cambridge University Press.

Appell, P., and Kampé de Fériet, J. 1926. *Fonctions Hypergéométriques et Hypersphériques. Polynômes d'Hermite*. Gauthier-Villars.

Ariznabarreta, G., and Mañas, M. 2016. Multivariate orthogonal polynomials and integrable systems. *Adv. Math.*, **302**, 628–739.

Atkinson, K., Chien, D., and Hansen, O. 2014. Evaluating polynomials over the unit disk and the unit ball. *Numer. Algorithms*, **67**, 691–711.

Barrio, R., Peña, J. M., and Sauer, T. 2010. Three term recurrence for the evaluation of multivariate orthogonal polynomials. *J. Approx. Theory*, **162**, 407–420.

Beerends, R. J. 1991. Chebyshev polynomials in several variables and the radial part of the Laplace–Beltrami operator. *Trans. Amer. Math. Soc.*, **328**, 779–814.

Beerends, R. J., and Opdam, E. M. 1993. Certain hypergeometric series related to the root system *BC*. *Trans. Amer. Math. Soc.*, **339**, 581–609.

Berens, H., Schmid, H. J., and Xu, Y. 1995a. Multivariate Gaussian cubature formulae. *Arch. Math. (Basel)*, **64**, 26–32.

Berens, H., Schmid, H. J., and Xu, Y. 1995b. On two-dimensional definite orthogonal systems and a lower bound for the number of nodes of associated cubature formulae. *SIAM J. Math. Anal.*, **26**, 468–487.

Berg, C. 1987. The multidimensional moment problem and semigroups. Pages 110–124 of: *Moments in Mathematics*. Proc. Sympos. Appl. Math., vol. 37. Amer. Math. Soc.

Bonami, A., and Clerc, J.-L. 1973. Sommes de Cesàro et multiplicateurs des développements en harmoniques sphériques. *Trans. Amer. Math. Soc.*, **183**, 223–263.

zu Castell, W., Filbir, F., and Xu, Y. 2009. Cesàro means of Jacobi expansions on the parabolic biangle. *J. Approx. Theory*, **159**, 167–179.

Cichoń, D., Stochel, J., and Szafraniec, F. H. 2005. Three term recurrence relation modulo ideal and orthogonality of polynomials of several variables. *J. Approx. Theory*, **134**, 11–64.

Connett, W. C., and Schwartz, A. L. 1995. Continuous 2-variable polynomial hypergroups. Pages 89–109 of: *Applications of Hypergroups and Related Measure Algebras*. Contemp. Math., vol. 183. Amer. Math. Soc.

Dai, F., and Xu, Y. 2009a. Boundedness of projection operators and Cesàro means in weighted L^p space on the unit sphere. *Trans. Amer. Math. Soc.*, **361**, 3189–3221.

Dai, F., and Xu, Y. 2009b. Cesàro means of orthogonal expansions in several variables. *Constr. Approx.*, **29**, 129–155.

Dai, F., and Xu, Y. 2013. *Approximation Theory and Harmonic Analysis on Spheres and Balls*. Springer Monographs in Mathematics. Springer.

Delgado, A. M., Geronimo, J. S., Iliev, P., and Xu, Y. 2009. On a two-variable class of Bernstein–Szegő measures. *Constr. Approx.*, **30**, 71–91.

Delgado, A. M., Fernández, L., Pérez, T. E., Piñar, M. A., and Xu, Y. 2010. Orthogonal polynomials in several variables for measures with mass points. *Numer. Algorithms*, **55**, 245–264.

Diaconis, P., and Griffiths, R. 2014. An introduction to multivariate Krawtchouk polynomials and their applications. *J. Statist. Plann. Inference*, **154**, 39–53.

Dubiner, M. 1991. Spectral methods on triangles and other domains. *J. Sci. Comput.*, **6**, 345–390.

Dunkl, C. F. 1987. Orthogonal polynomials on the hexagon. *SIAM J. Appl. Math.*, **47**, 343–351.

Dunkl, C. F. 1988. Reflection groups and orthogonal polynomials on the sphere. *Math. Z.*, **197**, 33–60.

Dunkl, C. F. 1989. Differential-difference operators associated to reflection groups. *Trans. Amer. Math. Soc.*, **311**, 167–183.

Dunkl, C. F., and Xu, Y. 2014. *Orthogonal Polynomials of Several Variables*. Second edn. Encyclopedia of Mathematics and Its Applications, vol. 155. Cambridge University Press.

Dunn, K. B., and Lidl, R. 1982. Generalizations of the classical Chebyshev polynomials to polynomials in two variables. *Czechoslovak Math. J.*, **32**, 516–528.

Eier, R., and Lidl, R. 1974. Tschebyscheffpolynome in einer und zwei Variablen. *Abh. Math. Sem. Univ. Hamburg*, **41**, 17–27.

Eier, R., and Lidl, R. 1982. A class of orthogonal polynomials in k variables. *Math. Ann.*, **260**.

Engélis, G. K. 1974. Certain two-dimensional analogues of the classical orthogonal polynomials (in Russian). Pages 169–202, 235 of: *Latvian Mathematical Yearbook, 15*. Izdat. "Zinatne", Riga.

Erdélyi, A. et al. 1953. *Higher Transcendental Functions, Vol. II*. McGraw-Hill.

Exton, H. 1976. *Multiple Hypergeometric Functions and Applications*. Ellis Horwood, Chichester.

Fackerell, E. D., and Littler, R. A. 1974. Polynomials biorthogonal to Appell's polynomials. *Bull. Austral. Math. Soc.*, **11**, 181–195.

Fernández, L., Pérez, T. E., and Piñar, M. A. 2005. Classical orthogonal polynomials in two variables: a matrix approach. *Numer. Algorithms*, **39**, 131–142.

Folland, G. B. 1975. Spherical harmonic expansion of the Poisson–Szegő kernel for the ball. *Proc. Amer. Math. Soc.*, **47**, 401–408.

Fuglede, B. 1983. The multidimensional moment problem. *Exposition. Math.*, **1**, 47–65.

Gekhtman, M. I., and Kalyuzhny, A. A. 1994. On the orthogonal polynomials in several variables. *Integral Equations Operator Theory*, **19**, 404–418.

Geronimo, J. S., and Iliev, P. 2010. Bispectrality of multivariable Racah–Wilson polynomials. *Constr. Approx.*, **31**, 417–457.

Ghanmi, A. 2008. A class of generalized complex Hermite polynomials. *J. Math. Anal. Appl.*, **340**, 1395–1406.

Griffiths, R. C. 1971. Orthogonal polynomials on the multinomial distribution. *Austral. J. Statist.*, **13**, 27–35.

Griffiths, R. C. 1975. Orthogonal polynomials on the negative multinomial distribution. *J. Multivariate Anal.*, **5**, 271–277.

Groemer, H. 1996. *Geometric Applications of Fourier Series and Spherical Harmonics*. Encyclopedia of Mathematics and Its Applications, vol. 61. Cambridge University Press.

Grundmann, A., and Möller, H. M. 1978. Invariant integration formulas for the n-simplex by combinatorial methods. *SIAM J. Numer. Anal.*, **15**, 282–290.

Helgason, S. 2000. *Groups and Geometric Analysis*. Amer. Math. Soc. Corrected reprint of the 1984 original.

Hoare, M. R., and Rahman, M. 2008. A probabilistic origin for a new class of bivariate polynomials. *SIGMA*, **4**, Paper 089, 18 pp.

Hobson, E. W. 1931. *The Theory of Spherical and Ellipsoidal Harmonics*. Cambridge University Press. Reprinted by Chelsea, 1955.

Ikeda, M. 1967. On spherical functions for the unitary group. I, II, III. *Mem. Fac. Engrg. Hiroshima Univ.*, **3**, 17–29, 31–53, 55–75.

Iliev, P. 2012a. A Lie-theoretic interpretation of multivariate hypergeometric polynomials. *Compos. Math.*, **148**, 991–1002.

Iliev, P. 2012b. Meixner polynomials in several variables satisfying bispectral difference equations. *Adv. in Appl. Math.*, **49**, 15–23.

Iliev, P., and Terwilliger, P. 2012. The Rahman polynomials and the Lie algebra $\mathfrak{sl}_3(\mathbb{C})$. *Trans. Amer. Math. Soc.*, **364**, 4225–4238.

Iliev, P., and Xu, Y. 2007. Discrete orthogonal polynomials and difference equations of several variables. *Adv. Math.*, **212**, 1–36.

Intissar, A., and Intissar, A. 2006. Spectral properties of the Cauchy transform on $L_2(\mathbb{C}, e^{-|z|^2}\lambda(z))$. *J. Math. Anal. Appl.*, **313**, 400–418.

Ismail, M. E. H. 2016. Analytic properties of complex Hermite polynomials. *Trans. Amer. Math. Soc.*, **368**, 1189–1210.

Itô, K. 1952. Complex multiple Wiener integral. *Japan. J. Math.*, **22**, 63–86.

Jackson, D. 1936. Formal properties of orthogonal polynomials in two variables. *Duke Math. J.*, **2**, 423–434.

Kanjin, Y. 1985. Banach algebra related to disk polynomials. *Tohoku Math. J. (2)*, **37**, 395–404.

Karlin, S., and McGregor, J. 1975. Linear growth models with many types and multidimensional Hahn polynomials. Pages 261–288 of: *Theory and Application of Special Functions*. Academic Press.

Koornwinder, T. H. 1974a. Orthogonal polynomials in two variables which are eigenfunctions of two algebraically independent partial differential operators. I, II. *Indag. Math.*, **36**, 48–58, 59–66.

Koornwinder, T. H. 1974b. Orthogonal polynomials in two variables which are eigenfunctions of two algebraically independent partial differential operators. III, IV. *Indag. Math.*, **36**, 357–369, 370–381.

Koornwinder, T. 1975. Two-variable analogues of the classical orthogonal polynomials. Pages 435–495 of: *Theory and Application of Special Functions*. Academic Press.

Koornwinder, T. H., and Schwartz, A. L. 1997. Product formulas and associated hypergroups for orthogonal polynomials on the simplex and on a parabolic biangle. *Constr. Approx.*, **13**, 537–567.

Koornwinder, T., and Sprinkhuizen-Kuyper, I. 1978. Generalized power series expansions for a class of orthogonal polynomials in two variables. *SIAM J. Math. Anal.*, **9**, 457–483.

Kowalski, M. A. 1982a. Orthogonality and recursion formulas for polynomials in n variables. *SIAM J. Math. Anal.*, **13**(2), 316–323.

Kowalski, M. A. 1982b. The recursion formulas for orthogonal polynomials in n variables. *SIAM J. Math. Anal.*, **13**, 309–315.

Krall, H. L., and Sheffer, I. M. 1967. Orthogonal polynomials in two variables. *Ann. Mat. Pura Appl. (4)*, **76**, 325–376.

Kwon, K. H., Lee, J. K., and Littlejohn, L. L. 2001. Orthogonal polynomial eigenfunctions of second-order partial differential equations. *Trans. Amer. Math. Soc.*, **353**, 3629–3647.

Larcher, H. 1959. Notes on orthogonal polynomials in two variables. *Proc. Amer. Math. Soc.*, **10**, 417–423.

Lee, J. K., and Littlejohn, L. L. 2006. Sobolev orthogonal polynomials in two variables and second order partial differential equations. *J. Math. Anal. Appl.*, **322**, 1001–1017.

Li, H., and Xu, Y. 2010. Discrete Fourier analysis on fundamental domain and simplex of A_d lattice in d-variables. *J. Fourier Anal. Appl.*, **16**(3), 383–433.

Li, H., and Xu, Y. 2014. Spectral approximation on the unit ball. *SIAM J. Numer. Anal.*, **52**, 2647–2675.

Li, H., Sun, J., and Xu, Y. 2008. Discrete Fourier analysis, cubature, and interpolation on a hexagon and a triangle. *SIAM J. Numer. Anal.*, **46**, 1653–1681.

Li, Z., and Xu, Y. 2000. Summability of product Jacobi expansions. *J. Approx. Theory*, **104**, 287–301.

Li, Z., and Xu, Y. 2003. Summability of orthogonal expansions of several variables. *J. Approx. Theory*, **122**, 267–333.

Littlejohn, L. L. 1988. Orthogonal polynomial solutions to ordinary and partial differential equations. Pages 98–124 of: *Orthogonal Polynomials and their Applications*. Lecture Notes in Math., vol. 1329. Springer.

Logan, B. F., and Shepp, L. A. 1975. Optimal reconstruction of a function from its projections. *Duke Math. J.*, **42**, 645–659.

Lyskova, A. S. 1991. Orthogonal polynomials of several variables. *Dokl. Akad. Nauk SSSR*, **316**, 1301–1306. Translation in *Soviet Math. Dokl.* **43** (1991), 264–268.

Marr, R. B. 1974. On the reconstruction of a function on a circular domain from a sampling of its line integrals. *J. Math. Anal. Appl.*, **45**, 357–374.

Milch, P. R. 1968. A multi-dimensional linear growth birth and death process. *Ann. Math. Statist.*, **39**, 727–754.

Mizukawa, H., and Tanaka, H. 2004. $(n+1, m+1)$-hypergeometric functions associated to character algebras. *Proc. Amer. Math. Soc.*, **132**, 2613–2618.

Möller, H. M. 1973. *Polynomideale und Kubaturformeln*. Ph.D. thesis, Universität Dortmund.

Möller, H. M. 1976. Kubaturformeln mit minimaler Knotenzahl. *Numer. Math.*, **25**, 185–200.

Möller, H. M. 1979. Lower bounds for the number of nodes in cubature formulae. Pages 221–230 of: *Numerische Integration*. Internat. Ser. Numer. Math., vol. 45. Birkhäuser.

Morrow, C. R., and Patterson, T. N. L. 1978. Construction of algebraic cubature rules using polynomial ideal theory. *SIAM J. Numer. Anal.*, **15**, 953–976.

Müller, C. 1966. *Spherical Harmonics*. Lecture Notes in Math., vol. 17. Springer.

Müller, C. 1998. *Analysis of Spherical Symmetries in Euclidean Spaces*. Springer.

Mysovskikh, I. P. 1976. Numerical characteristics of orthogonal polynomials in two variables. *Vestnik Leningrad. Univ. Math.*, **3**, 323–332. Translated from the 1970 Russian original.

Mysovskikh, I. P. 1981. *Interpolatory Cubature Formulas* (in Russian). "Nauka", Moscow.

Pérez, T. E., Piñar, M. A., and Xu, Y. 2013. Weighted Sobolev orthogonal polynomials on the unit ball. *J. Approx. Theory*, **171**, 84–104.

Petrushev, P. P. 1999. Approximation by ridge functions and neural networks. *SIAM J. Math. Anal.*, **30**, 155–189.

Piñar, M., and Xu, Y. 2009. Orthogonal polynomials and partial differential equations on the unit ball. *Proc. Amer. Math. Soc.*, **137**, 2979–2987.

Proriol, J. 1957. Sur une famille de polynomes à deux variables orthogonaux dans un triangle. *C. R. Acad. Sci. Paris*, **245**, 2459–2461.

Radon, J. 1948. Zur mechanischen Kubatur. *Monatsh. Math.*, **52**, 286–300.

Rösler, M. 1998. Generalized Hermite polynomials and the heat equation for Dunkl operators. *Comm. Math. Phys.*, **192**, 519–542.

Schmid, H. J., and Xu, Y. 1994. On bivariate Gaussian cubature formulae. *Proc. Amer. Math. Soc.*, **122**, 833–841.

Schmüdgen, K. 1990. *Unbounded Operator Algebras and Representation Theory*. Birkhäuser.

Shishkin, A. D. 1997. Some properties of special classes of orthogonal polynomials in two variables. *Integral Transform. Spec. Funct.*, **5**, 261–272.

Sogge, C. D. 1986. Oscillatory integrals and spherical harmonics. *Duke Math. J.*, **53**, 43–65.

Sprinkhuizen-Kuyper, I. G. 1976. Orthogonal polynomials in two variables. A further analysis of the polynomials orthogonal over a region bounded by two lines and a parabola. *SIAM J. Math. Anal.*, **7**, 501–518.

Stein, E. M., and Weiss, G. 1971. *Introduction to Fourier Analysis on Euclidean Spaces*. Princeton University Press.

Stroud, A. H. 1971. *Approximate Calculation of Multiple Integrals*. Prentice-Hall.

Suetin, P. K. 1999. *Orthogonal Polynomials in Two Variables*. Gordon and Breach. Translated from the 1988 Russian original.

Thangavelu, S. 1993. *Lectures on Hermite and Laguerre Expansions*. Princeton University Press.

Tratnik, M. V. 1989a. Multivariable biorthogonal Hahn polynomials. *J. Math. Phys.*, **30**, 627–634.

Tratnik, M. V. 1989b. Multivariable Meixner, Krawtchouk, and Meixner–Pollaczek polynomials. *J. Math. Phys.*, **30**, 2740–2749.

Tratnik, M. V. 1991. Some multivariable orthogonal polynomials of the Askey tableau—discrete families. *J. Math. Phys.*, **32**, 2337–2342.

Vilenkin, N. Ja. 1968. *Special Functions and the Theory of Group Representations*. Translated from the 1965 Russian original. Amer. Math. Soc.

Vilenkin, N. Ja., and Klimyk, A. U. 1993. *Representation of Lie Groups and Special Functions. Vol. 2*. Kluwer.

Volkmer, H. 1999. Expansions in products of Heine–Stieltjes polynomials. *Constr. Approx.*, **15**, 467–480.

Vretare, L. 1984. Formulas for elementary spherical functions and generalized Jacobi polynomials. *SIAM J. Math. Anal.*, **15**, 805–833.

Waldron, S. 2008. Orthogonal polynomials on the disc. *J. Approx. Theory*, **150**, 117–131.

Waldron, S. 2009. Continuous and discrete tight frames of orthogonal polynomials for a radially symmetric weight. *Constr. Approx.*, **30**, 33–52.

Wünsche, A. 2005. Generalized Zernike or disc polynomials. *J. Comput. Appl. Math.*, **174**, 135–163.

Xu, Y. 1993. On multivariate orthogonal polynomials. *SIAM J. Math. Anal.*, **24**, 783–794.

Xu, Y. 1994a. *Common Zeros of Polynomials in Several Variables and Higher-dimensional Quadrature*. Pitman Research Notes in Math., vol. 312. Longman.

Xu, Y. 1994b. Multivariate orthogonal polynomials and operator theory. *Trans. Amer. Math. Soc.*, **343**, 193–202.

Xu, Y. 1997a. Integration of the intertwining operator for h-harmonic polynomials associated to reflection groups. *Proc. Amer. Math. Soc.*, **125**, 2963–2973.

Xu, Y. 1997b. Orthogonal polynomials for a family of product weight functions on the spheres. *Canad. J. Math.*, **49**, 175–192.

Xu, Y. 1998a. Orthogonal polynomials and cubature formulae on spheres and on balls. *SIAM J. Math. Anal.*, **29**, 779–793.

Xu, Y. 1998b. Orthogonal polynomials and cubature formulae on spheres and on simplices. *Methods Appl. Anal.*, **5**, 169–184.

Xu, Y. 1999. Summability of Fourier orthogonal series for Jacobi weight on a ball in $\mathbf{R}^d$. *Trans. Amer. Math. Soc.*, **351**, 2439–2458.

Xu, Y. 2000. Funk–Hecke formula for orthogonal polynomials on spheres and on balls. *Bull. London Math. Soc.*, **32**, 447–457.

Xu, Y. 2001a. Orthogonal polynomials and summability in Fourier orthogonal series on spheres and on balls. *Math. Proc. Cambridge Philos. Soc.*, **131**, 139–155.

Xu, Y. 2001b. Orthogonal polynomials on the ball and the simplex for weight functions with reflection symmetries. *Constr. Approx.*, **17**, 383–412.

Xu, Y. 2004. On discrete orthogonal polynomials of several variables. *Adv. in Appl. Math.*, **33**, 615–632.

Xu, Y. 2005a. Lecture notes on orthogonal polynomials of several variables. Pages 141–196 of: *Inzell Lectures on Orthogonal Polynomials*. Nova Sci. Publ., Hauppauge, NY.

Xu, Y. 2005b. Monomial orthogonal polynomials of several variables. *J. Approx. Theory*, **133**, 1–37.

Xu, Y. 2005c. Second-order difference equations and discrete orthogonal polynomials of two variables. *Int. Math. Res. Not.*, 449–475.

Xu, Y. 2006a. A family of Sobolev orthogonal polynomials on the unit ball. *J. Approx. Theory*, **138**, 232–241.

Xu, Y. 2006b. A new approach to the reconstruction of images from Radon projections. *Adv. in Appl. Math.*, **36**, 388–420.

Xu, Y. 2007. Reconstruction from Radon projections and orthogonal expansion on a ball. *J. Phys. A*, **40**, 7239–7253.

Xu, Y. 2008. Sobolev orthogonal polynomials defined via gradient on the unit ball. *J. Approx. Theory*, **152**, 52–65.

Xu, Y. 2012. Orthogonal polynomials and expansions for a family of weight functions in two variables. *Constr. Approx.*, **36**, 161–190.

Xu, Y. 2015a. Complex versus real orthogonal polynomials of two variables. *Integral Transforms Spec. Funct.*, **26**, 134–151.

Xu, Y. 2015b. Hahn, Jacobi, and Krawtchouk polynomials of several variables. *J. Approx. Theory*, **195**, 19–42.

Zernike, F. 1934. Beugungstheorie des Schneidenverfahrens und seiner verbesserten Form, der Phasenkontrastmethode. *Physica*, **1**, 689–704.

Zernike, F., and Brinkman, H. C. 1935. Hypersphärische Funktionen und die in sphärischen Bereichen orthogonalen Polynome. *Proc. Akad. Wetensch. Amsterdam*, **38**, 161–170.

3

Appell and Lauricella Hypergeometric Functions

Keiji Matsumoto

3.1 Introduction

Appell (1880) introduced four kinds of hypergeometric series in two variables as extensions of the hypergeometric series $F(a, b, c; x)$. Lauricella (1893) generalized them to hypergeometric series in m variables, and he considered systems of partial differential equations satisfied by them. See Erdélyi (1953, §§5.7, 5.14) for the early history of hypergeometric functions in several variables.

In this chapter, we give definitions of Appell's and Lauricella's hypergeometric series and we state fundamental properties of them such as domains of convergence, integral representations, systems of partial differential equations, fundamental systems of solutions and transformation formulas. Here we follow Appell and Kampé de Fériet (1926, §§XXXII–XXXIX) and Erdélyi (1953, §§5.7–5.13). We define the rank and the singular locus of a system of partial differential equations, and list them for Appell's and Lauricella's systems. We describe Pfaffian systems, contiguity relations, monodromy representations and twisted period relations for the systems. We give their explicit forms for Lauricella's E_D, which is the simplest among Lauricella's systems. We also mention the uniformization of the complement of the singular locus of E_D by the projectivization of its fundamental system of solutions.

Formulas involving Appell's and Lauricella's functions are so numerous that it is impossible to cover all of them here. In general we present formulas in Lauricella's m-variable case. We also give a few examples of reduction formulas. For further formulas see Aomoto and Kita (2011), Appell and Kampé de Fériet (1926), Bailey (1935), Exton (1976), Schlosser (2013), Srivastava and Karlsson (1985), Yoshida (1997) and references given there.

There are systematic ways to obtain the rank, integral representations and contiguity relations of a hypergeometric system if one considers such systems as A-hypergeometric systems, defined by Gel'fand, Zelevinsky and Kapranov; see Chapter 4.

Notation $\mathrm{i} := \sqrt{-1}$, $\mathbb{N}_0 := \{0, 1, 2, \ldots\}$, ${}^tA$ denotes matrix transpose.

3.2 Appell's Hypergeometric Series

The four kinds of hypergeometric series F_1, F_2, F_3, F_4 of two variables were defined by Appell (1880, 1882) and Appell and Kampé de Fériet (1926, §III).

Definition 3.2.1 (Appell's hypergeometric series)

$$F_1(a, b_1, b_2, c; x_1, x_2) := \sum_{n_1,n_2=0}^{\infty} \frac{(a)_{n_1+n_2}(b_1)_{n_1}(b_2)_{n_2}}{(c)_{n_1+n_2} n_1! n_2!} x_1^{n_1} x_2^{n_2},$$

$$F_2(a, b_1, b_2, c_1, c_2; x_1, x_2) := \sum_{n_1,n_2=0}^{\infty} \frac{(a)_{n_1+n_2}(b_1)_{n_1}(b_2)_{n_2}}{(c_1)_{n_1}(c_2)_{n_2} n_1! n_2!} x_1^{n_1} x_2^{n_2},$$

$$F_3(a_1, a_2, b_1, b_2, c; x_1, x_2) := \sum_{n_1,n_2=0}^{\infty} \frac{(a_1)_{n_1}(a_2)_{n_2}(b_1)_{n_1}(b_2)_{n_2}}{(c)_{n_1+n_2} n_1! n_2!} x_1^{n_1} x_2^{n_2},$$

$$F_4(a, b, c_1, c_2; x_1, x_2) := \sum_{n_1,n_2=0}^{\infty} \frac{(a)_{n_1+n_2}(b)_{n_1+n_2}}{(c_1)_{n_1}(c_2)_{n_2} n_1! n_2!} x_1^{n_1} x_2^{n_2},$$

where x_1, x_2 are variables, a, b, c and a_i, b_i, c_i $(i = 1, 2)$ are complex parameters with $c, c_1, c_2 \notin -\mathbb{N}_0$. The series F_1, F_2, F_3, F_4 respectively converge in the domains

$$D_1 := \{(x_1, x_2) \in \mathbb{C}^2 \mid |x_1|, |x_2| < 1\}, \quad D_2 := \{(x_1, x_2) \in \mathbb{C}^2 \mid |x_1| + |x_2| < 1\},$$
$$D_3 := \{(x_1, x_2) \in \mathbb{C}^2 \mid |x_1|, |x_2| < 1\}, \quad D_4 := \{(x_1, x_2) \in \mathbb{C}^2 \mid \sqrt{|x_1|} + \sqrt{|x_2|} < 1\}.$$

Horn (1889) generalized Appell's hypergeometric series to series $\sum_{n_1,n_2=0}^{\infty} A_{n_1,n_2} x_1^{n_1} x_2^{n_2}$ whose ratios $A_{n_1+1,n_2}/A_{n_1,n_2}$ and $A_{n_1,n_2+1}/A_{n_1,n_2}$ of coefficients are ratios of polynomials of degree 2 in n_1 and n_2, and he studied the systems of partial differential equations satisfied by them. See Erdélyi (1953, §§5.7–5.13) for the formulas of this generalization and Srivastava and Karlsson (1985) for other kinds of hypergeometric series in two variables.

3.3 Lauricella's Hypergeometric Series

The hypergeometric series F_2, F_3, F_4 and F_1 were generalized, respectively, to series F_A, F_B, F_C and F_D of m variables by Lauricella (1893); see also Appell and Kampé de Fériet (1926, §XXXVII).

Definition 3.3.1 (Lauricella's hypergeometric series)

$$F_A(a, (b), (c); (x)) := \sum_{(n)\in\mathbb{N}_0^m} \frac{(a)_{n_1+\cdots+n_m} \prod_{i=1}^m (b_i)_{n_i}}{\prod_{i=1}^m (c_i)_{n_i} \prod_{i=1}^m n_i!} \prod_{i=1}^m x_i^{n_i},$$

$$F_B((a), (b), c; (x)) := \sum_{(n)\in\mathbb{N}_0^m} \frac{\prod_{i=1}^m (a_i)_{n_i} \prod_{i=1}^m (b_i)_{n_i}}{(c)_{n_1+\cdots+n_m} \prod_{i=1}^m n_i!} \prod_{i=1}^m x_i^{n_i},$$

$$F_C(a, b, (c); (x)) := \sum_{(n)\in\mathbb{N}_0^m} \frac{(a)_{n_1+\cdots+n_m}(b)_{n_1+\cdots+n_m}}{\prod_{i=1}^m (c_i)_{n_i} \prod_{i=1}^m n_i!} \prod_{i=1}^m x_i^{n_i},$$

$$F_D(a,(b),c;(x)) := \sum_{(n)\in\mathbb{N}_0^m} \frac{(a)_{n_1+\cdots+n_m}\prod_{i=1}^m (b_i)_{n_i}}{(c)_{n_1+\cdots+n_m}\prod_{i=1}^m n_i!}\prod_{i=1}^m x_i^{n_i},$$

where $(x) = (x_1,\dots,x_m)$, $(a) = (a_1,\dots,a_m)$, $(b) = (b_1,\dots,b_m)$, $(c) = (c_1,\dots,c_m)$, $(n) = (n_1,\dots,n_m)$, $x_1,\dots,x_m$ are variables, a, b, c and a_i, b_i, c_i $(i = 1,\dots,m)$ are complex parameters with $c, c_1,\dots,c_m \notin -\mathbb{N}_0$ and each n_i runs over the set $\mathbb{N}_0$. The series F_A, F_B, F_C, F_D respectively converge in the domains

$$D_A := \left\{(x)\in\mathbb{C}^m \;\middle|\; \textstyle\sum_{1\le i\le m}|x_i| < 1\right\},\qquad D_B := \left\{(x)\in\mathbb{C}^m \;\middle|\; \max_{1\le i\le m}|x_i| < 1\right\},$$
$$D_C := \left\{(x)\in\mathbb{C}^m \;\middle|\; \textstyle\sum_{1\le i\le m}\sqrt{|x_i|} < 1\right\},\quad D_D := \left\{(x)\in\mathbb{C}^m \;\middle|\; \max_{1\le i\le m}|x_i| < 1\right\}.$$

Each of Lauricella's hypergeometric series can be written in succinct form as

$$\sum_{(n)\in\mathbb{N}_0^m} \frac{((a),(n))((b),(n))}{((c),(n))(n)!}(x)^{(n)},$$

where we use the notation $(n)! := n_1!\cdots n_m!$, $(x)^{(n)} := x_1^{n_1}\cdots x_m^{n_m}$ and

$$((a),(n)) := \begin{cases} (a)_{n_1+\cdots+n_m} & \text{if } a\in\mathbb{C},\\ \prod_{i=1}^m (a_i)_{n_i} & \text{if } a = (a_1,\dots,a_m)\in\mathbb{C}^m.\end{cases}$$

By Definition 3.3.1, we have

$$F_B((a),(b),c;(z)) = F_B((b),(a),c;(z)),\qquad F_C(a,b,(c);(z)) = F_C(b,a,(c);(z)).$$

Furthermore, each of the functions in Definition 3.3.1 is invariant under the symmetric group S_m by simultaneous permutation of the variables and the parameters in $\mathbb{C}^m$.

See Srivastava and Karlsson (1985) for other kinds of hypergeometric series in three variables, and see Chapter 4 for the A-hypergeometric series.

3.4 Integral Representations

We consider five kinds of Euler-type kernel $P((a),(t),(x))$ as in Aomoto and Kita (2011, §3.3), Kita (1992) and Yoshida (1987, §6.4):

$$\prod(1-x_it_i)^{-a_i},\quad \prod(1-x_it)^{-a_i},\quad \prod(1-\tfrac{x_i}{t_i})^{-a_i},\quad (1-\textstyle\sum x_it_i)^{-a},\quad (1-\sum \tfrac{x_i}{t_i})^{-a},$$

where $\sum = \sum_{i=1}^m$ and $\prod = \prod_{i=1}^m$. Each of them admits Taylor's expansion

$$P((a),(t),(x)) = \sum_{n\in\mathbb{N}_0^m} p_{(n)}((a),(t))(x)^{(n)}$$

around $(x) = (0,\dots,0)$. By selecting a suitable function $f((a),(t))$ and a suitable cycle Δ, and by interchanging the order of the integration and the summation we obtain

$$\int_\Delta f((a),(t))P((a),(t),(x))\,dt = \sum_{n\in\mathbb{N}_0^m}(x)^{(n)}\int_\Delta f((a),(t))p_{(n)}((a),(t))\,dt, \tag{3.4.1}$$

where dt is the volume form on Δ. Note that the resulting integral is a generalized beta integral which can be expressed in terms of the gamma function. Thus we have the following integral representations of Euler type for the hypergeometric series F_A, F_B, F_C, F_D.

Theorem 3.4.1 (Integral representations of Euler type)

$$\begin{aligned}
&F_A(a, b_1, \ldots, b_m, c_1, \ldots, c_m; x_1, \ldots, x_m)\\
&= \left(\prod \frac{\Gamma(c_i)}{\Gamma(b_i)\Gamma(c_i-b_i)}\right) \int_{(0,1)^m} \left(\prod t_i^{b_i-1}(1-t_i)^{c_i-b_i-1}\right)(1-\textstyle\sum x_i t_i)^{-a}\, dt_1 \cdots dt_m\\
&\hspace{22em} (\operatorname{Re} b_i, \operatorname{Re}(c_i-b_i) > 0)\\
&= \frac{(\prod \Gamma(c_i))\Gamma(a-\sum c_i+m)}{(2\pi \mathrm{i})^m \Gamma(a)} \int_{\mathrm{i}\mathbb{R}^m_\xi} \left(\prod t_i^{-c_i}\right)(1-\textstyle\sum t_i)^{\sum c_i - a - m} \prod\left(1-\dfrac{x_i}{t_i}\right)^{-b_i} dt_1 \cdots dt_m\\
&\hspace{16em} (\operatorname{Re}(\textstyle\sum_{j\neq i} c_j - a) < m-1, \sum |x_i| < m^{-1}),\\
&F_B(a_1, \ldots, a_m, b_1, \ldots, b_m, c; x_1, \ldots, x_m)\\
&= \frac{\Gamma(c)}{(\prod \Gamma(a_i))\Gamma(c-\sum a_i)} \int_{\Delta^m} \left(\prod t_i^{a_i-1}\right)(1-\textstyle\sum t_i)^{c-\sum a_i - 1} \prod (1-x_i t_i)^{-b_i}\, dt_1 \cdots dt_m\\
&\hspace{22em} (\operatorname{Re} a_i, \operatorname{Re}(c-\textstyle\sum a_i) > 0)\\
&= (\text{idem with the exchange } (a) \leftrightarrow (b)),\\
&F_C(a, b, c_1, \ldots, c_m; x_1, \ldots, x_m)\\
&= \frac{(\prod \Gamma(c_i))\Gamma(a-\sum c_i+m)}{(2\pi \mathrm{i})^m \Gamma(a)} \int_{\mathrm{i}\mathbb{R}^m_\xi} \left(\prod t_i^{-c_i}\right)(1-\textstyle\sum t_i)^{\sum c_i - a - m} \left(1-\sum\dfrac{x_i}{t_i}\right)^{-b} dt_1 \cdots dt_m\\
&\hspace{16em} (\operatorname{Re}(\textstyle\sum_{j\neq i} c_j - a) < m-1, \sum |x_i| < m^{-1})\\
&= (\text{idem with the exchange } (a) \leftrightarrow (b)),\\
&F_D(a, b_1, \ldots, b_m, c; x_1, \ldots, x_m)\\
&= \frac{\Gamma(c)}{(\prod \Gamma(b_i))\Gamma(c-\sum b_i)} \int_{\Delta^m} \left(\prod t_i^{b_i-1}\right)(1-\textstyle\sum t_i)^{c-\sum b_i - 1} (1-\sum x_i t_i)^{-a}\, dt_1 \cdots dt_m\\
&\hspace{22em} (\operatorname{Re} b_i, \operatorname{Re}(c-\textstyle\sum b_i) > 0)\\
&= \frac{\Gamma(c)}{\Gamma(a)\Gamma(c-a)} \int_0^1 t^{a-1}(1-t)^{c-a-1}\left(\prod (1-x_i t)^{-b_i}\right) dt \qquad (\operatorname{Re} a, \operatorname{Re}(c-a) > 0),
\end{aligned}$$

where $\Delta^m := \{(t_1, \ldots, t_m) \in \mathbb{R}^m \mid t_1, \ldots, t_m > 0,\ t_1 + \cdots + t_m < 1\}$, *the m-dimensional simplex, and* $\mathrm{i}\mathbb{R}^m_\xi := \{(\xi, \ldots, \xi) + \mathrm{i}(s_1, \ldots, s_m) \mid (s_1, \ldots, s_m) \in \mathbb{R}^m\}$, *a translation of the pure imaginary space* $\mathrm{i}\mathbb{R}^m$ *by a positive real number* ξ *satisfying* $\sum_{i=1}^m |x_i| < \xi < 1/m$.

The first integral representations of F_A, F_B above and the integral representations of F_D were given by Appell and Kampé de Fériet (1926, §XXXVIII) and Lauricella (1893). The integrals over $\mathrm{i}\mathbb{R}^m_\xi$ were introduced by Aomoto, as mentioned in Oshima (2012, §13.10.2).

If we regularize the regions $(0, 1)$, $(0, 1)^m$ and Δ^m of the integrations in Theorem 3.4.1 then, in the convergence conditions, positivity of real parts of exponents is replaced by non-integrality of exponents. Let us explain their construction according to Yoshida (1997, Chap. IV). The *regularized cycle* reg(0, 1) of the open interval (0, 1) with respect to $t^{\alpha_0}(1-t)^{\alpha_1}$ ($\alpha_0, \alpha_1 \in \mathbb{C} \setminus \mathbb{Z}$) is the formal sum

$$I_\varepsilon - \frac{1}{1-e^{2\pi i \alpha_0}} S_0 + \frac{1}{1-e^{2\pi i \alpha_1}} S_1,$$

where ε is a small real number, I_ε is the closed interval $[\varepsilon, 1-\varepsilon]$ and S_0 and S_1 are oriented circles given by

$$S_0\colon [0, 2\pi] \ni \theta \mapsto \varepsilon e^{i\theta}, \quad S_1\colon [0, 2\pi] \ni \theta \mapsto 1 - \varepsilon e^{i\theta}.$$

We set

$$\begin{aligned}\int_{\mathrm{reg}(0,1)} t^{\alpha_0}(1-t)^{\alpha_1} P(a,t,x)\,dt &:= \int_{I_\varepsilon} t^{\alpha_0}(1-t)^{\alpha_1} P(a,t,x)\,dt \\ - \frac{1}{1-e^{2\pi i \alpha_0}} \int_{S_0} t^{\alpha_0}(1-t)^{\alpha_1} P(a,t,x)\,dt &+ \frac{1}{1-e^{2\pi i \alpha_1}} \int_{S_1} t^{\alpha_0}(1-t)^{\alpha_1} P(a,t,x)\,dt,\end{aligned} \tag{3.4.2}$$

where for the moment $P(a,t,x)$ is an Euler-type kernel with $m = 1$ and its branch on each path is fixed by putting $P(a,t,x) = 1$ at $x = 0$. The branch of

$$t^{\alpha_0}(1-t)^{\alpha_1} = e^{\alpha_0 \log t + \alpha_1 \log(1-t)}$$

is given by $\arg t = \arg(1-t) = 0$ on I_ε and at the initial points of S_0, S_1. The integral over reg(0, 1) is defined whenever $\alpha_0, \alpha_1 \in \mathbb{C} \setminus \mathbb{Z}$, and it is independent of ε by Cauchy's integral theorem. Moreover, it is equal to the original integral along (0, 1) under the convergence condition $\mathrm{Re}(\alpha_0), \mathrm{Re}(\alpha_1) > -1$. Note that the integral over reg(0, 1) on the left-hand side of (3.4.2), when multiplied by $(1-e^{2\pi i \alpha_0})(1-e^{2\pi i \alpha_1})$, is equal to that integral along the double contour loop around $t = 0, 1$, which is also known as the *Jordan–Pochhammer double loop*.

The regularized cycle $\mathrm{reg}(0,1)^m$ is defined by the m-fold direct product of reg(0, 1). For the construction of the regularized cycle $\mathrm{reg}(\Delta^m)$ ($m \geq 2$), see Kita (1992).

Another integral representation for F_4 is given by Chaundy (1954):

$$\begin{aligned}F_4(a,b,c_1,c_2;x_1,x_2) &= \frac{\Gamma(c_1)\Gamma(c_2)}{\Gamma(a)\Gamma(b)\Gamma(c_1-a)\Gamma(c_2-b)} \\ &\times \iint_S t_1^{a-1} t_2^{b-1} (1-t_1+t_1t_2x_2)^{c_1-a-1}(1-t_2+t_1t_2x_1)^{c_2-b-1}\,dt_1\,dt_2,\end{aligned}$$

where S is the bounded connected component of

$$\left\{(t_1,t_2) \mid t_1 > 0,\ t_2 > 0,\ 1-t_1+t_1t_2x_2 > 0,\ 1-t_2+t_1t_2x_1 > 0\right\},$$

with $\sqrt{|x_1|} + \sqrt{|x_2|} < 1$ and $\mathrm{Re}(c_1) > \mathrm{Re}(a) > 0$, $\mathrm{Re}(c_2) > \mathrm{Re}(b) > 0$. This formula is obtained by a change of variables in

$$\begin{aligned}F_4(a,b,c_1,c_2;x_1(1-x_2),x_2(1-x_1)) &= \frac{\Gamma(c_1)\Gamma(c_2)}{\Gamma(a)\Gamma(b)\Gamma(c_1-a)\Gamma(c_2-b)} \\ \times \int_0^1\int_0^1 t_1^{a-1}t_2^{b-1}(1-t_1)^{c_1-a-1}(1-t_2)^{c_2-b-1}&(1-t_1x_1)^{a-c_1-c_2+1}(1-t_2x_2)^{b-c_1-c_2+1}\end{aligned}$$

$$\cdot (1 - t_1 x_1 - t_2 x_2)^{c_1+c_2-a-b-1}\, dt_1\, dt_2 \quad (\mathrm{Re}(c_1) > \mathrm{Re}(a) > 0,\ \mathrm{Re}(c_2) > \mathrm{Re}(b) > 0),$$

given by Burchnall and Chaundy (1940, equation (68)) and reproduced in Erdélyi (1953, §5.8(4)).

Expressions of F_A and F_C over the regularized cycle $\mathrm{reg}(\Delta^m)$ are due to Kita (1992). Another expression of F_C is given by Pastro (1989). See Chapter 4, for a systematic way to find kernel functions.

The Mellin–Barnes integral representation (Andrews et al., 1999, Theorem 2.4.1) of the Gauss hypergeometric function has been generalized for Appell's functions; see Erdélyi (1953, §5.8.3). It is easy to extend this to the Lauricella case.

Theorem 3.4.2 (Integral representations of Barnes type) *Suppose that each entry of* $\mathrm{Re}(a)$ *and* $\mathrm{Re}(b)$ *is positive. Put*

$$F(x) := \frac{1}{(2\pi \mathrm{i})^m} \int_{\mathrm{i}\mathbb{R}^m_{-\varepsilon}} \Psi(t) \prod \Gamma(-t_i)(-x_i)^{t_i}\, dt_1 \ldots dt_m,$$

where $\arg(-x_i) \in (-\pi, \pi)$, $0 < \varepsilon < \min(\mathrm{Re}(a), \mathrm{Re}(b))$ *and*

$$\mathrm{i}\mathbb{R}^m_{-\varepsilon} := \{-(\varepsilon, \ldots, \varepsilon) + \mathrm{i}(s_1, \ldots, s_m) \mid (s_1, \ldots, s_m) \in \mathbb{R}^m\}.$$

If $\Psi(t)$ *equals one of the following* $\Psi_A(t), \ldots, \Psi_D(t)$, *then* $F(x)$ *equals* $F_A(a, (b), (c); (x)), \ldots, F_D(a, (b), c; (x))$, *respectively*:

$$\Psi_A = \frac{\Gamma(a + \sum t_i)(\prod \Gamma(b_i + t_i)\Gamma(c_i))}{\Gamma(a)(\prod \Gamma(b_i)\Gamma(c_i + t_i))}, \qquad \Psi_B = \frac{(\prod \Gamma(a + t_i))(\prod \Gamma(b_i + t_i))\Gamma(c)}{(\prod \Gamma(a_i)\Gamma(b_i))\Gamma(c + \sum t_i)},$$

$$\Psi_C = \frac{\Gamma(a + \sum t_i)\Gamma(b + \sum t_i)(\prod \Gamma(c_i))}{\Gamma(a)\Gamma(b)(\prod \Gamma(c_i + t_i))}, \qquad \Psi_D = \frac{\Gamma(a + \sum t_i)(\prod \Gamma(b_i + t_i))\Gamma(c)}{\Gamma(a)(\prod \Gamma(b_i))\Gamma(c + \sum t_i)}.$$

This theorem is still valid if no entry of (a) and (b) is in $-\mathbb{N}_0$ after a slight modification of $\mathrm{i}\mathbb{R}^m_{-\varepsilon}$.

3.5 Systems of Hypergeometric Differential Equations

Let $\mathbb{C}[x, \partial]$ be the ring generated by the polynomial ring $\mathbb{C}[x]$ in m variables $x_1, \ldots, x_m$ and the partial differential operators

$$\partial_1 = \frac{\partial}{\partial x_1}, \ldots, \partial_m = \frac{\partial}{\partial x_m}$$

with relations $\partial_i \partial_j = \partial_j \partial_i$ $(1 \le i, j \le m)$ and

$$\partial_i f(x) = \frac{\partial f(x)}{\partial x_i} + f(x)\partial_i, \quad f(x) \in \mathbb{C}[x].$$

This ring naturally acts from the left on a holomorphic function $F(x)$ in an open set in $\mathbb{C}^m$; for example,

$$f(x)\cdot F(x) = f(x)F(x), \quad \partial_i \cdot F(x) = \frac{\partial F(x)}{\partial x_i}, \quad (\partial_i\partial_j)\cdot F(x) = \frac{\partial^2 F(x)}{\partial x_i\,\partial x_j},$$

where $f(x) \in \mathbb{C}[x] \subset \mathbb{C}[x,\partial]$. We can consider a left ideal E of $\mathbb{C}[x,\partial]$ as a system of linear differential equations.

Definition 3.5.1 (Appell and Kampé de Fériet, 1926, §IV) The elements of $\mathbb{C}[x,\partial]$ in Table 3.1 annihilate Appell's hypergeometric series F_1, F_2, F_3, F_4, respectively. They are called *Appell's hypergeometric differential equations*. The left ideals E_1, E_2, E_3, E_4 of $\mathbb{C}[x,\partial]$ generated by them are called *Appell's systems* of hypergeometric differential equations.

Definition 3.5.2 (Appell and Kampé de Fériet, 1926, §XXXIX; Lauricella, 1893) The elements of $\mathbb{C}[x,\partial]$ in Table 3.2 ($i = 1,\dots,m$) annihilate Lauricella's hypergeometric series F_A, F_B, F_C, F_D, respectively. They are called *Lauricella's hypergeometric differential equations*. The left ideals E_A, E_B, E_C, E_D of $\mathbb{C}[x,\partial]$ generated by them are called *Lauricella's systems* of hypergeometric differential equations.

Remark 3.5.3
1. Although the hypergeometric series F_A, F_B, F_C, F_D are defined under the conditions that $c, c_1, \dots, c_m \neq 0, -1, -2, \dots$, the systems E_A, E_B, E_C, E_D are defined for any parameter values $c \in \mathbb{C}$ and $(c) = (c_1, \dots, c_m) \in \mathbb{C}^m$.
2. We can see the behavior of systems E_A, E_B, E_C, E_D or their solutions around $x_i = \infty$ ($i = 1,\dots,m$) as that around $x'_i = 0$ by the variable change $x'_i = 1/x_i$ and the relation $x_i\partial/\partial x_i = -x'_i\partial/\partial x'_i$. The systems E_A, E_B, E_C, E_D are regarded as defined on $(\mathbb{P}^1)^m$, where $\mathbb{P}^1$ denotes the complex projective line.
3. Lauricella's system $E_B((a),(b),c)$ is obtained by the left ideal of $\mathbb{C}[x,\partial]$ annihilating the function around $(x_1,\dots,x_m) = (\infty,\dots,\infty)$:

$$\left(\prod x_i^{-a_i}\right) F_A(1 - c + \textstyle\sum a_i, (a), (a) - (b) + (1,\dots,1); x_1^{-1},\dots,x_m^{-1}).$$

   In particular, Appell's system E_3 can be obtained from Appell's system E_2 in this way.

Let E be a left ideal of $\mathbb{C}[x,\partial]$. The *rank* of the system E of linear differential equations is defined by

$$\dim_{\mathbb{C}(x)}[\mathbb{C}(x) \otimes_{\mathbb{C}[x]} (\mathbb{C}[x,\partial]/E)],$$

where the tensor product $\mathbb{C}(x) \otimes_{\mathbb{C}[x]} (\mathbb{C}[x,\partial]/E)$ is regarded as a vector space over the field $\mathbb{C}(x)$ of rational functions in $x_1,\dots,x_m$. For a left ideal E of $\mathbb{C}[x,\partial]$ and the $\mathbb{C}$-algebra $\mathcal{O}(U_x)$ of (single-valued) holomorphic functions in a neighborhood U_x of $x \in \mathbb{C}^m$, we have the $\mathbb{C}$-vector space

$$\mathrm{Sol}_E(U_x) = \{F(x) \in \mathcal{O}(U_x) \mid P(x,\partial)\cdot F(x) = 0 \ \ \forall\, P(x,\partial) \in E\},$$

which is called the *local solution space* of the system E in U_x. It is known that

$$\dim(\mathrm{Sol}_E(U_x)) \le r = (\text{the rank of } E).$$

The maximum over x and U_x of $\dim(\mathrm{Sol}_E(U_x))$ is equal to r (which is an alternative definition of the rank), and $\dim(\mathrm{Sol}_E(U_x)) = r$ for generic values of x. A point $x \in \mathbb{C}^m$ satisfying $\mathrm{Sol}_E(U_x) < r$ for any neighborhood U_x is called a *singular point* of the system E, and the set $S(E)$ of singular points of E is called the *singular locus* of the system E. Note that this is a divisor contained in the D-module theoretical singular locus; for its definition, see textbooks on D-modules (e.g., Ōaku, 1994 and Saito et al., 2000). The singular loci of $E_1, \ldots, E_4$ and $E_A, \ldots, E_D$ are denoted by $S_1, \ldots, S_4$ and $S_A, \ldots, S_D$, respectively.

Table 3.1 *Appell's systems*

$E_1(a, b_1, b_2, c)$	$x_1(1-x_1)\partial_1^2 + x_2(1-x_1)\partial_1\partial_2 + [c-(a+b_1+1)x_1]\partial_1 - b_1x_2\partial_2 - ab_1,$
	$x_2(1-x_2)\partial_2^2 + x_1(1-x_2)\partial_1\partial_2 + [c-(a+b_2+1)x_2]\partial_2 - b_2x_1\partial_1 - ab_2,$
	$(x_1-x_2)\partial_1\partial_2 - b_2\partial_1 + b_1\partial_2$
$E_2(a, b_1, b_2, c_1, c_2)$	$x_1(1-x_1)\partial_1^2 - x_1x_2\partial_1\partial_2 + [c_1-(a+b_1+1)x_1]\partial_1 - b_1x_2\partial_2 - ab_1,$
	$x_2(1-x_2)\partial_2^2 - x_1x_2\partial_1\partial_2 + [c_2-(a+b_2+1)x_2]\partial_2 - b_2x_1\partial_1 - ab_2$
$E_3(a_1, a_2, b_1, b_2, c)$	$x_1(1-x_1)\partial_1^2 + x_2\partial_1\partial_2 + [c-(a_1+b_1+1)x_1]\partial_1 - a_1b_1,$
	$x_2(1-x_2)\partial_2^2 + x_1\partial_1\partial_2 + [c-(a_2+b_2+1)x_2]\partial_2 - a_2b_2$
$E_4(a, b, c_1, c_2)$	$x_1(1-x_1)\partial_1^2 - x_2^2\partial_2^2 - 2x_1x_2\partial_1\partial_2 + [c_1-(a+b+1)x_1]\partial_1$
	$-(a+b+1)x_2\partial_2 - ab,$
	$x_2(1-x_2)\partial_2^2 - x_1^2\partial_1^2 - 2x_1x_2\partial_1\partial_2 + [c_2-(a+b+1)x_2]\partial_2$
	$-(a+b+1)x_1\partial_1 - ab$

Table 3.2 *Lauricella's systems*

$E_A(a,(b),(c))$	$x_i(1-x_i)\partial_i^2 - x_i\sum_{1\le j\le m, j\ne i} x_j\partial_i\partial_j + [c_i-(a+b_i+1)x_i]\partial_i - b_i\sum_{1\le j\le m, j\ne i} x_j\partial_j - ab_i$
$E_B((a),(b),c)$	$x_i(1-x_i)\partial_i^2 + \sum_{1\le j\le m, j\ne i} x_j\partial_i\partial_j + [c-(a_i+b_i+1)x_i]\partial_i - a_ib_i$
$E_C(a,b,(c))$	$x_i(1-x_i)\partial_i^2 - x_i\sum_{1\le j\le m, j\ne i} x_j\partial_i\partial_j - \sum_{1\le j_1,j_2\le m, j_1\ne i} x_{j_1}x_{j_2}\partial_{j_1}\partial_{j_2}$
	$+[c_i-(a+b+1)x_i]\partial_i - (a+b+1)\sum_{1\le j\le m, j\ne i} x_j\partial_j - ab$
$E_D(a,(b),c)$	$x_i(1-x_i)\partial_i^2 + (1-x_i)\sum_{1\le j\le m, j\ne i} x_j\partial_i\partial_j + [c-(a+b_i+1)x_i]\partial_i - b_i\sum_{1\le j\le m, j\ne i} x_j\partial_j - ab_i,$
	$(x_i-x_j)\partial_i\partial_j - b_j\partial_i + b_i\partial_j \quad (1\le i<j\le m)$

Theorem 3.5.4 *The rank of Lauricella's system of hypergeometric differential equations with any parameters and the singular locus in $\mathbb{C}^m$ of that with generic parameters are in Table 3.3. Here $x_i = \infty$ $(i = 1, \ldots, m)$ belong to the singular locus of each system.*

Table 3.3 *Rank and singular locus*

System	Rank	Singular locus
E_1	3	$x_1(1-x_1)x_2(1-x_2)(x_1-x_2)=0$
E_2	4	$x_1(1-x_1)x_2(1-x_2)(1-x_1-x_2)=0$
E_3	4	$x_1(1-x_1)x_2(1-x_2)(x_1+x_2-x_1x_2)=0$
E_4	4	$x_1x_2(1-2x_1-2x_2-2x_1x_2+x_1^2+x_2^2)=0$
E_D	$m+1$	$(\prod x_i(1-x_i))\prod_{i<j}(x_i-x_j)=0$
E_A	2^m	$(\prod x_i)\prod_{r=1}^m \prod_{i_1<i_2<\cdots<i_r}(1-x_{i_1}-\cdots-x_{i_r})=0$
E_B	2^m	$(\prod x_i)\prod_{r=1}^m \prod_{i_1<i_2<\cdots<i_r} x_{i_1}\cdots x_{i_r}(1-x_{i_1}^{-1}-\cdots-x_{i_r}^{-1})=0$
E_C	2^m	$(\prod x_i)\prod_{\varepsilon_1,\ldots,\varepsilon_m=\pm 1}(1+\varepsilon_1\sqrt{x_1}+\cdots+\varepsilon_m\sqrt{x_m})=0$

Lauricella (1893) gave the ranks of the systems now named after him (see also Appell and Kampé de Fériet, 1926, §XXXIX), and he essentially gave the singular loci S_A, S_B and S_D by Proposition 3.6.1, Corollary 3.7.3 and Remark 3.5.3. For the singular locus S_C, see Hattori and Takayama (2014).

Remark 3.5.5

1. Lauricella (1893) showed that, if parameters of $E_D(a,(b),c)$ satisfy $a+1-c \neq 0$, then the operators $(x_i-x_j)\partial_i\partial_j - b_j\partial_i + b_i\partial_j$ $(1 \le i < j \le m)$ can be derived from the operators in the first line in $E_D(a,(b),c)$. Otherwise, we cannot omit them. For example, the rank of $E_1(1,(3,4),2)$ without the operator $(x_1-x_2)\partial_1\partial_2 - b_2\partial_1 + b_1\partial_2$ becomes infinite-dimensional.
2. It is shown by Hattori and Takayama (2014) that the local solution space of $E_4(-\frac{1}{2},-2,\frac{1}{2},\frac{1}{2})$ around a point near to $(0,0)$ is spanned by the four functions
$$1+2(x_1+x_2-x_1x_2)-\frac{x_1^2+x_2^2}{3},\quad \sqrt{x_1},\quad \sqrt{x_2},\quad \sqrt{x_1x_2}\Big(1-\frac{x_1+x_2}{3}\Big).$$
By their analytic continuations, the singular locus of $E_4(-\frac{1}{2},-2,\frac{1}{2},\frac{1}{2})$ consists of divisors $x_1=0$, $x_2=0$, $x_1=\infty$ and $x_2=\infty$.

Each system $E_A,\ldots,E_D$ can be transformed into the *Pfaffian system*
$$df(x)=\Omega(x)f(x),\quad \text{i.e., } \partial_k f_i(x)=\sum_j \Omega_{ij,k}(x)f_j(x)\quad (k=1,\ldots,m)$$

with the *integrability condition* $d\Omega(x) = \Omega(x) \wedge \Omega(x)$, i.e.,

$$\partial_l \Omega_{ij,k}(x) - \partial_k \Omega_{ij,l}(x) = \textstyle\sum_t (\Omega_{it,l}\Omega_{tj,k} - \Omega_{it,k}\Omega_{tj,l}) \quad (1 \le k < l \le m),$$

where the entries of the *connection matrix* $\Omega(x)$ are rational 1-forms in x, and

$$f = \begin{cases} {}^t(F, \partial_1 F, \ldots, \partial_m F), & E_D, \\ {}^t\Big(F, (\partial_i F), \ldots, (\partial_{I_1} \cdots \partial_{i_r} F)_{i_1 < i_2 < \cdots < i_r}, \ldots, \partial_{i_1} \cdots \partial_{i_m} F\Big), & E_A, E_B, E_C. \end{cases}$$

Below we explicitly give the Pfaffian system (Matsumoto, 2013) of E_D, which admits a simple expression. In order to simplify the connection matrix $\Omega(x)$, we use the vector-valued function $f = {}^t(f_0, \ldots, f_m)$, where $f_i = \int_{\mathrm{reg}(1,\infty)} u(t_1, x)\varphi_i$ $(i = 0, 1, \ldots, m)$, $a, c - a \notin \mathbb{Z}$,

$$u(t, x) = t^{\alpha_0}(t-1)^{\alpha_{m+1}} \prod_{j=1}^{m} (t - x_j)^{\alpha_j}, \tag{3.5.1}$$

$$\alpha_0 = -c + \sum_{j=1}^{m} b_j, \quad \alpha_i = -b_i \; (i = 1, \ldots, m), \quad \alpha_{m+1} = c - a, \quad \alpha_{m+2} = a,$$

$$\varphi_0 = \frac{dt}{t-1}, \qquad \varphi_i = \frac{(1 - x_i)\, dt}{(t - x_i)(t-1)} \quad (i = 1, \ldots, m).$$

Note that if $b_i \neq 0$ $(i = 1, \ldots, m)$ then

$$f_0 = \frac{\Gamma(a)\Gamma(c-a)}{\Gamma(c)} F_D(a, (b), c; x), \quad f_i = \frac{1 - x_i}{b_i} \partial_i f_0.$$

Proposition 3.5.6 *Suppose that $\alpha_i \neq 0$ $(i = 0, \ldots, m+2)$. The system E_D is equivalent to the integrable Pfaffian system*

$$df = \Omega(x) f, \quad \Omega(x) = \sum_{0 \le i < j \le m+1} A_{ij}\, d\log(x_i - x_j)$$

for an unknown vector-valued function f constructed from f_0 as above, where the $(m+1) \times (m+1)$ matrices (A_{ij}) $(i = 0, 1, \ldots, m,\ j = 1, 2, \ldots, m+1)$ are given as

$$A_{ij} = (\alpha_i + \alpha_j) C\, {}^t v_{ij} (v_{ij} C\, {}^t v_{ij})^{-1} v_{ij}$$

with

$$v_{ij} = \begin{cases} \tilde{e}_i - \tilde{e}_j & \text{if } 1 \le i < j \le m, \\ \alpha_0 \tilde{e}_j + \alpha_{m+2}\tilde{e}_0 + \sum_{k=1}^{m} \alpha_k \tilde{e}_k & \text{if } i = 0,\ 1 \le j \le m, \\ \tilde{e}_i & \text{if } 1 \le i \le m,\ j = m+1, \end{cases}$$

$$\tilde{e}_i = (\delta_{i0}, \delta_{i1}, \ldots, \delta_{im}),$$

$$C = \frac{1}{c-a} \begin{pmatrix} 1 & \cdots & 1 \\ \vdots & \ddots & \vdots \\ 1 & \cdots & 1 \end{pmatrix} + \mathrm{diag}(a^{-1}, -b_1^{-1}, \ldots, -b_m^{-1}) \in \mathrm{GL}_{m+1}(\mathbb{C}).$$

Remark 3.5.7 The matrix $2\pi \mathrm{i} C$ is given by Cho and Matsumoto (1995) as the intersection matrix for the forms φ_i $(i = 0, \ldots, m)$. It satisfies a relation $\Omega(x)C - C\,{}^t\Omega(x) = 0$. The non-zero eigenspace of A_{ij} is spanned by the vector v_{ij} and the zero eigenspace of A_{ij} is given by

$\{v \in \mathbb{C}^{m+1} \mid vC\,{}^t v_{ij} = 0\}$. The form $v_{ij}\,{}^t(\varphi_0, \varphi_1, \dots, \varphi_m)$ vanishes when x_i approaches x_j. These properties yield the expression of A_{ij}; see Matsumoto (2013).

3.6 Local Solution Spaces

Lauricella (1893) gives bases of the local solution spaces of E_A, E_B and E_C; see also Appell and Kampé de Fériet (1926, §XXXIX).

Proposition 3.6.1 *Suppose that $c_i \notin \mathbb{Z}$ for $i = 1, \dots, m$ and that U is a small neighborhood of $(x_1, \dots, x_m)$ near $(0, \dots, 0)$ with $\prod x_i \neq 0$. Bases of the local solution spaces of $E_A(a, (b), (c))$ and $E_C(a, b, (c))$ in U are given by*

$$F_A\colon \quad \Big(\prod_{i\in I_r} x_i^{\lambda_i}\Big) F_A\Big(a + \sum_{i\in I_r} \lambda_i, (b) + \sum_{i\in I_r} \lambda_i e_i, (c) + 2\sum_{i\in I_r} \lambda_i e_i; x\Big), \quad r = 1, \dots, m, \tag{3.6.1}$$

$$F_C\colon \quad \Big(\prod_{i\in I_r} x_i^{\lambda_i}\Big) F_C\Big(a + \sum_{i\in I_r} \lambda_i, b + \sum_{i\in I_r} \lambda_i, (c) + 2\sum_{i\in I_r} \lambda_i e_i; x\Big), \quad r = 1, \dots, m. \tag{3.6.2}$$

Here $\lambda_i = 1 - c_i$ $(i = 1, \dots, m)$, e_i is the ith unit row vector and $I_r = \{i_1, \dots, i_r\}$ with $1 \leq i_1 < i_2 < \cdots < i_r \leq m$. A basis of the local solution spaces of $E_B((a), (b), c)$ on a small neighborhood of $x \in \mathbb{C}^m$ near $(\infty, \dots, \infty)$ is obtained from that of E_A on U by Remark 3.5.3.

The cases $r = 1$ and $r = m$ of (3.6.1) simplify to

$$x_i^{\lambda_i} F_A(a + \lambda_i, (b) + \lambda_i e_i, (c) + 2\lambda_i e_i; x),$$
$$\Big(\prod_{i=1}^m x_i^{\lambda_i}\Big) F_A\Big(a + \sum_{i=1}^m \lambda_i, (b) + \sum_{i=1}^m \lambda_i e_i, (c) + 2\sum_{i=1}^m \lambda_i e_i; x\Big),$$

and similarly for these cases of (3.6.2).

For local solution spaces of Appell's E_2 (resp. E_3) around points (x) in $\mathbb{C}^2 - S_2$ (resp. $\mathbb{C}^2 - S_3$), bases expressed by series are given by Olsson (1977). See Diekema and Koornwinder (2019) for their integral representations.

We do not have a simple basis of the local solution space of E_D in a neighborhood U $(\subset \mathbb{C}^m - S_D)$ of a point (x) near (0) as in Proposition 3.6.1, since the singular locus of E_D around (0) is complicated. See Chapter 4 for a basis in terms of A-hypergeometric series. Here we give a basis in terms of integrals following Yoshida (1997, Ch. IV, §2).

Proposition 3.6.2 *Let U be a small neighborhood of $(\dot{x}) \in \mathbb{C}^m - S_D$. The local solution space $\mathrm{Sol}_{E_D}(U)$ of E_D is spanned by the integrals*

$$f_i(x) = \int_{\mathrm{reg}(x_i, x_{i+1})} u(t, x) \frac{dt}{t-1} \quad (i = 0, 1, \dots, m),$$

where $(x_0, x_1, \dots, x_m, x_{m+1}) = (0, x_1, \dots, x_m, 1)$, $u(t, x)$ is given in (3.5.1),

$$a, b_1, \dots, b_m, c, c - a, c - \textstyle\sum b_i \notin \mathbb{Z}$$

and $\mathrm{reg}(x_i, x_{i+1})$ *is the regularized cycle of the open interval* (x_i, x_{i+1}) *defined in §3.4. By deforming regularized cycles, we can make analytic continuation of this basis along a path in* $\mathbb{C}^m - S_D$ *connecting* $(\dot{x})$ *to any point* $(x) \in \mathbb{C}^m - S_D$.

Let $(\dot{x})$ be a point in $\mathbb{R}^m$ satisfying $0 < \dot{x}_1 < \dot{x}_2 < \cdots < \dot{x}_m < 1$, $1 - \dot{x}_1 - \cdots - \dot{x}_m > 0$. A basis of the local solution space of Lauricella's system E_A around $(\dot{x})$ is given in terms of integrals of Euler type by Matsumoto and Yoshida (2014). By deforming areas of integrals, we can make analytic continuation of this basis along a path in $\mathbb{C}^m - S_A$ connecting $(\dot{x})$ to any point $(x) \in \mathbb{C}^m - S_A$. In this way, we obtain a basis of the local solution space of Lauricella's system E_A around any point (x) in $\mathbb{C}^m - S_A$. Similarly, by continuations of integrals of Euler type expressing solutions in Proposition 3.6.1, we have a basis of the local solution space of Lauricella's E_C around any point $(x) \in \mathbb{C}^m - S_C$; see Goto (2013).

3.7 Transformation Formulas

By the first integral representation of F_A and the second one of F_D in Theorem 3.4.1, we have the following formulas; see Appell and Kampé de Fériet (1926, §XXXVIII) and Lauricella (1893).

Proposition 3.7.1

1. *For each* $I_r = \{i_1, \ldots, i_m\}$ $(r = 1, \ldots, m,\ 1 \le i_1 < \cdots < i_r \le m)$,

$$F_A(a, (b), (c); (x)) = \Big(1 - \textstyle\sum_{i \in I_r} x_i\Big)^{-a} F_A\big(a, (c - b)_{[I_r]}, (c); \chi_{I_r}(x)\big),$$

where $(c - b)_{[I_r]}$ *and* $\chi_{I_r}(x) \in \mathbb{C}^m$ *are row vectors with entries respectively given by*

$$\big((c - b)_{[I_r]}\big)_i = \begin{cases} c_i - b_i & \text{if } i \in I_r, \\ b_i & \text{if } i \notin I_r, \end{cases} \quad \text{and} \quad (\chi_{I_r}(x))_i = \begin{cases} \dfrac{-x_i}{1 - \sum_{j \in I_r} x_j} & \text{if } i \in I_r, \\[2ex] \dfrac{x_i}{1 - \sum_{j \in I_r} x_j} & \text{if } i \notin I_r. \end{cases}$$

2. *For each* i $(1 \le i \le m)$,

$$\begin{aligned} F_D(a, (b), c; (x)) &= \Big(\textstyle\prod_{j=1}^m (1 - x_j)^{-\beta_j}\Big) F_D\big(c - a, (b), c; \chi_{(m+1,m+2)}(x)\big) \\ &= (1 - x_i)^{-a} F_D\big(a, (b)_{[i]}, c; \chi_{(0,i)}(x)\big) \\ &= (1 - x_i)^{c-a} \Big(\textstyle\prod_{j=1}^m (1 - x_j)^{-\beta_j}\Big) F_D\big(c - a, (b)_{[i]}, c; \chi_{(0,i)(m+1,m+2)}(x)\big), \end{aligned}$$

where

$$\begin{aligned} (b)_{[i]} &= \Big(b_1, \ldots, b_{i-1}, c - \textstyle\sum_{j=1}^m b_j, b_{i+1}, \ldots, b_m\Big), \\ \chi_{(m+1,m+2)}(x) &= \left(\frac{-x_1}{1 - x_1}, \frac{-x_2}{1 - x_2}, \ldots, \frac{-x_m}{1 - x_m}\right), \\ \chi_{(0,i)}(x) &= \left(\frac{x_1 - x_i}{1 - x_i}, \ldots, \frac{x_{i-1} - x_i}{1 - x_i}, \frac{-x_i}{1 - x_i}, \frac{x_{i+1} - x_i}{1 - x_i}, \ldots, \frac{x_m - x_i}{1 - x_i}\right), \\ \chi_{(0,i)(m+1,m+2)}(x) &= \left(\frac{x_i - x_1}{1 - x_1}, \ldots, \frac{x_i - x_{i-1}}{1 - x_{i-1}}, x_i, \frac{x_i - x_{i+1}}{1 - x_{i+1}}, \ldots, \frac{x_i - x_m}{1 - x_m}\right). \end{aligned}$$

Remark 3.7.2

1. Every transformation χ_I ($\emptyset \neq I \subset \{1,\dots,m\}$) in Proposition 3.7.1(1) is an automorphism of $\mathbb{C}^m - S_A$ of order 2, while $\chi_\emptyset$ is the identity. Furthermore, $\chi_I \circ \chi_J = \chi_{I \ominus J}$, where $I \ominus J := (I \backslash J) \cup (J \backslash I)$ is the symmetric difference. Hence the set $\chi_A := \{\chi_I \mid I \subset \{1,\dots,m\}\}$ admits the structure of an abelian group, isomorphic to $(\mathbb{Z}/(2\mathbb{Z}))^m$.
2. The transformations $\chi_{(m+1,m+2)}, \chi_{(0,i)}, \chi_{(0,i)(m+1,m+2)}$ ($1 \leq i \leq m$) in Proposition 3.7.1(2) are automorphisms of $\mathbb{C}^m - S_D$ of order 2. They satisfy

$$\chi_{(0,i)}\chi_{(m+1,m+2)} = \chi_{(m+1,m+2)}\chi_{(0,i)} = \chi_{(0,i)(m+1,m+2)} \qquad (1 \leq i \leq m),$$
$$\chi_{(0,i)}\chi_{(0,j)}\chi_{(0,i)} = \chi_{(0,j)}\chi_{(0,i)}\chi_{(0,j)} = \text{the transposition } x_i \leftrightarrow x_j \quad (1 \leq i < j \leq m).$$

Thus the group χ'_D generated by $\chi_{(0,i)}$ ($1 \leq i \leq m$) and $\chi_{(m+1,m+2)}$ is isomorphic to the direct product $S_{m+1} \times S_2$ of the symmetric groups S_{m+1} and S_2. An involution

$$\chi_{(0,m+1)} : \mathbb{C}^m - S_D \in (x) \mapsto (1-x_1, 1-x_2, \dots, 1-x_m) \in \mathbb{C}^m - S_D$$

satisfies

$$\chi_{(0,m+1)}\chi_{(m+1,m+2)}\chi_{(0,m+1)} = \chi_{(m+1,m+2)}\chi_{(0,m+1)}\chi_{(m+1,m+2)},$$
$$\chi_{(0,i)}\chi_{(0,m+1)}\chi_{(0,i)} = \chi_{(0,m+1)}\chi_{(0,i)}\chi_{(0,m+1)} \quad (1 \leq i \leq m).$$

The group χ_D generated by χ'_D and $\chi_{(0,m+1)}$ is isomorphic to the symmetric group S_{m+3}.

Corollary 3.7.3 *Let $F(x)$ be a local solution of $E_A(a, (c-b)_{[I_r]}, (c))$ around a point in $\mathbb{C}^m - S_A$ for $I_r = \{i_1, \dots, i_r\}$. Then the product of $(1 - \sum_{i \in I_r} x_i)^{-a}$ and the pull-back $F(\chi_{I_r}(x))$ under χ_{I_r} becomes a solution to $E_A(a,(b),(c))$. The same statement holds for E_D and a transformation in χ_D under coordination of parameters and the product of some powers of x_i, $1-x_i$ and $x_j - x_i$ (with $1 \leq i < j \leq m$).*

For other transformation formulas, see Carlson (1976), Exton (1976), Matsumoto and Ohara (2009), Srivastava and Karlsson (1985) and Erdélyi (1953, §5.11).

3.8 Contiguity Relations

If we add ± 1 to an entry of parameters a, b, c of Lauricella's hypergeometric series then it becomes a linear combination of the original series and its partial derivatives. For F_D, the following simple formulas, with an underlying Lie-algebraic structure of $\mathfrak{sl}_{m+3}(\mathbb{C})$, are given by Miller Jr. (1972).

Theorem 3.8.1 (Contiguity relations for F_D) *Suppose that $a, b_1, \dots, b_m, c-1, c-a, c-\sum_{j=1}^m b_j \neq 0$. We have*

$$F_D(a+1,(b),c;(x)) = a^{-1}\Big(\sum_{j=1}^m x_j \partial_j + a\Big) F_D,$$
$$F_D(a-1,(b),c;(x)) = (c-a)^{-1}\Big(\sum_{j=1}^m x_j(1-x_j)\partial_j - \sum_{j=1}^m b_j x_j + c - a\Big) F_D,$$

$$F_D(a,(b)+e_i,c;(x)) = b_i^{-1}(x_i\partial_i + b_i)F_D,$$
$$F_D(a,(b)-e_i,c;(x)) = \frac{1}{c-\sum b_j}\Big(x_i\textstyle\sum_{j=1}^m(1-x_j)\partial_j - ax_i + c - \sum_{j=1}^m b_j\Big)F_D,$$
$$F_D(a,(b),c+1;(x)) = \frac{c}{(c-a)(c-\sum b_j)}\Big(\textstyle\sum_{j=1}^m(1-x_j)\partial_j + c - a - \sum_{j=1}^m b_j\Big)F_D,$$
$$F_D(a,(b),c-1;(x)) = (c-1)^{-1}\Big(\textstyle\sum_{j=1}^m x_j\partial_j + c - 1\Big)F_D,$$

where $F_D = F_D(a,(b),c;(x))$ *and* e_i *is the ith unit row vector.*

See contiguity relations of this type for F_A, F_B, F_C in Saito (1995), and for A-hypergeometric series in Chapter 4.

By using the convention $((a),(k))$ defined in §3.3, we can combine formulas given by Exton (1976, §2.8), as follows.

Theorem 3.8.2 *Lauricella's* F_A, F_B, F_C *and* F_D *satisfy functional equations*

$$\partial^{(k)}F((a),(b),(c);(x)) = \frac{((a),(k))((b),(k))}{((c),(k))}F((a)+(k),(b)+(k),(c)+(k);(x)),$$

where $(k) = (k_1,\ldots,k_m) \in \mathbb{N}_0^m$, $\partial^{(k)} = \partial_1^{k_1}\cdots\partial_m^{k_m}$,

$$(a)+(k) = \begin{cases} a + \sum_{i=1}^m k_i & \text{if } a \in \mathbb{C}, \\ (a_1+k_1,\ldots,a_m+k_m) & \text{if } a = (a_1,\ldots,a_m) \in \mathbb{C}^m. \end{cases}$$

For contiguity relations of this type for Appell's F_1, F_2, F_3, F_4 (the specialization $m = 2$ of the above results) see Schlosser (2013, §2.1) and references given there.

Remark 3.8.3 While the above contiguity relations involve first-order derivatives, the notion of *contiguous relations* can also be found in the literature, with prototypes considered by Gauss for the Gauss hypergeometric series; see Andrews et al. (1999, §2.5). These are linear relations of hypergeometric series of the same type, possibly multiplied by an independent variable, and where in each pair of terms the corresponding parameters differ by 0 or ±1. See Buschman (1987) and Schlosser (2013, §2.1) for contiguous relations for Appell's hypergeometric series.

3.9 Monodromy Representations

Let E be one of Lauricella's systems of differential equations and let S be its singular locus. For an element $x \in \mathbb{C}^m - S$, we have the local solution space $\mathrm{Sol}_E(U)$ of the system E in a simply connected neighborhood $U(\subset \mathbb{C}^m - S)$ of x, and the fundamental group $\pi_1(\mathbb{C}^m - S)$ of the complement of the singular locus S with the initial point x. We can make the analytic continuation $\mathcal{M}_\rho(f_0)$ of an element f_0 of $\mathrm{Sol}_E(U)$ along a loop $\rho \in \pi_1(\mathbb{C}^m - S)$. Note that $\mathcal{M}_\rho(f_0) \in \mathrm{Sol}_E(U)$ and that the map $\mathcal{M}_\rho\colon \mathrm{Sol}_E(U) \to \mathrm{Sol}_E(U)$ belongs to the general linear group $\mathrm{GL}(\mathrm{Sol}_E(U))$. In this way, we have a homomorphism

$$\mathcal{M}: \pi_1(\mathbb{C}^m - S) \ni \rho \mapsto \mathcal{M}_\rho \in \mathrm{GL}(\mathrm{Sol}_E(U)),$$

which is called the *monodromy representation* of E. If we choose a basis $f = {}^t(f_1, \dots f_r)$ of $\mathrm{Sol}_E(U)$ with rank r, then the linear transformation $\mathcal{M}_\rho$ is represented by an element $M_\rho \in \mathrm{GL}_r(\mathbb{C})$, which is called the *monodromy matrix* or *circuit matrix* along ρ with respect to this basis. The image

$$\{M_\rho \in \mathrm{GL}_r(\mathbb{C}) \mid \rho \in \pi_1(\mathbb{C}^m - S)\}$$

of the map $\pi_1(\mathbb{C}^m - S) \ni \rho \mapsto M_\rho \in \mathrm{GL}_r(\mathbb{C})$ is called the *monodromy group* of E with respect to the basis f.

For the monodromy groups of Appell's system, see Kaneko (1981), Kato (1995, 1997, 2000), Yoshida (1987, §10.6) and references therein. For Lauricella's system E_A, monodromy matrices along loops generating the fundamental group of $\mathbb{C}^m - S_A$ are explicitly given by Matsumoto and Yoshida (2014).

The singular locus S_C in $\mathbb{C}^m$ consists of the m coordinate hyperplanes and a hypersurface of degree 2^{m-1}. Goto (2016) shows that the fundamental group of $\mathbb{C}^m - S_C$ is generated by $m+1$ loops, and the monodromy matrices of Lauricella's system E_C along these loops are explicitly given.

For A-hypergeometric systems admitting a fundamental system of solutions in terms of Mellin–Barnes integrals, a method to give expressions of several monodromy matrices is described by Beukers (2016). It is possible to apply this method to Lauricella's E_A, E_B and E_D.

Here we give expressions for the monodromy matrices for Lauricella's system E_D. For this purpose, we introduce the intersection form on the $\mathbb{C}(\gamma)$-vector space V spanned by regularized cycles used in Proposition 3.6.2, where $\alpha_0 = \sum b_i - c$, $\alpha_i = -b_i$ $(i = 1, \dots, m)$, $\alpha_{m+1} = c - a$ are regarded as indeterminants, and $\mathbb{C}(\gamma)$ is the field of rational functions in $\gamma_i = \exp(2\pi \mathrm{i} \alpha_i)$ $(i = 0, 1, \dots, m+1)$. Let

$$J = \sum_j n_j(\gamma) \cdot J_j, \quad J' = \sum_k n'_k(\gamma) \cdot J'_k$$

be elements in V, where branches of $u(t) = u(t, x) = t^{\alpha_0}(t-1)^{\alpha_{m+1}} \prod_{i=1}^m (t - x_i)^{\alpha_i}$ are assigned on 1-chains J_j and J'_k in the t-space of the integral, the coefficients $n_j(\gamma)$ and $n'_k(\gamma)$ belong to $\mathbb{C}(\gamma)$ and the sums are finite. The *intersection number* $J \cdot J'$ is defined by

$$J \cdot J' = \sum_{j,k} \sum_{p_{jk} \in J_j \cap J'_k} (J_j \cdot J'_k)_{p_{jk}} \cdot n_j(\gamma) \cdot n'_k(\gamma)^\vee \cdot u(p_{jk})|_{J_j} \cdot u(p_{jk})^\vee|_{J'_k},$$

where $(J_j \cdot J_k)_{p_{jk}}$ is the topological intersection number of 1-chains J_j and J'_k at their intersection point p_{jk}, $n'_k(\gamma)^\vee$ is an element of $\mathbb{C}(\gamma)$ given by the substitutions $\alpha_i \mapsto -\alpha_i$ $(i = 0, 1, \dots, m+1)$ into $n'_k(\gamma)$, $u(p_{jk})|_{J_j}$ is the value of the branch of $u(t)$ on J_j at p_{jk} and $u(p_{jk})^\vee|_{J'_k} = 1/u(p_{jk})|_{J'_k}$. The map

$$V \times V \ni (J, J') \mapsto J \cdot J' \in \mathbb{C}(\gamma)$$

can be regarded as an inner product on V, which is called the *intersection form*. By using the intersection matrix $H = (h_{ij})_{i,j=0}^m = (\mathrm{reg}(x_i, x_{i+1}) \cdot \mathrm{reg}(x_j, x_{j+1}))_{i,j=0}^m$ given in Yoshida (1997, Ch. IV, §7) as

$$h_{ij} = \begin{cases} \frac{1-\gamma_i\gamma_{i+1}}{(1-\gamma_i)(1-\gamma_{i+1})} & \text{if } j = i, \\ \frac{-1}{1-\gamma_j} & \text{if } j = i+1, \\ \frac{-\gamma_i}{1-\gamma_i} & \text{if } j = i-1, \\ 0 & \text{otherwise,} \end{cases} \tag{3.9.1}$$

we can express the intersection form as

$$(n_0(\gamma), n_1(\gamma), \ldots, n_m(\gamma))H\,{}^t(n'_0(\gamma)^\vee, n'_1(\gamma)^\vee, \ldots, n'_m(\gamma)^\vee),$$

where $J, J' \in V$ are given by linear combinations

$$\sum_{i=0}^{m} n_i(\gamma) \cdot \mathrm{reg}(x_i, x_{i+1}), \quad \sum_{i=0}^{m} n'_i(\gamma) \cdot \mathrm{reg}(x_i, x_{i+1}).$$

Since the intersection matrix H is independent of (x), any monodromy matrix M_ρ satisfies

$$M_\rho H\,{}^t M_\rho^\vee = H,$$

where $M_\rho^\vee$ denotes the matrix with $^\vee$ acting on the entries of M_ρ. When the parameters $\alpha_0, \ldots, \alpha_{m+1}$ are in $\mathbb{R}$, this identity is equivalent to $M_\rho H\,{}^t\overline{M_\rho} = H$ since $\gamma_i^\vee = \exp(-2\pi \mathrm{i}\alpha_i) = \overline{\gamma_i}$ for $\alpha_i \in \mathbb{R}$.

To obtain a monodromy matrix, we use the following lemma from Matsumoto (2013), Matsumoto and Yoshida (2014). The properties in this lemma correspond to the fact that each eigenvalue of a unitary matrix is a complex number with absolute value 1, and that its eigenspaces of different eigenvalues are orthogonal to each other.

Lemma 3.9.1 *Suppose that a basis of a local solution space* $\mathrm{Sol}_E(U)$ *is given by integrals of Euler type. Let H be the intersection matrix of regularized cycles of regions of integration.*

1. *If an α-eigenvector v of a monodromy matrix M_ρ satisfies $vHs\,{}^tv^\vee \neq 0$ then the eigenvalue α satisfies $\alpha \cdot \alpha^\vee = 1$.*
2. *If α and β are eigenvalues of a monodromy matrix M_ρ satisfying $\alpha \cdot \beta^\vee \neq 1$ then the eigenspace of eigenvalue α is orthogonal to that of β with respect to the inner product by the intersection matrix H.*

We define a loop ρ_{ij} $(1 \le i < j \le m)$ in $\mathbb{C}^m - S_D$ with initial point $\dot{x}$ $(0 < \dot{x}_1 < \cdots < \dot{x}_m < 1)$ by a path in the x_i-space starting from $\dot{x}_i$, approaching $\dot{x}_j$ via the lower half-space, turning around $\dot{x}_j$ counterclockwise and tracing back. Define loops $\rho_{0,i}$ and $\rho_{i,m+1}$ $(1 \le i \le m)$ by loops in the x_i-space regarding $\dot{x}_j$ as 0 and 1, respectively. The fundamental group $\pi_1(\mathbb{C}^m - S_D)$ is generated by ρ_{ij} $(0 \le i < j \le m+1, (i,j) \neq (0, m+1))$.

Proposition 3.9.2 (Monodromy representation of E_D) *Let $M_{\rho_{ij}}$ be the monodromy matrix along the loop ρ_{ij} with respect to the fundamental system f to $E_D(a,(b),c)$ given in Proposition 3.6.2. If $\gamma_i\gamma_j \neq 1$ then $M_{\rho_{ij}}$ is conjugate to*

$$\mathrm{diag}(1, \ldots, 1, \gamma_i\gamma_j).$$

Its eigenvector of the eigenvalue $\gamma_i\gamma_j$ *is* $\tilde{e}_{i,\dots,j-1} = \tilde{e}_i + \cdots + \tilde{e}_{j-1}$ *corresponding to* $f_i + \cdots + f_{j-1} \in \mathrm{Sol}_{E_D}(U)$*, where* $\tilde{e}_i$ *is given in Proposition 3.5.6. This monodromy matrix is expressed as*

$$M_{\rho_{ij}} = \mathrm{id}_{m+1} - (1-\gamma_i\gamma_j)H\,{}^t\tilde{e}^{\vee}_{i,\dots,j-1}(\tilde{e}_{i,\dots,j-1}H\,{}^t\tilde{e}^{\vee}_{i,\dots,j-1})^{-1}\tilde{e}_{i,\dots,j-1},$$

where id_{m+1} *is the unit matrix of size* $m+1$.

Remark 3.9.3

1. The $\gamma_i\gamma_j$-eigenvector of $M_{\rho_{ij}}$ in Proposition 3.9.2 corresponds to the regularized cycle of a path from x_i to x_j via the lower half-space, which vanishes when x_i approaches x_j. Lemma 3.9.1 implies that the 1-eigenspace is the orthogonal complement of this vanishing cycle with respect to the intersection form. Thus the monodromy matrix $M_{\rho_{ij}}$ is characterized by a complex reflection with respect to the intersection form.
2. If $\gamma_i\gamma_j = 1$ then its eigenvector $\tilde{e}_{i,\dots,j-1}$ satisfies $\tilde{e}_{i,\dots,j-1}H\,{}^t\tilde{e}^{\vee}_{i,\dots,j-1} = 0$. The factor $(1-\gamma_i\gamma_j)$ in the expression of $M_{\rho_{ij}}$ in Proposition 3.9.2 is canceled by that in $(\tilde{e}_{i,\dots,j-1}H\,{}^t\tilde{e}^{\vee}_{i,\dots,j-1})^{-1}$; this expression is valid even in the case $\gamma_i\gamma_j = 1$. In this case, the monodromy matrix $M_{\rho_{ij}}$ is not diagonalizable.

Mimachi and Sasaki (2012a) made a detailed study of the (ir)reducibility of E_D. For the monodromy group of E_D (resp. E_2 and E_4) to be finite and irreducible, see Cohen and Wolfart (1993), Mimachi and Sasaki (2012b), Sasaki (1977) (resp. Kato, 2000, 1997).

3.10 Twisted Period Relations

We can regard integral representations of hypergeometric functions as pairings of twisted cohomology and homology groups defined by the integrand. There are the intersection pairings between twisted homology groups and between twisted cohomology groups. See Aomoto and Kita (2011, Ch. 2), Cho and Matsumoto (1995) and Yoshida (1997, Ch. IV) for the twisted (co)homology groups and the intersection pairings. The compatibility among these pairings implies *twisted period relations*, which are analogies of period relations among period integrals on a compact Riemann surface. For example, defining period matrices by

$$P_+(x) := \Big(\int_{\mathrm{reg}(x_j,x_{j+1})} u\varphi_i\Big)_{0\le i,j\le m}, \quad P_-(x) := \Big(\int_{\mathrm{reg}(x_j,x_{j+1})} u^{-1}\varphi_i\Big)_{0\le i,j\le m},$$

they satisfy twisted period relations

$$P_+(x)\,{}^tH^{-1}\,{}^tP_-(x) = 2\pi\mathrm{i}C, \quad \text{i.e., } {}^tP_-(x)C^{-1}P_+(x) = 2\pi\mathrm{i}\,{}^tH,$$

where $u = u(t,x)$, φ_i, C, $\mathrm{reg}(x_i, x_{i+1})$ and H are as in Propositions 3.5.6, 3.6.2 and 3.9.2. Below we give some relations among hypergeometric series derived from twisted period relations.

Proposition 3.10.1 (Cho and Matsumoto, 1995; Goto, 2013, 2015; Goto and Matsumoto, 2015) *Suppose that no entry of* (a), (b), (c), $(c)-(a)$, $(c)-(b)$ *belongs to* $\mathbb{Z}$. *We have the following identities*:

$$(1+b_1-c_1)(1+b_2-c_2)F_2(a,(b),(c);x)F_2(1-a,-(b),2e_{12}-(c);x)$$

$$
\begin{aligned}
&\quad - b_1(1+b_2-c_2)(1-x_1)F_2(a, e_1+(b), (c); x)F_2(1-a, e_1-(b), 2e_{12}-(c); x)\\
&\quad - b_2(1+b_1-c_1)(1-x_2)F_2(a, e_2+(b), (c); x)F_2(1-a, e_2-(b), 2e_{12}-(c); x)\\
&\quad + b_1b_2(1-x_1-x_2)F_2(a, e_{12}+(b), (c); x)F_2(1-a, e_{12}-(b), 2e_{12}-(c); x)\\
&= (1-c_1)(1-c_2),
\end{aligned}
$$

$$
\begin{aligned}
&F_3((a), (b), c; x)F_3(-(a), -(b), 1-c; x) - 1\\
&= \sum_{i=1}^{2} \frac{a_ib_ix_i(x_i-1)}{c(c-1)} F_3(e_i+(a), e_i+(b), 1+c; x)F_3(e_i-(a), e_i-(b), 2-c; x)\\
&\quad - \frac{a_1a_2b_1b_2}{(c+1)c(c-1)(c-2)} x_1x_2(x_1x_2-x_1-x_2)F_3(e_{12}+(a), e_{12}+(b), 2+c; x)\\
&\quad \times F_3(e_{12}-(a), e_{12}-(b), 3-c; x),
\end{aligned}
$$

$$
\begin{aligned}
&(1-a)F_4(a, b, (c_1, c_2); x)F_4(2-a, 1-b, (2-c_1, 2-c_2); x)\\
&\quad + (c_1+c_2-1-a)F_4(a-c_1-c_2+2, b-c_1-c_2+2, (2-c_1, 2-c_2); x)\\
&\quad \times F_4(c_1+c_2-a, c_1+c_2-b-1, (c_1, c_2); x)\\
&= (c_1-a)F_4(a-c_1+1, b-c_1+1, (2-c_1, c_2); x)\\
&\quad \times F_4(1-a+c_1, c_1-b, (c_1, 2-c_2); x)\\
&\quad + (c_2-a)F_4(a-c_2+1, b-c_2+1, (c_1, 2-c_2); x)\\
&\quad \times F_4(1-a+c_2, c_2-b, (2-c_1, c_2); x),
\end{aligned}
$$

$$
\begin{aligned}
&F_D(a, (b), c; x)F_D(-a, -(b), 1-c; x) - 1\\
&= \frac{a}{c(c-1)} \sum_{i=1}^{m} b_ix_i(x_i-1)F_D(1+a, e_i+(b), 1+c; x)F_D(1-a, e_i-(b), 2-c; x),
\end{aligned}
$$

where e_i is the ith unit row vector and $e_{ij} = e_i + e_j$.

3.11 The Schwarz Map for Lauricella's F_D

Let S_D be the singular locus of Lauricella's system E_D and let $f = {}^t(f_0, f_1, \dots, f_m)$ be the basis of the local solution space of E_D given in Proposition 3.6.2. By the analytic continuation of f we have the *Schwarz map* Φ from the universal covering of $\mathbb{C}^m - S_D$ to the complex projective space $\mathbb{P}^m$. Set $\nu_i = -\alpha_i$ $(0 \le i \le m)$ and $\nu_j = 1 - \alpha_j$ $(j = m+1, m+2)$, i.e.,

$$\nu_0 = c - \sum b_i, \quad \nu_i = b_i \ (1 \le i \le m), \quad \nu_{m+1} = 1 + a - c, \quad \nu_{m+2} = 1 - a.$$

Theorem 3.11.1 (Deligne and Mostow, 1986; Terada, 1983, 1985) *If the parameters ν_i $(0 \le i \le m+2)$ satisfy*

$$0 < \nu_i < 1 \quad \textit{and} \quad |1 - \nu_i - \nu_j|^{-1} \in \{2, 3, 4, \dots, \infty\} \quad (0 \le i < j \le m+2) \tag{3.11.1}$$

then

1. *the matrix* $\mathrm{i}H^{-1}$ *is hermitian with signature* $(1, m)$*, where* H *is the intersection matrix given in* (3.9.1),
2. *the image of* Φ *is open dense in the complex ball*

$$\mathbb{B}_H := \{z = {}^t(z_0, \dots, z_m) \in \mathbb{P}^m \mid \mathrm{i}\,{}^t\overline{z} H^{-1} z > 0\},$$

3. *the inverse of the Schwarz map* Φ *induces a single-valued map from the quotient space* $\mathbb{B}_H/\Gamma$ *to* $\mathbb{C}^m - S_D$*, where* Γ *is the monodromy group of* E_D.

For a complete list of parameters ν_i $(0 \le i \le m+2)$ satisfying (3.11.1) see Couwenberg et al. (2005), Deligne and Mostow (1986) and Terada (1983, 1985).

3.12 Reduction Formulas

For special cases of the parameters, it can occur that a hypergeometric series is expressible in terms of a more simple series of that type. For example,

$$F_1(a, (b_1, b_2), b_1 + b_2; x_1, x_2) = (1 - x_2)^{-a}\, {}_2F_1\left(\begin{matrix} a, b_1 \\ b_1 + b_2 \end{matrix}; \frac{x_1 - x_2}{1 - x_2}\right),$$

$$F_2(a, (b_1, b_2), (b_1, c_2); x_1, x_2) = (1 - x_1)^{-a}\, {}_2F_1\left(\begin{matrix} a, b_2 \\ c_2 \end{matrix}; \frac{x_2}{1 - x_1}\right),$$

$$F_2(a, (b_1, b_2), (a, a); x_1, x_2) = (1 - x_1)^{-b_1}(1 - x_2)^{-b_2}\, {}_2F_1\left(\begin{matrix} b_1, b_2 \\ a \end{matrix}; \frac{x_1 x_2}{(1 - x_1)(1 - x_2)}\right),$$

$$F_3((a, c - a), (b, c - b), c; x_1, x_2) = (1 - x_2)^{a+b-c}\, {}_2F_1\left(\begin{matrix} a, b \\ c \end{matrix}; x_1 + x_2 - x_1 x_2\right),$$

$$\begin{aligned} &F_4(a, c_1 + c_2 - a - 1, (c_1, c_2); x_1(1 - x_2), x_2(1 - x_1)) \\ &\quad = {}_2F_1\left(\begin{matrix} a, c_1 + c_2 - a - 1 \\ c_1 \end{matrix}; x_1\right) {}_2F_1\left(\begin{matrix} a, c_1 + c_2 - a - 1 \\ c_2 \end{matrix}; x_2\right), \end{aligned}$$

where ${}_2F_1\left(\begin{smallmatrix} a,b \\ c \end{smallmatrix}; x\right)$ is the Gauss hypergeometric function. Some of these cases are collected in Erdélyi (1953, §5.10), Schlosser (2013, §2.5), Exton (1976), Murley and Saad (2008) and Srivastava and Karlsson (1985). See also Vidūnas (2009, 2010), who induces some kinds of reduction formulas such as

$$F_1(a, 2b, a - b, 1 + b; x, x^2) = (1 - x)^{-2a}\, {}_2F_1\left(\begin{matrix} a, \frac{1}{2} \\ 1 + b \end{matrix}; \frac{-4x}{(x - 1)^2}\right).$$

References

Andrews, G. E., Askey, R., and Roy, R. 1999. *Special Functions*. Encyclopedia of Mathematics and Its Applications, vol. 71. Cambridge University Press.

Aomoto, K., and Kita, M. 2011. *Theory of Hypergeometric Functions*. Springer. With an appendix by T. Kohno. Translated from the Japanese by K. Iohara.

Appell, P. 1880. Sur les séries hypergéométriques de deux variables, et sur des équations différentielles linéaires aux dérivées partielles. *C. R. Acad. Sci. Paris*, **90**, 296–298.

Appell, P. 1882. Sur les fonctions hypergéométriques de deux variables. *J. Math. Pures Appl.(3)*, **8**, 173–216.

Appell, P., and Kampé de Fériet, J. 1926. *Fonctions Hypergéométriques et Hypersphériques. Polynômes d'Hermite*. Gauthier-Villars.

Bailey, W. N. 1935. *Generalized Hypergeometric Series*. Cambridge University Press.

Beukers, F. 2016. Monodromy of A-hypergeometric functions. *J. Reine Angew. Math.*, **718**, 183–206.

Burchnall, J. L., and Chaundy, T. W. 1940. Expansions of Appell's double hypergeometric functions. *Quart. J. Math., Oxford Ser.*, **11**, 249–270.

Buschman, R. G. 1987. Contiguous relations for Appell functions. *Indian J. Math.*, **29**, 165–171.

Carlson, B. C. 1976. Quadratic transformations of Appell functions. *SIAM J. Math. Anal.*, **7**, 291–304.

Chaundy, T. W. 1954. An integral for Appell's hypergeometric function $F^{(4)}$. *Ganita*, **5**, 231–235.

Cho, K., and Matsumoto, K. 1995. Intersection theory for twisted cohomologies and twisted Riemann's period relations. I. *Nagoya Math. J.*, **139**, 67–86.

Cohen, P. B., and Wolfart, J. 1993. Fonctions hypergéométriques en plusieurs variables et espaces des modules de variétés abéliennes. *Ann. Sci. École Norm. Sup. Ser (4)*, **26**, 665–690.

Couwenberg, W., Heckman, G., and Looijenga, E. 2005. Geometric structures on the complement of a projective arrangement. *Publ. Math. Inst. Hautes Études Sci.*, **101**, 69–161.

Deligne, P., and Mostow, G. D. 1986. Monodromy of hypergeometric functions and non-lattice integral monodromy. *Publ. Math. Inst. Hautes Études Sci.*, **63**, 5–89.

Diekema, E., and Koornwinder, T. H. 2019. Integral representations for Horn's H_2 function and Olsson's F_P function. *Kyushu J. Math.*, **73**, 1–24. Also available as arXiv:1607.07349.

Erdélyi, A. et al. 1953. *Higher Transcendental Functions, Vol. I*. McGraw-Hill.

Exton, H. 1976. *Multiple Hypergeometric Functions and Applications*. Ellis Horwood, Chichester.

Goto, Y. 2013. Twisted cycles and twisted period relations for Lauricella's hypergeometric function F_C. *Internat. J. Math.*, **24**, 1350094, 19 pp.

Goto, Y. 2015. Twisted period relations for Lauricella's hypergeometric functions F_A. *Osaka J. Math.*, **52**, 861–877.

Goto, Y. 2016. The monodromy representation of Lauricella's hypergeometric function F_C. *Ann. Sc. Norm. Super. Pisa Cl. Sci. (5)*, **16**, 1409–1445.

Goto, Y., and Matsumoto, K. 2015. The monodromy representation and twisted period relations for Appell's hypergeometric function F_4. *Nagoya Math. J.*, **217**, 61–94.

Hattori, R., and Takayama, N. 2014. The singular locus and the holonomic rank of Lauricella's F_C. *J. Math. Soc. Japan*, **66**, 981–995.

Horn, J. 1889. Ueber die Convergenz der hypergeometrischen Reihen zweier und dreier Veränderlichen. *Math. Ann.*, **34**, 544–600.

Kaneko, J. 1981. Monodromy group of Appell's system (F_4). *Tokyo J. Math.*, **4**, 35–54.

Kato, M. 1995. Connection formulas for Appell's system F_4 and some applications. *Funkcial. Ekvac.*, **38**, 243–266.

Kato, M. 1997. Appell's F_4 with finite irreducible monodromy group. *Kyushu J. Math.*, **51**, 125–147.

Kato, M. 2000. Appell's hypergeometric systems F_2 with finite irreducible monodromy groups. *Kyushu J. Math.*, **54**, 279–305.

Kita, M. 1992. On hypergeometric functions in several variables. I. New integral representations of Euler type. *Japan. J. Math. (N.S.)*, **18**, 25–74.

Lauricella, G. 1893. Sulle funzioni ipergeometriche a più variabili. *Rend. Circ. Mat. Palermo*, **7**, 111–158.

Matsumoto, K. 2013. Monodromy and Pfaffian of Lauricella's F_D in terms of the intersection forms of twisted (co)homology groups. *Kyushu J. Math.*, **67**, 367–387.

Matsumoto, K., and Ohara, K. 2009. Some transformation formulas for Lauricella's hypergeometric functions F_D. *Funkcial. Ekvac.*, **52**, 203–212.

Matsumoto, K., and Yoshida, M. 2014. Monodromy of Lauricella's hypergeometric F_A-system. *Ann. Sc. Norm. Super. Pisa Cl. Sci. (5)*, **13**, 551–577.

Miller Jr., W. 1972. Lie theory and Lauricella functions F_D. *J. Math. Phys.*, **13**, 1393–1399.

Mimachi, K., and Sasaki, T. 2012a. Irreducibility and reducibility of Lauricella's system of differential equations E_D and the Jordan–Pochhammer differential equation E_{JP}. *Kyushu J. Math.*, **66**, 61–87.

Mimachi, K., and Sasaki, T. 2012b. Monodromy representations associated with Appell's hypergeometric function F_1 using integrals of a multivalued function. *Kyushu J. Math.*, **66**, 89–114.

Murley, J., and Saad, N. 2008. *Tables of the Appell hypergeometric functions F_2*. arXiv:0809.5203.

Ōaku, T. 1994. Computation of the characteristic variety and the singular locus of a system of differential equations with polynomial coefficients. *Japan J. Industr. Appl. Math.*, **11**, 485–497.

Olsson, P. O. M. 1977. On the integration of the differential equations of five-parametric double-hypergeometric functions of second order. *J. Math. Phys.*, **18**, 1285–1294.

Oshima, T. 2012. *Fractional Calculus of Weyl Algebra and Fuchsian Differential Equations*. MSJ Memoirs, vol. 28. Mathematical Society of Japan.

Pastro, P. I. 1989. On the integral representation of F_4 of Appell and its Lauricella generalization. *Bull. Sci. Math.*, **113**, 119–124.

Saito, M. 1995. Contiguity relations for the Lauricella functions. *Funkcial. Ekvac.*, **38**, 37–58.

Saito, M., Sturmfels, B., and Takayama, N. 2000. *Gröbner Deformations of Hypergeometric Differential Equations*. Algorithms and Computation in Math., vol. 6. Springer.

Sasaki, T. 1977. On the finiteness of the monodromy group of the system of hypergeometric differential equations (F_D). *J. Fac. Sci. Univ. Tokyo Sect. IA Math.*, **24**, 565–573.

Schlosser, M. J. 2013. Multiple hypergeometric series. Appell series and beyond. Pages 305–324 of: *Computer Algebra in Quantum Field Theory*. Springer. Also available as arXiv:1305.1966.

Srivastava, H. M., and Karlsson, P. W. 1985. *Multiple Gaussian Hypergeometric Series*. Ellis Horwood, Chichester.

Terada, T. 1983. Fonctions hypergéométriques F_1 et fonctions automorphes I. *J. Math. Soc. Japan*, **35**, 451–475.

Terada, T. 1985. Fonctions hypergéométriques F_1 et fonctions automorphes II. *J. Math. Soc. Japan*, **37**, 173–185.
Vidūnas, R. 2009. Specialization of Appell's functions to univariate hypergeometric functions. *J. Math. Anal. Appl.*, **355**, 145–163.
Vidūnas, R. 2010. On singular univariate specializations of bivariate hypergeometric functions. *J. Math. Anal. Appl.*, **365**, 135–141.
Yoshida, M. 1987. *Fuchsian Differential Equations*. Vieweg, Braunschweig.
Yoshida, M. 1997. *Hypergeometric Functions, My Love. Modular Interpretations of Configuration Spaces*. Vieweg, Braunschweig.

4

A-Hypergeometric Functions

Nobuki Takayama

4.1 Introduction

The A-hypergeometric differential equations in the present form were introduced by Gel'fand et al. (1989). Series solutions of these equations are multivariable hypergeometric series defined by a matrix A. Although there were analogous approaches before their work, they brought the new insight that affine toric ideals and their algebraic and combinatorial properties describe solution spaces of the A-hypergeometric differential equations. This also opened new research areas in commutative algebra, combinatorics, polyhedral geometry and algebraic statistics. Several textbooks describe aspects of these new research areas: see Hibi (2013), Sturmfels (1995), Ziegler (1995) and references therein. The book by Saito et al. (2000) and its references give a comprehensive presentation on A-hypergeometric equations as far as the field had developed before 2000, but substantial progress has been made since. While the present chapter aims to give directions for these new advances, it will also describe the fundamental facts. Applications of A-hypergeometric functions are getting broader. Early applications mainly concerned period maps and algebraic geometry. For these two fields, and also for the theory of hypergeometric functions, the interplay with commutative algebra and combinatorics has been a source of new ideas. There are recent new applications to multivariate analysis in statistics.

This chapter starts with systems of differential equations and examples of matrices A which define A-hypergeometric functions. We briefly describe an interplay with combinatorics, with Gröbner bases (Cox et al., 2007) and with software systems. There follows a discussion of series solutions, including some important examples. In §§4.6 and 4.7 we illustrate that contiguity relations, isomorphisms, holonomic ranks and reducibility conditions have simple descriptions in terms of algebra and combinatorics. Recent new applications to statistics will be briefly discussed in the final section.

4.2 A-Hypergeometric Equations

Let A be a $d \times n$ matrix with integer entries a_{ij}. We denote by the point a_j in $\mathbf{Z}^d$ the jth column vector of A. We suppose that the a_j generate the lattice $\mathbf{Z}^d$, i.e., $\sum_{j=1}^{n} \mathbf{Z} a_j = \mathbf{Z}^d$. Let $\beta = (\beta_1, \ldots, \beta_d)^T \in \mathbf{C}^d$ be a vector of parameters. The ring of differential operators

$$\mathbf{C}\langle x_1,\dots,x_n,\partial_1,\dots,\partial_n\rangle \quad \text{with } x_ix_j = x_jx_i,\ \partial_i\partial_j = \partial_j\partial_i,\ \partial_i x_j = x_j\partial_i + \delta_{ij}$$

is denoted by D or by D_n. We use the multi-index notation $x^p = x_1^{p_1}\cdots x_n^{p_n}$ and $\partial^q = \partial_1^{q_1}\cdots\partial_n^{q_n}$. The action of $x^p\partial^q$ to a function $f(x)$ is defined by $x^p\partial^q \bullet f(x) := x^p \frac{\partial^{|q|} f(x)}{\partial x_1^{q_1}\cdots\partial x_n^{q_n}}$.

Definition 4.2.1 (Gel'fand et al., 1989) For $i = 1,\dots,d$ and for $0 \neq u \in \mathbf{Z}^n$ such that $Au = 0$, let

$$E_i := \sum_{j=1}^n a_{ij}x_j\partial_j, \qquad \Box_u := \prod_{i=1,\dots,n;u_i>0} \partial_i^{u_i} - \prod_{j=1,\dots,n;u_j<0} \partial_j^{-u_j}.$$

An *A-hypergeometric system* or a *GKZ hypergeometric system* is a system of differential equations given by

$$\begin{aligned} (E_i-\beta_i)\bullet f = 0, &\quad i = 1,\dots,d,\\ \Box_u \bullet f = 0, &\quad 0 \neq u \in A^{-1}(0). \end{aligned}$$

We denote by I_A the ideal in $S_n = \mathbf{C}[\partial_1,\dots,\partial_n]$ generated by the $\Box_u$ ($u \in A^{-1}(0)$). This is an *affine toric ideal*; see Sturmfels (1995). The left ideal in D generated by $E_i - \beta_i$ ($i = 1,\dots,d$) and I_A is denoted by $H_A(\beta)$ and is called the *A-hypergeometric ideal*. The quotient left D-module $D/H_A(\beta)$ is denoted by $M_A(\beta)$ and called the *A-hypergeometric D-module*. When the points a_i lie on a hyperplane which does not pass through the origin, the D-module $M_A(\beta)$ is regular holonomic (Hotta, 1998). Then the matrix A is called a *configuration matrix*.

Several invariants of the D-module can be described in terms of the set $\{a_1,\dots,a_n\}$, as we will see later. In the following, A will not only denote the matrix A but also the set $\{a_1,\dots,a_n\}$. Usually the meaning of A will be clear from the context. Some further notation:

$$\mathbf{N}_0 = \mathbf{Z}_{\geq 0} := \{0,1,2,\dots\}; \text{ then } \mathbf{N}_0A := \textstyle\sum_{i=1}^n \mathbf{N}_0 a_i \text{ and } \mathbf{Z}A := \sum_{i=1}^n \mathbf{Z}a_i.$$

Although the A-hypergeometric system can be defined for any matrix A, there are nice classes of matrices A (or sets of points a_i) which lead to systems having well-known special functions as solutions. Let us introduce some of them. Take integers k and k' satisfying $1 \leq k \leq k'$. Let $e_1,\dots,e_{k+1}$ and $e'_1,\dots,e'_{k'+1}$ be the standard bases of $\mathbf{Z}^{k+1}$ and $\mathbf{Z}^{k'+1}$, respectively. Let $A(k,k')$ be a $(k+k'+1)\times(k+1)(k'+1)$ matrix of which columns consist of $p(e_i \oplus e'_j)$ where p is the projection to the first $k+k'+1$ coordinates (the projection which removes the last coordinate). In Table 4.1, $A(1,1)$, $A(1,2)$, $A(2,2)$ are given.

The columns of $A(k,k')$ generate $\mathbf{Z}^{k+k'+1}$ and they lie on the hyperplane $\sum_{j=1}^{k+1} y_j = 1$ in $\mathbf{R}^{k+k'+1} = \{(y_1,\dots,y_{k+k'+1})\}$. Since the convex hull of $e_1,\dots,e_{k+1}$ is the simplex Δ_k and that of $e'_1,\dots,e'_{k'+1}$ is the simplex $\Delta_{k'}$, we call this A-hypergeometric system a $\Delta_k \times \Delta_{k'}$-*hypergeometric system*. It is also called the hypergeometric system $E'(k+1,k+k'+2)$ because it is related to the hypergeometric system $E(k,n)$ (see §4.5). For this system, we often denote the variables x_p by $x_{ij} := x_{(i-1)(k'+1)+(j-1)+1}$ ($1 \leq i \leq k+1$, $1 \leq j \leq k'+1$). This double index notation is convenient. We also regard a vector of length $(k+1)(k'+1)$ as a matrix under this double index notation. For example, for a vector e, the condition $A(k,k')e = \beta$ means that the

Table 4.1 *The matrix A*

$$A(1,1) = \begin{array}{c} \begin{matrix} x_{11} & x_{12} & x_{21} & x_{22} \end{matrix} \\ \begin{pmatrix} 1 & 1 & 0 & 0 \\ 0 & 0 & 1 & 1 \\ 1 & 0 & 1 & 0 \end{pmatrix} \end{array}, \quad A(1,2) = \begin{array}{c} \begin{matrix} x_{11} & x_{12} & x_{13} & x_{21} & x_{22} & x_{23} \end{matrix} \\ \begin{pmatrix} 1 & 1 & 1 & 0 & 0 & 0 \\ 0 & 0 & 0 & 1 & 1 & 1 \\ 1 & 0 & 0 & 1 & 0 & 0 \\ 0 & 1 & 0 & 0 & 1 & 0 \end{pmatrix} \end{array}$$

$$A(2,2) = \begin{pmatrix} 1 & 1 & 1 & 0 & 0 & 0 & 0 & 0 & 0 \\ 0 & 0 & 0 & 1 & 1 & 1 & 0 & 0 & 0 \\ 0 & 0 & 0 & 0 & 0 & 0 & 1 & 1 & 1 \\ 1 & 0 & 0 & 1 & 0 & 0 & 1 & 0 & 0 \\ 0 & 1 & 0 & 0 & 1 & 0 & 0 & 1 & 0 \end{pmatrix}$$

$$A(F_A,2) = \begin{pmatrix} 1 & 0 & 0 & 0 & 0 & 1 & 1 \\ 0 & 1 & 0 & 0 & 0 & 1 & 0 \\ 0 & 0 & 1 & 0 & 0 & 0 & 1 \\ 0 & 0 & 0 & 1 & 0 & -1 & 0 \\ 0 & 0 & 0 & 0 & 1 & 0 & -1 \end{pmatrix}, \quad A(F_C,2) = \begin{pmatrix} 1 & 0 & 0 & -1 & 0 & 0 \\ 0 & 1 & 0 & 0 & -1 & 0 \\ 0 & 0 & 1 & 0 & 0 & -1 \\ 1 & 1 & 1 & 1 & 1 & 1 \end{pmatrix}$$

$$A(0134) = \begin{pmatrix} 1 & 1 & 1 & 1 \\ 0 & 1 & 3 & 4 \end{pmatrix}, \quad A_s = \begin{pmatrix} 1 & 1 & 1 & 1 & 1 & 1 \\ 0 & 2 & 3 & 0 & 2 & 3 \\ 0 & 0 & 0 & 1 & 1 & 1 \end{pmatrix}, \quad A(P_4) = \begin{pmatrix} 1 & 1 & 1 & 1 & 1 \\ 1 & 0 & 0 & 0 & -1 \\ 0 & 1 & 0 & 0 & -1 \\ 0 & 0 & 1 & -1 & -2 \end{pmatrix}$$

row sums and the column sums of e, expressed in terms of the $(k+1) \times (k'+1)$ matrix, are $(\beta_1, \ldots, \beta_{k+1})$ and $(\beta_{k+2}, \ldots, \beta_{k+k'+1}, \sum_{i=1}^{k+1} \beta_i - \sum_{j=k+2}^{k+k'+1} \beta_j)$, respectively.

The ideal I_A for $A = A(k,k')$ is generated by

$$\partial_{iq}\partial_{jp} - \partial_{ip}\partial_{jq}, \quad 1 \leq i < j \leq k+1, \; 1 \leq p < q \leq k'+1.$$

More precisely, it is the *reduced Gröbner basis* with respect to the *graded reverse lexicographic order* $>$, i.e., $\partial_{1,1} > \partial_{1,2} > \cdots > \partial_{1,k} > \partial_{2,1} > \cdots$; see Sturmfels (1995, Prop. 5.4). For any A, generators of I_A can be obtained by a Gröbner basis computation (Sturmfels, 1995, Alg. 4.5). In algebraic statistics the generators of I_A are called the *Markov basis*. There are theoretical and computational efforts to find explicit Markov bases. We have a database of Markov bases for several matrices A (see 4ti2, 2015 or Kahle and Rauh, 2011).

The matrix $A(1,k')$ stands for the Lauricella hypergeometric function F_D of k' variables (see Example 4.4.4 for the correspondence). In particular, when $k' = 1$, it stands for the Gauss hypergeometric function. As we will see later, the correspondence can be described in terms of series solutions or integral representations. In a more sophisticated context, a categorical correspondence is given in Zamaere et al. (2013). Here we explain an elementary correspondence in the case of the Gauss hypergeometric equation (case $A(1,1)$) as an introduction. Put $A = A(1,1)$ and suppose that $f(x) := x^v F(z)$, where $x^v := \prod_{i,j=1}^{2} x_{ij}^{v_{ij}}$ and $z := x_{11}x_{22}/(x_{12}x_{21})$, is a solution of the A-hypergeometric system $H_A(\beta)$. We denote $x_{ij}\partial_{ij}$ by θ_{ij}. Then we have

$$
\begin{aligned}
(\theta_{11}+\theta_{12}-\beta_1)\bullet x^v F(z) &= (v_{11}+v_{12}-\beta_1)x^v F(z),\\
(\theta_{21}+\theta_{22}-\beta_2)\bullet x^v F(z) &= (v_{21}+v_{22}-\beta_2)x^v F(z),\\
(\theta_{11}+\theta_{21}-\beta_3)\bullet x^v F(z) &= (v_{11}+v_{21}-\beta_3)x^v F(z),
\end{aligned}
\tag{4.2.1}
$$

by the relation $\theta_{ij}x^v = x^v(\theta_{ij}+v_{ij})$ in the ring of differential operators, and because $\theta_{ii}\bullet F(z) = zF'(z)$, $\theta_{ij}\bullet F(z) = -zF'(z)$ $(i \neq j)$. The three expressions in (4.2.1) are equal to 0 by $(E_i-\beta_i)\bullet f = 0$, which implies $Av = \beta$ where $v = (v_{11}, v_{12}, v_{21}, v_{22})^T$ and $\beta = (\beta_1, \beta_2, \beta_3)^T$. Since $\ker(A\colon \mathbf{Z}^4 \to \mathbf{Z}^3) = \mathbf{Z}(1,-1,-1,1)^T$, we can show that the toric ideal I_A is generated by $\partial_{12}\partial_{21} - \partial_{11}\partial_{22}$. Note that, in the case of $A(1,k')$ with $k' > 1$, the basis of the kernel is not sufficient for generating the toric ideal I_A. By letting the operator $x_{11}x_{22}(\partial_{11}\partial_{22} - \partial_{12}\partial_{21}) = \theta_{11}\theta_{22} - z\theta_{12}\theta_{21}$ act on the function $x^v F(z)$, we obtain

$$
x^v((\theta_z+v_{11})(\theta_z+v_{22}) - z(\theta_z - v_{12})(\theta_z - v_{21}))\bullet F(z) = 0,
$$

where $\theta_z = z\partial_z$. When $v_{11} = 0$ or $v_{22} = 0$, this is the Gauss hypergeometric equation. The other v_{ij} are determined by $Av = \beta$.

We next give matrices A for other Lauricella functions (see Chapter 3 for details of these functions). For Lauricella's F_A let $e_0, e_1, \ldots, e_{2m}$ be the standard basis of $\mathbf{Z}^{2m+1}$. Put

$$
A := \{e_0, e_1, \ldots, e_{2m}, e_0 + e_1 - e_{m+1}, e_0 + e_2 - e_{m+2}, \ldots, e_0 + e_m - e_{2m}\}.
$$

Then A is a $(2m+1)\times(3m+1)$ matrix, which stands for the Lauricella function F_A of m variables (Saito, 1995). The column vectors in A lie on the hyperplane $y_0 + y_1 + \cdots + y_{2m} = 1$ in $\mathbf{R}^{2m+1} = \{(y_0, \ldots, y_{2m})\}$. We denote the matrix by $A(F_A, m)$. The associated toric ideal I_A is generated by $\partial_0\partial_j - \partial_{m+j}\partial_{2m+j}$, $j = 1, \ldots, m$. Here we use the variables $u_0, u_1, \ldots, u_{3m}$ as independent variables instead of $x_1, \ldots, x_n$. When $m = 2$, we get the Appell function F_2; the matrix is given in Table 4.1.

For Lauricella's F_C let $e_1, \ldots, e_{m+1}, e_{m+2}$ be the standard basis of $\mathbf{Z}^{m+2}$. Put

$$
A := \{e_1 + e_{m+2}, e_2 + e_{m+2}, \ldots, e_{m+1} + e_{m+2}, -e_1 + e_{m+2}, -e_2 + e_{m+2}, \ldots, -e_{m+1} + e_{m+2}\}.
$$

Then A is an $(m+2)\times 2(m+1)$ matrix, which stands for the Lauricella function F_C of m variables (Saito, 1995). The column vectors in A lie on the hyperplane $z_{m+2} = 1$ in $\mathbf{R}^{m+2}$. We denote the matrix by $A(F_C, m)$. Note that the lattice generated by the columns of $A(F_C, m)$ is a proper sublattice of $\mathbf{Z}^{m+2}$. Then we need to regard the sublattice as $\mathbf{Z}^{m+2}$. The associated toric ideal I_A is generated by $\partial_j\partial_{-j} - \partial_{m+1}\partial_{-(m+1)}$, $j = 1, \ldots, m$. Here we use the variables $u_1, \ldots, u_{m+1}, u_{-1}, \ldots, u_{-(m+1)}$ as independent variables. When $m = 2$, we get the Appell function F_4; the matrix is given in Table 4.1.

The notion of binomial D-modules is proposed and studied by Dickenstein et al. (2010). Binomial D-modules are generalizations of A-hypergeometric equations and they fit the study of Appell–Horn equations and their generalizations to several variables by algebraic methods.

A-hypergeometric systems associated with smooth Fano polytopes or better behaved *GKZ* systems have importance in studies of period maps for $K3$ and Calabi–Yau varieties (see e.g., Batyrev, 1993; Borisov and Horja, 2013; Hosono et al., 1996, 1997; Stienstra, 1998 and references therein). For example, the matrix $A(P_4)$ (Nagano, 2012) appears in this context.

Next we discuss integral representations of solutions of A-hypergeometric equations. Suppose that we are given n_k points $a_i \in \mathbf{Z}^m$. We divide these points into k groups and construct $m \times n_i$ matrices $A_1 = (a_1, \dots, a_{n_1}), \dots, A_k = (a_{n_{k-1}+1}, \dots, a_{n_k})$. For each group, define the polynomial $f_j(x, t) = \sum_{i=n_{j-1}+1}^{n_j} x_i t^{a_i}$ where $t^{a_i} = \prod_{j=1}^m t_j^{(a_i)_j}$. Note that we use the multi-index notation for $t = (t_1, \dots, t_m)$. Take complex numbers α_j, $\gamma = (\gamma_1, \dots, \gamma_m)$. We consider the integral $\Phi(\alpha, \gamma; x) := \int_C \left(\prod_{j=1}^k f_j(x, t)^{\alpha_j} \right) t^\gamma \, dt_1 \cdots dt_m$, where C is any twisted m-cycle defined for $\left(\prod_{j=1}^k f_j(x, t)^{\alpha_j} \right) t^\gamma$. Define the $(k + m) \times n_k$ matrix

$$A(A_1, \dots, A_k) := \begin{pmatrix} 1 & \cdots & 1 & 0 & \cdots & 0 & & 0 & \cdots & 0 \\ 0 & \cdots & 0 & 1 & \cdots & 1 & & 0 & \cdots & 0 \\ 0 & \cdots & 0 & 0 & \cdots & 0 & & 0 & \cdots & 0 \\ \cdot & \cdots & \cdot & \cdot & \cdots & \cdot & & \cdot & \cdots & \cdot \\ \cdot & \cdots & \cdot & \cdot & \cdots & \cdot & \cdots & \cdot & \cdots & \cdot \\ \cdot & \cdots & \cdot & \cdot & \cdots & \cdot & & \cdot & \cdots & \cdot \\ 0 & \cdots & 0 & 0 & \cdots & 0 & & 1 & \cdots & 1 \\ a_1 & \cdots & a_{n_1} & a_{n_1+1} & \cdots & a_{n_2} & & a_{n_{k-1}+1} & \cdots & a_{n_k} \end{pmatrix},$$

which is called the *Cayley matrix*. The function $\Phi(\alpha, \gamma; x)$ satisfies the A-hypergeometric system $H_A(\beta)$ for $A = A(A_1, \dots, A_m)$ and $\beta = (\alpha_1, \dots, \alpha_k, -\gamma_1 - 1, \dots, -\gamma_m - 1)^T$. When the f_j are linear with respect to the variable t, we call the function Φ a *hypergeometric function for hyperplane arrangements*. Note that when $A_1 = \cdots = A_k = \Delta_{k'}$, i.e., when all A_i are equal to the $k' \times (k' + 1)$ matrix

$$E_{k'} \oplus 0 := \begin{pmatrix} 1 & 0 & \cdots & 0 & 0 \\ 0 & 1 & \cdots & 0 & 0 \\ \cdot & \cdots & \cdot & \cdot & \cdot \\ \cdot & \cdots & \cdot & \cdot & \cdot \\ 0 & 0 & \cdots & 1 & 0 \end{pmatrix},$$

we have $A(A_1, \dots, A_k) = A(k, k')$. For studies of these hypergeometric functions in terms of twisted cohomology groups see Adolphson and Sperber (2012), Aomoto (1977), Aomoto and Kita (2011) and Orlik and Terao (2007).

When the toric ideal I_A is not a homogeneous ideal (the case that the a_i do not lie on an affine hyperplane), the integral

$$\Phi(\gamma; x) := \int_C \exp\Big(\sum_{i=1}^n x_i t^{a_i}\Big) t^\gamma \, dt_1 \cdots dt_d$$

satisfies $H_A(\beta)$ with $\beta = (-\gamma_1 - 1, \dots, -\gamma_d - 1)^T$ for any rapid decay d-cycle under some conditions (Esterov and Takeuchi, 2015). A more constructive method to construct a cycle C was recently given by Matsubara-Heo (2019).

4.3 Combinatorics, Polytopes and Gröbner Basis

The matrix A is said to be *pointed* when $a_1, \dots, a_n$ lie in a single open half-space. For example, $A = (-1, 1)$ is not pointed and all A's in Table 4.1 are pointed. The set of points A is called *normal* if A satisfies $(\sum \mathbf{R}_{\geq 0} a_k) \cap \mathbf{Z}^n = \sum \mathbf{Z}_{\geq 0} a_k$.

For a facet σ (i.e., a face of codimension one) of the cone $\text{pos}(A) := \mathbf{R}_{\geq 0} A$, let F_σ be the linear function on $\mathbf{R}A = \mathbf{R}^d$ which is uniquely determined by the conditions

(i) $F_\sigma(\mathbf{Z}A) = \mathbf{Z}$; (ii) $F_\sigma(a_i) \geq 0$ for all $i = 1, \dots, n$; (iii) $F_\sigma(a_i) = 0$ for all $a_i \in \sigma$.

We call F_σ the *primitive integral support function* of σ.

For $\Delta_k \times \Delta_{k'}$ embedded in $\mathbf{R}^{k+1} \times \mathbf{R}^{k'+1} = \{(x_1, \dots, x_{k+1}; y_1, \dots, y_{k'+1})\}$, the support functions are x_i and y_j. When we project the points to $\mathbf{R}^{k+1} \times \mathbf{R}^{k'}$, the primitive integral support functions are x_i $(i = 1, \dots, k+1)$ and y_j $(j = 1, \dots, k')$, and $1 - \sum_{j=1}^{k'} y_j$.

The support functions for $A(F_A, m)$ are s_j, $s_j + s_{m+j}$ $(1 \leq j \leq m)$ and $s_0 + \sum_{j \in J} s_{m+j}$ $(J \subseteq [1, m])$, where $\{s_i\}$ is the dual basis of $\{e_i\}$. Those for $A(F_C, m)$ are $\frac{1}{2}(s_{m+2} + \sum_{j \in J} s_j - \sum_{j \notin J} s_j)$ $(J \subseteq [1, m+1])$; see Saito (1995).

Let $\mathbf{Z}A$ be the lattice generated by the columns of A. Set the volume of the convex hull U of the lattice base and the origin to 1. The volume of polytopes in $\mathbf{R}A$ normalized with U is called the *normalized volume*. The normalized volume of the convex hull of A and the origin is denoted by $\text{vol}(A)$. The normalized volume of $A(k, k')$ is known to be equal to $\binom{k+k'}{k}$. For given A it can be evaluated by geometry software systems like *polymake*, or by computer algebra systems which use a formula $\deg(I_A) = \text{vol}(A)$.

An interplay of theory and computation has been indispensable in the study of A-hypergeometric systems. A lot of algorithms and software systems have been developed for studying these systems and related areas. The textbook by Hibi (2013) gives an introduction, including mathematical software systems. Here we give a few examples using the computer algebra system *Macaulay2* (Macaulay 2, 2016).

Example 4.3.1 Commands to evaluate the volume (the degree) of $A(0134)$. Here o5 is I_A.

```
loadPackage "FourTiTwo"
M=matrix "1,1,1,1; 0,1,3,4"
R=QQ[a..d]
I=toricGroebner(M,R)
   o5 = ideal (b^3 - a^2*c, b*c - a*d, - a*c^2 + b^2*d, c^3 - b*d^2 )
degree(I)
   o6 = 4
```

Example 4.3.2 For a given weight vector $w \in \mathbf{R}^n$ (`Weights` in the code below) consider points $\{(a_i, w_i)\}$ in $\mathbf{R}^{d+1}$ and take their convex hull. The projection of the convex hull to the first d coordinates naturally induces a triangulation of the set of points A for a generic weight w, which is called a *regular triangulation* (Gel'fand et al., 1994; Sturmfels, 1995, Chapter 8; Hibi, 2013, §5.5.2). We compute a regular triangulation of $\Delta_1 \times \Delta_2$ for $w = (4, 2, 0, 10, 8, 6)$.

```
i1 : loadPackage "FourTiTwo"
i2 : M=matrix "1,1,1,0,0,0; 0,0,0,1,1,1; 1,0,0,1,0,0; 0,1,0,0,1,0"
i3 : R=QQ[x11,x12,x13,x21,x22,x23, MonomialOrder=>{Weights=>{4,2,0,10,8,6}}]
```

```
i4 : I=toricGroebner(M,R)
  o4 = ideal (x13*x21 - x11*x23, x12*x21 - x11*x22, x13*x22 - x12*x23)
i5 : J=leadTerm(I)
  o8 = | x13x22 x13x21 x12x21 |
i6 : associatedPrimes(ideal(J))
  o12 = {ideal (x22, x21), ideal (x13, x12), ideal (x13, x21)}
```

By taking the complements of the indices of each of the associated primes, we get a regular triangulation (11, 12, 13, 23), (11, 21, 22, 23), (11, 12, 22, 23).

4.4 A-Hypergeometric Series

We introduce A-hypergeometric series following Gel'fand et al. (1989) and Saito et al. (2000, §3.4). Let $v = (v_1, \ldots, v_n)^T$ be a vector in $\mathbf{C}^n$ and $u = (u_1, \ldots, u_n)^T$ a vector in $\mathbf{Z}^n$. In the sequel we omit the transpose sign ^{T}, for changing a row vector into a column vector, as long as no confusion arises. We decompose u into positive and negative parts, $u = u_+ - u_-$, where u_+ and u_- are non-negative vectors with disjoint support. Consider the following two scalars in $\mathbf{C}$, which can be expressed by falling factorials:

$$[v]_{u_-} = \prod_{i:\, u_i<0} \prod_{j=1}^{-u_i} (v_i - j + 1),$$

$$[u+v]_{u_+} = \prod_{i:\, u_i>0} \prod_{j=1}^{u_i} (u_i + v_i - j + 1) = \prod_{i:\, u_i>0} \prod_{j=1}^{u_i} (v_i + j).$$

For example,

$$\frac{[v]_{u_-}}{[v+u]_{u_+}} = \frac{v_1(v_1-1)v_4(v_4-1)}{2(v_2+2)(v_2+1)} \quad \text{if } v = (v_1, v_2, 0, v_4) \text{ and } u = (-2, 2, 2, -2).$$

Note that when $v \in (\mathbf{C} \setminus \mathbf{Z}_{<0})^n$ we have $[u+v]_{u_+} \neq 0$. We set $L = \ker(\mathbf{Z}^n \xrightarrow{A} \mathbf{Z}^d)$.

Theorem 4.4.1 (Saito et al., 2000, Prop. 3.4.1) *Suppose that $v \in (\mathbf{C} \setminus \mathbf{Z}_{<0})^n$ and $Av = \beta$. Then the formal series*

$$\phi_v := \sum_{u \in L} \frac{[v]_{u_-}}{[v+u]_{u_+}} x^{v+u} \tag{4.4.1}$$

is well defined and is a formal solution of $H_A(\beta)$.

We call the formal series (4.4.1) the *A-hypergeometric series in falling factorial form*. Let us introduce another expression for the series. We set $\Gamma(u+v+1) := \prod_{i=1}^n \Gamma(u_i + v_i + 1)$ and when $u_i + v_i \in \mathbf{Z}_{<0}$ for an i, we define $1/\Gamma(u+v+1) := 0$. Under this convention, we have

$$\frac{1}{\Gamma(v+u+1)} = \frac{[v]_{u_-}}{[v+u]_{u_+}} \frac{1}{\Gamma(v+1)}, \qquad u \in L,\ v \in (\mathbf{C} \setminus \mathbf{Z}_{<0})^n$$

(use $\Gamma(\alpha+m) = \Gamma(\alpha)(\alpha)_m$, $\Gamma(\alpha - m + 1) = \Gamma(\alpha+1)(-1)^m/(-\alpha)_m$). Define

$$\Phi_v := \sum_{u \in L} \frac{1}{\Gamma(u+v+1)} x^{v+u}. \tag{4.4.2}$$

Then we have $\Phi_v = \frac{1}{\Gamma(v+1)}\phi_v$ when none of v_i is negative integer. We call the formal series the *A-hypergeometric series in the gamma function form*. Note that when v_i is a negative integer, the two series are different. For example, if $v_i = -1$ and $u_i = 1$, then we have $[u_i + v_i]_{u_i} = 0$ and ϕ_v is not well defined, but $\Gamma(u_i + v_i + 1) = 1$. When $v = (1, -2, 3, 0)$ and $L = \mathbf{Z}(1, -1, -1, 1)$, ϕ_v is a non-zero polynomial, but Φ_v is identically 0.

For a given weight vector $w \in \mathbf{Z}^n$ and $\ell \in I_A$, define $\mathrm{in}_w(\ell)$ as the sum of the highest w-order terms in ℓ. The ideal in S_n generated by $\mathrm{in}_w(\ell)$ $(\ell \in I_A)$ is denoted by $\mathrm{in}_w(I_A)$ and is called the *initial ideal* of I_A (Sturmfels, 1995). Two weight vectors w and w' are equivalent with respect to the ideal I when $\mathrm{in}_w(I) = \mathrm{in}_{w'}(I)$. Fix w. The closure of the equivalence class of the weight vector w is called the *Gröbner cone* for w (Sturmfels, 1995, Chapter 1; Saito et al., 2000, §2.1; Hibi, 2013, §5.3.2). Let C be the Gröbner cone of I_A for a generic weight vector w. The initial ideal $\mathrm{in}_{w'}(I_A)$ does not change by definition when w' runs over the relative interior of C (Sturmfels, 1995; Saito et al., 2000, Ch. 2). For a series f with support on a translate of the dual cone C^*, for which we may assume $(w, C^* \setminus \{0\}) > 0$, the starting term of f is the sum of the lowest-weight terms in f with respect to w. If f is a solution of $\ell \bullet f = 0$, $\ell \in D$, then the starting term of f is a solution of $\mathrm{in}_{(-w,w)}(\ell)$, which is the sum of the highest-order terms in ℓ with respect to the weight $(-w, w)$ where $-w$ (resp. w) stands for x (resp. ∂). This observation gives us the following method (Saito et al., 2000, Ch. 2) for finding series solutions of $H_A(\beta)$:

1. determine the initial ideal $\mathrm{in}_{(-w,w)}(H_A(\beta))$;
2. solve it to determine the starting terms;
3. extend the starting terms to series solutions.

Theorem 4.4.2 *For generic β, the initial ideal* $\mathrm{in}_{(-w,w)}(H_A(\beta))$ *is generated by $E_i - \beta_i$ $(1 \le i \le d)$ and* $\mathrm{in}_w(I_A)$.

This theorem is proved in (Saito et al., 2000, Theorem 3.1.3), but the proof given there needs to be corrected to utilize the homogenized Weyl algebra.

Suppose that I_A is a homogeneous ideal and take a generic weight vector w such that $\mathrm{in}_w(I_A)$ is a monomial ideal. Let G be the reduced Gröbner basis of I_A with respect to the order $<_w$ (Sturmfels, 1995). We consider the system of differential equations

$$(E_i - \beta_i) \bullet s = 0, \quad i = 1, \ldots, d \text{ and } \ell \bullet s = 0, \ \ell \in \mathrm{in}_w(G). \tag{4.4.3}$$

The solutions v of the algebraic equations

$$Av = \beta, \quad \prod_{i=1}^{n} v_i(v_i - 1) \cdots (v_i - e_i + 1) = 0 \quad \text{for } \partial^e \in \mathrm{in}_w(G) \tag{4.4.4}$$

are called *fake exponents*. Note that the fake exponents can be expressed in terms of standard pairs of the monomial ideal $\mathrm{in}_w(I_A)$ (Saito et al., 2000, 3.2). When the β_i are generic, there are $\mathrm{vol}(A)$ linearly independent solutions of (4.4.3) of the form $s = x^v = \prod_{i=1}^n x_i^{v_i}$ such that v is a fake exponent, and they span the solution space over $\mathbf{C}$ when v runs over the fake exponents.

Theorem 4.4.3 (Gel'fand et al., 1989; Saito et al., 2000, Th. 3.4.2) *If v is a fake exponent and $v \in (\mathbf{C} \setminus \mathbf{Z}_{<0})^n$, then ϕ_v is a formal solution of $H_A(\beta)$ with support in $v + (C^* \cap L)$.*

Originally Gel'fand et al. (1989) constructed series solutions by regular triangulations of A. Our construction differs from their construction, but it is still related to it in view of Sturmfels (1995, Theorem 8.3), who states that $\sqrt{\mathrm{in}_w(I_A)}$ is the Stanley–Reisner ideal for the regular triangulation by w. The function ϕ_v converges when $(-\log|x_1|, \ldots, -\log|x_n|)$ lies in a translate of the secondary cone attached to the regular triangulation.

For a special class of A-hypergeometric functions a more explicit form of A-hypergeometric series is known, as we will describe now. For $A = A(p-1, q-1)$, the staircase Gröbner basis in Sturmfels (1995, Proposition 5.4) gives series solutions. A sequence of index pairs starting with $(1, 1)$ and ending with (p, q) is called a *stair* if any non-concluding element (i, j) in the sequence is followed by $(i + 1, j)$ or $(i, j + 1)$ (see Table 4.2).

The initial ideal of I_A for the reverse lexicographic order is generated by $\partial_{i\ell}\partial_{jk}$, $1 \le i < j \le p$, $1 \le k < \ell \le q$ (Sturmfels, 1995, Proposition 5.4). We can obtain the fake exponents from this initial ideal by solving (4.4.4). It is known that there is a one-to-one correspondence between the roots of the system of equations and the stairs. For a given stair S, the system has a unique solution such that $v_{ij} = 0$ for $(i, j) \notin S$. In other words, the support of each exponent has the form of the stair for generic β. In the sequel, we use e rather than v to denote exponents. The support of the series solution standing for the exponent e has the form

$$e + L', \quad L' = \sum_{(i,j)\in\overline{\mathrm{supp}(e)}} \mathbf{Z}_{\ge 0} b_e^{(i,j)},$$

where $b_e^{(i,j)}$ is an element of $\ker A$ such that the (i, j)th element of $b_e^{(i,j)}$ is 1 for $(i, j) \in \overline{\mathrm{supp}(e)}$ and the (i', j')th element is 0 for $(i', j') \in \overline{\mathrm{supp}(e)} \setminus \{(i, j)\}$. Here $\overline{S}$ denotes the complement of the set S.

We next give some choices of A for which series solutions can be written in terms of solutions of Lauricella systems (§3.4).

Example 4.4.4 Put $A = A(1, N - 1)$. Let a, $b_1, \ldots, b_{N-1}$, c be (generic) constants. Put $b_N := a + 1 - c$ and

$$e(k) := \begin{pmatrix} -b_1 & \cdots & -b_{k-1} & -\sum_{j=k}^N b_j + a & 0 & \cdots & 0 & 0 \\ 0 & \cdots & 0 & \sum_{j=k+1}^N b_j - a & -b_{k+1} & \cdots & -b_{N-1} & -b_N \end{pmatrix},$$

which is the fake exponent standing for the kth stair.

Put $m = (m_1, \ldots, m_{k-1}, m_{k+1}, \ldots, m_N)$, $m_k = -\sum_{j=1}^{k-1} m_j + \sum_{j=k+1}^N m_j$ and $z_j = \frac{x_{2j}x_{1N}}{x_{1j}x_{2N}}$ for $1 \le j \le N$. Note that $z_N = 1$. Define a series

$$\phi_k(e; z) := \sum_{m\in\mathbf{Z}_{\ge 0}^{N-1}} \frac{\left(\prod_{j=1}^{k-1}[e_{1j}]_{m_j}\right)\left(\prod_{j=k+1}^{N}[e_{2j}]_{m_j}\right)}{m_1! \cdots m_{k-1}!\, m_{k+1}! \cdots m_N!} c_m \left(\prod_{j=1}^{k-1}(z_j z_k^{-1})^{m_j}\right)\left(\prod_{j=k+1}^{n}(z_k z_j^{-1})^{m_j}\right), \quad (4.4.5)$$

where $e = e(k)$, $c_m = [e_{1k}]_{m_k}/[e_{2k} + m_k]_{m_k}$ when $m_k > 0$ and $c_m = [e_{2k}]_{-m_k}/[e_{1k} - m_k]_{-m_k}$ when $m_k < 0$, and $c_m = 1$ when $m_k = 0$. For $\beta = (-\sum b_i + c - 1, -a, -b_1, \ldots, -b_{N-1}, c - 1 - a)$, the function $x^{e(k)}\phi_k(e(k); z)$, $1 \le k \le N$ is a solution of $H_A(\beta)$ and $x^{e(k)-e(N)}\phi_k(e(k); z)$ is a solution of the Lauricella system $E_D(a, (b), c)$. The series $\phi_N(e(N); z)$ is Lauricella's F_D. All series ϕ_k have a common domain of convergence $|z_1| < \cdots < |z_{N-1}| < 1$.

Example 4.4.5 *The function*

$$u_0^{-a}\left(\prod_{j=1}^{m} u_j^{-b_j}\right)\left(\prod_{j=1}^{m} u_{m+j}^{c_j-1}\right) f_A\left(a, b_1, \ldots, b_m, c_1, \ldots, c_m; \frac{u_{m+1}u_{2m+j}}{u_0 u_1}, \ldots, \frac{u_{m+m}u_{2m+m}}{u_0 u_m}\right) \tag{4.4.6}$$

is a solution of $H_{A(F_A,m)}(\beta)$, $\beta^T = (-a, -b_1, \ldots, -b_m, c_1 - 1, \ldots, c_m - 1)$ *when* f_A *is a solution of Lauricella's* $E_A(a, (b), (c))$. *Any classical solution of* $H_{A(F_A,m)}(\beta)$ *can be expressed as* (4.4.6).

Example 4.4.6 The function

$$u_{m+1}^{-a} u_{-m}^{-b}\left(\prod_{j=1}^{m} u_{-j}^{c_j-1}\right) f_C\left(a, b, c_1, \ldots, c_m; \frac{u_1 u_{-1}}{u_{m+1}u_{-(m+1)}}, \ldots, \frac{u_m u_{-m}}{u_{m+1}u_{-(m+1)}}\right) \tag{4.4.7}$$

is a solution of $H_{A(F_C,m)}(\beta)$, $\beta^T = (1 - c_1, \ldots, 1 - c_m, b - a, \sum_{j=1}^{m} c_j - a - b - m)$ when f_c is a solution of Lauricella's $E_C(a, b, (c))$. Any classical solution of $H_{A(F_C,m)}(\beta)$ can be expressed as (4.4.7).

Table 4.2 *Exponents*

Stair	e: exponent
$\begin{pmatrix} * & * & * \\ 0 & 0 & * \\ 0 & 0 & * \end{pmatrix}$	$e(1) = \begin{pmatrix} \gamma_1 & \gamma_2 & \alpha_1 - \gamma_1 - \gamma_2 \\ 0 & 0 & \alpha_2 \\ 0 & 0 & \alpha_3 \end{pmatrix}$
$\begin{pmatrix} * & * & 0 \\ 0 & * & * \\ 0 & 0 & * \end{pmatrix}$	$e(2) = \begin{pmatrix} \gamma_1 & \alpha_1 - \gamma_1 & 0 \\ 0 & -\alpha_1 + \gamma_1 + \gamma_2 & \alpha_1 + \alpha_2 - \gamma_1 - \gamma_2 \\ 0 & 0 & \alpha_3 \end{pmatrix}$
$\begin{pmatrix} * & * & 0 \\ 0 & * & 0 \\ 0 & * & * \end{pmatrix}$	$e(3) = \begin{pmatrix} \gamma_1 & \alpha_1 - \gamma_1 & 0 \\ 0 & \alpha_2 & 0 \\ 0 & -\alpha_1 - \alpha_2 + \gamma_1 + \gamma_2 & \alpha_1 + \alpha_2 + \alpha_3 - \gamma_1 - \gamma_2 \end{pmatrix}$
$\begin{pmatrix} * & 0 & 0 \\ * & * & * \\ 0 & 0 & * \end{pmatrix}$	$e(4) = \begin{pmatrix} \alpha_1 & 0 & 0 \\ -\alpha_1 + \gamma_1 & \gamma_2 & \alpha_1 + \alpha_2 - \gamma_1 - \gamma_2 \\ 0 & 0 & \alpha_3 \end{pmatrix}$
$\begin{pmatrix} * & 0 & 0 \\ * & * & 0 \\ 0 & * & * \end{pmatrix}$	$e(5) = \begin{pmatrix} \alpha_1 & 0 & 0 \\ -\alpha_1 + \gamma_1 & \alpha_1 + \alpha_2 - \gamma_1 & 0 \\ 0 & -\alpha_1 - \alpha_2 + \gamma_1 + \gamma_2 & \alpha_1 + \alpha_2 + \alpha_3 - \gamma_1 - \gamma_2 \end{pmatrix}$
$\begin{pmatrix} * & 0 & 0 \\ * & 0 & 0 \\ * & * & * \end{pmatrix}$	$e(6) = \begin{pmatrix} \alpha_1 & 0 & 0 \\ \alpha_2 & 0 & 0 \\ -\alpha_1 - \alpha_2 + \gamma_1 & \gamma_2 & \alpha_1 + \alpha_2 + \alpha_3 - \gamma_1 - \gamma_2 \end{pmatrix}$

Example 4.4.7 Series solutions for $A(2, 2)$ and $\beta^T = (\alpha_1, \alpha_2, \alpha_3, \gamma_1, \gamma_2)$ $(E'(3, 6))$ have attracted special interest (Matsumoto et al., 1992; Sekiguchi and Takayama, 1997). We present a set of series solutions of this system. When we express an exponent as a 3×3 matrix under the

Table 4.3 *Bases of* $\ker A$

Stair	b_e^1	b_e^2	b_e^3	b_e^4
$\begin{pmatrix} * & * & * \\ 0 & 0 & * \\ 0 & 0 & * \end{pmatrix}$	$\begin{pmatrix} -1 & 0 & 1 \\ 1 & 0 & -1 \\ 0 & 0 & 0 \end{pmatrix}$	$\begin{pmatrix} 0 & -1 & 1 \\ 0 & 1 & -1 \\ 0 & 0 & 0 \end{pmatrix}$	$\begin{pmatrix} -1 & 0 & 1 \\ 0 & 0 & 0 \\ 1 & 0 & -1 \end{pmatrix}$	$\begin{pmatrix} 0 & -1 & 1 \\ 0 & 0 & 0 \\ 0 & 1 & -1 \end{pmatrix}$
$\begin{pmatrix} * & * & 0 \\ 0 & * & * \\ 0 & 0 & * \end{pmatrix}$	$\begin{pmatrix} 0 & -1 & 1 \\ 0 & 1 & -1 \\ 0 & 0 & 0 \end{pmatrix}$	$\begin{pmatrix} -1 & 1 & 0 \\ 1 & -1 & 0 \\ 0 & 0 & 0 \end{pmatrix}$	$\begin{pmatrix} -1 & 1 & 0 \\ 0 & -1 & 1 \\ 1 & 0 & -1 \end{pmatrix}$	$\begin{pmatrix} 0 & 0 & 0 \\ 0 & -1 & 1 \\ 0 & 1 & -1 \end{pmatrix}$
$\begin{pmatrix} * & * & 0 \\ 0 & * & 0 \\ 0 & * & * \end{pmatrix}$	$\begin{pmatrix} 0 & -1 & 1 \\ 0 & 0 & 0 \\ 0 & 1 & -1 \end{pmatrix}$	$\begin{pmatrix} -1 & 1 & 0 \\ 1 & -1 & 0 \\ 0 & 0 & 0 \end{pmatrix}$	$\begin{pmatrix} 0 & 0 & 0 \\ 0 & -1 & 1 \\ 0 & 1 & -1 \end{pmatrix}$	$\begin{pmatrix} -1 & 1 & 0 \\ 0 & 0 & 0 \\ 1 & -1 & 0 \end{pmatrix}$
$\begin{pmatrix} * & 0 & 0 \\ * & * & * \\ 0 & 0 & * \end{pmatrix}$	$\begin{pmatrix} -1 & 1 & 0 \\ 1 & -1 & 0 \\ 0 & 0 & 0 \end{pmatrix}$	$\begin{pmatrix} -1 & 0 & 1 \\ 1 & 0 & -1 \\ 0 & 0 & 0 \end{pmatrix}$	$\begin{pmatrix} 0 & 0 & 0 \\ -1 & 0 & 1 \\ 1 & 0 & -1 \end{pmatrix}$	$\begin{pmatrix} 0 & 0 & 0 \\ 0 & -1 & 1 \\ 0 & 1 & -1 \end{pmatrix}$
$\begin{pmatrix} * & 0 & 0 \\ * & * & 0 \\ 0 & * & * \end{pmatrix}$	$\begin{pmatrix} -1 & 1 & 0 \\ 1 & -1 & 0 \\ 0 & 0 & 0 \end{pmatrix}$	$\begin{pmatrix} -1 & 0 & 1 \\ 1 & -1 & 0 \\ 0 & 1 & -1 \end{pmatrix}$	$\begin{pmatrix} 0 & 0 & 0 \\ 0 & -1 & 1 \\ 0 & 1 & -1 \end{pmatrix}$	$\begin{pmatrix} 0 & 0 & 0 \\ -1 & 1 & 0 \\ 1 & -1 & 0 \end{pmatrix}$
$\begin{pmatrix} * & 0 & 0 \\ * & 0 & 0 \\ * & * & * \end{pmatrix}$	$\begin{pmatrix} -1 & 1 & 0 \\ 0 & 0 & 0 \\ 1 & -1 & 0 \end{pmatrix}$	$\begin{pmatrix} -1 & 0 & 1 \\ 0 & 0 & 0 \\ 1 & 0 & -1 \end{pmatrix}$	$\begin{pmatrix} 0 & 0 & 0 \\ -1 & 1 & 0 \\ 1 & -1 & 0 \end{pmatrix}$	$\begin{pmatrix} 0 & 0 & 0 \\ -1 & 0 & 1 \\ 1 & 0 & -1 \end{pmatrix}$

double index notation, α_i is the ith row sum and γ_j is the jth column sum. The hypergeometric series associated to the exponent $e(i)$ is written as

$$\phi_{e(i)}(x) = x^{e(i)} \sum_{m \in \mathbf{N}_0^4} \frac{[e(i)]_{u_-}}{[e(i)+u]_{u_+}} x^u, \qquad u = \sum_{j=1}^{4} b^j_{e(i)} m_j. \tag{4.4.8}$$

A choice of $e(i)$ and $b^j_{e(i)}$ is given in Tables 4.2 and 4.3. For other series solutions see Sekiguchi and Takayama (1997) and references therein. An interesting series solution of $E'(3,6)$, which is not obtained by the method in this section, is studied by Matsumoto and Terasoma (2012) in terms of arithmetic and geometric means.

In the case of non-generic parameters we have series solutions containing logarithmic functions. If I_A is homogeneous then we can construct $\mathrm{vol}(A)$ linearly independent solutions by introducing a perturbation parameter ε in the parameters and expanding the series solution in terms of ε (Saito et al., 2000, §3.5, Theorem 3.5.1). We will explain the procedure by an example.

Example 4.4.8 We consider the case of $\alpha_i = \gamma_i = \frac{1}{2}$ for $E'(3,6)$ (Tables 4.2, 4.3). The system with this parameter has special importance in algebraic geometry (Matsumoto et al., 1992; Yoshida, 1997). Let us construct a set of series solutions for this case. The exponents $e(1)$ and $e(6)$ are not degenerate and give two linearly independent solutions. The exponents

$e(i)$, $i = 2, \ldots, 5$ are degenerate: $e(2) = e(3) = e(4) = e(5) = \mathrm{diag}\,(\frac{1}{2}, \frac{1}{2}, \frac{1}{2})$. We will construct four linearly independent solutions for the degenerate exponent. We set

$$\alpha_1 = \tfrac{1}{2} + 3\varepsilon, \quad \alpha_2 = \tfrac{1}{2} + 2\varepsilon, \quad \alpha_3 = \tfrac{1}{2} + \varepsilon, \quad \gamma_1 = \tfrac{1}{2} + \varepsilon, \quad \gamma_2 = \tfrac{1}{2} + 2\varepsilon, \quad \gamma_3 = \tfrac{1}{2} + 3\varepsilon.$$

Also put $y_i = x^{b^i_{e(2)}}$. Then we have the following series containing the parameter ε:

$$\begin{aligned}
\phi_{e(2)} &= x^{e(2)} f_2(\varepsilon; y_1, y_2, y_3, y_4),\\
\phi_{e(3)} &= x^{e(2)}(1 - 2\varepsilon \log y_4 + 2\varepsilon^2(\log y_4)^2 + O(\varepsilon^3)) f_3(\varepsilon; y_1y_4, y_2, y_4, y_3/y_4),\\
\phi_{e(4)} &= x^{e(2)}(1 - 2\varepsilon \log y_2 + 2\varepsilon^2(\log y_2)^2 + O(\varepsilon^3)) f_4(\varepsilon; y_2, y_2y_2, y_3/y_2, y_4),\\
\phi_{e(5)} &= x^{e(2)}(1 - 2\varepsilon \log(y_2y_4) + 2\varepsilon^2(\log(y_2y_4))^2 + O(\varepsilon^3)) f_5(\varepsilon; y_2, y_1y_2y_4, y_4, y_3/(y_2y_4)),
\end{aligned}$$

where $f_i(\varepsilon; z_1, z_2, z_3, z_4) = \sum_{m \in \mathbf{N}_0^4} \frac{[e(i)]_{u-}}{[e(i)+u]_{u+}} z^m$ and $u = \sum_{j=1}^4 m_j b^j_{e(i)}$. We expand f_i in ε as $f_i^{(0)} + \varepsilon f_i^{(1)} + \varepsilon^2 f_i^{(2)} + O(\varepsilon^2)$. Note that all $\phi_{e(i)}$ $(i = 2, 3, 4, 5)$ give the same series when $\varepsilon = 0$, which implies that all $f_i^{(0)}$ $(i = 2, 3, 4, 5)$ are equal. Hence

$$\begin{aligned}
\phi_{e(3)} - \phi_{e(2)} &= \varepsilon(x_{11}x_{22}x_{33})^{1/2}(\underline{-2f_2^{(0)} \log y_4} + f_3^{(1)} - f_2^{(1)}) + O(\varepsilon^2),\\
\phi_{e(4)} - \phi_{e(2)} &= \varepsilon(x_{11}x_{22}x_{33})^{1/2}(\underline{-2f_2^{(0)} \log y_2} + f_4^{(1)} - f_2^{(1)}) + O(\varepsilon^2),\\
\phi_{e(5)} - \phi_{e(2)} &= \varepsilon(x_{11}x_{22}x_{33})^{1/2}(\underline{-2f_2^{(0)} \log(y_2y_4)} + f_5^{(1)} - f_2^{(1)}) + O(\varepsilon^2).
\end{aligned}$$

Here the underlined terms denote the starting monomials (Saito et al., 2000, p. 92). The coefficients of ε are solutions. Let us find the fourth solution. We have $\lim_{\varepsilon \to 0} \frac{1}{\varepsilon} f_{2345} = 0$, $f_{2345} = (\phi_{e(5)} - \phi_{e(2)}) - (\phi_{e(3)} - \phi_{e(2)}) - (\phi_{e(4)} - \phi_{e(2)})$. Therefore the series f_{2345} starts with ε^2 and the coefficient ε^2 of f_{2345} is the fourth solution. It is

$$\begin{aligned}
(x_{11}x_{22}x_{33})^{1/2}\big(&\underline{2(\log y_2)(\log y_4) f_2^{(0)}} - 2\log(y_2y_4) f_5^{(1)} + f_5^{(2)}\\
&- 2\log(y_2) f_3^{(1)} + f_3^{(2)} - 2\log(y_4) f_4^{(1)} + f_4^{(2)} + f_2^{(2)}\big).
\end{aligned}$$

Example 4.4.9 Let $\beta = (1, 2)$ and $A = A(0134)$. Set $w = (0, 1, 2, 0)$. Then the Gröbner basis of I_A with respect to this order is

$$\underline{\partial_2\partial_3} - \partial_1\partial_4, \quad \underline{\partial_1\partial_3^2} - \partial_2^2\partial_4, \quad \underline{\partial_2^3} - \partial_1^2\partial_3, \quad \underline{\partial_3^3} - \partial_2\partial_4^2$$

where the underlined terms are the leading monomials. Therefore, the fake exponents are

$$v^{(1)} = (\tfrac{1}{2}, 0, 0, \tfrac{1}{2}), \quad v^{(2)} = (\tfrac{1}{4}, 1, 0, \tfrac{1}{4}), \quad v^{(3)} = (\tfrac{1}{4}, 0, 1, -\tfrac{1}{4}), \quad v^{(4)} = (-1, 2, 0, 0),$$

$\phi^{(1)}$, $\phi^{(2)}$ and $\phi^{(3)}$ are convergent series solutions, but $\phi^{(4)} \equiv 0$. By examining $\mathrm{in}_{(-w,w)}(I_A)$ we can find two more solutions: x_2^2/x_1 and x_3^2/x_4; see Cattani et al. (1999), Sturmfels and Takayama (1998).

Series solutions with logarithms have been constructed for a class of non-generic β in view of applications to mirror symmetry (Hosono et al., 1996, 1997; Stienstra, 1998). For non-homogeneous I_A, series solutions are divergent in most cases, but there is a class of series solutions which are convergent. These were studied by Ohara and Takayama (2009) and Dickenstein et al. (2012). The Gevrey order of divergent series solutions was studied by Fernández-Fernández (2010). The notion of fully supported series solutions was introduced by Matusevich (2009) and Schulze and Walther (2008). See Cattani et al. (2001) for rational solutions of $H_A(\beta)$ and Beukers (2010) for algebraic solutions.

4.5 Hypergeometric Function of Type $E(k,n)$

We fix two numbers k and n satisfying $n \geq 2k \geq 4$. Let α_j be generic parameters satisfying $\sum_{j=1}^{n} \alpha_j = n - k$. The *hypergeometric function of type* $E(k,n)$ or the *Aomoto–Gel'fand hypergeometric function* is defined by the integral

$$\Psi(\alpha;u) := \int_C \prod_{j=1}^{n} \Big(\sum_{i=1}^{k} u_{ij}s_i \Big)^{\alpha_j} ds_2 \cdots ds_k,$$

where we put $s_1 = 1$ and u is a $k \times n$ matrix and C is a bounded $(k-1)$-cell in the hyperplane arrangement defined by $\prod_{j=1}^{n} \sum_{i=1}^{k} u_{ij}s_i = 0$ in the $(s_2, \ldots, s_k)$-space (Gel'fand, 1986).

The hypergeometric function of type $E(k,n)$ is quasi-invariant under the action of complex torus $(\mathbf{C}^*)^n$ and the general linear group $\mathrm{GL}(k) = \mathrm{GL}(k,\mathbf{C})$. In fact, for $h = \mathrm{diag}(h_1, \ldots, h_n) \in (\mathbf{C}^*)^n$ and $g \in \mathrm{GL}(k)$, we have

$$\Psi(\alpha;uh) = \Big(\prod_j h_j^{\alpha_j}\Big)\Psi(\alpha;u), \qquad \Psi(\alpha;gu) = |g|^{-1}\Psi(\alpha;u).$$

It follows from the quasi-invariance property and the integral representation that the function $\Psi(\alpha;u)$ satisfies both a system of first-order equations and a system of second-order equations:

Theorem 4.5.1 (Gel'fand, 1986) *The function* $f = \Psi(\alpha;u)$ *satisfies*

$$\Big(\sum_{i=1}^{k} u_{ip}\frac{\partial}{\partial u_{ip}} - \alpha_p\Big)f = 0, \quad p = 1, \ldots, n, \qquad \Big(\sum_{p=1}^{n} u_{ip}\frac{\partial}{\partial u_{jp}} + \delta_{ij}\Big)f = 0, \quad i,j = 1, \ldots, k,$$

$$\Big(\frac{\partial^2}{\partial u_{ip}\partial u_{jq}} - \frac{\partial^2}{\partial u_{iq}\partial u_{jp}}\Big)f = 0, \quad i,j = 1, \ldots, k, \ p,q = 1, \ldots, n.$$

We call this system of equations $E(k,n)$.

When we restrict the hypergeometric system $E(k,n)$ to $u_{ij} = \delta_{ij}$ for $1 \leq i \leq k$, $1 \leq j \leq k$, we obtain the A-hypergeometric system associated to $A(k-1, n-k-1)$ and

$$\beta = (-\alpha_1 - 1, \ldots, -\alpha_k - 1, \alpha_{k+1} - 1, \ldots, \alpha_{n-1} - 1).$$

We denote it by $E'(k,n)$. Here $u_{i,j+k}$ stands for the variable x_{ij} in §4.2.

Let $\mathfrak{S}_n$ be the permutation group of n letters. If $\Psi(\alpha;u)$ is a solution of $E(k,n)$, then $\Psi(\alpha^s;u^s)$, $s \in \mathfrak{S}_n$ is also a solution. This $\mathfrak{S}_n$ symmetry leads to Kummer-type relations (Takayama, 2003). The confluent $E(k,n)$ has been geometrically studied. Furthermore, a general framework to derive Kummer-type relations has been given (see Kimura and Takano, 2006 and references therein).

4.6 Contiguity Relations

Note in the Weyl algebra D the relation $(\sum_{j=1}^n a_{ij}\theta_j - \beta_i)\partial_k = \partial_k(\sum_{j=1}^n a_{ij}\theta_j - \beta_i - a_{ik})$. Since ∂_k commutes with $\square_u$, we can see that if f is a solution of $H_A(\beta + a_k)$, then $\partial_k \bullet f$ is a solution of $H_A(\beta)$.

Let B_k be the ideal which is the intersection of $\mathbf{C}[s_1, \ldots, s_d]$ and the left ideal generated by ∂_k and $H_A(s)$ in $D[s_1, \ldots, s_d]$. If A is normal and I_A is homogeneous then B_k can be expressed in terms of primitive support functions.

Theorem 4.6.1 (Saito, 1992) *The ideal B_k is the principal ideal generated by*

$$\prod_{\sigma \in S} \prod_{i=0}^{F_\sigma(a_k)-1} (F_\sigma(s) - i),$$

where S is the set of facets σ of the convex hull of A for which $F_\sigma(a_k) > 0$.

It follows from Theorem 4.6.1 that, if $\beta \notin V(B_k)$, then there exists an operator $Q_k \in D$ such that $Q_k\partial_k = 1 \bmod H_A(\beta)$. The operators ∂_k and Q_k give contiguity relations for A-hypergeometric series.

The symmetry algebra introduced by Saito (2001) gives contiguity relations of A-hypergeometric systems in a general framework. The ideal B_k is a special case of the b-ideal introduced in Saito (2001).

4.6.1 Contiguity Relations for $E'(k, n)$

We will give a contiguity relation for $E'(k, n)$ following Sasaki (1991). Instead of x_{ij} we will use the variable u_{ij} as in §4.5. Put

$$X_{pa} = -u_{ap} - \sum_{q=k+1}^{n} u_{aq} \sum_{i=1}^{k} u_{ip}\partial_{iq}. \tag{4.6.1}$$

Let $\varphi(\alpha; u)$ be a solution of the system $E'(k, n)$ with the set of parameters α and with 1_a being the ath unit vector in $\mathbf{Z}^n$.

Theorem 4.6.2 (Sasaki, 1991) *We have*

$$\partial_{ap}\varphi(\alpha; u) = \varphi(\alpha + 1_a - 1_p; u), \quad X_{pa}\varphi(\alpha; u) = \varphi(\alpha - 1_a + 1_p; u), \quad X_{pa}\partial_{ap} - (\alpha_p - 1)\alpha_a \in H_A(\beta).$$

The introduction of extra variables in hypergeometric series of several variables in order to study contiguity relations was done in the pioneering work of Kalnins et al. (1980). Contiguity relations for the Lauricella functions F_A, F_B and F_C were derived in Saito (1995) by following this idea and by utilizing the b-ideal B_k for them. See also §3.5 for some explicit contiguity relations of Lauricella functions.

4.6.2 Isomorphism Between A-Hypergeometric D-Modules $M_A(\beta)$

Recall the definition of the left D-module $M_A(\beta)$ (after Definition 4.2.1). Above we gave contiguity operators ∂_k and Q_k. If they exist, there is an isomorphism $\partial_k\colon M_A(\beta-a_k) \to M_A(\beta)$.

The question of whether $M_A(\beta)$ and $M_A(\beta')$ are isomorphic as left D-modules is quite fundamental. It was studied by Saito et al. (2000, §§4.4, 4.5). A final answer was given by Saito (2001). Let τ be a face of $\mathrm{pos}(A)$. Define

$$E_\tau(\beta) := \{\lambda \in \mathbf{C}(A \cap \tau)/\mathbf{Z}(A \cap \tau) \mid \beta - \lambda \in \mathbf{N}_0 A + \mathbf{Z}(A \cap \tau)\}. \tag{4.6.2}$$

Theorem 4.6.3 (Saito, 2001; Saito and Traves, 2001, Theorem 3.4.4) *The left D-modules $M_A(\beta)$ and $M_A(\beta')$ are isomorphic if and only if $E_\tau(\beta) = E_\tau(\beta')$ for all faces τ of* $\mathrm{pos}(A)$.

The condition in the above theorem can be rewritten as a condition on the primitive integral supporting function if A is normal.

Theorem 4.6.4 (Saito, 2001, Theorem 5.2) *Assume A is normal and I_A is homogeneous. The left D-module $M_A(\beta)$ is isomorphic to $M_A(\beta')$ if and only if $\beta - \beta' \in \mathbf{Z}A$ and*

$$\{\text{facets } \sigma \mid F_\sigma(\beta) \in \mathbf{N}_0\} = \{\text{facets } \sigma \mid F_\sigma(\beta') \in \mathbf{N}_0\}. \tag{4.6.3}$$

4.7 Properties of A-Hypergeometric Equations

4.7.1 Rank Formula and the Euler–Koszul Complex

The *holonomic rank* $H_A(\beta)$ is the dimension of $R/(RH_A(\beta))$ considered as the vector space over the field of rational functions $\mathbf{C}(x_1, \dots, x_n)$. Here R is the ring of differential operators with rational function coefficients. For generic β the rank of $H_A(\beta)$ is equal to the normalized volume of A. Furthermore, rank $H_A(\beta) \geq \mathrm{vol}(A)$ (see Adolphson, 1994; Gel'fand et al., 1989; Saito et al., 2000). More precise discussion requires the Euler–Koszul complex (Berkesch, 2011; Matusevich et al., 2005).

We assume in this subsection that A is pointed. For $\partial^v \in S_n := \mathbf{C}[\partial_1, \dots, \partial_n]$, we define the *A-multidegree* $\deg(\partial^v) := -Av \in \mathbf{Z}^d$. Its ith component is denoted by $\deg_i(\partial^v)$. This definition can be naturally extended to the Weyl algebra D by putting $\deg(x^u\partial^v) := Au - Av$. Set $E_i := \sum_{j=1}^n a_{ij}\theta_j$. Then $\deg(E_i) = \mathbf{0}$. The following identity is fundamental:

$$\partial^v E_i = E_i\,\partial^v - \deg_i(\partial^v)\,\partial^v = (E_i - \deg_i(\partial^v))\,\partial^v.$$

Let S_A be the ring $\mathbf{C}[\partial_1, \dots, \partial_n]/I_A$, which is isomorphic to $\mathbf{C}[t^{a_1}, \dots, t^{a_n}] = \mathbf{C}[\mathbf{N}_0 A]$, and denote $D_n \otimes_{S_n} S_A \simeq D_n/(D_n I_A)$ by D_A. We consider the complex

$$\mathcal{K}_\bullet : 0 \xleftarrow{d_0} D_A^{\binom{n}{0}} \xleftarrow{d_1} D_A^{\binom{n}{1}} \xleftarrow{d_2} \cdots \xleftarrow{d_{n-1}} D_A^{\binom{n}{n-1}} \xleftarrow{d_n} D_A^{\binom{n}{n}} \longleftarrow 0.$$

For A-homogeneous $a \otimes b \in D_A$ we define

$$(E_i - \beta_i) \circ (a \otimes b) := (E_i - \beta_i - \deg_i(a \otimes b))(a \otimes b).$$

Denote the basis of $D_A^{\binom{d}{k}}$ by $e_{i_1,\dots,i_k}$, $1 \le i_1 < \cdots < i_k \le d$. The *boundary map* d_k is defined by

$$D_A^{\binom{d}{k}} \ni (a \otimes b)e_{i_1,\dots,i_k} \mapsto \sum_{i_j \in \{i_1,\dots,i_k\}} (E_{i_j} - \beta_{i_j}) \circ (a \otimes b)(-1)^{j-1} e_{\{i_1,\dots,i_k\}\setminus\{i_j\}} \in D_A^{\binom{d}{k-1}}. \qquad (4.7.1)$$

The complex is called the *Euler–Koszul complex* over D_A. It is well defined because we have

$$(E_i - \beta_i) \circ (a \otimes \Box_u) = (a\Box_u(E_i - \beta_i)) \otimes 1 = (a(E_i - \beta_i - \deg_i(\partial^{u_+}))\Box_u) \otimes 1 \equiv 0.$$

The *homology group* $\mathcal{H}_i(E - \beta; S_A) := H_i(\ker d_i / \operatorname{im} d_{i-1})$ of the Euler–Koszul complex has a natural A-grading by the A-multidegree. The 0th homology group is just $M_A(\beta)$. This leads us to a more functorial object for the study of A-hypergeometric systems, namely the *Euler–Koszul homology* for toric modules (Matusevich et al., 2005). For this purpose fix $E - \beta$ and replace S_A by (A-)toric modules. We will explain such modules by an example. Let A be $A(0134)$ and let $\tilde{A}$ be its saturation. Note that $n = 4$ and that the multigrading is determined by A. We may suppose that $\tilde{A} = \left(\begin{smallmatrix} 1 & 1 & 1 & 1 & 1 \\ 0 & 1 & 3 & 4 & 2 \end{smallmatrix}\right)$ and $S_{\tilde{A}} = D_5/I_{\tilde{A}}$. Then we have a short exact sequence

$$0 \longrightarrow D_4 \otimes_{S_4} S_A \longrightarrow D_4 \otimes_{S_4} S_{\tilde{A}} \longrightarrow D_4 \otimes_{S_4} S_{\tilde{A}}/S_A \longrightarrow 0.$$

All modules are A-graded and toric. The module $C := D_4 \otimes S_{\tilde{A}}/S_A$ has its support only at degree $(1, 2)$. We have

$$\mathcal{H}_0(E - \beta; D_4 \otimes S_{\tilde{A}}) \simeq D_5/x_5 D_5 \otimes_{D_5} M_{\tilde{A}}(\beta) \simeq M_{\tilde{A}}(\beta)$$

and

$$H_0(E - \beta; C) = \begin{cases} 0, & \beta \neq (1, 2), \\ D_4 \otimes [\partial_5], & \beta = (1, 2). \end{cases}$$

Theorem 4.7.1 (Matusevich et al., 2005) *Put* $\mathbf{m} := \langle \partial_1, \dots, \partial_n \rangle$, *a maximal ideal in* $S_n := \mathbf{C}[\partial_1, \dots, \partial_n]$.

1. *If k equals the smallest homological degree i for which $-\beta$ is a quasi-degree of $H^i_{\mathbf{m}}(S_A)$, then the Euler–Koszul homology $\mathcal{H}_{d-k}(E - \beta; S_A)$ is of non-zero rank and $\mathcal{H}_i = 0$ for $i > d - k$. Here γ is called the* quasi-degree *when γ is contained in the Zariski closure of the non-zero degrees of the homology group.*
2. *$H^i_{\mathbf{m}}(S_A) = 0$ holds for $0 \le i < d$ if and only if S_A is Cohen–Macaulay.*
3. *The rank of $H_A(\beta)$ is equal to the normalized volume of A if and only if β is not a quasi-degree of $H^i_{\mathbf{m}}(S_A)$.*

Put $\varepsilon_A := \sum a_i$. The part of the local cohomology group which has degree $-\alpha + \varepsilon_A$ is $H^{n-i}_{\mathbf{m}}(S_A)_{-\alpha+\varepsilon_A} = \operatorname{Hom}_{\mathbf{C}}(\operatorname{Ext}^i_{S_n}(S_A, S_n)_\alpha, \mathbf{C})$.

Example 4.7.2 Consider the case $A = A(0134)$, $\varepsilon_A = (4, 8)^T$. Construct an A-graded resolution of R/I_A by Schreyer's method (see, e.g., Oaku and Takayama, 2001). Then we have $\operatorname{Ext}^4 = 0$ and $\operatorname{Ext}^3 = \mathbf{C}$ at degree $(5, 10)$, which implies that $H^{4-3}_{\mathbf{m}} \neq 0$ at degree $-(1, 2)$. In fact, the rank of the system is 5 when $\beta = (1, 2)$ and it is 4 when $\beta \neq (1, 2)$.

4.7.2 Characteristic Variety and Principal A-Determinant

Let I be a left ideal in D. The initial ideal $\mathrm{in}_{(0,1)}(I)$ is the ideal in $\mathbf{C}[x_1,\ldots,x_n,\xi_1,\ldots,\xi_n]$ generated by the principal symbols of I. The ideal is called the *characteristic ideal* of I, and the zero set of the ideal in $\mathbf{C}^{2n}$ is called the *characteristic variety* of D/I and is denoted by $\mathrm{Ch}(D/I)$. The projection of $\mathrm{Ch}(D/I)\setminus V(\xi_1,\ldots,\xi_n)$ to $\mathbf{C}^n=\{x\}$ is called the *singular locus* of D/I and is denoted by $\mathrm{Sing}(D/I)$ (see, e.g., Saito et al., 2000, p. 36).

Theorem 4.7.3 (Gel'fand et al., 1989) *1. If $H_1(\mathrm{gr}_{(0,1)}\mathcal{K}_\bullet)=0$, then the characteristic ideal of $H_A(\beta)$ is generated by $Ax\xi$ and by $I'_A := I_A|_{\partial\to\xi}$. Here $Ax\xi$ is the ideal generated by $\sum_{j=1}^n a_{ij}x_j\xi_j$ $(i=1,\ldots,d)$.*
2. If I_A is Cohen–Macaulay then the first homology above vanishes.

Characteristic varieties and micro-characteristic varieties of $M_A(\beta)$ were combinatorially studied by Gel'fand et al. (1989) and by Schulze and Walther (2008).

Let E_A be the *principal A-determinant* (Gel'fand et al., 1994). The projection of $V(\langle Ax\xi, I'_A\rangle)\setminus V(\xi_1,\ldots,\xi_n)$ to $\mathbf{C}^n$ is expressed as $V(E_A)$.

Theorem 4.7.4 (Gel'fand et al., 1994, p. 300) *The principal A-determinant for $A(k,k')$ $(k\le k')$ is the product of the determinants of all $p\times p$ minors $(1\le p\le k)$ of the matrix (x_{ij}).*

Example 4.7.5 For $A=A(1,k'-1)$ we have $E_A=\left(\prod_{i=1}^2\prod_{j=1}^{k'}x_{ij}\right)\prod_{1\le j<j'\le k'}\begin{vmatrix}x_{1j} & x_{1j'}\\ x_{2j} & x_{2j'}\end{vmatrix}$. The variety $V(E_A)$ is the singular locus of $H_A(\beta)$.

4.7.3 Reducibility and Monodromy Groups

Consider the set $\mathrm{Res}(A):=\cup_\tau(\mathbf{Z}A+\tau)$ with union taken over all linear subspaces τ of $\mathbf{C}^d$ that form a boundary component of $\mathrm{pos}(A)$. The elements of $\mathrm{Res}(A)$ are called *resonant parameters*.

Let R be the ring of differential operators with rational function coefficients. Consider the left R-module $\mathbf{C}(x_1,\ldots,x_n)\otimes_{D_n}M_A(\beta)=R/(RH_A(\beta))$. If this module has a non-zero proper R-submodule, it is called *reducible*.

Theorem 4.7.6 (Beukers, 2011) *Suppose that I_A is homogeneous and A is not a pyramid. Then $\mathbf{C}(x_1,\ldots,x_n)\otimes M_A(\beta)$ is reducible if and only if $\beta\notin$* Res(A).

An analogue of this theorem holds without the homogeneity condition; see Schulze and Walther (2012). The irreducible quotients of $M_A(\beta)$ as D-modules were combinatorially discussed by Saito (2011).

The reducibility of a Lauricella system can be described by the reducibility of the corresponding A-hypergeometric system. See Hattori and Takayama (2014, §6) for the case F_C. Zamaere et al. (2013) described a systematic approach to study reducibilities for Appell–Horn or Mellin-type systems.

Connection formulas are studied for $A(1,n)$ by restrictions (Saito and Takayama, 1994). The global monodromy groups are calculated for some interesting cases of A. See Matsumoto et al. (1992, 1993), Yoshida (1997) for the case of $A(2,2)$. See Nagano (2012) for some of 3-dimensional Fano polytopes related to families of $K3$ surfaces. See §3.7 for monodromy groups for Lauricella functions. Monodromies at infinity are also discussed; see Ando et al. (2015) and its references. Recently, a general method to compute a subgroup of monodromy groups was proposed by Beukers (2016).

4.8 *A*-Hypergeometric Polynomials and Statistics

Assume that A is a configuration matrix. The *A-hypergeometric polynomial* for the configuration A and the parameter vector $\beta \in \mathbf{N}_0^d$ (see Saito et al., 1999) is defined by

$$Z(\beta; p) := \sum_{Au=\beta, u\in \mathbf{N}_0^n} \frac{p^u}{u!}, \tag{4.8.1}$$

where $p^u := \prod_{i=1}^n p_i^{u_i}$ and $u! := \prod_{i=1}^n u_i!$. It is a solution of the A-hypergeometric system $H_A(\beta)$. Set $p_i := \exp \xi_i$ and let $\exp \xi$ denote the vector $(\exp \xi_1, \ldots, \exp \xi_n)$. We fix $\beta \neq 0$ such that $\beta \in \mathbf{N}_0 A := \sum_{i=1}^n \mathbf{N}_0 a_i$. A random variable $U \in \mathbf{N}_0^n$ of the (A,β) *hypergeometric distribution with parameter* $p \in \mathbf{R}_{>0}^n$ (or $\xi \in \mathbf{R}^n$) is defined by the following probability that U takes the value u:

$$P(U = u \mid Au = \beta) := \frac{p(\xi)^u}{u!\, Z(\beta; p(\xi))} = \frac{\exp(u\cdot\xi)}{u!\, Z(\beta; p(\xi))}, \quad u\cdot\xi := \sum_{i=1}^n u_i \xi_i. \tag{4.8.2}$$

It is the conditional distribution of u given by $\beta = Au$ under the *Poisson distribution*

$$P(U = u) := \frac{p^u}{u!} \exp(-\mathbf{1}\cdot p), \quad \mathbf{1} := (1, \ldots, 1). \tag{4.8.3}$$

The polynomial Z is the *normalizing constant* or *partition function* of the (A,β) hypergeometric distribution. In statistics (Hibi, 2013, 4.1) the $(A(k,k'),\beta)$ hypergeometric distribution has been called the *generalized hypergeometric distribution for* $(k+1)\times(k'+1)$ *contingency tables with marginal sum* β.

Let $\overline{A}$ be an $n \times (n-d)$ matrix with integer entries satisfying the conditions that $A\overline{A} = 0$ and that the rank of $\overline{A}$ as a $\mathbf{Q}$-matrix is $n-d$. Denote by $\overline{a}_i$ the ith column vector of $\overline{A}$. An asymptotic study of the probability distribution (4.8.2) yields the following theorem, which gives the approximate behaviour of the A-hypergeometric polynomial with scaled parameter vector $\kappa\beta$ ($\kappa \to +\infty$).

Theorem 4.8.1 (Takayama et al., 2018) *Fix* $p \in \mathbf{R}_{>0}^n$ *and* $\beta \in \mathbf{N}_0 A \cap \mathrm{int}(\mathbf{R}_{\geq 0} A)$. *There exists a unique* $m \in \mathbf{R}_{>0}^n$ *such that* $Am = \beta$, $m^{\overline{a}_i} = p^{\overline{a}_i}$. *Then, with* $M := \mathrm{diag}(m)$,

$$Z(\kappa\beta; p) \sim \frac{\left(\prod p_i^{m_i}\right)^\kappa}{\Gamma(\kappa m + 1)} \frac{(2\pi\kappa)^{n-d}}{\det\left(\overline{A} M^{-1} \overline{A}^T\right)^{1/2}} \quad \text{as } \kappa \to +\infty.$$

Conversely, applications of A-hypergeometric equations to statistics are given by Hibi et al. (2017) and Takayama et al. (2018).

Acknowledgements The author is grateful to Uli Walther for providing Example 4.7.2. He also thanks Frits Beukers, Alicia Dickenstein, Yoshiaki Goto, Keiji Matsumoto, Mutsumi Saito and Uli Walther for several comments.

References

4ti2. 2015. Available at `www.4ti2.de`. A software package for algebraic, geometric and combinatorial problems on linear spaces (version 1.6.7).

Adolphson, A. 1994. Hypergeometric functions and rings generated by monomials. *Duke Math. J.*, **73**, 269–290.

Adolphson, A., and Sperber, S. 2012. A-hypergeometric systems that come from geometry. *Proc. Amer. Math. Soc.*, **140**, 2033–2042.

Ando, K., Esterov, A., and Takeuchi, K. 2015. Monodromies at infinity of confluent A-hypergeometric functions. *Adv. Math.*, **272**, 1–19.

Aomoto, K. 1977. On the structure of integrals of power products of linear functions. *Sci. Papers College Gen. Ed. Univ. Tokyo*, **27**, 49–61.

Aomoto, K., and Kita, M. 2011. *Theory of Hypergeometric Functions*. Springer. With an appendix by T. Kohno. Translated from the Japanese by K. Iohara.

Batyrev, V. 1993. Variations of the mixed Hodge structure of affine hypersurfaces in algebraic tori. *Duke Math. J.*, **69**, 349–409.

Berkesch, C. 2011. The rank of a hypergeometric system. *Compos. Math.*, **147**, 284–318.

Beukers, F. 2010. Algebraic A-hypergeometric functions. *Invent. Math.*, **180**, 589–610.

Beukers, F. 2011. Irreducibility of A-hypergeometric systems. *Indag. Math. (N.S.)*, **221**, 30–39.

Beukers, F. 2016. Monodromy of A-hypergeometric functions. *J. Reine Angew. Math.*, **718**, 183–206.

Borisov, L. A., and Horja, R. P. 2013. On the better behaved version of the GKZ hypergeometric system. *Math. Ann.*, **357**, 585–603.

Cattani, E., D'Andrea, C., and Dickenstein, A. 1999. The $\mathcal{A}$-hypergeometric system associated with a monomial curve. *Duke Math. J.*, **99**, 179–207.

Cattani, E., Dickenstein, A., and Sturmfels, B. 2001. Rational hypergeometric functions. *Compos. Math.*, **128**, 217–239.

Cox, D., Little, J., and O'Shea, D. 2007. *Ideals, Varieties and Algorithms*. Third edn. Springer.

Dickenstein, A., Matusevich, L., and Miller, E. 2010. Binomial D-modules. *Duke Math. J.*, **151**, 385–429.

Dickenstein, A., Martinez, F., and Matusevich, L. 2012. Nilsson solutions for irregular A-hypergeometric systems. *Rev. Mat. Iberoam.*, **28**, 723–758.

Esterov, A., and Takeuchi, K. 2015. Confluent A-hypergeometric functions and rapid decay homology cycles. *Amer. J. Math.*, **137**, 365–409.

Fernández-Fernández, M. 2010. Irregular hypergeometric $\mathcal{D}$-modules. *Adv. Math.*, **224**, 1735–1764.

Gel'fand, I. M. 1986. General theory of hypergeometric functions. *Soviet Math. Dokl.*, **33**, 573–577.

Gel'fand, I. M., Zelevinsky, A. V., and Kapranov, M. M. 1989. Hypergeometric functions and toral manifolds. *Funct. Anal. Appl.*, **23**, 94–106. Correction in *Funct. Anal. Appl.* **27** (1993), p. 295.

Gel'fand, I. M., Kapranov, M. M., and Zelevinsky, A. V. 1994. *Discriminants, Resultants and Multidimensional Determinants*. Birkhäuser.

Hattori, R., and Takayama, N. 2014. The singular locus and the holonomic rank of Lauricella's F_C. *J. Math. Soc. Japan*, **66**, 981–995.

Hibi, T. (ed). 2013. *Gröbner Bases: Statistics and Software Systems*. Springer.

Hibi, T., Nishiyama, K., and Takayama, N. 2017. Pfaffian systems of A-hypergeometric equations I: bases of twisted cohomology groups. *Adv. Math.*, **306**, 303–327.

Hosono, S., Lian, B. H., and Yau, S.-T. 1996. GKZ-generalized hypergeometric systems in mirror symmetry of Calabi–Yau hypersurfaces. *Comm. Math. Phys.*, **182**, 535–577.

Hosono, S., Lian, B. H., and Yau, S.-T. 1997. Maximal degeneracy points of GKZ systems. *J. Amer. Math. Soc.*, **10**, 427–443.

Hotta, R. 1998. *Equivariant D-modules*. arXiv:math/9805021.

Kahle, T., and Rauh, J. 2011. *The Markov bases database*. Available at `http://markov-bases.de`.

Kalnins, E. G., Manocha, H. L., and Miller, Jr., W. 1980. The Lie theory of two-variable hypergeometric functions. *Stud. Appl. Math.*, **62**, 143–173.

Kimura, H., and Takano, K. 2006. On confluences of general hypergeometric systems. *Tohoku Math. J. (2)*, **58**, 1–31.

Macaulay 2. 2016. Available at `www.math.uiuc.edu/Macaulay2`. A computer algebra system for algebraic geometry (version 1.9.2).

Matsubara-Heo, S. J. 2019. *Euler Laplace integral representations of GKZ hypergeometric functions*. arXiv:1904.00565.

Matsumoto, K., and Terasoma, T. 2012. Thomae type formula for $K3$ surfaces given by double covers of the projective plane branching along six lines. *J. Reine Angew. Math.*, **669**, 121–149.

Matsumoto, K., Sasaki, T., and Yoshida, M. 1992. The monodromy of the period map of a 4-parameter family of $K3$ surfaces and the hypergeometric function of type $(3, 6)$. *Internat. J. Math.*, **3**, 1–164.

Matsumoto, K., Sasaki, T., Takayama, N., and Yoshida, M. 1993. Monodromy of the hypergeometric differential equation of type (3,6), I. *Duke Math. J.*, **71**, 403–426.

Matusevich, L. 2009. Weyl closure of hypergeometric systems. *Collect. Math.*, **60**, 147–158.

Matusevich, L., Miller, E., and Walther, U. 2005. Homological methods for hypergeometric families. *J. Amer. Math. Soc.*, **18**, 919–941.

Nagano, A. 2012. Period differential equations for families of $K3$ surfaces derived from some 3 dimensional reflexive polytopes. *Kyushu J. Math.*, **66**, 193–244.

Oaku, T., and Takayama, N. 2001. Algorithms for D-modules — restriction, tensor product, localization, and local cohomology groups. *J. Pure Appl. Alg.*, **156**, 267–308.

Ohara, K., and Takayama, N. 2009. Holonomic rank of $\mathcal{A}$-hypergeometric differential-difference equations. *J. Pure Appl. Algebra*, **213**, 1536–1544.

Orlik, P., and Terao, H. 2007. *Arrangements and Hypergeometric Integrals*. Second edn. MSJ Memoirs, vol. 9. Mathematical Society of Japan.

Saito, M. 1992. Parameter shift in normal generalized hypergeometric systems. *Tohoku Math. J. (2)*, **44**, 523–534.

Saito, M. 1995. Contiguity relations for the Lauricella functions. *Funkcial. Ekvac.*, **38**, 37–58.

Saito, M. 2001. Isomorphism classes of A-hypergeometric systems. *Compos. Math.*, **128**, 323–338.

Saito, M. 2011. Irreducible quotients of A-hypergeometric systems. *Compos. Math.*, **147**, 613–632.

Saito, M., and Takayama, N. 1994. $\mathcal{A}$-hypergeometric systems and connection formula of the $\Delta_1 \times \Delta_{n-1}$-hypergeometric function. *Internat. J. Math.*, **5**, 537–560.
Saito, M., and Traves, W. 2001. Differential algebras on semigroup algebras. Pages 207–226 of: *Symbolic Computation: Solving Equations in Algebra, Geometry, and Engineering*. Contemp. Math., vol. 286. Amer. Math. Soc.
Saito, M., Sturmfels, B., and Takayama, N. 1999. Hypergeometric polynomials and integer programming. *Compos. Math.*, **115**, 185–204.
Saito, M., Sturmfels, B., and Takayama, N. 2000. *Gröbner Deformations of Hypergeometric Differential Equations*. Algorithms and Computation in Math., vol. 6. Springer.
Sasaki, T. 1991. Contiguity relations of Aomoto-Gel'fand hypergeometric functions and applications to Appell's system F_3 and Goursat's system ${}_3F_2$. *SIAM J. Math. Anal.*, **22**, 821–846.
Schulze, M., and Walther, U. 2008. Irregularity of hypergeometric systems via slopes along coordinate subspaces. *Duke Math. J.*, **142**, 465–509.
Schulze, M., and Walther, U. 2012. Resonance equals reducibility for A-hypergeometric systems. *Algebra Number Theory*, **6**, 527–537.
Sekiguchi, J., and Takayama, N. 1997. Compactifications of the configuration space of six points of the projective plane and fundamental solutions of the hypergeometric system of type (3,6). *Tohoku Math. J. (2)*, **49**, 379–413.
Stienstra, J. 1998. Resonant hypergeometric systems and mirror symmetry. Pages 412–452 of: *Integrable Systems and Algebraic Geometry*. World Scientific.
Sturmfels, B. 1995. *Gröbner Bases and Convex Polytopes*. University Lecture Notes, vol. 8. Amer. Math. Soc.
Sturmfels, B., and Takayama, N. 1998. Gröbner bases and hypergeometric functions. Pages 246–258 of: *Gröbner Bases and Applications*. London Math. Soc. Lecture Note Ser., vol. 251. Cambridge University Press.
Takayama, N. 2003. Generating Kummer type formulas for hypergeometric functions. Pages 131–145 of: *Algebra, Geometry, and Software Systems*. Springer.
Takayama, N., Kuriki, S., and Takemura, A. 2018. A-Hypergeometric distributions and Newton polytopes. *Adv. Appl. Math.*, **99**, 109–133.
Yoshida, M. 1997. *Hypergeometric Functions, My Love. Modular Interpretations of Configuration Spaces*. Vieweg, Braunschweig.
Zamaere, C. B., Matusevich, L. F., and Walther, U. 2013. *Torus equivariant D-modules and hypergeometric systems*. arXiv:1308.5901.
Ziegler, G. 1995. *Lectures on Polytopes*. Graduate Texts in Math., vol. 152. Springer.

5

Hypergeometric and Basic Hypergeometric Series and Integrals Associated with Root Systems

Michael J. Schlosser

5.1 Introduction

Hypergeometric series associated with root systems first appeared implicitly in the work of Ališauskas et al. (1972) and Chacón et al. (1972) in the context of the representation theory of unitary groups, more precisely, as the multiplicity-free Wigner and Racah coefficients ($3j$- and $6j$-symbols) of the group $\mathrm{SU}(n+1)$. A few years later, Holman et al. (1976) investigated these coefficients more explicitly as generalized hypergeometric series and obtained a first summation theorem for them. The series in question have explicit summands and contain the Weyl denominator of the root system A_n, and can thus be considered as hypergeometric series associated with this root system. (These series are not to be confused with the hypergeometric functions associated with root systems considered in Chapter 8, which generalize the spherical functions on noncompact Riemannian symmetric spaces.) Subsequently, A_n hypergeometric series were shown to satisfy various extensions of well-known identities for classical hypergeometric series (Gustafson, 1987a,b; Holman, 1980; Milne, 1980). For example, the A_n extension (Holman, 1980) of the terminating balanced Pfaff–Saalschütz ${}_3F_2$ summation is

$$\sum_{k_1,\dots,k_n=0}^{N_1,\dots,N_n} \left(\prod_{1\le i<j\le n} \frac{x_i+k_i-x_j-x_k}{x_i-x_j} \prod_{i,j=1}^{n} \frac{(-N_j+x_i-x_j)_{k_i}}{(1+x_i-x_j)_{k_i}} \right.$$
$$\left. \times \prod_{i=1}^{n} \frac{(a+x_i)_{k_i}(b+x_i)_{k_i}}{(c+x_i)_{k_i}(a+b-c+1-|N|+x_i)_{k_i}} \right) = \frac{(c-a)_{|N|}(c-b)_{|N|}}{\prod_{i=1}^{n}(c+x_i)_{N_i}(c-a-b+|N|-N_i-x_i)_{N_i}}, \tag{5.1.1}$$

where, throughout this chapter, $|N| := N_1 + \cdots + N_n$ and $(a)_k$ is the usual notation (5.2.1) for the shifted factorial.

An extensive study of the *basic* or q-analogue of A_n series was initiated in a series of papers by Milne (1985a, 1985b, 1985c). The following application of the "fundamental theorem of A_n series" (Milne, 1985a, Theorem 1.49),

$$\sum_{\substack{k_1,\dots,k_n\ge 0\\ |k|=N}} \left(\prod_{1\le i<j\le n} \frac{x_i q^{k_i} - x_j q^{k_j}}{x_i-x_j} \prod_{i,j=1}^{n} \frac{(a_j x_i/x_j;q)_{k_i}}{(q x_i/x_j;q)_{k_i}} \right) = \frac{(a_1\cdots a_n;q)_N}{(q;q)_N}, \tag{5.1.2}$$

where we are using the usual notation (5.2.2) for the q-shifted factorial, demonstrates a phenomenon which is typical for the A_n theory: In (5.1.2), let $n \mapsto n+1$ and replace k_{n+1} by $N-(k_1+\cdots+k_n)$. Then, after further replacing the variables a_i by c_i $(i=1,\ldots,n)$, a_{n+1} by $q^{-N}b/a$ and x_{n+1} by q^{-N}/a, the following terminating A_n ${}_6\phi_5$ summation is obtained:

$$\sum_{\substack{k_1,\ldots,k_n\geq 0\\ |k|\leq N}} \left(\frac{(q^{-N};q)_{|k|}}{(aq/b;q)_{|k|}} \left(\frac{aq^{1+N}}{bc_1\cdots c_n}\right)^{|k|} \prod_{1\leq i<j\leq n} \frac{x_iq^{k_i}-x_jq^{k_j}}{x_i-x_j} \prod_{i=1}^{n} \frac{1-ax_iq^{k_i+|k|}}{1-ax_i} \prod_{i,j=1}^{n} \frac{(c_jx_i/x_j;q)_{k_i}}{(qx_i/x_j;q)_{k_i}} \right.$$
$$\left. \times \prod_{i=1}^{n} \frac{(ax_i;q)_{|k|}(bx_i;q)_{k_i}}{(ax_iq/c_i;q)_{|k|}(ax_iq^{1+N};q)_{k_i}} \right) = \frac{(aq/bc_1\cdots c_n;q)_N}{(aq/b;q)_N} \prod_{i=1}^{n} \frac{(ax_iq;q)_N}{(ax_iq/c_i;q)_N}. \tag{5.1.3}$$

By application of the one-variable q-binomial theorem, it follows that another consequence of (5.1.2) is the following A_n extension of the nonterminating q-binomial theorem:

$$\sum_{k_1,\ldots,k_n\geq 0} \left(\prod_{1\leq i<j\leq n} \frac{x_iq^{k_i}-x_jq^{k_j}}{x_i-x_j} \prod_{i,j=1}^{n} \frac{(a_jx_i/x_j;q)_{k_i}}{(qx_i/x_j;q)_{k_i}} \cdot z^{|k|} \right) = \frac{(a_1\cdots a_nz;q)_\infty}{(z;q)_\infty}, \tag{5.1.4}$$

valid for $|q|<1$ and $|z|<1$.

While A_n (basic) hypergeometric series have also been referred to as SU(n) or U(n) series, the terminology *(basic) hypergeometric series associated to the root system* A_n, or simply A_n *(basic) hypergeometric series*, is preferred by most authors nowadays.

A further important development was the introduction by Gustafson (1989, 1990b) of very-well-poised series for other root systems. He also introduced related multivariate integrals associated with root systems (Denis and Gustafson, 1992; Gustafson, 1990a, 1992, 1994a,b). In this setting the multiple series or integrals are classified according to the type of specific factors (such as a Weyl denominator) appearing in the summand or integrand.

Most of the known results for multivariate (basic) hypergeometric series and integrals associated with root systems indeed concern classical root systems, while only sporadically summations or transformations for series or integrals associated with exceptional root systems have been obtained (van de Bult, 2011; Gustafson, 1990a; Ito, 2002, 2003; Ito and Tsubouchi, 2010).

The root system classification appears to be very useful and one would hope that the various relations satisfied by the series or integrals can be interpreted in terms of root systems or even Lie theory. Although the type of series in question first arose (in the limit $q\to 1$) in the representation theory of compact Lie groups, many questions remain open about this connection. While a (quantum) group interpretation for the A_n-type series has been given by Rosengren (2011) – even for the elliptic extension of the series, surveyed in Chapter 6 – no analogous interpretations for the other root systems have yet been revealed.

In many instances various types (still referring to the root system classification) of series/integrals can be combined with each other after, which one obtains series/integrals of some "mixed type" for which the correct classification is not really clear. The conclusion is that the root system classification of the series/integrals considered here is only rough and not always precise.

In terms of the rough classification by Bhatnagar (1999, §2), Bhatnagar and Schlosser (1998, §1), and Milne (2001, §5), a multivariate series $\sum_{k_1,\dots,k_{n+1}} S_{k_1,\dots,k_{n+1}}$ is considered to be an A_n *hypergeometric series* if the summand $S_{k_1,\dots,k_{n+1}}$ contains the factor

$$\prod_{1\le i<j\le n+1} (x_i + k_i - x_j - k_j). \tag{5.1.5a}$$

It is considered to be an A_n *basic hypergeometric series* if the summand contains the factor

$$\prod_{1\le i<j\le n+1} (x_i q^{k_i} - x_j q^{k_j}). \tag{5.1.5b}$$

If we take the sum over $k_1 + \cdots + k_{n+1} = N$, we may replace k_{n+1} by $N - |k|$ (where, as before, $|k| = k_1 + \cdots + k_n$). Also substitute $x_{n+1} = -a - N$ in (5.1.5a) and $x_{n+1} = q^{-N}/a$ in (5.1.5b). Then the two products can be respectively written as

$$\prod_{1\le i<j\le n} (x_i + k_i - x_j - k_j) \prod_{i=1}^{n} (a + x_i + k_i + |k|) \tag{5.1.6a}$$

and

$$\left(-aq^{|k|}\right)^{-n} \prod_{1\le i<j\le n} (x_i q^{k_i} - x_j q^{k_j}) \prod_{i=1}^{n} (1 - a x_i q^{k_i+|k|}). \tag{5.1.6b}$$

Likewise, a C_n *hypergeometric series* contains the factor

$$\prod_{1\le i<j\le n} (x_i + k_i - x_j - k_j) \prod_{1\le i\le j\le n} (x_i + k_i + x_j + k_j) \tag{5.1.7a}$$

and a C_n *basic hypergeometric series* the factor

$$\prod_{1\le i<j\le n} (x_i q^{k_i} - x_j q^{k_j}) \prod_{1\le i\le j\le n} (1 - x_i x_j q^{k_i+k_j}). \tag{5.1.7b}$$

We omit giving similar definitions for other root systems. The above factors may be associated with the *Weyl denominators* $\prod_{\alpha>0}(1 - e^{-\alpha})$ with the product taken over the positive roots in the root system. (Weyl denominators similarly appear in §8.4.2.)

A very similar classification applies to hypergeometric *integrals* associated with root systems, by considering certain factors of the integrand. For specific examples, see §5.3.

A special feature of the theory of hypergeometric or basic hypergeometric functions associated with root systems is that often there exist several different identities for one and the same root system that extend a particular one-variable identity. See the various A_n ${}_3\phi_2$ and A_n ${}_2\phi_1$ summations given by Milne (1997). At this point we would also like to mention that various special A_n hypergeometric series possess rich structures of symmetry; these were made explicit by Kajihara (2014).

This chapter deals to a large extent with the multivariate *basic* hypergeometric theory. The reason is that most results for ordinary hypergeometric series have basic hypergeometric analogues,[1] so it does not make sense to treat the hypergeometric case separately.

[1] Identities for basic hypergeometric series reduce to those for ordinary hypergeometric series by taking the limit $q \to 1$ in an appropriate manner (Gasper and Rahman, 2004, Section 1.2); see for instance the derivation of (5.2.31) from (5.2.30).

A good amount of the theory of basic hypergeometric series associated with root systems has recently been generalized to the *elliptic* level. For an introduction to elliptic hypergeometric functions associated with root systems, see Chapter 6. In the present chapter we emphasize some general facts about series associated with root systems which are important for understanding the nature of the series, but otherwise, to avoid overlap, mainly focus on parts of the theory which do *not* directly extend to the elliptic level. This in particular concerns identities obtained as confluent limits of more general identities and identities for nonterminating and/or multilateral series.

The following sections are devoted to multivariate identities, ranging from very simple identities to more complicated ones. In particular, in §§5.2 and 5.3 various summations, transformations, and integral evaluations are reviewed. Then §5.4 surveys the theory of basic hypergeometric series with Macdonald polynomial argument, and the chapter concludes with §5.5 which contains a brief discussion on applications of basic hypergeometric series associated with root systems.

Notation. In this chapter, we follow the convention that in lines with slashed fractions, multiplication has priority over division (i.e., $c/ab = c/(ab)$ etc.). This convention is also silently used in the book by Gasper and Rahman (2004), which we often refer to. We also follow the convention that a product symbol $\prod$ with its multiplication indices governs the subsequent expressions until the next product symbol $\prod$ or until a multiplication sign given by $\times$ or a dot $(\cdot)$. Finally,

$$\sum_{k_1,\dots,k_n=0}^{N_1,\dots,N_n} := \sum_{k_1=0}^{N_1} \cdots \sum_{k_n=0}^{N_n}.$$

5.2 Some Identities for (Basic) Hypergeometric Series Associated with Root Systems

A large number of identities for hypergeometric and basic hypergeometric series associated with root systems have appeared in the literature. Due to space limitations, we only provide a small representative selection of identities. Nevertheless, they are meant to give a flavor of the expressions which typically occur in the multivariate theory. For more details the reader is pointed to specific literature.

We use the following notation for *shifted* and *q-shifted factorials* (which are also referred to as *Pochhammer* and *q-Pochhammer symbols*, respectively):

$$(a)_k := \begin{cases} 1, & k = 0, \\ a(a+1)\cdots(a+k-1), & k = 1, 2, \dots, \\ ((a-1)(a-2)\cdots(a+k))^{-1}, & k = -1, -2, \dots, \end{cases} \tag{5.2.1}$$

$$(a;q)_k := \begin{cases} 1, & k=0, \\ (1-a)(1-aq)\cdots(1-aq^{k-1}), & k=1,2,\ldots, \\ ((1-aq^{-1})(1-aq^{-2})\cdots(1-aq^{k}))^{-1}, & k=-1,-2,\ldots, \end{cases} \tag{5.2.2a}$$

$$(a;q)_\infty := \prod_{i\geq 0}(1-aq^i). \tag{5.2.2b}$$

When dealing with products of shifted and q-shifted factorials, we frequently use the short-hand notation

$$(a_1,\ldots,a_n)_j := (a_1)_j\cdots(a_n)_j \quad \text{and} \quad (a_1,\ldots,a_n;q)_k := (a_1;q)_k\cdots(a_n;q)_k,$$

where j is an integer, and k is an integer or ∞.

This chapter reviews results for multivariate extensions associated with root systems of the following univariate series, whose definitions we give for reasons of self-containedness.

Hypergeometric ${}_rF_s$ series and *bilateral hypergeometric ${}_rH_s$ series* are defined as

$${}_rF_s\left(\begin{matrix}a_1,\ldots,a_r\\ b_1,\ldots,b_s\end{matrix};z\right) := \sum_{k=0}^{\infty}\frac{(a_1,\ldots,a_r)_k}{(1,b_1,\ldots,b_s)_k}z^k, \tag{5.2.3a}$$

$${}_rH_s\left(\begin{matrix}a_1,\ldots,a_r\\ b_1,\ldots,b_s\end{matrix};z\right) := \sum_{k=-\infty}^{\infty}\frac{(a_1,\ldots,a_r)_k}{(b_1,\ldots,b_s)_k}z^k. \tag{5.2.3b}$$

Similarly, *basic hypergeometric ${}_r\phi_s$ series* and *bilateral basic hypergeometric ${}_r\psi_s$ series* are defined as

$${}_r\phi_s\left(\begin{matrix}a_1,\ldots,a_r\\ b_1,\ldots,b_s\end{matrix};q,z\right) := \sum_{k=0}^{\infty}\frac{(a_1,\ldots,a_r;q)_k}{(q,b_1,\ldots,b_s;q)_k}\left((-1)^k q^{\binom{k}{2}}\right)^{1+s-r}z^k, \tag{5.2.4a}$$

$${}_r\psi_s\left(\begin{matrix}a_1,\ldots,a_r\\ b_1,\ldots,b_s\end{matrix};q,z\right) := \sum_{k=-\infty}^{\infty}\frac{(a_1,\ldots,a_r;q)_k}{(b_1,\ldots,b_s;q)_k}\left((-1)^k q^{\binom{k}{2}}\right)^{s-r}z^k. \tag{5.2.4b}$$

See the books by Slater (1966) and by Gasper and Rahman (2004) for the conditions of these series to terminate, to converge, to be balanced, or to be (very) well-poised, and for various identities satisfied by these series. Instead of "basic hypergeometric" one may also write "q-hypergeometric".

5.2.1 Some Useful Elementary Facts

We start with a few elementary ingredients which are useful for manipulating basic hypergeometric series associated with root systems.

(i) A fundamental ingredient (for inductive proofs and functional equations, etc.) is the following partial fraction decomposition (Milne, 1988b, Section 7):

$$\prod_{i=1}^{n}\frac{1-tx_iy_i}{1-tx_i} = y_1y_2\cdots y_n + \sum_{k=1}^{n}\frac{\prod_{i=1}^{n}(1-y_ix_i/x_k)}{(1-tx_k)\prod_{i\in\{1,\ldots,n\}\setminus\{k\}}(1-x_i/x_k)}. \tag{5.2.5}$$

In particular, this identity can be used to prove the fundamental theorem of A_n series in (5.1.2).

A slightly more general partial fraction decomposition was derived by Schlosser (2003, Lemma 3.2). The identity (5.2.5) can be obtained as a limiting case of an elliptic partial fraction decomposition of type A; cf. Whittaker and Watson (1927, Example 3, p. 451). A related partial fraction decomposition of type D was established by Gustafson (1987a, Lemma 4.14).

(ii) For the simplification of products the following identity is useful (Milne, 1997, Lemma 4.3):

$$\prod_{1\le i<j\le n} \frac{x_i q^{k_i} - x_j q^{k_j}}{x_i q^{m_i} - x_j q^{m_j}} \prod_{i,j=1}^{n} \frac{(q^{m_i-k_j} x_i/x_j;q)_{k_i-m_i}}{(q^{1+m_i-m_j} x_i/x_j;q)_{k_i-m_i}} = (-1)^{|k|-|m|} q^{-\binom{|k|-|m|+1}{2}}. \tag{5.2.6}$$

(iii) When one wants to reverse the order of the summations

$$\sum_{k_1,\dots,k_n=0}^{N_1,\dots,N_n} S_{k_1,\dots,k_n} = \sum_{k_1,\dots,k_n=0}^{N_1,\dots,N_n} S_{N_1-k_1,\dots,N_n-k_n}$$

it is convenient to use the fact that the following variant of an "A_n q-binomial coefficient"

$$\prod_{i,j=1}^{n} \frac{(qx_i/x_j;q)_{N_i}}{(qx_i/x_j;q)_{k_i}(q^{1+k_i-k_j} x_i/x_j;q)_{N_i-k_i}} \tag{5.2.7}$$

(the usual q-binomial coefficient is given in (5.2.9)) remains unchanged after performing the simultaneous substitutions $k_i \mapsto N_i - k_i$ and $x_i \mapsto q^{-N_i}/x_i$ for $i = 1,\dots,n$; see Schlosser (1999, Remark B.3).

5.2.2 *Some Terminating* A_n *q-Binomial Theorems*

The *terminating q-binomial theorem* can be written in the form (cf. Gasper and Rahman, 2004, Ex. 1.2(vi))

$$(z;q)_N = \sum_{k=0}^{N} \begin{bmatrix} N \\ k \end{bmatrix}_q (-1)^k q^{\binom{k}{2}} z^k, \tag{5.2.8}$$

where

$$\begin{bmatrix} N \\ k \end{bmatrix}_q := \frac{(q;q)_N}{(q;q)_k (q;q)_{N-k}} \tag{5.2.9}$$

is the *q-binomial coefficient*. (Here N denotes a nonnegative integer.) In basic hypergeometric notation this identity corresponds to a terminating ${}_1\phi_0$ summation. It can be immediately obtained from the *nonterminating q-binomial theorem* (or ${}_1\phi_0$ summation; cf. Gasper and Rahman, 2004, Equation (II.3)),

$${}_1\phi_0\left(\begin{matrix} a \\ - \end{matrix}; q, z\right) = \frac{(az;q)_\infty}{(z;q)_\infty}, \qquad |q|, |z| < 1. \tag{5.2.10}$$

To obtain (5.2.8) from (5.2.10), replace a and z by q^{-n} and zq^n, respectively.

We have the following three multisum identities (for equivalent forms, where the Vandermonde determinant of type A (5.1.5b) explicitly appears in the summands, see Milne, 1997, Theorems 5.44, 5.46, and 5.48), each involving the A_n q-binomial coefficient in (5.2.7):

$$(z;q)_{|N|} = \sum_{k_1,\dots,k_n=0}^{N_1,\dots,N_n} \left(\prod_{i,j=1}^{n} \frac{(qx_i/x_j;q)_{N_i}}{(qx_i/x_j;q)_{k_i}(q^{1+k_i-k_j}x_i/x_j;q)_{N_i-k_i}} \cdot (-1)^{|k|} q^{\binom{|k|}{2}} z^{|k|} \right), \tag{5.2.11a}$$

$$\prod_{i=1}^{n}(zx_i;q)_{N_i} = \sum_{k_1,\dots,k_n=0}^{N_1,\dots,N_n} \left(\prod_{i,j=1}^{n} \frac{(qx_i/x_j;q)_{N_i}}{(qx_i/x_j;q)_{k_i}(q^{1+k_i-k_j}x_i/x_j;q)_{N_i-k_i}} \cdot (-1)^{|k|} q^{\sum_{i=1}^{n}\binom{k_i}{2}} z^{|k|} \prod_{i=1}^{n} x_i^{k_i} \right), \tag{5.2.11b}$$

$$\prod_{i=1}^{n}(zq^{|N|-N_i}/x_i;q)_{N_i} = \sum_{k_1,\dots,k_n=0}^{N_1,\dots,N_n} \left(\prod_{i,j=1}^{n} \frac{(qx_i/x_j;q)_{N_i}}{(qx_i/x_j;q)_{k_i}(q^{1+k_i-k_j}x_i/x_j;q)_{N_i-k_i}} \right.$$
$$\left. \times (-1)^{|k|} q^{\binom{|k|}{2}+\sum_{1\le i<j\le n} k_i k_j} z^{|k|} \prod_{i=1}^{n} x_i^{-k_i} \right). \tag{5.2.11c}$$

The summations in (5.2.11b) and (5.2.11c) are related by reversal of sums as explained in §5.2.1(iii). In the last two identities the variable z is redundant. Inclusion turns both sides into polynomials in z.

Here are two other terminating A_n q-binomial theorems (Milne, 1997, Theorems 5.52 and 5.50):

$$(z;q)_N = \sum_{\substack{k_1,\dots,k_n\ge 0\\ |k|\le N}} \left(\prod_{1\le i<j\le n} \frac{x_iq^{k_i}-x_jq^{k_j}}{x_i-x_j} \prod_{i,j=1}^{n} (qx_i/x_j;q)_{k_i}^{-1} \cdot (q^{-N};q)_{|k|} q^{N|k|} z^{|k|} \right), \tag{5.2.11d}$$

$$(z;q)_N = \sum_{\substack{k_1,\dots,k_n\ge 0\\ |k|\le N}} \left(\prod_{1\le i<j\le n} \frac{x_iq^{k_i}-x_jq^{k_j}}{x_i-x_j} \prod_{i,j=1}^{n} (qx_i/x_j;q)_{k_i}^{-1} \cdot (q^{-N};q)_{|k|} \right.$$
$$\left. \times (-1)^{(n-1)|k|} q^{N|k|-\binom{|k|}{2}+n\sum_{i=1}^{n}\binom{k_i}{2}} z^{|k|} \prod_{i=1}^{n} x_i^{nk_i-|k|} \right). \tag{5.2.11e}$$

Note that (5.2.11d) and (5.2.11e) are equivalent under base inversion $q \to q^{-1}$.

Yet another terminating A_n q-binomial theorem, implicit from Bhatnagar and Schlosser (1998), is

$$\prod_{i=1}^{n}(z/x_i;q)_N = \sum_{\substack{k_1,\dots,k_n\ge 0\\ |k|\le N}} \left(\prod_{1\le i<j\le n} \frac{x_iq^{k_i}-x_jq^{k_j}}{x_i-x_j} \prod_{i,j=1}^{n} (qx_i/x_j;q)_{k_i}^{-1} \cdot (q^{-N};q)_{|k|} \right.$$
$$\left. \times q^{N|k|+\sum_{1\le i<j\le n} k_i k_j} z^{|k|} \prod_{i=1}^{n} x_i^{-k_i} (z/x_i;q)_{|k|-k_i} \right). \tag{5.2.11f}$$

Four of the terminating A_n q-binomial theorems of this subsection are special cases of nonterminating A_n q-binomial theorems. In particular, (5.2.11a) is a special case of (5.1.4). Similarly, (5.2.11b) is a special case of Lilly and Milne (1993, Theorem 4.7). Furthermore, (5.2.11f) is a special case of Milne (1997, Theorem 5.42), (which is the $b_1 = \cdots = b_n = q$ special case of (5.2.26)). Finally, (5.2.11e) is a special case of Bhatnagar and Schlosser (1998, Theorem 5.19).

5.2.3 Other Terminating Summations

All the A_n q-binomial theorems (i.e., terminating ${}_1\phi_0$ summations) listed in the previous subsection (and others not listed here, such as those implicit in Rosengren and Schlosser, 2003, §7.7) admit generalizations to summations involving more parameters. These include, in particular, various A_n ${}_2\phi_1$, ${}_3\phi_2$, ${}_6\phi_5$, or ${}_8\phi_7$ summations (see e.g. Milne, 1997; Rosengren and Schlosser, 2003; Schlosser, 2008).

The terminating balanced ${}_3\phi_2$ summation (or *q-Pfaff–Saalschütz summation*) is (cf. Gasper and Rahman, 2004, (II.12))

$$ {}_3\phi_2\left(\begin{matrix} a, b, q^{-N} \\ c, abq^{1-N}/c \end{matrix}; q, q\right) = \frac{(c/a, c/b; q)_N}{(c, c/ab; q)_N}. \tag{5.2.12} $$

Here are two A_n ${}_3\phi_2$ summations (cf. Milne, 1997):

$$ \sum_{k_1,\dots,k_n=0}^{N_1,\dots,N_n} \left(\prod_{1\le i<j\le n} \frac{x_i q^{k_i} - x_j q^{k_j}}{x_i - x_j} \prod_{i,j=1}^{n} \frac{(q^{-N_j} x_i/x_j; q)_{k_i}}{(q x_i/x_j; q)_{k_i}} \prod_{i=1}^{n} \frac{(a x_i; q)_{k_i}}{(c x_i; q)_{k_i}} \cdot \frac{(b; q)_{|k|}}{(abq^{1-|N|}/c; q)_{|k|}} q^{|k|} \right) $$
$$ = \frac{(c/a; q)_{|N|}}{(c/ab; q)_{|N|}} \prod_{i=1}^{n} \frac{(c x_i/b; q)_{N_i}}{(c x_i; q)_{N_i}}, \tag{5.2.13} $$

$$ \sum_{k_1,\dots,k_n=0}^{N_1,\dots,N_n} \left(\prod_{1\le i<j\le n} \frac{x_i q^{k_i} - x_j q^{k_j}}{x_i - x_j} \prod_{i,j=1}^{n} \frac{(q^{-N_j} x_i/x_j; q)_{k_i}}{(q x_i/x_j; q)_{k_i}} \prod_{i=1}^{n} \frac{(a x_i, b x_i; q)_{k_i}}{(c x_i, ab x_i q^{1-|N|}/c; q)_{k_i}} q^{|k|} \right) $$
$$ = (c/a, c/b; q)_{|N|} \prod_{i=1}^{n} (c x_i, c x_i q^{|N|-N_i}/ab; q)_{N_i}^{-1}. \tag{5.2.14} $$

Several other A_n ${}_3\phi_2$ summations are given by Milne (1997). For instance, a simple polynomial argument applied to (5.2.13) yields

$$ \sum_{\substack{k_1,\dots,k_n\ge 0 \\ |k|\le N}} \left(\prod_{1\le i<j\le n} \frac{x_i q^{k_i} - x_j q^{k_j}}{x_i - x_j} \prod_{i,j=1}^{n} \frac{(a_j x_i/x_j; q)_{k_i}}{(q x_i/x_j; q)_{k_i}} \prod_{i=1}^{n} \frac{(b x_i; q)_{k_i}}{(c x_i; q)_{k_i}} \cdot \frac{(q^{-N}; q)_{|k|}}{(a_1 \cdots a_n b q^{1-N}/c; q)_{|k|}} q^{|k|} \right) $$
$$ = \frac{(c/b; q)_N}{(c/a_1 \cdots a_n b; q)_N} \prod_{i=1}^{n} \frac{(c x_i/a_i; q)_N}{(c x_i; q)_N}. \tag{5.2.15} $$

Here is another terminating balanced ${}_3\phi_2$ summation (Milne and Lilly, 1995, Theorem 4.3 rewritten) which may be considered of "mixed type":

$$\sum_{\substack{k_1,\dots,k_n\geq 0\\ |k|\leq N}}\left(\prod_{1\leq i<j\leq n}\frac{x_iq^{k_i}-x_jq^{k_j}}{x_i-x_j}\frac{1}{(x_ix_j;q)_{k_i+k_j}}\prod_{i,j=1}^{n}\frac{(a_jx_i/x_j,x_ix_j/a_j;q)_{k_i}}{(qx_i/x_j;q)_{k_i}}\right.$$
$$\left.\times\frac{(q^{-N};q)_{|k|}}{\prod_{i=1}^n(bx_iq^{-N},qx_i/b;q)_{k_i}}q^{|k|}\right)=\prod_{i=1}^{n}\frac{(qa_i/bx_i,qx_i/a_ib;q)_N}{(q/bx_i,qx_i/b;q)_N}. \tag{5.2.16}$$

Again, a simple polynomial argument, which we do not state here explicitly, can be applied to transform this summation to another one, in this case to a sum over a rectangular region (see Milne and Lilly, 1995, Theorem 4.2).

Among the most general summations for basic hypergeometric series associated with root systems are various multivariate ${}_8\phi_7$ Jackson summations. In the univariate case, *Jackson's terminating balanced very-well-poised ${}_8\phi_7$ summation* is (cf. Gasper and Rahman, 2004, (II.22))

$${}_8\phi_7\left(\begin{matrix}a,qa^{\frac12},-qa^{\frac12},b,c,d,e,q^{-N}\\ a^{\frac12},-a^{\frac12},aq/b,aq/c,aq/d,aq/e,aq^{N+1}\end{matrix};q,q\right)=\frac{(aq,aq/bc,aq/bd,aq/cd;q)_N}{(aq/b,aq/c,aq/d,aq/bcd;q)_N}, \tag{5.2.17}$$

where $a^2q=bcdeq^{-N}$. Some of the multivariate Jackson summations have been extended to the level of elliptic hypergeometric series and are partly covered in §6.3. One of the most important is the following A_n Jackson summation (Milne, 1988a, Theorem 6.14):

$$\sum_{k_1,\dots,k_n=0}^{N_1,\dots,N_n}\left(\prod_{1\leq i<j\leq n}\frac{x_iq^{k_i}-x_jq^{k_j}}{x_i-x_j}\prod_{i=1}^{n}\frac{1-ax_iq^{k_i+|k|}}{1-ax_i}\prod_{i,j=1}^{n}\frac{(q^{-N_j}x_i/x_j;q)_{k_i}}{(qx_i/x_j;q)_{k_i}}\right.$$
$$\left.\times\prod_{i=1}^{n}\frac{(ax_i;q)_{|k|}(dx_i,a^2x_iq^{1+N_i}/bcd;q)_{k_i}}{(ax_iq^{1+|N|};q)_{|k|}(ax_iq/b,ax_iq/c;q)_{k_i}}\cdot\frac{(b,c;q)_{|k|}}{(aq/d,bcdq^{-|N|}/a;q)_{|k|}}q^{|k|}\right)$$
$$=\frac{(aq/bd,aq/cd;q)_{|N|}}{(aq/d,aq/bcd;q)_{|N|}}\prod_{i=1}^{n}\frac{(ax_iq,ax_iq/bc;q)_{N_i}}{(ax_iq/b,ax_iq/c;q)_{N_i}}. \tag{5.2.18}$$

It was initially proved by partial fraction decompositions and functional equations; a more direct proof (which extends to the elliptic level) utilizes partial fraction decompositions and induction (Rosengren, 2004a). For an elliptic extension of (5.2.18), see Chapter 6, (6.3.1c).

From (5.2.18), by multivariable matrix inversion, the following A_n Jackson summation was deduced by Schlosser (2008, Theorem 4.1):

$$\sum_{k_1,\dots,k_n=0}^{N_1,\dots,N_n}\left(\prod_{1\leq i<j\leq n}\frac{x_iq^{k_i}-x_jq^{k_j}}{x_i-x_j}\prod_{i=1}^{n}\frac{(bcd/ax_i;q)_{|k|-k_i}(d/x_i;q)_{|k|}(a^2x_iq^{1+|N|}/bcd;q)_{k_i}}{(d/x_i;q)_{|k|-k_i}(bcdq^{-N_i}/ax_i;q)_{|k|}(ax_iq/d;q)_{k_i}}\right.$$
$$\left.\times\prod_{i,j=1}^{n}\frac{(q^{-N_j}x_i/x_j;q)_{k_i}}{(qx_i/x_j;q)_{k_i}}\cdot\frac{(1-aq^{2|k|})}{(1-a)}\frac{(a,b,c;q)_{|k|}}{(aq^{1+|N|},aq/b,aq/c;q)_{|k|}}q^{|k|}\right)$$
$$=\frac{(aq,aq/bc;q)_{|N|}}{(aq/b,aq/c;q)_{|N|}}\prod_{i=1}^{n}\frac{(ax_iq/bd,ax_iq/bc;q)_{N_i}}{(ax_iq/d,ax_iq/bcd;q)_{N_i}}. \tag{5.2.19}$$

(Its elliptic extension is deduced by Rosengren, 2011, (7.7).) Both summations (5.2.18) and (5.2.19), which are summed over rectangular regions, can be turned to summations over a tetrahedral region (or simplex) $\{k_1, \ldots, k_n \geq 0,\ |k| \leq N\}$ by a polynomial argument, a standard procedure in the multivariate theory, used extensively by Milne (1997). Other Jackson summations which have been extended to the elliptic level are the C_n Jackson summations in Denis and Gustafson (1992, Theorem 4.1) (independently derived by Milne and Lilly, 1995, Theorem 6.13) and Schlosser (2000, Theorem 4.3) (see Chapter 6, (6.3.4) and (6.3.8)), the D_n Jackson summations (also referred to as A_n Jackson summations by some authors) of Bhatnagar (1999, Theorem 7) and Schlosser (1997, Theorem 5.14) (see also Chapter 6, (6.3.2)), the A_n and D_n Jackson summations by Bhatnagar and Schlosser (2018, Section 11), and the BC_n Jackson summation by van Diejen and Spiridonov (2000, Theorem 3) (also derived by Rosengren, 2001, $p \to 0$ in Theorem 2.1) which was originally conjectured by Warnaar (2002).

The following A_n Jackson summation is due to Gustafson and Rakha (2000, Theorem 1.2) (but stated here as in Rosengren, 2017 where it has been extended to the elliptic level) – see also Chapter 6, (6.3.3):

$$\sum_{\substack{k_1,\ldots,k_n \geq 0 \\ |k| \leq N}} \Bigg(\prod_{1 \leq i < j \leq n} \frac{x_i q^{k_i} - x_j q^{k_j}}{x_i - x_j} (x_i x_j; q)_{k_i + k_j} \prod_{1 \leq i,j \leq n} (q x_i / x_j; q)_{k_i}^{-1} \prod_{i=1}^{n} \frac{1 - a x_i q^{k_i + |k|}}{1 - a x_i}$$

$$\times \prod_{i=1}^{n} \frac{(a x_i; q)_{|k|}}{(aq/x_i; q)_{|k| - k_i} (a x_i q^{1+N_i}; q)_{k_i}} \prod_{j=1}^{4} \frac{\prod_{i=1}^{n} (x_i b_j; q)_{k_i}}{(aq/b_j; q)_{|k|}} \cdot (q^{-N}; q)_{|k|} q^{|k|} \Bigg)$$

$$= (aq/b_1, aq/b_2, aq/b_3, aq/b_1 b_2 b_3 X^2; q)_N^{-1} \prod_{i=1}^{n} \frac{(a x_i q; q)_N}{(aq/x_i; q)_N}$$

$$\times \begin{cases} (aq/X, aq/b_1 b_2 X, aq/b_1 b_3 X, aq/b_2 b_3 X; q)_N & \text{if } n \text{ is odd}, \\ (aq/b_1 X, aq/b_2 X, aq/b_3 X, aq/b_1 b_2 b_3 X; q)_N & \text{if } n \text{ is even}, \end{cases} \tag{5.2.20}$$

where $X = x_1 \cdots x_n$, under the assumption that $a^2 q^{N+1} = b_1 b_2 b_3 b_4 X^2$.

Two similar multivariate ${}_8\phi_7$ summations (at the elliptic level) are established by Rosengren (2017), and two others are conjectured by Spiridonov and Warnaar (2011, Conjectures 6.2 and 6.5). The latter actually look more complicated, the sums running over pairs of partitions whose Ferrers diagrams differ by a horizontal strip (cf. Macdonald, 1995). Those summations may play a role in the construction of Askey–Wilson polynomials of type A.

An A_n Jackson summation of a quite different type, intimately related to Macdonald polynomials (Macdonald, 1995), has been derived by Schlosser (2007, Theorem 4.1):

$$\sum_{\substack{k_1,\ldots,k_n \geq 0 \\ |k| \leq N}} \Bigg(\prod_{i,j=1}^{n} \frac{(q x_i / t_i x_j; q)_{k_i}}{(q x_i / x_j; q)_{k_i}} \prod_{1 \leq i < j \leq n} \frac{(t_j x_i / x_j; q)_{k_i - k_j}}{(q x_i / t_i x_j; q)_{k_i - k_j}} \frac{1}{x_i - x_j}$$

$$\times \det_{1 \leq i,j \leq n} \left[(x_i q^{k_i})^{n-j} \left(1 - t_i^{j-n-1} \frac{1 - t_0 x_i q^{k_i}}{1 - t_0 x_i q^{k_i} / t_i} \prod_{s=1}^{n} \frac{x_i q^{k_i} - x_s}{x_i q^{k_i} / t_i - x_s} \right) \right]$$

$$\times \prod_{i=1}^{n} \frac{(dq^{-N}/t_0x_i;q)_{|k|-k_i}(t_0x_iq/t_i, bx_i, t_0^2x_iq^{1+N}/bdt_1\cdots t_n;q)_{k_i}}{(dt_iq^{-N}/t_0x_i;q)_{|k|-k_i}(t_0x_iq, t_0x_iq/dt_i, t_0x_iq^{1+N}/t_i;q)_{k_i}}$$

$$\times \frac{(d,q^{-N};q)_{|k|}}{(bdq^{-N}/t_0, t_0q/bt_1\cdots t_n;q)_{|k|}} q^{\sum_{i=1}^n (2-i)k_i} \prod_{i=1}^{n} t_i^{(i-1)k_i+\sum_{j=i+1}^n k_j}\Bigg)$$

$$= \frac{(t_0q/b, t_0q/bdt_1\cdots t_n;q)_N}{(t_0q/bd, t_0q/bt_1\cdots t_n;q)_N} \prod_{i=1}^{n} \frac{(t_0x_iq/t_i, t_0x_iq/d;q)_N}{(t_0x_iq, t_0x_iq/dt_i;q)_N}. \tag{5.2.21}$$

A similar Jackson sum of type C_n has been conjectured (Schlosser, 2007, Conjecture 4.5). For the A_n identity (5.2.21) and the similar C_n identity from Schlosser (2007) (conjectured) elliptic extensions have not yet been established. The difficulty stems from the special type of determinants (which do not allow termwise elliptic extension) appearing in the respective summands of the series.

The A_n Jackson summation in (5.2.21) is also remarkable in the sense that no corresponding multivariate Bailey transformation has yet been found or conjectured. (The same applies to Schlosser (2007), Conjecture 4.5.) The other Jackson summations which we have discussed in this subsection all can be generalized to transformations. Since in this chapter we are mainly concerned with identities that do not directly extend to the elliptic setting we are not reproducing any of the multivariate Bailey transformations here. (The only exception is the C_n nonterminating Bailey transformation in (5.2.48), as nonterminating series do not admit a direct elliptic extension, for the reason of convergence.) For a discussion on multivariate extensions of Bailey's ${}_{10}\phi_9$ transformation, see §§6.3.3, 6.3.4.

5.2.4 Some Multilateral Summations

Dougall's bilateral ${}_2H_2$ summation (Dougall, 1907, §13) is

$${}_2H_2\left(\begin{matrix} a, b \\ c, d \end{matrix}; 1\right) = \frac{\Gamma(1-a)\Gamma(1-b)\Gamma(c)\Gamma(d)\Gamma(c+d-a-b-1)}{\Gamma(c-a)\Gamma(c-b)\Gamma(d-a)\Gamma(d-b)}, \tag{5.2.22}$$

where the series either terminates, or $\mathrm{Re}(c+d-a-b-1) > 0$ for convergence. This identity does not admit a direct basic extension (as a closed form ${}_2\psi_2$ summation with general parameters does not exist). A related, similar looking identity is *Ramanujan's ${}_1\psi_1$ summation theorem* (cf. Gasper and Rahman, 2004, (II.29)),

$${}_1\psi_1\left(\begin{matrix} a \\ b \end{matrix}; q, z\right) = \frac{(q, b/a, az, q/az;q)_\infty}{(b, q/a, z, b/az;q)_\infty}, \quad |q|<1,\ |b/a|<|z|<1. \tag{5.2.23}$$

An A_n extension of Dougall's ${}_2H_2$ summation theorem was proved in Gustafson (1987a), Theorem 1.11, by induction and residue calculus:

$$\sum_{k_1,\dots,k_n=-\infty}^{\infty} \left(\prod_{1\le i<j\le n} \frac{x_i+k_i-x_j-k_j}{x_i-x_j} \prod_{i=1}^{n}\prod_{j=1}^{n+1} \frac{(a_j+x_i)_{k_i}}{(b_j+x_i)_{k_i}} \right)$$

$$= \frac{\Gamma\Big(-n + \sum_{j=1}^{n+1}(b_j - a_j)\Big)\prod_{i=1}^{n}\prod_{j=1}^{n+1}\Gamma(1 - a_j - x_i)\Gamma(b_j + x_i)}{\prod_{i,j=1}^{n+1}\Gamma(b_j - a_i)\prod_{1\le i<j\le n}\Gamma(1 - x_i + x_j)\Gamma(1 + x_i - x_j)}, \tag{5.2.24}$$

provided $\mathrm{Re}\,\big(\sum_{j=1}^{n+1}(b_j - a_j)\big) > n$.

Similarly, an A_n extension of Ramanujan's ${}_1\psi_1$ summation theorem was proved by Gustafson (1987a, Theorem 1.17):

$$\sum_{k_1,\ldots,k_n=-\infty}^{\infty}\left(\prod_{1\le i<j\le n}\frac{x_iq^{k_i} - x_jq^{k_j}}{x_i - x_j}\prod_{i,j=1}^{n}\frac{(a_jx_i/x_j;q)_{k_i}}{(b_jx_i/x_j;q)_{k_i}}\cdot z^{|k|}\right)$$

$$= \frac{(a_1\cdots a_nz, q/a_1\cdots a_nz;q)_\infty}{(z, b_1\cdots b_nq^{1-n}/a_1\cdots a_nz;q)_\infty}\prod_{i,j=1}^{n}\frac{(b_jx_i/a_ix_j, qx_i/x_j;q)_\infty}{(qx_i/a_ix_j, b_jx_i/x_j;q)_\infty}, \tag{5.2.25}$$

where $|q| < 1$ and $|b_1\cdots b_nq^{1-n}/a_1\cdots a_n| < |z| < 1$.

The special case of (5.2.25), in which $b_1 = \cdots = b_n = b$, was previously obtained by Milne (1986, Theorem 1.15). See Milne (1986, 1988b, 1997) for further application of this first multilateral ${}_1\psi_1$ summation.

Another A_n ${}_1\psi_1$ summation theorem was found by Milne and Schlosser (2002, Theorem 3.2):

$$\sum_{k_1,\ldots,k_n=-\infty}^{\infty}\left(\prod_{1\le i<j\le n}\frac{x_iq^{k_i} - x_jq^{k_j}}{x_i - x_j}\prod_{i,j=1}^{n}(b_jx_i/x_j;q)_{k_i}^{-1}\prod_{i=1}^{n}x_i^{nk_i-|k|}\cdot(a;q)_{|k|}(-1)^{(n-1)|k|}\right.$$

$$\left.\times\, q^{-\binom{|k|}{2}+n\sum_{i=1}^{n}\binom{k_i}{2}}z^{|k|}\right) = \frac{(az, q/az, b_1\cdots b_nq^{1-n}/a;q)_\infty}{(z, b_1\cdots b_nq^{1-n}/az, q/a;q)_\infty}\prod_{i,j=1}^{n}\frac{(qx_i/x_j;q)_\infty}{(b_jx_i/x_j;q)_\infty}, \tag{5.2.26}$$

where $|q| < 1$ and $|b_1\cdots b_nq^{1-n}/a| < |z| < 1$. (The specified region of convergence can be determined by an analysis as carried out by Schlosser, 2005, Appendix A.)

Another A_n extension of Ramanujan's ${}_1\psi_1$ summation is implicitly contained in Macdonald (2003b). Written out in explicit terms, it reads as the last identity in Warnaar (2010):

$$\sum_{k_1,\ldots,k_n=-\infty}^{\infty}\left(\prod_{1\le i<j\le n}\frac{x_iq^{k_i} - x_jq^{k_j}}{x_i - x_j}\frac{(x_i/tx_j;q)_{k_i-k_j}}{(qtx_i/x_j;q)_{k_i-k_j}}q^{-k_j}t^{k_i-k_j}\cdot\frac{(a;q)_{|k|}}{(b;q)_{|k|}}z^{|k|}\right)$$

$$= \frac{(az, q/az, b/a, qt;q)_\infty}{(z, b/az, q/a, b;q)_\infty}\prod_{i=1}^{n-1}\frac{(qt^{i+1};q)_\infty}{(t^i;q)_\infty}\prod_{i,j=1}^{n}\frac{(qx_i/x_j;q)_\infty}{(qtx_i/x_j;q)_\infty}, \tag{5.2.27}$$

where $|q| < 1$, $|t| < 1$, and $|b/a| < |z| < 1$.

Taking coefficients of z^N on both sides of (5.2.27) while appealing to (the univariate version of) Ramanujan's ${}_1\psi_1$ summation, we obtain the interesting identity

$$\sum_{\substack{k_1,\ldots,k_n=-\infty\\|k|=N}}^{\infty}\left(\prod_{1\le i<j\le n}\frac{x_iq^{k_i} - x_jq^{k_j}}{x_i - x_j}\frac{(x_i/tx_j;q)_{k_i-k_j}}{(qtx_i/x_j;q)_{k_i-k_j}}q^{-k_j}t^{k_i-k_j}\right)$$

$$= \frac{(qt;q)_\infty}{(q;q)_\infty}\prod_{i=1}^{n-1}\frac{(qt^{i+1};q)_\infty}{(t^i;q)_\infty}\prod_{i,j=1}^{n}\frac{(qx_i/x_j;q)_\infty}{(qtx_i/x_j;q)_\infty}, \tag{5.2.28}$$

subject to $|q| < 1$. Observe that the right-hand side is independent of N. Some additional (simpler) A_n ${}_1\psi_1$ summations are given by Gustafson and Krattenthaler (1996), Schlosser (2000, §2), and Rosengren and Schlosser (2003, §6).

Bailey's very-well-poised ${}_6\psi_6$ summation is (cf. Gasper and Rahman, 2004, (II.33))

$$
{}_6\psi_6\left(\begin{matrix} qa^{\frac{1}{2}}, -qa^{\frac{1}{2}}, b, c, d, e \\ a^{\frac{1}{2}}, -a^{\frac{1}{2}}, aq/b, aq/c, aq/d, aq/e \end{matrix}; q, q\right)
= \frac{(q, aq, q/a, aq/bc, aq/bd, aq/be, aq/cd, aq/ce, aq/de; q)_\infty}{(aq/b, aq/c, aq/d, aq/e, q/b, q/c, q/d, q/e, a^2q/bcde; q)_\infty}, \qquad |q| < 1,\ |a^2q/bcde| < 1.
$$

Several root system extensions of Bailey's ${}_6\psi_6$ summation formula exist. Due to the fundamental importance of the ${}_6\psi_6$ summation, we will review several of these summations.

We start with an A_n extension of the ${}_6\psi_6$ summation (Gustafson, 1987a, Theorem 1.15):

$$
\sum_{k_1,\dots,k_n=-\infty}^{\infty}\left(\frac{(d;q)_{|k|}}{(aq/c;q)_{|k|}}\prod_{1\le i<j\le n}\frac{x_iq^{k_i}-x_jq^{k_j}}{x_i-x_j}\prod_{i=1}^{n}\frac{1-ax_iq^{k_i+|k|}}{1-ax_i}\prod_{i,j=1}^{n}\frac{(b_jx_i/x_j;q)_{k_i}}{(ax_iq/e_jx_j;q)_{k_i}}\right.
$$
$$
\left.\times\prod_{i=1}^{n}\frac{(e_ix_i;q)_{|k|}(cx_i;q)_{k_i}}{(ax_iq/b_i;q)_{|k|}(ax_iq/d;q)_{k_i}}\cdot\left(\frac{a^{n+1}q}{BcdE}\right)^{|k|}\right) = \frac{(aq/Bc, a^nq/dE, aq/cd;q)_\infty}{(a^{n+1}q/BcdE, aq/c, q/d;q)_\infty}
$$
$$
\times\prod_{i,j=1}^{n}\frac{(ax_iq/b_ie_jx_j, qx_i/x_j;q)_\infty}{(qx_i/b_ix_j, ax_iq/e_jx_j;q)_\infty}\prod_{i=1}^{n}\frac{(aq/ce_ix_i, ax_iq/b_id, ax_iq, q/ax_i;q)_\infty}{(ax_iq/b_i, q/e_ix_i, q/cx_i, ax_iq/d;q)_\infty}, \tag{5.2.29}
$$

where $B = b_1\cdots b_n$ and $E = e_1\cdots e_n$, provided $|q| < 1$ and $|a^{n+1}q/BcdE| < 1$.

The multilateral identity above can also be written in a more compact form. We then have the A_n ${}_6\psi_6$ summation from Gustafson (1987a, Theorem 1.15):

$$
\sum_{\substack{k_1,\dots,k_{n+1}=-\infty\\ k_1+\cdots+k_{n+1}=0}}^{\infty}\left(\prod_{1\le i<j\le n+1}\frac{x_iq^{k_i}-x_jq^{k_j}}{x_i-x_j}\prod_{i,j=1}^{n+1}\frac{(a_jx_i/x_j;q)_{k_i}}{(b_jx_i/x_j;q)_{k_i}}\right)
$$
$$
= \frac{(b_1\cdots b_{n+1}q^{-n}, q/a_1\cdots a_{n+1};q)_\infty}{(q, b_1\cdots b_{n+1}q^{-n}/a_1\cdots a_{n+1};q)_\infty}\prod_{i,j=1}^{n+1}\frac{(qx_i/x_j, b_jx_i/a_ix_j;q)_\infty}{(b_jx_i/x_j, x_iq/a_ix_j;q)_\infty}, \tag{5.2.30}
$$

provided $|q| < 1$ and $|b_1\cdots b_{n+1}q^{-n}/a_1\cdots a_{n+1}| < 1$. It is not difficult to see that (5.2.29) and (5.2.30) are equivalent by a change of variables.

If in (5.2.30) one replaces the parameters x_i, a_i, and b_i by q^{x_i}, q^{a_i}, and q^{b_i}, respectively, and formally lets $q \to 1$, one obtains the following A_n ${}_5H_5$ summation from Gustafson (1987a, Theorem 1.13) (where a direct proof is given by functional equations without appealing to a $q \to 1$ limit):

$$
\sum_{\substack{k_1,\dots,k_{n+1}=-\infty\\ k_1+\cdots+k_{n+1}=0}}^{\infty}\left(\prod_{1\le i<j\le n+1}\frac{x_i+k_i-x_j-k_j}{x_i-x_j}\prod_{i,j=1}^{n+1}\frac{(a_j+x_i-x_j)_{k_i}}{(b_j+x_i-x_j)_{k_i}}\right)
$$
$$
= \frac{\Gamma\left(-n+\sum_{i=1}^{n+1}(b_i-a_i)\right)}{\Gamma\left(1-\sum_{i=1}^{n+1}a_i\right)\Gamma\left(-n+\sum_{i=1}^{n+1}b_i\right)}\prod_{i,j=1}^{n+1}\frac{\Gamma(b_j+x_i-x_j)\Gamma(1-a_i+x_i-x_j)}{\Gamma(1+x_i-x_j)\Gamma(b_j-a_i+x_i-x_j)}, \tag{5.2.31}
$$

provided $\mathrm{Re}\left(\sum_{i=1}^{n+1}(b_i - a_i)\right) > n$.

Another A_n very-well-poised ${}_6\psi_6$ summation was derived by Schlosser (2008):

$$\sum_{k_1,\ldots,k_n=-\infty}^{\infty}\left(\frac{1-aq^{2|k|}}{1-a}\frac{(E/a^{n-1};q)_{|k|}}{(aq/C;q)_{|k|}}\prod_{1\le i<j\le n}\frac{x_iq^{k_i}-x_jq^{k_j}}{x_i-x_j}\prod_{i,j=1}^{n}\frac{(c_jx_i/x_j;q)_{k_i}}{(ax_iq/e_jx_j;q)_{k_i}}\right.$$
$$\left.\times\prod_{i=1}^{n}\frac{(aq/bCx_i;q)_{|k|-k_i}(dE/a^{n-1}e_ix_i;q)_{|k|}(bx_i;q)_{k_i}}{(dE/a^nx_i;q)_{|k|-k_i}(ac_iq/bCx_i;q)_{|k|}(ax_iq/d;q)_{k_i}}\cdot\left(\frac{a^{n+1}q}{bCdE}\right)^{|k|}\right)$$
$$=\frac{(aq,q/a,aq/bd;q)_\infty}{(aq/C,a^{n+1}q/bCdE,a^{n-1}q/E;q)_\infty}\prod_{i,j=1}^{n}\frac{(qx_i/x_j,ax_iq/c_ie_jx_j;q)_\infty}{(qx_i/c_ix_j,ax_iq/e_jx_j;q)_\infty}$$
$$\times\prod_{i=1}^{n}\frac{(a^nx_iq/dE,aq/be_ix_i,aq/bCx_i,ax_iq/c_id;q)_\infty}{(a^{n-1}e_ix_iq/dE,q/bx_i,ax_iq/d,ac_iq/bCx_i;q)_\infty}, \tag{5.2.32}$$

where $C = c_1\cdots c_n$ and $E = e_1\cdots e_n$, provided $|q|<1$ and $|a^{n+1}q/BcdE|<1$.

A C_n very-well-poised ${}_6\psi_6$ summation was derived by Gustafson (1989, Theorem 5.1):

$$\sum_{k_1,\ldots,k_n=-\infty}^{\infty}\left(\prod_{1\le i<j\le n}\frac{x_iq^{k_i}-x_jq^{k_j}}{x_i-x_j}\prod_{1\le i\le j\le n}\frac{1-ax_ix_jq^{k_i+k_j}}{1-ax_ix_j}\prod_{i,j=1}^{n}\frac{(c_jx_i/x_j,e_jx_ix_j;q)_{k_i}}{(ax_ix_jq/c_j,ax_iq/e_jx_j;q)_{k_i}}\right.$$
$$\left.\times\prod_{i=1}^{n}\frac{(bx_i,dx_i;q)_{k_i}}{(ax_iq/b,ax_iq/d;q)_{k_i}}\cdot\left(\frac{a^{n+1}q}{bCdE}\right)^{|k|}\right)=\prod_{1\le i<j\le n}(ax_ix_jq/c_ic_j,aq/e_ie_jx_ix_j;q)_\infty$$
$$\times\prod_{1\le i\le j\le n}(ax_ix_jq,q/ax_ix_j;q)_\infty\prod_{i,j=1}^{n}\frac{(ax_iq/c_ie_jx_j,qx_i/x_j;q)_\infty}{(ax_iq/e_jx_j,q/e_jx_ix_j,ax_ix_jq/c_i,qx_i/c_ix_j;q)_\infty}$$
$$\times\prod_{i=1}^{n}\frac{(ax_iq/bc_i,aq/be_ix_i,ax_iq/c_id,aq/de_ix_i;q)_\infty}{(ax_iq/b,q/bx_i,ax_iq/d,q/dx_i;q)_\infty}\cdot\frac{(aq/bd;q)_\infty}{(a^{n+1}q/bCdE;q)_\infty}, \tag{5.2.33}$$

where $C = c_1\cdots c_n$ and $E = e_1\cdots e_n$, provided $|q|<1$ and $|a^{n+1}q/bCdE|<1$.

Here is a $B_n^\vee$ (using terminology for affine root systems in Macdonald, 1972; or labeled $A_{2n-1}^{(2)}$ by Kac, 1990) very-well-poised ${}_6\psi_6$ summation, obtained by Gustafson (1989, Theorem 6.1):

$$\sum_{\substack{k_1,\ldots,k_n=-\infty\\|k|\equiv\sigma \bmod 2}}^{\infty}\left(\prod_{1\le i<j\le n}\frac{x_iq^{k_i}-x_jq^{k_j}}{x_i-x_j}\frac{1-ax_ix_jq^{k_i+k_j}}{1-ax_ix_j}\prod_{i,j=1}^{n}\frac{(c_jx_i/x_j,e_jx_ix_j;q)_{k_i}}{(ax_ix_jq/c_j,ax_iq/e_jx_j;q)_{k_i}}\cdot\left(-\frac{a^n}{bCdE}\right)^{|k|}\right)$$
$$=\frac{(-q;q)_\infty}{(-a^n/CE;q)_\infty}\prod_{1\le i<j\le n}(ax_ix_jq/c_ic_j,aq/e_ie_jx_ix_j;q)_\infty(ax_ix_jq,q/ax_ix_j;q)_\infty$$
$$\times\prod_{i,j=1}^{n}\frac{(ax_iq/c_ie_jx_j,qx_i/x_j;q)_\infty}{(ax_iq/e_jx_j,q/e_jx_ix_j,ax_ix_jq/c_i,qx_i/c_ix_j;q)_\infty}\prod_{i=1}^{n}\frac{(aqx_i^2/c_i^2,aq/e_i^2u_i^2;q^2)_\infty}{(aqx_i^2,q/ax_i^2;q^2)_\infty}, \tag{5.2.34}$$

where $C = c_1\cdots c_n$ and $E = e_1\cdots e_n$, and where $\sigma = 0, 1$, provided $|q|<1$ and $|a^n/CE|<1$.

As observed by Spiridonov and Warnaar (2011), the identity (5.2.34) is closely connected to the $b = \sqrt{aq}$, $d = -\sqrt{aq}$ case of the identity (5.2.33), where the sum evaluates to twice the product on the right-hand side of (5.2.34) (the latter being independent of σ).

Another C_n very-well-poised ${}_6\psi_6$ summation was established by van Diejen (1997, (2.22)):

$$\sum_{k_1,\dots,k_n=-\infty}^{\infty}\Bigg(\prod_{1\le i<j\le n}\frac{x_iq^{k_i}-x_jq^{k_j}}{x_i-x_j}\frac{1-ax_ix_jq^{k_i+k_j}}{1-ax_ix_j}\frac{(tax_ix_j;q)_{k_i+k_j}(tx_i/x_j;q)_{k_i-k_j}}{(ax_ix_jq/t;q)_{k_i+k_j}(qx_i/tx_j;q)_{k_i-k_j}}$$
$$\times\prod_{i=1}^{n}\frac{(bx_i,cx_i,dx_i,ex_i;q)_{k_i}}{(ax_iq/b,ax_iq/c,ax_iq/d,ax_iq/e;q)_{k_i}}\cdot\left(\frac{t^2}{q}\right)^{\sum_{i=1}^n(i-1)k_i}\left(\frac{t^{2-2n}a^2q}{bcde}\right)^{|k|}\Bigg)$$
$$=\prod_{i,j=1}^{n}\frac{(qx_i/x_j;q)_\infty}{(qx_i/tx_j;q)_\infty}\cdot\frac{\prod_{1\le i\le j\le n}(ax_ix_jq,q/ax_ix_j;q)_\infty}{\prod_{1\le i<j\le n}(ax_ix_jq/t,q/tax_ix_j;q)_\infty}\prod_{i=1}^{n}\frac{(qt^{-i};q)_\infty}{(qt^{2-i-n}a^2/bcde;q)_\infty}$$
$$\times\prod_{i=1}^{n}\frac{(at^{1-i}q/bc,at^{1-i}q/bd,at^{1-i}q/be,at^{1-i}q/cd,at^{1-i}q/ce,at^{1-i}q/de;q)_\infty}{(q/bx_i,q/cx_i,q/dx_i,q/ex_i,ax_iq/b,ax_iq/c,ax_iq/d,ax_iq/e;q)_\infty},\quad(5.2.35)$$

provided $|q|<1$, $|a^2q^{2-n}/bcde|<1$, and $|t^{2-2n}a^2q/bcde|<1$.

The next $B_n^\vee$ (or $A_{2n-1}^{(2)}$) ${}_6\psi_6$ summation (compare with (5.2.34)) was derived by Spiridonov and Warnaar (2011, Theorem 4.1):

$$\sum_{\substack{k_1,\dots,k_n=-\infty\\|k|\equiv\sigma \bmod 2}}^{\infty}\Bigg(\prod_{i=1}^{n}\frac{(bx_i,cx_i;q)_{k_i}}{(ax_iq/b,ax_iq/c;q)_{k_i}}\prod_{1\le i<j\le n}\frac{x_iq^{k_i}-x_jq^{k_j}}{x_i-x_j}\prod_{1\le i\le j\le n}\frac{1-ax_ix_jq^{k_i+k_j}}{1-ax_ix_j}$$
$$\times\prod_{1\le i<j\le n}\frac{(tax_ix_j;q)_{k_i+k_j}(tx_i/x_j;q)_{k_i-k_j}}{(ax_ix_jq/t;q)_{k_i+k_j}(qx_i/tx_j;q)_{k_i-k_j}}\cdot\left(\frac{t^2}{q}\right)^{\sum_{i=1}^n(i-1)k_i}\left(-\frac{t^{2-2n}a}{bc}\right)^{|k|}\Bigg)$$
$$=\frac{1}{2}\prod_{i,j=1}^{n}\frac{(qx_i/x_j;q)_\infty}{(qx_i/tx_j;q)_\infty}\frac{\prod_{1\le i\le j\le n}(ax_ix_jq,q/ax_ix_j;q)_\infty}{\prod_{1\le i<j\le n}(ax_ix_jq/t,q/tax_ix_j;q)_\infty}\prod_{i=1}^{n}\frac{(qt^{-i};q)_\infty}{(-t^{2-i-n}a/bc;q)_\infty}$$
$$\times\prod_{i=1}^{n}\frac{(at^{1-i}q/bc,-t^{1-i};q)_\infty(at^{2-2i}q/b,at^{2-2i}q/c;q^2)_\infty}{(q/bx_i,q/cx_i,ax_iq/b,ax_iq/c;q)_\infty(q/ax_i^2,aqx_i^2;q^2)_\infty},\quad(5.2.36)$$

where $\sigma=0,1$, provided $|q|<1$, $|aq^{1-n}/bc|<1$, and $|t^{2-2n}a/bc|<1$.

Similar to the relation between (5.2.34) and (5.2.33), the identity (5.2.36) is closely connected to the $d=\sqrt{aq}$, $e=-\sqrt{aq}$ case of the identity (5.2.35), where the sum evaluates to twice the product on the right-hand side of (5.2.36) (the latter being independent of σ). Two other (simpler) C_n ${}_6\psi_6$ summations are given by Schlosser (2000, Theorem 3.4).

Multivariate analogues of Bailey's ${}_6\psi_6$ summation for *exceptional* root systems were derived by Gustafson (1990b) (summation for G_2), Ito (2002) (summation for F_4; see also Ito, 2003), and Ito and Tsubouchi (2010) (further summations for G_2).

5.2.5 Watson Transformations

The *Watson transformation* (cf. Gasper and Rahman, 2004, (III.18)),

$$
{}_8\phi_7\left(\begin{matrix} a, qa^{\frac{1}{2}}, -qa^{\frac{1}{2}}, b, c, d, e, q^{-N} \\ a^{\frac{1}{2}}, -a^{\frac{1}{2}}, aq/b, aq/c, aq/d, aq/e, aq^{N+1} \end{matrix}; q, \frac{a^2q^{N+2}}{bcde}\right)
= \frac{(aq, aq/de; q)_N}{(aq/d, aq/e; q)_N}\, {}_4\phi_3\left(\begin{matrix} aq/bc, d, e, q^{-N} \\ aq/b, aq/c, deq^{-N}/a \end{matrix}; q, q\right), \tag{5.2.37}
$$

is very useful. For instance, it can be used for a quick proof of the Rogers–Ramanujan identities, see Gasper and Rahman (2004, §2.7).

A number of Watson transformations for basic hypergeometric series associated with root systems exist. We reproduce only a few of them here.

Gustafson (1987b, Theorem 2.24) applied the representation theory of U(n) to derive the first multivariable generalization of Whipple's classical transformation of an ordinary ($q = 1$) terminating well-poised ${}_7F_6(1)$ into a terminating balanced ${}_4F_3(1)$. Its q-analogue, the first multivariable Watson transformation, was obtained by Milne (1988a, Theorems 6.1 and 6.4; 1989, Theorems 6.1 and 6.4) by a direct, elementary proof utilizing q-difference equations and induction. Further details and applications are given by Milne (1988a, 1989, 1994). A more symmetrical A_n Watson transformation was derived by Milne (2000, Theorem 2.1) by means of the summation theorems and analysis from Milne (1997), which also provides an A_n generalization of much of the analysis in Gasper and Rahman (2004, Chapters 1 and 2).

The following A_n Watson transformation was derived by Milne and Newcomb (1996, Theorem A.3):

$$
\sum_{k_1,\dots,k_n=0}^{N_1,\dots,N_n}\left(\prod_{1\le i<j\le n}\frac{x_iq^{k_i}-x_jq^{k_j}}{x_i-x_j}\prod_{i,j=1}^{n}\frac{(q^{-N_j}x_i/x_j;q)_{k_i}}{(qx_i/x_j;q)_{k_i}}\prod_{i=1}^{n}\frac{1-ax_iq^{k_i+|k|}}{1-ax_i}\frac{(ax_i;q)_{|k|}}{(ax_iq^{1+N_i};q)_{|k|}}\right.
$$
$$
\left.\times\prod_{i=1}^{n}\frac{(bx_i,cx_i;q)_{k_i}}{(ax_iq/d,ax_iq/e;q)_{k_i}}\cdot\frac{(d,e;q)_{|k|}}{(aq/b,aq/c;q)_{|k|}}\left(\frac{a^2q^{|N|+2}}{bcde}\right)^{|k|}\right)
$$
$$
=\frac{(aq/ce;q)_{|N|}}{(aq/c;q)_{|N|}}\prod_{i=1}^{n}\frac{(ax_iq;q)_{N_i}}{(ax_iq/e;q)_{N_i}}\sum_{k_1,\dots,k_n=0}^{N_1,\dots,N_n}\left(\prod_{1\le i<j\le n}\frac{x_iq^{k_i}-x_jq^{k_j}}{x_i-x_j}\prod_{i,j=1}^{n}\frac{(q^{-N_j}x_i/x_j;q)_{k_i}}{(qx_i/x_j;q)_{k_i}}\right.
$$
$$
\left.\times\prod_{i=1}^{n}\frac{(cx_i;q)_{k_i}}{(ax_iq/d;q)_{k_i}}\cdot\frac{(aq/bd,e;q)_{|k|}}{(aq/b,ceq^{-|N|}/a;q)_{|k|}}q^{|k|}\right). \tag{5.2.38}
$$

For a very similar but different A_n Watson transformation, see the $f_i = q^{-N_i}$, $i = 1, \dots, n$ case of (5.2.50).

The following $\mathrm{C}_n \leftrightarrow \mathrm{A}_{n-1}$ Watson transformation was first derived by Milne and Lilly (1995, Theorem 6.6):

$$
\sum_{k_1,\dots,k_n=0}^{N_1,\dots,N_n}\left(\prod_{1\le i<j\le n}\frac{x_iq^{k_i}-x_jq^{k_j}}{x_i-x_j}\prod_{1\le i\le j\le n}\frac{1-x_ix_jq^{k_i+k_j}}{1-x_ix_j}\prod_{i,j=1}^{n}\frac{(q^{-N_j}x_i/x_j,x_ix_j;q)_{k_i}}{(qx_i/x_j,q^{1+N_i}x_ix_j;q)_{k_i}}\right.
$$

$$\times \prod_{i=1}^{n} \frac{(bx_i, cx_i, dx_i, ex_i; q)_{k_i}}{(qx_i/b, qx_i/c, qx_i/d, qx_i/e; q)_{k_i}} \cdot \left(\frac{q^{|N|+2}}{bcde}\right)^{|k|}\Bigg) = (q/bc;q)_{|N|} \prod_{i=1}^{n} \frac{1}{(qx_i/b, qx_i/c; q)_{N_i}}$$
$$\times \prod_{i,j=1}^{n} (qx_ix_j;q)_{N_i} \prod_{1\le i<j\le n} \frac{1}{(qx_ix_j;q)_{N_i+N_j}} \sum_{k_1,\dots,k_n=0}^{N_1,\dots,N_n} \Bigg(\prod_{1\le i<j\le n} \frac{x_iq^{k_i} - x_jq^{k_j}}{x_i - x_j}$$
$$\times \prod_{i,j=1}^{n} \frac{(q^{-N_j}x_i/x_j;q)_{k_i}}{(qx_i/x_j;q)_{k_i}} \prod_{i=1}^{n} \frac{(bx_i, cx_i; q)_{k_i}}{(qx_i/d, qx_i/e; q)_{k_i}} \cdot \frac{(q/de;q)_{|k|}}{(bcq^{-|N|};q)_{|k|}} q^{|k|}\Bigg). \tag{5.2.39}$$

This identity was utilized by Bartlett (2013) and Bartlett and Warnaar (2015) to obtain identities for characters of affine Lie algebras.

Several other Watson transformations are given by Bhatnagar (1999), Bhatnagar and Schlosser (1998), Bhatnagar and Schlosser (2018), and Coskun (2008). One of them is the following (cf. Bhatnagar and Schlosser, 1998, Theorem 4.10):

$$\sum_{k_1,\dots,k_n=0}^{N_1,\dots,N_n} \Bigg(\prod_{1\le i<j\le n} \frac{x_iq^{k_i} - x_jq^{k_j}}{x_i - x_j} \frac{(ax_ix_jq/c;q)_{k_i+k_j}}{(ex_ix_j;q)_{k_i+k_j}} \prod_{i=1}^{n} \frac{1 - ax_iq^{k_i+|k|}}{1 - ax_i} \frac{(aq/ex_i;q)_{|k|-k_i}}{(c/x_i;q)_{|k|-k_i}}$$
$$\times \prod_{i=1}^{n} \frac{(ax_i, c/x_i; q)_{|k|}(bx_i;q)_{k_i}}{(ax_iq^{1+N_i}, aq^{1-N_i}/ex_i;q)_{|k|}(ax_iq/d;q)_{k_i}} \prod_{i,j=1}^{n} \frac{(q^{-N_j}x_i/x_j, ex_ix_jq^{N_j};q)_{k_i}}{(qx_i/x_j, ax_ix_jq/c;q)_{k_i}}$$
$$\times \frac{(d;q)_{|k|}}{(aq/b;q)_{|k|}} \left(\frac{q^2a^2}{bcde}\right)^{|k|}\Bigg) = d^{-|N|} \prod_{i=1}^{n} \frac{(ax_iq, dex_i/a;q)_{N_i}}{(ex_i/a, ax_iq/d;q)_{N_i}} \sum_{k_1,\dots,k_n=0}^{N_1,\dots,N_n} \Bigg(\prod_{i=1}^{n} \frac{(ax_iq/bc;q)_{k_i}}{(dex_i/a;q)_{k_i}}$$
$$\times \prod_{1\le i<j\le n} \frac{x_iq^{k_i} - x_jq^{k_j}}{x_i - x_j} \frac{(ax_ix_jq/c;q)_{k_i+k_j}}{(ex_ix_j;q)_{k_i+k_j}} \prod_{i,j=1}^{n} \frac{(q^{-N_j}x_i/x_j, ex_ix_jq^{N_j};q)_{k_i}}{(qx_i/x_j, ax_ix_jq/c;q)_{k_i}} \cdot \frac{(d;q)_{|k|}}{(aq/b;q)_{|k|}} q^{|k|}\Bigg). \tag{5.2.40}$$

This multivariate Watson transformation cannot be simplified to any multivariate Jackson summation as a special case.

5.2.6 Dimension-Changing Transformations

Heine's q-analogue of the classical Euler transformation of ${}_2F_1$ series is (cf. Gasper and Rahman, 2004, (III.3))

$${}_2\phi_1\left(\begin{matrix} a, b \\ c \end{matrix}; q, z\right) = \frac{(abz/c;q)_\infty}{(z;q)_\infty}\, {}_2\phi_1\left(\begin{matrix} c/a, c/b \\ c \end{matrix}; q, \frac{abz}{c}\right), \quad |q| < 1,\ |z| < 1,\ |abz/c| < 1. \tag{5.2.41}$$

The following result, which was first derived by Kajihara (2004), connects A_n and A_m basic hypergeometric series and reduces, for $n = m = 1$, to the q-Euler transformation:

$$\sum_{k_1,\dots,k_n\ge 0} \Bigg(\prod_{1\le i<j\le n} \frac{x_iq^{k_i} - x_jq^{k_j}}{x_i - x_j} \prod_{1\le i,j\le n} \frac{(a_jx_i/x_j;q)_{k_i}}{(qx_i/x_j;q)_{k_i}} \prod_{\substack{1\le i\le n\\ 1\le l\le m}} \frac{(b_lx_iy_l;q)_{k_i}}{(cx_iy_l;q)_{k_i}} \cdot z^{|k|}\Bigg) = \frac{(ABz/c^m;q)_\infty}{(z;q)_\infty}$$

$$\times \sum_{\kappa_1,\dots,\kappa_m\geq 0}\left(\prod_{1\leq j<l\leq m}\frac{y_jq^{\kappa_j}-y_lq^{\kappa_l}}{y_j-y_l}\prod_{1\leq j,l\leq m}\frac{(cy_j/b_ly_l;q)_{\kappa_j}}{(qy_j/y_l;q)_{\kappa_j}}\prod_{\substack{1\leq i\leq n\\1\leq l\leq m}}\frac{(cx_iy_l/a_i;q)_{\kappa_l}}{(cx_iy_l;q)_{\kappa_l}}\cdot\left(\frac{ABz}{c^m}\right)^{|\kappa|}\right), \tag{5.2.42}$$

where $A=a_1\cdots a_n$ and $B=b_1\cdots b_m$, provided $|q|<1$, $|z|<1$, and $|ABz/c^m|<1$.

Now let

$$\Phi_N^{n,m}\left(\begin{matrix}\{a_i\}_n\\\{x_i\}_n\end{matrix}\middle|\begin{matrix}\{b_ly_l\}_m\\\{cy_l\}_m\end{matrix}\right)
:=\sum_{\substack{k_1,\dots,k_n\geq 0\\|k|=N}}\left(\prod_{1\leq i<j\leq n}\frac{x_iq^{k_i}-x_jq^{k_j}}{x_i-x_j}\prod_{1\leq i,j\leq n}\frac{(a_jx_i/x_j;q)_{k_i}}{(qx_i/x_j;q)_{k_i}}\prod_{\substack{1\leq i\leq n\\1\leq l\leq m}}\frac{(b_lx_iy_l;q)_{k_i}}{(cx_iy_l;q)_{k_i}}\right). \tag{5.2.43}$$

The transformation (5.2.42) was used to derive the following identity in Kajihara (2016, Theorem 3.1) (which we state here in corrected form):

$$\sum_{K=0}^{N}\Phi_K^{n_2,m_2}\left(\begin{matrix}\{f/e_t\}_{n_2}\\\{v_t\}_{n_2}\end{matrix}\middle|\begin{matrix}\{fw_r/d_r\}_{m_2}\\\{fw_r\}_{m_2}\end{matrix}\right)\Phi_{N-K}^{n_1,m_1}\left(\begin{matrix}\{a_i\}_{n_1}\\\{x_i\}_{n_1}\end{matrix}\middle|\begin{matrix}\{b_ly_l\}_{m_1}\\\{cy_l\}_{m_1}\end{matrix}\right)\left(\frac{d_1\cdots d_{m_2}e_1\cdots e_{n_2}}{f^{n_2}}\right)^K$$
$$=\sum_{L=0}^{N}\Phi_L^{m_1,n_1}\left(\begin{matrix}\{c/b_l\}_{m_1}\\\{y_l\}_{m_1}\end{matrix}\middle|\begin{matrix}\{cx_i/a_i\}_{n_1}\\\{cx_i\}_{n_1}\end{matrix}\right)\Phi_{N-L}^{m_2,n_2}\left(\begin{matrix}\{d_r\}_{m_2}\\\{w_r\}_{m_2}\end{matrix}\middle|\begin{matrix}\{e_tv_t\}_{n_2}\\\{fv_t\}_{n_2}\end{matrix}\right)\left(\frac{a_1\cdots a_{n_1}b_1\cdots b_{m_1}}{c^{m_1}}\right)^L, \tag{5.2.44}$$

where $a_1\cdots a_{n_1}b_1\cdots b_{m_1}/c^{m_1}=d_1\cdots d_{m_2}e_1\cdots e_{n_2}/f^{n_2}$. This identity can be viewed as a multivariate extension of the *Sears transformation* (Gasper and Rahman, 2004, Equation (III.16)) (to which it reduces for $n=m=1$ after some elementary manipulations). A transformation similar to (5.2.42) but connecting the C_n and C_m basic hypergeometric series has been given by Komori et al. (2016).

Several other transformations connecting sums of different dimensions exist. For instance, Rosengren (2004b) derived the following reduction formula for a multilateral Karlsson–Minton-type basic hypergeometric series associated with the root system A_n (a basic hypergeometric series is said to be of *Karlsson–Minton type* if the quotient of corresponding upper and lower parameters is a nonnegative integer power of q):

$$\sum_{\substack{k_1,\dots,k_n=-\infty\\k_1+\cdots+k_n=0}}^{\infty}\left(\prod_{1\leq i<j\leq n}\frac{x_iq^{k_i}-x_jq^{k_j}}{x_i-x_j}\prod_{\substack{1\leq i\leq n\\1\leq j\leq p}}\frac{(x_iy_jq^{m_j};q)_{k_i}}{(x_iy_j;q)_{k_i}}\prod_{i,j=1}^{n}\frac{(x_ia_j;q)_{k_i}}{(x_ib_j;q)_{k_i}}\right)$$
$$=\frac{(q^{1-|m|}/AX,q^{1-n}BX;q)_\infty}{(q,q^{1-|m|-n}B/A;q)_\infty}\prod_{i,j=1}^{n}\frac{(b_i/a_j,qx_i/x_j;q)_\infty}{(q/x_ia_j,x_ib_j;q)_\infty}\prod_{\substack{1\leq i\leq n\\1\leq j\leq p}}\frac{(q^{-m_j}b_i/y_j;q)_{m_j}}{(q^{1-m_j}/x_iy_j;q)_{m_j}}$$
$$\times\sum_{\kappa_1,\dots,\kappa_p=0}^{m_1,\dots,m_p}\left(q^{|\kappa|}\frac{(q^n/BX;q)_{|\kappa|}}{(q^{1-|m|}/AX;q)_{|\kappa|}}\prod_{1\leq i<j\leq n}\frac{y_iq^{\kappa_i}-y_jq^{\kappa_j}}{y_i-y_j}\prod_{\substack{1\leq i\leq n\\1\leq j\leq p}}\frac{(y_j/a_i;q)_{\kappa_j}}{(qy_j/b_i;q)_{\kappa_j}}\prod_{i,j=1}^{p}\frac{(q^{-m_i}y_j/y_i;q)_{\kappa_j}}{(qy_j/y_i;q)_{\kappa_j}}\right), \tag{5.2.45}$$

where $A=a_1\cdots a_n$, $B=b_1\cdots b_n$, $X=x_1\cdots x_n$, provided $|q|<1$ and $|q^{1-|m|-n}B/A|<1$.

Similarly, Rosengren (2003) derived the following reduction formula for a multilateral Karlsson–Minton-type basic hypergeometric series associated with the root system C_n:

$$\sum_{k_1,\dots,k_n=-\infty}^{\infty}\Bigg(\prod_{1\le i<j\le n}\frac{x_iq^{k_i}-x_jq^{k_j}}{x_i-x_j}\prod_{1\le i\le j\le n}\frac{1-x_ix_jq^{k_i+k_j}}{1-x_ix_j}\prod_{\substack{1\le i\le n\\1\le j\le p}}\frac{(x_iy_jq^{m_j},qx_i/y_j;q)_{k_i}}{(x_iy_j,q^{1-m_j}x_i/y_j;q)_{k_i}}$$

$$\times\prod_{\substack{1\le i\le n\\1\le j\le 2n+2}}\frac{(x_ia_j;q)_{k_i}}{(qx_i/a_j;q)_{k_i}}\cdot\left(\frac{q^{1-|m|}}{A}\right)^{|k|}\Bigg)=\frac{\prod_{1\le i\le j\le n}(qx_ix_j,q/x_ix_j;q)_\infty\prod_{i,j=1}^{n}(qx_i/x_j;q)_\infty}{\prod_{1\le i\le n,1\le j\le 2n+2}(qx_i/a_j,q/x_ia_j;q)_\infty}$$

$$\times\frac{\prod_{1\le i<j\le 2n+2}(q/a_ia_j;q)_\infty}{(q/A;q)_\infty}\frac{\prod_{1\le i\le 2n+2,1\le j\le p}(y_ja_i;q)_{m_j}}{\prod_{1\le i\le n,1\le j\le p}(y_jx_i,y_j/x_i;q)_{m_j}}\frac{\prod_{1\le i<j\le p}(y_iy_j;q)_{m_i+m_j}}{\prod_{i,j=1}^{p}(y_iy_j;q)_{m_i}}$$

$$\times\frac{1}{(A;q)_{|m|}}\sum_{\kappa_1,\dots,\kappa_p=0}^{m_1,\dots,m_p}\Bigg(\prod_{1\le i<j\le n}\frac{y_iq^{\kappa_i}-y_jq^{\kappa_j}}{y_i-y_j}\prod_{1\le i\le j\le n}\frac{1-y_iy_jq^{\kappa_i+\kappa_j-1}}{1-y_iy_jq^{-1}}$$

$$\times\prod_{\substack{1\le i\le 2n+2\\1\le j\le p}}\frac{(y_j/a_i;q)_{\kappa_j}}{(y_ja_i;q)_{\kappa_j}}\prod_{i,j=1}^{p}\frac{(q^{-1}y_iy_j,q^{-m_j}y_i/y_j;q)_{\kappa_i}}{(qy_i/y_j,q^{m_j}y_iy_j;q)_{\kappa_i}}\cdot\left(Aq^{|m|}\right)^{|\kappa|}\Bigg),\tag{5.2.46}$$

where $A=a_1\cdots a_{2n+2}$, provided $|q|<1$ and $|q^{1-|m|}/A|<1$. A substantially more general transformation (involving fourfold multiple sums) was given by Masuda (2013, Theorem 3).

Both (5.2.45) and (5.2.46) have many interesting consequences. In particular, these transformations form bridges between the one-variable and the multivariable theory and can be used to prove various summations and transformations for A_n and C_n basic hypergeometric series. For details we refer the reader to Rosengren (2003, 2004b). Other dimension-changing transformations have been given (or conjectured) by Bhatnagar (2019), Gessel and Krattenthaler (1997), Krattenthaler (2001), Rains (2010), and Rosengren (2006).

5.2.7 Multiterm Transformations

Bailey's nonterminating balanced very-well-poised ${}_{10}\phi_9$ transformation is (cf. Gasper and Rahman, 2004, (III.39))

$${}_{10}\phi_9\left(\begin{matrix}a,qa^{\frac12},-qa^{\frac12},b,c,d,e,f,g,h\\a^{\frac12},-a^{\frac12},aq/b,aq/c,aq/d,aq/e,aq/f,aq/g,aq/h\end{matrix};q,q\right)$$

$$+\frac{(aq,b/a,c,d,e,f,g,h,bq/c,bq/d,bq/e,bq/f,bq/g,bq/h;q)_\infty}{(b^2q/a,a/b,aq/c,aq/d,aq/e,aq/f,aq/g,aq/h,bc/a,bd/a,be/a,bf/a,bg/a,bh/a;q)_\infty}$$

$$\times\,{}_{10}\phi_9\left(\begin{matrix}b^2/a,qba^{-\frac12},-qba^{-\frac12},b,bc/a,bd/a,be/a,bf/a,bg/a,bh/a\\ba^{-\frac12},-ba^{-\frac12},bq/a,bq/c,bq/d,bq/e,bq/f,bq/g,bq/h\end{matrix};q,q\right)$$

$$=\frac{(aq,b/a,\lambda q/f,\lambda q/g,\lambda q/h,bf/\lambda,bg/\lambda,bh/\lambda;q)_\infty}{(\lambda q,b/\lambda,aq/f,aq/g,aq/h,bf/a,bg/a,bh/a;q)_\infty}$$

$$\times\,{}_{10}\phi_9\left(\begin{matrix}\lambda,q\lambda^{\frac12},-q\lambda^{\frac12},b,\lambda c/a,\lambda d/a,\lambda e/a,f,g,h\\\lambda^{\frac12},-\lambda^{\frac12},\lambda q/b,aq/c,aq/d,aq/e,\lambda q/f,\lambda q/g,\lambda q/h\end{matrix};q,q\right)$$

$$+\frac{(aq,b/a,f,g,h,bq/f,bq/g,bq/h,\lambda c/a,\lambda d/a,\lambda e/a,abq/\lambda c,abq/\lambda d,abq/\lambda e;q)_\infty}{(b^2q/\lambda,\lambda/b,aq/c,aq/d,aq/e,aq/f,aq/g,aq/h,bc/a,bd/a,be/a,bf/a,bg/a,bh/a;q)_\infty}$$
$$\times\,{}_{10}\phi_9\left(\begin{matrix}b^2/\lambda,qb\lambda^{-\frac12},-qb\lambda^{-\frac12},b,bc/a,bd/a,be/a,bf/\lambda,bg/\lambda,bh/\lambda\\ b\lambda^{-\frac12},-b\lambda^{-\frac12},bq/\lambda,abq/c\lambda,abq/d\lambda,abq/e\lambda,bq/f,bq/g,bq/h\end{matrix};q,q\right), \tag{5.2.47}$$

where $\lambda=a^2q/cde$, $a^3q^2=bcdefgh$, and $|q|<1$. This identity is at the top of the classical hierarchy of identities for basic hypergeometric series.

The following identity from Rosengren and Schlosser (2003, Corollary 4.1) was first derived by determinant evaluations, following a method first used by Gustafson and Krattenthaler (1996, 1997) to derive A_n extensions of Heine's ${}_2\phi_1$ transformations, and subsequently used in a systematic manner by Schlosser (2000) and Rosengren and Schlosser (2003). The identity concerns a C_n extension of Bailey's four-term transformation (5.2.47), where both sides of the identity involve 2^n nonterminating C_n basic hypergeometric series. Let $a^3q^{3-n}=bc_id_ie_ix_ifgh$ and $\lambda=a^2q/c_id_ie_ix_i$ for $i=1,\dots,n$. Then

$$\sum_{S\subseteq\{1,2,\dots,n\}}\left[\left(\frac{b}{a}\right)^{\binom{n-|S|}{2}}\prod_{i\notin S}\frac{(ax_i^2q,c_ix_i,d_ix_i,e_ix_i;q)_\infty}{(ax_i/b,ax_iq/c_i,ax_iq/d_i,ax_iq/e_i;q)_\infty}\right.$$
$$\times\prod_{i\notin S}\frac{(fx_i,gx_i,hx_i,b/ax_i,bq/c_i,bq/d_i,bq/e_i,bq/f,bq/g,bq/h;q)_\infty}{(ax_iq/f,ax_iq/g,ax_iq/h,b^2q/a,bc_i/a,bd_i/a,be_i/a,bf/a,bg/a,bh/a;q)_\infty}$$
$$\times\sum_{k_1,\dots,k_n=0}^{\infty}\left(\prod_{\substack{1\le i<j\le n\\ i,j\in S}}\frac{(x_iq^{k_i}-x_jq^{k_j})(1-ax_ix_jq^{k_i+k_j})}{(x_i-x_j)(1-ax_ix_j)}\prod_{i\in S}\frac{1-ax_i^2q^{2k_i}}{1-ax_i^2}\right.$$
$$\times\prod_{\substack{1\le i<j\le n\\ i,j\notin S}}\frac{(q^{k_i}-q^{k_j})(1-b^2q^{k_i+k_j}/a)}{(x_i-x_j)(1-ax_ix_j)}\prod_{i\notin S}\frac{1-b^2q^{2k_s}/a}{1-b^2/a}\prod_{i\in S,j\notin S}\frac{(x_iq^{k_i}-bq^{y_j}/a)(1-bx_jq^{k_i+k_j})}{(x_i-x_j)(1-ax_ix_j)}$$
$$\times\prod_{i\in S}\frac{(ax_i^2,bx_i,c_ix_i,d_ix_i,e_ix_i,fx_i,gx_i,hx_i;q)_{k_i}}{(q,ax_iq/b,ax_iq/c_i,ax_iq/d_i,ax_iq/e_i,ax_iq/f,ax_iq/g,ax_iq/h;q)_{k_i}}$$
$$\left.\left.\times\prod_{i\notin S}\frac{(b^2/a,bx_i,bc_i/a,bd_i/a,be_i/a,bf/a,bg/a,bh/a;q)_{k_i}}{(q,bq/ax_i,bq/c_i,bq/d_i,bq/e_i,bq/f,bq/g,bq/h;q)_{k_i}}\cdot q^{|k|}\right)\right]$$
$$=\prod_{i=1}^{n}\frac{(ax_i^2q,b/ax_i,\lambda x_iq/f,\lambda x_iq/g,\lambda x_iq/h,bfq^{i-1}/\lambda,bgq^{i-1}/\lambda,bhq^{i-1}/\lambda;q)_\infty}{(\lambda x_i^2q,b/\lambda x_i,ax_iq/f,ax_iq/g,ax_iq/h,bfq^{i-1}/a,bgq^{i-1}/a,bhq^{i-1}/a;q)_\infty}\prod_{1\le i<j\le n}\frac{1-\lambda x_ix_j}{1-ax_ix_j}$$
$$\times\sum_{S\subseteq\{1,2,\dots,n\}}\left[\left(\frac{b}{\lambda}\right)^{\binom{n-|S|}{2}}\prod_{i\notin S}\frac{(\lambda x_i^2q,\lambda c_ix_i/a,\lambda d_ix_i/a,\lambda e_ix_i/a,fx_i,gx_i,hx_i;q)_\infty}{(\lambda x_i/b,ax_iq/c_i,ax_iq/d_i,ax_iq/e_i,\lambda x_iq/f,\lambda x_iq/g,\lambda x_iq/h;q)_\infty}\right.$$
$$\times\prod_{i\notin S}\frac{(b/\lambda x_i,abq/c_i\lambda,abq/d_i\lambda,abq/e_i\lambda,bq/f,bq/g,bq/h;q)_\infty}{(b^2q/\lambda,bc_i/a,bd_i/a,be_i/a,bf/\lambda,bg/\lambda,bh/\lambda;q)_\infty}$$
$$\times\sum_{k_1,\dots,k_n=0}^{\infty}\left(\prod_{\substack{1\le i<j\le n\\ i,j\in S}}\frac{(x_iq^{k_i}-x_jq^{k_j})(1-\lambda x_ix_jq^{k_i+k_j})}{(x_i-x_j)(1-\lambda x_ix_j)}\prod_{i\in S}\frac{1-\lambda x_i^2q^{2k_i}}{1-\lambda x_i^2}\right.$$

$$\times \prod_{\substack{1\le i<j\le n\\ i,j\notin S}} \frac{(q^{k_i}-q^{k_j})(1-b^2q^{k_i+k_j}/\lambda)}{(x_i-x_j)(1-\lambda x_ix_j)} \prod_{i\notin S} \frac{1-b^2q^{2k_i}/\lambda}{1-b^2/\lambda} \prod_{i\in S,\, j\notin S} \frac{(x_iq^{k_i}-bq^{k_j}/\lambda)(1-bx_iq^{k_i+k_j})}{(x_i-x_j)(1-\lambda x_ix_j)}$$

$$\times \prod_{i\in S} \frac{(\lambda x_i^2, bx_i, \lambda c_ix_i/a, \lambda d_ix_i/a, \lambda e_ix_i/a, fx_i, gx_i, hx_i;q)_{k_i}}{(q, \lambda x_iq/b, ax_iq/c_i, ax_iq/d_i, ax_iq/e_i, \lambda x_iq/f, \lambda x_iq/g, \lambda x_iq/h;q)_{k_i}}$$

$$\times \prod_{i\notin S} \frac{(b^2/\lambda, bx_i, bc_i/a, bd_i/a, be_i/a, bf/\lambda, bg/\lambda, bh/\lambda;q)_{k_i}}{(q, bq/\lambda x_i, abq/c_i\lambda, abq/d_i\lambda, abq/e_i\lambda, bq/f, bq/g, bq/h;q)_{k_i}} \cdot q^{|k|}\Bigg)\Bigg], \tag{5.2.48}$$

where $|q|<1$.

The identity (5.2.50) (see below) from Milne and Newcomb (2012) concerns an A_n extension of the *nonterminating Watson transformation* (cf. Gasper and Rahman, 2004, (III.36)),

$$\begin{aligned} {}_8\phi_7\left(\begin{matrix} a, qa^{\frac12}, -qa^{\frac12}, b, c, d, e, f \\ a^{\frac12}, -a^{\frac12}, aq/b, aq/c, aq/d, aq/e, aq/f \end{matrix}; q, \frac{a^2q^2}{bcdef}\right) &= \frac{(aq, aq/de, aq/df, aq/ef;q)_\infty}{(aq/d, aq/e, aq/f, aq/def;q)_\infty} \\ \times {}_4\phi_3\left(\begin{matrix} aq/bc, d, e, f \\ aq/b, aq/c, def/a \end{matrix}; q, q\right) &+ \frac{(aq, aq/bc, d, e, f, a^2q^2/bdef, a^2q^2/cdef;q)_\infty}{(aq/b, aq/c, aq/d, aq/e, aq/f, a^2q^2/bcdef, def/aq;q)_\infty} \\ \times {}_4\phi_3\left(\begin{matrix} aq/de, aq/df, aq/ef, a^2q^2/bcdefa \\ a^2q^2/bdef, a^2q^2/cdef, aq^2/def \end{matrix}; q, q\right), \end{aligned} \tag{5.2.49}$$

where $|q|<1$ and $|a^2q^2/bcdef|<1$. In the multivariate case (5.2.50), this is a transformation of a nonterminating very-well-poised A_n basic hypergeometric series into $n+1$ multiples of nonterminating balanced A_n basic hypergeometric series. It is interesting to point out that although the A_n ${}_8\phi_7$ series on the left-hand side is of the same type as that on the left-hand side of (5.2.38), this nonterminating A_n Watson transformation does *not* reduce to the terminating A_n Watson transformation (5.2.38) as the A_n ${}_4\phi_3$ series on the respective right-hand sides are of different type. Specifically,

$$\sum_{k_1,\dots,k_n\ge 0}\Bigg(\prod_{1\le i<j\le n} \frac{x_iq^{k_i}-x_jq^{k_j}}{x_i-x_j} \prod_{i,j=1}^n \frac{(f_jx_i/x_j;q)_{k_i}}{(qx_i/x_j;q)_{k_i}} \prod_{i=1}^n \frac{1-ax_iq^{k_i+|k|}}{1-ax_i} \frac{(ax_i;q)_{|k|}}{(ax_iq/f_i;q)_{|k|}}$$

$$\times \prod_{i=1}^n \frac{(bx_i, cx_i;q)_{k_i}}{(ax_iq/d, ax_iq/e;q)_{k_i}} \cdot \frac{(d,e;q)_{|k|}}{(aq/b, aq/c;q)_{|k|}} \left(\frac{a^2q^2}{bcdef_1\cdots f_n}\right)^{|k|}\Bigg)$$

$$= \frac{(aq/bf_1\cdots f_n, aq/cf_1\cdots f_n;q)_\infty}{(aq/b, aq/c;q)_\infty} \prod_{i=1}^n \frac{(ax_iq, af_iq/bcf_1\cdots f_nx_i;q)_\infty}{(aq/bcf_1\cdots f_nx_i, ax_iq/f_i;q)_\infty}$$

$$\times \sum_{k_1,\dots,k_n\ge 0}\Bigg(q^{|k|} \prod_{1\le i<j\le n} \frac{x_iq^{k_i}-x_jq^{k_j}}{x_i-x_j} \prod_{i,j=1}^n \frac{(f_jx_i/x_j;q)_{k_i}}{(qx_i/x_j;q)_{k_i}} \prod_{i=1}^n \frac{(ax_iq/de, bx_i, cx_i;q)_{k_i}}{(ax_iq/d, ax_iq/e, bcf_1\cdots f_nx_i/a;q)_{k_i}}\Bigg)$$

$$+ \frac{(q, a^2q^2/bcdf_1\cdots f_n, a^2q^2/bcef_1\cdots f_n;q)_\infty}{(a^2q^2/bcdef_1\cdots f_n, aq/b, aq/c;q)_\infty} \prod_{i=1}^n \frac{(ax_iq;q)_\infty}{(ax_iq/f_i;q)_\infty}$$

$$\times \sum_{s=1}^{n} \Bigg[q^{(n-1)k_s} \frac{(ax_sq/de, bx_s, cx_s; q)_\infty}{(bcf_1 \cdots f_n x_s/aq, ax_sq/d, ax_sq/e; q)_\infty} \prod_{i=1}^{n} \frac{(f_i x_s/x_i; q)_\infty}{(qx_s/x_i; q)_\infty} \prod_{\substack{1\le i\le n\\ i\ne s}} \frac{x_i}{x_i - x_s}$$

$$\times \sum_{k_1,\ldots,k_n\ge 0} \Bigg(\prod_{\substack{1\le i<j\le n\\ i,j\ne s}} \frac{x_i q^{k_i} - x_j q^{k_j}}{x_i - x_j} \prod_{\substack{1\le i,j\le n\\ i\ne s}} \frac{(f_j x_i/x_j; q)_{k_i}}{(qx_i/x_j; q)_{k_i}}$$

$$\times \prod_{\substack{1\le i\le n\\ i\ne s}} \frac{1 - bcf_1 \cdots f_n x_i q^{k_i - y_s - 1}/a}{1 - bcf_1 \cdots f_n x_i/aq} \frac{(ax_iq/de, bx_i, cx_i; q)_{k_i}}{(bcf_1 \cdots f_n x_i/a, ax_iq/d, ax_iq/e; q)_{k_i}}$$

$$\times \frac{(a^2q^2/bcdef_1 \cdots f_n, aq/bf_1 \cdots f_n, aq/cf_1 \cdots f_n; q)_{k_s}}{(q, a^2q^2/bcdf_1 \cdots f_n, a^2q^2/bcef_1 \cdots f_n; q)_{k_s}} \prod_{i=1}^{n} \frac{(af_iq/bcf_1 \cdots f_n x_i; q)_{k_s}}{(aq^2/bcf_1 \cdots f_n x_i; q)_{k_s}} \cdot q^{|k|} \Bigg) \Bigg], \tag{5.2.50}$$

where $|q| < 1$ and $|a^2q^2/bcdef_1 \cdots f_n| < 1$.

The $f_i = q^{-N_i}$, $i = 1, \ldots, n$ case of (5.2.50) gives a terminating A_n Watson transformation which is different from the one in (5.2.38). Milne and Newcomb (2012) obtained yet another nonterminating A_n Watson transformation.

Ito (2008) derived a BC_n extension of the Slater (1952) general transformation formula for very-well-poised balanced ${}_{2r}\psi_{2r}$ series. Some interesting and potentially useful transformations for A_n basic hypergeometric series involving nested sums were recently given by Fang (2016).

For further references to summations and transformations for basic hypergeometric series associated with root systems, see the survey by Milne (2001) and the paper by Milne and Newcomb (2012) and the references therein.

5.3 Hypergeometric and Basic Hypergeometric Integrals Associated with Root Systems

There exist a number of hypergeometric integral evaluations associated with root systems. Several of them can be viewed as extensions of Selberg's multivariate extension in 1944 of the classical beta integral evaluation,

$$\int_{[0,1]^n} \prod_{1\le i<j\le n} |z_i - z_j|^{2\gamma} \prod_{i=1}^{n} z_i^{\alpha-1}(1 - z_i)^{\beta-1} \mathrm{d}z_i = \prod_{i=1}^{n} \frac{\Gamma(\alpha + (i-1)\gamma)\Gamma(\beta + (i-1)\gamma)\Gamma(1 + i\gamma)}{\Gamma(\alpha + \beta + (n+i-2)\gamma)\Gamma(1+\gamma)}, \tag{5.3.1}$$

provided $\mathrm{Re}(\alpha) > 0$, $\mathrm{Re}(\beta) > 0$, and $\mathrm{Re}(\gamma) + \max\left(\frac{1}{n}, \mathrm{Re}\,\frac{\alpha}{n-1}, \mathrm{Re}\,\frac{\beta}{n-1}\right) > 0$. The Selberg integral is used in many areas; see Chapter 11 and Forrester and Warnaar (2008).

In 1982 Macdonald conjectured related constant term identities associated with root systems together with q-analogues. Assume R to be a reduced root system of rank n with set of positive roots R^+ and with basis of simple roots $\{\alpha_1, \ldots, \alpha_n\}$. Furthermore, let e^α ($\alpha \in R$) be the formal exponentials, which form the group ring of the lattice generated by R, and let $d_1, \ldots, d_n$

be the degrees of the fundamental invariants of the Weyl group $W(R)$. Then Macdonald (1982, Conjecture 3.1) conjectured that, for any nonnegative integer k,

$$\prod_{\alpha\in R^+}\prod_{i=1}^{k}(1-q^{i-1}e^{-\alpha})(1-q^ie^{\alpha}) \tag{5.3.2}$$

has *constant term* (i.e., the term not containing any e^{α})

$$\prod_{i=1}^{n}\frac{(q;q)_{kd_i}}{(q;q)_k(q;q)_{k(d_i-1)}}. \tag{5.3.3}$$

(For the root system A_{n-1}, this exactly corresponds to the $t=q^k$ case of the squared norm evaluation of Macdonald polynomials indexed by $\lambda=(0,\dots,0)$, the empty partition, in (5.4.1).) This conjecture can be reformulated in terms of reduced affine root systems and further strengthened. Generalizations with an extra parameter were proposed by Morris (1982). A thorough account of the historic development of q-Selberg integrals and corresponding constant term identities is provided by Forrester and Warnaar (2008, §2.3).

Of particular interest are those multiple integrals which in the univariate case reduce to the *Askey–Wilson integral* (Askey and Wilson, 1985)

$$\frac{1}{2\pi\mathrm{i}}\int_{\mathbb{T}}\frac{(z^2,1/z^2;q)_\infty}{(az,a/z,bz,b/z,cz,c/z,dz,d/z;q)_\infty}\frac{\mathrm{d}z}{z}=\frac{2(abcd;q)_\infty}{(q,ab,ac,ad,bc,bd,cd;q)_\infty}, \tag{5.3.4}$$

where $|q|<1$, $|a|<1$, $|b|<1$, $|c|<1$, $|d|<1$, and $\mathbb{T}$ is the positively oriented unit circle. The Askey–Wilson integral is responsible for the orthogonality of the Askey–Wilson polynomials, which sit at the top of the q-Askey scheme of q-orthogonal polynomials. Such multivariate integral evaluations were first obtained by Gustafson in the early 1990s. In the following we list some of the Askey–Wilson integral evaluations associated with root systems. Many of these or related integrals arise as constant term identities for (extensions of) Macdonald polynomials. This provides a natural link between the material presented here and in Chapter 9.

All these multivariate Askey–Wilson integral evaluations can be further generalized to multivariate extensions of the Nassrallah–Rahman integral evaluation (Gasper and Rahman, 2004, (6.4.1)) (which has one more parameter than the Askey–Wilson integral evaluation). The latter admit elliptic extensions. They are treated in §6.2 together with some further extensions to integral transformations.

In the following, let $\mathbb{T}^n$ be the positively oriented n-dimensional complex torus. Gustafson (1994a, Theorem 6.1) derived the following A_n Askey–Wilson integral evaluation:

$$\frac{1}{(2\pi\mathrm{i})^n}\int_{\mathbb{T}^n}\frac{\prod_{1\le i<j\le n+1}(z_i/z_j,z_j/z_i;q)_\infty}{\prod_{i,j=1}^{n+1}(a_i/z_j,b_iz_j;q)_\infty}\prod_{i=1}^{n}\frac{\mathrm{d}z_i}{z_i}$$
$$=\frac{(n+1)!\left(\prod_{i=1}^{n+1}a_ib_i;q\right)_\infty}{(q;q)_\infty^n\left(\prod_{i=1}^{n+1}a_i,\prod_{i=1}^{n+1}b_i;q\right)_\infty\prod_{i,j=1}^{n+1}(a_ib_j;q)_\infty},\qquad \prod_{i=1}^{n+1}z_i=1, \tag{5.3.5}$$

provided $|q|<1$ and $|a_i|,|b_i|<1$ $(1\le i\le n+1)$.

A considerably more complicated A_n Askey–Wilson integral evaluation, depending on the parity of n, was given by Gustafson and Rakha (2000, Theorem 1.1):

$$\frac{1}{(2\pi \mathrm{i})^n}\int_{\mathbb{T}^n}\frac{\prod_{1\le i<j\le n+1}(z_i/z_j, z_j/z_i;q)_\infty}{\prod_{i,j=1}^{n+1}(a_i/z_j;q)_\infty\prod_{1\le i<j\le n+1}(bz_iz_j;q)_\infty}\prod_{i=1}^{n+1}\frac{(S/z_i;q)_\infty}{\prod_{j=1}^{3}(bc_jz_i;q)_\infty}\frac{dz_i}{z_i}$$
$$=\begin{cases}\dfrac{(n+1)!\left(b^{(n+4)/2}\prod_{i=1}^{n+1}a_i\prod_{j=1}^{3}c_j;q\right)_\infty\prod_{i=1}^{n+1}(S/a_i;q)_\infty}{(q;q)_\infty^n\left(\prod_{i=1}^{n+1}a_i, b^{(n+4)/2}\prod_{j=1}^{3}c_j;q\right)_\infty\prod_{j=1}^{3}(b^{(m+2)/2}c_j;q)_\infty} \\ \quad\times\dfrac{\prod_{j=1}^{3}\left(b^{(n+2)/2}c_j\prod_{i=1}^{n+1}a_i;q\right)_\infty}{\prod_{i=1}^{n+1}\prod_{j=1}^{3}(ba_ic_j;q)_\infty\prod_{1\le i<j\le n+1}(ba_ia_j;q)_\infty} & \text{for } n \text{ even,} \\[2ex] \dfrac{(n+1)!\left(b^{(n+1)/2}\prod_{i=1}^{n+1}a_i;q\right)_\infty\prod_{i=1}^{n+1}(S/a_i;q)_\infty}{(q;q)_\infty^n(b^{(n+1)/2},\prod_{i=1}^{n+1}a_i;q)_\infty\prod_{i=1}^{n+1}\prod_{j=1}^{3}(ba_ic_j;q)_\infty} \\ \quad\times\dfrac{\prod_{j=1}^{3}\left(b^{(n+3)/2}\prod_{i=1}^{n+1}a_i\prod_{\substack{1\le k\le 3\\ k\ne j}}c_k;q\right)_\infty}{\prod_{1\le i<j\le 3}(b^{(n+3)/2}c_ic_j;q)_\infty\prod_{1\le i<j\le n+1}(ba_ia_j;q)_\infty} & \text{for } n \text{ odd.}\end{cases} \tag{5.3.6}$$

where $\prod_{i=1}^{n+1} z_i = 1$ and $S = b^{n+2}\prod_{i=1}^{n+1} a_i \prod_{j=1}^{3} c_j$, provided $|q| < 1$, $|b| < 1$, $|a_i| < 1$ $(1 \le i \le n+1)$, and $|c_j| < 1$ $(1 \le j \le n)$.

The following C_n Askey–Wilson integral evaluation was derived by Gustafson (1994a, Theorem 7.1):

$$\frac{1}{(2\pi \mathrm{i})^n}\int_{\mathbb{T}^n}\frac{\prod_{1\le i<j\le n}(z_i/z_j, z_j/z_i, z_iz_j, 1/z_iz_j;q)_\infty}{\prod_{i=1}^{2n+2}\prod_{j=1}^{n}(a_iz_j, a_i/z_j;q)_\infty}\prod_{i=1}^{n}(z_i^2, 1/z_i^2;q)_\infty\frac{dz_i}{z_i}$$
$$=\frac{2^n n!\left(\prod_{i=1}^{2n+2}a_i;q\right)_\infty}{(q;q)_\infty^n\prod_{1\le i<j\le 2n+2}(a_ia_j;q)_\infty},\qquad |q|<1,\ |a_i|<1\ (1\le i\le n). \tag{5.3.7}$$

Another C_n Askey–Wilson integral evaluation was given by Gustafson (1990a, (2)):

$$\frac{1}{(2\pi \mathrm{i})^n}\int_{\mathbb{T}^n}\prod_{1\le i<j\le n}\frac{(z_i/z_j, z_j/z_i, z_iz_j, 1/z_iz_j;q)_\infty}{(bz_i/z_j, bz_j/z_i, bz_iz_j, b/z_iz_j;q)_\infty}\prod_{i=1}^{n}\frac{(z_i^2, 1/z_i^2;q)_\infty}{\prod_{j=1}^{4}(a_jz_i, a_j/z_j;q)_\infty}\frac{dz_i}{z_i}$$
$$=\frac{2^n n!\,(b;q)_\infty^n}{(q;q)_\infty^n}\prod_{i=1}^{n}\frac{\left(b^{n+i-2}\prod_{j=1}^{4}a_j;q\right)_\infty}{(b^i;q)_\infty\prod_{1\le j<k\le 4}(a_ja_kb^{i-1};q)_\infty}, \tag{5.3.8}$$

provided $|q| < 1$, $|b_i| < 1$ $(1 \le i \le n)$, and $|a_j| < 1$ $(1 \le j \le 4)$. By suitably specializing the variables a_j for $1 \le j \le 4$, this multivariate integral evaluation can be used (see Gustafson, 1990a) to prove generalizations of the Macdonald conjectures due to Morris (1982) for the affine root systems C_n, $C_n^\vee$, BC_n, B_n, $B_n^\vee$, and D_n (using the classification in Macdonald, 1972).

The multivariate integral evaluation (5.3.8) explicitly describes the normalization factor of the orthogonality measure for the Macdonald–Koornwinder polynomials (see Koornwinder, 1992 and Chapter 9), the BC_n generalization of the Askey–Wilson polynomials.

Gustafson (1994a, Theorem 8.1) gave an Askey–Wilson integral evaluation for the root system G_2:

$$\frac{1}{(2\pi\mathrm{i})^2}\int_{\mathbb{T}^2}\frac{\prod_{1\le i,j\le 3, i\neq j}(z_i/z_j;q)_\infty\prod_{j=1}^{3}(z_j,1/z_j;q)_\infty}{\prod_{i=1}^{4}\prod_{j=1}^{3}(a_iz_j,a_i/z_j;q)_\infty}\frac{\mathrm{d}z_1}{z_1}\frac{\mathrm{d}z_2}{z_2}$$
$$=\frac{12\big(\prod_{i=1}^{4}a_i^2;q\big)_\infty\prod_{i=1}^{4}(a_i;q)_\infty}{(q;q)_\infty^2\big(\prod_{i=1}^{4}a_i;q\big)_\infty\prod_{1\le i\le j\le 4}(a_ia_j;q)_\infty\prod_{1\le i<j<k\le 4}(a_ia_ja_k;q)_\infty}, \tag{5.3.9}$$

where $\prod_{j=1}^{3}z_j=1$ and $|a_i|<1$ for $1\le i\le 4$.

All these basic hypergeometric integral evaluations can be specialized to ordinary hypergeometric integral evaluations by taking suitable limits. In particular, if in (5.3.5) one replaces the parameters a_i by q^{a_i} and b_i by q^{b_i}, for $1\le i\le n+1$, and then takes the limit as $q\to 1^-$, one obtains the following *multidimensional Mellin–Barnes integral* (Gustafson, 1990a, Theorem 9.1):

$$\frac{1}{(2\pi\mathrm{i})^n}\int_{-\mathrm{i}\infty}^{\mathrm{i}\infty}\cdots\int_{-\mathrm{i}\infty}^{\mathrm{i}\infty}\frac{\prod_{i,j=1}^{n+1}\Gamma(a_i-z_j)\Gamma(b_i+z_j)}{\prod_{1\le i,j\le n+1, i\neq j}\Gamma(z_i-z_j)}\prod_{i=1}^{n}\frac{\mathrm{d}z_i}{z_i}$$
$$=\frac{(n+1)!\,\Gamma(a_1+\cdots+a_{n+1})\Gamma(b_1+\cdots+b_{n+1})\prod_{i,j=1}^{n+1}\Gamma(a_i+b_j)}{\Gamma(a_1+\cdots+a_{n+1}+b_1+\cdots+b_{n+1})},\qquad \sum_{i=1}^{n+1}z_i=0, \tag{5.3.10}$$

provided $\mathrm{Re}(a_i),\mathrm{Re}(b_i)>0$ $(1\le i\le n+1)$. (For a generalization of (5.3.10), obtained by taking a suitable $q\to 1^-$ limit from an A_n Nassrallah–Rahman integral that extends (5.3.5), see Gustafson, 1992, Theorem 5.1).

We have reproduced here only a few of the many existing integral evaluations. More can be found in papers by Gustafson (1994a,b), Rains (2010), and Spiridonov and Warnaar (2011) (to list just a few relevant ones). An interesting integral transformation with F_4 symmetry has been given by van de Bult (2011). For further discussion of integral identities (evaluations and transformations) associated with root systems, where such identities are considered at the elliptic level, see §6.2.

5.4 Basic Hypergeometric Series with Macdonald Polynomial Argument

The series considered here were first introduced by Macdonald (in unpublished work Macdonald, 1989), and by Kaneko (1996). Important special cases were considered earlier. Basic hypergeometric series with Schur polynomial argument (the Schur polynomials correspond to the $q=t$ case of the Macdonald polynomials) were studied by Milne (1992) who derived ${}_1\phi_0$, ${}_2\phi_1$, and ${}_1\psi_1$ summations and several transformations for such series. Hypergeometric series with Jack polynomial argument (the Jack polynomials indexed by α correspond to the $q=t^\alpha$, $t\to 1$ specialization of the Macdonald polynomials) were studied by Constantine (1963),

Herz (1955), and Muirhead (1970) for $\alpha = 2$ (the zonal polynomial case) and for arbitrary α by Kaneko (1993), Korányi (1991), and Yan (1990, 1992).

By *Macdonald polynomials* we mean the GL_n-type symmetric Macdonald polynomials in the terminology of Chapter 9, where §§9.1.1 and 9.3.7 give a thorough treatment (for the general root system case see Macdonald, 2003a and again Chapter 9). The standard reference for Macdonald polynomials is Macdonald (1995, Chapter VI). This book also deals thoroughly with important special cases of the Macdonald polynomials, including in particular the aforementioned Schur, zonal, and Jack polynomials. See Chapter 10 of the current volume for a survey on combinatorial aspects of these multivariate polynomials.

Let Λ_n denote the *ring of symmetric functions* in the variables $z = (z_1, \ldots, z_n)$ over $\mathbb{C}$. Furthermore, we assume two nonzero generic parameters q, t satisfying $|q|, |t| < 1$. The Macdonald polynomials $P_\lambda(z; q, t)$ (often shortened to P_λ or $P_\lambda(z)$ as long as no ambiguity arises), indexed by partitions λ of length $l(\lambda) \leq n$, form an orthogonal basis of Λ_n. They can be defined as the unique family of symmetric polynomials whose expansion in terms of the monomial symmetric functions $m_\lambda(z)$ is uni-upper-triangular with respect to the dominance order $<$ of partitions,

$$P_\lambda(z; q, t) = m_\lambda(z) + \sum_{\mu<\lambda} c_{\lambda\mu}(q, t) m_\mu(z)$$

(with $c_{\lambda\mu}(q, t)$ being a rational function in q and t), being orthogonal with respect to the scalar product (Macdonald, 1995, Chapter VI, §9)

$$\langle f, g \rangle'' = \frac{1}{n!(2\pi \mathrm{i})^n} \int_{\mathbb{T}^n} f(z)\overline{g(z)}\, \Delta_{q,t}(z) \prod_{i=1}^{n} \frac{\mathrm{d}z_i}{z_i}, \qquad f, g \in \Lambda_n,$$

where

$$\Delta_{q,t}(z) := \prod_{\substack{1\leq i,j\leq n \\ i\neq j}} \frac{(z_i/z_j; q)_\infty}{(tz_i/z_j; q)_\infty}.$$

As in §5.3, $\mathbb{T}^n$ is the positively oriented n-dimensional complex torus. The squared norm evaluation of P_λ is (Macdonald, 1995, Chapter VI, §9, Example 1(d)):

$$\langle P_\lambda, P_\lambda \rangle = \prod_{1\leq i<j\leq n} \frac{(q^{\lambda_i-\lambda_j} t^{j-i}, q^{\lambda_i-\lambda_j+1} t^{j-i}; q)_\infty}{(q^{\lambda_i-\lambda_j} t^{j-i+1}, q^{\lambda_i-\lambda_j+1} t^{j-i-1}; q)_\infty}. \tag{5.4.1}$$

Macdonald (1995, Chapter VI) develops most of the theory for the polynomials $P_\lambda(z; q, t)$ using a different (albeit, up to normalization, equivalent) scalar product (which we are not displaying here) that is more algebraic in nature and does not require the conditions $|q| < 1$ and $|t| < 1$. Rather than considering symmetric functions over $\mathbb{C}$, Macdonald assumes q and t to be indeterminate and considers symmetric functions over $\mathbb{Q}(q, t)$. The above scalar product has the advantage that the structure of the root system A_{n-1} is clearly visible. This aspect of the theory generalizes to other root systems; see Macdonald (2003a) and Chapter 9.

The $P_\lambda(z;q,t)$ are homogeneous in $z=(z_1,\dots,z_n)$ of degree $|\lambda|$. They satisfy the stability property

$$P_\lambda(z_1,\dots,z_n;q,t)=P_\lambda(z_1,\dots,z_n,0;q,t).$$

Furthermore, they satisfy (Macdonald, 1995, Chapter VI, (4.17))

$$P_\lambda(z;q,t)=(z_1\cdots z_n)^{\lambda_n}P_{\lambda-\lambda_n}(z;q,t), \tag{5.4.2}$$

where $\lambda-\lambda_n:=(\lambda_1-\lambda_n,\dots,\lambda_{n-1}-\lambda_n,0)$ for any partition λ with $l(\lambda)\le n$.

For any partition λ, and $f\in\Lambda_n$ let $u_\lambda\colon\Lambda_n\to\mathbb{C}$ be the evaluation homomorphism defined by

$$u_\lambda(f(z)):=\begin{cases} f(z)\big|_{z_i=q^{\lambda_i}t^{n-i},1\le i\le n} & \text{for } l(\lambda)\le n,\\ 0 & \text{otherwise.}\end{cases}$$

The following evaluation symmetry (cf. Macdonald, 1995, Chapter VI, (6.6)), first proved by Koornwinder in unpublished work, is very useful (in particular, for interchanging summations in the process of deriving transformations):

$$u_0(P_\lambda)u_\lambda(P_\mu)=u_0(P_\mu)u_\mu(P_\lambda). \tag{5.4.3}$$

For any partition λ, let

$$(a;q,t)_\lambda:=\prod_{i\ge1}(at^{1-i};q)_{\lambda_i}\quad\text{and}\quad n(\lambda):=\sum_{i\ge1}(i-1)\lambda_i=\sum_{i\ge1}\binom{\lambda_i'}{2},$$

where λ' denotes the conjugate partition of λ. We also use the shorthand notation

$$(a_1,\dots,a_n;q,t)_\lambda:=(a_1;q,t)_\lambda\cdots(a_n;q,t)_\lambda.$$

Furthermore, for $l(\lambda)\le n$ we define

$$c_\lambda(q,t):=\prod_{i=1}^n(t^{n-i+1};q)_{\lambda_i}\prod_{1\le i<j\le n}\frac{(t^{j-i};q)_{\lambda_i-\lambda_j}}{(t^{j-i+1};q)_{\lambda_i-\lambda_j}}, \tag{5.4.4a}$$

$$c'_\lambda(q,t):=\prod_{i=1}^n(qt^{n-i};q)_{\lambda_i}\prod_{1\le i<j\le n}\frac{(qt^{j-i-1};q)_{\lambda_i-\lambda_j}}{(qt^{j-i};q)_{\lambda_i-\lambda_j}}. \tag{5.4.4b}$$

An important normalization of the Macdonald polynomials is given by

$$Q_\lambda(z;q,t):=b_\lambda(q,t)P_\lambda(z;q,t),\qquad\text{where } b_\lambda(q,t):=\frac{h_\lambda(q,t)}{c'_\lambda(q,t)}.$$

The Q_λ are exactly the polynomials dual to P_λ with respect to the scalar product mentioned right after Equation (5.4.1). The two normalizations of Macdonald polynomials appear jointly in the *Cauchy identity*

$$\sum_\lambda P_\lambda(z;q,t)Q_\lambda(y;q,t)=\prod_{i,j=1}^n\frac{(tz_iy_j;q)_\infty}{(z_iy_j;q)_\infty}. \tag{5.4.5}$$

Let a be an indeterminate and define the homomorphism $\epsilon_{a;t}\colon \Lambda_n \to \mathbb{C}[a]$ by its action on the *power sum symmetric functions* $p_r = p_r(z_1, \ldots, z_n) := \sum_{i=1}^n z_i^r$ for $r \geq 1$ (which algebraically generate Λ_n):

$$\epsilon_{a;t}(p_r) := \frac{1-a^r}{1-t^r}, \qquad r \geq 1.$$

For $a = t^n$, we have $\epsilon_{t^n;t}(f) = f(1, t, \ldots, t^{n-1})$ for any $f \in \Lambda_n$.

The following evaluations are useful (cf. Kaneko, 1996, Theorem 3.3):

$$\epsilon_{a;t}(P_\lambda(z;q,t)) = t^{n(\lambda)} \frac{(a;q,t)_\lambda}{c_\lambda(q,t)}, \qquad \epsilon_{a;t}(Q_\lambda(z;q,t)) = t^{n(\lambda)} \frac{(a;q,t)_\lambda}{c'_\lambda(q,t)}. \tag{5.4.6}$$

Basic hypergeometric series with Macdonald polynomial argument are defined as

$${}_r\Phi_s\left(\begin{matrix} a_1, \ldots, a_r \\ b_1, \ldots, b_s \end{matrix}; q, t, z\right) := \sum_\lambda \left((-1)^{|\lambda|} q^{n(\lambda')} t^{-n(\lambda)}\right)^{s+1-r} \frac{t^{n(\lambda)}}{c'_\lambda(q,t)} \frac{(a_1, \ldots, a_r; q,t)_\lambda}{(b_1, \ldots, b_s; q,t)_\lambda} P_\lambda(z;q,t), \tag{5.4.7}$$

provided that the series converges.

Application of the homomorphism $\epsilon_{a;t}$ with respect to y to both sides of the Cauchy identity (5.4.5) immediately gives the following *q-binomial theorem for Macdonald polynomials*:

$${}_1\Phi_0\left(\begin{matrix} a \\ - \end{matrix}; q, t, z\right) = \prod_{i=1}^n \frac{(az_i;q)_\infty}{(z_i;q)_\infty}, \qquad |z_i| < 1 \ (1 \leq i \leq n). \tag{5.4.8}$$

It is interesting that the right-hand side is independent of t.

Baker and Forrester (1999) made use of the q-binomial theorem for Macdonald polynomials and the evaluation symmetry (5.4.3) to derive the following *multivariate generalization of the Heine transformation*:

$${}_2\Phi_1\left(\begin{matrix} a, b \\ c \end{matrix}; q, t, xt^\delta\right) = \prod_{i=1}^n \frac{(bt^{1-i}, axt^{n-i};q)_\infty}{(ct^{1-i}, xt^{n-i};q)_\infty} \cdot {}_2\Phi_1\left(\begin{matrix} c/b, xt^{n-1} \\ axt^{n-1} \end{matrix}; q, t, bt^{1-n}t^\delta\right), \tag{5.4.9}$$

valid for $|x| < 1$ and $|bt^{1-n}| < 1$, where xt^δ stands for the argument $(x, xt, \ldots, xt^{n-1})$. Notice that this transformation involves specialized Macdonald polynomials on both sides (these factorize since $P_\lambda(xt^\delta) = x^{|\lambda|} t^{n(\lambda)} (t^n;q)_\lambda / c_\lambda(q,t)$ due to homogeneity, and the specialization (5.4.6)). A multivariate generalization of the first iterate of the Heine transformation involving unspecialized interpolation Macdonald polynomials was given by Lascoux et al. (2009, Corollary 10.2). A further extension was obtained by Lascoux and Warnaar (2011, Corollary 6.3) as a special case of multivariate extension of the q-Kummer–Thomae–Whipple transformation (Lascoux and Warnaar, 2011, Corollary 6.2).

For $x = ct^{1-n}/ab$, the right-hand side of (5.4.9) reduces to a ${}_1\Phi_0$ series which can be summed using (5.4.8). This gives a multivariate extension of the *q-Gauss summation*

$${}_2\Phi_1\left(\begin{matrix} a, b \\ c \end{matrix}; q, t, \frac{ct^{1-n}}{ab} t^\delta\right) = \prod_{i=1}^n \frac{(ct^{1-i}/b, ct^{1-i}/a;q)_\infty}{(ct^{1-i}/ab, ct^{1-i};q)_\infty}, \qquad |ct^{1-n}/ab|. \tag{5.4.10}$$

More general q-Gauss summations involving unspecialized (non)symmetric Macdonald polynomials were given by Lascoux et al. (2009), and by Lascoux and Warnaar (2011, Corollary 5.4).

For general unspecialized argument $z=(z_1,\dots,z_n)$, Baker and Forrester (1999), building on work of Kaneko (1996), proved the following *multivariate extension of the Euler transformation* (or equivalently, of the second iterate of the Heine transformation):

$$ {}_2\Phi_1\left(\begin{matrix}a,b\\ c\end{matrix};q,t,z\right)=\prod_{i=1}^{n}\frac{(abz_i/c;q)_\infty}{(z_i;q)_\infty}\cdot {}_2\Phi_1\left(\begin{matrix}c/a,c/b\\ c\end{matrix};q,t,abz/c\right), \tag{5.4.11}$$

valid for $|z_i|<1$ and $|abz_i/c|<1$ $(1\le i\le n)$. A nonsymmetric extension was given by Lascoux et al. (2009, Corollary 10.3).

We list two other results from Baker and Forrester (1999) for N a nonnegative integer. The *q-Pfaff–Saalschütz summation* for basic hypergeometric series with specialized Macdonald polynomial argument is

$$ {}_3\Phi_2\left(\begin{matrix}a,b,q^{-N}\\ c,abq^{1-N}t^{n-1}/c\end{matrix};q,t,qt^{\delta}\right)=\prod_{i=1}^{n}\frac{(ct^{1-i}/a,ct^{1-i}/b;q)_N}{(ct^{1-i},ct^{1-i}/ab;q)_N}. \tag{5.4.12}$$

This can be generalized to a *multivariate Sears' transformation with specialized Macdonald polynomial arguments*,

$$ \begin{aligned}&{}_4\Phi_3\left(\begin{matrix}a,b,c,q^{-N}\\ d,e,ft^{n-1}\end{matrix};q,t,qt^{\delta}\right)\\ &=a^{nN}\prod_{i=1}^{n}\frac{(et^{1-i}/a,ft^{n-i}/a;q)_N}{(et^{1-i},ft^{n-i};q)_N}\cdot {}_4\Phi_3\left(\begin{matrix}a,d/b,d/c,q^{-N}\\ d,aq^{1-N}t^{n-1}/e,aq^{1-N}/f\end{matrix};q,t,qt^{\delta}\right),\end{aligned} \tag{5.4.13}$$

where $def=abcq^{1-N}$. Rains (2005, §4) proved extensions of (5.4.12) and (5.4.13) for Macdonald polynomials indexed by partitions of skew shape. Extensions of (5.4.12) and (5.4.13) to nonsymmetric Macdonald polynomials were given by Lascoux et al. (2009, Theorem 6.6 and Proposition 6.8).

Kaneko (1996) developed q-difference equations for basic hypergeometric series with Macdonald polynomial argument and related them to q-Selberg integrals. Warnaar (2005) proved various generalizations of q-Selberg integral evaluations and constant term identities, including a q-analogue of the Hua–Kadell formula for Jack polynomials (cf. Hua, 1979, Theorem 5.2.1 and Kadell, 1993, Theorem 2). Rains and Warnaar (2015, §5.3) includes further multivariate ${}_4\Phi_3$ transformations.

We end this section with a multivariate extension of Ramanujan's ${}_1\psi_1$ summation formula due to Kaneko (1998), which is a t-extension of an earlier result by Milne (1992) (namely, for basic hypergeometric series with Schur function argument).

Let $\mathbb{Z}^n_{\geq}:=\{(\lambda_1,\dots,\lambda_n)\in\mathbb{Z}^n\mid \lambda_1\ge\cdots\ge\lambda_n\}$. By (5.4.2), the Macdonald polynomials $P_\lambda(z;q,t)$ can be defined for any $\lambda\in\mathbb{Z}^n_{\geq}$. Bilateral basic hypergeometric series with Macdonald polynomial argument are defined as

$$ {}_r\Psi_{s+1}\left(\begin{matrix}a_1,\dots,a_r\\ b,b_1,\dots,b_s\end{matrix};q,t,z\right):=\sum_{\lambda\in\mathbb{Z}^n_{\geq}}\left(\left((-1)^{|\lambda|}q^{n(\lambda')}t^{-n(\lambda)}\right)^{s+1-r}\frac{(q;q)^n_\infty}{(b;q)^n_\infty}\prod_{i=1}^{n}\frac{(bq^{\lambda_i}t^{n-i};q)_\infty}{(q^{\lambda_i+1}t^{n-i};q)_\infty}\right.$$

$$\times \frac{t^{n(\lambda)}}{c'_\lambda(q,t)} \frac{(a_1,\ldots,a_r;q,t)_\lambda}{(b_1,\ldots,b_s;q,t)_\lambda} P_\lambda(z;q,t)\Bigg), \tag{5.4.14}$$

provided that the series converges. With this notation, Kaneko's ${}_1\psi_1$ summation for Macdonald polynomials is

$${}_1\Psi_1\left(\begin{matrix}a\\b\end{matrix};q,t,z\right) = \prod_{i=1}^{n} \frac{(qt^{n-i}, bt^{1-i}/a, az_i, q/az_i;q)_\infty}{(bqt^{1-i}, qt^{n-i}/a, z_i, bt^{1-n}/az_i;q)_\infty}, \quad |bt^{1-n}/a| < |z_i| < 1 \ (1 \le i \le n). \tag{5.4.15}$$

The result by Warnaar (2013, Theorem 2.6) is a generalization of (5.4.15) which involves a pair of Macdonald polynomials in two independent sets of variables. Other identities of this type were obtained by Warnaar (2008).

5.5 Remarks on Applications

As mentioned in the introduction, hypergeometric series associated with root systems first arose in the context of $3j$- and $6j$-symbols for the unitary groups (Ališauskas et al., 1972; Chacón et al., 1972; Holman et al., 1976). This initiated their study and that of their basic analogues from a pure mathematics point of view.

Basic hypergeometric series associated with root systems have found applications in various areas. We list a few occurrences, making no claim about completeness. First of all, such series, in particular multivariate ${}_6\psi_6$ summations associated with root systems, were used to give elementary proofs of the Macdonald identities (Gustafson, 1989; Milne, 1985a). More generally, these series were used for deriving expansions of various special powers of the eta function (Bartlett and Warnaar, 2015; Leininger and Milne, 1999a,b; Milne, 2000; Warnaar and Zudilin, 2012) and for establishing infinite families of exact formulae for sums of squares and of triangular numbers (Milne, 2002; Rosengren, 2007, 2008b). Basic hypergeometric series associated with root systems were also employed in the enumeration of plane partitions (Gessel and Krattenthaler, 1997; Krattenthaler and Schlosser, 2014; Rosengren, 2008a). Applications to Macdonald polynomials were given by Kajihara and Noumi (2000) and Schlosser (2007). Basic hypergeometric integrals associated with root systems were used in the construction of BC_n orthogonal polynomials and BC_n biorthogonal rational functions that generalize the Macdonald polynomials; see Koornwinder (1992) and Rains (2005). Watson transformations (and related transformations) associated to root systems were used by Bartlett (2013), Bartlett and Warnaar (2015), Coskun (2008), Griffin et al. (2016), and Zhang and Wu (2015) to derive multiple Rogers–Ramanujan identities and characters for affine Lie algebras. For applications to quantum groups, see Rosengren (2011). Basic hypergeometric series of Macdonald polynomial argument were used to construct Selberg-type integrals for A_{n-1} (Warnaar, 2009). Also, hypergeometric series with Jack and zonal polynomial argument appeared in studies on random matrices (Desrosiers and Liu, 2015; Forrester and Rains, 2009) and Selberg integrals (Kaneko, 1993; Korányi, 1991).

Very recently the subject has gained growing attention from physicists working in spin models and in quantum field theory. In particular, it was shown by Derkachov and Manashov (2017) and Derkachov et al. (2017) that Gustafson's multivariate hypergeometric integrals appear naturally in the integrable spin models. Furthermore, it was shown (Hama et al., 2011; Jafferis, 2012) that the partition functions in $3d$ field theories can be expressed in terms of specific basic hypergeometric integrals. As made explicit by Dolan et al. (2011), these partition functions can also be obtained by reduction from the $4d$ superconformal indices which, according to Dolan and Osborn (2009), can be identified with elliptic hypergeometric integrals. (The latter are reviewed in Chapter 6.) Accordingly, multivariate basic hypergeometric integrals and series associated with various symmetry groups (or gauge groups, in the terminology of quantum field theory) appear as explicit expressions for the respective partition functions (Dolan et al., 2011; Spiridonov and Vartanov, 2011, 2012). Several of these are new and await further mathematical study. These partition functions can also be interpreted as solutions of the Yang–Baxter equation; see Gahramanov (2015) and Gahramanov and Spiridonov (2015).

Acknowledgements I would like to thank Gaurav Bhatnagar, Tom Koornwinder, Stephen Milne, Hjalmar Rosengren, Jasper Stokman, and Ole Warnaar for careful reading and many valuable comments. The author's research was partially supported by Austrian Science Fund grant F50-08.

References

Ališauskas, S. J., Jucys, A.-A. A., and Jucys, A. P. 1972. On the symmetric tensor operators of the unitary groups. *J. Math. Phys.*, **13**, 1329–1333.

Askey, R., and Wilson, J. 1985. *Some Basic Hypergeometric Orthogonal Polynomials that Generalize Jacobi Polynomials.* Mem. Amer. Math. Soc., vol. 54, no. 319.

Baker, T. H., and Forrester, P. J. 1999. Transformation formulas for multivariable basic hypergeometric series. *Methods Appl. Anal.*, **6**, 147–164.

Bartlett, N. 2013. *Hall–Littlewood polynomials and characters of affine Lie algebras.* Ph.D. thesis, University of Queensland.

Bartlett, N., and Warnaar, S. O. 2015. Hall–Littlewood polynomials and characters of affine Lie algebras. *Adv. Math.*, **285**, 1066–1105.

Bhatnagar, G. 1999. D_n basic hypergeometric series. *Ramanujan J.*, **3**, 175–203.

Bhatnagar, G. 2019. Heine's method and A_n to A_m transformation formulas. *Ramanujan J.*, **48**, 191–215.

Bhatnagar, G., and Schlosser, M. 1998. C_n and D_n very-well-poised ${}_{10}\phi_9$ transformations. *Constr. Approx.*, **14**, 531–567.

Bhatnagar, G., and Schlosser, M. J. 2018. Elliptic well-poised Bailey transforms and lemmas on root systems. *SIGMA*, **14**, 025, 44 pp.

van de Bult, F. J. 2011. An elliptic hypergeometric integral with $W(F_4)$ symmetry. *Ramanujan J.*, **25**, 1–20.

Chacón, E., Ciftan, M., and Biedenharn, L. C. 1972. On the evaluation of the multiplicity-free Wigner coefficients of $U(n)$. *J. Math. Phys.*, **13**, 577–590.

Constantine, A. G. 1963. Some non-central distribution problems in multivariate analysis. *Ann. Math. Statist.*, **34**, 1270–1285.

Coskun, H. 2008. An elliptic BC_n Bailey lemma, multiple Rogers–Ramanujan identities and Euler's pentagonal number theorems. *Trans. Amer. Math. Soc.*, **360**, 5397–5433.

Denis, R. Y., and Gustafson, R. A. 1992. An SU(n) q-beta integral transformation and multiple hypergeometric series identities. *SIAM J. Math. Anal.*, **23**, 552–561.

Derkachov, S. É., and Manashov, A. N. 2017. Spin chains and Gustafson's integrals. *J. Phys. A*, **50**, Paper 294006, 20 pp.

Derkachov, S. É., Manashov, A. N., and Valinevich, P. A. 2017. Gustafson integrals for $SL(2, \mathbb{C})$ spin magnet. *J. Phys. A*, **50**, Paper 294007, 12 pp.

Desrosiers, P., and Liu, D.-Z. 2015. Selberg integrals, super-hypergeometric functions and applications to β-ensembles of random matrices. *Random Matrices Theory Appl.*, **4**, Paper 1550007, 59 pp.

van Diejen, J. F. 1997. On certain multiple Bailey, Rogers and Dougall type summation formulas. *Publ. Res. Inst. Math. Sci.*, **33**, 483–508.

van Diejen, J. F., and Spiridonov, V. P. 2000. An elliptic Macdonald–Morris conjecture and multiple modular hypergeometric sums. *Math. Res. Lett.*, **7**, 729–746.

Dolan, F. A., and Osborn, H. 2009. Applications of the superconformal index for protected operators and q-hypergeometric identities to $\mathcal{N} = 1$ dual theories. *Nuclear Phys. B*, **818**, 137–178.

Dolan, F. A. H., Spiridonov, V. P., and Vartanov, G. S. 2011. From 4d superconformal indices to 3d partition functions. *Phys. Lett. B*, **704**, 234–241.

Dougall, J. 1907. On Vandermonde's theorem, and some more general expansions. *Proc. Edinb. Math. Soc.*, **25**, 114–132.

Fang, J.-P. 2016. Generalizations of Milne's $U(n + 1)$ q-Chu-Vandermonde summation. *Czechoslovak Math. J.*, **66**, 395–407.

Forrester, P. J., and Rains, E. M. 2009. Matrix averages relating to Ginibre ensembles. *J. Phys. A*, **42**, Paper 385205, 13 pp.

Forrester, P. J., and Warnaar, S. O. 2008. The importance of the Selberg integral. *Bull. Amer. Math. Soc. (N.S.)*, **45**, 489–534.

Gahramanov, I. 2015. Mathematical structures behind supersymmetric dualities. *Arch. Math. (Brno)*, **51**, 273–286.

Gahramanov, I., and Spiridonov, V. P. 2015. The star-triangle relation and $3d$ superconformal indices. *J. High Energy Phys.*, Paper 040, 22 pp.

Gasper, G., and Rahman, M. 2004. *Basic Hypergeometric Series*. Second edn. Encyclopedia of Mathematics and Its Applications, vol. 96. Cambridge University Press.

Gessel, I. M., and Krattenthaler, C. 1997. Cylindric partitions. *Trans. Amer. Math. Soc.*, **349**, 429–479.

Griffin, M. J., Ono, K., and Warnaar, S. O. 2016. A framework of Rogers–Ramanujan identities and their arithmetic properties. *Duke Math. J.*, **165**, 1475–1527.

Gustafson, R. A. 1987a. Multilateral summation theorems for ordinary and basic hypergeometric series in U(n). *SIAM J. Math. Anal.*, **18**, 1576–1596.

Gustafson, R. A. 1987b. A Whipple's transformation for hypergeometric series in U(n) and multivariable hypergeometric orthogonal polynomials. *SIAM J. Math. Anal.*, **18**, 495–530.

Gustafson, R. A. 1989. The Macdonald identities for affine root systems of classical type and hypergeometric series very-well-poised on semisimple Lie algebras. Pages 185–224 of: *Ramanujan International Symposium on Analysis (Pune, 1987)*. Macmillan of India, New Delhi.

Gustafson, R. A. 1990a. A generalization of Selberg's beta integral. *Bull. Amer. Math. Soc. (N.S.)*, **22**, 97–105.

Gustafson, R. A. 1990b. A summation theorem for hypergeometric series very-well-poised on G_2. *SIAM J. Math. Anal.*, **21**, 510–522.

Gustafson, R. A. 1992. Some q-beta and Mellin–Barnes integrals with many parameters associated to the classical groups. *SIAM J. Math. Anal.*, **23**, 525–551.

Gustafson, R. A. 1994a. Some q-beta and Mellin–Barnes integrals on compact Lie groups and Lie algebras. *Trans. Amer. Math. Soc.*, **341**, 69–119.

Gustafson, R. A. 1994b. Some q-beta integrals on SU(n) and Sp(n) that generalize the Askey-Wilson and Nasrallah-Rahman integrals. *SIAM J. Math. Anal.*, **25**, 441–449.

Gustafson, R. A., and Krattenthaler, C. 1996. Heine transformations for a new kind of basic hypergeometric series in $U(n)$. *J. Comput. Appl. Math.*, **68**, 151–158.

Gustafson, R. A., and Krattenthaler, C. 1997. Determinant evaluations and U(n) extensions of Heine's ${}_2\phi_1$-transformations. Pages 83–89 of: *Special Functions, q-Series and Related Topics*. Fields Inst. Commun., vol. 14. Amer. Math. Soc.

Gustafson, R. A., and Rakha, M. A. 2000. q-Beta integrals and multivariate basic hypergeometric series associated to root systems of type A_m. *Ann. Comb.*, **4**, 347–373.

Hama, N., Hosomichi, K., and Lee, S. 2011. Notes on SUSY gauge theories on three-sphere. *J. High Energy Phys.*, Paper 127, 14 pp.

Herz, C. S. 1955. Bessel functions of matrix argument. *Ann. of Math. (2)*, **61**, 474–523.

Holman, III, W. J. 1980. Summation theorems for hypergeometric series in U(n). *SIAM J. Math. Anal.*, **11**, 523–532.

Holman, III, W. J., Biedenharn, L. C., and Louck, J. D. 1976. On hypergeometric series well-poised in SU(n). *SIAM J. Math. Anal.*, **7**, 529–541.

Hua, L. K. 1979. *Harmonic Analysis of Functions of Several Complex Variables in the Classical Domains*. Translations of Mathematical Monographs, vol. 6. Amer. Math. Soc. Translated from the Russian, which was a translation of the Chinese original.

Ito, M. 2002. A product formula for Jackson integral associated with the root system F_4. *Ramanujan J.*, **6**, 279–293.

Ito, M. 2003. Symmetry classification for Jackson integrals associated with the root system BC_n. *Compos. Math.*, **136**, 209–216.

Ito, M. 2008. A multiple generalization of Slater's transformation formula for a very-well-poised-balanced ${}_{2r}\psi_{2r}$ series. *Quart. J. Math.*, **59**, 221–235.

Ito, M., and Tsubouchi, A. 2010. Bailey type summation formulas associated with the root system $G_2^\vee$. *Ramanujan J.*, **22**, 231–248.

Jafferis, D. L. 2012. The exact superconformal R-symmetry extremizes Z. *J. High Energy Phys.*, Paper 159, 20 pp.

Kac, V. G. 1990. *Infinite-dimensional Lie Algebras*. Third edn. Cambridge University Press.

Kadell, K. W. J. 1993. An integral for the product of two Selberg-Jack symmetric polynomials. *Compos. Math.*, **87**, 5–43.

Kajihara, Y. 2004. Euler transformation formula for multiple basic hypergeometric series of type A and some applications. *Adv. Math.*, **187**, 53–97.

Kajihara, Y. 2014. Symmetry groups of A_n hypergeometric series. *SIGMA*, **10**, Paper 026, 29 pp.

Kajihara, Y. 2016. Transformation formulas for bilinear sums of basic hypergeometric series. *Canad. Math. Bull.*, **59**, 136–143.

Kajihara, Y., and Noumi, M. 2000. Raising operators of row type for Macdonald polynomials. *Compos. Math.*, **120**, 119–136.

Kaneko, J. 1993. Selberg integrals and hypergeometric functions associated with Jack polynomials. *SIAM J. Math. Anal.*, **24**, 1086–1110.

Kaneko, J. 1996. q-Selberg integrals and Macdonald polynomials. *Ann. Sci. École Norm. Sup. (4)*, **29**, 583–637.

Kaneko, J. 1998. A ${}_1\Psi_1$ summation theorem for Macdonald polynomials. *Ramanujan J.*, **2**, 379–386.

Komori, Y., Masuda, Y., and Noumi, M. 2016. Duality transformation formulas for multiple elliptic hypergeometric series of type *BC*. *Constr. Approx.*, **44**, 483–516.

Koornwinder, T. H. 1992. Askey–Wilson polynomials for root systems of type *BC*. Pages 189–204 of: *Hypergeometric Functions on Domains of Positivity, Jack Polynomials, and Applications*. Contemp. Math., vol. 138. Amer. Math. Soc.

Korányi, A. 1991. Hua-type integrals, hypergeometric functions and symmetric polynomials. Pages 169–180 of: *International Symposium in Memory of Hua Loo Keng, Vol. II*. Springer.

Krattenthaler, C. 2001. Proof of a summation formula for an $\tilde{A}_n$ basic hypergeometric series conjectured by Warnaar. Pages 153–161 of: *q-Series with Applications to Combinatorics, Number Theory, and Physics*. Contemp. Math., vol. 291. Amer. Math. Soc.

Krattenthaler, C., and Schlosser, M. J. 2014. The major index generating function of standard Young tableaux of shapes of the form "staircase minus rectangle". Pages 111–122 of: *Ramanujan 125*. Contemp. Math., vol. 627. Amer. Math. Soc.

Lascoux, A., and Warnaar, S. O. 2011. Branching rules for symmetric functions and $\mathfrak{sl}_n$ basic hypergeometric series. *Adv. in Appl. Math.*, **46**, 424–456.

Lascoux, A., Rains, E. M., and Warnaar, S. O. 2009. Nonsymmetric interpolation Macdonald polynomials and $\mathfrak{gl}_n$ basic hypergeometric series. *Transform. Groups*, **14**, 613–647.

Leininger, V. E., and Milne, S. C. 1999a. Expansions for $(q)_\infty^{n^2+2n}$ and basic hypergeometric series in $U(n)$. *Discrete Math.*, **204**, 281–317.

Leininger, V. E., and Milne, S. C. 1999b. Some new infinite families of η-function identities. *Methods Appl. Anal.*, **6**, 225–248.

Lilly, G. M., and Milne, S. C. 1993. The C_l Bailey transform and Bailey lemma. *Constr. Approx.*, **9**, 473–500.

Macdonald, I. G. 1972. Affine root systems and Dedekind's η-function. *Invent. Math.*, **15**, 91–143.

Macdonald, I. G. 1982. Some conjectures for root systems. *SIAM J. Math. Anal.*, **13**, 988–1007.

Macdonald, I. G. 1989. *Hypergeometric functions II (q-analogues)*. Unpublished manuscript, available at arXiv:1309.5208 (2013).

Macdonald, I. G. 1995. *Symmetric Functions and Hall Polynomials*. Second edn. Oxford University Press.

Macdonald, I. G. 2003a. *Affine Hecke Algebras and Orthogonal Polynomials*. Cambridge Tracts in Mathematics, vol. 157. Cambridge University Press.

Macdonald, I. G. 2003b. A formal identity for affine root systems. Pages 195–211 of: *Lie Groups and Symmetric Spaces*. Amer. Math. Soc. Transl. Ser. 2, vol. 210. Amer. Math. Soc.

Masuda, Y. 2013. Kernel identities for van Diejen's q-difference operators and transformation formulas for multiple basic hypergeometric series. *Ramanujan J.*, **32**, 281–314.

Milne, S. C. 1980. Hypergeometric series well-poised in SU(n) and a generalization of Biedenharn's G-functions. *Adv. Math.*, **36**, 169–211.

Milne, S. C. 1985a. An elementary proof of the Macdonald identities for $A_l^{(1)}$. *Adv. Math.*, **57**, 34–70.

Milne, S. C. 1985b. A q-analog of hypergeometric series well-poised in SU(n) and invariant G-functions. *Adv. Math.*, **58**, 1–60.

Milne, S. C. 1985c. A q-analog of the ${}_5F_4(1)$ summation theorem for hypergeometric series well-poised in SU(n). *Adv. Math.*, **57**, 14–33.

Milne, S. C. 1986. A U(n) generalization of Ramanujan's ${}_1\Psi_1$ summation. *J. Math. Anal. Appl.*, **118**, 263–277.

Milne, S. C. 1988a. Multiple q-series and U(n) generalizations of Ramanujan's ${}_1\Psi_1$ sum. Pages 473–524 of: *Ramanujan Revisited*. Academic Press.

Milne, S. C. 1988b. A q-analog of the Gauss summation theorem for hypergeometric series in U(n). *Adv. Math.*, **72**, 59–131.

Milne, S. C. 1989. The multidimensional ${}_1\Psi_1$ sum and Macdonald identities for $A_l^{(1)}$. Pages 323–359 of: *Theta Functions*. Proc. Sympos. Pure Math., vol. 49, Part 2. Amer. Math. Soc.

Milne, S. C. 1992. Summation theorems for basic hypergeometric series of Schur function argument. Pages 51–77 of: *Progress in Approximation Theory*. Springer.

Milne, S. C. 1994. A q-analog of a Whipple's transformation for hypergeometric series in U(n). *Adv. Math.*, **108**, 1–76.

Milne, S. C. 1997. Balanced ${}_3\phi_2$ summation theorems for U(n) basic hypergeometric series. *Adv. Math.*, **131**, 93–187.

Milne, S. C. 2000. A new U(n) generalization of the Jacobi triple product identity. Pages 351–370 of: *q-Series From a Contemporary Perspective*. Contemp. Math., vol. 254. Amer. Math. Soc.

Milne, S. C. 2001. Transformations of U($n+1$) multiple basic hypergeometric series. Pages 201–243 of: *Physics and Combinatorics 1999*. World Scientific.

Milne, S. C. 2002. Infinite families of exact sums of squares formulas, Jacobi elliptic functions, continued fractions, and Schur functions. *Ramanujan J.*, **6**, 7–149.

Milne, S. C., and Lilly, G. M. 1995. Consequences of the A_l and C_l Bailey transform and Bailey lemma. *Discrete Math.*, **139**, 319–346.

Milne, S. C., and Newcomb, J. W. 1996. U(n) very-well-poised ${}_{10}\phi_9$ transformations. *J. Comput. Appl. Math.*, **68**, 239–285.

Milne, S. C., and Newcomb, J. W. 2012. Nonterminating q-Whipple transformations for basic hypergeometric series in $U(n)$. Pages 181–224 of: *Partitions, q-Series, and Modular Forms*. Dev. Math., vol. 23. Springer.

Milne, S. C., and Schlosser, M. 2002. A new A_n extension of Ramanujan's ${}_1\psi_1$ summation with applications to multilateral A_n series. *Rocky Mountain J. Math.*, **32**, 759–792.

Morris, W. G. 1982. *Constant term identities for finite and affine root systems: conjectures and theorems*. Ph.D. thesis, Univ. of Wisconsin–Madison.

Muirhead, R. J. 1970. Systems of partial differential equations for hypergeometric functions of matrix argument. *Ann. Math. Statist.*, **41**, 991–1001.

Rains, E. M. 2005. BC$_n$-symmetric polynomials. *Transform. Groups*, **10**, 63–132.

Rains, E. M. 2010. Transformations of elliptic hypergeometric integrals. *Ann. of Math. (2)*, **171**, 169–243.

Rains, E. M., and Warnaar, S. O. 2015. *Bounded Littlewood identities*. arXiv:1506.02755; Mem. Amer. Math. Soc., in press.

Rosengren, H. 2001. A proof of a multivariable elliptic summation formula conjectured by Warnaar. Pages 193–202 of: *q-Series with Applications to Combinatorics, Number Theory, and Physics*. Contemp. Math., vol. 291. Amer. Math. Soc.

Rosengren, H. 2003. Karlsson–Minton type hypergeometric functions on the root system C_n. *J. Math. Anal. Appl.*, **281**, 332–345.

Rosengren, H. 2004a. Elliptic hypergeometric series on root systems. *Adv. Math.*, **181**, 417–447.

Rosengren, H. 2004b. Reduction formulas for Karlsson–Minton-type hypergeometric functions. *Constr. Approx.*, **20**, 525–548.

Rosengren, H. 2006. New transformations for elliptic hypergeometric series on the root system A_n. *Ramanujan J.*, **12**, 155–166.

Rosengren, H. 2007. Sums of triangular numbers from the Frobenius determinant. *Adv. Math.*, **208**, 935–961.

Rosengren, H. 2008a. Schur Q-polynomials, multiple hypergeometrical series and enumeration of marked shifted tableaux. *J. Combin. Theory Ser. A*, **115**, 376–406.

Rosengren, H. 2008b. Sums of squares from elliptic Pfaffians. *Int. J. Number Theory*, **4**, 873–902.

Rosengren, H. 2011. Felder's elliptic quantum group and elliptic hypergeometric series on the root system A_n. *Int. Math. Res. Not.*, 2861–2920.

Rosengren, H. 2017. Gustafson–Rakha-type elliptic hypergeometric series. *SIGMA*, **13**, Paper 037, 11 pp.

Rosengren, H., and Schlosser, M. 2003. Summations and transformations for multiple basic and elliptic hypergeometric series by determinant evaluations. *Indag. Math. (N.S.)*, **14**, 483–513.

Schlosser, M. 1997. Multidimensional matrix inversions and A_r and D_r basic hypergeometric series. *Ramanujan J.*, **1**, 243–274.

Schlosser, M. 1999. Some new applications of matrix inversions in A_r. *Ramanujan J.*, **3**, 405–461.

Schlosser, M. 2000. Summation theorems for multidimensional basic hypergeometric series by determinant evaluations. *Discrete Math.*, **210**, 151–169.

Schlosser, M. 2003. A multidimensional generalization of Shukla's ${}_8\psi_8$ summation. *Constr. Approx.*, **19**, 163–178.

Schlosser, M. 2005. Abel–Rothe type generalizations of Jacobi's triple product identity. Pages 383–400 of: *Theory and Applications of Special Functions*. Dev. Math., vol. 13. Springer.

Schlosser, M. J. 2007. Macdonald polynomials and multivariable basic hypergeometric series. *SIGMA*, **3**, 056, 30 pp.

Schlosser, M. 2008. A new multivariable ${}_6\psi_6$ summation formula. *Ramanujan J.*, **17**, 305–319.

Selberg, A. 1944. Bemerkinger om et multipelt integral (in Norwegian). *Norsk Mat. Tidsskr.*, **26**, 71–78.

Slater, L. J. 1952. General transformations of bilateral series. *Quart. J. Math., Oxford Ser. (2)*, **3**, 73–80.

Slater, L. J. 1966. *Generalized Hypergeometric Functions*. Cambridge University Press.

Spiridonov, V. P., and Vartanov, G. S. 2011. Elliptic hypergeometry of supersymmetric dualities. *Comm. Math. Phys.*, **304**, 797–874.

Spiridonov, V. P., and Vartanov, G. S. 2012. Superconformal indices of $\mathcal{N} = 4$ SYM field theories. *Lett. Math. Phys.*, **100**, 97–118.

Spiridonov, V. P., and Warnaar, S. O. 2011. New multiple ${}_6\psi_6$ summation formulas and related conjectures. *Ramanujan J.*, **25**, 319–342.

Warnaar, S. O. 2002. Summation and transformation formulas for elliptic hypergeometric series. *Constr. Approx.*, **18**, 479–502.

Warnaar, S. O. 2005. q-Selberg integrals and Macdonald polynomials. *Ramanujan J.*, **10**, 237–268.

Warnaar, S. O. 2008. Bisymmetric functions, Macdonald polynomials and $\mathfrak{sl}_3$ basic hypergeometric series. *Compos. Math.*, **144**, 271–303.

Warnaar, S. O. 2009. A Selberg integral for the Lie algebra A_n. *Acta Math.*, **203**, 269–304.

Warnaar, S. O. 2010. The $\mathfrak{sl}_3$ Selberg integral. *Adv. Math.*, **224**, 499–524.

Warnaar, S. O. 2013. Ramanujan's ${}_1\psi_1$ summation. *Notices Amer. Math. Soc.*, **60**(1), 18–22.

Warnaar, S. O., and Zudilin, W. 2012. Dedekind's η-function and Rogers-Ramanujan identities. *Bull. London Math. Soc.*, **44**, 1–11.

Whittaker, E. T., and Watson, G. N. 1927. *A Course of Modern Analysis*. Fourth edn. Cambridge University Press.

Yan, Z. M. 1990. Generalized hypergeometric functions. *C. R. Acad. Sci. Paris Sér. I Math.*, **310**, 349–354.

Yan, Z. M. 1992. A class of generalized hypergeometric functions in several variables. *Canad. J. Math.*, **44**, 1317–1338.

Zhang, Z., and Wu, Y. 2015. A $U(n+1)$ Bailey lattice. *J. Math. Anal. Appl.*, **426**, 747–764.

6

Elliptic Hypergeometric Functions Associated with Root Systems

Hjalmar Rosengren and S. Ole Warnaar

6.1 Introduction

Let $f = \sum_{n\geq 0} c_n$. The series f is called *hypergeometric* if the ratio c_{n+1}/c_n, viewed as a function of n, is rational. A simple example is the Taylor series $\exp(z) = \sum_{n=0}^{\infty} z^n/n!$. Similarly, if the ratio of consecutive terms of f is a rational function of q^n for some fixed q – known as the *base* – then f is called a *basic hypergeometric series* or *q-hypergeometric series*. An early example of a basic hypergeometric series is Euler's q-exponential function $e_q(z) = \sum_{n\geq 0} z^n/((1-q)\cdots(1-q^n))$. If we express the base as $q = \exp(2\pi\mathrm{i}/\omega)$ then c_{n+1}/c_n becomes a trigonometric function in n, with period ω. This motivates the more general definition of an *elliptic hypergeometric series* as a series f for which c_{n+1}/c_n is a doubly periodic meromorphic function of n.

Elliptic hypergeometric series first appeared in 1988 in the work of Date et al. on exactly solvable lattice models in statistical mechanics (Date et al., 1988). They were formally defined and identified as mathematical objects of interest in their own right by Frenkel and Turaev (1997). Subsequently, Spiridonov (2001) introduced the elliptic beta integral, initiating a parallel theory of elliptic hypergeometric integrals. Together with Zhedanov he also showed (Spiridonov, 2004; Spiridonov and Zhedanov, 2000a,b) that the theory of biorthogonal rational functions (Rahman, 1986; Wilson, 1991) – itself a generalization of the Askey scheme (Koekoek et al., 2010) of classical orthogonal polynomials – can be lifted to the elliptic level.

All three aspects of the theory of elliptic hypergeometric functions (series, integrals and biorthogonal functions) have been generalized to higher dimensions, connecting them to root systems and Macdonald–Koornwinder theory. Warnaar (2002) introduced elliptic hypergeometric series associated with root systems, including a conjectural series evaluation of type C_n. This was recognized by van Diejen and Spiridonov (2000, 2001a) as a discrete analogue of a multiple elliptic beta integral (or elliptic Selberg integral). They formulated the corresponding integral evaluation, again as a conjecture. This in turn led Rains (2006, 2010) to develop an elliptic analogue of Macdonald–Koornwinder theory, resulting in continuous as well as discrete biorthogonal elliptic functions attached to the non-reduced root system BC_n. In this theory, the elliptic multiple beta integral and its discrete analogue give the total mass of the biorthogonality measure.

Although a relatively young field, the theory of elliptic hypergeometric functions has already seen some remarkable applications. Many of these involve the multivariable theory. Dolan and Osborn (2009) showed that supersymmetric indices of four-dimensional supersymmetric quantum field theories are expressible in terms of elliptic hypergeometric integrals. Conjecturally, such field theories admit electric–magnetic dualities known as Seiberg dualities, such that dual theories have the same index. This leads to non-trivial identities between elliptic hypergeometric integrals (or, for so-called confining theories, to integral evaluations). In some cases these are known identities, which thus gives a partial confirmation of the underlying Seiberg duality. However, in many cases it leads to new identities that are yet to be rigorously proved; see e.g. Gadde et al. (2010a,b), Spiridonov and Vartanov (2011, 2012, 2014) and the recent survey Rastelli and Razamat (2017). Another application of elliptic hypergeometric functions is to exactly solvable lattice models in statistical mechanics. We already mentioned the occurrence of elliptic hypergeometric series in the work of Date et al., but more recently it was shown that elliptic hypergeometric integrals are related to solvable lattice models with continuous spin parameters (Bazhanov and Sergeev, 2012b; Spiridonov, 2012). In the one-variable case, this leads to a generalization of many well-known discrete models such as the two-dimensional Ising model and the chiral Potts model. This relation to solvable lattice models has been extended to multivariable elliptic hypergeometric integrals by Bazhanov and Sergeev (2012a), Bazhanov et al. (2013) and Spiridonov (2012). Further applications of multivariable elliptic hypergeometric functions pertain to elliptic Calogero–Sutherland-type systems (Razamat, 2014; Spiridonov, 2007a) and the representation theory of elliptic quantum groups (Rosengren, 2011).

In the current chapter we give a survey of elliptic hypergeometric functions associated with root systems, comprised of three main parts. The first two form in essence an annotated table of the main evaluation and transformation formulas for elliptic hypergeometric integrals and series on root systems. The third and final part gives an introduction to Rains' elliptic Macdonald–Koornwinder theory (in part also developed by Coskun and Gustafson, 2006). Due to space limitations, applications will not be covered here and we refer the interested reader to the above-mentioned papers and references therein.

Rather than throughout the text, references for the main results are given in the form of separate notes at the end of each section. These notes also contain some brief historical comments and further pointers to the literature.

6.1.1 Preliminaries

Elliptic functions are doubly periodic meromorphic functions on $\mathbb{C}$. That is, a meromorphic function $g\colon \mathbb{C} \to \mathbb{C}$ is *elliptic* if there exist ω_1, ω_2 with $\mathrm{Im}(\omega_1/\omega_2) > 0$ such that $g(z + \omega_1) = g(z + \omega_2) = g(z)$ for all $z \in \mathbb{C}$. If we define the *elliptic nome* p by $p = \mathrm{e}^{2\pi \mathrm{i}\omega_1/\omega_2}$ (so that $|p| < 1$) then $z \mapsto \mathrm{e}^{2\pi \mathrm{i}z/\omega_2}$ maps the period parallelogram spanned by ω_1, ω_2 to an annulus with radii $|p|$ and 1. Given an elliptic function g with periods ω_1 and ω_2, the function $f\colon \mathbb{C}^* \to \mathbb{C}$ defined by $g(z) = f(\mathrm{e}^{2\pi \mathrm{i}z/\omega_2})$ is thus periodic in an annulus: $f(pz) = f(z)$. By mild abuse of terminology we will also refer to such an f as an elliptic function. A more precise description would be that f is an elliptic function in multiplicative form.

The basic building blocks for elliptic hypergeometric functions are

$$\theta(z) = \theta(z;p) := \prod_{i=0}^{\infty}(1-zp^i)(1-p^{i+1}/z),$$

$$(z)_k = (z;q,p)_k := \prod_{i=0}^{k-1}\theta(zq^i;p),$$

$$\Gamma(z) = \Gamma(z;p,q) := \prod_{i,j=0}^{\infty}\frac{1-p^{i+1}q^{j+1}/z}{1-zp^iq^j},$$

known as the *modified theta function*, *elliptic shifted factorial* and *elliptic gamma function*, respectively. Note that the dependence on the elliptic nome p and base q will mostly be suppressed from our notation. One exception is the q-shifted factorial $(z;q)_\infty := \prod_{i\geq 0}(1-zq^i)$ which, to avoid possible confusion, will never be shortened to $(z)_\infty$.

For simple relations satisfied by the above three functions we refer the reader to Gasper and Rahman (2004). Here we only note that the elliptic gamma function is symmetric in p and q and satisfies

$$\Gamma(pq/z)\Gamma(z) = 1 \quad \text{and} \quad \Gamma(qz) = \theta(z)\Gamma(z).$$

For each of the functions $\theta(z)$, $(z)_k$ and $\Gamma(z)$, we employ condensed notation as exemplified by

$$\begin{aligned}\theta(z_1,\dots,z_m) &= \theta(z_1)\cdots\theta(z_m),\\ (az^{\pm})_k &= (az)_k(a/z)_k,\\ \Gamma(tz^{\pm}w^{\pm}) &= \Gamma(tzw)\Gamma(tz/w)\Gamma(tw/z)\Gamma(t/zw).\end{aligned}$$

In the trigonometric case $p=0$ we have $\theta(z) = 1-z$, so that $(z)_k$ becomes a standard q-shifted factorial and $\Gamma(z)$ a rescaled version of the q-gamma function.

We also need elliptic shifted factorials indexed by partitions. A *partition* $\lambda = (\lambda_1,\lambda_2,\dots)$ is a weakly decreasing sequence of non-negative integers such that only finitely many λ_i are non-zero. The number of positive λ_i is called the *length* of λ and denoted by $l(\lambda)$. The sum of the λ_i will be denoted by $|\lambda|$. The *diagram* of λ consists of the points $(i,j)\in\mathbb{Z}^2$ such that $1\leq i\leq l(\lambda)$ and $1\leq j\leq\lambda_i$. If these inequalities hold for $(i,j)\in\mathbb{Z}^2$ we write $(i,j)\in\lambda$. Reflecting the diagram in the main diagonal yields the *conjugate partition* λ'. In other words, the rows of λ are the columns of λ' and vice versa. A standard statistic on partitions is

$$n(\lambda) := \sum_{i\geq 1}(i-1)\lambda_i = \sum_{i\geq 1}\binom{\lambda_i'}{2}.$$

For a pair of partitions λ, μ we write $\mu\subset\lambda$ if $\mu_i\leq\lambda_i$ for all $i\geq 1$. In particular, when $l(\lambda)\leq n$ and $\lambda_i\leq N$ for all $1\leq i\leq n$ we write $\lambda\subset(N^n)$. Similarly, we write $\mu\prec\lambda$ if the interlacing conditions $\lambda_1\geq\mu_1\geq\lambda_2\geq\mu_2\geq\cdots$ hold.

With t an additional fixed parameter, we will need the following three types of *elliptic shifted factorials indexed by partitions*:

$$(z)_\lambda = (z;q,t;p)_\lambda := \prod_{(i,j)\in\lambda} \theta(zq^{j-1}t^{1-i}) = \prod_{i\geq 1} (zt^{1-i})_{\lambda_i},$$
$$C^-_\lambda(z) = C^-_\lambda(z;q,t;p) := \prod_{(i,j)\in\lambda} \theta(zq^{\lambda_i-j}t^{\lambda'_j-i}),$$
$$C^+_\lambda(z) = C^+_\lambda(z;q,t;p) := \prod_{(i,j)\in\lambda} \theta(zq^{\lambda_i+j-1}t^{2-\lambda'_j-i}).$$

By $\theta(pz) = -z^{-1}\theta(z)$ it follows that $(a)_\lambda$ is quasi-periodic:

$$(p^k z)_\lambda = \left[(-z)^{-|\lambda|} q^{-n(\lambda')} t^{n(\lambda)}\right]^k p^{-\binom{k}{2}|\lambda|} (z)_\lambda, \quad k \in \mathbb{Z}. \tag{6.1.1}$$

Again we use condensed notation so that, for example, $(a_1,\dots,a_k)_\lambda = (a_1)_\lambda \cdots (a_k)_\lambda$.

6.1.2 Elliptic Weyl Denominators

Suppressing their p-dependence we define

$$\Delta^{\mathrm{A}}(x_1,\dots,x_{n+1}) := \prod_{1\leq i<j\leq n+1} x_j\,\theta(x_i/x_j),$$
$$\Delta^{\mathrm{C}}(x_1,\dots,x_n) := \prod_{j=1}^{n} \theta(x_j^2) \prod_{1\leq i<j\leq n} x_j\,\theta(x_i x_j^{\pm}),$$

which are essentially the *Weyl denominators* of the affine root systems $\mathrm{A}_n^{(1)}$ and $\mathrm{C}_n^{(1)}$ (Kac, 1990; Macdonald, 1972). Although we have no need for the theory of affine root systems here, it may be instructive to explain the connection to the root system $\mathrm{C}_n^{(1)}$ (the case of $\mathrm{A}_n^{(1)}$ is similar). The Weyl denominator of an affine root system R is the formal product $\prod_{\alpha\in R_+}(1-\mathrm{e}^{-\alpha})^{m(\alpha)}$, where R_+ denotes the set of positive roots and m is a multiplicity function. For $\mathrm{C}_n^{(1)}$, the positive roots are

$$\begin{aligned}
&m\delta, && m\geq 1,\\
&m\delta+2\varepsilon_i, && m\geq 0,\ 1\leq i\leq n,\\
&m\delta-2\varepsilon_i, && m\geq 1,\ 1\leq i\leq n,\\
&m\delta+\varepsilon_i\pm\varepsilon_j, && m\geq 0,\ 1\leq i<j\leq n,\\
&m\delta-\varepsilon_i\pm\varepsilon_j, && m\geq 1,\ 1\leq i<j\leq n,
\end{aligned}$$

where $\varepsilon_1,\dots,\varepsilon_n$ are the coordinate functions on $\mathbb{R}^n$ and δ is the constant function 1. The roots $m\delta$ have multiplicity n, while all other roots have multiplicity 1. Thus, the Weyl denominator for $\mathrm{C}_n^{(1)}$ is

$$\prod_{m=0}^{\infty}\Bigg((1-\mathrm{e}^{-(m+1)\delta})^n\prod_{i=1}^{n}(1-\mathrm{e}^{-m\delta-2\varepsilon_i})(1-\mathrm{e}^{-(m+1)\delta+2\varepsilon_i})$$
$$\times\prod_{1\le i<j\le n}(1-\mathrm{e}^{-m\delta-\varepsilon_i-\varepsilon_j})(1-\mathrm{e}^{-(m+1)\delta+\varepsilon_i+\varepsilon_j})(1-\mathrm{e}^{-m\delta-\varepsilon_i+\varepsilon_j})(1-\mathrm{e}^{-(m+1)\delta+\varepsilon_i-\varepsilon_j})\Bigg).$$

It is easy to check that this equals $(p;p)_\infty^n x_1^0x_2^{-1}\cdots x_n^{1-n}\Delta^{\mathrm{C}}(x_1,\dots,x_n)$, where $p=\mathrm{e}^{-\delta}$ and $x_i=\mathrm{e}^{-\varepsilon_i}$.

We will consider elliptic hypergeometric series containing the factor $\Delta^{\mathrm{A}}(xq^k)$ or $\Delta^{\mathrm{C}}(xq^k)$, where $xq^k:=(x_1q^{k_1},x_2q^{k_2},\dots,x_rq^{k_r})$, the $k_i\in\mathbb{Z}$ being summation indices, and $r=n+1$ in the case of A_n and $r=n$ in the case of C_n. We refer to these as A_n *series* and C_n *series*, respectively. In the case of A_n, the summation variables typically satisfy a restriction of the form $k_1+\dots+k_{n+1}=N$. Eliminating k_{n+1} gives series containing the $\mathrm{A}_{n-1}^{(1)}$ Weyl denominator times $\prod_{i=1}^n\theta(ax_iq^{k_i+|k|})$, where $a=q^{-N}/x_{n+1}$; these will also be viewed as A_n series.

Similarly, A_n *integrals* contain the factor

$$\frac{1}{\prod_{1\le i<j\le n+1}\Gamma(z_i/z_j,z_j/z_i)},\tag{6.1.2a}$$

where $z_1\cdots z_{n+1}=1$, while C_n *integrals* contain

$$\frac{1}{\prod_{i=1}^n\Gamma(z_i^{\pm 2})\prod_{1\le i<j\le n}\Gamma(z_i^{\pm}z_j^{\pm})}.\tag{6.1.2b}$$

If we denote the expression (6.1.2b) by $g(z)$ then it is easy to verify that, for $k\in\mathbb{Z}^n$,

$$\frac{g(zq^k)}{g(z)}=\left(\prod_{i=1}^n q^{-nk_i-(n+1)k_i^2}z_i^{-2(n+1)k_i}\right)\frac{\Delta^{\mathrm{C}}(zq^k)}{\Delta^{\mathrm{C}}(z)}.\tag{6.1.3}$$

A similar relation holds for the A-type factors. This shows that the series can be considered as discrete analogues of the integrals. In fact, in many instances the series can be obtained from the integrals via residue calculus.

It is customary to attach a "type" to hypergeometric integrals associated with root systems, although different authors have used slightly different definitions of type. As the terminology will be used here, in *type-I integrals* the only factors containing more than one integration variable are (6.1.2), while *type-II integrals* contain twice the number of such factors. For example, $\mathrm{C}_n^{(\mathrm{II})}$ integrals contain the factor $\prod_{i<j}\Gamma(tz_i^{\pm}z_j^{\pm})/\Gamma(z_i^{\pm}z_j^{\pm})$. It may be noted that, under appropriate assumptions on the parameters,

$$\lim_{q\to1}\lim_{p\to0}\prod_{i<j}\frac{\Gamma(q^{t\pm z_i\pm z_j})}{\Gamma(q^{\pm z_i\pm z_j})}=\lim_{q\to1}\prod_{i<j}\frac{(q^{\pm z_i\pm z_j};q)_\infty}{(q^{t\pm z_i\pm z_j};q)_\infty}$$
$$=\prod_{i<j}(1-z_i^{\pm}z_j^{\pm})^t=\prod_{i<j}\left((z_i+z_i^{-1})-(z_j+z_j^{-1})\right)^{2t}.$$

For this reason, C_n beta integrals of type II are sometimes referred to as *elliptic Selberg integrals*. There are also integrals containing an intermediate number of factors. We will refer to these as *integrals of mixed type*.

6.2 Integrals

Throughout this section we assume that $|q| < 1$. Whenever possible, we have restricted the parameters in such a way that the integrals may be taken over the n-dimensional complex torus $\mathbb{T}^n$. However, all results can be extended to more general parameter domains by appropriately deforming $\mathbb{T}^n$. When $n = 1$ all the stated A_n and C_n beta integral evaluations reduce to Spiridonov's elliptic beta integral (Spiridonov, 2001).

6.2.1 A_n *Beta Integrals*

We will present four A_n beta integrals. In each of these the integrand contains a variable z_{n+1} which is determined from the integration variables $z_1, \dots, z_n$ by the relation $z_1 \cdots z_{n+1} = 1$. To shorten the expressions we define the constant κ_n^{A} by

$$\kappa_n^{\mathrm{A}} := \frac{(p;p)_\infty^n (q;q)_\infty^n}{(n+1)!\,(2\pi \mathrm{i})^n}.$$

For $1 \le i \le n+2$, let $|s_i| < 1$ and $|t_i| < 1$, such that $ST = pq$, where $S = s_1 \cdots s_{n+2}$ and $T = t_1 \cdots t_{n+2}$. Then we have the type-I integral

$$\kappa_n^{\mathrm{A}} \int_{\mathbb{T}^n} \frac{\prod_{i=1}^{n+2} \prod_{j=1}^{n+1} \Gamma(s_i z_j, t_i/z_j)}{\prod_{1 \le i < j \le n+1} \Gamma(z_i/z_j, z_j/z_i)} \frac{\mathrm{d}z_1}{z_1} \cdots \frac{\mathrm{d}z_n}{z_n} = \prod_{i=1}^{n+2} \Gamma(S/s_i, T/t_i) \prod_{i,j=1}^{n+2} \Gamma(s_i t_j). \tag{6.2.1}$$

Next, let $|s| < 1$, $|t| < 1$, $|s_i| < 1$ and $|t_i| < 1$ for $1 \le i \le 3$, such that $s^{n-1} t^{n-1} s_1 s_2 s_3 t_1 t_2 t_3 = pq$. Then we have the type-II integral

$$\kappa_n^{\mathrm{A}} \int_{\mathbb{T}^n} \prod_{1 \le i < j \le n+1} \frac{\Gamma(s z_i z_j, t/z_i z_j)}{\Gamma(z_i/z_j, z_j/z_i)} \prod_{i=1}^{3} \prod_{j=1}^{n+1} \Gamma(s_i z_j, t_i/z_j) \frac{\mathrm{d}z_1}{z_1} \cdots \frac{\mathrm{d}z_n}{z_n}$$

$$= \begin{cases} \displaystyle \prod_{m=1}^{N} \Big(\Gamma(s^m t^m) \prod_{1 \le i < j \le 3} \Gamma(s^{m-1} t^m s_i s_j, s^m t^{m-1} t_i t_j) \prod_{i,j=1}^{3} \Gamma(s^{m-1} t^{m-1} s_i t_j) \Big) \\ \displaystyle \quad \times \Gamma(s^{N-1} s_1 s_2 s_3, t^{N-1} t_1 t_2 t_3) \prod_{i=1}^{3} \Gamma(s^N s_i, t^N t_i), & n = 2N, \\ \displaystyle \prod_{m=1}^{N} \Big(\Gamma(s^m t^m) \prod_{1 \le i < j \le 3} \Gamma(s^{m-1} t^m s_i s_j, s^m t^{m-1} t_i t_j) \Big) \prod_{m=1}^{N+1} \prod_{i,j=1}^{3} \Gamma(s^{m-1} t^{m-1} s_i t_j) \\ \displaystyle \quad \times \Gamma(s^{N+1}, t^{N+1}) \prod_{1 \le i < j \le 3} \Gamma(s^N s_i s_j, t^N t_i t_j), & n = 2N+1. \end{cases} \tag{6.2.2}$$

Let $|t| < 1$, $|t_i| < 1$ for $1 \le i \le n+3$ and $|t| < |s_i| < |t|^{-1}$ for $1 \le i \le n$, where $t^2 t_1 \cdots t_{n+3} = pq$. Then

$$\kappa_n^{\mathrm{A}} \int_{\mathbb{T}^n} \prod_{1\le i<j\le n+1} \frac{1}{\Gamma(z_i/z_j, z_j/z_i, t^2 z_i z_j)} \prod_{j=1}^{n+1}\Big(\prod_{i=1}^{n}\Gamma(ts_i^{\pm}z_j)\prod_{i=1}^{n+3}\Gamma(t_i/z_j)\Big)\frac{\mathrm{d}z_1}{z_1}\cdots\frac{\mathrm{d}z_n}{z_n}$$
$$= \prod_{i=1}^{n}\prod_{j=1}^{n+3}\Gamma(ts_i^{\pm}t_j)\prod_{1\le i<j\le n+3}\frac{1}{\Gamma(t^2 t_i t_j)}, \tag{6.2.3}$$

which is an integral of mixed type.

Finally, let $|t| < 1$, $|s_i| < 1$ for $1 \le i \le 4$ and $|t_i| < 1$ for $1 \le i \le n+1$ such that $t^{n-1}s_1\cdots s_4 T = pq$, where $T = t_1\cdots t_{n+1}$. Then we have a second mixed-type integral:

$$\kappa_n^{\mathrm{A}} \int_{\mathbb{T}^n} \prod_{1\le i<j\le n+1} \frac{\Gamma(tz_i z_j)}{\Gamma(z_i/z_j, z_j/z_i)} \prod_{j=1}^{n+1}\Big(\prod_{i=1}^{4}\Gamma(s_i z_j)\prod_{i=1}^{n+1}\Gamma(t_i/z_j)\Big)\frac{\mathrm{d}z_1}{z_1}\cdots\frac{\mathrm{d}z_n}{z_n}$$
$$= \begin{cases} \Gamma(T)\displaystyle\prod_{i=1}^{4}\frac{\Gamma(t^N s_i)}{\Gamma(t^N T s_i)}\prod_{1\le i<j\le n+1}\Gamma(tt_i t_j)\prod_{i=1}^{4}\prod_{j=1}^{n+1}\Gamma(s_i t_j), & n = 2N,\\ \displaystyle\frac{\Gamma(t^{N+1}, T)}{\Gamma(t^{N+1}T)}\prod_{1\le i<j\le 4}\Gamma(t^N s_i s_j)\prod_{1\le i<j\le n+1}\Gamma(tt_i t_j)\prod_{i=1}^{4}\prod_{j=1}^{n+1}\Gamma(s_i t_j), & n = 2N+1. \end{cases} \tag{6.2.4}$$

6.2.2 C_n *Beta Integrals*

We will give three C_n beta integrals. They all involve the constant

$$\kappa_n^{\mathrm{C}} := \frac{(p;p)_\infty^n (q;q)_\infty^n}{n!\,2^n(2\pi \mathrm{i})^n}.$$

Let $|t_i| < 1$ for $1 \le i \le 2n+4$ such that $t_1\cdots t_{2n+4} = pq$. We then have the following C_n beta integral of type I:

$$\kappa_n^{\mathrm{C}} \int_{\mathbb{T}^n} \prod_{1\le i<j\le n}\frac{1}{\Gamma(z_i^{\pm}z_j^{\pm})}\prod_{j=1}^{n}\frac{\prod_{i=1}^{2n+4}\Gamma(t_i z_j^{\pm})}{\Gamma(z_j^{\pm 2})}\frac{\mathrm{d}z_1}{z_1}\cdots\frac{\mathrm{d}z_n}{z_n} = \prod_{1\le i<j\le 2n+4}\Gamma(t_i t_j). \tag{6.2.5}$$

Next, let $|t| < 1$ and $|t_i| < 1$ for $1 \le i \le 6$ such that $t^{2n-2}t_1\cdots t_6 = pq$. We then have the type-II C_n beta integral

$$\kappa_n^{\mathrm{C}} \int_{\mathbb{T}^n} \prod_{1\le i<j\le n}\frac{\Gamma(tz_i^{\pm}z_j^{\pm})}{\Gamma(z_i^{\pm}z_j^{\pm})}\prod_{j=1}^{n}\frac{\prod_{i=1}^{6}\Gamma(t_i z_j^{\pm})}{\Gamma(z_j^{\pm 2})}\frac{\mathrm{d}z_1}{z_1}\cdots\frac{\mathrm{d}z_n}{z_n} = \prod_{m=1}^{n}\Big(\frac{\Gamma(t^m)}{\Gamma(t)}\prod_{1\le i<j\le 6}\Gamma(t^{m-1}t_i t_j)\Big). \tag{6.2.6}$$

This is the elliptic Selberg integral mentioned in the introduction.

At this point it is convenient to introduce notation for more general C_n integrals of type II. For m a non-negative integer, let $|t| < 1$ and $|t_i| < 1$ for $1 \le i \le 2m+6$ such that

$$t^{2n-2}t_1\cdots t_{2m+6} = (pq)^{m+1}. \tag{6.2.7}$$

We then define

$$J^{(m)}_{\mathrm{C}_n}(t_1,\dots,t_{2m+6};t) := \kappa^{\mathrm{C}}_n \int_{\mathbb{T}^n} \prod_{1\le i<j\le n} \frac{\Gamma(tz_i^{\pm}z_j^{\pm})}{\Gamma(z_i^{\pm}z_j^{\pm})} \prod_{j=1}^{n} \frac{\prod_{i=1}^{2m+6}\Gamma(t_iz_j^{\pm})}{\Gamma(z_j^{\pm 2})} \frac{\mathrm{d}z_1}{z_1}\cdots\frac{\mathrm{d}z_n}{z_n}. \tag{6.2.8}$$

Note that (6.2.6) gives a closed-form evaluation for the integral $J^{(0)}_{\mathrm{C}_n}$. As outlined in the appendix to Rains' paper (2010), $J^{(m)}_{\mathrm{C}_n}$ can be continued to a single-valued meromorphic function in the parameters t_i and t subject to the constraint (6.2.7). For generic values of the parameters this continuation is obtained by replacing the integration domain with an appropriate deformation of $\mathbb{T}^n$. We can now state the second C_n beta integral of type II as

$$\begin{aligned} &J^{(n-1)}_{\mathrm{C}_n}(t_1,\dots,t_4,s_1,\dots,s_n,pq/ts_1,\dots,pq/ts_n;t) \\ &\quad = \Gamma(t)^n \prod_{l=1}^{n} \prod_{1\le i<j\le 4} \Gamma(t^{l-1}t_it_j) \prod_{i=1}^{n}\prod_{j=1}^{4} \frac{\Gamma(s_it_j)}{\Gamma(ts_i/t_j)}, \end{aligned} \tag{6.2.9}$$

where $t^{n-2}t_1t_2t_3t_4 = 1$. In this identity it is necessary to work with an analytic continuation of (6.2.8) since the inequalities $|t_i|, |t| < 1$ are incompatible with $t^{n-2}t_1t_2t_3t_4 = 1$ for $n \ge 2$.

6.2.3 Integral Transformations

We now turn to integral transformations, starting with integrals of type I. For m a non-negative integer we introduce the notation

$$I^{(m)}_{\mathrm{A}_n}(s_1,\dots,s_{m+n+2};t_1,\dots,t_{m+n+2}) := \kappa^{\mathrm{A}}_n \int_{\mathbb{T}^n} \frac{\prod_{i=1}^{m+n+2}\prod_{j=1}^{n+1}\Gamma(s_iz_j, t_i/z_j)}{\prod_{1\le i<j\le n+1}\Gamma(z_i/z_j, z_j/z_i)} \frac{\mathrm{d}z_1}{z_1}\cdots\frac{\mathrm{d}z_n}{z_n},$$

where $|s_i| < 1$ and $|t_i| < 1$ for all i, $\prod_{i=1}^{m+n+2} s_it_i = (pq)^{m+1}$ and $z_1\cdots z_{n+1} = 1$. We also define

$$I^{(m)}_{\mathrm{C}_n}(t_1,\dots,t_{2m+2n+4}) := \kappa^{\mathrm{C}}_n \int_{\mathbb{T}^n} \prod_{1\le i<j\le n} \frac{1}{\Gamma(z_i^{\pm}z_j^{\pm})} \prod_{j=1}^{n} \frac{\prod_{i=1}^{2m+2n+4}\Gamma(t_iz_j^{\pm})}{\Gamma(z_j^{\pm 2})} \frac{\mathrm{d}z_1}{z_1}\cdots\frac{\mathrm{d}z_n}{z_n},$$

where $|t_i| < 1$ for all i and $t_1\cdots t_{2m+2n+4} = (pq)^{m+1}$. The A_n integral satisfies

$$I^{(m)}_{\mathrm{A}_n}(s_1,\dots,s_{m+n+2};t_1,\dots,t_{m+n+2}) = I^{(m)}_{\mathrm{A}_n}(s_1\zeta,\dots,s_{m+n+2}\zeta;t_1/\zeta,\dots,t_{m+n+2}/\zeta)$$

for ζ any $(n+1)$th root of unity, whereas the C_n integral is invariant under simultaneous negation of all of the t_i. We further note that (6.2.1) and (6.2.5) provide closed-form evaluations of $I^{(0)}_{\mathrm{A}_n}$ and $I^{(0)}_{\mathrm{C}_n}$, respectively.

For the integral $I^{(m)}_{\mathrm{A}_n}$, the following transformation reverses the roles of m and n:

$$\begin{aligned} &I^{(m)}_{\mathrm{A}_n}(s_1,\dots,s_{m+n+2};t_1,\dots,t_{m+n+2}) \\ &\quad = \prod_{i,j=1}^{m+n+2} \Gamma(s_it_j) \cdot I^{(n)}_{\mathrm{A}_m}\Big(\frac{\lambda}{s_1},\dots,\frac{\lambda}{s_{m+n+2}};\frac{pq}{\lambda t_1},\dots,\frac{pq}{\lambda t_{m+n+2}}\Big), \end{aligned} \tag{6.2.10}$$

where $\lambda^{m+1} = s_1 \cdots s_{m+n+2}$, $(pq/\lambda)^{m+1} = t_1 \cdots t_{m+n+2}$. Moreover, for $t_1 \cdots t_{2m+2n+4} = (pq)^{m+1}$, there is an analogous transformation of type C:

$$I^{(m)}_{C_n}(t_1, \dots, t_{2m+2n+4}) = \prod_{1\le i<j\le 2m+2n+4} \Gamma(t_it_j) \cdot I^{(n)}_{C_m}\left(\frac{\sqrt{pq}}{t_1}, \dots, \frac{\sqrt{pq}}{t_{2m+2n+4}}\right). \tag{6.2.11}$$

It is easy to check that $I^{(m)}_{A_1}(t_1, \dots, t_{m+3}; t_{m+4}, \dots, t_{2m+6}) = I^{(m)}_{C_1}(t_1, \dots, t_{2m+6})$. Thus, combining (6.2.10) and (6.2.11) leads to

$$\begin{aligned} I^{(1)}_{A_n}(s_1, \dots, s_{n+3}; t_1, \dots, t_{n+3}) = &\prod_{1\le i<j\le n+3} \Gamma(S/s_is_j, T/t_it_j) \\ &\times I^{(1)}_{C_n}(s_1/v, \dots, s_{n+3}/v, t_1v, \dots, t_{n+3}v), \end{aligned} \tag{6.2.12}$$

where $S = s_1 \cdots s_{n+3}$, $T = t_1 \cdots t_{n+3}$ and $v^2 = S/pq = pq/T$. Since $I^{(1)}_{C_n}$ is symmetric, (6.2.12) implies non-trivial symmetries of $I^{(1)}_{A_n}$, such as

$$\begin{aligned} I^{(1)}_{A_n}(s_1, \dots, s_{n+3}; t_1, \dots, t_{n+3}) = &\prod_{i=1}^{n+2} \Gamma(s_it_{n+3}, t_is_{n+3}, S/s_is_{n+3}, T/t_it_{n+3}) \\ &\times I^{(1)}_{A_n}(s_1/v, \dots, s_{n+2}/v, s_{n+3}v^n; t_1v, \dots, t_{n+2}v, t_{n+3}/v^n), \end{aligned} \tag{6.2.13}$$

where, with the same definitions of S and T as above, $ST = (pq)^2$ and $v^{n+1} = St_{n+3}/pqs_{n+3}$.

Recalling the notation (6.2.8), we have the type-II C_n integral transformation

$$\begin{aligned} J^{(1)}_{C_n}(t_1, \dots, t_8; t) = &\prod_{m=1}^{n}\Bigg(\prod_{1\le i<j\le 4} \Gamma(t^{m-1}t_it_j) \prod_{5\le i<j\le 8} \Gamma(t^{m-1}t_it_j)\Bigg) \\ &\times J^{(1)}_{C_n}(t_1v, \dots, t_4v, t_5/v, \dots, t_8/v; t), \end{aligned} \tag{6.2.14}$$

where $t^{2n-2}t_1 \cdots t_8 = (pq)^2$ and $v^2 = pqt^{1-n}/t_1t_2t_3t_4 = t^{n-1}t_5t_6t_7t_8/pq$. Iterating this transformation yields a symmetry of $J^{(1)}_{C_n}$ under the Weyl group of type E_7 (Rains, 2010).

We conclude with a transformation between C_n and C_m integrals of type II:

$$\begin{aligned} &J^{(m+n-1)}_{C_n}(t_1, \dots, t_4, s_1, \dots, s_{m+n}, pq/ts_1, \dots, pq/ts_{m+n}; t) \\ &\quad = \Gamma(t)^{n-m} \prod_{1\le i<j\le 4} \frac{\prod_{l=1}^{n} \Gamma(t^{l-1}t_it_j)}{\prod_{l=1}^{m} \Gamma(t^{l+n-m-1}t_it_j)} \prod_{i=1}^{m+n}\prod_{j=1}^{4} \frac{\Gamma(s_it_j)}{\Gamma(ts_i/t_j)} \\ &\quad\quad \times J^{(m+n-1)}_{C_m}(t/t_1, \dots, t/t_4, s_1, \dots, s_{m+n}, pq/ts_1, \dots, pq/ts_{m+n}; t), \end{aligned} \tag{6.2.15}$$

where $t_1t_2t_3t_4 = t^{m-n+2}$.

6.2.4 Notes

For $p = 0$ the integrals (6.2.1), (6.2.2), (6.2.5) and (6.2.6) are due to Gustafson (1992, 1994), the integral (6.2.4) to Gustafson and Rakha (2000) and the transformation (6.2.13) to Denis and Gustafson (1992). None of the $p = 0$ instances of (6.2.3), (6.2.9)–(6.2.12), (6.2.14) and (6.2.15) were known prior to the elliptic case.

For general p, van Diejen and Spiridonov (2000, 2001a) conjectured the type-I C_n beta integral (6.2.5) and showed that it implies the elliptic Selberg integral (6.2.6). A rigorous derivation of the classical Selberg integral as a special limit of (6.2.6) is due to Rains (2009). Spiridonov (2004) conjectured the type-I A_n beta integral (6.2.1) and showed that, combined with (6.2.5), it implies the type-II A_n beta integral (6.2.2), as well as the integral (6.2.4) of mixed type. He also showed that (6.2.1) implies (6.2.13). The first proofs of the fundamental type-I integrals (6.2.1) and (6.2.5) were obtained by Rains (2010). For subsequent proofs of (6.2.1), (6.2.5) and (6.2.6), see Spiridonov (2007b), Rains and Spiridonov (2009)/Spiridonov (2007b) and Ito and Noumi (2017b), respectively. Rains (2010) also proved the integral transformations (6.2.10), (6.2.11) and (6.2.14), and gave further transformations analogous to (6.2.14). The integral (6.2.3) of mixed type is due to Spiridonov and Warnaar (2006). The transformation (6.2.15), which includes (6.2.9) as its $m = 0$ case, was conjectured by Rains (2012) and also appears in Spiridonov and Vartanov (2011). It was first proved by van de Bult (2009) and subsequently proved and generalized to an identity for the "interpolation kernel" (an analytic continuation of the elliptic interpolation functions R^*_λ of §6.4) by Rains (2018).

Several of the integral identities surveyed here have analogues for $|q| = 1$. The cases of (6.2.1), (6.2.3) and (6.2.5) were found by van Diejen and Spiridonov (2005), who also stated the unit-circle analogue of (6.2.6).

Spiridonov (2004) gave one more C_n beta integral, which lacks the $p \leftrightarrow q$ symmetry present in all the integrals considered here, and is more elementary in that it follows as a determinant of one-variable beta integrals.

Rains (2012) conjectured several quadratic integral transformations involving the interpolation functions R^*_λ. These conjectures were proved in van de Bult (2011) and Rains (2018). In special cases, they simplify to transformations for the function $J^{(2)}_{\mathrm{C}_n}$.

Motivated by quantum field theories on lens spaces, Spiridonov (2018) evaluated certain finite sums of C_n integrals, for both type I and type II. In closely related work, Kels and Yamazaki (2018) obtained transformation formulas for finite sums of A_n and C_n integrals of type I.

As mentioned in the introduction, the recent identification of elliptic hypergeometric integrals as indices in supersymmetric quantum field theory by Dolan and Osborn (2009) has led to a large number of conjectured integral evaluations and transformations (Gadde et al., 2010a,b; Spiridonov and Vartanov, 2011, 2012, 2014). It is too early to give a survey of the emerging picture, but it is clear that the identities stated in this section are a small sample from a much larger collection of identities.

6.3 Series

In this section we give the most important summation and transformation formulas for elliptic hypergeometric series associated to A_n and C_n. In the $n = 1$ case all summations except for (6.3.7) simplify to the elliptic Jackson summation of Frenkel and Turaev (1997). Similarly, most transformations may be viewed as generalizations of the elliptic Bailey transformation.

6.3.1 A_n *Summations*

The following A_n *elliptic Jackson summation* is a discrete analogue of the multiple beta integral (6.2.1):

$$\sum_{\substack{k_1,\dots,k_{n+1}\geq 0\\ k_1+\cdots+k_{n+1}=N}} \frac{\Delta^{\mathrm{A}}(xq^k)}{\Delta^{\mathrm{A}}(x)} \prod_{i=1}^{n+1} \frac{\prod_{j=1}^{n+2}(x_i a_j)_{k_i}}{(bx_i)_{k_i}\prod_{j=1}^{n+1}(qx_i/x_j)_{k_i}} = \frac{(b/a_1,\dots,b/a_{n+2})_N}{(q,bx_1,\dots,bx_{n+1})_N}, \tag{6.3.1a}$$

where $b = a_1\cdots a_{n+2}x_1\cdots x_{n+1}$. Using the constraint on the summation indices to eliminate k_{n+1}, this identity can be written in less symmetric form as

$$\sum_{\substack{k_1,\dots,k_n\geq 0\\ |k|\leq N}} \frac{\Delta^{\mathrm{A}}(xq^k)}{\Delta^{\mathrm{A}}(x)} \prod_{i=1}^{n}\left(\frac{\theta(ax_iq^{k_i+|k|})}{\theta(ax_i)} \frac{(ax_i)_{|k|}\prod_{j=1}^{n+2}(x_ib_j)_{k_i}}{(aq^{N+1}x_i, aqx_i/c)_{k_i}\prod_{j=1}^{n}(qx_i/x_j)_{k_i}}\right) \frac{(q^{-N},c)_{|k|}}{\prod_{i=1}^{n+2}(aq/b_i)_{|k|}} q^{|k|}$$
$$= c^N \prod_{i=1}^{n}\frac{(aqx_i)_N}{(aqx_i/c)_N}\prod_{i=1}^{n+2}\frac{(aq/cb_i)_N}{(aq/b_i)_N}, \tag{6.3.1b}$$

where $b_1\cdots b_{n+2}cx_1\cdots x_n = a^2q^{N+1}$. By analytic continuation one can then deduce the companion identity

$$\sum_{k_1,\dots,k_n=0}^{N_1,\dots,N_n} \left(\frac{\Delta^{\mathrm{A}}(xq^k)}{\Delta^{\mathrm{A}}(x)} \prod_{i=1}^{n}\left(\frac{\theta(ax_iq^{k_i+|k|})}{\theta(ax_i)} \frac{(ax_i)_{|k|}(dx_i,ex_i)_{k_i}}{(aq^{N_i+1}x_i)_{|k|}(aqx_i/b,aqx_i/c)_{k_i}}\right)\right.$$
$$\left.\times\frac{(b,c)_{|k|}q^{|k|}}{(aq/d,aq/e)_{|k|}}\prod_{i,j=1}^{n}\frac{(q^{-N_j}x_i/x_j)_{k_i}}{(qx_i/x_j)_{k_i}}\right) = \frac{(aq/cd,aq/bd)_{|N|}}{(aq/d,aq/bcd)_{|N|}}\prod_{i=1}^{n}\frac{(aqx_i,aqx_i/bc)_{N_i}}{(aqx_i/b,aqx_i/c)_{N_i}}, \tag{6.3.1c}$$

where $bcde = a^2q^{|N|+1}$. For the discrete analogue of the type-II integral (6.2.2) we refer the reader to Notes at the end of this section.

Our next result corresponds to a discretization of (6.2.3):

$$\sum_{\substack{k_1,\dots,k_{n+1}\geq 0\\ k_1+\cdots+k_{n+1}=N}} \frac{\Delta^{\mathrm{A}}(xq^k)}{\Delta^{\mathrm{A}}(x)} \prod_{1\leq i<j\leq n+1}\frac{1}{(x_ix_j)_{k_i+k_j}} \prod_{i=1}^{n+1}\frac{q^{\binom{k_i}{2}}x_i^{k_i}\prod_{j=1}^{n}(x_ia_j^{\pm})_{k_i}}{(bx_i,q^{1-N}x_i/b)_{k_i}\prod_{j=1}^{n+1}(qx_i/x_j)_{k_i}}$$
$$= (-bq^{N-1})^N \frac{\prod_{i=1}^{n}(ba_i^{\pm})_N}{(q)_N\prod_{i=1}^{n+1}(bx_i^{\pm})_N}. \tag{6.3.2}$$

Mimicking the steps that led from (6.3.1a) to (6.3.1c), identity (6.3.2) can be rewritten as a sum over an n-dimensional rectangle; see Rosengren (2004). Some authors have associated (6.3.2) and related results with the root system D_n rather than A_n.

Finally, the following summation is a discrete analogue of (6.2.4):

$$\sum_{\substack{k_1,\dots,k_{n+1}\geq 0\\ k_1+\cdots+k_{n+1}=N}} \frac{\Delta^A(xq^k)}{\Delta^A(x)} \prod_{1\leq i<j\leq n+1} q^{k_ik_j}(x_ix_j)_{k_i+k_j} \prod_{i=1}^{n+1} \frac{\prod_{j=1}^4 (x_ib_j)_{k_i}}{x_i^{k_i}\prod_{j=1}^{n+1}(qx_i/x_j)_{k_i}}$$
$$= \begin{cases} \dfrac{(Xb_1,Xb_2,Xb_3,Xb_4)_N}{X^N(q)_N}, & n \text{ odd},\\[2ex] \dfrac{(X,Xb_1b_2,Xb_1b_3,Xb_1b_4)_N}{(Xb_1)^N(q)_N}, & n \text{ even}, \end{cases} \tag{6.3.3}$$

where $X = x_1\cdots x_{n+1}$ and $q^{N-1}b_1\cdots b_4X^2 = 1$.

6.3.2 C_n Summations

The following C_n *elliptic Jackson summation* is a discrete analogue of (6.2.5):

$$\sum_{k_1,\dots,k_n=0}^{N_1,\dots,N_n} \frac{\Delta^{\mathrm{C}}(xq^k)}{\Delta^{\mathrm{C}}(x)} \prod_{i=1}^n \frac{(bx_i,cx_i,dx_i,ex_i)_{k_i}q^{k_i}}{(qx_i/b,qx_i/c,qx_i/d,qx_i/e)_{k_i}} \prod_{i,j=1}^n \frac{(q^{-N_j}x_i/x_j,x_ix_j)_{k_i}}{(qx_i/x_j,q^{N_j+1}x_ix_j)_{k_i}}$$
$$= \frac{\prod_{i,j=1}^n (qx_ix_j)_{N_i}}{\prod_{1\leq i<j\leq n}(qx_ix_j)_{N_i+N_j}} \frac{(q/bc,q/bd,q/cd)_{|N|}}{\prod_{i=1}^n (qx_i/b,qx_i/c,qx_i/d,q^{-N_i}e/x_i)_{N_i}}, \tag{6.3.4}$$

where $bcde = q^{|N|+1}$.

The discrete analogues of the type-II integrals (6.2.6) and (6.2.9) are most conveniently expressed in terms of the series

$${}_{r+1}V_r^{(n)}(a;b_1,\dots,b_{r-4}) := \sum_\lambda \Bigg(\prod_{i=1}^n \frac{\theta(at^{2-2i}q^{2\lambda_i})}{\theta(at^{2-2i})} \frac{(at^{1-n},b_1,\dots,b_{r-4})_\lambda}{(qt^{n-1},aq/b_1,\dots,aq/b_{r-4})_\lambda}$$
$$\times \prod_{1\leq i<j\leq n} \left(\frac{\theta(t^{j-i}q^{\lambda_i-\lambda_j},at^{2-i-j}q^{\lambda_i+\lambda_j})}{\theta(t^{j-i},at^{2-i-j})} \frac{(t^{j-i+1})_{\lambda_i-\lambda_j}(at^{3-i-j})_{\lambda_i+\lambda_j}}{(qt^{j-i-1})_{\lambda_i-\lambda_j}(aqt^{1-i-j})_{\lambda_i+\lambda_j}} \right) q^{|\lambda|}t^{2n(\lambda)} \Bigg), \tag{6.3.5}$$

where the summation is over partitions $\lambda = (\lambda_1,\dots,\lambda_n)$ of length at most n. Note that this implicitly depends on t as well as p and q. When $b_{r-4} = q^{-N}$ with $N \in \mathbb{Z}_{\geq 0}$, this becomes a terminating series, with sum ranging over partitions $\lambda \subset (N^n)$. The series (6.3.5) is associated with C_n since

$$\prod_{i=1}^n \frac{\theta(at^{2-2i}q^{2\lambda_i})}{\theta(at^{2-2i})} \prod_{1\leq i<j\leq n} \frac{\theta(t^{j-i}q^{\lambda_i-\lambda_j},at^{2-i-j}q^{\lambda_i+\lambda_j})}{\theta(t^{j-i},at^{2-i-j})} = q^{-n(\lambda)}\frac{\Delta^{\mathrm{C}}(xq^\lambda)}{\Delta^{\mathrm{C}}(x)},$$

with $x_i = \sqrt{a}\,t^{1-i}$.

Using the above notation, the discrete analogue of (6.2.6) is

$$ {}_{10}V_9^{(n)}(a;b,c,d,e,q^{-N}) = \frac{(aq,aq/bc,aq/bd,aq/cd)_{(N^n)}}{(aq/b,aq/c,aq/d,aq/bcd)_{(N^n)}}, \tag{6.3.6} $$

where $bcdet^{n-1} = a^2q^{N+1}$. This is the C_n summation mentioned in the introduction.

Next we give a discrete analogue of (6.2.9):

$$ \begin{aligned} &{}_{2r+8}V_{2r+7}^{(n)}\Big(a;t^{1-n}b,\frac{a}{b},c_1q^{k_1},\dots,c_rq^{k_r},\frac{aq}{c_1},\dots,\frac{aq}{c_r},q^{-N}\Big) \\ &\quad = \frac{(aq,qt^{n-1})_{(N^n)}}{(bq,aqt^{n-1}/b)_{(N^n)}} \prod_{i=1}^{r} \frac{(c_ib/a,c_it^{n-1}/b)_{(k_i^n)}}{(c_i,c_it^{n-1}/a)_{(k_i^n)}}, \end{aligned} \tag{6.3.7} $$

where the k_i are non-negative integers such that $k_1 + \cdots + k_r = N$. As the summand contains the factors $(c_iq^{k_i})_\lambda/(c_i)_\lambda$, this is a so-called *Karlsson–Minton-type summation*.

Finally, we have the C_n summation

$$ \begin{aligned} &\sum_{k_1,\dots,k_n=0}^{N} \frac{\Delta^{\mathrm{C}}(xq^k)}{\Delta^{\mathrm{C}}(x)} \prod_{i=1}^{n} \frac{(x_i^2,bx_i,cx_i,dx_i,ex_i,q^{-N})_{k_i}}{(q,qx_i/b,qx_i/c,qx_i/d,qx_i/e,q^{N+1}x_i^2)_{k_i}} q^{k_i} \\ &\quad = \prod_{1\le i<j\le n} \frac{\theta(q^Nx_ix_j)}{\theta(x_ix_j)} \prod_{i=1}^{n} \frac{(qx_i^2,q^{2-i}/bc,q^{2-i}/bd,q^{2-i}/cd)_N}{(qx_i/b,qx_i/c,qx_i/d,q^{-N}e/x_i)_N}, \end{aligned} \tag{6.3.8} $$

where $bcde = q^{N-n+2}$. There is a corresponding integral evaluation (Spiridonov, 2004), which was mentioned in §6.2.4.

6.3.3 Series Transformations

Several of the transformations stated below have companion identities (similar to the different versions of (6.3.1)) which will not be stated explicitly.

The following A_n *Bailey transformation* is a discrete analogue of (6.2.13):

$$ \begin{aligned} &\sum_{k_1,\dots,k_n=0}^{N_1,\dots,N_n} \Bigg(\frac{\Delta^{\mathrm{A}}(xq^k)}{\Delta^{\mathrm{A}}(x)} \prod_{i=1}^{n} \Bigg(\frac{\theta(ax_iq^{k_i+|k|})}{\theta(ax_i)} \frac{(ax_i)_{|k|}(ex_i,fx_i,gx_i)_{k_i}}{(aq^{N_i+1}x_i)_{|k|}(aqx_i/b,aqx_i/c,aqx_i/d)_{k_i}} \Bigg) \\ &\qquad \times \frac{(b,c,d)_{|k|}q^{|k|}}{(aq/e,aq/f,aq/g)_{|k|}} \prod_{i,j=1}^{n} \frac{(q^{-N_j}x_i/x_j)_{k_i}}{(qx_i/x_j)_{k_i}} \Bigg) \\ &\quad = \Big(\frac{a}{\lambda}\Big)^{|N|} \frac{(\lambda q/f,\lambda q/g)_{|N|}}{(aq/f,aq/g)_{|N|}} \prod_{i=1}^{n} \frac{(aqx_i,\lambda qx_i/d)_{N_i}}{(\lambda qx_i,aqx_i/d)_{N_i}} \\ &\qquad \times \sum_{k_1,\dots,k_n=0}^{N_1,\dots,N_n} \Bigg(\frac{\Delta^{\mathrm{A}}(xq^k)}{\Delta^{\mathrm{A}}(x)} \prod_{i=1}^{n} \Bigg(\frac{\theta(\lambda x_iq^{k_i+|k|})}{\theta(\lambda x_i)} \frac{(\lambda x_i)_{|k|}(\lambda ex_i/a,fx_i,gx_i)_{k_i}}{(\lambda q^{N_i+1}x_i)_{|k|}(aqx_i/b,aqx_i/c,\lambda qx_i/d)_{k_i}} \Bigg) \\ &\qquad \times \frac{(\lambda b/a,\lambda c/a,d)_{|k|}q^{|k|}}{(aq/e,\lambda q/f,\lambda q/g)_{|k|}} \prod_{i,j=1}^{n} \frac{(q^{-N_j}x_i/x_j)_{k_i}}{(qx_i/x_j)_{k_i}} \Bigg), \end{aligned} \tag{6.3.9} $$

where $bcdefg = a^3q^{|N|+2}$ and $\lambda = a^2q/bce$. For $be = aq$ the sum on the right trivializes and the transformation simplifies to (6.3.1c).

The next transformation, which relates an A_n and a C_n series, is a discrete analogue of (6.2.12):

$$\begin{aligned}
&\sum_{k_1,\dots,k_n=0}^{N_1,\dots,N_n} \frac{\Delta^{C}(xq^k)}{\Delta^{C}(x)} \prod_{i=1}^{n} \frac{(bx_i,cx_i,dx_i,ex_i,fx_i,gx_i)_{k_i}q^{k_i}}{(qx_i/b,qx_i/c,qx_i/d,qx_i/e,qx_i/f,qx_i/g)_{k_i}} \prod_{i,j=1}^{n} \frac{(q^{-N_j}x_i/x_j,x_ix_j)_{k_i}}{(qx_i/x_j,q^{N_j+1}x_ix_j)_{k_i}} \\
&= \frac{\prod_{i,j=1}^{n}(qx_ix_j)_{N_i}}{\prod_{1\le i<j\le n}(qx_ix_j)_{N_i+N_j}} \frac{(\lambda q/e,\lambda q/f,q/ef)_{|N|}}{\prod_{i=1}^{n}(\lambda qx_i,qx_i/e,qx_i/f,q^{-N_i}g/x_i)_{N_i}} \\
&\quad\times \sum_{k_1,\dots,k_n=0}^{N_1,\dots,N_n} \Bigg(\frac{\Delta^{A}(xq^k)}{\Delta^{A}(x)} \prod_{i=1}^{n} \bigg(\frac{\theta(\lambda x_iq^{k_i+|k|})}{\theta(\lambda x_i)} \frac{(\lambda x_i)_{|k|}(ex_i,fx_i,gx_i)_{k_i}}{(\lambda q^{N_i+1}x_i)_{|k|}(qx_i/b,qx_i/c,qx_i/d)_{k_i}} \bigg) \\
&\quad\times \frac{(\lambda b,\lambda c,\lambda d)_{|k|}q^{|k|}}{(\lambda q/e,\lambda q/f,\lambda q/g)_{|k|}} \prod_{i,j=1}^{n} \frac{(q^{-N_j}x_i/x_j)_{k_i}}{(qx_i/x_j)_{k_i}} \Bigg),
\end{aligned} \tag{6.3.10}$$

where $bcdefg = q^{|N|+2}$ and $\lambda = q/bcd$. For $bc = q$ this reduces to (6.3.4).

The discrete analogue of (6.2.10) provides a duality between A_n and A_m elliptic hypergeometric series:

$$\begin{aligned}
&\sum_{\substack{k_1,\dots,k_{n+1}\ge 0\\ k_1+\dots+k_{n+1}=N}} \frac{\Delta^{A}(xq^k)}{\Delta^{A}(x)} \prod_{i=1}^{n+1} \frac{\prod_{j=1}^{m+n+2}(x_ia_j)_{k_i}}{\prod_{j=1}^{m+1}(x_iy_j)_{k_i}\prod_{j=1}^{n+1}(qx_i/x_j)_{k_i}} \\
&= \sum_{\substack{k_1,\dots,k_{m+1}\ge 0\\ k_1+\dots+k_{m+1}=N}} \frac{\Delta^{A}(yq^k)}{\Delta^{A}(y)} \prod_{i=1}^{m+1} \frac{\prod_{j=1}^{m+n+2}(y_i/a_j)_{k_i}}{\prod_{j=1}^{n+1}(y_ix_j)_{k_i}\prod_{j=1}^{m+1}(qy_i/y_j)_{k_i}},
\end{aligned} \tag{6.3.11}$$

where $w_1\cdots w_{m+1} = x_1\cdots x_{n+1}a_1\cdots a_{m+n+2}$. For $m = 0$ this reduces to (6.3.1a).

We next give a discrete analogue of (6.2.11). If M_i $(i = 1,\dots,m)$ and N_j $(j = 1,\dots,n)$ are non-negative integers and $bcde = q^{|N|-|M|+1}$, then

$$\begin{aligned}
&\sum_{k_1,\dots,k_n=0}^{N_1,\dots,N_n} \Bigg(\frac{\Delta^{C}(xq^k)}{\Delta^{C}(x)} \prod_{i=1}^{n} \frac{(bx_i,cx_i,dx_i,ex_i)_{k_i}q^{k_i}}{(qx_i/b,qx_i/c,qx_i/d,qx_i/e)_{k_i}} \\
&\quad\times \prod_{i=1}^{n}\prod_{j=1}^{m} \frac{(q^{M_j}x_iy_j,qx_i/y_j)_{k_i}}{(x_iy_j,q^{1-M_j}x_i/y_j)_{k_i}} \prod_{i,j=1}^{n} \frac{(q^{-N_j}x_i/x_j,x_ix_j)_{k_i}}{(qx_i/x_j,q^{N_j+1}x_ix_j)_{k_i}} \Bigg) \\
&= q^{-|N||M|} \frac{(q/bc,q/bd,q/cd)_{|N|}}{(q^{-|N|}bc,q^{-|N|}bd,q^{-|N|}cd)_{|M|}} \prod_{i=1}^{m}\prod_{j=1}^{n} \frac{(q^{-N_j}y_i/x_j)_{M_i}}{(y_i/x_j)_{M_i}} \\
&\quad\times \frac{\prod_{i,j=1}^{n}(qx_ix_j)_{N_i}\prod_{1\le i<j\le m}(y_iy_j)_{M_i+M_j}}{\prod_{i,j=1}^{m}(y_iy_j)_{M_i}\prod_{1\le i<j\le n}(qx_ix_j)_{N_i+N_j}} \frac{\prod_{i=1}^{m}(by_i,cy_i,dy_i,q^{1-M_i}/y_ie)_{M_i}}{\prod_{i=1}^{n}(qx_i/b,qx_i/c,qx_i/d,q^{-N_i}e/x_i)_{N_i}} \\
&\quad\times \sum_{k_1,\dots,k_m=0}^{M_1,\dots,M_m} \Bigg(\frac{\Delta^{C}(q^{-1/2}yq^k)}{\Delta^{C}(q^{-1/2}y)} \prod_{i=1}^{m} \frac{(y_i/b,y_i/c,y_i/d,y_i/e)_{k_i}q^{k_i}}{(by_i,cy_i,dy_i,ey_i)_{k_i}}
\end{aligned}$$

$$\times \prod_{i=1}^{m}\prod_{j=1}^{n}\frac{(q^{N_j}y_ix_j, y_i/x_j)_{k_i}}{(y_ix_j, q^{-N_j}y_i/x_j)_{k_i}}\prod_{i,j=1}^{m}\frac{(q^{-M_j}y_i/y_j, q^{-1}y_iy_j)_{k_i}}{(qy_i/y_j, q^{M_j}y_iy_j)_{k_i}}\Bigg). \tag{6.3.12}$$

Recalling the notation (6.3.5), we have the following discrete analogue of (6.2.14):

$$\begin{aligned} &{}_{12}V_{11}^{(n)}(a;b,c,d,e,f,g,q^{-N}) \\ &\quad = \frac{(aq, aq/ef, \lambda q/e, \lambda q/f)_{(N^n)}}{(\lambda q, \lambda q/ef, aq/e, aq/f)_{(N^n)}}\,{}_{12}V_{11}^{(n)}\Big(\lambda; \frac{\lambda b}{a}, \frac{\lambda c}{a}, \frac{\lambda d}{a}, e, f, g, q^{-N}\Big), \end{aligned} \tag{6.3.13}$$

where $bcdefgt^{n-1} = a^3q^{N+2}$ and $\lambda = a^2q/bcd$.

Finally, the following *Karlsson–Minton-type transformation* is an analogue of (6.2.15):

$$\begin{aligned} &{}_{2r+8}V_{2r+7}^{(n)}\Big(a; bt^{1-n}, \frac{aq^{-M}}{b}, c_1q^{k_1}, \dots, c_rq^{k_r}, \frac{aq}{c_1}, \dots, \frac{aq}{c_r}, q^{-N}\Big) \\ &\quad = \frac{(aq, t^{n-1}q)_{(N^n)}}{(bq, t^{n-1}aq/b)_{(N^n)}}\frac{(bq, t^{n-1}bq/a)_{(M^n)}}{(b^2q/a, t^{n-1}q)_{(M^n)}}\prod_{i=1}^{r}\frac{(bc_i/a, t^{n-1}c_i/b)_{(k_i^n)}}{(c_i, t^{n-1}c_i/a)_{(k_i^n)}} \\ &\quad \times {}_{2r+8}V_{2r+7}^{(n)}\Big(\frac{b^2}{a}; bt^{1-n}, \frac{bq^{-N}}{a}, \frac{bc_1q^{k_1}}{a}, \dots, \frac{bc_rq^{k_r}}{a}, \frac{bq}{c_1}, \dots, \frac{bq}{c_r}, q^{-M}\Big), \end{aligned} \tag{6.3.14}$$

where the k_i are non-negative integers such that $k_1 + \cdots + k_r = M + N$.

6.3.4 Notes

For $p = 0$ the A_n summations (6.3.1), (6.3.2) and (6.3.3) are due to Milne (1997) (see also (5.2.18)), Schlosser (1997) (see also Bhatnagar, 1999), and Gustafson and Rakha (2000) (see also (5.2.20)), respectively. The $p = 0$ case of the C_n summation (6.3.4) was found independently by Denis and Gustafson (1992) and Milne and Lilly (1995). The $p = 0$ case of the C_n summation (6.3.8) is due to Schlosser (2000). The $p = 0$ case of the transformation (6.3.9) was obtained, again independently, by Denis and Gustafson (1992) and Milne and Newcomb (1996). The $p = 0$ cases of (6.3.10) and (6.3.11) are due to Bhatnagar and Schlosser (1998) and Kajihara (2004), respectively. The $p = 0$ instances of (6.3.6), (6.3.7), (6.3.12), (6.3.13) and (6.3.14) were not known prior to the elliptic cases.

For general p, the A_n summations (6.3.1) and (6.3.2) were first obtained by Rosengren (2004) using an elementary inductive argument. A derivation of (6.3.1) from (6.2.1) using residue calculus is given by Spiridonov (2004) and a similar derivation of (6.3.2) from (6.2.3) is due to Spiridonov and Warnaar (2006). The summation (6.3.3) was conjectured by Spiridonov (2004) and proved, independently, by Ito and Noumi (personal communication, 2017) and by Rosengren (2017).

As mentioned in the introduction, Warnaar (2002) conjectured the C_n summation (6.3.6). He also proved the more elementary C_n summation (6.3.8). Van Diejen and Spiridonov (2000, 2001b) showed that the C_n summations (6.3.4) and (6.3.6) follow from the (at that time conjectural) integral identities (6.2.5) and (6.2.6). This in particular implied the first proof of (6.3.6) for $p = 0$. For general p, the summations (6.3.4) and (6.3.6) were proved by Rosengren (2001, 2004) using the case $N = 1$ of Warnaar's identity (6.3.8). Subsequent proofs of

(6.3.6) were given by Coskun and Gustafson (2006), Ito and Noumi (2017a) and Rains (2006, 2010). The proofs by Coskun and Gustafson (2006) and Rains (2006) establish the more general sum (6.4.14) for elliptic binomial coefficients. Rains (2010) obtains the identity (6.3.6) as a special case of the discrete biorthogonality relation (6.4.27) for the elliptic biorthogonal functions $\tilde{R}_\lambda$.

The transformations (6.3.9) and (6.3.10) were obtained by Rosengren (2004), together with two more A_n transformations that are not surveyed here. The transformation (6.3.11) was obtained independently by Kajihara and Noumi (2003) and Rosengren (2006). Both these papers contain further transformations that can be obtained by iterating (6.3.11). The transformation (6.3.12) was proved by Rains (personal communication, 2003) by specializing the parameters of Theorem 7.9 in Rains' paper (2010) to a union of geometric progressions. It appeared explicitly in subsequent work by Komori et al. (2016, Theorem 4.2) using a similar approach to that of Rains. The transformation (6.3.13) was conjectured by Warnaar (2002) and established by Rains (2006) using the symmetry of the expression (6.4.16) below. The transformation (6.3.14) is stated somewhat implicitly by Rains (2012); it includes (6.3.7) as a special case.

A discrete analogue of the type-II A_n beta integral (6.2.2) has been conjectured by Spiridonov and Warnaar (2011). Surprisingly, this conjecture contains the C_n identity (6.3.6) as a special case.

The summation formula (6.3.8) can be obtained as a determinant of one-dimensional summations. Further summations and transformations of determinantal type are given by Rosengren and Schlosser (2003). The special case $t = q$ of (6.3.6) and (6.3.13) is also closely related to determinants; see Schlosser (2007).

Transformations related to the sum (6.3.3) are discussed in Rosengren (2017). In their work on elliptic Bailey lemmas on root systems, Bhatnagar and Schlosser (2018) discovered two further elliptic Jackson summations for A_n, as well as corresponding transformation formulas. For none of these is an integral analogue known. Langer et al. (2009) proved a curious A_n transformation formula, which was unknown even in the one-variable case.

6.4 Elliptic Macdonald–Koornwinder Theory

A function f on $(\mathbb{C}^*)^n$ is said to be BC_n-*symmetric* if it is invariant under the action of the hyperoctahedral group $(\mathbb{Z}/2\mathbb{Z}) \wr S_n$. Here the symmetric group S_n acts by permuting the variables and $\mathbb{Z}/2\mathbb{Z}$ acts by replacing a variable with its reciprocal. The *interpolation functions*

$$R^*_\lambda(x_1, \ldots, x_n; a, b; q, t; p), \tag{6.4.1}$$

introduced independently by Rains (2006, 2010) and by Coskun and Gustafson (2006), are BC_n-symmetric elliptic functions that generalize Okounkov's BC_n interpolation Macdonald polynomials (Okounkov, 1998) as well as the Macdonald polynomials of type A (Macdonald, 1995) (see also Chapter 9). They form the building blocks of Rains' more general BC_n-symmetric functions (Rains, 2006, 2010):

$$\tilde{R}_\lambda(x_1, \ldots, x_n; a\!:\!b, c, d; u, v; q, t; p).$$

The $\tilde{R}_\lambda$ are an elliptic generalization of the Koornwinder polynomials (Koornwinder, 1992) (see also Chapter 9), themselves a generalization to BC_n of the Askey–Wilson polynomials (Askey and Wilson, 1979). The price one pays for ellipticity is that the functions R^*_λ and $\tilde{R}_\lambda$ are neither polynomial nor orthogonal. The latter do however form a biorthogonal family, and for $n = 1$ they reduce to the continuous biorthogonal functions of Spiridonov (2004) (in the elliptic case) and Rahman (1986) (the $p = 0$ case) and, appropriately specialized, to the discrete biorthogonal functions of Spiridonov and Zhedanov (2000a,b) (in the elliptic case) and Wilson (1991) (the $p = 0$ case).

There are a number of ways to define the elliptic interpolation functions. Here we will describe them via a branching rule. The *branching coefficient* $c_{\lambda\mu}$ is a complex function on $(\mathbb{C}^*)^7$, indexed by a pair of partitions λ, μ. It is defined to be zero unless $\lambda \succ \mu$, in which case

$$\begin{aligned} c_{\lambda\mu}(z;a,b;q,t,T;p) :=& \frac{(aTz^{\pm},pqa/bt)_\lambda}{(aTz^{\pm},pqa/bt)_\mu}\frac{(pqz^{\pm}/bt,T)_\mu}{(pqz^{\pm}/b,tT)_\lambda} \\ &\times \prod_{\substack{(i,j)\in\lambda \\ \lambda'_j=\mu'_j}} \frac{\theta(q^{\lambda_i+j-1}t^{2-i-\lambda'_j}aT/b)}{\theta(pq^{\mu_i-j+1}t^{\mu'_j-i})} \prod_{\substack{(i,j)\in\lambda \\ \lambda'_j\neq\mu'_j}} \frac{\theta(q^{\lambda_i-j}t^{\lambda'_j-i+1})}{\theta(pq^{\mu_i+j}t^{-i-\mu'_j}aT/b)} \\ &\times \prod_{\substack{(i,j)\in\mu \\ \lambda'_j=\mu'_j}} \frac{\theta(pq^{\lambda_i-j+1}t^{\lambda'_j-i})}{\theta(q^{\mu_i+j-1}t^{1-i-\mu'_j}aT/b)} \prod_{\substack{(i,j)\in\mu \\ \lambda'_j\neq\mu'_j}} \frac{\theta(pq^{\lambda_i+j}t^{1-i-\lambda'_j}aT/b)}{\theta(q^{\mu_i-j}t^{\mu'_j-i+1})}. \end{aligned} \tag{6.4.2}$$

From (6.1.1) and the invariance under the substitution $z \mapsto z^{-1}$ it follows that $c_{\lambda\mu}$ is a BC_1-symmetric elliptic function of z. The elliptic interpolation functions are uniquely determined by the branching rule

$$R^*_\lambda(x_1,\ldots,x_{n+1};a,b;q,t;p) = \sum_\mu c_{\lambda\mu}(x_{n+1};a,b;q,t,t^n;p)R^*_\mu(x_1,\ldots,x_n;a,b;q,t;p), \tag{6.4.3}$$

subject to the initial condition $R^*_\lambda(-;a,b;q,t;p) = \delta_{\lambda,0}$. It immediately follows that the interpolation function (6.4.1) vanishes if $l(\lambda) > n$. From the symmetry and ellipticity of the branching coefficient it also follows that the interpolation functions are BC_1-symmetric and elliptic in each of the x_i. However, S_n-symmetry (and thus BC_n-symmetry) is not manifest and is a consequence of the non-trivial fact that

$$\sum_\mu c_{\lambda\mu}(z;a,b;q,t,T;p)c_{\mu\nu}(w;a,b;q,t,T/t;p) \tag{6.4.4}$$

is a symmetric function in z and w; see also the discussion around (6.4.17) below.

In the remainder of this section $x = (x_1,\ldots,x_n)$. Comparison of their respective branching rules shows that Okounkov's BC_n *interpolation Macdonald polynomials* $P^*_\lambda(x;q,t,s)$ and the ordinary *Macdonald polynomials* $P_\lambda(x;q,t)$ arise in the limit as

$$P^*_\lambda(x;q,t,s) = \lim_{p\to 0}(-s^2t^{2n-2})^{-|\lambda|}q^{-n(\lambda')}t^{2n(\lambda)}\frac{(t^n)_\lambda}{C^-_\lambda(t)}R^*_\lambda(st^\delta x;s,p^{1/2}b;q,t;p)$$

and

$$P_\lambda(x;q,t) = \lim_{z\to\infty} z^{-|\lambda|} \lim_{p\to 0} (-at^{n-1})^{-|\lambda|} q^{-n(\lambda')} t^{2n(\lambda)} \frac{(t^n)_\lambda}{C^-_\lambda(t)} R^*_\lambda(zx; a, p^{1/2}b; q,t;p),$$

where $\delta := (n-1,\dots,1,0)$ is the *staircase partition* of length $n-1$, $st^\delta x = (st^{n-1}x_1,\dots,st^0x_n)$ and $zx = (zx_1,\dots,zx_n)$.

Many standard properties of $P_\lambda(x;q,t)$ and $P^*_\lambda(x;q,t,s)$ have counterparts for the elliptic interpolation functions. Here we have space for only a small selection. Up to normalization, Okounkov's BC_n interpolation Macdonald polynomials are uniquely determined by symmetry and vanishing properties. The latter carry over to the elliptic case as follows:

$$R^*_\mu(aq^\lambda t^\delta; a,b;q,t;p) = 0 \tag{6.4.5}$$

if $\mu \not\subset \lambda$. For $q,t,a,b,c,d \in \mathbb{C}^*$ the elliptic difference operator $D^{(n)}(a,b,c,d;q,t;p)$, acting on BC_n-symmetric functions, is given by

$$(D^{(n)}(a,b,c,d;q,t;p)f)(x) := \sum_{\sigma\in\{\pm1\}^n} f(q^{\sigma/2}x) \prod_{i=1}^n \frac{\theta(ax_i^{\sigma_i}, bx_i^{\sigma_i}, cx_i^{\sigma_i}, dx_i^{\sigma_i})}{\theta(x_i^{2\sigma_i})} \prod_{1\le i<j\le n} \frac{\theta(tx_i^{\sigma_i}x_j^{\sigma_j})}{\theta(x_i^{\sigma_i}x_j^{\sigma_j})},$$

where $q^{\sigma/2}x = (q^{\sigma_1/2}x_1,\dots,q^{\sigma_n/2}x_n)$. Then

$$\begin{aligned} &D^{(n)}(a,b,c,d;q,t;p)R^*_\lambda(x;aq^{1/2},bq^{1/2};q,t;p) \\ &\quad = \prod_{i=1}^n \theta(abt^{n-i}, acq^{\lambda_i}t^{n-i}, bcq^{-\lambda_i}t^{i-1}) \cdot R^*_\lambda(x;a,b;q,t;p) \end{aligned} \tag{6.4.6}$$

provided that $t^{n-1}abcd = p$. Like for the Macdonald polynomials, there is no simple closed-form expression for the elliptic interpolation functions. When indexed by rectangular partitions of length n, however, they do admit a simple form, viz.

$$R^*_{(N^n)}(x;a,b;q,t;p) = \prod_{i=1}^n \frac{(ax_i^\pm)_N}{(pqx_i^\pm/b)_N}. \tag{6.4.7}$$

The *principal specialization formula* for the elliptic interpolation functions is

$$R^*_\lambda(zt^\delta;a,b;q,t;p) = \frac{(t^{n-1}az, a/z)_\lambda}{(pqt^{n-1}z/b, pq/bz)_\lambda}. \tag{6.4.8}$$

The R^*_λ satisfy numerous symmetries, all direct consequence of symmetries of the branching coefficients $c_{\lambda\mu}$. Two of the most notable ones are

$$R^*_\lambda(x;a,b;q,t;p) = R^*_\lambda(-x;-a,-b;q,t;p) \tag{6.4.9a}$$

$$= \left(\frac{qt^{n-1}a}{b}\right)^{2|\lambda|} q^{4n(\lambda')} t^{-4n(\lambda)} R^*_\lambda(x;1/a,1/b;1/q,1/t;p). \tag{6.4.9b}$$

Specializations of R^*_μ give rise to elliptic binomial coefficients. Before defining these we introduce the function

$$\Delta_\lambda(a|b_1,\dots,b_k) := \frac{(pqa)_{2\lambda^2}}{C^-_\lambda(t,pq)C^+_\lambda(a,pqa/t)} \frac{(b_1,\dots,b_k)_\lambda}{(pqa/b_1,\dots,pqa/b_k)_\lambda},$$

where the dependence on q, t and p has been suppressed and where $2\lambda^2$ is shorthand for the partition $(2\lambda_1, 2\lambda_1, 2\lambda_2, 2\lambda_2, \dots)$. Explicitly, for λ such that $l(\lambda) \le n$,

$$\Delta_\lambda(a|b_1,\dots,b_k) = \left(\frac{(-1)^k a^{k-3} q^{k-3} t}{b_1\cdots b_k}\right)^{|\lambda|} q^{(k-4)n(\lambda')} t^{-(k-6)n(\lambda)} \frac{(at^{1-n}, aqt^{-n}, b_1,\dots,b_k)_\lambda}{(qt^{n-1}, t^n, aq/b_1,\dots,aq/b_k)_\lambda}$$
$$\times \prod_{i=1}^n \frac{\theta(at^{2-2i}q^{2\lambda_i})}{\theta(at^{2-2i})} \prod_{1\le i<j\le n}\left(\frac{\theta(t^{j-i}q^{\lambda_i-\lambda_j}, at^{2-i-j}q^{\lambda_i+\lambda_j})}{\theta(t^{j-i}, at^{2-i-j})} \frac{(t^{j-i+1})_{\lambda_i-\lambda_j}(at^{3-i-j})_{\lambda_i+\lambda_j}}{(qt^{j-i-1})_{\lambda_i-\lambda_j}(aqt^{1-i-j})_{\lambda_i+\lambda_j}}\right),$$

so that (cf. (6.3.5))

$$ {}_{r+1}V_r^{(n)}(a;b_1,\dots,b_{r-4}) = \sum_\lambda \frac{(b_3,\dots,b_{r-4}, qt^{n-1}b_1b_2)_\lambda}{(aq/b_3,\dots,aq/b_{r-4}, at^{1-n}/b_1b_2)_\lambda} \Delta_\lambda\!\left(a\middle| t^n, b_1, b_2, \frac{at^{1-n}}{b_1b_2}\right). \tag{6.4.10}$$

The *elliptic binomial coefficients* $\binom{\lambda}{\mu}_{[a,b]} = \binom{\lambda}{\mu}_{[a,b];q,t;p}$ may now be defined as

$$\binom{\lambda}{\mu}_{[a,b]} := \Delta_\mu(a/b|t^n, 1/b) R_\mu^*(x_1,\dots,x_n; a^{1/2}t^{1-n}, ba^{-1/2}; q,t;p)\big|_{x_i=a^{1/2}q^{\lambda_i}t^{1-i}}, \tag{6.4.11}$$

where on the right n can be chosen arbitrarily provided that $n \ge l(\lambda), l(\mu)$. Apart from their n-independence, the elliptic binomial coefficients are also independent of the choice of square root of a. Although $\binom{\lambda}{0}_{[a,b]} = 1$, they are not normalized like ordinary binomial coefficients, and

$$\binom{\lambda}{\lambda}_{[a,b]} = \frac{(1/b, pqa/b)_\lambda}{(b, pqa)_\lambda} \frac{C_\lambda^+(a)}{C_\lambda^+(a/b)}. \tag{6.4.12}$$

The elliptic binomial coefficients vanish unless $\mu \subset \lambda$, are elliptic in both a and b, and invariant under the simultaneous substitution $(a,b,q,t) \mapsto (1/a, 1/b, 1/q, 1/t)$. They are also conjugation symmetric:

$$\binom{\lambda}{\mu}_{[a,b];q,t;p} = \binom{\lambda'}{\mu'}_{[aq/t,b];1/t,1/q;p}. \tag{6.4.13}$$

A key identity is

$$\binom{\lambda}{\nu}_{[a,c]} = \frac{(b,ce,cd,bde)_\lambda}{(cde,bd,be,c)_\lambda} \frac{(1/c,bd,be,cde)_\nu}{(bcde,e,d,b/c)_\nu} \sum_\mu \frac{(c/b,d,e,bcde)_\mu}{(bde,ce,cd,1/b)_\mu} \binom{\lambda}{\mu}_{[a,b]} \binom{\mu}{\nu}_{[a/b,c/b]} \tag{6.4.14}$$

for generic parameters such that $bcde = aq$. The $c \to 1$ limit of $\binom{\lambda}{\nu}_{[a,c]}(c)_\lambda/(1/c)_\nu$ exists and is given by $\delta_{\lambda\nu}$. Multiplying both sides of (6.4.14) by $(c)_\lambda/(1/c)_\nu$ and then letting c tend to 1 thus yields the orthogonality relation

$$\sum_\mu \binom{\lambda}{\mu}_{[a,b]} \binom{\mu}{\nu}_{[a/b,1/b]} = \delta_{\lambda\mu}. \tag{6.4.15}$$

For another important application of (6.4.14) we note that by (6.4.8),

$$\binom{(N^n)}{\mu}_{[a,b]} = \Delta_\mu(a/b|t^n, 1/b, aq^Nt^{1-n}, q^{-N}).$$

Setting $\lambda = (N^n)$ and $\nu = 0$ in (6.4.14) and recalling (6.4.10) yields the C_n Jackson summation (6.3.6). Also (6.3.8) may be obtained as a special case of (6.4.14) but the details of the derivation are more intricate; see Rains (2006). As a final application of (6.4.14) it can be shown that

$$\frac{(b,b'e)_\lambda}{(b'de,bd)_\lambda}\frac{(b'de,bf/c)_\nu}{(b/c,b'de/f)_\nu}\sum_\mu \frac{(c/b,d,e,bb'de,b'g/c,b'f/c)_\mu}{(bb'de/c,b'e,b'd,1/b,f,g)_\mu}\binom{\lambda}{\mu}_{[a,b]}\binom{\mu}{\nu}_{[a/b,c/b]} \tag{6.4.16}$$

is symmetric in b and b', where $bb'de = aq$ and $cde = fg$. Setting $\lambda = (N^n)$ and $\nu = 0$ results in the transformation formula (6.3.13). Now assume that $bcd = b'c'd'$. Twice using the symmetry of (6.4.16) it follows that

$$\begin{aligned}\sum_\mu &\frac{(c,d,aq/c',aq/d')_\lambda}{(c,d,aq/c',aq/d')_\mu}\frac{(c'/b,d'/b,aq/bc,aq/bd)_\mu}{(c'/b,d'/b,aq/bc,aq/bd)_\nu}\\ &\times\frac{(1/b,aq/b)_\nu}{(1/b,aq/bb')_\mu}\frac{(b',aq)_\mu}{(b',aq/b)_\lambda}\binom{\lambda}{\mu}_{[a,b]}\binom{\mu}{\nu}_{[a/b,b']}\end{aligned} \tag{6.4.17}$$

is invariant under the simultaneous substitution $(b,c,d)\leftrightarrow(b',c',d')$. The branching coefficient (6.4.2) may be expressed as an elliptic binomial coefficient as

$$c_{\lambda\mu}(z;a,b;q,t,T;p) = \frac{(aTz^\pm,pqa/bt,t)_\lambda}{(aTz^\pm,pqa/bt,1/t)_\mu}\frac{(pqz^\pm/bt,T,pqaT/b)_\mu}{(pqz^\pm/b,tT,pqaT/bt)_\lambda}\binom{\lambda}{\mu}_{[aT/b,t]}, \tag{6.4.18}$$

so that, up to a simple change of variables and the use of (6.1.1), the z,w-symmetry of (6.4.4) corresponds to the $b = b'$ case of the symmetry of (6.4.17). To conclude our discussion of the elliptic binomial coefficients we remark that they also arise as connection coefficients between the interpolation functions. Specifically,

$$\begin{aligned}&R^*_\lambda(x;a,b;q,t;p)\\ &\quad=\sum_\mu\binom{\lambda}{\mu}_{[t^{n-1}a/b,a/a']}\frac{(a/a',t^{n-1}aa')_\lambda}{(a'/a,t^{n-1}aa')_\mu}\frac{(pqt^{n-1}a/b,pq/ab)_\mu}{(pqt^{n-1}a'/b,pq/a'b)_\lambda}R^*_\mu(x;a',b;q,t;p).\end{aligned} \tag{6.4.19}$$

Let a,b,c,d,u,v,q,t be complex parameters such that $t^{2n-2}abcduv = pq$, and λ a partition of length at most n. Then the BC_n*-symmetric biorthogonal functions* $\tilde{R}_\lambda$ are defined as

$$\begin{aligned}&\tilde{R}_\lambda(x;a\!:\!b,c,d;u,v;q,t;p)\\ &\quad:=\sum_{\mu\subset\lambda}\binom{\lambda}{\mu}_{[1/uv,t^{1-n}/av]}\frac{(pq/bu,pq/cu,pq/du,pq/uv)_\mu}{(t^{n-1}ab,t^{n-1}ac,t^{n-1}ad,t^{n-1}av)_\mu}R^*_\mu(x;a,u;q,t;p).\end{aligned} \tag{6.4.20}$$

By (6.4.15) this relation between the two families of BC_n-symmetric elliptic functions can be inverted. We also note that from (6.4.19) it follows that

$$\tilde{R}_\lambda(x;a\!:\!b,c,d;u,t^{1-n}/b;q,t;p) = \frac{(pq/au,pqt^{n-1}a/u)_\lambda}{(b/a,t^{n-1}ab)_\lambda}R^*_\lambda(x;b,u;q,t;p), \tag{6.4.21}$$

so that the interpolation functions are a special case of the biorthogonal functions. Finally, from Okounkov's binomial formula for Koornwinder polynomials (Okounkov, 1998) it follows that in the $p \to 0$ limit the $\tilde{R}_\lambda$ simplify to the *Koornwinder polynomials* $K_\lambda(x; a, b, c, d; q, t)$:

$$K_\lambda(x; a, b, c, d; q, t) = \lim_{p\to 0} (at^{n-1})^{-|\lambda|} t^{n(\lambda)} \frac{(t^n, t^{n-1}ab, t^{n-1}ac, t^{n-1}ad)_\lambda}{C_\lambda^-(t)C_\lambda^+(abcdt^{2n-2}/q)} \times \tilde{R}_\lambda(x; a{:}b, c, d; up^{1/2}, vp^{1/2}, q, t; p).$$

Most of the previously listed properties of the interpolation functions have implications for the biorthogonal functions. For example, using (6.4.8) and (6.4.14) one can prove the *principal specialization formula*

$$\tilde{R}_\lambda(bt^\delta; a{:}b, c, d; u, v; q, t; p) = \frac{(t^{n-1}bc, t^{n-1}bd, t^{1-n}/bv, pqt^{n-1}a/u)_\lambda}{(t^{n-1}ac, t^{n-1}ad, t^{1-n}/av, pqt^{n-1}b/u)_\lambda}. \tag{6.4.22}$$

Another result that carries over is the elliptic difference equation (6.4.6). Combined with (6.4.20) it yields

$$\begin{aligned} &D^{(n)}(a, u, b, pt^{1-n}/uab; q, t; p)\tilde{R}_\lambda(x; aq^{1/2}{:}bq^{1/2}, cq^{-1/2}, dq^{-1/2}; uq^{1/2}, vq^{-1/2}, q, t; p) \\ &\quad = \prod_{i=1}^n \theta(abt^{n-i}, aut^{n-i}, but^{n-i}) \cdot \tilde{R}_\lambda(x; a{:}b, c, d; u, v, q, t; p). \end{aligned} \tag{6.4.23}$$

The Koornwinder polynomials are symmetric in the parameters a, b, c, d. From (6.4.20) it follows that $\tilde{R}_\lambda$ is symmetric in b, c, d, but the choice of normalization breaks the full S_4 symmetry. Instead,

$$\tilde{R}_\lambda(x; a{:}b, c, d; u, v; q, t; p) = \tilde{R}_\lambda(x; b{:}a, c, d; u, v; q, t; p)\tilde{R}_\lambda(bt^\delta; a{:}b, c, d; u, v; q, t; p). \tag{6.4.24}$$

For partitions λ, μ such that $l(\lambda), l(\mu) \le n$ the biorthogonal functions satisfy *evaluation symmetry*:

$$\tilde{R}_\lambda(at^\delta q^\mu; a{:}b, c, d; u, v; q, t; p) = \tilde{R}_\mu(\hat{a}t^\delta q^\lambda; \hat{a}{:}\hat{b}, \hat{c}, \hat{d}; \hat{u}, \hat{v}; q, t; p), \tag{6.4.25}$$

where

$$\hat{a} = \sqrt{abcd/pq}, \quad \hat{a}\hat{b} = ab, \quad \hat{a}\hat{c} = ac, \quad \hat{a}\hat{d} = ad, \quad a\hat{u} = \hat{a}u, \quad a\hat{v} = v\hat{a}.$$

Given a pair of partitions λ, μ such that $l(\lambda), l(\mu) \le n$, define

$$\tilde{R}_{\lambda\mu}(x; a{:}b, c, d; u, v; t; p, q) := \tilde{R}_\lambda(x; a{:}b, c, d; u, v; p, t; q)\tilde{R}_\mu(x; a{:}b, c, d; u, v; q, t; p).$$

Note that $\tilde{R}_{\lambda\mu}(x; a{:}b, c, d; u, v; t; p, q)$ is invariant under the simultaneous substitutions $\lambda \leftrightarrow \mu$ and $p \leftrightarrow q$. The functions $\tilde{R}_{\lambda\mu}(x; a{:}b, c, d; u, v; t; p, q)$ form a biorthogonal family, with continuous biorthogonality relation

$$\kappa_n^C \int_{C_{\lambda\nu,\mu\omega}} \tilde{R}_{\lambda\mu}(z_1, \dots, z_n; t_1{:}t_2, t_3, t_4; t_5, t_6; t; p, q)\tilde{R}_{\nu\omega}(z_1, \dots, z_n; t_1{:}t_2, t_3, t_4; t_6, t_5; t; p, q)$$

$$\times \prod_{1\le i<j\le n} \frac{\Gamma(tz_i^{\pm}z_j^{\pm})}{\Gamma(z_i^{\pm}z_j^{\pm})} \prod_{j=1}^{n} \frac{\prod_{i=1}^{6}\Gamma(t_iz_j^{\pm})}{\Gamma(z_j^{\pm 2})} \frac{\mathrm{d}z_1}{z_1}\cdots\frac{\mathrm{d}z_n}{z_n}$$

$$= \delta_{\lambda\nu}\delta_{\mu\omega} \prod_{m=1}^{n}\Bigl(\frac{\Gamma(t^m)}{\Gamma(t)} \prod_{1\le i<j\le 6}\Gamma(t^{m-1}t_it_j)\Bigr)$$

$$\times \frac{1}{\Delta_\lambda(1/t_5t_6|t^n, t^{n-1}t_1t_2, t^{n-1}t_1t_3, t^{n-1}t_1t_4, t^{1-n}/t_1t_5, t^{1-n}/t_1t_6; p,t;q)}$$

$$\times \frac{1}{\Delta_\mu(1/t_5t_6|t^n, t^{n-1}t_1t_2, t^{n-1}t_1t_3, t^{n-1}t_1t_4, t^{1-n}/t_1t_5, t^{1-n}/t_1t_6; q,t;p)}. \tag{6.4.26}$$

Here, $C_{\lambda\nu,\mu\omega}$ is a deformation of $\mathbb{T}^n$ which separates sequences of poles of the integrand tending to zero from sequences tending to infinity. The location of these poles depends on the choice of partitions; see Rains (2010) for details. Provided $|t| < 1$ and $|t_i| < 1$ for $1 \le i \le 6$ we can take $C_{00,00} = \mathbb{T}^n$ so that for $\lambda = \mu = \nu = \omega = 0$ one recovers the type-$\mathrm{C}_n^{(\mathrm{II})}$ integral (6.2.6). The summation (6.3.6), which is the discrete analogue of (6.2.6), follows in a similar manner from the discrete biorthogonality relation

$$\sum_{\mu\subset(N^n)} \Delta_\mu(t^{2n-2}a^2|t^n, t^{n-1}ac, t^{n-1}ad, t^{n-1}au, t^{n-1}av, q^{-N})$$

$$\times \tilde{R}_\lambda(aq^\mu t^\delta; a\!:\!b,c,d;u,v;t;p,q)\tilde{R}_\nu(aq^\mu t^\delta; a\!:\!b,c,d;v,u;t;p,q)$$

$$= \frac{\delta_{\lambda\nu}}{\Delta_\lambda(1/uv|t^n, t^{n-1}ab, t^{n-1}ac, t^{n-1}ad, t^{1-n}/au, t^{1-n}/av)}$$

$$\times \frac{(b/a, pq/uc, pq/ud, pq/uv)_{(N^n)}}{(pqt^{n-1}a/u, t^{n-1}bc, t^{n-1}bd, t^{n-1}bv)_{(N^n)}}, \tag{6.4.27}$$

where $t^{2n-2}abcduv = pq$ and $q^N t^{n-1}ab = 1$. The discrete biorthogonality can also be lifted to the functions $\tilde{R}_{\lambda\mu}$ but since the resulting identity factors into two copies of (6.4.27) – the second copy with q replaced by p and N by a second discrete parameter M – this is no more general than the above.

The final result listed here is a (dual) *Cauchy identity* which incorporates the Cauchy identities for the Koornwinder polynomials, BC_n interpolation Macdonald polynomials and ordinary Macdonald polynomials:

$$\sum_{\lambda\subset(N^n)} \Delta_\lambda(q^{1-2N}/uv|t^n, q^{-N}, q^{1-N}t^{1-n}/av, a/u)$$

$$\times \tilde{R}_\lambda(x; a\!:\!b,c,d; q^N u, q^{N-1}v; q,t;p)\tilde{R}_{\hat{\lambda}}(y; a\!:\!b,c,d; t^n u, t^{n-1}v; t,q;p)$$

$$= \frac{(a/u, pq^{1-N}/au, pq^{1-N}/bu, pq^{1-N}/cu, pq^{1-N}/du, pq^{2-2N}/uv)_{(N^n)}}{(t^{n-1}ab, t^{n-1}ac, t^{n-1}ad, q^{N-1}t^{n-1}av)_{(N^n)}}$$

$$\times \prod_{i=1}^{n}\prod_{j=1}^{N}\theta(x_i^{\pm}y_j)\prod_{i=1}^{n}\frac{1}{(ux_i^{\pm})_m}\prod_{j=1}^{N}\frac{1}{(p/uy_j, y_j/u; 1/t,p)_n}, \tag{6.4.28}$$

where $x = (x_1, \ldots, x_n)$, $y = (y_1, \ldots, y_N)$, $\hat{\lambda} = (n-\lambda'_m, \ldots, n-\lambda'_1)$ and $abcduvq^{2m-2}t^{2n-2} = p$.

Notes

Instead of $R^*_\lambda(x_1,\ldots,x_n,a,b;q,t;p)$, Rains (2006, 2010, 2012) denotes the BC_n-symmetric interpolation functions as $R^{*(n)}_\lambda(x_1,\ldots,x_n,a,b;q,t;p)$. An equivalent family of functions is defined by Coskun and Gustafson (2006) (see also Coskun, 2008). They refer to these as *well-poised Macdonald functions*, denoted by $W_\lambda(x_1,\ldots,x_n;q,p,t,a,b)$. The precise relation between the two families is given by

$$W_\lambda(x_1/a,\ldots,x_n/a;q,p,t,a^2,a/b)$$
$$=\left(\frac{t^{1-n}b^2}{q^2}\right)^{|\lambda|}q^{-2n(\lambda')}t^{2n(\lambda)}\frac{(t^n)_\lambda}{C^-_\lambda(t)}\frac{(qt^{n-2}a/b;q,t^2;p)_{2\lambda}}{(qa/tb)_\lambda C^+_\lambda(qt^{n-2}a/b)}R^*_\lambda(x_1,\ldots,x_n;a,b;q,t;p).$$

Similarly, Rains (2006, 2010, 2012) writes $\tilde{R}^{(n)}_\lambda(x_1,\ldots,x_n;a{:}b,c,d;u,v;q,t;p)$ for the biorthogonal functions instead of $\tilde{R}_\lambda(x_1,\ldots,x_n;a{:}b,c,d;u,v;q,t;p)$.

The branching rule (6.4.3) is the $k=1$ instance of (Rains, 2006, (4.40)) or the $\mu=0$ case of (Coskun and Gustafson, 2006, (2.14)). The vanishing property (6.4.5) follows by combining (6.4.21) with (Rains, 2010, Corollary 8.12) combined with (6.4.21), or from (Coskun and Gustafson, 2006, Theorem 2.6). The elliptic difference equation (6.4.6) is (Rains, 2006, (3.34)). Formula (6.4.7) for the interpolation function indexed by a rectangular partition of length n is the $\lambda=0$ case of (Rains, 2006, (3.42)) or (Coskun and Gustafson, 2006, Corollary 2.4). The principal specialization formula (6.4.8) is (Rains, 2006, (3.35)). The symmetry (6.4.9a) is (Rains, 2006, (3.39)) and the symmetry (6.4.9b) is (Rains, 2006, (3.38)) or (Coskun and Gustafson, 2006, Proposition 2.8). The definition of the elliptic binomial coefficients, (6.4.11), is due to Rains: see (Rains, 2006, (4.1)). Coskun and Gustafson (2006, (2.38)) define so-called *elliptic Jackson coefficients*

$$\omega_{\lambda/\mu}(z;r;a,b)=\omega_{\lambda/\mu}(z;r,q,p;a,b).$$

Up to normalization these are the elliptic binomial coefficients:

$$\omega_{\lambda/\mu}(z;r;a,b)=\frac{(1/z,az)_\lambda}{(qbz,qb/az)_\lambda}\frac{(qbz/r,qb/azr,bq,r)_\mu}{(1/z,az,qb/r^2,1/r)_\mu}\binom{\lambda}{\mu}_{[b,r]}.$$

The value of the elliptic binomials (6.4.12) is (Rains, 2006, (4.8)). It is equivalent to (Coskun and Gustafson, 2006, (2.9)) and also (Coskun and Gustafson, 2006, (4.23)). The conjugation symmetry (6.4.13) of the elliptic binomial coefficients is (Rains, 2006, Corollary 4.4). The summation (6.4.14) is (Rains, 2006, Theorem 4.1), and is equivalent to the "cocycle identity" (Coskun and Gustafson, 2006, (3.7)) for the elliptic Jackson coefficients. The orthogonality relation (6.4.15) is (Rains, 2006, Corollary 4.3) or (Coskun and Gustafson, 2006, (4.16)). The symmetry of (6.4.16) is (Rains, 2006, Theorem 4.9) or (Coskun and Gustafson, 2006, (3.8)), and the symmetry of (6.4.17) is (Rains, 2006, Corollary 4.11). The expression (6.4.18) for the branching coefficients is a consequence of (Rains, 2006, Corollary 4.5) or (Coskun and Gustafson, 2006, Lemma 3.11). The connection coefficient identity (6.4.19) is (Rains, 2006, Corollary 4.14). It is equivalent to the "Jackson sum" (Coskun and Gustafson, 2006, (3.6)) for the well-poised Macdonald functions W_λ.

Definition (6.4.20) of the biorthogonal functions is (Rains, 2006, (5.1)), its specialization (6.4.21) to interpolation functions is (Rains, 2006, (5.2)), its principal specialization (6.4.22) is (Rains, 2006, (5.4)) and the difference equation (6.4.23) is (Rains, 2006, Lemma 5.2). The parameter and evaluation symmetries (6.4.24) and (6.4.25) are, respectively, (Rains, 2006, Theorems 5.1 and 5.4). The important biorthogonality relation (6.4.26) is a combination of Theorems 8.4 and 8.10 in (Rains, 2010). Its discrete analogue (6.4.27) is (Rains, 2006, Theorem 5.8); see also (Rains, 2006, Theorem 8.11). Finally, the Cauchy identity (6.4.28) is (Rains, 2006, Theorem 5.11).

The BC_n-symmetric interpolation functions satisfy several further important identities not covered in the main text, such as a "bulk" branching rule, which extends (6.4.3), and a generalized Pieri rule; they are, respectively, (Rains, 2006, Theorems 4.16 and 4.17). Coskun (2008) applies the elliptic binomial coefficients (elliptic Jackson coefficients in his language) to formulate an elliptic Bailey lemma of type BC_n. The interpolation functions further admit a generalization to skew interpolation functions $\mathcal{R}_{\lambda/\mu}([v_1,\dots,v_{2n}];a,b;q,t;p)$ for $\mu\subseteq\lambda$; see Rains (2012). These are elliptic functions, symmetric in the variables $v_1,\dots,v_{2n}$, such that (Rains, 2012, Theorem 2.5)

$$R^*_\lambda(x_1,\dots,x_n;a,b;q,t;p)=\frac{(pqa/tb)_\lambda}{(t^n)_\lambda}\mathcal{R}^*_{\lambda/0}([t^{1/2}x_1^\pm,\dots,t^{1/2}x_n^\pm];t^{n-1/2}a,t^{1/2}b;q,t;p).$$

They also generalize the n-variable skew elliptic Jackson coefficients (Coskun and Gustafson, 2006, (2.43)) denoted by $\omega_{\lambda/\mu}(x_1,\dots,x_n;r,q,p;a,b)$:

$$\begin{aligned}\omega_{\lambda/\mu}(r^{n-1/2}x_1/a,\dots,r^{n-1/2}x_n/a;r;a^2r^{1-2n},ar^{1-n}/b)&=\Big(-\frac{b^3}{q^3a}\Big)^{|\lambda|-|\mu|}\\ \times q^{3n(\mu')-3n(\lambda')}t^{3n(\lambda)-3n(\mu)}r^{-n|\mu|}\frac{(aq/br)_\lambda}{(aqr^{-n-1}/b)_\mu}\frac{(r)_\mu}{(r)_\lambda}&\mathcal{R}^*_{\lambda/\mu}([r^{1/2}x_1^\pm,\dots,r^{1/2}x_n^\pm];a,b;q,t;p).\end{aligned}$$

A very different generalization of the interpolation functions is given by Rains (2018) in the form of an interpolation kernel $\mathcal{K}_c(x_1,\dots,x_n;y_1,\dots,y_n;q,t;p)$. By specializing $y_i=q^{\lambda_i}t^{n-i}a/c$ with $c=\sqrt{t^{n-1}ab}$ for all $1\le i\le n$ one recovers, up to a simple normalizing factor, $R^*_\lambda(x_1,\dots,x_n;a,b;q,t;p)$. In the same paper Rains uses this kernel to prove quadratic transformation formulas for elliptic Selberg integrals.

Also for the biorthogonal functions we have omitted a number of further results, such as a quasi-Pieri formula (Rains, 2006, Theorem 5.10) and a connection coefficient formula of Askey–Wilson type (Rains, 2006, Theorem 5.6), generalizing (6.4.19).

Acknowledgements We thank Ilmar Gahramanov, Eric Rains, Michael Schlosser and Vyacheslav Spiridonov for valuable comments.

References

Askey, R., and Wilson, J. 1979. A set of orthogonal polynomials that generalize the Racah coefficients or 6-j symbols. *SIAM J. Math. Anal.*, **10**, 1008–1016.

Bazhanov, V. V., and Sergeev, S. M. 2012a. Elliptic gamma-function and multi-spin solutions of the Yang–Baxter equation. *Nuclear Phys. B*, **856**, 475–496.

Bazhanov, V. V., and Sergeev, S. M. 2012b. A master solution of the quantum Yang–Baxter equation and classical discrete integrable equations. *Adv. Theor. Math. Phys.*, **16**, 65–95.

Bazhanov, V. V., Kels, A. P., and Sergeev, S. M. 2013. Comment on star-star relations in statistical mechanics and elliptic gamma-function identities. *J. Phys. A*, **46**, 152001, 7 pp.

Bhatnagar, G. 1999. D_n basic hypergeometric series. *Ramanujan J.*, **3**, 175–203.

Bhatnagar, G., and Schlosser, M. 1998. C_n and D_n very-well-poised ${}_{10}\phi_9$ transformations. *Constr. Approx.*, **14**, 531–567.

Bhatnagar, G., and Schlosser, M. J. 2018. Elliptic well-poised Bailey transforms and lemmas on root systems. *SIGMA*, **14**, 025, 44 pp.

van de Bult, F. J. 2009. *An elliptic hypergeometric beta integral transformation*. arXiv:0912.3812.

van de Bult, F. J. 2011. *Two multivariate quadratic transformations of elliptic hypergeometric integrals*. arXiv:1109.1123.

Coskun, H. 2008. An elliptic BC_n Bailey lemma, multiple Rogers–Ramanujan identities and Euler's pentagonal number theorems. *Trans. Amer. Math. Soc.*, **360**, 5397–5433.

Coskun, H., and Gustafson, R. A. 2006. Well-poised Macdonald functions W_λ and Jackson coefficients ω_λ on BC_n. Pages 127–155 of: *Jack, Hall–Littlewood and Macdonald Polynomials*. Contemp. Math., vol. 417. Amer. Math. Soc.

Date, E., Jimbo, M., Kuniba, A., Miwa, T., and Okado, M. 1988. Exactly solvable SOS models. II. Proof of the star-triangle relation and combinatorial identities. Pages 17–122 of: *Conformal Field theory and Solvable Lattice Models*. Adv. Stud. Pure Math., vol. 16. Academic Press.

Denis, R. Y., and Gustafson, R. A. 1992. An SU(n) q-beta integral transformation and multiple hypergeometric series identities. *SIAM J. Math. Anal.*, **23**, 552–561.

van Diejen, J. F., and Spiridonov, V. P. 2000. An elliptic Macdonald–Morris conjecture and multiple modular hypergeometric sums. *Math. Res. Lett.*, **7**, 729–746.

van Diejen, J. F., and Spiridonov, V. P. 2001a. Elliptic Selberg integrals. *Int. Math. Res. Not.*, 1083–1110.

van Diejen, J. F., and Spiridonov, V. P. 2001b. Modular hypergeometric residue sums of elliptic Selberg integrals. *Lett. Math. Phys.*, **58**(3), 223–238.

van Diejen, J. F., and Spiridonov, V. P. 2005. Unit circle elliptic beta integrals. *Ramanujan J.*, **10**, 187–204.

Dolan, F. A., and Osborn, H. 2009. Applications of the superconformal index for protected operators and q-hypergeometric identities to $\mathcal{N} = 1$ dual theories. *Nuclear Phys. B*, **818**, 137–178.

Frenkel, I. B., and Turaev, V. G. 1997. Elliptic solutions of the Yang–Baxter equation and modular hypergeometric functions. Pages 171–204 of: *The Arnold–Gelfand Mathematical Seminars*. Birkhäuser.

Gadde, A., Pomoni, E., Rastelli, L., and Razamat, S. S. 2010a. S-duality and 2d topological QFT. *J. High Energy Phys.*, 032, 22 pp.

Gadde, A., Rastelli, L., Razamat, S. S., and Yan, W. 2010b. The superconformal index of the E_6 SCFT. *J. High Energy Phys.*, 107, 27 pp.

Gasper, G., and Rahman, M. 2004. *Basic Hypergeometric Series*. Second edn. Encyclopedia of Mathematics and Its Applications, vol. 96. Cambridge University Press.

Gustafson, R. A. 1992. Some q-beta and Mellin–Barnes integrals with many parameters associated to the classical groups. *SIAM J. Math. Anal.*, **23**, 525–551.

Gustafson, R. A. 1994. Some q-beta integrals on SU(n) and Sp(n) that generalize the Askey-Wilson and Nasrallah-Rahman integrals. *SIAM J. Math. Anal.*, **25**, 441–449.

Gustafson, R. A., and Rakha, M. A. 2000. q-Beta integrals and multivariate basic hypergeometric series associated to root systems of type A_m. *Ann. Comb.*, **4**, 347–373.

Ito, M., and Noumi, M. 2017a. Derivation of a BC_n elliptic summation formula via the fundamental invariants. *Constr. Approx.*, **45**, 33–46.

Ito, M., and Noumi, M. 2017b. Evaluation of the BC_n elliptic Selberg integral via the fundamental invariants. *Proc. Amer. Math. Soc.*, **145**, 689–703.

Kac, V. G. 1990. *Infinite-dimensional Lie Algebras*. Third edn. Cambridge University Press.

Kajihara, Y. 2004. Euler transformation formula for multiple basic hypergeometric series of type A and some applications. *Adv. Math.*, **187**, 53–97.

Kajihara, Y., and Noumi, M. 2003. Multiple elliptic hypergeometric series. An approach from the Cauchy determinant. *Indag. Math. (N.S.)*, **14**, 395–421.

Kels, A. P., and Yamazaki, M. 2018. Elliptic hypergeometric sum/integral transformations and supersymmetric lens index. *SIGMA*, **14**, 013, 29 pp.

Koekoek, R., Lesky, P. A., and Swarttouw, R. F. 2010. *Hypergeometric Orthogonal Polynomials and their q-Analogues*. Springer.

Komori, Y., Masuda, Y., and Noumi, M. 2016. Duality transformation formulas for multiple elliptic hypergeometric series of type BC. *Constr. Approx.*, **44**, 483–516.

Koornwinder, T. H. 1992. Askey–Wilson polynomials for root systems of type BC. Pages 189–204 of: *Hypergeometric Functions on Domains of Positivity, Jack Polynomials, and Applications*. Contemp. Math., vol. 138. Amer. Math. Soc.

Langer, R., Schlosser, M. J., and Warnaar, S. O. 2009. Theta functions, elliptic hypergeometric series, and Kawanaka's Macdonald polynomial conjecture. *SIGMA*, **5**, Paper 055, 20 pp.

Macdonald, I. G. 1972. Affine root systems and Dedekind's η-function. *Invent. Math.*, **15**, 91–143.

Macdonald, I. G. 1995. *Symmetric Functions and Hall Polynomials*. Second edn. Oxford University Press.

Milne, S. C. 1997. Balanced ${}_3\phi_2$ summation theorems for U(n) basic hypergeometric series. *Adv. Math.*, **131**, 93–187.

Milne, S. C., and Lilly, G. M. 1995. Consequences of the A_l and C_l Bailey transform and Bailey lemma. *Discrete Math.*, **139**, 319–346.

Milne, S. C., and Newcomb, J. W. 1996. U(n) very-well-poised ${}_{10}\phi_9$ transformations. *J. Comput. Appl. Math.*, **68**, 239–285.

Okounkov, A. 1998. BC-type interpolation Macdonald polynomials and binomial formula for Koornwinder polynomials. *Transform. Groups*, **3**, 181–207.

Rahman, M. 1986. An integral representation of a ${}_{10}\varphi_9$ and continuous bi-orthogonal ${}_{10}\varphi_9$ rational functions. *Canad. J. Math.*, **38**, 605–618.

Rains, E. M. 2006. BC_n-symmetric Abelian functions. *Duke Math. J.*, **135**, 99–180.

Rains, E. M. 2009. Limits of elliptic hypergeometric integrals. *Ramanujan J.*, **18**, 257–306.

Rains, E. M. 2010. Transformations of elliptic hypergeometric integrals. *Ann. of Math. (2)*, **171**, 169–243.

Rains, E. M. 2012. Elliptic Littlewood identities. *J. Combin. Theory Ser. A*, **119**, 1558–1609.

Rains, E. M. 2018. Multivariate quadratic transformations and the interpolation kernel. *SIGMA*, **14**, 019, 69 pp.

Rains, E. M., and Spiridonov, V. P. 2009. Determinants of elliptic hypergeometric integrals. *Funct. Anal. Appl.*, **43**, 297–311.

Rastelli, L., and Razamat, S. S. 2017. The supersymmetric index in four dimensions. *J. Phys. A*, **50**, 443013, 34 pp.

Razamat, S. S. 2014. On the $\mathcal{N} = 2$ superconformal index and eigenfunctions of the elliptic RS model. *Lett. Math. Phys.*, **104**, 673–690.

Rosengren, H. 2001. A proof of a multivariable elliptic summation formula conjectured by Warnaar. Pages 193–202 of: *q-Series with Applications to Combinatorics, Number Theory, and Physics*. Contemp. Math., vol. 291. Amer. Math. Soc.

Rosengren, H. 2004. Elliptic hypergeometric series on root systems. *Adv. Math.*, **181**, 417–447.

Rosengren, H. 2006. New transformations for elliptic hypergeometric series on the root system A_n. *Ramanujan J.*, **12**, 155–166.

Rosengren, H. 2011. Felder's elliptic quantum group and elliptic hypergeometric series on the root system A_n. *Int. Math. Res. Not.*, 2861–2920.

Rosengren, H. 2017. Gustafson–Rakha-type elliptic hypergeometric series. *SIGMA*, **13**, Paper 037, 11 pp.

Rosengren, H., and Schlosser, M. 2003. Summations and transformations for multiple basic and elliptic hypergeometric series by determinant evaluations. *Indag. Math. (N.S.)*, **14**, 483–513.

Schlosser, M. 1997. Multidimensional matrix inversions and A_r and D_r basic hypergeometric series. *Ramanujan J.*, **1**, 243–274.

Schlosser, M. 2000. Summation theorems for multidimensional basic hypergeometric series by determinant evaluations. *Discrete Math.*, **210**, 151–169.

Schlosser, M. 2007. Elliptic enumeration of nonintersecting lattice paths. *J. Combin. Theory Ser. A*, **114**, 505–521.

Spiridonov, V. P. 2001. On the elliptic beta function. *Russian Math. Surveys*, **56**, 185–186.

Spiridonov, V. P. 2004. Theta hypergeometric integrals. *St. Petersburg Math. J.*, **15**, 929–967.

Spiridonov, V. P. 2007a. Elliptic hypergeometric functions and models of Calogero–Sutherland type. *Theoret. Math. Phys.*, **150**, 266–277.

Spiridonov, V. P. 2007b. Short proofs of the elliptic beta integrals. *Ramanujan J.*, **13**, 265–283.

Spiridonov, V. P. 2012. Elliptic beta integrals and solvable models of statistical mechanics. Pages 181–211 of: *Algebraic Aspects of Darboux Transformations, Quantum Integrable Systems and Supersymmetric Quantum Mechanics*. Contemp. Math., vol. 563. Amer. Math. Soc.

Spiridonov, V. P. 2018. Rarefied elliptic hypergeometric functions. *Adv. Math.*, **331**, 830–873.

Spiridonov, V. P., and Vartanov, G. S. 2011. Elliptic hypergeometry of supersymmetric dualities. *Comm. Math. Phys.*, **304**, 797–874.

Spiridonov, V. P., and Vartanov, G. S. 2012. Superconformal indices of $\mathcal{N} = 4$ SYM field theories. *Lett. Math. Phys.*, **100**, 97–118.

Spiridonov, V. P., and Vartanov, G. S. 2014. Elliptic hypergeometry of supersymmetric dualities II. Orthogonal groups, knots, and vortices. *Comm. Math. Phys.*, **325**, 421–486.

Spiridonov, V. P., and Warnaar, S. O. 2006. Inversions of integral operators and elliptic beta integrals on root systems. *Adv. Math.*, **207**, 91–132.

Spiridonov, V. P., and Warnaar, S. O. 2011. New multiple ${}_6\psi_6$ summation formulas and related conjectures. *Ramanujan J.*, **25**, 319–342.

Spiridonov, V., and Zhedanov, A. 2000a. Classical biorthogonal rational functions on elliptic grids. *C. R. Math. Acad. Sci. Soc. R. Can.*, **22**, 70–76.

Spiridonov, V., and Zhedanov, A. 2000b. Spectral transformation chains and some new biorthogonal rational functions. *Comm. Math. Phys.*, **210**, 49–83.

Warnaar, S. O. 2002. Summation and transformation formulas for elliptic hypergeometric series. *Constr. Approx.*, **18**, 479–502.

Wilson, J. A. 1991. Orthogonal functions from Gram determinants. *SIAM J. Math. Anal.*, **22**, 1147–1155.

7

Dunkl Operators and Related Special Functions

Charles F. Dunkl

7.1 Introduction

Functions like the exponential, the Chebyshev polynomials and the monomial symmetric polynomials are pre-eminent among all special functions. They have simple definitions and can be expressed using easily specified integers like $n!$. Families of functions like Bessel functions, Gegenbauer, Jacobi and Jack symmetric polynomials are labeled by parameters. These could be unspecified transcendental numbers or drawn from large sets of real numbers, for example the complement of $\{-\frac{1}{2}, -\frac{3}{2}, -\frac{5}{2}, \dots\}$. It is well known that the Gegenbauer polynomials C_n^λ are realized as zonal spherical harmonics for the unit sphere in $\mathbb{R}^d$ for $\lambda = \frac{1}{2}d - 1$. One aim of this chapter is to provide a harmonic analysis setting for any parameter value (generally restricted to be positive) for such situations. As an example consider the weight function $|\sin\theta|^{2\lambda}$ on the unit circle $\{(\cos\theta, \sin\theta) \mid -\pi < \theta \le \pi\}$ in $\mathbb{R}^2$. Then $\{C_n^\lambda(\cos\theta), \sin\theta\, C_n^{\lambda+1}(\cos\theta)\}_{n\ge 0}$ is an orthogonal basis for the associated L^2-space. There is a second-order differential operator which has the polynomials $C_n^\lambda(\cos\theta)$ as eigenfunctions but not the other polynomials in the basis. By introducing difference terms one can define an operator which has each basis element as an eigenfunction. In this situation the difference term comes from the reflection $(x_1, x_2) \mapsto (x_1, -x_2)$. This is the simplest example of the structure to be described in this chapter. The basic objects are finite reflection (Coxeter) groups and algebras of operators on polynomials which generalize the algebra of partial differential operators. These algebras have as many parameters as the number of conjugacy classes of reflections in the associated groups. The first-order operators have acquired the name *Dunkl operators*.

Coxeter groups are related to root systems. This chapter begins with a presentation of these systems, the definition of a Coxeter group and the classification of the indecomposable systems. Then the theory of the operators is developed in detail with an emphasis on inner products and their relation to Macdonald–Opdam integrals, a generalization of the Fourier transform, the Laplacian operator and harmonic polynomials, and applications to particular groups to study Gegenbauer, Jacobi and nonsymmetric Jack polynomials.

For $x, y \in \mathbb{R}^d$ the *inner product* is $\langle x, y\rangle := \sum_{j=1}^d x_j y_j$ and the *norm* is $|x| := \langle x, x\rangle^{1/2}$. A matrix $w = (w_{ij})_{i,j=1}^d$ is called *orthogonal* if $ww^T = I_d$. The group of orthogonal $d \times d$ matrices is denoted by $O(d)$. The standard unit basis vectors of $\mathbb{R}^d$ are denoted by ε_i, $1 \le i \le d$. The nonnegative integers $\{0, 1, 2, \ldots\}$ are denoted by $\mathbb{N}_0$ and the cardinality of a set S is denoted by $\#S$.

Definition 7.1.1 For $a \in \mathbb{R}^d \backslash \{0\}$, the *reflection along* a, denoted by s_a, is defined by

$$s_a x := x - 2\frac{\langle x, a\rangle}{|a|^2} a.$$

Writing $s_a = I_d - 2(a^T a)^{-1} a a^T$ shows that $s_a = s_a^T$ and $s_a^2 = I_d$, i.e., $s_a \in O(d)$. The matrix entries of s_a are $(s_a)_{ij} = \delta_{ij} - 2a_i a_j/|a|^2$. The hyperplane $a^\perp := \{x \mid \langle x, a\rangle = 0\}$ is the invariant set for s_a. Also $s_a a = -a$, and any nonzero multiple of a determines the same reflection. The reflection s_a has one eigenvector for the eigenvalue -1 and $d-1$ independent eigenvectors for the eigenvalue $+1$ and $\det s_a = -1$.

For $a, b \in \mathbb{R}^d \backslash \{0\}$, let $\cos \measuredangle(a, b) := \langle a, b\rangle / (|a|\,|b|)$; if a, b are linearly independent in $\mathbb{R}^d$ then $s_a s_b$ is a plane rotation in $\mathrm{span}_{\mathbb{R}}\{a, b\}$ through an angle $2\measuredangle(a, b)$. Consequently, for given $m = 1, 2, 3, \ldots$, $(s_a s_b)^m = I_d$ if and only if $\cos \measuredangle(a, b) = \cos(\pi j/m)$ for some integer j. Two reflections s_a and s_b commute if and only if $\langle a, b\rangle = 0$, since $(s_a s_b)^{-1} = s_b s_a$, in general. The conjugate of a reflection is again a reflection: suppose $w \in O(d)$ and $a \in \mathbb{R}^d \backslash \{0\}$ then $w s_a w^{-1} = s_{wa}$.

7.2 Root Systems

Definition 7.2.1 A *root system* is a finite set $R \subset \mathbb{R}^d \backslash \{0\}$ such that $a, b \in R$ implies $s_b a \in R$. If, additionally, $\mathbb{R}a \cap R = \{\pm a\}$ for each $a \in R$ then R is said to be *reduced*. The *rank* of R is defined to be $\dim(\mathrm{span}_{\mathbb{R}}(R))$.

If the root system is not reduced then generally the crystallographic condition is imposed (see Definition 7.2.5 below). Note that $a \in R$ implies $-a = s_a a \in R$ for any root system. By choosing some $a_0 \in \mathbb{R}^d$ such that $\langle a, a_0\rangle \neq 0$ for all $a \in R$ one defines the *set of positive roots* to be $R_+ = \{a \in R \mid \langle a, a_0\rangle > 0\}$. Also set $R_- := -R_+$ so that $R = R_+ \cup R_-$, the disjoint union of the sets of positive and negative roots.

Definition 7.2.2 The *Coxeter group* or *finite reflection group* $W(R)$ is defined to be the subgroup of $O(d)$ generated by $\{s_a\}_{a \in R_+}$.

The definition is independent of the choice of R_+ since $s_{-a} = s_a$. Thus R_+ is a convenient index set for the reflections in the reduced case. The group $W(R)$ is finite because each $w \in W(R)$ fixes the orthogonal complement of $\mathrm{span}_{\mathbb{R}}(R)$ pointwise and s_a permutes the finite set R for each $a \in R$.

There is a distinguished set $S = \{r_1, \ldots, r_n\}$, $n = \mathrm{rank}(R)$, of positive roots, called the *simple roots*, such that S is a basis for $\mathrm{span}_{\mathbb{R}}(R)$ and $a \in R_+$ implies $a = \sum_{i=1}^n c_i r_i$ with $c_i \geq 0$ (see Humphreys, 1990, Theorem 1.3). The corresponding reflections $s_i := s_{r_i}$ $(i = 1, \ldots, n)$ are called *simple reflections* and $\{s_1, \ldots, s_n\}$ generates $W(R)$.

Definition 7.2.3 For a reduced root system R the *length* of $w \in W(R)$ is $\ell(w) := \#(wR_+ \cap R_-)$.

Equivalently (see Humphreys, 1990, Corollary 1.7), $\ell(w)$ equals the number of factors in the shortest product $w = s_{i_1} s_{i_2} \cdots s_{i_m}$ for expressing w in terms of simple reflections. In particular, w has length 1 if and only if it is a simple reflection.

For the purpose of studying the group $W(R)$ one can replace R by a reduced root system (example: $\{2^{1/2}a/|a|\}_{a\in R}$ is a commonly used normalization).

Definition 7.2.4 The *discriminant*, or *alternating polynomial*, of the reduced root system R is

$$a_R(x) := \prod\nolimits_{a\in R_+} \langle x, a\rangle.$$

If R is reduced and $b \in R_+$ then $\langle s_b x, b\rangle = -\langle x, b\rangle$ and $\#\{a \in R_+\backslash\{b\} \mid s_b a \in -R_+\}$ is even. It follows that $a_R(s_b x) = -a_R(x)$ and furthermore that $a_R(wx) = \det(w) a_R(x)$, $w \in W(R)$.

The set of reflections in $W(R)$ is exactly $\{s_a\}_{a\in R_+}$. This is a consequence of a divisibility property: if $b \in \mathbb{R}^d\backslash\{0\}$ and p is a polynomial such that $p(s_b x) = -p(x)$ for all $x \in \mathbb{R}^d$, then $p(x)$ is divisible by $\langle x, b\rangle$. Indeed, without loss of generality assume that $b = \varepsilon_1$. Let $p(x) = \sum_{j=0}^n x_1^j p_j(x_2,\ldots,x_d)$; then $p(s_b x) = \sum_{j=0}^n (-x_1)^j p_j(x_2,\ldots,x_d)$ and $p(s_b x) = -p(x)$ implies $p_j = 0$ unless j is odd. If $w = s_b \in W(R)$ is a reflection then $\det w = -1$ and so $a_R(wx) = -a_R(x)$ and hence $a_R(x)$ is divisible by $\langle x, b\rangle$. Linear factors are irreducible and the unique factorization theorem shows that some multiple of b is an element of R_+.

Definition 7.2.5 The root system R is called *crystallographic* if $2\langle a, b\rangle/|b|^2 \in \mathbb{Z}$ for all $a, b \in R$.

In the crystallographic case R is in the $\mathbb{Z}$-lattice generated by the simple roots and $W(R)$ acts on this lattice (see Humphreys, 1972, §10.1). In this case $W(R)$ is a *Weyl group*. In some contexts, for example where the root system itself is the basic object, the crystallographic condition is assumed to hold.

If R can be expressed as a disjoint union $R_1 \cup R_2$ of nonempty sets with $\langle a, b\rangle = 0$ for every $a \in R_1$, $b \in R_2$ then each R_i $(i = 1, 2)$ is itself a root system and $W(R) = W(R_1) \times W(R_2)$, a direct product. Furthermore, $W(R_1)$ and $W(R_2)$ act on the orthogonal subspaces $\mathrm{span}_{\mathbb{R}}(R_1)$ and $\mathrm{span}_{\mathbb{R}}(R_2)$, respectively. In this case, the root system R and the reflection group $W(R)$ are called *decomposable*. Otherwise the system and group are *indecomposable* (also called *irreducible*). There is a complete classification of indecomposable finite reflection groups.

Assume that the rank of R is d, i.e., $\mathrm{span}_{\mathbb{R}}(R) = \mathbb{R}^d$. The complement of the union $H := \bigcup_{a\in R_+} a^\perp$ of hyperplanes consists of connected (open) components called *chambers*. The order of the group equals the number of chambers (see Humphreys, 1990, Theorems 1.4, 1.8). Recall that $R_+ = \{a \in R \mid \langle a, a_0\rangle > 0\}$ for some $a_0 \in \mathbb{R}^d\backslash H$. The connected component of $\mathbb{R}^d\backslash H$ which contains a_0 is called the *fundamental chamber*. The simple roots correspond to the bounding hyperplanes of this chamber and they form a basis of $\mathbb{R}^d$ (note that the definitions of chamber and fundamental chamber extend to the situation where $\mathrm{span}_{\mathbb{R}}(R)$ is embedded in a higher-dimensional space). The simple reflections s_i, $i = 1, \ldots, d$ correspond to the simple roots. Let m_{ij} be the order of $s_i s_j$ (clearly $m_{ii} = 1$ and $m_{ij} = 2$ if and only if $s_i s_j = s_j s_i$, for $i \neq j$). The group $W(R)$ is isomorphic to the abstract group generated by $\{s_i\}_{i=1}^d$ subject to the relations $(s_i s_j)^{m_{ij}} = 1$ (see Humphreys, 1990, Theorem 1.9).

The *Coxeter diagram* is a graphical way of displaying the relations: it is a graph with d nodes corresponding to the simple reflections, the nodes i and j are joined with an edge when $m_{ij} > 2$; the edge is labeled by m_{ij} when $m_{ij} > 3$. The root system is indecomposable if and only if the Coxeter diagram is connected. Below are given brief descriptions of the indecomposable root systems and the corresponding groups. The rank is indicated by the subscript. The systems are crystallographic except for H_3, H_4 and $I_2(m)$, $m \notin \{2, 3, 4, 6\}$.

7.2.1 Type A_{d-1}

The root system is $R = \{\varepsilon_i - \varepsilon_j\}_{i \neq j} \subset \mathbb{R}^d$. The span is $(\sum_{i=1}^d \varepsilon_i)^{\perp}$, thus the rank is $d - 1$. The reflection $s_{ij} = s_{\varepsilon_i - \varepsilon_j}$ interchanges the components x_i and x_j of each $x \in \mathbb{R}^d$ and is called a *transposition*, often denoted by (ij). Thus $W(R)$ is the symmetric (or permutation) group $\mathcal{S}_d$ of d objects. Choose $a_0 := \sum_{i=1}^d (d + 1 - i)\varepsilon_i$; then $R_+ = \{\varepsilon_i - \varepsilon_j\}_{i<j}$ and the simple roots are $\{\varepsilon_i - \varepsilon_{i+1}\}_{i=1}^{d-1}$. The corresponding reflections are the adjacent transpositions $s_i = (i, i + 1)$. The structure constants satisfy $m_{i,i+1} = 3$ and $m_{ij} < 3$ otherwise. The Coxeter diagram is $\circ - \circ - \circ - \cdots - \circ$. The alternating polynomial is $a_R(x) = \prod_{1 \le i < j \le d}(x_i - x_j)$. The fundamental chamber is $\{x \mid x_1 > x_2 > \cdots > x_d\}$.

7.2.2 Type B_d

The root system is $R = \{\pm \varepsilon_i \pm \varepsilon_j \mid 1 \le i < j \le d\} \cup \{\pm \varepsilon_i \mid 1 \le i \le d\}$. For $d = 1$, $R = \{\pm\varepsilon_1\}$, which is essentially the same as A_1. The group $W(B_d)$ is the full symmetry group of the hyperoctahedron $\{\pm\varepsilon_1, \pm\varepsilon_2, \ldots, \pm\varepsilon_d\} \subset \mathbb{R}^d$ (also of the hypercube) and is thus called the *hyperoctahedral group*. Its elements are the $d \times d$ generalized permutation matrices with entries ± 1 (i.e., each row and each column has exactly one nonzero element ± 1). With the same a_0 as used for A_{d-1}, the positive root system is $R_+ = \{\varepsilon_i - \varepsilon_j, \varepsilon_i + \varepsilon_j\}_{i<j} \cup \{\varepsilon_i\}_{1 \le i \le d}$ and the simple roots are $\{\varepsilon_i - \varepsilon_{i+1}\}_{i<d} \cup \{\varepsilon_d\}$. The order of $s_{\varepsilon_{d-1} - \varepsilon_d} s_{\varepsilon_d}$ is 4. The Coxeter diagram is $\circ - \circ - \circ - \cdots - \circ \overset{4}{-} \circ$. The alternating polynomial is $a_R(x) = \prod_{i=1}^d x_i \cdot \prod_{1 \le i < j \le d}(x_i^2 - x_j^2)$. The fundamental chamber is $\{x \mid x_1 > x_2 > \cdots > x_d > 0\}$.

7.2.3 Types C_d and BC_d

The root system C_d is $\{\pm \varepsilon_i \pm \varepsilon_j \mid 1 \le i < j \le d\} \cup \{2\varepsilon_i \mid 1 \le i \le d\}$ and the system BC_d is $\{\pm \varepsilon_i \pm \varepsilon_j \mid 1 \le i < j \le d\} \cup \{2\varepsilon_i, \varepsilon_i \mid 1 \le i \le d\}$. The latter system is not reduced. Both of these systems generate the same group as $W(B_d)$. The system C_d has to be distinguished from B_d because of the crystallographic condition in Definition 7.2.5. For $d = 1, 2$ there is no essential distinction.

7.2.4 Type D_d

For $d \ge 4$ the root system is $R = \{\pm \varepsilon_i \pm \varepsilon_j \mid 1 \le i < j \le d\}$, a subset of B_d. The group $W(D_d)$ is the subgroup of $W(B_d)$ fixing the polynomial $\prod_{j=1}^d x_j$. The simple roots are given by $\{\varepsilon_i - \varepsilon_{i+1}\}_{i=1}^{d-1} \cup \{\varepsilon_{d-1} + \varepsilon_d\}$ and the Coxeter diagram is $\circ - \circ - \circ - \cdots - \circ <^{\circ}_{\circ}$. The alternating polynomial is $a_R(x) = \prod_{1 \le i < j \le d}(x_i^2 - x_j^2)$. The fundamental chamber is $\{x \mid x_1 > x_2 > \cdots > x_{d-1} > |x_d|\}$.

7.2.5 Type F_4

The root system is $R_1 \cup R_2$ where $R_1 = \{\pm\varepsilon_i \pm \varepsilon_j \mid 1 \le i < j \le 4\}$ and $R_2 = \{\pm\varepsilon_i\}_{i=1}^{4} \cup \{\frac{1}{2}(\pm\varepsilon_1 \pm \varepsilon_2 \pm \varepsilon_3 \pm \varepsilon_4)\}$. Each set contains 24 roots. The group $W(F_4)$ contains $W(B_4)$ as a subgroup of index 3. The simple roots are $\varepsilon_2 - \varepsilon_3$, $\varepsilon_3 - \varepsilon_4$, ε_4, $\frac{1}{2}(\varepsilon_1 - \varepsilon_2 - \varepsilon_3 - \varepsilon_4)$ and the Coxeter diagram is $\circ - \circ \overset{4}{-} \circ - \circ$. With the orthogonal coordinates $y_1 = 2^{-1/2}(-x_1 + x_2)$, $y_2 = 2^{-1/2}(x_1 + x_2)$, $y_3 = 2^{-1/2}(-x_3 + x_4)$, $y_4 = 2^{-1/2}(x_3 + x_4)$, the alternating polynomial is $a_R(x) = 2^{-6}\prod_{1 \le i < j \le 4}(x_i^2 - x_j^2)(y_i^2 - y_j^2)$.

7.2.6 Type G_2

This system is a subset $R_1 \cup R_2$ of $\{\sum_{i=1}^{3} x_i\varepsilon_i \mid \sum_{i=1}^{3} x_i = 0\} \subset \mathbb{R}^3$, where $R_1 = \{\pm(\varepsilon_i - \varepsilon_j) \mid 1 \le i < j \le 3\}$ and $R_2 = \{\pm(2\varepsilon_1 - \varepsilon_2 - \varepsilon_3), \pm(2\varepsilon_2 - \varepsilon_3 - \varepsilon_1), \pm(2\varepsilon_3 - \varepsilon_1 - \varepsilon_2)\}$. The simple roots are $\varepsilon_1 - \varepsilon_2$, $-2\varepsilon_1 + \varepsilon_2 + \varepsilon_3$ and the Coxeter diagram is $\circ \overset{6}{-} \circ$.

7.2.7 Types E_6, E_7, E_8

The root system E_8 equals $R_1 \cup R_2$ where

$$R_1 = \{\pm\varepsilon_i \pm \varepsilon_j \mid 1 \le i < j \le 8\}, \quad R_2 = \left\{\tfrac{1}{2}\textstyle\sum_{i=1}^{8}(-1)^{n_i}\varepsilon_i \mid n_i \in \{0,1\},\ \sum_{i=1}^{8} n_i = 0 \bmod 2\right\},$$

with $\#R_1 = 112$ and $\#R_2 = 128$. The simple roots are $r_1 = \frac{1}{2}(\varepsilon_1 - \sum_{i=2}^{7}\varepsilon_i + \varepsilon_8)$, $r_2 = \varepsilon_1 + \varepsilon_2$, $r_i = \varepsilon_{i-1} - \varepsilon_{i-2}$ for $3 \le i \le 8$. The systems E_6 and E_7 consist of the elements of R which lie in $\mathrm{span}_{i=1}^{6}\{r_i\}$ and $\mathrm{span}_{i=1}^{7}\{r_i\}$, respectively. (Because these systems are crystallographic, the spans are over $\mathbb{Z}$.) For more details on these systems see Humphreys (1990, pp. 42–43).

7.2.8 Type $I_2(m)$

These are the dihedral systems corresponding to symmetry groups of regular m-gons in $\mathbb{R}^2$ for $m \ge 3$. Using a complex coordinate system $z = x_1 + \mathrm{i}x_2$ and $\bar{z} = x_1 - \mathrm{i}x_2$, the map $z \mapsto ze^{\mathrm{i}\theta}$ is a rotation through the angle θ and the reflection along $(\sin\phi, -\cos\phi)$ is $z \mapsto \bar{z}e^{2\mathrm{i}\phi}$. The reflection along $v_{(j)} = (\sin\frac{\pi j}{m}, -\cos\frac{\pi j}{m})$ corresponds to $s_j\colon z \mapsto \bar{z}e^{2\pi\mathrm{i}j/m}$ $(1 \le j \le 2m)$; note that $v_{(m+j)} = -v_{(j)}$. For $a_0 = (\cos\frac{\pi}{2m}, \sin\frac{\pi}{2m})$, the positive roots are $\{v_{(j)}\}_{j=1}^{m}$ and the simple roots are $v_{(1)}, v_{(m)}$. Then $s_1 s_m$ maps z to $ze^{2\pi\mathrm{i}/m}$ and has period m. The Coxeter diagram is $\circ \overset{m}{-} \circ$. Since $s_j s_n s_j = s_{2j-n}$ for any n, j, there are two conjugacy classes of reflections $\{s_{2i}\}$, $\{s_{2i+1}\}$ when m is even and one class when m is odd. There are three special cases: the groups $W(I_2(3))$, $W(I_2(4))$, $W(I_2(6))$ are isomorphic to $W(A_2)$, $W(B_2)$, $W(G_2)$, respectively. The alternating polynomial is a multiple of $(z^m - \bar{z}^m)/\mathrm{i}$.

7.2.9 *Type* H_3

Let $\tau := \frac{1}{2}(1+\sqrt{5})$ (so $\tau^2 = \tau + 1$). Take the positive root system to be $R_+ = \{(2,0,0), (0,2,0), (0,0,2), (\tau, \pm\tau^{-1}, \pm 1), (\pm 1, \tau, \pm\tau^{-1}), (\tau^{-1}, \pm 1, \tau), (-\tau^{-1}, 1, \tau), (\tau^{-1}, 1, -\tau)\}$, thus $\#R_+ = 15$. The root system $R = R_+ \cup (-R_+)$ as a configuration in $\mathbb{R}^3$ is called the *icosidodecahedron*. The group $W(H_3)$ is the symmetry group of the icosahedron Q_{12} (12 vertices, 20 triangular faces) and of the dodecahedron Q_{20} (20 vertices, 12 pentagonal faces), which are given by

$$Q_{12} := \{(0, \pm\tau, \pm 1), (\pm 1, 0, \pm\tau), (\pm\tau, \pm 1, 0)\},$$
$$Q_{20} := \{(0, \pm\tau^{-1}, \pm\tau), (\pm\tau, 0, \pm\tau^{-1}), (\pm\tau^{-1}, \pm\tau, 0), (\pm 1, \pm 1, \pm 1)\}.$$

The simple roots are $(\tau, -\tau^{-1}, -1)$, $(-1, \tau, -\tau^{-1})$, $(\tau^{-1}, -1, \tau)$ and the Coxeter diagram is $\circ - \circ \overset{5}{-} \circ$.

7.2.10 *Type* H_4

This root system has 60 positive roots and the Coxeter diagram is $\circ - \circ - \circ \overset{5}{-} \circ$. The group $W(H_4)$ is the symmetry group of the 600-cell and is sometimes called the *hecatonicosahedroidal group*. See Humphreys (1990, §2.13).

7.2.11 *Miscellaneous Results*

For any root system, the subgroup generated by a subset of simple reflections (i.e., the result of deleting one or more nodes from the Coxeter diagram) is called a *parabolic subgroup* of $W(R)$. More generally, any conjugate of such a subgroup is also called *parabolic*. Therefore, for any $a \in \mathbb{R}^d$ the stabilizer $W_a := \{w \in W(R) \mid wa = a\}$ is parabolic (see Humphreys, 1990, §1.10). The number of conjugacy classes of reflections equals the number of connected components of the Coxeter diagram after all edges with an even label have been removed (see Dunkl and Xu, 2014, p. 183).

7.3 Invariant Polynomials

For $w \in O(d)$ and $p \in \Pi^d$, the space of polynomials on $\mathbb{R}^d$, let $(wp)(x) := p(w^{-1}x)$ (thus $((w_1 w_2)p)(x) = (w_1(w_2 p))(x)$, $w_1, w_2 \in O(d)$). For a finite subgroup G of $O(d)$ let Π^G denote the space of G-invariant polynomials $\{p \in \Pi^d \mid wp = p \text{ for all } w \in G\}$.

When G is a finite reflection group $W(R)$, Π^G has an elegant structure: there is a set of algebraically independent homogeneous generators, whose degrees are fundamental constants associated with R.

Theorem 7.3.1 (Humphreys, 1990, Theorems 3.5, 3.9) *For R a root system in $\mathbb{R}^d$, there exist d algebraically independent $W(R)$-invariant homogeneous polynomials q_j of degree n_j ($j = 1, \dots, d$) such that $\Pi^{W(R)}$ is the ring of polynomials generated by $\{q_j\}$. Furthermore, $\#W(R) = n_1 n_2 \cdots n_d$ and the number of reflections in $W(R)$ is $\sum_{j=1}^{d}(n_j - 1)$.*

For an indecomposable root system R of rank d the numbers n_j ($j - 1, \ldots, d$) are called the *fundamental degrees* of $W(R)$ (or sometimes the *primitive degrees*). The coefficient of t^k in the product $\prod_{j=1}^{d}(1 + (n_j - 1)t)$ is the number of elements of $W(R)$ whose fixed point set is of codimension k (see Humphreys, 1990, Remark 3.9). The structural constants (see Humphreys, 1990, Table 3.1) are shown below.

Type	$\#R_+$	$\#W(R)$	$n_1, \ldots, n_d$
A_d	$\frac{1}{2}d(d+1)$	$(d+1)!$	$2, 3, \ldots, d+1$
B_d	d^2	$2^d d!$	$2, 4, 6, \ldots, 2d$
D_d	$d(d-1)$	$2^{d-1}d!$	$2, 4, 6, \ldots, 2(d-1), d$
G_2	6	12	2, 6
H_3	15	120	2, 6, 10
F_4	24	1152	2, 6, 8, 12
H_4	60	14400	2, 12, 20, 30
E_6	36	$2^7 \times 3^4 \times 5$	2, 5, 6, 8, 9, 12
E_7	63	$2^{10} \times 3^4 \times 5 \times 7$	2, 6, 8, 10, 12, 14, 18
E_8	120	$2^{14} \times 3^4 \times 5^2 \times 7$	2, 8, 12, 14, 18, 20, 24, 30
$I_2(m)$	m	$2m$	$2, m$

7.4 Dunkl Operators

Throughout this section R denotes a reduced root system contained in $\mathbb{R}^d$ and $G = W(R)$, a subgroup of $O(d)$. For $\alpha \in \mathbb{N}_0^d$ and $x \in \mathbb{R}^d$ let $|\alpha| := \sum_{i=1}^{d} \alpha_i$ and $x^\alpha := \prod_{i=1}^{d} x_i^{\alpha_i}$, a monomial of degree $|\alpha|$. Let $\Pi_n^d := \operatorname{span}_{\mathbb{F}}\{x^\alpha \mid \alpha \in \mathbb{N}^d,\ |\alpha| = n\}$, the space of homogeneous polynomials of degree n, $n \in \mathbb{N}_0$, where $\mathbb{F}$ is some extension field of $\mathbb{R}$ containing the parameter values. Generally these values are positive real numbers so that $\mathbb{F} = \mathbb{R}$, but at other times it is convenient to use transcendental values (for example in symbolic computations).

Definition 7.4.1 A *multiplicity function* on R is a G-invariant function κ with values in $\mathbb{R}$ or a transcendental extension of $\mathbb{Q}$, i.e., $a \in R, w \in G$ implies $\kappa(wa) = \kappa(a)$. Note that $\kappa(-a) = \kappa(a)$ since $s_a a = -a$.

Suppose $a \in \mathbb{R}^d \backslash \{0\}$ and $w \in O(d)$; then $w s_a w^{-1} = s_{wa}$. Thus, κ can be considered as a function on the set of reflections $\{s_a\}_{a \in R_+}$ which is constant on conjugacy classes.

For any reflection s_a and polynomial $p \in \Pi^d$, the polynomial $p(x) - s_a p(x)$ vanishes on $a^\perp$. Hence it is divisible by $\langle x, a \rangle$. The *gradient* is denoted by ∇.

Definition 7.4.2 For $p \in \Pi^d$ and $x \in \mathbb{R}^d$, let

$$\nabla_\kappa p(x) := \nabla p(x) + \sum_{v \in R_+} \kappa(v) \frac{p(x) - s_v p(x)}{\langle x, v \rangle} v,$$

and for $a \in \mathbb{R}^d$ let $\mathcal{D}_a p(x) := \langle \nabla_\kappa p(x), a \rangle$, where $\mathcal{D}_a$ is a *Dunkl operator*. For $1 \le i \le d$ denote $\mathcal{D}_{\varepsilon_i}$ by $\mathcal{D}_i$.

Thus each $\mathcal{D}_a$ is an operator on polynomials and maps Π_n^d into Π_{n-1}^d, $n \in \mathbb{N}_0$ (i.e., $\mathcal{D}_a$ is homogeneous of degree -1). The important properties of $\{\mathcal{D}_a\}_{a\in\mathbb{R}^d}$ are G-covariance and commutativity.

Proposition 7.4.3 *Let $w \in G$ and $a \in \mathbb{R}^d$; then (as operators on Π^d) $w\mathcal{D}_a w^{-1} = \mathcal{D}_{wa}$.*

Proof Let $p \in \Pi^d$; then

$$\begin{aligned} w\mathcal{D}_a w^{-1} p(x) &= w\langle \nabla w^{-1} p(x), a\rangle + \tfrac{1}{2} w \sum_{v\in R} \kappa(v) \frac{p(wx) - p(ws_v x)}{\langle x, v\rangle} \langle v, a\rangle \\ &= \langle \nabla p(x), wa\rangle + \tfrac{1}{2} \sum_{v\in R} \kappa(v) \frac{p(x) - p(ws_v w^{-1} x)}{\langle w^{-1}x, v\rangle} \langle v, a\rangle \\ &= \langle \nabla p(x), wa\rangle + \tfrac{1}{2} \sum_{u\in R} \kappa(w^{-1}u) \frac{p(x) - p(s_u x)}{\langle x, u\rangle} \langle u, wa\rangle = \mathcal{D}_{wa} p(x). \end{aligned}$$

In the reflection part of $\mathcal{D}_a$ the sum over R_+ can be replaced by $\frac{1}{2}$ of the sum over R. Then the summation variable $v \in R$ is replaced by $v = w^{-1}u$. □

For $t \in \mathbb{R}^d$ let m_t denote the multiplier operator on Π^d given by $(m_t p)(x) := \langle t, x\rangle p(x)$. The *commutator* of two operators A, B on Π^d is $[A, B] := AB - BA$. The identities $[[A, B], C] = [[A, C], B] - [[B, C], A]$ and $[A^2, B] = A[A, B] + [A, B]A$ are used below.

Proposition 7.4.4 *For $a, t \in \mathbb{R}^d$, $[\mathcal{D}_a, m_t] = \langle a, t\rangle + 2\sum_{v\in R_+} \kappa(v) \frac{\langle a,v\rangle\langle t,v\rangle}{|v|^2} s_v$.*

Proof Let $p \in \Pi^d$. Then $\langle \nabla(\langle t, x\rangle p(x)), a\rangle - \langle t, x\rangle\langle \nabla p(x), a\rangle = \langle t, a\rangle p(x)$. Next, for any $v \in R_+$, we have

$$\begin{aligned} \frac{\langle t, x\rangle p(x) - \langle t, s_v x\rangle p(s_v x)}{\langle x, v\rangle} - \langle t, x\rangle \frac{p(x) - p(s_v x)}{\langle x, v\rangle} &= \frac{p(s_v x)}{\langle x, v\rangle}(\langle t, x\rangle - \langle s_v t, x\rangle) \\ &= \frac{2p(s_v x)\langle x, v\rangle}{\langle x, v\rangle |v|^2} \langle t, v\rangle. \end{aligned}$$

□

Lemma 7.4.5 *For $b \in \mathbb{R}^d$ and $v \in R$, we have $[s_v, \mathcal{D}_b] = 2\frac{\langle b,v\rangle}{|v|^2} s_v \mathcal{D}_v$.*

Proof Indeed, by Proposition 7.4.3 we have

$$[s_v, \mathcal{D}_b] = s_v\mathcal{D}_b - \mathcal{D}_b s_v = s_v(\mathcal{D}_b - \mathcal{D}_{s_v b}) = s_v(2\langle b, v\rangle |v|^{-2} \mathcal{D}_v).$$ □

Theorem 7.4.6 *If $a, b \in \mathbb{R}^d$ then $[\mathcal{D}_a, \mathcal{D}_b] = 0$.*

Proof Let $t \in \mathbb{R}^d$; we will show $[[\mathcal{D}_a, \mathcal{D}_b], m_t] = 0$. Indeed,

$$[[\mathcal{D}_a, \mathcal{D}_b], m_t] = [[\mathcal{D}_a, m_t], \mathcal{D}_b] - [[\mathcal{D}_b, m_t], \mathcal{D}_a]$$

and

$$[[\mathcal{D}_a, m_t], \mathcal{D}_b] = [\langle a, t\rangle, \mathcal{D}_b] + 2 \sum_{v\in R_+} \kappa(v) \frac{\langle a, v\rangle\langle t, v\rangle}{|v|^2} [s_v, \mathcal{D}_b]$$
$$= 4 \sum_{v\in R_+} \kappa(v) \frac{\langle a, v\rangle\langle t, v\rangle\langle b, v\rangle}{|v|^4} s_v \mathcal{D}_v,$$

which is symmetric in a, b. The algebra generated by $\{m_t\}_{t\in\mathbb{R}^d}$ is Π^d. For any $p \in \Pi^d$ we have $[\mathcal{D}_a, \mathcal{D}_b]p = p[\mathcal{D}_a, \mathcal{D}_b]1 = 0$. □

Note that Π^d is implicitly used in three different ways: as the abstract algebra $\mathbb{R}[x_1, \ldots, x_d]$, as a linear space of polynomial functions on $\mathbb{R}^d$ and as an algebra of multiplicative operators on the space of functions on $\mathbb{R}^d$.

The method of proof of Theorem 7.4.6 is due to Etingof and Ma (2010, Theorem 2.15) (the result was first established by Dunkl, 1989). One important consequence is that the operators $\mathcal{D}_a$ generate a commutative algebra.

Definition 7.4.7 Let $\mathcal{A}_\kappa$ denote the algebra of operators on Π^d generated by $\{\mathcal{D}_i\}_{i=1}^d$. Let ρ denote the homomorphism $\Pi^d \to \mathcal{A}_\kappa$ given by $\rho p(x_1, \ldots, x_d) := p(\mathcal{D}_1, \ldots, \mathcal{D}_d)$, $p \in \Pi^d$.

Proposition 7.4.8 *If $w \in G$ and $p \in \Pi^d$ then $\rho(wp) = w\rho(p)w^{-1}$.*

Proof It suffices to show this for first-degree polynomials. For $t \in \mathbb{R}^d$ let $p_t(x) := \langle t, x\rangle$; then $\rho p_t = \sum_{i=1}^d t_i \mathcal{D}_i = \mathcal{D}_t$. Also $wp_t(x) = \langle t, w^{-1}x\rangle = \langle wt, x\rangle = p_{wt}(x)$. Then $\rho(wp_t) = \rho(p_{wt}) = \mathcal{D}_{wt} = w\mathcal{D}_t w^{-1} = w\rho(p)w^{-1}$ by Proposition 7.4.3. □

There is a Laplacian-type operator in the algebra $\mathcal{A}_\kappa$. This operator is an important part of the analysis of the L^2-theory associated to the G-invariant weight functions.

Definition 7.4.9 For $p \in \Pi^d$ the *Dunkl Laplacian* Δ_κ is given by

$$\Delta_\kappa p(x) := \Big(\Delta + 2 \sum_{v\in R_+} \kappa(v) T_v\Big) p(x), \quad \text{where } T_v p(x) := \frac{\langle \nabla p(x), v\rangle}{\langle x, v\rangle} - \frac{|v|^2}{2} \frac{p(x) - p(s_v x)}{\langle x, v\rangle^2}.$$

Sometimes Δ_κ is called the *h-Laplacian*, corresponding to a weight function denoted by h, but here denoted by w_κ; see Definition 7.4.18.

Theorem 7.4.10 $\Delta_\kappa = \sum_{i=1}^d \mathcal{D}_i^2$, *and* $[\Delta_\kappa, m_t] = 2\mathcal{D}_t$ *for each* $t \in \mathbb{R}^d$.

Proof We use the same method as in the proof of Theorem 7.4.6, i.e., we will show that $[\Delta_\kappa, m_t] = \sum_{i=1}^d [\mathcal{D}_i^2, m_t]$ for $t \in \mathbb{R}^d$. Clearly $\Delta_\kappa 1 = 0 = \sum_{i=1}^d \mathcal{D}_i^2 1$. We have

$$\sum_{i=1}^d [\mathcal{D}_i^2, m_t] = \sum_{i=1}^d \big(\mathcal{D}_i[\mathcal{D}_i, m_t] + [\mathcal{D}_i, m_t]\mathcal{D}_i\big) = \sum_{i=1}^d \Big(2t_i \mathcal{D}_i + 2 \sum_{v\in R_+} \kappa(v) \frac{\langle t, v\rangle v_i}{|v|^2} (\mathcal{D}_i s_v + s_v \mathcal{D}_i)\Big)$$
$$= 2\mathcal{D}_t + 2 \sum_{v\in R_+} \kappa(v) \frac{\langle t, v\rangle}{|v|^2} (\mathcal{D}_v s_v + s_v \mathcal{D}_v) = 2\mathcal{D}_t,$$

because $\mathcal{D}_v s_v + s_v \mathcal{D}_v = \mathcal{D}_v s_v + \mathcal{D}_{s_v v} s_v$ and $\mathcal{D}_{s_v v} = -\mathcal{D}_v$. Next $[\Delta, m_t] = 2\sum_{i=1}^d t_i \frac{\partial}{\partial x_i}$. For $v \in R_+$ let T_v be the operator defined in Definition 7.4.9; then

$$[T_v, m_t]p(x) = \frac{\langle t, v\rangle}{\langle x, v\rangle} p(x) - \frac{|v|^2}{2} \frac{\langle x, t\rangle - \langle s_v x, t\rangle}{\langle x, v\rangle^2} p(s_v x) = \frac{\langle t, v\rangle}{\langle x, v\rangle}(p(x) - p(s_v x)),$$

since $\langle x, t\rangle - \langle s_v x, t\rangle = 2\langle x, v\rangle\langle t, v\rangle/|v|^2$ for any $p \in \Pi^d$. Thus $[\Delta_\kappa, m_t]p = 2\mathcal{D}_t p$. □

Corollary 7.4.11 $\Delta_\kappa \in \mathcal{A}_\kappa$ *and* $[\Delta_\kappa, w] = 0$ *for* $w \in G$.

Proof For any reflection group, $p_2(x) = |x|^2$ is G-invariant. Then Theorem 7.4.10 shows that $\Delta_\kappa = \rho(p_2) \in \mathcal{A}_\kappa$ and thus $w\Delta_\kappa w^{-1} = \Delta_\kappa$. □

There is a natural bilinear G-invariant form on Π^d associated with Dunkl operators. We will show that the form is symmetric, and positive definite when $\kappa \geq 0$, i.e., $\kappa(v) \geq 0$ for all $v \in R$. The proof involves a number of ingredients.

Lemma 7.4.12 *For each* $n \in \mathbb{N}_0$ *the space* Π_n^d *is a direct sum of eigenvectors of* $\sum_{i=1}^d x_i \mathcal{D}_i$.

Proof If $p \in \Pi^d$ then $\sum_{i=1}^d x_i \mathcal{D}_i p(x) = \sum_{i=1}^d x_i \frac{\partial}{\partial x_i} p(x) + \sum_{v \in R_+} \kappa(v)(p(x) - s_v p(x))$. The space Π_n^d is a G-module under the action $w \mapsto (p \mapsto wp)$ for $w \in G$, $p \in \Pi_n^d$. So $\Pi_n^d = \sum_{j=1}^m \oplus M_j$ where each M_j is an irreducible G-submodule. There are constants $c_j(\kappa)$ such that $\sum_{v \in R_+} \kappa(v)(1 - s_v)p = c_j(\kappa)p$ for each $p \in M_j$, $1 \leq j \leq m$, because $\sum_{v \in R_+} \kappa(v)(1 - s_v)$ is in the center of the group algebra of G. (Indeed, it is a sum over conjugacy classes and $c_j(\kappa) = \sum_{v \in R_+} \kappa(v) c_{j,v}$ where $c_{j,v} \in \mathbb{Q}$ and the map $v \mapsto c_{j,v}$ is constant on G-orbits in R.) Thus $\sum_{i=1}^d x_i \mathcal{D}_i p = (n + c_j(\kappa))p$ for $p \in M_j$. □

Definition 7.4.13 For $p, q \in \Pi^d$, let $\langle p, q\rangle_\kappa := \rho(p)q(x)|_{x=0}$ ($\rho(p)q$ evaluated at $x = 0$).

Theorem 7.4.14 *The pairing* $\langle\cdot,\cdot\rangle_\kappa$ *has the following properties*:
1. *If* $p \in \Pi_n^d$, $q \in \Pi_m^d$ *and* $m \neq n$ *then* $\langle p, q\rangle_\kappa = 0$.
2. *If* $w \in G$ *and* $p, q, r \in \Pi^d$ *then* $\langle wp, wq\rangle_\kappa = \langle p, q\rangle_\kappa$ *and* $\langle rp, q\rangle_\kappa = \langle p, \rho(r)q\rangle_\kappa$.
3. *The form is bilinear and symmetric.*

Proof For part 1, if $p \in \Pi_n^d$, $q \in \Pi_m^d$ and $m \geq n$ then $\rho(p)q \in \Pi_{m-n}^d$ and it vanishes at $x = 0$ if $m > n$; if $m < n$ then $\rho(p)q = 0$. A nonzero value at $x = 0$ is possible only if $m = n$ and then $\rho(p)q$ is a constant.

For part 2 we may assume $p, q \in \Pi_n^d$ for some n and $\langle p, q\rangle_\kappa = \rho(p)q$ (a constant). By Proposition 7.4.3 $\rho(wp) = w\rho(p)w^{-1}$, thus $\langle wp, wq\rangle_\kappa = w\rho(p)w^{-1}wq = w\rho(p)q = \langle p, q\rangle_\kappa$ (because $w1 = 1$). That $\langle rp, q\rangle_\kappa = \langle p, \rho(r)q\rangle_\kappa$ follows easily from the definition.

Use induction on part 3: for constants p_1, p_2 the form equals $\langle p_1 1, p_2\rangle_\kappa = p_1 p_2$; assume the form is symmetric on $\sum_{m=0}^n \Pi_m^d$ for some n and let $p, q \in \Pi_{n+1}^d$. Using Lemma 7.4.12 suppose p and q are eigenfunctions of $\sum_{v \in R_+} \kappa(v)(1 - s_v)$ with eigenvalues $c_1(\kappa)$ and $c_2(\kappa)$ respectively; then

$$c_1(\kappa)\langle p, q\rangle_\kappa = \sum_{v \in R_+} \kappa(v)\langle(1 - s_v)p, q\rangle_\kappa = \sum_{v \in R_+} \kappa(v)\langle p, (1 - s_v)q\rangle = c_2(\kappa)\langle p, q\rangle_\kappa.$$

So $c_1(\kappa) \neq c_2(\kappa)$ implies $\langle p, q\rangle_\kappa = 0$ (a symmetric relation). Now assume $c_1(\kappa) = c_2(\kappa)$; then

$$\begin{aligned}(n + c_1(\kappa))\langle p, q\rangle_\kappa &= \langle \textstyle\sum_{i=1}^d x_i \mathcal{D}_i p, q\rangle = \sum_{i=1}^d \langle \mathcal{D}_i p, \mathcal{D}_i q\rangle_\kappa \\ &= \sum_{i=1}^d \langle \mathcal{D}_i q, \mathcal{D}_i p\rangle_\kappa = \langle \textstyle\sum_{i=1}^d x_i \mathcal{D}_i q, p\rangle = (n + c_1(\kappa))\langle q, p\rangle_\kappa,\end{aligned}$$

where the inductive hypothesis was used to imply $\langle \mathcal{D}_i p, \mathcal{D}_i q\rangle_\kappa = \langle \mathcal{D}_i q, \mathcal{D}_i p\rangle_\kappa$. For fixed p, q, $\langle p, q\rangle_\kappa - \langle q, p\rangle_\kappa$ is a polynomial in the values of κ and $n + c_1(\kappa) = 0$ defines a hyperplane in the parameter space (the dimension equals the number of G-orbits in R). Thus $\langle p, q\rangle_\kappa = \langle q, p\rangle_\kappa$ for all κ. □

Corollary 7.4.15 *Suppose $\langle \cdot, \cdot\rangle_1$ is a bilinear symmetric form on Π^d such that $\langle 1, 1\rangle_1 = 1$ and $\langle x_i p, q\rangle_1 = \langle p, \mathcal{D}_i q\rangle_1$ for all $p, q \in \Pi^d$ and $1 \le i \le d$; then $\langle \cdot, \cdot\rangle_1 = \langle \cdot, \cdot\rangle_\kappa$.*

Proof Let $\alpha \in \mathbb{N}_0^d$ and $|\alpha| = n$ for some $n \ge 0$. Let $q \in \sum_{m=0}^{n-1} \Pi_m^d$. By hypothesis, $\langle x^\alpha, q\rangle_1 = \langle 1, \rho(x^\alpha)q\rangle_1 = 0$. If $q \in \Pi_n^d$ then $\langle x^\alpha, q\rangle_1 = \langle 1, \rho(x^\alpha)q\rangle_1 = \langle 1, \rho(x^\alpha)q\rangle_\kappa = \langle x^\alpha, q\rangle_\kappa$. By linearity and symmetry the proof is complete. □

7.4.1 The Gaussian Form

The operator $e^{\Delta_\kappa/2}$ maps Π_n^d into $\sum_{m=0}^n \Pi_m^d$ and its inverse is $e^{-\Delta_\kappa/2}$.

Definition 7.4.16 The *Gaussian form* on Π^d is given by $\langle p, q\rangle_g := \langle e^{\Delta_\kappa/2} p, e^{\Delta_\kappa/2} q\rangle_\kappa$.

Proposition 7.4.17 *The Gaussian form is symmetric and bilinear, and if $p, q \in \Pi^d$ then $\langle wp, wq\rangle_g = \langle p, q\rangle_g$ ($w \in G$), $\langle \mathcal{D}_i p, q\rangle_g = \langle p, (x_i - \mathcal{D}_i)q\rangle_g$ ($i = 1, \ldots, d$) and $\langle x_i p, q\rangle_g = \langle p, x_i q\rangle_g$ ($i = 1, \ldots, d$).*

Proof The first claim follows from the property $w\Delta_\kappa = \Delta_\kappa w$. By Theorem 7.4.10 we have $[\Delta_\kappa, m_{\varepsilon_i}] = 2\mathcal{D}_i$ (m_{ε_i} denotes multiplication by x_i). Repeated use of the general formula $[A^n, B] = A[A^{n-1}, B] + [A, B]A^{n-1}$ shows that $[\Delta_\kappa^n, m_{\varepsilon_i}] = 2n\mathcal{D}_i\Delta_\kappa^{n-1}$. This implies that, for any $p \in \Pi^d$, we have $e^{\Delta_\kappa/2} x_i p(x) - x_i e^{\Delta_\kappa/2} p(x) = \mathcal{D}_i e^{\Delta_\kappa/2} p(x)$. Thus

$$\begin{aligned}\langle \mathcal{D}_i p, q\rangle_g &= \langle \mathcal{D}_i e^{\Delta_\kappa/2} p, e^{\Delta_\kappa/2} q\rangle_\kappa = \langle e^{\Delta_\kappa/2} p, x_i e^{\Delta_\kappa/2} q\rangle_\kappa \\ &= \langle e^{\Delta_\kappa/2} p, e^{\Delta_\kappa/2}(x_i - \mathcal{D}_i) q\rangle_\kappa = \langle p, (x_i - \mathcal{D}_i) q\rangle_g.\end{aligned}$$

Finally $\langle p, x_i q\rangle_g = \langle \mathcal{D}_i p, q\rangle_g + \langle p, \mathcal{D}_i q\rangle_g$, which is symmetric in p, q. □

Thus the Gaussian form satisfies $\langle p, q\rangle_g = \langle 1, pq\rangle_g$. This suggests that there may be an integral formula; this is indeed the situation when $\kappa \ge 0$. The properties in Proposition 7.4.17 imply a uniqueness result by "reading the proof backwards": set $\langle p, q\rangle_1 := \langle e^{-\Delta_\kappa/2} p, e^{-\Delta_\kappa/2} q\rangle_g$ and use Corollary 7.4.15.

Definition 7.4.18 For $\kappa \ge 0$ the fundamental *G-invariant weight function* is

$$w_\kappa(x) := \prod_{v \in R_+} |\langle x, v\rangle|^{2\kappa(v)}, \quad x \in \mathbb{R}^d.$$

The G-invariance is a consequence of the definition of multiplicity functions.

Definition 7.4.19 For $p, q \in \Pi^d$ and $\kappa \geq 0$, let

$$\langle p, q\rangle_2 := \frac{c_\kappa}{(2\pi)^{d/2}} \int_{\mathbb{R}^d} p(x)q(x)w_\kappa(x)e^{-|x|^2/2}\,dx,$$

where c_κ is the normalizing constant resulting in $\langle 1, 1\rangle_2 = 1$.

The constant c_κ is related to the Macdonald–Mehta integral. This will be discussed below in Theorem 7.4.21.

Theorem 7.4.20 *If $p, q \in \Pi^d$ and $1 \leq i \leq d$ then $\langle \mathcal{D}_i p, q\rangle_2 = \langle p, (x_i - \mathcal{D}_i)q\rangle_2$. The forms $\langle \cdot,\cdot\rangle_2$ and $\langle \cdot,\cdot\rangle_g$ are equal when $\kappa \geq 0$.*

Proof Let $p, q \in \Pi^d$, $1 \leq i \leq d$. Note that

$$\frac{\partial}{\partial x_i}(w_\kappa(x)e^{-|x|^2/2}) = \left(- x_i + 2\sum_{v\in R_+} \frac{\kappa(v)v_i}{\langle x, v\rangle}\right) w_\kappa(x)e^{-|x|^2/2}$$

(in the special case $0 < \kappa(v) < \frac{1}{2}$ the formula is valid provided $\langle x, v\rangle \neq 0$).
It suffices to show that the following integral vanishes:

$$\begin{aligned}\int_{\mathbb{R}^d} (q\mathcal{D}_i p + p\mathcal{D}_i q - x_i pq)w_\kappa e^{-|x|^2/2}\,dx &= \int_{\mathbb{R}^d} \frac{\partial}{\partial x_i}(pq)w_\kappa e^{-|x|^2/2}\,dx \\ - \int_{\mathbb{R}^d} x_i pq w_\kappa e^{-|x|^2/2}dx &+ \sum_{v\in R_+} \kappa(v)v_i \int_{\mathbb{R}^d} \frac{2p(x)q(x) - p(s_v x)q(x) - p(x)q(s_v x)}{\langle x, v\rangle} w_\kappa e^{-|x|^2/2}dx \\ = - \sum_{v\in R_+} \kappa(v)v_i \int_{\mathbb{R}^d} &\frac{p(s_v x)q(x) + p(x)q(s_v x)}{\langle x, v\rangle} w_\kappa e^{-|x|^2/2}\,dx = 0,\end{aligned}$$

because in each term the integrand is odd under the action of a reflection s_v. Integration by parts and exponential decay shows $\int_{\mathbb{R}^d} \frac{\partial}{\partial x_i}(pq)w_\kappa e^{-|x|^2/2}dx = -\int_{\mathbb{R}^d} pq\frac{\partial}{\partial x_i}(w_\kappa e^{-|x|^2/2})\,dx$. The terms in the sum over R_+ have the singularity $|\langle x, v\rangle|^{2\kappa(v)-1}$, which is integrable for $\kappa(v) > 0$ (the terms with $\kappa(v) = 0$ do not appear). □

As a consequence of Theorem 7.4.20 and of the invertibility of $e^{\Delta_\kappa/2}$ on polynomials it follows that the forms $\langle\cdot,\cdot\rangle_\kappa$ and $\langle\cdot,\cdot\rangle_g$ are positive definite when $\kappa \geq 0$.

There is an elegant formula for c_κ when there is only one G-orbit in R. Recall the discriminant $a_R(x) = \prod_{v\in R_+}\langle x, v\rangle$.

Theorem 7.4.21 (Macdonald–Mehta integral)
Suppose G has just one conjugacy class of reflections, $|v|^2 = 2$ for each $v \in R$ and $\kappa_0 > 0$; then

$$(2\pi)^{-d/2} \int_{\mathbb{R}^d} |a_R(x)|^{2\kappa_0} e^{-|x|^2/2}\,dx = \prod_{i=1}^{d} \frac{\Gamma(1 + n_i\kappa_0)}{\Gamma(1 + \kappa_0)},$$

where $\{n_i\}_{i=1}^d$ is the set of fundamental degrees of G (see Theorem 7.3.1).

(If the rank of G is less than d then some of the degrees equal 1.) The integral for the symmetric group can be deduced from Selberg's integral formula. Macdonald conjectured a formula for c_κ for any reflection group. Opdam (1993) proved the formula for all cases, except for a constant (independent of κ) multiple for H_3 and H_4; the constant was verified by Garvan with a computer-assisted proof. Etingof (2010) extended Opdam's method to a general proof for the one-class type.

Corollary 7.4.22 *With notation as in Theorem 7.4.21, $\kappa_0 \in \mathbb{C}$ and $\kappa(v) = \kappa_0$ ($v \in R$),*

$$\langle a_R, a_R \rangle_\kappa = \langle a_R, a_R \rangle_g = \#G \prod_{i=1}^{d} \prod_{j=1}^{n_i - 1} (j + n_i \kappa_0).$$

Proof $\Delta_\kappa a_R = 0$ because $\Delta_\kappa a_R$ is an alternating polynomial of degree $\#R_+ - 2$, while a_R is the minimal-degree nonzero alternating polynomial. Thus $e^{\Delta_\kappa/2} a_R = a_R$. First assume $\kappa_0 > 0$. From the definition of c_κ it follows that

$$\langle a_R, a_R \rangle_g = \frac{c_{\kappa_0}}{c_{\kappa_0+1}} = (1+\kappa_0)^{-d} \prod_{i=1}^{d} \frac{\Gamma(1 + n_i + n_i\kappa_0)}{\Gamma(1 + n_i\kappa_0)}$$

$$= (1+\kappa_0)^{-d} \prod_{i=1}^{d} \prod_{j=1}^{n_i} (j + n_i\kappa_0) = \prod_{i=1}^{d} n_i \prod_{j=1}^{n_i - 1} (j + n_i\kappa_0) = \#G \prod_{j=1}^{n_i - 1} (j + n_i\kappa_0).$$

This proves the formula for $\kappa_0 > 0$. It is valid for all $\kappa_0 \in \mathbb{C}$ because it is a polynomial identity. □

The indecomposable reflection groups with two classes of reflections consist of $I_2(2m)$ with $m \geq 2$, B_d and F_4. To get a convenient expression for the dihedral group $I_2(2m)$ let $z = x_1 + \mathrm{i}x_2$ and denote the values of κ by κ_0 and κ_1; write the weight function as $w_\kappa(x) = |z^m - \bar{z}^m|^{2\kappa_0} |z^m + \bar{z}^m|^{2\kappa_1}$. Then

$$\frac{1}{2\pi} \int_{\mathbb{R}^2} w_\kappa(x) e^{-|x|^2/2}\, dx = 2^{m(\kappa_0+\kappa_1)} \frac{\Gamma(1+2\kappa_0)\Gamma(1+2\kappa_1)\Gamma(1+m(\kappa_0+\kappa_1))}{\Gamma(1+\kappa_0)\Gamma(1+\kappa_1)\Gamma(1+\kappa_0+\kappa_1)}.$$

For the hyperoctahedral group B_d (see §7.2.2 for the root system) the integral is

$$(2\pi)^{-d/2} \int_{\mathbb{R}^d} \prod_{i=1}^{d} |x_i|^{2\kappa_1} \prod_{1 \le i < j \le d} |x_i^2 - x_j^2|^{2\kappa_0} e^{-|x|^2/2}\, dx$$

$$= 2^{d((d-1)\kappa_0 + \kappa_1)} \prod_{i=1}^{d} \frac{\Gamma(1 + i\kappa_0)\Gamma((i-1)\kappa_0 + \kappa_1 + \frac{1}{2})}{\Gamma(1+\kappa_0)\Gamma(\frac{1}{2})}.$$

In the formula, $\Gamma(1/2)$ is used in place of $\pi^{1/2}$ for the sake of appearance. The formula can be derived from Selberg's integral.

For F_4, in the notation of §7.2.5 for the y_i, we get (a special case of Opdam's result)

$$\frac{1}{4\pi^2} \int_{\mathbb{R}^4} \prod_{1 \le i < j \le 4} |x_i^2 - x_j^2|^{2\kappa_1} \prod_{1 \le i < j \le 4} |y_i^2 - y_j^2|^{2\kappa_2} e^{-|x|^2/2}\, dx$$

$$= 2^{12(\kappa_1+\kappa_2)} \frac{\Gamma(2\kappa_1 + \kappa_2 + \frac{1}{2})\Gamma(\kappa_1 + 2\kappa_2 + \frac{1}{2})\Gamma(3\kappa_1 + 3\kappa_2 + \frac{1}{2})}{\Gamma(\frac{1}{2})^3}$$

$$\times \frac{\Gamma(4\kappa_1 + 4\kappa_2 + 1)}{\Gamma(\kappa_1 + \kappa_2 + 1)} \prod_{i=1}^{2} \frac{\Gamma(2\kappa_i + 1)\Gamma(3\kappa_i + 1)}{\Gamma(\kappa_i + 1)^2}.$$

The formula agrees with the simpler single-class result when $\kappa_1 = \kappa_2$ and with the D_4 value when $\kappa_2 = 0$ (by use of the Gamma function duplication formula).

7.5 Harmonic Polynomials

The Gaussian form is an important part of our analysis. Accordingly the polynomials in the kernel of Δ_κ have properties relevant to the two forms $\langle\cdot,\cdot\rangle_\kappa$ and $\langle\cdot,\cdot\rangle_g$.

Definition 7.5.1 Let $\mathcal{H}_\kappa := \{p \in \Pi^d \mid \Delta_\kappa p = 0\}$ and $\mathcal{H}_{\kappa,n} := \mathcal{H}_\kappa \cap \Pi^d_n$ for $n = 0, 1, 2, \ldots$. These are the spaces of *harmonic and harmonic homogeneous polynomials*, respectively. Let $\gamma_\kappa := \sum_{v\in R_+} \kappa(v)$ (w_κ is positively homogeneous of degree $2\gamma_\kappa$).

In the original literature (Dunkl, 1989) these polynomials were called *h-harmonic* because the weight function $w_\kappa(x)$ was denoted by $h(x)$. In the sequel we use *harmonic* without specifying the Δ_κ aspect (i.e., not the traditional usage of $\Delta p = 0$). For convenience we let $|x|^2$ denote both the polynomial in Π^d_2 and the corresponding multiplier operator. We will show that $\Pi^d_n = \sum_{j=0}^{\lfloor n/2\rfloor} \oplus |x|^{2j} \mathcal{H}_{\kappa,n-2j}$ for each $n \geq 2$ provided that $\gamma_\kappa + \frac{d}{2} \notin -\mathbb{N}_0$. Trivially $\Pi^d_n = \mathcal{H}_{\kappa,n}$ for $n = 0, 1$. Note that the proof cannot use any nonsingularity property of the form $\langle\cdot,\cdot\rangle_\kappa$.

Lemma 7.5.2 $[\Delta_\kappa, |x|^{2m}] = 2m|x|^{2(m-1)}(2m - 2 + d + 2\gamma_\kappa + 2\sum_{i=1}^d x_i \frac{\partial}{\partial x_i})$, $m = 1, 2, 3, \ldots$.

Corollary 7.5.3 *If $m, n, k = 1, 2, 3, \ldots$ and $p \in \mathcal{H}_{\kappa,n}$ then*

1. $\Delta_\kappa(|x|^{2m} p(x)) = 2m(2m - 2 + d + 2\gamma_\kappa + 2n)|x|^{2m-2} p(x)$;
2. $\Delta^k_\kappa(|x|^{2m} p(x)) = 4^k(-m)_k(1 - m - d/2 - \gamma_\kappa - n)_k |x|^{2m-2k} p(x)$.

Part 2 implies that $\Delta^k_\kappa(|x|^{2m} p(x)) = 0$ if $k > m$. This leads to an orthogonality relation:

Proposition 7.5.4 *Suppose $m, k \leq \frac{n}{2}$, $n \geq 2$, $p \in \mathcal{H}_{\kappa,n-2k}$ and $q \in \mathcal{H}_{\kappa,n-2m}$. If $m \neq k$ then* $\langle |x|^{2k} p, |x|^{2m} q\rangle_\kappa = 0$.

Proof By the symmetry of the form we may assume $k > m$; then

$$\langle |x|^{2k} p, |x|^{2m} q\rangle_\kappa = \langle p, \Delta^k_\kappa |x|^{2m} q\rangle_\kappa = 0.$$

□

Definition 7.5.5 Suppose $\gamma_\kappa + \frac{1}{2}d \neq 0, -1, -2, \ldots$. Then let

$$\pi_{\kappa,n} := \begin{cases} \sum_{j=0}^{\lfloor n/2\rfloor} \dfrac{1}{4^j j!\,(-\gamma_\kappa - n + 2 - d/2)_j} |x|^{2j} \Delta^j_\kappa, & n = 2, 3, 4, \ldots, \\ I, & n = 0, 1. \end{cases}$$

The following is a version of *Dixon's summation theorem* (see Olver et al., 2010, 16.4.4).

Lemma 7.5.6 *Suppose that $k \in \mathbb{N}_0$ and $a+1, a-b+1 \notin -\mathbb{N}_0$. Then*

$$ {}_3F_2\left(\begin{matrix} -k, a, b \\ k+a+1, a-b+1 \end{matrix}; 1\right) = \frac{(a+1)_k(\frac{1}{2}a-b+1)_k}{(\frac{1}{2}a+1)_k(a-b+1)_k}. $$

Proposition 7.5.7 *If $\gamma_\kappa + \frac{1}{2}d \notin -\mathbb{N}_0$ and $p \in \Pi_n^d (n = 2, 3, 4, \ldots)$ then $\pi_{\kappa,n} p \in \mathcal{H}_{\kappa,n}$. If $p \in \mathcal{H}_{\kappa,n}$ then $p = \pi_{\kappa,n} p$, i.e., $\pi_{\kappa,n}$ is a projection. Furthermore,*

$$ p = \sum_{j=0}^{\lfloor n/2 \rfloor} \frac{1}{4^j\, j!\, (\gamma_\kappa + d/2 + n - 2j)_j} |x|^{2j} \pi_{\kappa,n-2j}(\Delta_\kappa^j p). $$

Proof The first part is a consequence of Lemma 7.5.2. The proof of the expansion formula depends on Lemma 7.5.6. Set $a = 1 - n - \gamma_\kappa - \frac{1}{2}d$ and $b = \frac{1}{2}a + 1$; then the coefficient of $(4^k k!)^{-1}|x|^{2k}\Delta_\kappa^k$ in the right-hand side is

$$ \begin{aligned} \sum_{j=0}^{k} \binom{k}{j} \frac{1}{(1-a-2j)_j (a+1+2j)_{k-j}} &= \sum_{j=0}^{k} \frac{(-k)_j (a+1)_{2j}}{j!\,(a+j)_j (a+1)_{k+j}} \\ = \frac{1}{(a+1)_k} \sum_{j=0}^{k} \frac{(-k)_j (a+1)_{2j} (a)_j}{j!\,(a)_{2j} (a+1+k)_j} &= \frac{1}{(a+1)_k} \sum_{j=0}^{k} \frac{(-k)_j (\frac{1}{2}a+1)_j (a)_j}{j!\,(\frac{1}{2}a)_j (a+1+k)_j}, \end{aligned} $$

which vanishes for $k \geq 1$ because of the term $(\frac{1}{2}a - b + 1)_k = (0)_k$ in the summation formula Lemma 7.5.6; the transformation $\frac{(a+1)_{2j}}{(a)_{2j}} = \frac{a+2j}{a} = \frac{(a/2+1)_j}{(a/2)_j}$ was used in the last step. The sum equals 1 when $k = 0$. The identity holds for generic γ_κ and the hypothesis $\gamma_\kappa + \frac{1}{2}d \notin -\mathbb{N}_0$ ensures that the terms in $\pi_{\kappa,n-2j}$ are well defined. □

This establishes the validity of $\Pi_n^d = \bigoplus_{j=0}^{\lfloor n/2 \rfloor} |x|^{2j} \mathcal{H}_{\kappa,n-2j}$, provided $\gamma_\kappa + \frac{1}{2}d \notin -\mathbb{N}_0$. (Argue by induction that $\mathcal{H}_{\kappa,n} \cap |x|^2 \Pi_{n-2}^d = \{0\}$.) To transfer these results to the Gaussian form, let $p \in \mathcal{H}_{\kappa,n}$ and $s \in \mathbb{R}$, and evaluate

$$ \begin{aligned} e^{s\Delta_\kappa}(|x|^{2m} p(x)) &= \sum_{j=0}^{m} \frac{1}{j!} s^j \Delta_\kappa^j (|x|^{2m} p(x)) \\ &= \sum_{j=0}^{m} \frac{1}{j!} (4s)^j (-m)_j (1 - m - \tfrac{1}{2}d - \gamma_\kappa - n)_j |x|^{2m-2j} p(x) = (4s)^m m!\, L_m^{(\alpha)}\left(-\frac{|x|^2}{4s}\right) p(x), \quad (7.5.1) \end{aligned} $$

where $\alpha = \gamma_\kappa + \frac{1}{2}d + n - 1$ and $L_m^{(\alpha)}$ denotes the *Laguerre polynomial* of degree m and index α; it is defined provided $\alpha + 1 \notin -\mathbb{N}_0$, and it is part of an orthogonal family of polynomials if $\alpha > -1$. Here we use $s = -\frac{1}{2}$.

We list the conditions on γ_κ specialized to some reflection groups:

$$ \begin{aligned} &I_2(2m): && m(\kappa_0 + \kappa_1) + 1 \notin -\mathbb{N}_0, \\ &A_{d-1} \subset \mathbb{R}^d: && \frac{d}{2}((d-1)\kappa + 1) \notin -\mathbb{N}_0, \\ &B_d: && d((d-1)\kappa_0 + \kappa_1 + \frac{1}{2}) \notin -\mathbb{N}_0. \end{aligned} $$

When $\kappa \geq 0$ the Gaussian form can be related to an integral over the unit sphere, and there is an analog of the spherical harmonics. Let ω denote the normalized rotation-invariant measure on the surface of the unit sphere $S^{d-1} := \{x \in \mathbb{R}^d \mid |x| = 1\}$. Suppose f is continuous and integrable over $\mathbb{R}^d$. Then

$$\int_{\mathbb{R}^d} f(x)\, dx = \frac{2\pi^{d/2}}{\Gamma(d/2)} \int_0^\infty r^{d-1} dr \int_{S^{d-1}} f(ru)\, d\omega(u).$$

The constant multiplier is evaluated by setting $f(x) = e^{-|x|^2/2}$. Now suppose f is positively homogeneous of degree β (i.e., $f(tx) = t^\beta f(x)$ for $t > 0$) and $\beta + d > 1$. Then

$$(2\pi)^{-d/2} \int_{\mathbb{R}^d} f(x) e^{-|x|^2/2}\, dx = 2^{\beta/2} \frac{\Gamma((\beta + d)/2)}{\Gamma(d/2)} \int_{S^{d-1}} f(u)\, d\omega(u).$$

To normalize the measure $w_\kappa\, d\omega$, set $f = w_\kappa$ (with $\beta = 2\gamma_\kappa$) and let

$$c_{\kappa,S}^{-1} := \int_{S^{d-1}} w_\kappa\, d\omega = 2^{-\gamma_\kappa} \frac{\Gamma(d/2)}{\Gamma(\gamma_\kappa + d/2)} c_\kappa^{-1}.$$

Observe that the condition $\gamma_\kappa + d/2 \notin -\mathbb{N}_0$ appears again.

Proposition 7.5.8 *Suppose $p \in \mathcal{H}_{\kappa,n}$ and $q \in \mathcal{H}_{\kappa,m}$.*
1. If $m \neq n$ then $c_{\kappa,S} \int_{S^{d-1}} pqw_\kappa\, d\omega = 0$.
2. If $m = n$ then

$$c_{\kappa,S} \int_{S^{d-1}} pqw_\kappa\, d\omega = \frac{c_\kappa (2\pi)^{-d/2}}{2^n (\gamma_\kappa + d/2)_n} \int_{\mathbb{R}^d} p(x) q(x) w_\kappa(x) e^{-|x|^2/2}\, dx = \frac{1}{2^n (\gamma_\kappa + d/2)_n} \langle p, q \rangle_\kappa.$$

That is, the spaces $\{\mathcal{H}_{\kappa,n}\}_{n \in \mathbb{N}_0}$ are pairwise orthogonal in $L^2(S^{d-1}, w_\kappa\, d\omega)$. By Proposition 7.5.7 each polynomial agrees with a harmonic one on S^{d-1} and so

$$L^2(S^{d-1}, w_\kappa\, d\omega) = \bigoplus_{n=0}^\infty \mathcal{H}_{\kappa,n}$$

by the density of polynomials. At this writing there are generally no known explicit orthogonal bases for $\mathcal{H}_{\kappa,n}$ but it is possible to define nonorthogonal bases by applying the projection $\pi_{\kappa,n}$ to the monomials x^α with $\alpha \in \mathbb{N}_0^d$ such that $|\alpha| = n$ and $\alpha_d = 0$ or 1. The argument for this is as follows: Let $\Pi_{n,0}^d := \operatorname{span}\{x^\alpha \mid |\alpha| = n,\ \alpha_d = 0, 1\}$; then $\dim \Pi_{n,0}^d = \binom{n+d-2}{d-2} + \binom{n+d-3}{d-2}$ and $p \in \Pi_{n,0}^d$, $\pi_{\kappa,n} p = 0$ implies $p = 0$ because any polynomial annihilated by $\pi_{\kappa,n}$ is divisible by $|x|^2$ (by Proposition 7.5.7). Thus $\pi_{\kappa,n}$ is one-to-one on $\Pi_{n,0}^d$ and $\dim \mathcal{H}_{\kappa,n} = \binom{n+d-1}{d-1} - \binom{n+d-3}{d-1} = \dim \Pi_{n,0}^d$. In the next section we consider the reproducing and Poisson kernels.

By a version of Hamburger's theorem Π^d is dense in $L^2(\mathbb{R}^d, w_\kappa(x) e^{-|x|^2/2} dx)$. Bases for $\mathcal{H}_{\kappa,n}$ can be used to produce bases for this L^2-space by forming products of harmonic polynomials and Laguerre polynomials with argument $|x|^2/2$.

Definition 7.5.9 For $n \in \mathbb{N}_0$ let $X_n := \operatorname{span}\{p(x) L_m^{(\alpha_n)}(|x|^2/2) \mid p \in \mathcal{H}_{\kappa,n},\ m \in \mathbb{N}_0\}$, where $\alpha_n := \gamma_\kappa + d/2 + n - 1$.

Suppose $k, l, m, n \in \mathbb{N}_0$ and $p \in \mathcal{H}_{\kappa,n}$, $q \in \mathcal{H}_{\kappa,l}$; then $\langle |x|^{2m} p(x), |x|^{2k} q(x) \rangle_\kappa = 0$ unless $n = l$ and $m = k$. If $2m + n \neq 2k + l$ this follows from part 1 of Theorem 7.4.14, while

Proposition 7.5.4 applies if $2m + n = 2k + l$ and $m \neq k$. By (7.5.1) with $s = -\frac{1}{2}$ we have $\langle L_m^{(\alpha_n)}(|x|^2/2)p(x), L_k^{(\alpha_l)}(|x|^2/2)q(x)\rangle_g = 0$ unless $n = l$ and $m = k$. Thus

$$L^2(\mathbb{R}^d, w_\kappa(x)e^{-|x|^2/2}dx) = \bigoplus_{n=0}^{\infty} X_n.$$

Suppose $p, q \in \mathcal{H}_{\kappa,n}$ and $k, m \in \mathbb{N}_0$; then

$$\langle L_m^{(\alpha_n)}(|x|^2/2)p(x), L_k^{(\alpha_l)}(|x|^2/2)q(x)\rangle_g = \delta_{mk}\frac{(\alpha_n + 1)_m}{m!}\langle p, q\rangle_g,$$

and $\langle p, q\rangle_g = 2^n(\gamma_\kappa + d/2)_n c_{\kappa,S} \int_{S^{d-1}} pqw_\kappa \, d\omega$. These formulae show how an orthogonal basis for $\mathcal{H}_{\kappa,n}$ can be used to produce such a basis for $L^2(\mathbb{R}^d, w_\kappa(x)e^{-|x|^2/2}dx)$.

7.6 The Intertwining Operator and the Dunkl Kernel

Several important objects can be defined when the form $\langle \cdot, \cdot \rangle_\kappa$ is nondegenerate for specific numerical values of κ. We refer to "generic" κ when κ has some transcendental value (formal parameter), and to "specific" κ when κ takes on real values. The form $\langle \cdot, \cdot \rangle_\kappa$ is defined for all κ, and so is the following operator.

Definition 7.6.1 The operator V_κ^0 on Π^d is given by $V_\kappa^0 p(y) := \langle e^{\langle y,x\rangle}, p(x)\rangle_\kappa$ $(p \in \Pi^d)$.

If this formal operator is applied to $p \in \Pi_n^d$ then $V_\kappa^0 p(y) = \sum_{\alpha\in\mathbb{N}_0^d, |\alpha|=n} y^\alpha \mathcal{D}^\alpha p/\alpha!$. Note that $e^{\langle y,x\rangle} = \sum_{n=0}^{\infty} \frac{1}{n!} \sum_{|\alpha|=n} \binom{n}{\alpha} x^\alpha y^\alpha$ (where $\alpha! := \prod_{i=1}^d \alpha_i!$ and $\binom{n}{\alpha} := n!/\alpha!$), and $\mathcal{D}^\alpha p = \langle x^\alpha, p(x)\rangle$ for $p \in \Pi_n^d$ ($|\alpha| = n \in \mathbb{N}_0$). Also $V_\kappa^0 1 = 1$.

Proposition 7.6.2 *If $1 \le i \le d$ and $p \in \Pi^d$ then $\frac{\partial}{\partial x_i} V_\kappa^0 p(x) = V_\kappa^0 \mathcal{D}_i p(x)$. If $w \in G$ then $V_\kappa^0 wp = wV_\kappa^0 p$.*

Proof We have $\frac{\partial}{\partial y_i} V_\kappa^0 p(y) = \langle x_i e^{\langle y,x\rangle}, p(x)\rangle_\kappa = \langle e^{\langle y,x\rangle}, \mathcal{D}_i p(x)\rangle_\kappa$. For $x, y \in \mathbb{R}^d$ let $f_x(y) := \langle x, y\rangle$. Suppose $p \in \Pi_n^d$; then $V_\kappa^0(wp)(x) = \frac{1}{n!}\langle f_x^n, wp\rangle_\kappa = \frac{1}{n!}\langle (w^{-1}f_x)^n, p\rangle_\kappa$, and $w^{-1}f_x(y) = \langle x, wy\rangle = \langle w^{-1}x, y\rangle$, so $V_\kappa^0(wp)(x) = V_\kappa^0 p(w^{-1}x)$. □

Definition 7.6.3 For specific κ define the *radical* $\mathrm{Rad}(\kappa) := \{p \in \Pi^d \mid \langle p, q\rangle_\kappa = 0 \; \forall \, q \in \Pi^d\}$.

Proposition 7.6.4 *The space* $\mathrm{Rad}(\kappa)$ *has the following properties*:
1. *$p \in \mathrm{Rad}(\kappa)$ $(1 \le i \le d)$ and $w \in G$ imply that $x_i p(x), \mathcal{D}_i p(x), wp(x) \in \mathrm{Rad}(\kappa)$.*
2. *$\mathrm{Rad}(\kappa) = \ker V_\kappa^0$.*
3. *$\mathrm{Rad}(\kappa) = \sum_{n=0}^{\infty}(\mathrm{Rad}(\kappa) \cap \Pi_n^d)$ (algebraic direct sum).*

Proof Part 1 follows directly from the properties of $\langle \cdot, \cdot \rangle_\kappa$ and the definition of the radical. For part 2 suppose the degree of p is n (i.e., $p \in \sum_{j=0}^n \Pi_j^d$) and $p \in \mathrm{Rad}(\kappa)$; then $V_\kappa^0 p(y) = \sum_{j=0}^n \frac{1}{j!}\langle \langle y, x\rangle^j, p(x)\rangle_\kappa = 0$. Conversely, suppose $V_\kappa^0 p(y) = 0$; then $\langle x^\alpha, p(x)\rangle_\kappa = 0$ for all $\alpha \in \mathbb{N}_0^d$ with $|\alpha| \le n$, and thus $p \in \mathrm{Rad}(\kappa)$. For part 3 use that the Euler operator satisfies

$$\sum_{i=1}^{d} x_i \frac{\partial}{\partial x_i} = \sum_{i=1}^{d} x_i \mathcal{D}_i - \sum_{v \in R_+} \kappa(v)(1 - s_v).$$

If $p \in \mathrm{Rad}(\kappa)$ then $\sum_{i=1}^{d} x_i \frac{\partial}{\partial x_i} p(x) \in \mathrm{Rad}(\kappa)$ by part 1, and hence $\mathrm{Rad}(\kappa)$ is the sum of its homogeneous subspaces. □

Part 1 of Proposition 7.6.4 implies that $\mathrm{Rad}(\kappa)$ is an ideal of the *rational Cherednik algebra* (an abstract algebra isomorphic to the algebra of operators on Π^d generated by the multipliers x_i, the operators $\mathcal{D}_i$ and the group G; this name was introduced in Etingof and Ginzburg, 2002, p. 251). We can now set up the key decomposition of the parameter space. Multiplicity functions can be identified with points in $\mathbb{R}^c$ where c is the number of G-orbits in R.

Definition 7.6.5 Let $\Lambda^0 := \{\kappa \mid \mathrm{Rad}(\kappa) \neq \{0\}\}$, the *singular set* and $\Lambda^{\mathrm{reg}} := \{\kappa \mid \mathrm{Rad}(\kappa) = \{0\}\}$, the *regular set*.

As a result of the papers of Opdam (1993) and Dunkl et al. (1994) there is a concise description of the singular set for indecomposable reflection groups: The value of the integral $\int_{\mathbb{R}^d} w_\kappa(x) e^{-|x|^2/2}\, dx$ is a meromorphic function of κ; the integral is defined for $\kappa \geq 0$ but the value extends analytically to $\mathbb{C}$. The poles coincide with the singular set. For the one-class type the singular set is $\{-j/n_m \mid j \in \mathbb{N}_0,\ 1 \leq m \leq d,\ j/n_m \notin \mathbb{Z}\}$, where the rank of G is d and the fundamental degrees are $n_1, \ldots, n_d$. The realization of the Gaussian form as an integral shows that $\kappa \geq 0$ implies $\kappa \in \Lambda^{\mathrm{reg}}$.

Below we use superscripts (x), (y) to indicate the variable on which an operator acts.

Definition 7.6.6 For $\kappa \in \Lambda^{\mathrm{reg}}$ let $V_\kappa := (V_\kappa^0)^{-1}$, the *intertwining operator*, and let

$$K_{\kappa,n}(x, y) := \frac{1}{n!} V_\kappa^{(y)} \langle x, y \rangle^n = \sum_{\alpha \in \mathbb{N}_0^d, |\alpha| = n} \frac{1}{\alpha!} x^\alpha V_\kappa(y^\alpha), \quad x, y \in \mathbb{R}^d,\ n \in \mathbb{N}_0.$$

The polynomial $K_{\kappa,n}$ is homogeneous of degree n in both x and y.

Theorem 7.6.7 *$K_{\kappa,n}$ and V_κ have the following properties:*

1. *If $p \in \Pi^d$ and $1 \leq i \leq d$ then $\mathcal{D}_i(V_\kappa p)(x) = V_\kappa(\frac{\partial}{\partial x_i} p(x))$; if $w \in G$ then $wV_\kappa = V_\kappa w$.*
2. *V_κ maps Π_n^d one-to-one onto Π_n^d for each n.*
3. *$\mathcal{D}_i^{(y)} K_{\kappa,n}(x, y) = x_i K_{\kappa,n-1}(x, y)$.*
4. *$\langle K_{\kappa,n}(x, \cdot), p \rangle_\kappa = p(x)$ for $p \in \Pi_n^d$.*
5. *$K_{\kappa,n}(x, y) = K_{\kappa,n}(y, x)$ for all $x, y \in \mathbb{R}^d$ and $K_{\kappa,n}(wx, wy) = K_{\kappa,n}(x, y)$ for each $w \in G$.*

Proof Parts 1 and 3 are straightforward. Part 2 holds because V_κ^0 maps Π_n^d into Π_n^d and its inverse exists. For part 4 let $\partial_y^\alpha := \prod_{i=1}^{d} (\frac{\partial}{\partial y_i})^{\alpha_i}$; if $p \in \Pi_n^d$ then $p(x) = \sum_{\alpha \in \mathbb{N}_0^d, |\alpha| = n} \frac{1}{\alpha!} x^\alpha p(\partial_y) y^\alpha$. Apply $V_\kappa^{(y)}$ to both sides (and the left-hand side is independent of y). Thus

$$p(x) = V_\kappa^{(y)} p(x) = \sum_{\alpha \in \mathbb{N}_0^d, |\alpha| = n} \frac{1}{\alpha!} x^\alpha V_\kappa^{(y)} p(\partial_y) y^\alpha = \sum_{\alpha \in \mathbb{N}_0^d, |\alpha| = n} \frac{1}{\alpha!} x^\alpha p(\mathcal{D}^{(y)}) V_\kappa^{(y)} y^\alpha = \langle p, K_{\kappa,n}(x, \cdot) \rangle_\kappa.$$

For part 5 use that the form $\langle\cdot,\cdot\rangle_\kappa$ is symmetric. Indeed, for any $p, q \in \Pi_n^d$, by part 4,

$$\langle p, q\rangle_\kappa = \langle K_{\kappa,n}(\cdot, \mathcal{D}^{(y)})p(y), q\rangle_\kappa = K_{\kappa,n}(\mathcal{D}^{(x)}, \mathcal{D}^{(y)})p(y)q(x) = \langle q, p\rangle_\kappa.$$

(This symmetry of $\langle\cdot,\cdot\rangle_\kappa$ was also stated in Theorem 7.4.14(3).) This implies $K_{\kappa,n}(x,y) = K_{\kappa,n}(y,x)$, because $\kappa \in \Lambda^{\text{reg}}$. Let $f_x(y) = K_{\kappa,n}(x,y)$, $w \in G$ and $p \in \Pi_n^d$; then $p(x) = \langle f_x, p\rangle_\kappa$ and $wp(x) = p(w^{-1}x) = \langle f_{w^{-1}x}, p\rangle_\kappa = \langle f_x, wp\rangle_\kappa = \langle w^{-1}f_x, p\rangle_\kappa$. Thus $K_{\kappa,n}(x, wy) = w^{-1}f_x = f_{w^{-1}x} = K_{\kappa,n}(w^{-1}x, y)$. □

Denote the formal sum $\sum_{n=0}^\infty K_{\kappa,n}(x,y)$ by $K_\kappa(x,y)$. The question now arises of whether the series converges in a useful way. There are some strong results for $\kappa \geq 0$. For $x, y \in \mathbb{R}^d$ let $\rho(x,y) := \max_{w\in G} |\langle x, wy\rangle|$.

Theorem 7.6.8 *Suppose $\kappa \geq 0$ and $x, y \in \mathbb{R}^d$. Then $|K_{\kappa,n}(x,y)| \leq \frac{1}{n!}\rho(x,y)^n$ for all $n \in \mathbb{N}_0$, the series for K_κ converges uniformly and absolutely on compact subsets of $\mathbb{R}^d \times \mathbb{R}^d$, and $|K_\kappa(x,y)| \leq e^{\rho(x,y)}$.*

Theorem 7.6.9 *Suppose $\kappa \geq 0$ and $x \in \mathbb{R}^d$. Then there exists a Baire probability measure μ_x with $\text{supp}(\mu_x) \subset \text{conv}\{wx\}_{w\in G}$ (the convex hull) such that $V_\kappa p(x) = \int_{\mathbb{R}^d} p\, d\mu_x$ for each $p \in \Pi^d$.*

Theorem 7.6.8 was shown by Dunkl (1991). There he constructed the *Dunkl kernel*. Later Rösler (1999) proved Theorem 7.6.9. The second inequality in the following corollary will be used in §7.7 on the Dunkl transform.

Corollary 7.6.10 *If $x, y \in \mathbb{R}^d$ then $K_\kappa(x,y) > 0$ and $|K_\kappa(x, \mathrm{i}y)| \leq 1$.*

Proof By Fubini's theorem, summation and integration can be interchanged in

$$K_\kappa(x,y) = \sum_{n=0}^\infty \frac{1}{n!} V_\kappa^{(y)} \langle x, y\rangle^n = \sum_{n=0}^\infty \frac{1}{n!} \int_{\mathbb{R}^d} \langle x, z\rangle^n \, d\mu_y(z) = \int_{\mathbb{R}^d} e^{\langle x,z\rangle}\, d\mu_y(z).$$

By homogeneity $K_\kappa(x, \mathrm{i}y) = \sum_{n=0}^\infty \mathrm{i}^n K_{\kappa,n}(x,y) = \int_{\mathbb{R}^d} e^{\mathrm{i}\langle x,z\rangle}\, d\mu_y(z)$. □

There is a mean-value-type result for V_κ:

Proposition 7.6.11 *Suppose $p \in \Pi^d$; then*

$$\frac{c_\kappa}{(2\pi)^{d/2}} \int_{\mathbb{R}^d} V_\kappa p(x) w_\kappa(x) e^{-|x|^2/2}\, dx = \frac{1}{(2\pi)^{d/2}} \int_{\mathbb{R}^d} p(x) e^{-|x|^2/2}\, dx.$$

Proof By Theorem 7.4.20 the left-hand side equals $\langle e^{\Delta_\kappa/2} V_\kappa p, e^{\Delta_\kappa/2} 1\rangle_\kappa = \langle V_\kappa e^{\Delta/2} p, 1\rangle_\kappa = \langle e^{\Delta/2} p, 1\rangle_0$, which equals the right-hand side (the subscript 0 indicates $\kappa = 0$). □

Corollary 7.6.12 *If $f \in C(\{x \in \mathbb{R}^d \mid |x| \leq 1\})$ and $\kappa > 0$ then*

$$c_{\kappa,S} \int_{S^{d-1}} (V_\kappa f) w_\kappa \, d\omega = 2\frac{\Gamma(\gamma_\kappa + d/2)}{\Gamma(\gamma_\kappa)\Gamma(d/2)} \int_{|x|\leq 1} f(x)(1 - |x|^2)^{\gamma_\kappa - 1}\, dx.$$

This is proven by applying the proposition to a homogeneous polynomial and then evaluating the two integrals in spherical polar coordinates (see Proposition 7.5.8; only the even-degree case needs to be computed). The formula extends to continuous functions by Theorem 7.6.9. This result is due to Xu (1997).

Opdam (1993) defined a *Bessel function* in the multiplicity function context. His approach was through G-invariant differential operators commuting with $\Delta + 2\sum_{v\in R_+}\kappa(v)\frac{\langle v,\nabla\rangle}{\langle x,v\rangle}$ (the differential part of Δ_κ). The result is that $J_G(x,y) := \frac{1}{\#G}\sum_{w\in G} K_\kappa(wx,y)$ is real-entire in x, y for each $\kappa \in \Lambda^{\mathrm{reg}}$, and $J_G(x,y)$ is meromorphic in κ with poles on Λ^0. Opdam (1993, Remark 6.12) observed that J_G can be interpreted as a spherical function on a Euclidean symmetric space, when G is a Weyl group and κ takes values in certain discrete sets.

The properties of $K_{\kappa,n}$ described in Theorem 7.6.7 extend to K_κ:

1. $\mathcal{D}_i^{(y)} K_\kappa(x,y) = x_i K_\kappa(x,y)$ for $1 \le i \le d$.
2. $\langle K_\kappa(x,\cdot), p\rangle_\kappa = p(x)$ for $p \in \Pi^d$.
3. $K_\kappa(x,y) = K_\kappa(y,x)$ for all $x, y \in \mathbb{R}^d$ and $K_\kappa(wx,wy) = K_\kappa(x,y)$ for each $w \in G$.

Property 3 shows that $J_G(wx,y) = J_G(x,wy) = J_G(x,y)$ for all $w \in G$.

7.6.1 Example: $\mathbb{Z}_2$

The objects described above can be stated explicitly for the smallest reflection group. We use κ for the value of the multiplicity function and suppress the subscript "1" (for example, in x_1). Throughout let $n \in \mathbb{N}_0$.

1. $\mathcal{D}p(x) = \partial_x p(x) + \kappa x^{-1}(p(x) - p(-x))$.
2. $\langle x^{2n}, x^{2n}\rangle_\kappa = 2^{2n} n!\,(\kappa + \frac{1}{2})_n$, $\langle x^{2n+1}, x^{2n+1}\rangle_\kappa = 2^{2n+1} n!\,(\kappa + \frac{1}{2})_{n+1}$.
3. $V_\kappa^0 x^{2n} = \frac{\kappa+1/2)_n}{(1/2)_n} x^{2n}$, $V_\kappa^0 x^{2n+1} = \frac{(\kappa+1/2)_{n+1}}{(1/2)_{n+1}} x^{2n+1}$.
4. $\Lambda^0 = \{-\frac{1}{2}, -\frac{3}{2}, \ldots\}$; if $\kappa = -\frac{1}{2} - n$ then $\mathrm{Rad}(\kappa) = \mathrm{span}\{x^m \mid m \ge 2n+1\}$.
5. $\langle p,q\rangle_g = 2^{-\kappa-1/2}\Gamma(\kappa+\frac{1}{2})^{-1}\int_{-\infty}^{\infty} p(x)q(x)|x|^{2\kappa}e^{-x^2/2}\,dx$, valid for $\kappa > -\frac{1}{2}$.
6. $e^{-\mathcal{D}^2/2}x^{2n} = (-1)^m n!\,2^n L_n^{(\kappa-1/2)}(x^2/2)$, $e^{-\mathcal{D}^2/2}x^{2n+1} = (-1)^n n!\,2^n x L_n^{(\kappa+1/2)}(x^2/2)$.
7. $V_\kappa p(x) = \frac{\Gamma(1/2)\Gamma(\kappa)}{\Gamma(\kappa+1/2)}\int_{-1}^{1} p(xt)(1+t)^\kappa(1-t)^{\kappa-1}\,dt$ $(\kappa > 0)$.
8. $K_\kappa(x,y) = \sum_{n=0}^{\infty}\frac{1}{(\kappa+1/2)_n\,n!}\left(\frac{xy}{2}\right)^{2n} + \frac{xy}{1+2\kappa}\sum_{n=0}^{\infty}\frac{1}{(\kappa+3/2)_n\,n!}\left(\frac{xy}{2}\right)^{2n}$.

Part 7 can be shown by substituting $p(x) = x^n$ in the integral and using part 3. In part 8 note that the *modified Bessel function* is given by $I_{\kappa-1/2}(x) := \frac{(x/2)^{\kappa-1/2}}{\Gamma(\kappa+1/2)}\sum_{n=0}^{\infty}\frac{1}{(\kappa+1/2)_n\,n!}\left(\frac{x}{2}\right)^{2n}$. Specializing Opdam's formula for the Bessel function yields

$$J_{\mathbb{Z}_2}(x,y) = \tfrac{1}{2}(K_\kappa(x,y) + K_\kappa(-x,y)) = \sum_{n=0}^{\infty}\frac{1}{(\kappa+1/2)_n\,n!}\left(\frac{xy}{2}\right)^{2n}.$$

This partly explains the use of "Bessel" in the name.

7.6.2 Asymptotic Properties of the Dunkl Kernel

Rösler and de Jeu (2002) proved the following results concerning the limiting behavior of $K_\kappa(x, y)$ as $x \to \infty$, for $\kappa \geq 0$. The variable x is restricted to the interior of a chamber. The fundamental chamber corresponds to the positive root system R_+. Let $\delta > 0$. Then define

$$\mathcal{C} := \{x \in \mathbb{R}^d \mid \langle x, v\rangle > 0 \;\forall\, v \in R_+\}, \quad \mathcal{C}_\delta := \{x \in \mathbb{R}^d \mid \langle x, v\rangle > \delta |x| \;\forall\, v \in R_+\}.$$

The walls of $\mathcal{C}$ are the hyperplanes $v^\perp$ for the simple roots v.

Theorem 7.6.13 *For each $w \in G$ there is a constant A_w such that for all $y \in \mathcal{C}$,*

$$\lim_{x \in \mathcal{C}_\delta, |x| \to \infty} \sqrt{w_\kappa(x) w_\kappa(y)}\, e^{-\mathrm{i}\langle x, wy\rangle} K_\kappa(\mathrm{i}x, wy) = A_w.$$

Recall $\gamma_\kappa = \sum_{v \in R_+} \kappa(v)$. For $z \in \mathbb{C}$ with $\operatorname{Re} z \geq 0$, let z^{γ_κ} denote the principal branch ($1^{\gamma_\kappa} = 1$). (See Definition 7.4.19 for c_κ.)

Theorem 7.6.14 *The constant A_1 equals $(\mathrm{i}^{\gamma_\kappa} c_\kappa)^{-1}$. Furthermore,*

$$\lim_{\operatorname{Re} z \geq 0, z \to \infty} z^{\gamma_\kappa} e^{-z\langle x, y\rangle} K_\kappa(zx, y) = \frac{1}{c_\kappa \sqrt{w_\kappa(x) w_\kappa(y)}}, \qquad x, y \in \mathcal{C}.$$

This limit is used in the context of a heat kernel.

7.6.3 The Heat Kernel

For functions defined on $\mathbb{R}^d \times (0, \infty)$ the *generalized heat equation* is

$$\Delta_\kappa u(x, t) - \frac{\partial}{\partial t} u(x, t) = 0.$$

The associated boundary-value problem is to find the solution u such that $u(x, 0) = f(x)$ where f is a given bounded continuous function on $\mathbb{R}^d$.

Definition 7.6.15 For $x, y \in \mathbb{R}^d$ and $t > 0$, the *generalized heat kernel* Γ_κ is given by

$$\Gamma_\kappa(t, x, y) := \frac{c_\kappa}{(2t)^{\gamma_\kappa + d/2} (2\pi)^{d/2}} \exp\left(-\frac{|x|^2 + |y|^2}{4t}\right) K_\kappa\left(\frac{x}{\sqrt{2t}}, \frac{y}{\sqrt{2t}}\right).$$

Definition 7.6.16 For a bounded continuous function f on $\mathbb{R}^d$ and $t > 0$, let

$$H(t)f(x) := \int_{\mathbb{R}^d} f(y) \Gamma_\kappa(t, x, y) w_\kappa(y)\, dy.$$

Theorem 7.6.17 *Suppose $f \in \mathcal{S}(\mathbb{R}^d)$ (the Schwartz space). Then $H(t)f \in \mathcal{S}(\mathbb{R}^d)$ $(t > 0)$, $H(s)H(t)f = H(s+t)f$ $(s, t > 0)$ and $\lim_{t \to 0_+} \sup_x |H(t)f(x) - f(x)| = 0$. Furthermore, the function $u(x, t) := H(t)f(x)$ $(t > 0)$, $u(x, 0) := f(x)$ solves the boundary-value problem.*

These results are due to Rösler (1998). Theorem 7.6.14 implies

$$\lim_{t\to 0_+} \frac{\sqrt{w_\kappa(x)w_\kappa(y)}\Gamma_\kappa(t,x,y)}{\Gamma_0(t,x,y)} = 1 \quad (x, y \in \mathcal{C}).$$

There is an associated *càdlàg Markov process* $X = (X_t)_{t\geq 0}$ with infinitesimal generator $\frac{1}{2}\Delta_\kappa$ ("càdlàg" is a French acronym for right-continuous and left limits). The semigroup densities are $p_t^{(\kappa)}(x,y) := \Gamma_\kappa(\frac{1}{2}t, x, y)w_\kappa(y)$. For further details see Rösler and Voit (1998), Gallardo and Yor (2006) and Graczyk et al. (2008).

7.7 The Dunkl Transform

The Dunkl kernel is used to define a generalization of the Fourier transform. The Fourier integral kernel $e^{-\mathrm{i}\langle x,y\rangle}$ is replaced by $K_\kappa(x, -\mathrm{i}y)w_\kappa(x)$. Throughout this section $\kappa \geq 0$. Recall $|K_\kappa(x, -\mathrm{i}y)| \leq 1$ for all $x, y \in \mathbb{R}^d$.

Definition 7.7.1 For $f \in L^1(\mathbb{R}^d, w_\kappa(x)\,dx)$, define the *Dunkl transform* by

$$\mathcal{F}f(y) := \frac{c_\kappa}{(2\pi)^{d/2}} \int_{\mathbb{R}^d} f(x)K_\kappa(x, -\mathrm{i}y)w_\kappa(x)\,dx.$$

By finding a set of eigenfunctions of $\mathcal{F}$ which is dense in $L^2(\mathbb{R}^d, w_\kappa(x)\,dx)$ (by Hamburger's theorem) we show that $\mathcal{F}$ is an L^2-isometry, has period 4 and $\mathcal{F}x_j = \mathrm{i}\mathcal{D}_j\mathcal{F}$ for $1 \leq j \leq d$. Convergence arguments, mostly depending on the dominated convergence theorem, are omitted, and appropriate smoothness restrictions on functions are implicitly assumed.

Theorem 7.7.2 *Let* $f(x) := p(x)L_m^{(\alpha)}(|x|^2)e^{-|x|^2/2}$, *where* $m, n \in \mathbb{N}_0$, $\alpha = n + \frac{1}{2}d + \gamma_\kappa - 1$ *and* $p \in \mathcal{H}_{\kappa,n}$; *then* $\mathcal{F}f(y) = (-\mathrm{i})^{n+2m}f(y)$ $(y \in \mathbb{R}^d)$.

Proof Suppose q is an arbitrary polynomial of degree n, for some n, and $N \geq n$. Then the formula $\langle q, \sum_{j=0}^N K_{\kappa,j}(\cdot, u)\rangle_\kappa = q(u)$ is valid for all $u \in \mathbb{C}^d$, since it is a polynomial relation. By Theorem 7.4.20 and by letting $N \to \infty$,

$$\frac{c_\kappa}{(2\pi)^{d/2}} \int_{\mathbb{R}^d} e^{-\Delta_\kappa/2}q(x)e^{-\Delta_\kappa^{(x)}/2}K_\kappa(x,u)e^{-|x|^2/2}w_\kappa(x)\,dx = q(u).$$

Since $e^{-\Delta_\kappa^{(x)}/2}K_\kappa(x,u) = \exp\big(-\frac{1}{2}\sum_{j=1}^d u_j^2\big)K_\kappa(x,u)$,

$$\frac{c_\kappa}{(2\pi)^{d/2}} \int_{\mathbb{R}^d} e^{-\Delta_\kappa/2}q(x)K_\kappa(x,u)e^{-|x|^2/2}w_\kappa(x)\,dx = \exp\big(\tfrac{1}{2}\textstyle\sum_{j=1}^d u_j^2\big)q(u). \tag{7.7.1}$$

In (7.7.1) set $q(x) := e^{\Delta_\kappa/4}(|x|^{2m}p(x)) = m!\,L_m^{(\alpha)}(-\sum_{j=1}^d x_j^2)p(x)$ (by (7.5.1)), and $u := -\mathrm{i}y$. In the left-hand side we have $e^{-\Delta_\kappa/4}(|x|^{2m}p(x))e^{-|x|^2/2} = (-1)^m m!\,f(x)$ and the right-hand side equals $m!\,e^{-|y|^2/2}L_m^{(\alpha)}(|y^2|)p(-\mathrm{i}y) = (-\mathrm{i})^n m!\,f(y)$. □

Corollary 7.7.3 *If* $f \in L^2(\mathbb{R}^d, w_\kappa(x)\,dx)$ *then* $\int_{\mathbb{R}^d} |\mathcal{F}f(y)|^2 w_\kappa(y)\,dy = \int_{\mathbb{R}^d} |f(y)|^2 w_\kappa(x)\,dx$, *and* $\mathcal{F}^2 f(x) = f(-x)$ *for almost all* $x \in \mathbb{R}^d$.

Suppose $f(x), |x|f(x) \in L^1(\mathbb{R}^d, w_\kappa(x)\,dx)$ and $1 \le j \le d$; then

$$\mathcal{D}_j \mathcal{F} f(y) = -\mathrm{i}\mathcal{F}(x_j f(x))(y), \quad y \in \mathbb{R}^d$$

since $K_\kappa(x, -\mathrm{i}y) = K_\kappa(-\mathrm{i}x, y)$. The G-invariance property of K_κ implies that $w\mathcal{F} = \mathcal{F}w$ for all $w \in G$.

The transform and the L^2-isometry result first appeared in Dunkl (1992). The uniform boundedness of $|K_\kappa(x, -\mathrm{i}y)|$ was shown by de Jeu (1993). For the special case $d = 1$, $G = Z_2$ and even functions f on $\mathbb{R}$, the transform $\mathcal{F}$ essentially coincides with the classical *Hankel transform*. For general d, when G is a Weyl group (crystallographic root system; see Definition 7.2.5), and for special values of κ, the transform $\mathcal{F}$ restricted to G-invariant functions is the spherical Fourier transform on a Cartan motion group.

7.8 The Poisson Kernel

In this section assume $\kappa \ge 0$. There is a natural boundary-value problem for harmonic functions. Given $f \in C(S^{d-1})$ find the function $P[f]$ which is smooth on $\{x \mid |x| < 1\}$ and which satisfies $\Delta_\kappa P[f] = 0$ and $\lim_{r\uparrow 1} P[f](rx) = f(x)$ $(x \in S^{d-1})$. We outline the argument for polynomial functions on S^{d-1}.

Let $n \in \mathbb{N}_0$ and $x, y \in \mathbb{R}^d$. By the definition of $\pi_{\kappa,n}$ and by Theorem 7.6.7(3),

$$\pi_{\kappa,n}^{(x)} K_{\kappa,n}(x,y) = \sum_{j=0}^{\lfloor n/2 \rfloor} \frac{1}{4^j j!\,(-\gamma_\kappa - n + 2 - d/2)_j} |x|^{2j}|y|^{2j} K_{\kappa,n-2j}(x,y).$$

Applying $\pi_{n,\kappa}^{(x)}$ to the reproducing equation $\langle p, K_{\kappa,n}(x,\cdot)\rangle_\kappa = p(x)$ for $p \in \Pi_n^d$, we obtain

$$\begin{aligned}\pi_{\kappa,n} p(x) = \langle p, \pi_{\kappa,n}^{(x)} K_{\kappa,n}(x,\cdot)\rangle_\kappa &= \frac{c_\kappa}{(2\pi)^{d/2}} \int_{\mathbb{R}^d} e^{-\Delta_\kappa/2} p(y) \pi_{\kappa,n}^{(x)} K_{\kappa,n}(x,y) e^{-|x|^2/2} w_\kappa(y)\,dy \\ &= 2^n(\gamma_\kappa + \tfrac{1}{2}d)_n c_{\kappa,S} \int_{S^{d-1}} p(y) \pi_{\kappa,n}^{(x)} K_{\kappa,n}(x,y) w_\kappa(y)\,d\omega(y),\end{aligned}$$

by Proposition 7.5.8, and because $e^{-\Delta_\kappa/2}p(y) = p(y) + p'(y)$ where p' is of degree $\le n-2$ and is thus orthogonal to $\pi_{\kappa,n}^{(x)} K_{\kappa,n}(x,y)$.

Definition 7.8.1 For $n \in \mathbb{N}_0$ and $x, y \in \mathbb{R}^d$, let

$$P_{\kappa,n}(x,y) := 2^n(\gamma_\kappa + \tfrac{1}{2}d)_n \sum_{j=0}^{\lfloor n/2 \rfloor} \frac{1}{4^j j!\,(-\gamma_\kappa - n + 2 - d/2)_j} |x|^{2j}|y|^{2j} K_{\kappa,n-2j}(x,y).$$

By the decomposition Proposition 7.5.7, any polynomial p satisfies $p(x) = \sum_{n,k\ge 0} |x|^{2k} p_{n,k}(x)$ where each $p_{n,k} \in \mathcal{H}_{\kappa,n}$, and so p agrees with the harmonic polynomial $q(x) = \sum_{n,k\ge 0} p_{n,k}(x)$ on S^{d-1}. Thus

$$q(x) = c_{\kappa,S} \int_{S^{d-1}} p(y) \sum_{j=0}^{N} P_{\kappa,j}(x,y) w_\kappa(y)\,d\omega(y),$$

where N is sufficiently large. The series converges for $|x| < 1$, $|y| = 1$.

Theorem 7.8.2 *For* $|x| < 1$, $|y| = 1$,

$$\sum_{j=0}^{\infty} P_{\kappa,j}(x, y) = V_\kappa^{(y)}\left(\frac{1 - |x|^2}{(1 - 2\langle x, y\rangle + |x|^2)^{\gamma_\kappa + d/2}}\right).$$

The result follows from expanding the right-hand side as a series in $V_\kappa^{(y)}(\langle x, y\rangle^j)$. The left-hand side of the equation is thus the *Poisson kernel* for harmonic functions in the unit ball. For fixed x with $|x| < 1$, the denominator in the $V_\kappa^{(y)}$-term does not vanish for $|y| \le 1$. There is a formula for $P_{\kappa,n}(x, y)$ restricted to $|x| = 1$:

$$V_\kappa^{(x)} \frac{n+\alpha}{\alpha} C_n^\alpha(\langle x, y/|y|\rangle)|y|^n = 2^n(\gamma_\kappa + \tfrac{1}{2}d)_n \sum_{j=0}^{\lfloor n/2\rfloor} \frac{1}{4^j j!\,(-\gamma_\kappa - n + 2 - d/2)_j} |y|^{2j} K_{\kappa,n-2j}(x, y),$$

where C_n^α is the *Gegenbauer polynomial* of degree n and index $\alpha = \gamma_\kappa + \frac{1}{2}d - 1$. (For the exceptional case $\alpha = 0$, replace the left-hand side by $V_\kappa^{(x)} 2T_n(\langle x, y/|y|\rangle)|y|^n$ for $n \ge 1$; T_n is the *Chebyshev polynomial of the first kind*.) The formula is suggested by a generating function for these polynomials. Maslouhi and Youssfi (2007) studied the properties of the Poisson kernel in connection with L^p-type convergence and with a generalized translation.

7.9 Harmonic Polynomials for $\mathbb{R}^2$

This section exhibits the classical Gegenbauer and Jacobi polynomials as spherical harmonics for the groups $\mathbb{Z}_2$ and $\mathbb{Z}_2 \times \mathbb{Z}_2$ with root systems $\{\pm\varepsilon_2\}$ and $\{\pm\varepsilon_1, \pm\varepsilon_2\}$ respectively. We mention here that the structure associated with the group $\mathbb{Z}_2^d$, $R = \{\pm\varepsilon_i\}_{i=1}^d$ and $\kappa(\varepsilon_i) = \kappa_i$ can be analyzed in a similar way and leads to multivariable Jacobi polynomials orthogonal on a simplex; see Chapter 2.

Polar coordinates will be used: $x_1 = r\cos\theta$, $x_2 = r\sin\theta$ $(r \ge 0, -\pi \le \theta \le \pi)$.

7.9.1 One Parameter

Let $R = \{\pm\varepsilon_2\}$ and $\kappa > -\frac{1}{2}$. Then

1. $w_\kappa(x) = |x_2|^{2\kappa} = r^{2\kappa}|\sin\theta|^{2\kappa}$;
2. $\mathcal{D}_1 p(x) = \frac{\partial}{\partial x_1} p(x)$, $\mathcal{D}_2 p(x) = \frac{\partial}{\partial x_2} p(x) + \kappa x_2^{-1}(p(x_1, x_2) - p(x_1 - x_2))$;
3. $\Delta_\kappa p(x) = \Delta p(x) + \kappa x_2^{-1}(2\frac{\partial}{\partial x_2} p(x) - x_2^{-1}(p(x_1, x_2) - p(x_1 - x_2)))$.

For $n \ge 1$ the space $\mathcal{H}_{\kappa,n}$ contains an orthogonal basis consisting of two polynomials, $p_{n,0}$ being even and $p_{n,1}$ being odd in x_2, and expressed in terms of Gegenbauer polynomials:

$$p_{n,0}(x) := r^n C_n^\kappa(\cos\theta), \quad p_{n,1}(x) := r^n \sin\theta\, C_{n-1}^{\kappa+1}(\cos\theta).$$

Let $a_\kappa := \frac{\Gamma(\kappa+1)}{2\sqrt{\pi}\,\Gamma(\kappa+1/2)}$; the normalizing constant such that $a_\kappa \int_{-\pi}^{\pi} |\sin\theta|^{2\kappa}\,d\theta = 1$. Then

$$a_\kappa \int_{-\pi}^{\pi} p_{n,0}(\cos\theta, \sin\theta)^2 |\sin\theta|^{2\kappa}\,d\theta = \frac{\kappa(2\kappa)_n}{(n+\kappa)\,n!},$$
$$a_\kappa \int_{-\pi}^{\pi} p_{n,1}(\cos\theta, \sin\theta)^2 |\sin\theta|^{2\kappa}\,d\theta = \frac{(\kappa+\frac{1}{2})(2\kappa+2)_{n-1}}{(n+\kappa)(n-1)!}.$$

If the Gegenbauer polynomials are replaced by $P_n^\kappa = \frac{n!}{(2\kappa)_n} C_n^\kappa$ (similarly for $C_{n-1}^{\kappa+1}$; these are normalized by $P_n^\kappa(1) = 1$) then at $\kappa = 0$ one obtains $p_{n,0} = r^n \cos n\theta$ and $p_{n,1} = r^n \sin n\theta$.

7.9.2 Two Parameters

Let $R = \{\pm\varepsilon_1, \pm\varepsilon_2\}$ and $\kappa_1, \kappa_2 > -\frac{1}{2}$. Then

1. $w_\kappa(x) = |x_1|^{2\kappa_1}|x_2|^{2\kappa_2} = r^{2\kappa_1+2\kappa_2}|\cos\theta|^{2\kappa_1}|\sin\theta|^{2\kappa_2}$;
2. $\mathcal{D}_1 p(x) = \frac{\partial}{\partial x_1}p(x) + \kappa_1 \frac{p(x)-p(-x_1,x_2)}{x_1}$, $\mathcal{D}_2 p(x) = \frac{\partial}{\partial x_2}p(x) + \kappa_2 \frac{p(x)-p(x_1,-x_2)}{x_2}$;
3. $\Delta_\kappa p(x) = \Delta p(x) + \frac{\kappa_1}{x_1}(2\frac{\partial}{\partial x_1}p(x) - \frac{p(x)-p(-x_1,x_2)}{x_1}) + \frac{\kappa_2}{x_2}(2\frac{\partial}{\partial x_2}p(x) - \frac{p(x)-p(x_1,-x_2)}{x_2})$.

There are four families of harmonic polynomials, expressed in terms of Jacobi polynomials $P_n^{(\alpha,\beta)}$. The first subscript indicates the degree and the second subscript is used to indicate the parity type; for example 01 denotes "even in x_1, odd in x_2":

$$p_{2n,00}(x) := r^{2n} P_n^{(\kappa_2-1/2,\kappa_1-1/2)}(\cos 2\theta),$$
$$p_{2n,11}(x) := r^{2n} \sin 2\theta\, P_{n-1}^{(\kappa_2+1/2,\kappa_1+1/2)}(\cos 2\theta),$$
$$p_{2n+1,10}(x) := r^{2n+1} \cos\theta\, P_n^{(\kappa_2-1/2,\kappa_1+1/2)}(\cos 2\theta),$$
$$p_{2n+1,01}(x) := r^{2n+1} \sin\theta\, P_n^{(\kappa_2+1/2,\kappa_1-1/2)}(\cos 2\theta).$$

The norms with respect to $L^2([-\pi,\pi], |\cos\theta|^{2\kappa_1}|\sin\theta|^{2\kappa_2}\,d\theta)$ can be computed from

$$\int_{-\pi}^{\pi} \left(P_n^{(\alpha,\beta)}(\cos 2\theta)\right)^2 |\sin\theta|^{2\alpha+1}|\cos\theta|^{2\beta+1}\,d\theta$$
$$= \frac{4\Gamma(\alpha+1)\Gamma(\beta+1)}{\Gamma(\alpha+\beta+2)} \frac{(\alpha+1)_n(\beta+1)_n}{n!\,(\alpha+\beta+2)_n} \frac{\alpha+\beta+n+1}{\alpha+\beta+2n+1}, \qquad n \in \mathbb{N}_0.$$

7.9.3 Dihedral Groups

The harmonic polynomials for the general dihedral groups can be expressed using the two previous types and complex coordinates $z = x_1 + \mathrm{i}x_2$, $\overline{z} = x_1 - \mathrm{i}x_2$. Interpret $p(z^m)$ as $p(\operatorname{Re} z^m, \operatorname{Im} z^m)$. For the one-parameter case $I_2(m)$ with m odd the harmonic polynomials are spanned by

1. $p_{n,0}(z^m)$, $p_{n,1}(z^m)$, of degree nm;
2. $\operatorname{Re} p_{nm+j}(z)$, $\operatorname{Im} p_{nm+j}(z)$ $(j = 1, \ldots, m-1)$, where $p_{nm+j}(z) := z^j(\frac{n+2\kappa}{2\kappa} p_{n,0}(z^m) + \mathrm{i}p_{n,1}(z^m))$.

For the two-parameter case $I_2(2m)$ with $w_\kappa(z) = |z^m + \overline{z}^m|^{2\kappa_1}|z^m - \overline{z}^m|^{2\kappa_2}$, let

$$q_{2n}(z) := p_{2n,00}(z) + \tfrac{1}{2}\mathrm{i}p_{2n,11}(z),$$

$$q_{2n+1}(z) := (n + \kappa_2 + \tfrac{1}{2})p_{2n+1,10}(z) + \mathrm{i}(n + \kappa_1 + \tfrac{1}{2})p_{2n+1,01}(z).$$

The harmonic polynomials are spanned by $z^j q_n(z^m)$ and $\overline{z^j q_n(z^m)}$ for $n \in \mathbb{N}_0$, $0 \le j \le m$ (note $q_0(z) = 1$). These results are from Dunkl (1989).

7.10 Nonsymmetric Jack Polynomials

For the symmetric group $\mathcal{S}_d$ acting naturally on $\mathbb{R}^d$ there is an elegant orthogonal basis for Π^d with respect to the form $\langle\cdot,\cdot\rangle_\kappa$. The basis consists of nonsymmetric Jack polynomials, so named because their symmetrization (summing over an $\mathcal{S}_d$-orbit) yields the Jack polynomials, with parameter $1/\kappa$. The construction of the basis depends on commuting self-adjoint "*Cherednik–Dunkl*" *operators* and an ordering of the monomial basis with respect to which the operators are represented by triangular matrices.

Recall the notation from §7.2.1: (i, j) denotes the transposition $s_{\varepsilon_i-\varepsilon_j}$ and $s_i = (i, i+1)$ for $1 \le i < d$. Interpret $\mathcal{S}_d$ as the set of bijections on $\{1, 2, \ldots, d\}$; then the action on $\mathbb{R}^d$ is given by $(w^{-1}x)_i = x_{w(i)}$ and the action on monomials is $w(x^\alpha) = x^{\alpha w^{-1}}$ where $(\alpha w^{-1})_i = \alpha_{w^{-1}(i)}$ for $w \in \mathcal{S}_d$, $1 \le i \le d$, $\alpha \in \mathbb{N}_0^d$.

The degree of the monomial x^α is $|\alpha| := \sum_{i=1}^d \alpha_i$. The elements of $\mathbb{N}_0^d$ are called *compositions* or *multi-indices*.

Definition 7.10.1 The set of *partitions* (of length $\le d$) is

$$\mathbb{N}_0^{d,+} := \{\lambda \in \mathbb{N}_0^d \mid \lambda_i \ge \lambda_{i+1},\ 1 \le i < d\}.$$

Also, for $\alpha \in \mathbb{N}_0^d$, let α^+ denote the unique partition such that $\alpha^+ = \alpha w$ for some $w \in \mathcal{S}_d$.

Definition 7.10.2 For $\alpha \in \mathbb{N}_0^d$ and $1 \le i \le d$, the *rank function* is

$$w_\alpha(i) := \#\{j \mid \alpha_j > \alpha_i\} + \#\{j \mid 1 \le j \le i,\ \alpha_j = \alpha_i\}.$$

Note that $w_\alpha(i) < w_\alpha(j)$ is equivalent to $\alpha_i > \alpha_j$, or $\alpha_i = \alpha_j$ and $i < j$. For any α the function w_α is one-to-one on $\{1, 2, \ldots, d\}$, hence $w_\alpha \in \mathcal{S}_d$. Also α is a partition if and only if $w_\alpha(i) = i$ for all i. In general $\alpha w_\alpha^{-1} = \alpha^+$ because $(\alpha w_\alpha^{-1})_i = \alpha_{w_\alpha^{-1}(i)}$ for $1 \le i \le d$.

There is one conjugacy class of reflections and we use κ for the value of the multiplicity function. For $p \in \Pi^d$ and $1 \le i \le d$,

$$\mathcal{D}_i p(x) = \frac{\partial}{\partial x_i} p(x) + \kappa \sum_{j \ne i} \frac{p(x) - p((i, j)x)}{x_i - x_j}.$$

The commutation relations from Proposition 7.4.4 (using x_i to denote the multiplication operator) become

$$[\mathcal{D}_j, x_i] = \begin{cases} -\kappa(i, j), & j \ne i, \\ 1 + \kappa \sum_{k \ne i}(i, k), & j = i. \end{cases} \tag{7.10.1}$$

A new order on compositions is defined in terms of the *dominance order*:

Definition 7.10.3 For $\alpha,\beta \in \mathbb{N}_0^d$ the partial order $\alpha > \beta$ (α dominates β) means that $\alpha \neq \beta$ and $\sum_{i=1}^j \alpha_i \geq \sum_{i=1}^j \beta_i$ for $1 \leq j \leq d$; and $\alpha \rhd \beta$ means that $|\alpha| = |\beta|$ and either $\alpha^+ > \beta^+$ or $\alpha^+ = \beta^+$ and $\alpha > \beta$.

For example $(5,1,4) \rhd (1,5,4) \rhd (4,3,3)$, while $(1,5,4)$ and $(6,2,2)$ are not comparable in $\rhd$. The following hold for $\alpha \in \mathbb{N}_0^d$:
1. $\alpha^+ \unrhd \alpha$.
2. If $\alpha_i > \alpha_j$ and $i < j$ then $\alpha \rhd \alpha(i,j)$.
3. If $1 \leq m < \alpha_i - \alpha_j$ then $\alpha^+ \rhd (\alpha - m(\varepsilon_i - \varepsilon_j))^+$.

The *Cherednik–Dunkl operators*, which are extensions of the Jucys–Murphy elements (see Murphy, 1981), are given by

$$\mathcal{U}_i := \mathcal{D}_i x_i - \kappa \sum_{1 \leq j < i} (i,j), \quad 1 \leq i \leq d.$$

Proposition 7.10.4 *The operators $\mathcal{U}_i$ satisfy*
1. $[\mathcal{U}_i, \mathcal{U}_j] = 0$ *for* $1 \leq i,j \leq d$;
2. $\langle \mathcal{U}_i p, q\rangle_\kappa = \langle p, \mathcal{U}_i q\rangle_\kappa$ *for* $p,q \in \Pi^d$;
3. $s_j \mathcal{U}_i s_j = \mathcal{U}_i$ *for* $j \neq i-1, i$ *and* $s_i \mathcal{U}_i s_i = \mathcal{U}_{i+1} + \kappa s_i$ *for* $1 \leq i \leq d$.

Proposition 7.10.5 *Let* $\alpha \in \mathbb{N}_0^d$ *and* $1 \leq i \leq d$; *then*

$$\mathcal{U}_i x^\alpha = ((d - w_\alpha(i))\kappa + \alpha_i + 1)x^\alpha + q_{\alpha,i}(x),$$

where $q_{\alpha,i}(x)$ *is a sum of terms* $\pm\kappa x^{\alpha(i,j)}$ *with* $\alpha \rhd \alpha(i,j)$ *and* $j \neq i$.

This shows that the matrix representing $\mathcal{U}_i$ on the monomial basis of Π_n^d for any $n \in \mathbb{N}_0$ is triangular (recall that any partial order can be embedded in a total order) and the eigenvalues of $\mathcal{U}_i$ are $\{(d - w_\alpha(i))\kappa + \alpha_i + 1 \mid \alpha \in \mathbb{N}_0^d,\ |\alpha| = n\}$.

For $\alpha \in \mathbb{N}_0^d$ and $1 \leq i \leq d$, let $\xi_i(\alpha) := (d - w_\alpha(i))\kappa + \alpha_i + 1$. To assert that a commuting collection of triangular matrices has a basis of joint eigenvectors, a separation property suffices: If $\alpha,\beta \in \mathbb{N}_0^d$ and $\alpha \neq \beta$ then $\xi_i(\alpha) \neq \xi_i(\beta)$ for some i. This condition is satisfied if κ is generic or $\kappa > 0$. In this case the eigenvalues $\xi_i(\alpha)$ $(1 \leq i \leq d)$ are pairwise distinct and the largest value is $(d-1)\kappa + \alpha_1^+$; if $\xi_i(\alpha) = \xi_i(\beta)$ for all i then $\alpha_j = \beta_j$ for $j = w_\alpha^{-1}(1) = w_\beta^{-1}(1)$ and so on.

Theorem 7.10.6 *Suppose κ is generic or $\kappa > 0$; then for each $\alpha \in \mathbb{N}_0^N$, there is a unique simultaneous eigenfunction ζ_α such that*

$$\mathcal{U}_i \zeta_\alpha = \xi_i(\alpha)\zeta_\alpha \ (1 \leq i \leq d) \quad \textit{and} \quad \zeta_\alpha = x^\alpha + \sum_{\alpha \rhd \beta} A_{\beta\alpha} x^\beta \ \textit{with coefficients}\ A_{\beta\alpha} \in \mathbb{Q}(\kappa).$$

These eigenfunctions are called *nonsymmetric Jack polynomials* and form a basis of Π^d by the triangularity property. There is a link to the *trigonometric* Jack polynomials which are polynomials in $\exp(2\pi \mathrm{i}\theta_j)$, $1 \leq j \leq d$, interpreted as periodic functions on $\mathbb{R}^d/\mathbb{Z}^d$. Essentially this comes from restricting the polynomials to the d-torus $\{x \in \mathbb{C}^d \mid |x_j| = 1,\ 1 \leq j \leq d\}$.

If $\alpha \neq \beta$ then $\xi_i(\alpha) \neq \xi_i(\beta)$ for some i and $\xi_i(\alpha)\langle\zeta_\alpha,\zeta_\beta\rangle_\kappa = \langle \mathcal{U}_i\zeta_\alpha,\zeta_\beta\rangle_\kappa = \langle\zeta_\alpha, \mathcal{U}_i\zeta_\beta\rangle_\kappa = \xi_i(\beta)\langle\zeta_\alpha,\zeta_\beta\rangle_\kappa$ and thus $\langle\zeta_\alpha,\zeta_\beta\rangle_\kappa = 0$. The formula for $\langle\zeta_\alpha,\zeta_\alpha\rangle_\kappa$ is more complicated.

Suppose $\alpha \in \mathbb{N}_0^d$ and $\alpha_i < \alpha_{i+1}$; then $\alpha s_i \rhd \alpha$ and $w_{\alpha s_i} = w_\alpha s_i$. Let $p := s_i\zeta_\alpha - c\zeta_\alpha$, where $c \in \mathbb{Q}(\kappa)$ is to be determined. By Proposition 7.10.4, $\mathcal{U}_j p = \xi_j(\alpha)p$ for $j \neq i, i+1$ and the leading term (with respect to $\rhd$) in p is $x^{\alpha s_i}$. Solve the equation $\mathcal{U}_i p = \xi_{i+1}(\alpha)p$ for c by using $\mathcal{U}_i s_i = s_i \mathcal{U}_{i+1} + \kappa$ to obtain $c = \frac{\kappa}{\xi_i(\kappa)-\xi_{i+1}(\kappa)}$. This implies that $\mathcal{U}_{i+1}p = \xi_i(\alpha)p$, $p = \zeta_{\alpha s_i}$ and

$$s_i\zeta_\alpha = c\zeta_\alpha + \zeta_{\alpha s_i}, \quad s_i\zeta_{\alpha s_i} = (1-c^2)\zeta_\alpha - c\zeta_{\alpha s_i}, \quad \langle \zeta_{\alpha s_i}, \zeta_{\alpha s_i}\rangle_\kappa = (1-c^2)\langle \zeta_\alpha, \zeta_\alpha\rangle_\kappa.$$

The last equation follows from $\langle \zeta_\alpha, \zeta_\alpha\rangle_\kappa = \langle s_i\zeta_\alpha, s_i\zeta_\alpha\rangle_\kappa = c^2\langle \zeta_\alpha, \zeta_\alpha\rangle_\kappa + \langle \zeta_{\alpha s_i}, \zeta_{\alpha s_i}\rangle_\kappa$. The other ingredient is a raising operator. From the commutations (7.10.1) we obtain

$$\mathcal{U}_i x_d = \begin{cases} x_d(\mathcal{U}_i - \kappa(i,d)), & 1 \le i < d, \\ x_d(1 + \mathcal{D}_d x_d), & i = d. \end{cases}$$

Let $\theta_d := s_1 s_2 \cdots s_{d-1}$; thus $\theta_d(d) = 1$ and $\theta_d(i) = i+1$ for $1 \le i < d$ (a cyclic shift). Then

$$\mathcal{U}_i x_d = \begin{cases} x_d(\theta_d^{-1}\mathcal{U}_{i+1}\theta_d), & 1 \le i < d, \\ x_d(1 + \theta_d^{-1}\mathcal{U}_1\theta_d), & i = d. \end{cases}$$

If p satisfies $\mathcal{U}_i p = \lambda_i p$ for $1 \le i \le d$ then $\mathcal{U}_i(x_d\theta_d^{-1}f) = \lambda_{i+1}(x_d\theta_d^{-1}f)$ for $1 \le i < d$ and $\mathcal{U}_d(x_d\theta_d^{-1}f) = (\lambda_1 + 1)(x_d\theta_d f)$. For $\alpha \in \mathbb{N}_0^d$ let $\phi(\alpha) := (\alpha_2, \alpha_3, \ldots, \alpha_d, \alpha_1 + 1)$; then $x_d\theta_d^{-1}x^\alpha = x^{\phi(\alpha)}$.

Proposition 7.10.7 *If $\alpha \in \mathbb{N}_0^d$ then $\zeta_{\phi(\alpha)} = x_d\theta_d^{-1}\zeta_\alpha$ and*

$$\langle \zeta_{\phi(\alpha)}, \zeta_{\phi(\alpha)}\rangle_\kappa = ((d - w_\alpha(1))\kappa + \alpha_1 + 1)\langle \zeta_\alpha, \zeta_\alpha\rangle_\kappa.$$

Proof The first part is shown by identifying the eigenvalues $\xi_i(\phi(\alpha))$. Also $\langle \zeta_{\phi(\alpha)}, \zeta_{\phi(\alpha)}\rangle_\kappa = \langle \theta_d^{-1}\zeta_\alpha, \mathcal{D}_d x_d \theta_d^{-1}\zeta_\alpha\rangle_\kappa = \langle \theta_d^{-1}\zeta_\alpha, \theta_d^{-1}\mathcal{D}_1 x_1 \zeta_\alpha\rangle_\kappa = \xi_1(\alpha)\langle \zeta_\alpha, \zeta_\alpha\rangle_\kappa$. □

The norm formula involves a hook-length product. For $\alpha \in \mathbb{N}_0^d$ let $\ell(\alpha) = \max\{j \mid \alpha_j > 0\}$, the *length* of α. For a point (i, j) in the *Ferrers diagram* $\{(k,l) \in \mathbb{N}_0^2 \mid 1 \le k \le \ell(\alpha),\ 1 \le l \le \alpha_k\}$ and $t \in \mathbb{Q}(\kappa)$ define the *generalized hook length* and the *hook-length product* by

$$h(\alpha, t; i, j) := \alpha_i - j + t + \kappa\#\{l \mid l > i,\ j \le \alpha_l \le \alpha_i\} + \kappa\#\{l \mid l < i,\ j \le \alpha_l + 1 \le \alpha_i\},$$

$$h(\alpha, t) := \prod_{i=1}^{\ell(\alpha)} \prod_{j=1}^{\alpha_i} h(\alpha, t; i, j).$$

For $\lambda \in \mathbb{N}_0^{d,+}$ let $(t)_\lambda := \prod_{i=1}^d (t - (i-1)\kappa)_{\lambda_i}$, the *generalized shifted factorial* (or *generalized Pochhammer symbol*).

Theorem 7.10.8 *For $\alpha \in \mathbb{N}_0^d$,*

$$\langle \zeta_\alpha, \zeta_\alpha\rangle_\kappa = (d\kappa + 1)_{\alpha^+}\frac{h(\alpha, 1)}{h(\alpha, \kappa + 1)}, \quad \zeta_\alpha(1, \ldots, 1) = \frac{(d\kappa + 1)_{\alpha^+}}{h(\alpha, \kappa + 1)}.$$

These formulae are proved by induction starting with $\alpha = 0$ and using the steps $\alpha \to \phi(\alpha)$, $\alpha \to \alpha s_i$ for $\alpha_i < \alpha_{i+1}$ (it suffices to use $\lambda \to \phi(\lambda)$ with $\lambda \in \mathbb{N}_0^{d,+}$ in the computation). The number of such steps in any sequence linking 0 to α equals

$$|\alpha| + \frac{1}{2} \sum_{1 \le i < j \le d} \left(|\alpha_i - \alpha_j| + |\alpha_i - \alpha_j + 1| - 1 \right).$$

There are two explicit results for $\mathcal{D}_i \zeta_\alpha$. Recall $\theta_m = s_1 s_2 \cdots s_{m-1}$ for $m \le d$.

Proposition 7.10.9 (Dunkl, 2005, Proposition 3.17) *Suppose $\alpha \in \mathbb{N}_0^d$ and $\ell(\alpha) = m$. Let $\widetilde{\alpha} := (\alpha_m - 1, \alpha_1, \ldots, \alpha_{m-1}, 0, \ldots)$ and $\beta_m := \alpha_m - w_\alpha(m)\kappa$. Then*

$$\mathcal{D}_i \zeta_\alpha = \begin{cases} 0, & m < i \le d, \\ \dfrac{(m\kappa + \beta_m)((d+1)\kappa + \beta_m)}{(m+1)\kappa + \beta_m} \theta_m^{-1} \zeta_{\widetilde{\alpha}}, & i = m. \end{cases}$$

Symmetric (i.e., $\mathcal{S}_d$-invariant) polynomials have bases labeled by partitions $\lambda \in \mathbb{N}_0^{d,+}$. We describe bases whose elements are mutually orthogonal in $\langle \cdot, \cdot \rangle_\kappa$. For a given $\lambda \in \mathbb{N}_0^{d,+}$ one can consider $\sum_{w \in \mathcal{S}_d} w\zeta_\lambda$ or a sum $\sum_{\alpha^+ = \lambda} b_\alpha \zeta_\alpha$ with suitable coefficients $\{b_\alpha\}$. Let $\lambda^R = (\lambda_d, \lambda_{d-1}, \ldots, \lambda_1)$; thus λ^R is the unique $\rhd$-minimum in $\{\alpha : \alpha^+ = \lambda\}$. No term x^α with $\alpha^+ = \lambda$ except for $\alpha = \lambda^R$ appears in ζ_{λ^R}. Also let $n_\lambda = \#\{w \in \mathcal{S}_d \mid \lambda w = \lambda\}$.

Definition 7.10.10 For $\lambda \in \mathbb{N}_0^{d,+}$ let $j_\lambda := h(\lambda, 1) \sum_{\alpha^+ = \lambda} \frac{\zeta_\alpha}{h(\alpha, 1)}$.

Theorem 7.10.11 *For $\lambda \in \mathbb{N}_0^{d,+}$, j_λ has the following properties:*

1. *j_λ is symmetric and the coefficient of x^λ in j_λ is* 1.
2. $j_\lambda = \dfrac{h(\lambda, \kappa+1)}{n_\lambda h(\lambda^R, \kappa+1)} \sum_{w \in \mathcal{S}_d} w\zeta_\lambda = \dfrac{1}{n_\lambda} \sum_{w \in \mathcal{S}_d} w\zeta_{\lambda^R}$.
3. $\langle j_\lambda, j_\lambda \rangle_\kappa = (d\kappa + 1)_\lambda \dfrac{d!\, h(\lambda, 1)}{n_\lambda h(\lambda^R, \kappa+1)}$ *and* $j_\lambda(1, 1, \ldots, 1) = \dfrac{d!\, (d\kappa+1)_\lambda}{n_\lambda h(\lambda^R, \kappa+1)}$.

Knop and Sahi (1997) proved the norm formulae and also showed that the coefficient of each monomial in $h(\alpha, \kappa+1)\zeta_\alpha$ is a polynomial in κ with nonnegative coefficients.

For $\lambda \in \mathbb{N}_0^{d,+}$, j_λ is a scalar multiple of the *Jack polynomial* $J_\lambda(x; \frac{1}{\kappa})$. For more details on the nonsymmetric Jack polynomials and their applications to Calogero–Moser–Sutherland systems and the groups of type B see Dunkl and Xu (2014, Chapters 10, 11).

References

Dunkl, C. F. 1989. Differential-difference operators associated to reflection groups. *Trans. Amer. Math. Soc.*, **311**, 167–183.

Dunkl, C. F. 1991. Integral kernels with reflection group invariance. *Canad. J. Math.*, **43**, 1213–1227.

Dunkl, C. F. 1992. Hankel transforms associated to finite reflection groups. Pages 123–138 of: *Hypergeometric Functions on Domains of Positivity, Jack Polynomials, and Applications*. Contemp. Math., vol. 138. Amer. Math. Soc.

Dunkl, C. F. 2005. Singular polynomials and modules for the symmetric groups. *Int. Math. Res. Not.*, 2409–2436.

Dunkl, C. F., and Xu, Y. 2014. *Orthogonal Polynomials of Several Variables*. Second edn. Encyclopedia of Mathematics and Its Applications, vol. 155. Cambridge University Press.

Dunkl, C. F., de Jeu, M. F. E., and Opdam, E. M. 1994. Singular polynomials for finite reflection groups. *Trans. Amer. Math. Soc.*, **346**, 237–256.

Etingof, P. 2010. A uniform proof of the Macdonald–Mehta-Opdam identity for finite Coxeter groups. *Math. Res. Lett.*, **17**, 275–282.

Etingof, P., and Ginzburg, V. 2002. Symplectic reflection algebras, Calogero–Moser space, and deformed Harish-Chandra homomorphism. *Invent. Math.*, **147**, 243–348.

Etingof, P., and Ma, X. 2010. *Lecture notes on Cherednik algebras*. arXiv:1001.0432v4.

Gallardo, L., and Yor, M. 2006. A chaotic representation property of the multidimensional Dunkl processes. *Ann. Probab.*, **34**, 1530–1549.

Graczyk, P., Rösler, M., and Yor, M. 2008. *Harmonic & Stochastic Analysis of Dunkl Processes*. Travaux en cours, vol. 71. Hermann.

Humphreys, J. E. 1972. *Introduction to Lie Algebras and Representation Theory*. Graduate Texts in Mathematics, vol. 9. Springer.

Humphreys, J. E. 1990. *Reflection Groups and Coxeter Groups*. Cambridge University Press.

de Jeu, M. F. E. 1993. The Dunkl transform. *Invent. Math.*, **113**, 147–162.

Knop, F., and Sahi, S. 1997. A recursion and a combinatorial formula for Jack polynomials. *Invent. Math.*, **128**, 9–22.

Maslouhi, M., and Youssfi, E. H. 2007. Harmonic functions associated to Dunkl operators. *Monatsh. Math.*, **152**, 337–345.

Murphy, G. E. 1981. A new construction of Young's seminormal representation of the symmetric groups. *J. Algebra*, **69**, 287–297.

Olver, F. W. J., Lozier, D. W., Boisvert, R. F., and Clark, C. W. (eds). 2010. *NIST Handbook of Mathematical Functions*. Cambridge University Press. Online available as *Digital Library of Mathematical Functions*, `https://dlmf.nist.gov`.

Opdam, E. M. 1993. Dunkl operators, Bessel functions and the discriminant of a finite Coxeter group. *Compos. Math.*, **85**, 333–373.

Rösler, M. 1998. Generalized Hermite polynomials and the heat equation for Dunkl operators. *Comm. Math. Phys.*, **192**, 519–542.

Rösler, M. 1999. Positivity of Dunkl's intertwining operator. *Duke Math. J.*, **98**, 445–463.

Rösler, M., and de Jeu, M. 2002. Asymptotic analysis for the Dunkl kernel. *J. Approx. Theory*, **119**, 110–126.

Rösler, M., and Voit, M. 1998. Markov processes related with Dunkl operators. *Adv. in Appl. Math.*, **21**(4), 575–643.

Xu, Y. 1997. Integration of the intertwining operator for h-harmonic polynomials associated to reflection groups. *Proc. Amer. Math. Soc.*, **125**, 2963–2973.

8

Jacobi Polynomials and Hypergeometric Functions Associated with Root Systems

Gert J. Heckman and Eric M. Opdam

8.1 The Gauss Hypergeometric Function

The *Gauss hypergeometric equation* is the second-order differential equation

$$(\theta(\theta + c - 1) - z(\theta + a)(\theta + b))f = 0$$

in the complex plane $\mathbb{C}$ with $\theta := z\,d/dz$ and a, b, c three complex parameters. It is regular outside $z = 0$, 1 and ∞. The singular points are regular singular with local exponents given by the *Riemann scheme*

$z = 0$	$z = 1$	$z = \infty$
0	0	a
$1 - c$	$c - (a + b)$	b

The first line contains the three singular points and the next two lines give the local exponents at these points. The *Gauss hypergeometric function*

$${}_2F_1\left(\begin{matrix} a, b \\ c \end{matrix}; z\right) = F(a, b; c; z) := \sum_{n=0}^{\infty} \frac{(a)_n (b)_n}{(c)_n\, n!} z^n = 1 + \frac{ab}{c\,1!} z + \frac{a(a+1)b(b+1)}{c(c+1)\,2!} z^2 + \cdots$$

is the holomorphic solution of the hypergeometric equation around $z = 0$ with exponent 0 and normalized by $F(a, b; c; 0) = 1$. It is well defined if $c \notin -\mathbb{N}$, is convergent for $|z| < 1$ and terminates if $a \in -\mathbb{N}$ or $b \in -\mathbb{N}$. The third way of defining the hypergeometric function is the *Euler integral representation*

$${}_2F_1\left(\begin{matrix} a, b \\ c \end{matrix}; z\right) = \frac{\Gamma(c)}{\Gamma(a)\Gamma(c-a)} \int_0^1 t^{a-1}(1-t)^{c-a-1}(1-tz)^{-b}\, dt, \quad \operatorname{Re} c > \operatorname{Re} a > 0, \quad (8.1.1)$$

for $|z| < 1$ or more generally, by analytic continuation, for z in the cut plane $\mathbb{C}\backslash[1, \infty)$. All three characterizations of the hypergeometric function, through the differential equation, the power series and the integral formula are in fact due to Euler.

In the next sections we shall sketch a multivariable generalization of the hypergeometric function in the context of root systems. It turns out that essentially all aspects of the one-variable case have suitable generalizations, with the exception of the Euler integral. In the general root system context, integral representations still remain a mystery, apart from a handful of isolated examples.

Consider the pull-back under the map $\mathbb{C}^\times \ni t \mapsto z \in \mathbb{C}$ given by

$$z = \left(\tfrac{1}{2} - \tfrac{1}{4}(t + 1/t)\right) = -\tfrac{1}{4}\left(t^{\frac{1}{2}} - t^{-\frac{1}{2}}\right)^2$$

of the hypergeometric equation. The transformation $t \mapsto z$ has degree 2 and ramification points $t = 1, -1$ lying above $z = 0, 1$ respectively. It is the quotient map for the action of the group $S_2 := \{\pm 1\}$ acting by $t \mapsto t^{\pm 1}$.

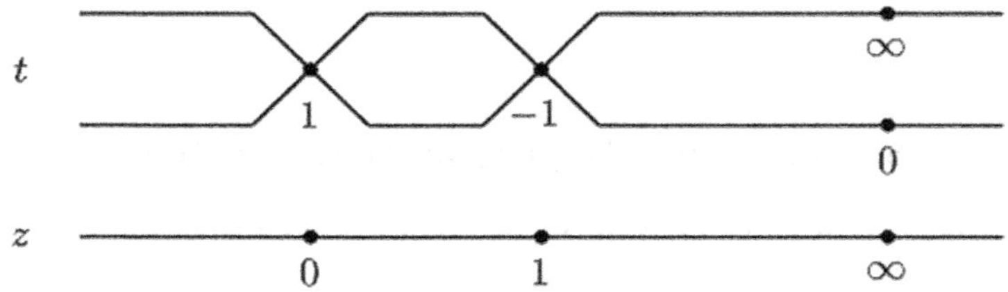

The pull-back of the hypergeometric equation under $t \mapsto z$ takes the form

$$\left(\vartheta^2 + k_1 \frac{1 + t^{-1}}{1 - t^{-1}} \vartheta + 2k_2 \frac{1 + t^{-2}}{1 - t^{-2}} \vartheta + \left(\tfrac{1}{2}k_1 + k_2\right)^2 - \lambda^2\right) f = 0$$

(with $\vartheta := t\, d/dt$) and with the linear relations

$$a = \lambda + \tfrac{1}{2}k_1 + k_2, \quad b = -\lambda + \tfrac{1}{2}k_1 + k_2, \quad c = \tfrac{1}{2} + k_1 + k_2$$

between the two parameter sets. Note the visible symmetry under $t \mapsto 1/t$. This equation has four regular singular points $t = 1, -1, 0, \infty$ with Riemann scheme

$t = 1$	$t = -1$	$t = 0$	$t = \infty$
0	0	a	a
$2 - 2c$	$2c - 2(a + b)$	b	b

as is clear from the ramification picture and the Riemann scheme of the Gauss hypergeometric equation.

The multiplicative group $\mathbb{C}^\times$ with the action of the group $S_2 = \{\pm 1\}$ by $z \mapsto z^{\pm 1}$, together with the pull-back of the hypergeometric equation, has a natural generalization. Let T be a maximal torus in a simply connected complex simple Lie group with Weyl group W. Instead of $\mathbb{C}^\times$ with the action of S_2 we consider the complex torus $T \cong (\mathbb{C}^\times)^n$ with the action of W. It turns out that on the quotient space $W \backslash T$ there is an integrable system, in fact the eigenvalue system for a commutative algebra of linear partial differential operators, which can be viewed as a natural multivariable generalization of the Gauss hypergeometric equation.

Initial steps in this direction for rank two were taken by Koornwinder (1974a,b). In general such a multivariable theory of hypergeometric functions associated with root systems was established by the authors Heckman and Opdam (1987), Heckman (1987) and Opdam (1988a,b, 1989, 1993). The original arguments used transcendental methods, but this all changed with the fundamental paper by Dunkl (1989). The extension of Dunkl operators from the rational to the trigonometric setting was obtained by Heckman (1991) with further simplifications by Cherednik (1991) and Opdam (1995). Dunkl operators are now the cornerstone for obtaining the hypergeometric equations associated with root systems. Additional survey articles on this subject were written by Heckman (1997) and Opdam (2000).

8.2 Root Systems

In this section we set up the notation and give a brief exposition of the theory of root systems. Standard references are Bourbaki (1968) and Humphreys (1990).

Let V be a finite-dimensional Euclidean vector space. The inner product of two vectors λ, μ in V will be denoted (λ, μ). For α a nonzero vector in V let $\alpha^\vee = 2\alpha/(\alpha, \alpha)$ be the covector of α and denote by

$$s_\alpha : \lambda \mapsto \lambda - (\lambda, \alpha^\vee)\alpha : V \to V$$

the orthogonal reflection with mirror the hyperplane V_α perpendicular to α. The transformation s_α is called the *reflection with root α*.

Definition 8.2.1 A *root system R* in V is a finite subset of nonzero vectors spanning V, such that $s_\alpha(\beta) \in R$ and $(\beta, \alpha^\vee) \in \mathbb{Z}$ for all $\alpha, \beta \in R$. The group W generated by the reflections s_α for $\alpha \in R$ is called the *Weyl group*. The second property is called the *crystallographic condition*.

In Dunkl's Chapter 7, Definition 7.2.1, the crystallographic condition is not included in the definition of a root system.. However, in the present chapter a root system will always be crystallographic, as is customary in semisimple Lie theory.

We do not require that $\mathbb{Q}\alpha \cap R = \{\pm\alpha\}$ for all $\alpha \in R$, and so R need not be *reduced*. It is obvious that $R^\vee := \{\alpha^\vee\}_{\alpha \in R}$ is again a root system, called the *coroot system*. The lattice $Q := \mathbb{Z}R$ is called the *root lattice* of R and the lattice P dual to the *coroot lattice* $Q^\vee := \mathbb{Z}R^\vee$ is called the *weight lattice* of R. Vectors in the weight lattice P are called *weights*. The root lattice is contained in the weight lattice by the crystallographic condition and both lattices are invariant under W. It is easy to see that for $\alpha \in R$ one has either $\mathbb{Q}\alpha \cap P = \mathbb{Z}\alpha$ or $\mathbb{Q}\alpha \cap P = \frac{1}{2}\mathbb{Z}\alpha$, and in the latter case we say that the root α is *twice a weight*.

Let $T := \mathrm{Hom}(P, \mathbb{C}^\times)$ be the *complex torus* with character lattice P. We have the polar decomposition

$$T = T_v T_u, \quad T_v := \mathrm{Hom}(P, \mathbb{R}_{>0}), \quad T_u := \mathrm{Hom}\left(P, \{z \in \mathbb{C}^\times \mid |z| = 1\}\right)$$

with T_v the real vector subgroup and T_u the *real compact torus*. Since the weight lattice P is equal to $\mathrm{Hom}(T, \mathbb{C}^\times)$ the group algebra $\mathbb{C}[P]$ gets identified with the algebra $\mathbb{C}[T]$ of regular functions (or Laurent polynomials) on T. This defines T as a complex algebraic torus.

As a complex manifold, $T = \mathfrak{t}/(2\pi i Q^\vee)$ with $\mathfrak{t} := \mathbb{C} \otimes V$ the Lie algebra of T. Extend the inner product $(\cdot, \cdot)$ on V to a symmetric bilinear form on $\mathfrak{t}$. For $\mu \in P$ the regular function t^μ on T, defined by $t^\mu := e^{(\mu, \log t)}$ with $\log t$ a representative in $\mathfrak{t}$ for $t \in \mathfrak{t}/(2\pi i Q^\vee)$, is called a

Laurent monomial. Addition on $\mathfrak{t}$ induces an Abelian group structure on T and so T becomes a complex torus. For a root α the submanifold $T_\alpha := \{t \in T \mid t^\alpha = 1\}$ is a subgroup of T, called a *toric mirror*. It consists of two or one connected components, depending on whether α is twice a weight or not.

Fix once and for all a decomposition $R = R_+ \cup R_-$ in positive and negative roots. The cone of *dominant weights* $P_+ = \{\mu \in P \mid (\mu, \alpha^\vee) \in \mathbb{N}\ \forall \alpha \in R_+\}$ has a basis over $\mathbb{N}$ of *fundamental weights* $\varpi_1, \ldots, \varpi_n$, which is just dual to the basis of *simple coroots* $\alpha_1^\vee, \ldots, \alpha_n^\vee$ of $R_+^\vee$. The corresponding *simple roots* $\alpha_1, \ldots, \alpha_n$ are a basis of simple roots for the root subsystem $R^0 := \{\alpha \in R\}_{2\alpha \notin R}$ of unmultipliable roots. The corresponding *simple reflections* $s_1, \ldots, s_n$ generate the Weyl group W as a Coxeter group.

For $\mu \in P_+$ the regular function $m_\mu(t) := \sum_{\nu \in W\mu} t^\nu$ is called the *monomial invariant function* with highest weight μ. Define a partial ordering $\leq$ on P by $\nu \leq \mu$ if $(\mu - \nu) \in \mathbb{N}R_+$, which explains the term *highest weight* μ for m_μ. It is easy to show that $m_\mu m_\nu = m_{\mu+\nu} + \cdots$ with $\cdots$ denoting a linear combination of terms m_λ ($\lambda \in P_+$, $\lambda < \mu + \nu$). Since the monomial invariant functions are a basis of $\mathbb{C}[T]^W$, one can derive that $\mathbb{C}[T]^W$ is equal to the polynomial algebra $\mathbb{C}[z_1, \ldots, z_n]$ with z_j the *fundamental monomial invariant function* with highest weight ϖ_j. In turn this implies that $W\backslash T$ is isomorphic to the linear space $\mathbb{C}^n$. The quotient map $T \to W\backslash T$ for the action of W on T has degree equal to the order of the Weyl group, and is ramified along the toric mirror arrangement $\cup_\alpha T_\alpha$. The hypergeometric system is an integrable system of linear partial differential equations with polynomial coefficients on $W\backslash T \cong \mathbb{C}^n$. Although in the rank-one case of the previous section, with W of order 2 acting on $\mathbb{C}^\times$ by $t \mapsto 1/t$ and with quotient map $t \mapsto (t + 1/t) \colon \mathbb{C}^\times \to \mathbb{C}$ a degree-2 covering, this might seem odd, for higher rank the only sensible approach is never to work on the quotient space $W\backslash T$, but to perform slick constructions on T with suitable equivariance under the Weyl group W.

8.3 The Hypergeometric System

The complex torus T, with character lattice equal to the weight lattice P of our given root system R, has Lie algebra $\mathfrak{t}$ with the trivial Lie bracket. We have natural isomorphisms $S\mathfrak{t} \cong P\mathfrak{t}^* \cong U\mathfrak{t}$ of the symmetric algebra $S\mathfrak{t}$ of $\mathfrak{t}$, the algebra $P\mathfrak{t}^*$ of polynomial functions on $\mathfrak{t}^*$ and the universal enveloping algebra $U\mathfrak{t}$ of invariant linear differential operators on T. For p a polynomial function on $\mathfrak{t}^*$ we denote by $\partial(p)$ the corresponding invariant linear differential operator on T. The characters $t \mapsto t^\mu = e^{\mu(\log t)}$ are eigenfunctions for $U\mathfrak{t}$, so $\partial(p)t^\mu = p(\mu)t^\mu$ for all $\mu \in P$. The root system R, the root lattice Q and the weight lattice P are naturally considered as subsets of the dual space $\mathfrak{t}^*$.

Let us denote by $T_{\text{reg}} := T \backslash (\cup_{\alpha \in R} T_\alpha)$ the complement of the toric mirror arrangement, and by $\mathbb{C}[T_{\text{reg}}]$ the algebra of regular functions on T_{reg} generated by $\mathbb{C}[T]$ and the functions $t \mapsto 1/(1 - t^{-\alpha})$ for $\alpha \in R$. Denote by $\mathbb{D}(T_{\text{reg}}) := \mathbb{C}[T_{\text{reg}}] \otimes U\mathfrak{t}$ the corresponding algebra of linear differential operators on T_{reg}. Clearly $\mathbb{C}[T_{\text{reg}}]$ is a natural left module for $\mathbb{D}[T_{\text{reg}}]$. The Weyl group algebra $\mathbb{C}[W]$ acts on $\mathbb{C}[T_{\text{reg}}]$ by left multiplication (so $w(e^\mu) = e^{w\mu}$ for all $\mu \in P$)

and on $\mathbb{D}[T_{\mathrm{reg}}]$ by conjugation in a compatible way. There is a unique associative algebra structure on $\mathbb{D}[T_{\mathrm{reg}}] \otimes \mathbb{C}[W]$ turning $\mathbb{C}[T_{\mathrm{reg}}]$ into a left module for $\mathbb{D}[T_{\mathrm{reg}}] \otimes \mathbb{C}[W]$.

Lemma 8.3.1   *The natural map* $\mathbb{D}[T_{\mathrm{reg}}] \otimes \mathbb{C}[W] \to \mathrm{Hom}\,(\mathbb{C}[T], \mathbb{C}[T_{\mathrm{reg}}])$ *is an injection.*

Definition 8.3.2   Put $\mathcal{K} := \{k = (k_\alpha)_{\alpha \in R} \in \mathbb{C}^R \mid k_{w\alpha} = k_\alpha \ \forall\, w \in W, \ \forall\, \alpha \in R\}$ and $\mathcal{K}_+ := \{k \in \mathcal{K} \mid k_\alpha \geq 0 \ \forall\, \alpha \in R\}$. The elements of $\mathcal{K}$ are called *multiplicity parameters* for R. For $\xi \in \mathfrak{t}$ and $k \in \mathcal{K}$ the expression

$$T(\xi, k) := \partial(\xi) - \rho(k)(\xi) + \sum_{\alpha > 0} k_\alpha \alpha(\xi)(1 - t^{-\alpha})^{-1} \otimes (1 - s_\alpha)$$

(viewed as an element of $\mathbb{D}[T_{\mathrm{reg}}] \otimes \mathbb{C}[W]$) is called the *Dunkl–Cherednik operator* or the *trigonometric Dunkl operator*, with

$$\rho(k) := \tfrac{1}{2} \sum_{\alpha > 0} k_\alpha \alpha \in \mathfrak{t}^* \tag{8.3.1}$$

the *Weyl vector* for the multiplicity parameter $k \in \mathcal{K}$.

The Dunkl operator acts as a linear operator on $\mathbb{C}[T_{\mathrm{reg}}]$ leaving the linear subspace $\mathbb{C}[T] \hookrightarrow \mathbb{C}[T_{\mathrm{reg}}]$ invariant. Indeed, for $\mu \in P$ with $\mu(\alpha^\vee) = m \in \mathbb{Z}$ we have

$$\frac{t^\mu - t^{s_\alpha \mu}}{1 - t^{-\alpha}} = t^\mu \frac{1 - t^{-m\alpha}}{1 - t^{-\alpha}} = \begin{cases} t^\mu(1 + e^{-\alpha} + \cdots + e^{-(m-1)\alpha}) & \text{if } m > 0, \\ 0 & \text{if } m = 0, \\ -t^\mu(e^\alpha + \cdots + e^{-m\alpha}) & \text{if } m < 0, \end{cases}$$

which in turn implies that $T(\xi, k)$ maps $\mathbb{C}[T]$ to $\mathbb{C}[T]$.

Lemma 8.3.3   *In the case* $k \in \mathcal{K}_+$ *we can define a Hermitian inner product* $\langle \cdot, \cdot \rangle_k$ *on* $\mathbb{C}[T]$ *by*

$$\langle f, g \rangle_k = |W|^{-1} \int_{T_u} f(t) \overline{g(t)} \prod_{\alpha > 0} \left| t^{\frac{1}{2}\alpha} - t^{-\frac{1}{2}\alpha} \right|^{2k_\alpha} d_u t$$

with $d_u t$ *the normalized Haar measure on the compact torus* T_u. *Moreover, the Dunkl operator satisfies*

$$\langle T(\xi, k) f, g \rangle_k = \langle f, T(\bar{\xi}, k) g \rangle_k$$

with the bar for complex conjugation on $\mathfrak{t}$ *with respect to the real form* $\mathfrak{t}_v$. *In particular* $T(\xi, k)$ *is self-adjoint on* $\mathbb{C}[T]$ *with respect to* $\langle \cdot, \cdot \rangle_k$ *for all* $\xi \in \mathfrak{t}_v$.

If $f(t) = \sum_{\mu \in P} c_\mu t^\mu$ is a Laurent polynomial on T then the constant term c_0 is equal to $\int_{T_u} f(t)\, d_u t$. So the Hermitian inner product $\langle \cdot, \cdot \rangle_k$ is defined in algebraic terms for $k \in \mathcal{K} \cap \mathbb{N}^R$, since

$$\delta(k; t) := \prod_{\alpha > 0} \left| t^{\frac{1}{2}\alpha} - t^{-\frac{1}{2}\alpha} \right|^{2k_\alpha} = \prod_{\alpha > 0} (2 - t^\alpha - t^{-\alpha})^{k_\alpha} \in \mathbb{C}[T]$$

for $t \in T_u$. In turn this implies that our proof of Theorem 8.3.5 below on the commutativity of the Dunkl operators is algebraic.

Lemma 8.3.4 *Recall the standard partial ordering $\leq$ on P defined by $\nu \leq \mu$ if $\mu - \nu \in \mathbb{N}R_+$. For $\mu \in P$ let $\mu_+ \in P_+$ be the unique dominant weight in the orbit $W\mu$. Define a new partial ordering $\trianglelefteq$ on P by*

$$\nu \trianglelefteq \mu \quad \text{if either } \nu_+ < \mu_+ \text{ or } \nu_+ = \mu_+ \wedge \mu \leq \nu.$$

So μ_+ is the smallest and $w_0\mu_+$ is the largest element in the orbit $W\mu$ in this new ordering $\trianglelefteq$. Here $w_0 \in W$ is the longest element. Then the Dunkl operators are upper triangular with respect to the basis t^μ of $\mathbb{C}[T]$ partially ordered by $\trianglelefteq$. More precisely, writing dots for lower-order terms with respect to $\trianglelefteq$, we have $T(\xi, k)t^\mu = \tilde{\mu}(\xi)t^\mu + \cdots$ for all $\mu \in P$, with

$$\tilde{\mu} := \mu + \tfrac{1}{2} \sum_{\alpha > 0} k_\alpha \epsilon(\mu(\alpha^\vee))\alpha$$

and $\epsilon \colon R \to \{\pm 1\}$ defined by $\epsilon(x) = +1$ if $x > 0$ and $\epsilon(x) = -1$ if $x \leq 0$.

For $k \in \mathcal{K}_+$ define a new basis $\{E(\mu, k)\}_{\mu \in P}$ of $\mathbb{C}[T]$ by the conditions

$$E(\mu, k) = t^\mu + \text{lower-order terms}, \quad \langle E(\mu, k), t^\nu \rangle_k = 0 \text{ for all } \nu \in P \text{ with } \nu \lhd \mu. \tag{8.3.2}$$

Here lower-order terms are taken in the partial ordering $\trianglelefteq$ on P. The Laurent polynomials $E(\mu, k)$ will be called *nonsymmetric Jacobi polynomials*; see after Definition 8.4.1. This new basis is obtained from the original monomial basis by an upper unitriangular transformation, so that the inverse transformation is again upper unitriangular. Therefore the Dunkl operators are also upper triangular with respect to this basis. Since $\langle E(\mu, k), E(\nu, k) \rangle_k = 0$ for all $\nu \in P$ with $\nu \lhd \mu$ it follows from Lemma 8.3.3 that the Dunkl operators with fixed multiplicity parameter $k \in \mathcal{K}$ are simultaneously diagonalized by the basis $\{E(\mu, k)\}_{\mu \in P}$. Hence Dunkl operators commute, and we have proven the following theorem.

Theorem 8.3.5 *We have $T(\xi, k)T(\eta, k) = T(\eta, k)T(\xi, k)$ for all $\xi, \eta \in \mathfrak{t}$ and all $k \in \mathcal{K}$.*

The equality in the theorem is polynomial in $k \in \mathcal{K}$, and so it follows for all $k \in \mathcal{K}$ once it is known on the Zariski dense subsets $\mathcal{K} \cap \mathbb{N}^R \subset \mathcal{K}_+$ of $\mathcal{K}$. Due to the commutativity of the Dunkl operators we can extend the linear map $\xi \mapsto T(\xi, k) \colon \mathfrak{t} \to \mathbb{D}[T_{\text{reg}}] \otimes \mathbb{C}[W]$ to an algebra homomorphism $p \mapsto T(p, k)$ from the symmetric algebra $S\mathfrak{t}$ into $\mathbb{D}[T_{\text{reg}}] \otimes \mathbb{C}[W]$, such that the induced natural action of $S\mathfrak{t}$ on $\mathbb{C}[T_{\text{reg}}]$ via higher-order Dunkl operators preserves the linear subspace $\mathbb{C}[T]$. It is clear that

$$T(p, k)E(\mu, k) = p(\tilde{\mu})E(\mu, k) \quad (\mu \in P,\ k \in \mathcal{K}_+).$$

The following definition goes back to Drinfel'd (1986) and Lusztig (1989).

Definition 8.3.6 The *degenerate affine Hecke algebra* $\mathbb{H} = \mathbb{H}(R_+, k)$ is the unique associative algebra satisfying

1. $\mathbb{H} = S\mathfrak{t} \otimes \mathbb{C}[W]$ as a vector space over $\mathbb{C}$;
2. The maps $p \mapsto p \otimes 1 \colon S\mathfrak{t} \to \mathbb{H}$ and $w \mapsto 1 \otimes w \colon \mathbb{C}[W] \to \mathbb{H}$ are algebra homomorphisms, and so we will identify $S\mathfrak{t}$ and $\mathbb{C}[W]$ with their images in $\mathbb{H}$ via these maps;
3. $p \cdot w = p \otimes w$ with $\cdot$ denoting the algebra multiplication in $\mathbb{H}$;

4. $s_i \cdot p - s_i(p) \cdot s_i = -k_i(p - s_i(p))/\alpha_i^\vee$ with $w(p)$ the natural transform of $p \in S\mathfrak{t}$ under $w \in W$, and with $k_i := \frac{1}{2}k_{\alpha_i/2} + k_{\alpha_i}$.

Note that the fourth item of this definition holds if and only if the item holds for all $p = \xi \in \mathfrak{t}$ homogeneous of degree 1, in which case it boils down to

$$s_i \cdot \xi - s_i(\xi) \cdot s_i = -k_i \alpha_i(\xi) \quad (\xi \in \mathfrak{t}).$$

If $W \ni w = s_{i_1} \cdots s_{i_p}$ is written as a shortest word in the simple reflections then $p = l(w)$ is called the *length* of this Weyl group element w. By induction on the length $l(w)$ one can show

$$w \cdot \xi \cdot w^{-1} = w(\xi) + \sum_{\alpha \in R_+ \cap wR_-} k_\alpha \alpha(w(\xi)) s_\alpha \quad (\xi \in \mathfrak{t},\ w \in W).$$

In turn this implies that the centralizer of $\mathfrak{t}$ in $\mathbb{H}$ is equal to $S\mathfrak{t}$. Using the last item of the above definition it is straightforward to describe the center of the degenerate affine Hecke algebra $\mathbb{H}$.

Proposition 8.3.7 *The center $Z(\mathbb{H})$ of $\mathbb{H}$ is equal to $S\mathfrak{t}^W$.*

An easy check from the definition of the Dunkl operator (just a rank-one computation) gives

$$s_i T(\xi, k) - T(s_i\xi, k)s_i = -k_i \alpha_i(\xi) \quad (\xi \in \mathfrak{t}).$$

Thus the conclusion is that the action of the Weyl group and the Dunkl operators on $\mathbb{C}[T]$ define a representation of the degenerate affine Hecke algebra $\mathbb{H}(R_+, k)$ on the function space $\mathbb{C}[T]$.

Definition 8.3.8 The representation via Dunkl operators

$$p \mapsto T(p, k), \quad w \mapsto w \colon \mathbb{H}(R_+, k) \to \mathrm{End}_{\mathbb{C}}(\mathbb{C}[T])$$

is called the *Dunkl representation* of the degenerate affine Hecke algebra.

Hence for $p \in S\mathfrak{t}^W$ the Dunkl operator

$$T(p, k) = \sum_w D(w, p, k) \otimes w \in \mathbb{D}[T_{\text{reg}}] \otimes \mathbb{C}[W]$$

commutes with all elements from W, and therefore the linear differential operator

$$D(p, k) := \sum_w D(w, p, k)$$

lies in $\mathbb{D}[T_{\text{reg}}]^W$. It is also clear that

$$D(p, k)D(q, k) = D(pq, k) \quad (p, q \in S\mathfrak{t}^W),$$

and so $\{D(p, k) \mid p \in S\mathfrak{t}^W\}$ is a commutative algebra of differential operators on T_{reg}. By definition $D(p, k)$ is the unique element of $\mathbb{D}[T_{\text{reg}}]^W$ which has the same restriction to $\mathbb{C}[T]^W$ as the Dunkl operator $T(p, k)$. In particular $D(p, k)$ preserves the space $\mathbb{C}[T]^W$.

Definition 8.3.9 Fix $k \in \mathcal{K}$ and $\lambda \in \mathfrak{t}^*$. The system of differential equations

$$D(p,k)f = p(\lambda)f \quad (p \in S\mathfrak{t}^W) \tag{8.3.3}$$

on $T_{\text{reg}} \subset T$ is called the *hypergeometric system* associated with the root system R with multiplicity parameter $k \in \mathcal{K}$ and spectral parameter $\lambda \in \mathfrak{t}^*$.

An explicit expression for the linear differential operator $D(p,k)$ is only manageable for p equal to the quadratic invariant.

Theorem 8.3.10 *If $\xi_1, \ldots, \xi_n$ is a real orthonormal basis of $\mathfrak{t}$ then*

$$D(\textstyle\sum_i \xi_i^2, k) = \displaystyle\sum_i \partial(\xi_i)^2 + \sum_{\alpha>0} k_\alpha \frac{1+t^{-\alpha}}{1-t^{-\alpha}} \partial(\alpha) + (\rho(k), \rho(k)) \tag{8.3.4}$$

with $\partial(p)t^\mu = p(\mu)t^\mu$ for $p \in S\mathfrak{t}$ and $\mu \in P \subset \mathfrak{t}^*$.

Proof For $p \in S\mathfrak{t}^W$ homogeneous the leading symbol of $D(p,k)$ is equal to $\partial(p)$ while the constant term equals $p(\rho(k))$. The intermediate linear terms require a small computation. □

Example 8.3.11 In the case $R = \{\pm 1, \pm 2\} \subset \mathbb{R}$ is a rank-one root system with $\mathbb{C}[T] = \mathbb{C}[t, t^{-1}]$ and $\vartheta := t\,d/dt$ the natural basis vector of $\mathfrak{t}$, the hypergeometric equation associated with R becomes

$$\left(\vartheta^2 + k_1 \frac{1+t^{-1}}{1-t^{-1}} \vartheta + 2k_2 \frac{1+t^{-2}}{1-t^{-2}} \vartheta + (\tfrac{1}{2}k_1 + k_2)^2 - \lambda^2\right) f = 0,$$

and after elimination of the symmetry $t \mapsto t^{\pm 1}$ it reduces in the new coordinate $z = \frac{1}{4} - \frac{1}{2}(t + t^{-1})$ to the Gauss hypergeometric equation, as discussed in §8.1.

The algebra of invariants $\mathbb{C}[T]^W$ is a polynomial algebra $\mathbb{C}[z_1, \ldots, z_n]$ in the monomial invariant functions z_i with highest weight the fundamental weight $\varpi_i \in P_+$. Hence we have constructed for each $k \in \mathcal{K}$ a commutative subalgebra of the Weyl algebra $\mathbb{C}[z_1, \ldots, z_n, \partial_1, \ldots, \partial_n]$ of maximal rank n. In these algebraic coordinates the hypergeometric system associated with a rank-one root system becomes the Gauss hypergeometric equation. However, in higher rank the algebraic coordinates become intractable (Koornwinder, 1974a,b; Opdam, 1988a) and it is best to work on the torus T in an equivariant way for W.

8.4 Jacobi Polynomials

Throughout this section we will assume that $k \in \mathcal{K}_+$, which implies that $\langle \cdot, \cdot \rangle_k$ is a Hermitian inner product on $\mathbb{C}[T]$. The monomial invariant functions m_μ for $\mu \in P_+$ form a basis of the vector space $\mathbb{C}[T]^W$.

Definition 8.4.1 The *Jacobi polynomials* $P(\mu,k)$ for $\mu \in P_+$ form a basis of $\mathbb{C}[T]^W$ satisfying

$$P(\mu,k) = m_\mu + \text{lower-order terms}, \quad \langle P(\mu,k), m_\nu \rangle_k = 0 \text{ for all } \nu \in P_+ \text{ with } \nu < \mu.$$

Here, lower-order terms are taken in the standard partial ordering $\leq$ on P_+.

This Gram–Schmidt-type definition is similar to that of the basis $E(\mu, k)$ for $\mu \in P$ of $\mathbb{C}[T]$; see (8.3.2). From this definition it follows that the Jacobi polynomials $P(\mu, k)$ are simultaneous eigenfunctions for the commutative algebra $\{T(p, k) \mid p \in S\mathrm{t}^W\}$. The $E(\mu, k) \in \mathbb{C}[T]$ are generally referred to as the *nonsymmetric Jacobi polynomials*. It is clear that $P(\mu, k) = E(w_0\mu, k) +$ lower-order terms, with $w_0 \in W$ the unique element interchanging positive and negative roots and the lower-order terms taken for the partial ordering $\trianglelefteq$ on P relative to the basis $E(\nu, k)$ of $\mathbb{C}[T]$. Hence, with $\rho(k)$ given by (8.3.1),

$$D(p, k)P(\mu, k) = p(\mu + \rho(k))P(\mu, k), \quad p \in S\mathrm{t}^W, \ \mu \in P_+. \tag{8.4.1}$$

Since real Dunkl operators acting on $\mathbb{C}[T]$ are self-adjoint with respect to $\langle\cdot,\cdot\rangle_k$ and the algebra $S\mathrm{t}^W$ separates the points of the real locus $P_+ + \rho(k)$ we find that

$$\langle P(\mu, k), P(\nu, k)\rangle_k = 0, \quad \mu, \nu \in P_+, \ \mu \neq \nu.$$

The conclusion is that the Jacobi polynomials are a set of orthogonal polynomials for $\mathbb{C}[T]^W$ with respect to $\langle\cdot,\cdot\rangle_k$. Our normalization of the Jacobi polynomials is by leading coefficient at infinity equal to 1.

8.4.1 Jacobi Polynomials and Zonal Spherical Functions

For special values of the multiplicity parameters the Jacobi polynomials have a group-theoretical meaning. We need some structure theory (for more details see Helgason, 2000, Ch. V; Helgason, 2001; Heckman and Schlichtkrull, 1994, Part II, Lecture 2). Let G be a noncompact real semisimple Lie group with Cartan subgroup A and Cartan decomposition $G = KAK$. Let $A_\mathbb{C} \subset G_\mathbb{C}$ be the complexification of A inside $G_\mathbb{C}$, with polar decomposition $A_\mathbb{C} = AA_u$, and let $U \subset G_\mathbb{C}$ be the corresponding compact real form of $G_\mathbb{C}$. The compact dual of the noncompact real Riemannian globally symmetric space $X = G/K$ is denoted $X_u = U/K$, and X and X_u are both real forms of the same complex symmetric space $X_\mathbb{C} = G_\mathbb{C}/K_\mathbb{C}$. We shall assume that $X_\mathbb{C}$ is both connected and simply connected.

If $T = A_\mathbb{C}K_\mathbb{C} \subset X_\mathbb{C}$ then $T \simeq A_\mathbb{C}/F$ where $F \subset A_\mathbb{C}$ is the 2-torsion subgroup. If $T = T_vT_u$ is the polar decomposition then $T_v \simeq AK \subset G/K$ is a maximal flat subspace of X, and $T_u \simeq A_uK \subset U/K$ is a maximal flat subspace of the compact dual X_u. The map $A_\mathbb{C}/F \ni aF \stackrel{\sim}{\longrightarrow} a^2 \in A_\mathbb{C}$ identifies $A_\mathbb{C}/F \subset X_\mathbb{C}$ with $A_\mathbb{C}$. Let $\Sigma \subset \mathfrak{a}^* = \mathrm{Lie}(A)$ be the restricted root system of G, and let $R \subset \mathfrak{a}^* = 2\Sigma$ be the corresponding set of characters of $T = A_\mathbb{C}/F$ via the above identification map. Choose $k_{2\alpha} \in \frac{1}{2}\mathbb{Z}_+$ such that for all $\alpha \in \Sigma$, $m_\alpha = 2k_{2\alpha}$ is equal to the multiplicity of the restricted root α of G. Then the density function $\delta_u(k)$ of the Weyl measure on T_u is the density function of the defining orthogonality measure of the Jacobi polynomials on T_u as in Lemma 8.3.3.

It follows that there exists a close relationship between the Jacobi polynomials $P(\mu, k)$ with the above parameters k_α and the zonal spherical functions of the Gelfand pair (U, K) on the compact Riemannian symmetric space X_u. The zonal spherical functions of X_u form a complete set of orthogonal polynomials on $L^2(K\backslash U/K) \simeq L^2(W\backslash T_u, \delta_u(k)\, d_ut)$ (cf. Helgason,

2000, Ch. V, §4.2). Therefore the restriction of a zonal spherical function on X_u to T_u is a Jacobi polynomial $P(\mu, k)$, up to normalization, and all Jacobi polynomials are obtained in this way. The leading coefficient at infinity of the zonal spherical function is by definition equal to the corresponding Harish-Chandra c-function. More precisely, the restrictions to T_u of the zonal spherical functions on X_u are the Laurent polynomials of the form

$$\phi(\mu, k)|_{T_u} = c(\mu + \rho(k), k) P(\mu, k), \quad \mu \in P^+.$$

It is a basic property of zonal spherical functions that their evaluation at eK is equal to 1. Hence we find

$$P(\mu, k; e) = \big(c(\mu + \rho(k), k)\big)^{-1}$$

for the normalization at the identity element of T.

Let $d(\mu, k)$ denote the dimension of the irreducible K-spherical representation of U associated to the zonal spherical function $\phi(\mu, k)$. Another basic aspect of the theory of zonal spherical functions is the following formula for their square L^2-norm:

$$(\phi(\mu, k), \phi(\mu, k)) = d(\mu, k)^{-1} \operatorname{Vol}(X_u).$$

According to a remarkable formula of Vretare (1984, p. 819) we can also express $d(\mu, k)$ in terms of the Harish-Chandra c-function:

$$d(\mu, k) = \lim_{\epsilon \to 0} \frac{c(-\rho(k) + \epsilon, k)}{c(\mu + \rho(k), k) c(-\mu - \rho(k) + \epsilon, k)}.$$

In addition, if we normalize the volume of T_u to be equal to 1, then the volume $\operatorname{Vol}(X_u)$ can also be expressed in terms of the close relatives of the Harish-Chandra c-function. By Weyl's integration formula we have

$$\operatorname{Vol}(X_u) = |W|^{-1} \int_{T_u} \delta_u(k)\, d_u t = |W|^{-1} \int_{T_u} \prod_{\alpha > 0} \left| e^{\alpha/2} - e^{-\alpha/2} \right|^{2k_\alpha} d_u t.$$

A closed formula for this integral was conjectured by I. G. Macdonald (1982), not only for k_α equal to half the restricted root multiplicities of a Riemannian symmetric space, but for arbitrary complex parameters k_α with $\operatorname{Re}(k_\alpha) > 0$, depending only on the length of the root α. This "constant term conjecture" of Macdonald stimulated much of the research on hypergeometric functions for root systems and double affine Hecke algebras. More generally, the evaluation and norm formulae for zonal spherical functions on X_u as discussed above extend holomorphically to arbitrary complex parameters k_α.

All these closed formulae are expressible in terms of generalizations of the Harish-Chandra c-function, as we will see in the next subsection.

8.4.2 Norm and Evaluation Formulae

The c-functions for the zonal spherical functions on a Riemannian symmetric space are expressible, by a famous formula of Gindikin and Karpelevič (1962), as a product over the positive roots of rank-one c-functions. This product formula is used to define the generalized

c-functions for arbitrary complex parameters k_α. In this section we will compute the square norms $\langle P(\mu,k), P(\mu,k)\rangle_k$ and their evaluation $P(\mu,k;e)$ at the identity element e of T, in terms of these generalized c-functions. In particular, these numbers are explicitly computable as a product taken over the positive roots of quotients of Γ-factors. This is one of the remarkable features of the theory of zonal spherical functions on semisimple symmetric spaces which generalizes to the theory of hypergeometric functions. In fact, this feature holds true even for the more general nonsymmetric Jacobi polynomials. Since the formulations as well as the proofs become somewhat easier when we exploit the extra freedom this generalization offers, we will discuss these results at this level of generality.

For $w \in W$ we define a function $\delta_w \colon R_+ \to \{0,1\}$ by $\delta_w(\alpha) := 0$ if $w(\alpha) > 0$ and $:= 1$ if $w(\alpha) < 0$, and we define

$$c^*_w(\lambda,k) := \prod_{\alpha\in R_+} \frac{\Gamma(-\lambda(\alpha^\vee) - \frac{1}{2}k_{\alpha/2} - k_\alpha + \delta_w(\alpha))}{\Gamma(-\lambda(\alpha^\vee) - \frac{1}{2}k_{\alpha/2} + \delta_w(\alpha))}$$

and

$$\tilde{c}_w(\lambda,k) := \prod_{\alpha\in R_+} \frac{\Gamma(\lambda(\alpha^\vee) + \frac{1}{2}k_{\alpha/2} + \delta_w(\alpha))}{\Gamma(\lambda(\alpha^\vee) + \frac{1}{2}k_{\alpha/2} + k_\alpha + \delta_w(\alpha))}. \tag{8.4.2}$$

For brevity we will write

$$\tilde{c}(\lambda,k) := \tilde{c}_e(\lambda,k), \quad c^*(\lambda,k) := c^*_{w_0}(\lambda,k).$$

The Harish-Chandra *c-function* for general multiplicity parameters is defined by

$$c(\lambda,k) := \frac{\tilde{c}(\lambda,k)}{\tilde{c}(\rho(k),k)}, \tag{8.4.3}$$

which in the Riemannian symmetric space case is just the celebrated Gindikin–Karpelevich (1962) evaluation of the original Harish-Chandra c-function.

Theorem 8.4.2 (Opdam, 1989, 1995) *Let $\lambda \in P_+$. We denote by W_λ the isotropy subgroup of λ in W, and by w_λ the longest element of W_λ. Let W^λ be the set of shortest length representatives for the left cosets of W_λ in W, and let $w \in W^\lambda$. Then*

(i) $$\|E(w\lambda,k)\|_k^2 = \frac{c^*_{ww_\lambda}(-\lambda-\rho(k),k)}{\tilde{c}_{ww_\lambda}(\lambda+\rho(k),k)},$$

(ii) $$E(w\lambda,k;e) = \frac{\tilde{c}_{w_0}(\rho(k),k)}{\tilde{c}_{ww_\lambda}(\lambda+\rho(k),k)},$$

(iii) $$\|P(\lambda,k)\|_k^2 = |W|\frac{c^*(-\lambda-\rho(k),k)}{\tilde{c}(\lambda+\rho(k),k)},$$

(iv) $$P(\lambda,k;e) = \frac{1}{c(\lambda+\rho(k),k)}.$$

Proof It is clearly enough to prove these assertions for a Zariski-dense subset of the parameter space. Therefore, without loss of generality, we may assume that all k_α are nonnegative integers. We may also assume without loss of generality that R is irreducible, since the general case easily reduces to this case.

Consider the subspace $\mathcal{E}(\lambda, k) \subset L^2(T_u, \delta(k)\, d_u t)$ spanned by the functions $E(w\lambda, k)$ with $w \in W$. Recall from Proposition 8.3.7 that the center $Z(\mathbb{H}) = S\mathfrak{t}^W$ of the degenerate affine Hecke algebra $\mathbb{H} := \mathbb{H}(R_+, k)$ acts on $\mathbb{C}[T]$ via the operators $T(p, k)$ (with $p \in S\mathfrak{t}^W$). By Lemma 8.3.4 we see that $\mathcal{E}(\lambda, k)$ is the $Z(\mathbb{H})$-eigenspace of the central character $S\mathfrak{t}^W \ni p \mapsto p(\lambda + \rho(k))$. By Lemma 8.3.3 it follows that the subspaces $\mathcal{E}(\lambda, k)$ are mutually orthogonal $\mathbb{H}$-submodules of $\mathbb{C}[T]$. Let $\mathbb{H}_\lambda \subset \mathbb{H}$ be the "parabolic subalgebra" $\mathbb{H}_\lambda := S\mathfrak{t} \otimes \mathbb{C}[W_\lambda] \subset \mathbb{H}$. The algebra $\mathbb{H}_\lambda$ has a one-dimensional trivial representation $\mathbb{C}_{\tilde{\lambda}}$ given by $\mathfrak{t} \ni \xi \mapsto \tilde{\lambda}(\xi)$ and $W_\lambda \ni w \mapsto \mathrm{id}_{\mathbb{C}}$. Consider the induced $\mathbb{H}$-module $V_{\lambda,k} := \mathrm{Ind}_{\mathbb{H}_\lambda}^{\mathbb{H}} \mathbb{C}_{\tilde{\lambda}} = \mathbb{H} \otimes_{\mathbb{H}_\lambda} \mathbb{C}_{\tilde{\lambda}}$. We show by induction on the length $l(w)$ of $w \in W^\lambda$ that $V_{\lambda,k}$ contains a nonzero $S\mathfrak{t}$-eigenvector v_w with eigenvalue $w\tilde{\lambda}$. The induction process starts with the eigenvector $v_e := 1 \otimes 1$ with eigenvalue $\tilde{\lambda}$. Now let $w \in W^\lambda$ and let s_i be a simple reflection such that $l(s_i w) < l(w)$. Then $s_i w \in W^\lambda$. By induction we may assume that there exists a nonzero eigenvector $v_{s_i w} \in V_{\lambda,k}$ with eigenvalue $s_i w\tilde{\lambda}$. Then it is easy to see that

$$v_w := \frac{s_i w\tilde{\lambda}(\alpha_i^\vee)}{s_i w\tilde{\lambda}(\alpha_i^\vee) + k_i}\left(s_i + \frac{k_i}{s_i w\tilde{\lambda}(\alpha_i^\vee)}\right) v_{s_i w} \in V_{\lambda,k}$$

is a *nonzero* eigenvector with eigenvalue $w\tilde{\lambda}$ for the action of $S\mathfrak{t}$. Since $V_{\lambda,k}$ obviously has dimension $|W^\lambda|$, it follows that $V_{\lambda,k}$ has a one-dimensional $S\mathfrak{t}$-eigenspace with eigenvalue $w\tilde{\lambda}$ for every $w \in W^\lambda$. In particular, by Frobenius reciprocity, it is clear that $V_{\lambda,k}$ is irreducible. By Frobenius reciprocity there exists a unique nonzero $\mathbb{H}$-module homomorphism

$$j = j_{\lambda,k} \colon V_{\lambda,k} \to L^2(T_u, \delta(k)\, d_u t) \quad \text{such that } j(v_e) = E(\lambda, k). \tag{8.4.4}$$

In particular, $V_{\lambda,k}$ admits a nondegenerate Hermitian form which turns $V_{\lambda,k}$ into a $*$-representation for $\mathbb{H}$ when we equip $\mathbb{H}$ with the $*$-structure $\mathfrak{t} \ni \xi \mapsto \xi^* := \overline{\xi}$ and $w^* := w^{-1}$. By the irreducibility of $V_{\lambda,k}$, this Hermitian inner product is unique up to normalization, and the basis $\{v_w\}_{w \in W^\lambda}$ of eigenvectors we constructed is orthogonal. It is clear that such a form is definite (since it comes from the inner product on $L^2(T_u, \delta(k)\, d_u t)$). It is an easy matter to prove that such an Hermitian inner product must be of the form

$$(v_w, v_{w'}) = a(\lambda, k)\delta_{w,w'} \prod_{\alpha \in R_+^0} \left(1 - \frac{k_\alpha + \frac{1}{2}k_{\alpha/2}}{w\tilde{\lambda}(\alpha^\vee)}\right)^{-1}$$

for some $a(\lambda, k) \neq 0$, where R^0 denotes the set of roots $\alpha \in R$ such that $2\alpha \notin R$. The eigenvector v_w is mapped via j to a multiple of $E(w\lambda, k)$. The constant of proportionality is easily determined inductively by comparing the normalization in the $E(\mu, k)$-polynomials and the basis v_w; we find

$$j(v_w) = \prod_{\alpha \in R_+ \cap w^{-1}R_-} \left(\frac{\tilde{\lambda}(\alpha^\vee) + \frac{1}{2}k_{\alpha/2} + k_\alpha}{\tilde{\lambda}(\alpha^\vee) + \frac{1}{2}k_{\alpha/2}}\right) E(w\lambda, k) = \prod_{\alpha \in R_+^0 \cap w^{-1}R_-^0} \left(1 + \frac{k_\alpha + \frac{1}{2}k_{\alpha/2}}{\tilde{\lambda}(\alpha^\vee)}\right) E(w\lambda, k).$$

Therefore the proof of (i) reduces to the determination of the constant $a(\lambda, k)$ such that j becomes an isometry. The following formula for $|W_\lambda|$ is a consequence of the well-known identity (Macdonald, 1972, Theorem 2.4 – substitute $t_\alpha = t^{k_\alpha + \frac{1}{2}k_{\alpha/2}}$ and take the limit $t \to 1$)

$$|W_\lambda| = \prod_{\alpha \in R_{\lambda,+}} \left(\frac{\rho(k)(\alpha^\vee) + k_\alpha + \frac{1}{2}k_{\alpha/2}}{\rho(k)(\alpha^\vee) + \frac{1}{2}k_{\alpha/2}} \right) = \prod_{\alpha \in R^0_{\lambda,+}} \left(1 + \frac{k_\alpha + \frac{1}{2}k_{\alpha/2}}{\rho(k)(\alpha^\vee)} \right). \tag{8.4.5}$$

An elementary calculation using (8.4.5) shows that (i) is equivalent to proving that

$$a(\lambda, k) = |W_\lambda|^2 \frac{c^*(-(\lambda + \rho(k)), k)}{\tilde{c}(\lambda + \rho(k), k)}. \tag{8.4.6}$$

The proof of (8.4.6) is an inductive argument on the parameter k, where the induction step is based on a generalization of Weyl's character formula. First we observe that (i) (hence (8.4.6)) holds if $k = 0$. Before discussing the induction step, let us point out a remarkable and extremely useful property of the polynomials $E(\lambda, k)$. Suppose that $R' \subset R$ is a subsystem of roots such that $k_\alpha = 0$ for $\alpha \in R \backslash R'$. It follows directly from the definitions that the Dunkl operators $T(\xi, k)$ for R and R' are the same for such parameters. As a consequence, the polynomials $E(\lambda, k)$ for R and R' are equal in this situation. This allows us to delete the roots from R on which k is zero. Therefore, to prove (8.4.6) it suffices to show that if (8.4.6) is true for a parameter k it is also true for $k + 1$, where 1 denotes the characteristic function of the set $R^0 \subset R$. Let $\Delta := e^\delta \prod_{\alpha \in R^0_+} (1 - e^{-\alpha})$ be the Weyl denominator, where δ is the Weyl vector of R^0_+. If $\lambda \in P_+$ is regular and $t \in T$ then we define

$$P^-(\lambda, k; t) := \sum_{w \in W} \epsilon(w) E(\lambda, k; wt)$$

where ϵ denotes the sign character of W. Then $P^-(\lambda, k)$ spans the subspace of W-skew invariant polynomials in $\mathcal{E}(\lambda, k)$. From the Gram–Schmidt-type definition of the $E(\mu, k)$ we obtain the following generalization of Weyl's character formula:

$$P(\lambda, k + 1) = \Delta^{-1} P^-(\lambda + \delta, k), \quad \lambda \in P_+. \tag{8.4.7}$$

In particular, we have

$$||P(\lambda, k + 1)||^2_{k+1} = ||P^-(\lambda + \delta, k)||^2_k. \tag{8.4.8}$$

On the other hand, via j (see (8.4.4)) the Jacobi polynomial $P(\lambda, k + 1)$ corresponds to the vector $|W_\lambda|^{-1} \sum_{w \in W} wv_e \in V_{\lambda,k+1}$. It is not very difficult to show that this vector has square norm equal to $|W||W_\lambda|^{-2} a(\lambda, k + 1)$. Similarly, $P^-(\lambda + \delta, k)$ corresponds via j to the vector $\sum_{w \in W} \epsilon(w) wv_e \in V_{\lambda+\delta,k}$, and the square norm of this expression can be shown to be

$$|W| a(\lambda + \delta, k) \prod_{\alpha \in R^0_+} \left(\frac{(\lambda + \rho(k) + \delta)(\alpha^\vee) + k_\alpha + \frac{1}{2}k_{\alpha/2}}{(\lambda + \rho(k) + \delta)(\alpha^\vee) - k_\alpha - \frac{1}{2}k_{\alpha/2}} \right).$$

Hence (8.4.8) implies that

$$a(\lambda, k + 1) = a(\lambda + \delta, k) |W_\lambda|^2 \prod_{\alpha \in R^0_+} \left(\frac{(\lambda + \rho(k) + \delta)(\alpha^\vee) + k_\alpha + \frac{1}{2}k_{\alpha/2}}{(\lambda + \rho(k) + \delta)(\alpha^\vee) - k_\alpha - \frac{1}{2}k_{\alpha/2}} \right). \tag{8.4.9}$$

By the induction hypothesis and by easy manipulations we see that the right-hand side of (8.4.9) is equal to the right-hand side of (8.4.6) (with k replaced by $k + 1$), as desired. This finishes the proof of (i) and (iii).

The proof of (ii) and (iv) uses a similar type of inductive argument, but we will skip the details. □

8.4.3 Hypergeometric Shift Operators

Let $\epsilon^+ \in \mathbb{C}[W]$ denote the central idempotent of the trivial character and $\epsilon^- \in \mathbb{C}[W]$ the central idempotent of the sign character. The relations of the degenerate affine Hecke algebra $\mathbb{H}$ (see Definition 8.3.6) easily imply that the elements

$$\pi^{\pm}(k) := \prod_{\alpha\in R_+^0} (\alpha^\vee \pm (k_\alpha + \tfrac{1}{2}k_{\alpha/2}))$$

satisfy the relations $\pi^\pm\epsilon^\pm = \epsilon^\mp\pi^\pm$ in $\mathbb{H}$. The Dunkl representation yields operators

$$T(\pi^\pm(k), k) = \sum_w D^\pm(w,k)\otimes w \in \mathbb{D}[T_{\mathrm{reg}}]\otimes\mathbb{C}[W]$$

on the algebra $\mathbb{C}[T]$. Similar to the construction of the W-invariant differential operators $D(p,k)$ (with $p \in S\mathfrak{t}^W$) we introduce the associated differential operators

$$D^\pm(\pi^\pm(k), k) := \sum_{w\in W} (\pm 1)^{l(w)} D^\pm(w,k) \in \mathbb{D}[T_{\mathrm{reg}}]^{-W}.$$

Here $\mathbb{D}[T_{\mathrm{reg}}]^{-W}$ denotes the space of W-skew invariant algebraic linear partial differential operators on T_{reg}. This gives rise to two W-invariant linear partial differential operators on T_{reg} as follows:

$$G^+(k) := \Delta^{-1}D^+(\pi^+(k),k), \quad G^-(k+1) := D^-(\pi^-(k),k)\Delta,$$

where, as in (8.4.7), 1 denotes the multiplicity parameter which is equal to 1 on R^0 and equal to 0 on $R\backslash R^0$. We have for $\lambda \in P_+$,

$$\begin{aligned} G^+(k)P(\lambda+\delta,k) &= \prod_{\alpha\in R_+^0}\big((k_\alpha + \tfrac{1}{2}k_{\alpha/2}) - (\lambda+\rho(k+1))(\alpha^\vee)\big)P(\lambda,k+1), \\ G^-(k+1)P(\lambda,k+1) &= \prod_{\alpha\in R_+^0}\big((k_\alpha + \tfrac{1}{2}k_{\alpha/2}) + (\lambda+\rho(k+1))(\alpha^\vee)\big)P(\lambda+\delta,k). \end{aligned} \tag{8.4.10}$$

Indeed, the generalized Weyl character formula (8.4.7) and the skew W-invariance of the $D^\pm(\pi^\pm(k),k)$ imply that these formulae hold, up to some constant factor. The upper unitriangularity of the action of the Dunkl operators on monomials t^λ of $\mathbb{C}[T]$ implies that we may easily determine the coefficients of the monomial $t^{w_0\lambda}$ in both the left-hand and the right-hand side, which determines the precise form of the constant. Because the Jacobi polynomials form a basis of the complex vector space $\mathbb{C}[T]^W$, we obtain the following corollary.

Corollary 8.4.3 *The differential operator $G^{\pm}(k)$ is the pull-back of a* polynomial *differential operator, i.e., an element (also denoted by $G^{\pm}(k)$) of the Weyl algebra of the affine space $W\backslash T$.*

From (8.4.1) and (8.4.10) we see that these operators satisfy for all $p \in S\mathfrak{t}^W$,

$$\begin{aligned} G^+(k)D(p,k) &= D(p,k+1)G^+(k), \\ G^-(k)D(p,k) &= D(p,k-1)G^-(k). \end{aligned} \tag{8.4.11}$$

The hypergeometric *shift operators* $G^{\pm}(k)$ derive their name from these relations.

These translations in the multiplicity parameters of the hypergeometric system along integer multiples of $1 \in \mathcal{K}$ can be further generalized to include the translations in $\mathcal{K}$ by vectors in the full lattice $\mathcal{K}^{\mathbb{Z}} \subset \mathcal{K}$ consisting of the elements $l \in \mathcal{K}$ such that $l_\alpha \in \mathbb{Z}$ for all $\alpha \in R$ and $l_{\alpha/2} \in 2\mathbb{Z}$ for all $\alpha \in R$. The translations in these lattice vectors can all be realized by similar hypergeometric shift operators.

Apart from these translations there exist fundamental reflection symmetries in the parameter space of the hypergeometric system. We again only discuss the principal instance of such a symmetry here. This symmetry originates from the following fundamental formula:

$$D(p, 1-k) = \delta(k - \tfrac{1}{2}) \circ D(p,k) \circ \delta(\tfrac{1}{2} - k), \quad p \in S\mathfrak{t}^W.$$

For $p = p_2$ equal to the quadratic invariant of W the formula can be established by a direct computation using Theorem 8.3.10 (checking this formula for p_2 involves computations similar to those worked out in the proof of Theorem 8.5.1). It then follows for arbitrary p because the commutation relation

$$D(p_2,k)D(p,k) - D(p,k)D(p_2,k) = 0,$$

together with the assertion $D(p,k) = \partial(p) + \sum_{\mu<0} e^\mu \partial(p_\mu)$, yields a recurrence relation on the p_μ which determines $D(p,k)$ completely. Similarly one proves the symmetry relation

$$G^+(-\tfrac{1}{2} - k) \circ \delta(k+1) = \delta(k) \circ G^-(\tfrac{3}{2} + k). \tag{8.4.12}$$

Let us introduce the rational *Dunkl operator* $T^{\mathrm{rat}}(\xi,k) \in \mathrm{End}_{\mathbb{C}}(\mathbb{C}[T])$ by the formula

$$T^{\mathrm{rat}}(\xi,k) := \partial(\xi) + \sum_{\alpha \in R_+^0} (k_\alpha + k_{\alpha/2}) \frac{\alpha(\xi)}{\alpha}(1 - s_\alpha),$$

which is discussed in Chapter 7 by Dunkl. These operators have homogeneous degree -1. If f is a holomorphic germ at $e \in T$ of vanishing order at least $\nu \geq 0$ we can canonically write $f = f_\nu +$ (terms of vanishing order greater than ν in $e \in T$), where f_ν is a homogeneous polynomial of degree ν on $\mathfrak{t}$. Then the relation between the trigonometric and rational Dunkl operators is expressed by the formula

$$(T(\xi,k)(f))_{\nu-1} = T^{\mathrm{rat}}(\xi,k)(f_\nu).$$

Observe that this implies that the rational Dunkl operators $T^{\mathrm{rat}}(\xi, k)$ mutually commute. Hence there exists a unique unital $\mathbb{C}$-algebra homomorphism $p \mapsto T^{\mathrm{rat}}(p,k): S\mathfrak{t} \to \mathrm{End}_{\mathbb{C}}(\mathbb{C}[T])$ extending the map $\xi \mapsto T^{\mathrm{rat}}(\xi, k)$. We see that for any holomorphic germ f at $e \in T$ we have

$$G^{-}(k)(f)(e) = f(e)T^{\mathrm{rat}}(\pi^{\vee}, k)(\pi), \tag{8.4.13}$$

where $\pi^{\vee} := \prod_{\alpha \in R_+^0} \alpha^{\vee} \in S\mathfrak{t}$ and $\pi := \prod_{\alpha \in R_+^0} \alpha \in S\mathfrak{t}$. Combining (8.4.13) with Theorem 8.4.2(iv) and with (8.4.10) gives as a corollary that

$$T^{\mathrm{rat}}(\pi^{\vee}, k)(\pi) = \frac{\tilde{c}(\rho(k), k)}{\tilde{c}(\rho(k+1), k+1)}. \tag{8.4.14}$$

This result has important consequences. By applying (8.4.12) to the constant function 1 and by using (8.4.14) one can prove the following result (Opdam, 1989) which was conjectured by Yano and Sekiguchi (1979).

Corollary 8.4.4 *Let $k = s.1 \in \mathcal{K}$, and let $\mathbb{A}$ be the Weyl algebra of polynomial linear partial differential operators on the affine space $W\backslash\mathfrak{t}$. Let $D \in \mathbb{A} \otimes \mathbb{C}[s]$ be defined by*

$$Df := \pi^{-1}T^{\mathrm{rat}}(\pi^{\vee}, -\tfrac{1}{2} - k)f, \quad f \in S\mathfrak{t}^{W}.$$

Then

$$D\pi^{2(s+1)} = |W| b(s) \pi^{2s},$$

where $b(s)$ is the Bernstein–Sato polynomial of the discriminant $\pi^2 \in S\mathfrak{t}^W$. Moreover, $b(s)$ is explicitly given by $b(s) := \prod_{i=1}^{n} \prod_{j=1}^{d_i-1} (d_i(s+\frac{1}{2}) + j)$, where the d_i denote the primitive degrees of W.

Also one may use (8.4.14) to prove the Macdonald–Mehta conjecture for crystallographic Weyl groups (Opdam, 1989):

Corollary 8.4.5 *Let γ denote the Gaussian measure on the Euclidean vector space $\mathfrak{a}$, i.e., $d\gamma(x) = (2\pi)^{-\frac{n}{2}} e^{-\frac{1}{2}|x|^2} dx$ where dx denotes the Lebesgue measure on $\mathfrak{a}$. Furthermore, let $\pi(x;k) := \prod_{\alpha \in R_+} (\tilde{\alpha}^2(x))^{k_\alpha}$, where $\tilde{\alpha} := \frac{\sqrt{2}}{|\alpha|}\alpha$. Then*

$$\int_{\mathfrak{a}} \pi(x;k)\, d\gamma(x) = \frac{|W|}{\tilde{c}(\rho(k), k)}, \quad k \in \mathcal{K},\ \mathrm{Re}(k_\alpha) \geq 0.$$

8.5 The Calogero–Moser System

In this section we shall view the weight function of the hypergeometric system

$$\delta(k;t) := \prod_{\alpha>0} (t^{\frac{1}{2}\alpha} - t^{-\frac{1}{2}\alpha})^{2k_\alpha} = \prod_{\alpha>0} (t^{\alpha} + t^{-\alpha} - 2)^{k_\alpha}$$

formally as a multivalued function of determination order 1 (i.e., a multivalued function for which any two of its branches have a locally constant ratio). Since we shall only conjugate linear differential operators by the square root of δ, the multivaluedness is of no concern.

Alternatively one could work on the regular part of the real vector subgroup T_v where $(t^\alpha + t^{-\alpha} - 2)$ is positive.

Theorem 8.5.1 *If $\xi_1, \ldots, \xi_n$ is an orthonormal basis of $\mathfrak{t}_v$ then conjugation of the quadratic operator $L(k) = D(\sum_i \xi_i^2, k)$ by the square root of the weight function $\delta(k)$ is given by*

$$\delta(k;t)^{\frac{1}{2}} \circ L(k) \circ \delta(k;t)^{-\frac{1}{2}} = \sum_i \partial(\xi_i)^2 + \sum_{\alpha>0} \frac{k_\alpha(1-k_\alpha-2k_{2\alpha})(\alpha,\alpha)}{(t^{\frac{1}{2}\alpha}-t^{-\frac{1}{2}\alpha})^2}$$

as equality in $\mathbb{D}[T_{\text{reg}}]$.

Proof For $\xi \in \mathfrak{t}$ we have

$$\delta(k)^{-\frac{1}{2}} \circ \partial(\xi) \circ \delta(k)^{\frac{1}{2}} = \partial(\xi) + \tfrac{1}{2}\partial(\xi)(\log\delta(k)),$$
$$\delta(k)^{-\frac{1}{2}} \circ \partial(\xi)^2 \circ \delta(k)^{\frac{1}{2}} = \partial(\xi)^2 + \partial(\xi)(\log\delta(k))\partial(\xi) + \delta(k)^{-\frac{1}{2}}\partial(\xi)^2(\delta(k)^{\frac{1}{2}})$$

with

$$\tfrac{1}{2}\partial(\xi)(\log\delta(k)) = \sum_{\alpha>0} \tfrac{1}{2}k_\alpha\alpha(\xi)\frac{t^{\frac{1}{2}\alpha}+t^{-\frac{1}{2}\alpha}}{t^{\frac{1}{2}\alpha}-t^{-\frac{1}{2}\alpha}}$$

the constant term of the first expression. In turn we get

$$\sum_i \partial(\xi_i)(\log\delta(k))\partial(\xi_i) = \sum_{\alpha>0} k_\alpha \frac{t^{\frac{1}{2}\alpha}+t^{-\frac{1}{2}\alpha}}{t^{\frac{1}{2}\alpha}-t^{-\frac{1}{2}\alpha}}\partial(\alpha),$$

which is precisely the first-order term of the differential operator $D(\sum \xi_i^2, k)$ in Theorem 8.3.10. If we write $\square := \sum \partial(\xi_i)^2$ then

$$\tfrac{1}{2}\square(\log\delta(k)) = -\sum_{\alpha>0} \frac{k_\alpha(\alpha,\alpha)}{(t^{\frac{1}{2}\alpha}-t^{-\frac{1}{2}\alpha})^2},$$

and so

$$\delta(k)^{-\frac{1}{2}}\square(\delta(k)^{\frac{1}{2}}) = -\sum_{\alpha>0}\frac{k_\alpha(\alpha,\alpha)}{(t^{\frac{1}{2}\alpha}-t^{-\frac{1}{2}\alpha})^2} + \sum_{\alpha,\beta>0}\tfrac{1}{4}k_\alpha k_\beta(\alpha,\beta)\frac{(t^{\frac{1}{2}\alpha}+t^{-\frac{1}{2}\alpha})(t^{\frac{1}{2}\beta}+t^{-\frac{1}{2}\beta})}{(t^{\frac{1}{2}\alpha}-t^{-\frac{1}{2}\alpha})(t^{\frac{1}{2}\beta}-t^{-\frac{1}{2}\beta})}.$$

We rewrite the second term on the right-hand side as

$$(\rho(k),\rho(k)) + \sum_{\alpha,\beta>0}\tfrac{1}{4}k_\alpha k_\beta(\alpha,\beta)\frac{2(t^{\frac{1}{2}(\alpha-\beta)}+t^{-\frac{1}{2}(\alpha-\beta)})}{(t^{\frac{1}{2}\alpha}-t^{-\frac{1}{2}\alpha})(t^{\frac{1}{2}\beta}-t^{-\frac{1}{2}\beta})},$$

or equivalently as

$$(\rho(k),\rho(k)) + \sum_{\alpha>0}\frac{k_\alpha(k_\alpha+2k_{2\alpha})(\alpha,\alpha)}{(t^{\frac{1}{2}\alpha}-t^{-\frac{1}{2}\alpha})^2} + \sum_{\alpha,\beta>0,\alpha\nsim\beta}\tfrac{1}{4}k_\alpha k_\beta(\alpha,\beta)\frac{2(t^{\frac{1}{2}(\alpha-\beta)}+t^{-\frac{1}{2}(\alpha-\beta)})}{(t^{\frac{1}{2}\alpha}-t^{-\frac{1}{2}\alpha})(t^{\frac{1}{2}\beta}t^{-\frac{1}{2}\beta})}$$

with $\alpha \nsim \beta$ meaning that α and β are not proportional.

We claim that the third term of this last expression vanishes identically. Indeed it is invariant under W and has simple poles along the zero locus $\cup_{\alpha\in R_+^0} T_\alpha$ of the Weyl denominator Δ. Hence its product with Δ becomes skew invariant under W and is a regular function on T. Since this product is of the form $\sum c_\mu t^\mu$ with $c_\mu = 0$ unless $\mu < \delta$, we conclude that the third term is zero. Here $\delta = \frac{1}{2}\sum \alpha$ is the Weyl vector of R^0_+. The theorem follows by collection of the various terms. □

After a switch from the coupling constant $k \in \mathcal{K}$ to $g \in \mathcal{K}$, by the substitution

$$g_\alpha^2 = \tfrac{1}{2}k_\alpha(k_\alpha + 2k_{2\alpha} - 1)(\alpha, \alpha)$$

the differential operator

$$S(g) := \tfrac{1}{2}\sum_i \partial(\xi_i)^2 - \sum_{\alpha>0} \frac{g_\alpha^2}{(t^{\frac{1}{2}\alpha} - t^{-\frac{1}{2}\alpha})^2}$$

is called the *Schrödinger operator* for the periodic *Calogero–Moser system*. If we denote by $\mathbb{D}[T_{\text{reg}}]^{W,L(k)}$ the commutant of $L(k)$ in $\mathbb{D}[T_{\text{reg}}]^W$ and likewise $\mathbb{D}[T_{\text{reg}}]^{W,S(g)}$ for the commutant of $S(g)$ then the conjugation map

$$D \mapsto \delta(k)^{\frac{1}{2}} \circ D \circ \delta(k)^{-\frac{1}{2}} : \ \mathbb{D}[T_{\text{reg}}]^{W,L(k)} \to \mathbb{D}[T_{\text{reg}}]^{W,S(g)}$$

is an isomorphism of algebras, which are both isomorphic to $S\mathfrak{t}^W$. The conclusion is that the Calogero–Moser system is a completely integrable quantum system, and the results of the previous section are just an exact solution of this integrable quantum system.

An element $D(p,k) \in \mathbb{D}[T_{\text{reg}}]^{W,L(k)}$ ($p \in S\mathfrak{t}^W$) has an asymptotic expansion of the form

$$D(p,k) = \sum_{\mu \leq 0} t^\mu \partial(p_\mu)$$

for $p_\mu \in S\mathfrak{t}$. In the previous section we have shown that the constant term $p_0 \in S\mathfrak{t} \cong P\mathfrak{t}^*$ is given by $p_0(\lambda) = p(\lambda + \rho(k))$ for all $\lambda \in \mathfrak{t}^*$. Likewise, after conjugation by $\delta^{\frac{1}{2}}$, the operator $\delta^{\frac{1}{2}} \circ D(p,k) \circ \delta^{-\frac{1}{2}}$ in $\mathbb{D}[T_{\text{reg}}]^{W,S(g)}$ has an asymptotic expansion

$$\delta^{\frac{1}{2}} \circ D(p,k) \circ \delta^{-\frac{1}{2}} = \sum_{\mu \leq 0} t^\mu \partial(q_\mu)$$

with $q_\mu \in S\mathfrak{t}$ and constant term given by $q_0(\lambda) = p_0(\lambda - \rho(k)) = p(\lambda)$ for $p \in P(\mathfrak{t}^*)^W$. These convergent asymptotic expansions are valid in the interior of the positive chamber in the vector group T_v. Because the constant term corresponds to the case $g = 0$ of a free particle, the Calogero–Moser system is called *asymptotically free*.

The commutation relation $[D, S(g)] = 0$ for $D \in \mathbb{D}[T_{\text{reg}}]^{W,S(g)}$ of the form $\sum_{\mu \leq 0} t^\mu \partial(q_\mu)$ amounts to a system of recurrence relations

$$(2\lambda + \mu, \mu) q_\mu(\lambda) = -2 \sum_{\alpha>0} g_\alpha^2 \sum_{j \geq 1} j(q_{\mu + j\alpha}(\lambda - j\alpha) - q_{\mu + j\alpha}(\lambda))$$

by a direct verification. Apparently one can pick the initial polynomial $q_0 \in P(\mathfrak{t}^*)^W$ freely for the constant term, and then solve the $q_\mu \in P(\mathfrak{t}^*)$ recurrently. Evidently these recurrence relations can be solved uniquely for chosen q_0 inside the algebra of rational functions on $\mathfrak{t}^*$

with poles on certain hyperplanes. The amazing fact of the integrability of the Calogero–Moser system is that all divisions can be carried out in the algebra $P(\mathfrak{t}^*)$, which is not at all clear from the recurrence relations. However there is one nontrivial conclusion we can draw from these recurrence relations, namely that all differential operators in the commutant $\mathbb{D}[T_{\text{reg}}]^{W,S(g)}$ of $S(g)$ must have polynomial dependence on the coupling constants $g^2 \in \mathcal{K}$. This was not clear before since the substitution $\mathcal{K} \ni k \mapsto g \in \mathcal{K}$ is algebraic.

Example 8.5.2 For the root system of type A_n given by

$$\mathfrak{t}_v = \{(x_0, \dots, x_n) \in \mathbb{R}^{n+1} \mid \textstyle\sum_j x_j = 0\}, \quad R = \{\alpha \in \mathfrak{t}_v \cap \mathbb{Z}^{n+1} \mid (\alpha, \alpha) = 2\},$$

all the roots are conjugated under the symmetric group $W = S_{n+1}$ and we have just one coupling parameter $g^2 = k(k-1)$. On the compact torus T_u the Schrödinger operator becomes

$$S(g) = -\tfrac{1}{2} \sum_j \partial(y_j)^2 + \sum_{j<k} \frac{g^2}{4 \sin^2\left(\frac{1}{2}(y_j - y_k)\right)}$$

using complex coordinates $z_j = x_j + iy_j$. Since $|e^{iy_j} - e^{iy_k}|^2 = 4\sin^2\left(\frac{1}{2}(y_j - y_k)\right)$, the periodic Calogero–Moser system describes a system of $n+1$ identical particles on the unit circle $\mathbb{R}/2\pi\mathbb{Z}$ in $\mathbb{C}$ with an inverse square potential. This was the original example studied by Calogero (1971) and Moser (1975).

Since T is a complex torus, the cotangent bundle T^*T_{reg} is canonically isomorphic to the direct product $T_{\text{reg}} \times \mathfrak{t}^*$. The Hamiltonian of the periodic Calogero–Moser system is defined by

$$H(g, t, \lambda) := -\tfrac{1}{2}(\lambda, \lambda) - \sum_{\alpha>0} \frac{g_\alpha^2}{(t^{\frac{1}{2}\alpha} - t^{-\frac{1}{2}\alpha})^2},$$

viewed as a function of $(g, t, \lambda) \in \mathcal{K} \times T_{\text{reg}} \times \mathfrak{t}^*$. The commutative algebra $\mathbb{C}[\mathcal{K}] \otimes \mathbb{C}[T_{\text{reg}}] \otimes S\mathfrak{t}$ of functions on $\mathcal{K} \times T_{\text{reg}} \times \mathfrak{t}^* \simeq \mathcal{K} \times T^*T_{\text{reg}}$ has a natural Poisson bracket (with K taken as the space of constant parameters). This Poisson bracket is derived from the filtration on the differential operator algebra $\mathbb{C}[\mathcal{K}] \otimes \mathbb{C}[T_{\text{reg}}] \otimes U\mathfrak{t}$ by taking the sum of the polynomial degrees in $\mathbb{C}[\mathcal{K}]$ and $U\mathfrak{t} \simeq S\mathfrak{t}$ as the total degree. The associated graded algebra of the commutative algebra of hypergeometric differential operators (twisted by conjugation with $\delta(k)^{\frac{1}{2}}$) yields a Poisson commutative algebra containing the Calogero–Moser Hamiltonian $H(g, t, \lambda)$ as homogeneous function of degree 2. In other words, the quantum integrability of the Calogero–Moser system gives the classical integrability of the Calogero–Moser system by taking the classical limit.

The original proof by Moser (1975) established the classical integrability for the root system of type A_n using a Lax pair representation. Olshanetsky and Perelomov (1976) extended this proof (under a linear parameter constraint) for the other classical root systems. However, for the exceptional root systems the only known proof of the classical integrability of the Calogero–Moser system is the above approach through quantum integrability.

8.6 The Hypergeometric Function

Recall the system of hypergeometric differential equation on T_{reg} of Definition 8.3.9. If $\text{Re}(k_\alpha) \geq 0$ for all $\alpha \in R$ and $\lambda = \mu + \rho(k) \in \mathfrak{t}^*$, then the Jacobi polynomial $P(\mu, k) \in \mathbb{C}[T]^W$ is a solution of (8.3.3). For other values of the spectral parameter $\lambda \in W\backslash\mathfrak{t}^*$ the solutions of (8.3.3) do not extend holomorphically to all of T. Remarkably, there always exists a unique solution $F(\lambda, k)$ of (8.3.3) which extends to a W-invariant holomorphic function on a tubular neighborhood of $T_v \subset T$, and which is normalized by $F(\lambda, k; e) = 1$. This solution of (8.3.3) is called the *hypergeometric function for root systems*. Before we look at the general theory establishing the existence and uniqueness of this function, let us consider its meaning in the context of Riemannian symmetric spaces. We saw in §8.4.1 that the Jacobi polynomials $P(\lambda, k)$ for a root system R with multiplicity parameters $k \in \mathcal{K}$ could be viewed as a generalization of the elementary zonal spherical functions on a compact Riemannian symmetric space $X_u = U/K$. In a similar manner, these hypergeometric functions can be thought of as a natural generalization of Harish-Chandra's spherical functions on the noncompact dual $X = G/K$ of X_u.

8.6.1 Hypergeometric Functions and Spherical Functions

Recall the setup of §8.4.1. In this subsection we shall discuss the relation between the hypergeometric functions for special multiplicity parameters and the theory of spherical functions on Riemannian symmetric spaces. A standard reference for the latter theory is Helgason (2000). Let G have *Iwasawa decomposition* $G = NAK$, and let $G \ni g \mapsto a(g)$ denote the associated Iwasawa projection onto the split maximal Abelian subgroup $A \subset G$. Let $\mathfrak{a}$ denote the Lie algebra of A, and let $\lambda \in \mathfrak{a}^*_{\mathbb{C}}$. In the harmonic analysis on X one defines the Harish-Chandra *spherical function* ϕ_λ on X by the integral formula

$$\phi_\lambda(gK) := \int_K a(kg)^{\lambda+\rho}\, dk,$$

where dk is the normalized Haar measure of K. Here $\rho := \frac{1}{2}\sum_{\alpha\in\Sigma_+} m_\alpha \alpha \in \mathfrak{a}^*$ is the Weyl vector for the minimal parabolic subgroup normalizing NA, where $\Sigma \subset \mathfrak{a}^*$ denotes the restricted root system of G and m_α the multiplicity of the restricted root α. This function is obviously bi-K-invariant as a function of $g \in G$ and satisfies $\phi_\lambda(eK) = 1$. It is well known that it is a joint eigenfunction of the algebra $\mathcal{D}(X)$ of G-invariant differential operators on X. More specifically, if $\gamma\colon \mathcal{D}(X) \to S\mathfrak{t}^W$ denotes the *Harish-Chandra isomorphism* then

$$\Delta\phi_\lambda = \gamma(\Delta)(\lambda)\phi_\lambda, \quad \Delta \in \mathcal{D}(X).$$

By the K-invariance of ϕ_λ and of the operators we can separate the variables in this system of differential equations and reduce to $A \cong AK \subset G/K$. The radial component of elements of $\mathcal{D}(X)$ yields an embedding of $\mathcal{D}(X)$ into the algebra of W-invariant partial linear differential operators on A_{reg}. Let us denote its image by $\mathcal{R}(X)$. We factor the Harish-Chandra isomorphism γ via the radial component isomorphism $\mathcal{D}(X) \to \mathcal{R}(X)$ and (by abuse of notation) denote the resulting algebra isomorphism also as $\gamma\colon \mathcal{R}(X) \to S\mathfrak{t}^W$. Hence

$$D(\phi_\lambda|_{A_{\mathrm{reg}}}) = \gamma(D)(\lambda)\phi_\lambda|_{A_{\mathrm{reg}}}, \quad D \in \mathcal{R}(X). \tag{8.6.1}$$

This is a system of eigenfunction equations for the commutative algebra $\mathcal{R}(X)$ of W-invariant linear partial differential operators on $T_{v,\mathrm{reg}} = A_{\mathrm{reg}}$. It follows from the material in §8.4.1 that the Jacobi polynomials $P(\mu, k)$ diagonalize the algebra $\mathcal{R}(X)$. It can be easily seen that $\mathcal{R}(X)$ is nothing but the algebra of the W-invariant differential operators $D(p, k)$ with $p \in S\mathfrak{t}^W$, where we make the same identifications as in §8.4.1, i.e., we take $R = 2\Sigma$ (with $\Sigma = \Sigma(\mathfrak{g}, \mathfrak{a})$ the restricted root system of $\mathfrak{g}$) and $m_\alpha = 2k_{2\alpha}$ for all $\alpha \in \Sigma$. The system of eigenfunction equations (8.6.1) identifies with the hypergeometric system (8.3.3) in this way, and we have therefore established the existence of a W-invariant solution $\phi_\lambda|_{T_v}$ on T_v of (8.3.3) in this special situation. It is easy to see that ϕ_λ actually extends holomorphically to the W-invariant tubular neighborhood $T_v \exp(\pi i \Omega) \subset T$, where $\Omega := \{X \in \mathfrak{a} \mid |\alpha(X)| < 1 \text{ for all } \alpha \in R\} \subset \mathfrak{a}$.

When we consider the hypergeometric system (8.3.3) for more general multiplicity parameters k_α we lose the group-theoretical techniques considered above to construct solutions. Yet it turns out that the essential features of the space of solutions of (8.3.3) remain intact.

8.6.2 Asymptotic Freedom and Monodromy

The first important observation is the generic asymptotic freedom of solutions of the system (8.3.3) as a function of $t \in T_{\mathrm{reg}}$ when $|t| \in T_v^+$ is deep in the positive chamber. Indeed, when we plug (following Harish-Chandra in the group case) an asymptotic series of the form

$$\Phi(\lambda - \rho(k), k; t) = \sum_{\kappa \in Q_-} \Gamma_\kappa(\lambda, k) t^{\lambda - \rho(k) + \kappa}$$

(with $\lambda \in \mathfrak{t}^*$, $|t| \in T_v^+$ and $\Gamma_0(\lambda, k) = 1$) into the single eigenfunction equation

$$D\big(\textstyle\sum_{i=1}^n \xi_i^2, k\big)\Phi(\lambda - \rho(k), k; t) = (\lambda, \lambda)\Phi(\lambda - \rho(k), k; t), \tag{8.6.2}$$

we obtain the recurrence relations

$$-(2\lambda + \kappa, \kappa)\Gamma_\kappa(\lambda, k) = 2\sum_{\alpha > 0} k_\alpha \sum_{j \geq 1} (\lambda - \rho(k) + \kappa + j\alpha, \alpha)\Gamma_{\kappa + j\alpha}(\lambda, k) \tag{8.6.3}$$

for the $\Gamma_\kappa(\lambda, k)$ which have a unique solution in the field of rational functions in λ and k. In view of (8.6.3) the $\Gamma_\kappa(\lambda, k)$ may have poles along hyperplanes of the form $(2\lambda + \kappa', \kappa') = 0$ for certain $\kappa' \in Q_- \setminus \{0\}$. These poles are removable except possibly when $\kappa' \in -\mathbb{N}R_+$ (Opdam, 1988b). This is a locally finite collection of affine hyperplanes in the spectral parameter space $\mathfrak{t}^*$. If λ is in the complement of the collection of hyperplanes then we can evaluate the coefficients of $\Phi(\lambda - \rho(k), k; t)$ at λ to obtain a formal solution of (8.6.2). It is not hard to show that such formal solutions are always convergent for all t with $|t| \in T_v^+$.

The fundamental group Π of the regular orbit space $W \backslash T_{\mathrm{reg}}$ has the following description. Consider the sequence of unramified covering maps

$$\mathfrak{t}_{\mathrm{reg}} \to T_{\mathrm{reg}} \to W \backslash T_{\mathrm{reg}}$$

with $\mathfrak{t}_{\text{reg}} := \{x \in \mathfrak{t} \mid \alpha(x) \notin 2\pi i\mathbb{Z}\ \forall\, \alpha \in R^0\}$ the regular points in $\mathfrak{t}$ for the action of the affine Weyl group with translation lattice $2\pi i Q^\vee$. Choose a base point $*$ inside a fundamental alcove (with the origin in its closure) in $\mathfrak{t}_{u,\text{reg}}$. The line segment $[0,1] \ni t \mapsto (1-t)* + tw*$ hits the singular locus in a finite number of points and, going around them through the complex upper half-plane (with coordinate t), we obtain elements $T_w \in \Pi$ with $T_{w_1}T_{w_2} = T_{w_3}$ if $w_3 = w_1 w_2 \in 2\pi i Q^\vee \rtimes W$ and their lengths add up. In fact, this gives a presentation of Π as the affine braid or affine Artin group. For translations over $2\pi i Q^{\vee,+}$ in the direction of the positive chamber containing the alcove the lengths add up, and so $2\pi i Q^{\vee,+}$ reproduces itself as an Abelian monoid inside Π. Another presentation of Π generated by this Abelian group, which can be thought of as the fundamental group of T_{reg}, and the Artin group for the finite Weyl group W was obtained by van der Lek (1983). Using the theory of torus compactifications (more specifically the fact that a mirror intersects a one-dimensional boundary stratum normally) it is in fact easy to show that

$$T_{s_i}T_x = T_x T_{s_i} \tag{8.6.4}$$

for all $x \in 2\pi i Q^{\vee,+}$ which are fixed by s_i, and the presentation of van der Lek and Looijenga is a further refinement of these relations.

Proposition 8.6.1 *For $\lambda \in \mathfrak{t}^*$ generic and $k \in \mathcal{K}$ such that* $\text{Re}(k) \in \mathcal{K}_+$, *the solution*

$$\tilde{c}(\lambda,k)\Phi(\lambda-\rho(k),k;t) + \tilde{c}(s_i\lambda,k)\Phi(s_i\lambda-\rho(k),k;t) \tag{8.6.5}$$

of the hypergeometric system (8.3.3), *which is a priori defined as a holomorphic function for $|t| \in T_v^+$, has a holomorphic continuation over the wall of T_v^+ corresponding to the simple reflection $s_i \in W$, which is invariant under the transformation s_i of T.*

Proof For $\lambda \in \mathfrak{t}^*$ generic the series $\Phi(w\lambda-\rho(k),k;t)$ for $|t| \in T_v^+$ are a basis of the solution space of the hypergeometric system (8.3.3) as w runs over the Weyl group W. The commutation relations (8.6.4) imply that the span of the two basis vectors with indices w and s_iw is invariant under the monodromy operator of T_{s_i}. By asymptotics along this wall, the monodromy calculation of T_{s_i} in these two basis vectors can be reduced to the monodromy calculation for the rank-one Gauss hypergeometric system. This ultimately follows from the Kummer continuation formula for the Gauss hypergeometric function

$$\begin{aligned}{}_2F_1\left(\begin{matrix}\alpha,\beta\\ \gamma\end{matrix};z\right) &= \frac{\Gamma(\gamma)\Gamma(\beta-\alpha)}{\Gamma(\beta)\Gamma(\gamma-\alpha)}(-z)^{-\alpha}\,{}_2F_1\left(\begin{matrix}\alpha,\alpha-\gamma+1\\ \alpha-\beta+1\end{matrix};\frac{1}{z}\right)\\ &\quad+\frac{\Gamma(\gamma)\Gamma(\alpha-\beta)}{\Gamma(\alpha)\Gamma(\gamma-\beta)}(-z)^{-\beta}\,{}_2F_1\left(\begin{matrix}\beta,\beta-\gamma+1\\ \beta-\alpha+1\end{matrix};\frac{1}{z}\right)\end{aligned}$$

by holomorphic continuation in z along the negative real axis $(-\infty,0)$. In turn, this shows that the given linear combination in (8.6.5) extends over the wall to a meromorphic function invariant under s_i. A computation of the local exponents of the hypergeometric system along the wall gives 0 and $1-\gamma_i$ (with $\gamma_i = \frac{1}{2} + k_{\alpha_i/2} + k_{\alpha_i}$) both with multiplicity $\frac{1}{2}|W|$. Therefore the expression (8.6.5) extends in fact holomorphically over the wall. □

Using Hartog's theorem the holomorphic extension over the walls in codimension one implies in fact a holomorphic continuation to a suitable tubular neighborhood of T_v in T.

Corollary 8.6.2 *The function $\tilde{F}(\lambda, k; t) := \sum_w \tilde{c}(w\lambda, k)\Phi(w\lambda - \rho(k); t)$ extends from $T_v^+ T_u$ to a holomorphic W-invariant function on the tubular neighborhood $T_v \exp(\pi i \Omega)$ of T_v in $T = T_v T_u$, where $\Omega := \{x \in \mathfrak{t}_v \mid |\alpha(X)| < 1 \; \forall\, \alpha \in R\}$.*

So far, $\lambda \in \mathfrak{t}^*$ has been a generic parameter, but the various functions have in fact meromorphic behavior in $\lambda \in \mathfrak{t}^*$ and k. A careful analysis of the loci of poles in λ of $\tilde{c}(\lambda, k)$ and of $\Phi(w\lambda - \rho(k), k)$ (via the recurrence relations (8.6.3)) shows (Opdam, 1988b) the following result.

Theorem 8.6.3 *The function $(\lambda, k, t) \mapsto \tilde{F}(\lambda, k; t)$ has a holomorphic extension to $\mathfrak{t}^* \times \mathcal{K} \times T_v \exp(\pi i \Omega)$ as a W-invariant function in both the spectral variable λ and the space variable t.*

For reasons which will become clear in §8.6.4 we renormalize this solution by the multiplying factor $\tilde{c}(\rho(k), k)^{-1}$ (see (8.4.3)) and denote this renormalized function by $F(\lambda, k; t)$. By (8.4.3) and Corollary 8.6.2 the asymptotic expansion formula for $F(\lambda, k)$ in terms of the asymptotically free solutions $\Phi(w\lambda - \rho(k), k; t)$ becomes

$$F(\lambda, k; t) = \sum_w c(w\lambda, k)\Phi(w\lambda - \rho(k), k; t). \tag{8.6.6}$$

Definition 8.6.4 The function $F(\lambda, k)$ is called the *hypergeometric function for the root system R* with multiplicity parameter $k \in \mathcal{K}$ and spectral parameter $\lambda \in \mathfrak{t}^*$.

We claim that the meromorphic function $\mathcal{K} \ni k \mapsto \tilde{c}(\rho(k), k)$ is entire. Indeed, from (8.4.2) it is clear that, for any $k_0 \in \mathcal{K}$, there exists an $n \in \mathbb{N}$ such that $\tilde{c}(\rho(k), k)$ is holomorphic in a neighborhood of $k = k_0 + n$. The claim follows by using that the left-hand side of (8.4.14) is a polynomial (namely the polynomial $T^{\mathrm{rat}}(\pi^\vee, k)(\pi)$ in the variables $k_\alpha + k_{\alpha/2}$, where α runs over R_+^0). By Corollary 8.4.5 it also follows that $\tilde{c}(\rho(k), k)$ is nonzero if $k \in \mathcal{K}^+$, where

$$\mathcal{K}^+ := \{k \in \mathcal{K} \mid \mathrm{Re}(k_\alpha + k_{\alpha/2}) \geq 0 \; \forall\, \alpha \in R^0\}.$$

Let $\mathcal{S} \subset \mathcal{K}$ denote the set of zeros of the entire function $\tilde{c}(\rho(k), k)$, and let $\mathcal{S}_0 \subset \mathcal{K}$ be the set of zeros of the polynomial $T^{\mathrm{rat}}(\pi^\vee, k)(\pi)$ as in (8.4.14). By the above and by (8.4.14) we have

$$\mathcal{S} = \cup_{n \geq 0}(S_0 - n.1) \tag{8.6.7}$$

(where $1 \in \mathcal{K}$ denotes the characteristic function of R^0). The set $\mathcal{S}_0$ is a finite union of hyperplanes in $\mathcal{K}$ which has been described explicitly in all cases (Dunkl et al., 1994).

From Theorem 8.6.3 one obtains the following proposition.

Proposition 8.6.5 *The function F extends to a meromorphic function on $\mathfrak{t}^* \times \mathcal{K} \times T_v \exp(\pi i b \Omega)$ which is holomorphic on $\mathfrak{t}^* \times (\mathcal{K} \backslash \mathcal{S}) \times T_v \exp(\pi i \Omega)$. The set $\mathcal{K} \backslash \mathcal{S}$ is the complement of a locally finite union of hyperplanes, and it contains the closed set $\mathcal{K}^+$.*

8.6.3 Knizhnik–Zamolodchikov and Matsuo's Isomorphism

The commuting Dunkl operators $T(\xi, k)$ of Definition 8.3.2 are deformations of the constant vector fields $\partial(\xi)$ on T_{reg}. This deformation satisfies the Leibniz rule

$$T(\xi, k)(fg) = (\partial(\xi)(f))g + f(T(\xi, k)(g))$$

if f is a W-invariant function on T_{reg}. This shows that we can think of the operators $T(\xi, k)$ as the covariant differentiations of a W-equivariant integrable connection on $W\backslash T_{\mathrm{reg}}$ on the free $\mathbb{C}[T_{\mathrm{reg}}]^W$-module $\mathbb{C}[T_{\mathrm{reg}}]$. If $\lambda \in \mathfrak{t}^*$ then exactly similar considerations apply to the operators $T(\xi, k) - \lambda(\xi)$.

For explicit computation of the connection form of such integrable connections, let us choose a point $Wt \in W\backslash T_{\mathrm{reg}}$. We define a tangent vector at Wt by fixing an element $t \in Wt$ and choosing a tangent vector $\xi \in \mathfrak{t}$. Here we view $\mathfrak{t}$ as the tangent space at $t \in T_{\mathrm{reg}}$ in the usual way, by identifying elements of $\mathfrak{t}$ as constant vector fields on T_{reg}. Let us denote this tangent vector at Wt by (t, ξ); then obviously $(wt, w\xi) = (t, \xi)$ for all $w \in W$. Let us denote by $\hat{\mathcal{O}}_t$ the completed local ring of $t \in T_{\mathrm{reg}}$, and by $\hat{\mathcal{O}}^W_{Wt}$ the completed local ring of $Wt \in W\backslash T_{\mathrm{reg}}$. Then $\hat{\mathcal{O}}_{Wt} := \hat{\mathcal{O}}^W_{Wt} \otimes_{\mathbb{C}[T_{\mathrm{reg}}]^W} \mathbb{C}[T_{\mathrm{reg}}]$ is isomorphic to $\oplus_{w\in W}\hat{\mathcal{O}}_{wt}$ as $\hat{\mathcal{O}}^W_{Wt}[W]$-algebra, by the Chinese remainder theorem. After fixing $t \in Wt$ (as we did above) we can thus identify a germ $\phi \in \hat{\mathcal{O}}_{Wt}$ with a collection of germs $(\phi_w)_{w\in W}$ with $\phi_w \in \hat{\mathcal{O}}_{wt}$ for all $w \in W$ via this isomorphism. With this notation we define an isomorphism

$$L_t\colon \phi = (\phi_w)_{w\in W} \mapsto \sum_{w\in W} {}^w\phi_{w^{-1}} \otimes w\colon \quad \hat{\mathcal{O}}_{Wt} \to \hat{\mathcal{O}}_t \otimes \mathbb{C}[W]$$

of $\hat{\mathcal{O}}^W_{Wt}[W]$-modules. Here ${}^w\phi_{w^{-1}} := \phi_{w^{-1}} \circ w^{-1} \in \hat{\mathcal{O}}_t$. The W-action on $\hat{\mathcal{O}}_{Wt}$ corresponds via L_t with the right regular action of W on the right tensor leg $\mathbb{C}[W]$ of $\hat{\mathcal{O}}_t \otimes \mathbb{C}[W]$, and multiplication by an element $(f_w)_{w\in W} \in \hat{\mathcal{O}}^W_{Wt}$ in $\hat{\mathcal{O}}_{Wt}$ corresponds with multiplication on the left by $f_e \in \hat{\mathcal{O}}_t$. The inverse of this isomorphism L_t maps $\psi \otimes w \in \hat{\mathcal{O}}_t \otimes \mathbb{C}[W]$ to ${}^{w^{-1}}\psi \in \hat{\mathcal{O}}_{w^{-1}t} \subset \hat{\mathcal{O}}_{Wt}$. Hence the integrable connection $\nabla(\lambda, k)$ on $\hat{\mathcal{O}}_t \otimes \mathbb{C}[W]$ whose covariant derivative with respect to the tangent vector (t, ξ) at Wt is given by the operator $\oplus_{w\in W}(T(w\xi, k) - \lambda(w\xi))$ on $\hat{\mathcal{O}}_{Wt} = \oplus_{w\in W}\hat{\mathcal{O}}_{wt}$ satisfies

$$\nabla_{(t,\xi)}(\lambda, k)(\psi \otimes w) = L_t\Big((T(w^{-1}\xi, k) - w\lambda(\xi))({}^{w^{-1}}\psi)\Big). \tag{8.6.8}$$

A straightforward computation shows that

$$\nabla_{(t,\xi)}(\lambda, k) = \partial_\xi\psi \otimes \mathrm{id} - \mathrm{id} \otimes D(\lambda, \xi) + \tfrac{1}{2}\sum_{\alpha\in R_+} k_\alpha \alpha(\xi)\left(\frac{1 + e^{-\alpha}}{1 - e^{-\alpha}} \otimes (1 - r_\alpha) + \mathrm{id} \otimes r_\alpha \epsilon_\alpha\right),$$

where $\epsilon_\alpha(w) := -\mathrm{sgn}(w^{-1}\alpha)w$ and $D(\lambda, \xi)(w) := w\lambda(\xi)w$.

It is easy to see that the connection $\nabla(\lambda, k)$ does not depend on the choice of $t \in Wt$, and that $\nabla(\lambda, k)$ is equivariant with respect to the diagonal action of W on $\mathcal{O}(T_{\mathrm{reg}}) \otimes \mathbb{C}[W]$ (where we act via the left regular action of W on the right tensor leg $\mathbb{C}[W]$).

The just-defined integrable, W-equivariant connection $\nabla(\lambda, k)$ on $\mathcal{O}(T_{\mathrm{reg}}) \otimes \mathbb{C}[W]$ is called the (trigonometric) *Knizhnik–Zamolodchikov connection* (*KZ* connection in the sequel). For a further discussion of *KZ* equations see Chapter 11.

Corollary 8.6.6 (Matsuo's isomorphism; see Matsuo, 1992; Cherednik, 1991; Opdam, 1995) *Assume that for all $\alpha \in R^0_+$ we have $\lambda(\alpha^\vee) \neq k_\alpha + \frac{1}{2}k_{\alpha/2}$. The $\mathbb{D}$-module on $W\backslash T_{\mathrm{reg}}$ defined by the W-equivariant integrable connection $\nabla(\lambda, k)$ on the trivial vector bundle $\mathcal{O}(T_{\mathrm{reg}}) \otimes \mathbb{C}[W]$ over T_{reg} is equivalent to the cyclic D-module on $W\backslash T_{\mathrm{reg}}$ defined by the hypergeometric system* (8.3.3) *via the map $\psi \otimes w \mapsto \psi$: $\hat{\mathcal{O}}_t \otimes \mathbb{C}[W] \to \hat{\mathcal{O}}_t$.*

Proof It is enough to show that this map restricts to an isomorphism between the sheaf of flat sections of $\nabla(\lambda, k)$ and the sheaf of solutions of (8.3.3). By (8.6.8) we see that a flat section of $\nabla(\lambda, k)$ is of the form $L_t(\psi)$, where $\psi \in \hat{\mathcal{O}}_{Wt}$ is a joint eigenfunction of the $T(\xi, k)$ satisfying $T(\xi, k)\psi = \lambda(\xi)\psi$. When we extend the corresponding image under the Matsuo map of this flat section in a W-invariant way we simply obtain $\overline{\psi} := \sum_{w\in W} {}^w\psi \in \hat{\mathcal{O}}^W_{Wt}$. Observe that $\hat{\mathcal{O}}_{Wt}$ is an $\mathbb{H}$-module via the Dunkl representation, and that ψ (and hence also $\overline{\psi}$) as above belongs to the submodule $S(\lambda, k)$ of $\hat{\mathcal{O}}_{Wt}$ on which $Z(\mathbb{H}) \simeq S\mathfrak{t}^W$ acts by the central character $W\lambda$. Recall the minimal principal series module $M(\lambda) = \mathbb{H} \otimes_{S\mathfrak{t}} \mathbb{C}_\lambda$ of the degenerate affine Hecke algebra $\mathbb{H}$. This module has central character $W\lambda$ and always contains a one-dimensional subspace $M(\lambda)^W$ of W-invariant vectors. If $\lambda(\xi) \neq \pm(k_\alpha + \frac{1}{2}k_{\alpha/2})$ for all $\alpha \in R^0$ then this W-dimensional module of $\mathbb{H}$ is known to be irreducible by a well-known theorem of Shinichi Kato (1981). If $M(\lambda)$ is irreducible then it is easy to see that the joint t-eigenspace $M(\lambda)_\lambda$ is one-dimensional, and that the symmetrization map $\sum_w w$ with respect to W defines an isomorphism from $M(\lambda)_\lambda$ to $M(\lambda)^W$. By a dimension count and by use of Definition 8.3.6 and Proposition 8.3.7, it follows that the quotient algebra $\mathbb{H}_\lambda$ by the maximal $Z(\mathbb{H})$-ideal of the point $W\lambda \in W\backslash\mathfrak{t}^*$ has dimension $|W|^2$. Hence, by Kato's theorem, if $\lambda(\xi) \neq \pm(k_\alpha + \frac{1}{2}k_{\alpha/2})$ for all $\alpha \in R^0$ then $\mathbb{H}_\lambda$ is isomorphic to the finite-dimensional simple $\mathbb{C}$-algebra $\mathrm{End}(M(\lambda))$. Hence $S(\lambda, k)$ is semisimple in this case, and isomorphic to $M(\lambda)^d$, where $d = \dim(S(\lambda, k)^W$. In particular, the symmetrization map $\sum_w w$ defines a linear isomorphism from the space $S(\lambda, k)_\lambda$ of joint eigenfunctions ψ of the $T(\xi, k)$ with joint eigenvalue λ onto $S(\lambda, k)^W$. Recall that $S(\lambda, k)^W$ is equal to the local solution space at Wt of the hypergeometric system (8.3.3), and via the map L_t defined above the space $S(\lambda, k)_\lambda$ is isomorphic to the space of flat sections of $\nabla(\lambda, k)$ locally at t. Via L_t the symmetrization map $\sum_w w$ corresponds to Matsuo's map on this space of flat sections. Therefore Matsuo's map defines an isomorphism onto $S(\lambda, k)^W$ if $\lambda(\xi) \neq \pm(k_\alpha + \frac{1}{2}k_{\alpha/2})$ for all $\alpha \in R^0$. Its inverse can be written down explicitly as a differential operator by using Dunkl's operators; this even shows that this Matsuo map is an isomorphism whenever $\lambda(\xi) \neq k_\alpha + \frac{1}{2}k_{\alpha/2}$ (see Opdam, 1995). □

Corollary 8.6.7 *The hypergeometric system* (8.3.3) *is holonomic of rank $|W|$ and is regular singular on $W\backslash T_{\mathrm{reg}}$.*

Proof It is not difficult to prove the claim that the system (8.3.3) is always holonomic of rank $|W|$ by using the fact that the commuting differential operators $D(p, k)$ with $p \in S\mathfrak{t}^W$ are of the form $\partial(p)$ + l.o.t. When $\lambda(\alpha^\vee) \neq k_\alpha + \frac{1}{2}k_{\alpha/2}$ then Matsuo's isomorphism proves that the system is equivalent to an algebraic integrable connection on the trivial vector bundle with fiber $\mathbb{C}[W]$ on the smooth quasi-projective variety $W\backslash T_{\mathrm{reg}}$, exhibiting simple poles only. Hence the system is clearly regular singular in that case. Since the holonomic rank of the system is equal to $|W|$, constant in the parameters, the regularity is detected by the regularity of the restrictions of a rank $|W|$ connection with holomorphic dependence on the parameters (λ, k) on punctured disks such that generically in (λ, k) the connection is regular singular at the puncture. By rewriting the first-order system of ordinary differential equations for the flat sections of this connection on the punctured disk as a higher-order ordinary differential equation with holomorphic coefficients and holomorphic dependence on (λ, k), it is clear that the generic regularity in (λ, k) of the singularity implies the regularity for all values of (λ, k). □

8.6.4 Normalization at e and Summation Formulae

We have introduced the hypergeometric function $F(\lambda, k)$ of Definition 8.6.4 via its asymptotically free expansion deep in the Weyl chamber. However, its normalization is motivated by the evaluation at e, as we will see in this section. The main theorem of this subsection is the following theorem.

Theorem 8.6.8 *We have $F(\lambda, k; e) = 1$ for all $\lambda \in \mathfrak{t}^*$ and $k \in \mathcal{K}$.*

As a first step we look at the special case $\lambda = \mu + \rho(k)$ for $\mu \in P^+$. Then we have $F(\lambda, k) = c(\mu + \rho(k), k)P(\mu, k)$, since both expressions are meromorphic in k (for fixed μ) and represent a W-invariant holomorphic solution of (8.3.3) in a tubular neighborhood of T_v. For generic k, and hence for all k, they must therefore be proportional, and we conclude that the asymptotically free expansion (8.6.6) of the left-hand side coincides with the right-hand side if $\lambda = \mu + \rho(k)$ with μ dominant and integral. By Theorem 8.4.2 we conclude that $F(\lambda, k; e) = 1$ if $\lambda = \mu + \rho(k)$ with μ dominant and integral.

By use of the theory of hypergeometric shift operators we see more generally that the meromorphic function $(\lambda, k) \mapsto F(\lambda, k; e)$ is periodic for translations of k in the integral lattice $\mathcal{K}^{\mathbb{Z}} \subset \mathcal{K}$ (for translations which are integral multiples of the constant multiplicity functions 1 this follows from (8.4.13) and (8.4.14), and this argument can be generalized to general integral translations in k). This statement is indeed more general than the former, since for the polynomial case, where $\mu = \lambda - \rho(k)$ is integral and dominant, the periodicity in k clearly implies that $F(\lambda, k; e)$ is constant in k, and hence constant equal to 1. We would now like to extend this result to arbitrary $\lambda \in \mathfrak{t}^*$ by using some kind of interpolation result like Carlson's theorem, although this is not literally possible. We begin with a bound on the growth order:

Lemma 8.6.9 *Let $k_0 \in \mathcal{K}^{\mathbb{Z}}$ and $\lambda \in \mathfrak{t}^*$. The entire function $\mathbb{C} \ni x \mapsto \epsilon^{k_0}(x) := F(\lambda, xk_0; e)$ is periodic with period 1 and has growth order at most 1.*

Proof The periodicity follows immediately from the above text. It is enough to prove that the growth order of the entire function $\tilde{\epsilon}^{k_0}(x) := \tilde{F}(\lambda, xk_0; e)$ is at most 1, using the (nontrivial) fact (Levin, 1996, §2.4, Theorem 1) that the growth order of an entire function that can be written as a quotient f/g of two entire functions of finite growth order has a finite growth order bounded by the maximum of the growth orders of f and g. By a well-known result of Lelong and Gruman (1986), the growth order $\rho(z)$ of a holomorphic family of entire functions

$w \mapsto f(z,w)$ has the property that its smallest upper-half continuous majorant $z \mapsto \rho^*(z)$ is plurisubharmonic, hence in particular satisfies the maximum principle. Applying this to $\mathfrak{t}^* \times T_v \exp(\pi i \Omega) \ni (\lambda, z) \mapsto \tilde{F}(\lambda, k; z)$ it suffices therefore, in view of Corollary 8.6.2, to prove that the order as an entire function of k of the holomorphic family $(\lambda, z) \mapsto \Phi(\lambda - \rho(k), k; z)$ is bounded by 1, for λ outside the locally finite set of hyperplanes $(\lambda, \kappa^\vee) + 1 = 0$ and $z \in T_v^+ \exp(\pi i \Omega)$ regular. This is a consequence of the recurrence relations (8.6.3); see Opdam (1993). □

The next two lemmas yield the nonvanishing of $F(\lambda, k; e)$ if $k \in \mathcal{K}^+$, which is an important intermediate result for the proof of Theorem 8.6.8.

Lemma 8.6.10 *There exists an open neighborhood* $\mathcal{U}$ *of* $\mathcal{K}^+$ *with the following property: if* F *is a nonzero holomorphic* W*-invariant solution of* (8.3.3) *defined in a neighborhood of* e *for some* $(\lambda, k) \in \mathfrak{t}^* \times \mathcal{U}$*, then* $F(e) \neq 0$.

Proof For some $k \in \mathcal{K}$ let G be a joint eigenfunction of the $T(\xi, k)$ with eigenvalue $\lambda \in \mathfrak{t}^*$. Suppose that G is nonzero and that $G(e) = 0$. Then the lowest homogeneous term g of the expansion of G at e is a nonzero homogeneous polynomial of positive degree d on $\mathfrak{t}$ which is killed by all operators $T^{\mathrm{rat}}(\xi, k)$, and hence by the degree-preserving operator $E(k) := \sum_i x_i T^{\mathrm{rat}}(\xi_i, k)$. We have

$$E(k) = \sum_i x_i \partial(\xi_i) + \sum_{\alpha \in R_+^0} (k_\alpha + k_{\alpha/2})(1 - s_\alpha). \tag{8.6.9}$$

Observe that $\sum_i x_i \partial(\xi_i)(g) = dg$ with $d > 0$, and that $\sum_{\alpha \in R_+^0} (k_\alpha + k_{\alpha/2})(1 - s_\alpha)$ is a scalar operator on any irreducible W-module by Schur's lemma. Let Harm denote the vector space of W-harmonic polynomials (cf. Kostant, 1963) and let $\mathrm{Harm}_+ \subset \mathrm{Harm}$ denote the space of W-harmonic polynomials on $\mathfrak{t}$ which vanish at 0. Define

$$\mathcal{U} := \{k \in \mathcal{K} \mid \mathrm{Re}(\epsilon(k)) > 0 \ \forall \text{ eigenvalues } \epsilon(k) \text{ of } E(k) \text{ on } \mathrm{Harm}_+\}. \tag{8.6.10}$$

Then $\mathcal{U} \subset \mathcal{K}$ is open, and nonempty since $0 \in \mathcal{U}$. Since the space of all polynomials on $\mathfrak{t}$ is a free $(S\mathfrak{t}^*)^W$-module of the form $(S\mathfrak{t}^*)^W \otimes_{\mathbb{C}} \mathrm{Harm}$ (cf. Kostant, 1963, Example 1), it follows that for all $k \in \mathcal{U}$ and any degree $d > 0$, every eigenvalue of $E(k)$ on the space of all homogeneous polynomials of degree d has a strictly positive real part. In particular, if $k \in \mathcal{U}$ then $E(k)$ is a degree-preserving linear isomorphism on the space of polynomials on $\mathfrak{t}$ of positive degree. We conclude that, if $k \in \mathcal{U}$, then $G \neq 0$ implies that $G(e) \neq 0$.

Now consider the $\mathbb{H}$-module $M = \mathbb{H}F$. It is obvious that $M = \mathbb{H}F$ has dimension $\leq |W|$, central character $W\lambda$, and that the trivial representation of W occurs in M with multiplicity 1. Let $G \in M$ be any nonzero simultaneous eigenvector of the $T(\xi, k)$. By the above we see that $G(e) \neq 0$. Hence we may and will assume that $G(e) = 1$. We conclude that $0 \neq \sum_{w \in W} {}^wG \in M^W$, and that $F = F(e)|W|^{-1} \sum_{w \in W} {}^wG$. Hence $F(e) \neq 0$.

Finally note that class functions of W of the form $\sum_\alpha (1 - s_\alpha)$, where α runs over the set of all positive roots of a fixed length, act as a nonnegative constant in any irreducible W-module. Hence $\mathcal{K}^+ \subset \mathcal{U}$, and combined with the above this finishes the proof. □

Lemma 8.6.11 *If* $(\lambda, k) \in \mathfrak{t}^* \times \mathcal{K}^+$ *then* $F(\lambda, k, e) \neq 0$.

Proof Let $\mathcal{U}, \mathcal{S} \subset \mathcal{K}$ be the subsets defined by (8.6.10) and (8.6.7), respectively, and put $\mathcal{V} := \mathcal{U} \backslash (\mathcal{S} \cap \mathcal{U})$. Notice that $\mathcal{V}$ is open and connected and that $\mathcal{K}^+ \subset \mathcal{V}$.

By Proposition 8.6.5, $F(\lambda, k)$ is a holomorphic family for $(\lambda, k) \in \mathfrak{t}^* \times \mathcal{V}$. Let $Z_{F,e}$ denote the zero locus of the holomorphic function $\mathfrak{t}^* \times \mathcal{V} \ni (\lambda, k) \mapsto F(\lambda, k; e)$. Thus $Z_{F,e} \subset \mathfrak{t}^* \times \mathcal{V}$ is an analytic hypersurface, and in particular a complex space. By Lemma 8.6.10 we see that $Z_{F,e} \times T_v \exp(\pi i \Omega) \subset Z_F$, where Z_F denotes the zero locus of F viewed as a holomorphic function on $\mathfrak{t}^* \times \mathcal{V} \times T_v \exp(\pi i \Omega)$.

On the other hand, observe that, if all summands of the asymptotic expansion (8.6.6) are well defined, nonvanishing and if the points $w\lambda$ are distinct modulo Q when w varies in W, then $F(\lambda, k) \neq 0$ (i.e., $F(\lambda, k; t)$ does not vanish identically for all $t \in T_v \exp(\pi i \Omega)$).

From (8.4.2), (8.6.6), (8.6.3) and the above we conclude that the analytic hypersurface $Z_{F,e}$ is contained in the locally finite union $U \subset \mathfrak{t}^* \times \mathcal{V}$ of hyperplanes $H_{\kappa,1}$ defined by the equation $(\lambda, \kappa^\vee) + 1 = 0$ for some $\kappa \in Q \backslash \{0\}$, and root hyperplanes $H_{\alpha,0}$ defined by the equation $(\lambda, \alpha^\vee) = 0$ for some $\alpha \in R$.

By the irreducible decomposition of $Z_{F,e}$ (Grauert and Remmert, 1984, Ch. 9, §9.2) and the inclusion $Z_{F,e} \subset U$ it follows that $Z_{F,e}$ is a union of a subset of the set of hyperplanes H contained in U. Suppose that $Z_{F,e}$ is nonempty, and let $(\lambda_0, k_0) \in Z_{F,e}$. Then there exists a hyperplane of the form $H_{\kappa,c}$ (given by an equation of the form $(\lambda, \kappa^\vee) + c = 0$) such that $(\lambda_0, k_0) \in H_{\kappa,c} \subset Z_{F,e}$. In particular, $(\lambda_0, 0) \in Z_{F,e}$ since $(\lambda_0, 0) \in H_{\kappa,c}$. But this is not possible since (as one easily checks) $F(\lambda, 0; e) = 1$ for all $\lambda \in \mathfrak{t}^*$. Hence $Z_{F,e}$ must be the empty set. □

Proof of Theorem 8.6.8 Lemma 8.6.11 and the 1-periodicity of the entire function $\mathbb{C} \ni x \mapsto \epsilon^{k_0}(x)$ imply that this function is nonvanishing if $k_0 \in \mathcal{K}^{\mathbb{Z},+} := \mathcal{K}^{\mathbb{Z}} \cap \mathcal{K}^+$, and by Lemma 8.6.9 it has growth order at most 1. This implies that $\epsilon^{k_0}(x) = \exp(p(x))$ where p is a polynomial of degree at most 1. Since this is a 1-periodic function we see that $p(x)$ has to be of the form $p_{n,c}(x) = 2\pi i n x + c$ for some $n \in \mathbb{Z}$ and $c \in \mathbb{C}$. It is also clear that for real λ we have $\epsilon^{k_0}(x) \in \mathbb{R}$ if $x \in \mathbb{R}$. Hence we have $n = 0$, proving that ϵ^{k_0} is a constant function. Hence, for λ real and $k \in \mathcal{K}^+$ rational, we see that $F(\lambda, k; e) = \epsilon^{k_0}(q) = \epsilon^{k_0}(0) = 1$, where we wrote $k = q k_0$ for some $k_0 \in \mathcal{K}^{\mathbb{Z},+}$ and $q \in \mathbb{Q}_+$. Since $F(\lambda, k; e)$ is meromorphic in λ and k the desired result follows. □

The following uniqueness result is based on the argument in the proof of Lemma 8.6.10.

Theorem 8.6.12 *The meromorphic family* $F(\lambda, k)$ *is holomorphic on* $\mathfrak{t}^* \times \mathcal{U}$. *If* $(\lambda, k) \in \mathfrak{t}^* \times \mathcal{U}$ *then* $F(\lambda, k)$ *is the* unique *holomorphic* W*-invariant solution of* (8.3.3) *defined in a neighborhood of* e, *up to scalar multiplication.*

Proof If the space of holomorphic W-invariant solutions of (8.3.3) defined in a neighborhood of e for a parameter $(\lambda, k) \in \mathfrak{t}^* \times \mathcal{U}$ has dimension higher than 1, then there also exists such a nonzero holomorphic W-invariant solution F_0 of (8.3.3) defined in a neighborhood of e such that $F_0(e) = 0$. This is impossible by Lemma 8.6.10, which proves the uniqueness claim.

Hence, by Theorem 8.6.8 we have $F(\lambda, k) = \tilde{F}(\lambda, k; e)^{-1}\tilde{F}(\lambda, k)$, a meromorphic family of solutions of (8.3.3), holomorphic and W-invariant on $T_v \exp(\pi i\Omega)$. If this meromorphic family had poles at some $(\lambda, k) \in \mathfrak{t}^* \times \mathfrak{U}$, then removing these poles would again yield a nonzero solution F_0 of (8.3.3) at the parameter (λ, k) such that $F_0(e) = 0$, in contradiction to Lemma 8.6.10. This finishes the proof. □

A more refined analysis than was outlined in the above proof of Theorem 8.6.8 makes it possible to evaluate the asymptotically free solutions $\Phi(\lambda, k; e)$ at e whenever this is possible. It is an analog of the Gauss summation formula for the classical hypergeometric function:

Theorem 8.6.13 (Opdam, 1993) *Let $k \in -\mathfrak{U}$ with $\mathfrak{U}$ as in* (8.6.10). *Then*

$$\lim_{A_+ \ni z \to e} \Phi(\lambda - \rho(k), k; z) = \frac{c^*(\rho(k), k)}{c^*(\lambda, k)}.$$

Such formulae also enable one to evaluate explicitly $F(\lambda, k; p)$ at special points $p \in \overline{\exp(\pi i\Omega)}$. Theorem 8.6.8 and the above summation formulae are multivariate examples of explicit (partial) solutions to a "connection problem" for a regular singular system of linear partial differential equations. Connection problems play a central role in the theory of ordinary linear differential equations with regular singularities on the projective line. There has been quite some progress in the theory for these one-dimensional connection problems, in particular in the case of rigid local systems, by the work of Crawley-Boevey (2003), Oshima and others (see Oshima, 2012 for an account of these developments). Oshima and Shimeno (2010) observed that the solution to the connection problem for rigid local systems in one dimension is relevant in the multivariate situation of Theorem 8.6.8 as well, by "restricting" the hypergeometric system to various one-dimensional strata of the singular locus. This yields a different approach than the one presented here. Though natural and possibly more elementary than the above proof, it seems inevitable that a case-by-case analysis will be part of such an approach. Perhaps the key step to a proof of Theorem 8.6.8 that is both uniform and satisfactory is the "global hypergeometric function" for root systems introduced by Cherednik. This function is a q-hypergeometric-function version of the $F(\lambda, k; x)$. Techniques of Stokman (2014) handle the normalization of this global hypergeometric function for root systems quite naturally, and Theorem 8.6.8 should follow by taking the limit when q tends to 1.

8.6.5 Noncompact Harmonic Analysis

In §8.6.1 (see also §8.4.1) we saw that the restriction of $F(\lambda, k)$ to T_v can be viewed as a generalization of the elementary zonal spherical function on a Riemannian symmetric space $X = G/K$ of noncompact type restricted to a maximally flat subspace $T_v = AK/K \subset G/K$. In the previous subsection we saw that $F(\lambda, k; e) = 1$, generalizing a basic property of the family of elementary zonal spherical functions. It turns out that the spherical Plancherel formula for G/K generalizes as well, as long as $k \in \mathcal{K}_+$. This is the "noncompact version" of the theory of the orthogonal basis of Jacobi polynomials for $\mathbb{C}[T]^W$ on $W\backslash T_u$ as discussed in §8.4.1, and can be viewed as a common generalization of Harish-Chandra's spherical Plancherel formula (Harish-Chandra, 1958a,b) for noncompact Riemannian symmetric spaces and the Jacobi function transform for even functions on the real line (Koornwinder, 1975).

We identify T_v with $\mathfrak{t}_v$ via the exponential mapping exp normalized such that for all $\alpha \in R$ we have $(\exp(\xi))^\alpha = \exp \alpha(\xi)$. Let dx denote a Haar measure on $\mathfrak{t}_v$ normalized such that the covolume of $2\pi Q^\vee$ equals 1, and let $d\lambda$ denote the Haar measure on $\mathfrak{t}^*$ defined by duality. Let da denote the Haar measure on T_v corresponding to dx via the identification. We equip $C_c^\infty(T_v)$ with a pre-Hilbertian structure by the Hermitian form

$$(f,g) := \int_{T_v} \overline{f(a)}\, g(a)\, d\mu(a), \quad \text{where } d\mu(a) := |W|^{-1} \prod_{\alpha \in R_+} \left|a^{\frac{1}{2}\alpha} - a^{-\frac{1}{2}\alpha}\right|^{2k_\alpha} da.$$

Let us define an absolutely continuous measure ν on $i\mathfrak{t}_v^*$ by the formula

$$d\nu(\lambda) := \frac{(2\pi)^{-n}}{\tilde{c}(\lambda,k)\tilde{c}_{w_0}(w_0\lambda,k)}\, d\lambda.$$

For $f \in C_c^\infty(T_v)^W$ we define its hypergeometric Fourier transform (with respect to the root system R and parameter function k) as the W-invariant function $\mathcal{H}(f)$ of $\lambda \in \mathfrak{t}^*$ defined by

$$\mathcal{H}(f)(\lambda) := \int_{a \in T_v} f(a) F(-\lambda, k; a)\, d\mu(a).$$

By Proposition 8.6.5 this is well defined, obviously W-invariant and holomorphic in λ. In the opposite direction we define a wave-packet operator $\mathcal{J}$. If h is a nice W-invariant function on $i\mathfrak{t}_v^*$ (say an integrable function with respect to the measure ν) then we define

$$\mathcal{J}(h)(a) := \int_{\lambda \in i\mathfrak{t}_v^*} h(\lambda) F(\lambda, k; a)\, d\nu(\lambda).$$

The transforms $\mathcal{H}$ and $\mathcal{J}$ can be extended to various, more general types of functions, and are in a formal sense adjoint to each other if we give these respective function spaces the Hermitian inner product structures associated to the measures μ and ν, respectively. By abuse of language we will not make any notational distinction between all these extensions of the transforms $\mathcal{H}$ and $\mathcal{J}$. The main results on these transforms state that $\mathcal{H}$ and $\mathcal{J}$ are inverse to each other, with important refinements describing the behavior of various important spaces of functions under these transforms. The proofs of such results are based on various types of estimates for the kernel $F(\lambda, k; a)$.

The following uniform (both in $\lambda \in \mathfrak{t}^*$ and in $a \in T_v$) estimate plays an important role.

Theorem 8.6.14 (Opdam, 1995) *We have*

$$|F(\lambda, k; a)| \leq |W|^{1/2} H_a(\mathrm{Re}(\lambda)),$$

where for $\lambda \in \mathfrak{t}_v^*$ and $a \in T_v$ one defines $H_a(\lambda) := \max_{w \in W} a^{w\lambda}$.

By use of this estimate one can prove the Paley–Wiener theorem. Define for $a \in T_v$ the (W-invariant) Paley–Wiener space $PW(a)^W$ consisting of all W-invariant entire complex functions h on $\mathfrak{t}^*$ with the property that for all $N \in \mathbb{N}$ there exists a constant C_N such that

$$|h(\lambda)| \leq C_N(1 + \|\lambda\|)^{-N} H_a(-\mathrm{Re}(\lambda)).$$

Let C_a denote the convex hull of the orbit Wa in T_v. By Theorem 8.6.14 one shows easily that $\mathcal{H}(C_c^\infty(C_a)^W) \subset PW(a)^W$, where $C_c^\infty(C_a)$ denotes the space of compactly supported smooth functions on T_v whose support is contained inside C_a. The converse statement can be proved by an argument going back to Rosenberg (1977) which uses the asymptotic expansion of Corollary 8.6.2 of the kernel $F(\lambda, k; a)$ in the positive chamber T_v^+. This yields the result $\mathcal{J}(PW(a)^W) \subset C_c^\infty(C_a)^W$. By an argument due to van den Ban and Schlichtkrull (1997) one can now prove the *Paley–Wiener theorem*:

Theorem 8.6.15 (Opdam, 1995) *The transforms* $\mathcal{H}\colon C_c^\infty(C_a)^W \to PW(a)^W$ *and* $\mathcal{J}\colon PW(a)^W \to C_c^\infty(C_a)^W$ *are inverse isomorphisms.*

In combination with the formal adjointness of $\mathcal{H}$ and $\mathcal{J}$ we immediately obtain the L^2 version of this result:

Theorem 8.6.16 (Opdam, 1995) *The transforms* $\mathcal{H}$ *and* $\mathcal{J}$ *admit a unique extension to inverse unitary isomorphisms between* $L^2(T_v, \mu)$ *and* $L^2(i\mathfrak{t}_v^*, \nu)$.

This result was further generalized by Opdam (1999) to include also the case of not necessarily positive root parameters k_α subject to the condition that μ is a locally integrable function on T_v. It is interesting that the spectrum is no longer continuous in this generality, but consists of series of various dimensions. This corresponds to the possible occurrence of spherical discrete series of graded affine Hecke algebras if the parameters are not necessarily positive. Opdam (1995) also refined the result by extending the transforms beyond W-invariant functions. In this version the transforms $\mathcal{H}$ and $\mathcal{J}$ extend to intertwining isomorphisms between $\mathbb{H}$-modules. A further refinement was provided by Delorme (1999), who defined the natural Schwartz spaces and proved that $\mathcal{H}$ and $\mathcal{J}$ (in the non-W-invariant version, and in the generality where we only require μ to be a locally integrable function) restrict to inverse topological isomorphisms between these Schwartz spaces. In the "repulsive" case where k_α is positive for all $\alpha > 0$, the argument of Delorme was simplified by Schapira (2008) by means of a beautiful sharpening of the uniform estimate Theorem 8.6.14. He proved the following striking results:

Theorem 8.6.17 (Schapira, 2008) *Let* $\lambda = \sigma + i\tau$ *with* $\sigma, \tau \in \mathfrak{t}_v^*$. *Let* $a = \exp x \in \overline{T_v^+}$, *and let* $k \in \mathcal{K}^+$. *Then*

(i) $|F(\lambda, k; a)| \leq F(\sigma, k; a)$,
(ii) $F(\sigma, k; a) \leq F(0, k; a) H_a(\sigma)$,
(iii) $F(0, k; a) \asymp \prod_{\alpha \in R_{0,+}} (1 + \alpha(x)) a^{-\rho(k)}$.

Following Harish-Chandra, Delorme defined the Schwartz space for $\mathcal{H}$ on T_v as the space $\mathcal{C}(T_v)$ consisting of smooth functions f on T_v such that for all constant coefficient differential operators D on T_v and all $N \in \mathbb{N}$ one has

$$\sup_{a \in T_v} (1 + \|\log(a)\|)^N F(0, k; a)^{-1} |Df(a)| < \infty. \tag{8.6.11}$$

The space $\mathcal{C}(T_v)$, equipped with its natural family of seminorms arising from (8.6.11), is a nuclear Fréchet space.

The results of Delorme and Schapira, restricted to the case at hand of W-invariant functions and positive root parameters k_α, can now be stated as follows:

Theorem 8.6.18 (Delorme, 1999; Schapira, 2008) *The transform $\mathcal{H}$ maps $\mathcal{C}(T_v)^W$ onto the space of W-invariant elements of the classical Schwartz space $\mathcal{S}(it_v^*)$. This yields an isomorphism $\mathcal{H}\colon \mathcal{C}(T_v)^W \to \mathcal{S}(it_v^*)^W$ of topological vector spaces, whose inverse is $\mathcal{J}$ (considered on the classical Fréchet space $\mathcal{S}(it_v^*)^W$).*

8.6.6 Further Comments

Schapira's estimates have been improved by Rösler et al. (2013, Theorem 3.3). The sharp asymptotics in Theorem 8.6.17(iii) have an analog at every λ in the closed positive Weyl chamber; see Voit (2015, Remark 3.1). Their proof requires some work; see Narayanan et al. (2014, Theorem 3.4).

In §8.6.5 we only gave an account of the L^2 harmonic analysis. For the L^p harmonic analysis with $p \geq 1$ much less is known: the characterization of the $F(\lambda, k)$ which are bounded is given by Narayanan et al. (2014, Theorem 4.2) but the product formulae and the L^1 convolution structure are longstanding open problems; see Flensted-Jensen and Koornwinder (1973) and some partial progress in papers by Rösler (2010) and Voit (2015).

8.7 Special Cases

8.7.1 Jack Polynomials

Let $\mathcal{P}$ denote the set of integer partitions and $\mathcal{P}_n \subset \mathcal{P}$ the subset of partitions in at most n parts. The Jack polynomials $J_\lambda(x;\alpha)$ in $x = (x_1, \ldots, x_n)$ (with $\lambda \in \mathcal{P}_n$ and $\alpha \in \mathbb{C}$) can be naturally considered as the GL_n-type extension of the Jacobi polynomials for the root system of type A_{n-1}. They form a $\mathbb{C}$-basis of the ring of symmetric polynomials in $x_1, \ldots, x_n$. If $\alpha = 1$ they reduce to the well-known Schur polynomials, up to normalization. The Jack polynomials (Jack, 1970) form a very important class of symmetric polynomials, with remarkably deep applications, interpretations and special properties. There are several elegant and meaningful definitions accordingly (see e.g. Macdonald, 1995). We will presently define the Jack polynomial via its relation with the type A_{n-1} Jacobi polynomial, and comment on the more conventional definitions afterwards.

We define the lower hook-length product

$$h_*(\lambda, \alpha) := \prod_{(i,j)\in\lambda} (\lambda'_j - i + 1 + \alpha(\lambda_i - j)), \quad \lambda \in \mathcal{P},\ \alpha \in \mathbb{C}, \tag{8.7.1}$$

where (i, j) runs over the set of coordinates of the boxes of λ when represented as a Young diagram in the usual way, and λ' denotes the conjugate partition (so that $\lambda'_j - i + 1 + \lambda_i - j$ equals the length of the "hook" inside λ with upper leftmost corner (i, j)).

Recall that an integer partition $\lambda \in \mathcal{P}_n$ determines canonically a dominant weight $\pi(\lambda)$ of the root system of type A_{n-1}.

Definition 8.7.1 Let λ be an integer partition with at most n parts. The n-variable *Jack polynomial* $J_\lambda(x;\alpha)$ is the unique symmetric polynomial of homogeneous degree $|\lambda|$ in $x_1,\ldots,x_n$ characterized by the property that its restriction to the complex algebraic torus T_A defined by $\{(x_1,\ldots,x_n)\mid x_1\cdots x_n=1\}$ equals $h_*(\lambda,\alpha)P_A(\pi(\lambda),\alpha^{-1};(x_1,\ldots,x_n))$, where $P_A(\mu,k;x)$ denotes the Jacobi polynomial of type A_{n-1} with highest weight μ.

We now recall some of the striking properties of the $J_\lambda(x;\alpha)$ which cannot be easily understood directly in terms of the Jacobi polynomials. First of all, they are *stable with respect to the number of variables*, i.e.,

$$J_\lambda(x_1,\ldots,x_m;\alpha)\Big|_{x_{n+1}=\cdots=x_m=0}=J_\lambda(x_1,\ldots,x_n;\alpha),\qquad m\geq n\geq l(\lambda),$$

where $l(\lambda)=\lambda'_1$ denotes the number of parts of λ. For this reason it is possible to view the $J_\lambda(x;\alpha)$ as restrictions to a finite set of variables of symmetric functions $J_\lambda(\alpha)$ in an infinite set of variables, which are called *Jack functions*.

The $J_\lambda(\alpha)$ ($\lambda\in\mathcal{P}$) form a basis of the ring of symmetric functions. The expansion of $J_\lambda(\alpha)$ in terms of the basis of monomial symmetric functions m_μ only involves partitions μ which are smaller than or equal to λ in the dominance ordering of integer partitions. In particular, it makes sense to speak about the coefficient of the monomial symmetric function $m_{(1^l)}$ where $l=|\lambda|$. This reveals a much more natural definition of the normalization of the family $J_\lambda(\alpha)$: the coefficient of $m_{(1^l)}$ equals $l!$. In fact, this normalization is part of the original definition of the Jack polynomials by Jack (1970) (see also Macdonald, 1995, Ch. VI, §10, Example 3(b)) and the equivalence with our normalization in a fixed number of variables can be derived from a nontrivial result due to Stanley (1989) (see Beerends and Opdam, 1993).

It follows in a straightforward way from our definition that, for $\alpha>0$, the n-variable Jack polynomials $J_\lambda(x;\alpha)$ are orthogonal with respect to the measure

$$\textstyle\prod_{i<j}|x_i-x_j|^{2/\alpha}\cdot x_1^{-1}\cdots x_n^{-1}\,dx_1\wedge\cdots\wedge dx_n$$

on the torus $T_B:=\{(x_1,\ldots,x_n)\mid |x_1|=\cdots=|x_n|=1\}$. In contrast, there exists a quite different inner product with respect to which the Jack functions $J_\lambda(\alpha)$ are orthogonal:

Theorem 8.7.2 *Define $z_\lambda:=\prod_{i\geq1}(i^{m_i}m_i!)$, where $m_i=m_i(\lambda)$ denotes the number of parts of λ that are equal to i. Let $(\cdot,\cdot)_\alpha$ denote the scalar product on the ring of symmetric functions such that $(p_\lambda,p_\mu)_\alpha=\delta_{\lambda,\mu}z_\lambda\alpha^{l(\lambda)}$. Then the Jack functions $J_\lambda(\alpha)$ are orthogonal with respect to the inner product $(\cdot,\cdot)_\alpha$.*

Another remarkable property of the $J_\lambda(\alpha)$ is the positivity and integrality of its coefficients when expressed with respect to the basis of normalized monomial symmetric functions $\tilde{m}_\lambda=n_\lambda m_\lambda$, where $n_\lambda:=\prod_{i\geq1}m_i!$.

Theorem 8.7.3 (Knop and Sahi, 1997) *The coefficients of $J_\lambda(\alpha)$ with respect to the basis $\{\tilde{m}_\mu\mid\mu\in\mathcal{P}\}$ of the ring of symmetric functions are polynomials in α with nonnegative integral coefficients.*

The Jack polynomials are also discussed in Chapter 7.

8.7.2 The Hypergeometric Function of Matrix Argument

Hypergeometric functions of matrix argument arose as certain zonal spherical functions on the cone of positive definite real symmetric $n \times n$ matrices in the work of Bochner (1952) and Herz (1955). This theory was generalized and further developed by Constantine (1963), James (1964), Muirhead (1970) and Takemura (1984). The hypergeometric functions of matrix argument find applications in random matrix models, number theory and quantum theory. The most general special functions of this type were introduced independently by Macdonald (1989) and Korányi (1991). Macdonald defined his functions as formal series in terms of the Jack functions $J_\lambda(\alpha)$, hence in infinitely many variables. When we restrict to n variables by setting $x_i = 0$ for all $i > n$ then one obtains a symmetric formal power series in $x_1, \ldots, x_n$. These formal power series are convergent if $|x_i| < 1$ for all $i = 1, \ldots, n$. When we interpret the x_i as the eigenvalues of an $n \times n$ matrix, we can think of these functions as functions of "matrix argument."

First define the dual Jack polynomials $J^*_\lambda(x; \alpha)$ by

$$J^*_\lambda(\alpha) := J_\lambda(\alpha)/(J_\lambda(\alpha), J_\lambda(\alpha))_\alpha.$$

We note in passing that it was shown by Stanley (1989) that

$$(J_\lambda(\alpha), J_\lambda(\alpha))_\alpha = h^*(\lambda, \alpha) h_*(\lambda, \alpha),$$

where $h_*(\lambda, \alpha)$ is the lower hook-length product (8.7.1) and

$$h^*(\lambda, \alpha) := \prod_{(i,j)\in\lambda} (\lambda'_j - i + \alpha(\lambda_i - j + 1)), \qquad \lambda \in \mathcal{P},\ \alpha \in \mathbb{C},$$

the *upper hook-length product*. Hence, when we restrict to n variables $x_1, \ldots, x_n$ we have

$$J^*_\lambda(x; \alpha) = P_A(\lambda, \alpha^{-1}; x)/h^*(\lambda, \alpha).$$

Finally we define the "C-normalization" of the Jack polynomials by

$$C_\lambda(\alpha) := \alpha^{|\lambda|} |\lambda|!\, J^*_\lambda(\alpha).$$

Recall the *shifted factorial* (or *Pochhammer symbol*) $(a)_s := a(a+1)\cdots(a+s-1)$ $(s \in \mathbb{Z}_{>0})$, $(a)_0 := 1$, where $a \in \mathbb{C}$. We define a *generalized shifted factorial* by

$$(a)^{(\alpha)}_\lambda := \prod_{i\geq 1} (a - \alpha^{-1}(i-1))_{\lambda_i}, \qquad \lambda \in \mathcal{P},\ \alpha \in \mathbb{C}\backslash\{0\}.$$

Definition 8.7.4 (Macdonald, 1989) Let $a_1, \ldots, a_p; b_1, \ldots, b_q$ and α be complex parameters. The *generalized hypergeometric function of matrix argument* is the formal series

$${}_pF_q(a_1, \ldots, a_p; b_1, \ldots, b_q; \alpha) := \sum_{\lambda\in\mathcal{P}} \frac{(a_1)^{(\alpha)}_\lambda \cdots (a_p)^{(\alpha)}_\lambda}{(b_1)^{(\alpha)}_\lambda \cdots (b_q)^{(\alpha)}_\lambda |\lambda|!} C_\lambda(\alpha).$$

Its restriction to n variables is obtained by setting $x_i = 0$ for all $i > n$ and is denoted by ${}_pF_q(a_1, \ldots, a_p; b_1, \ldots, b_q; x; \alpha)$ with $x = (x_1, \ldots, x_n)$. This symmetric power series in $x_1, \ldots, x_n$ is convergent on the polydisk defined by $|x_i| < 1$ for all i.

For $\alpha = \frac{1}{2}, 1$ or 2, the Jack polynomials occurring in the above power series can be interpreted as zonal polynomials on the cone of quaternionic, complex or real positive definite matrices, respectively, and in this way one can establish the link for these special parameter values between the functions ${}_pF_q$ defined by Macdonald and Korányi and the original functions of matrix argument studied by Constantine, James and Muirhead.

Let us now restrict our attention to the special case $p = 2, q = 1$ of the generalized hypergeometric functions ${}_pF_q$ of matrix argument. The power series ${}_2F_1$ is characterized uniquely by a system of n linear partial differential equations of order 2. Explicitly these equations are given as follows. Here and below we will always write $k = \alpha^{-1}$. Define

$$\begin{aligned}\Delta_i(a,b,c;k) :=\ & x_i(1-x_i)\partial_i^2 + (c - k(n-1) - (a+b+1-k(n-1))x_i)\partial_i \\ & + k \sum_{j=1;j\neq i}^{n} \frac{x_i(1-x_i)}{x_i - x_j}\partial_i - k \sum_{j=1;j\neq i}^{n} \frac{x_j(1-x_j)}{x_i - x_j}\partial_j.\end{aligned}$$

Theorem 8.7.5 (Korányi, 1991) *The hypergeometric function ${}_2F_1(a,b;c;x;\alpha)$ is the unique symmetric function in the n variables $x_1,\ldots,x_n$ that satisfies*

$$\Delta_i(a,b,c;\alpha^{-1})F = abF, \quad i = 1,2,\ldots,n, \tag{8.7.2}$$

is analytic at $x = 0$ and is normalized by $F(0) = 1$.

Observe that $\Delta_i(a,b,c;\alpha^{-1})$ depends only on $a+b$ and that (8.7.2) is symmetric in a and b. Note that if a symmetric function F is an eigenfunction of the operators $\Delta_i(a',b',c;\alpha^{-1})$ then the eigenvalues are independent of i and we can choose a and b such that $a+b = a'+b'$ and F satisfies (8.7.2).

Let us now turn to the relation between hypergeometric functions of matrix argument and hypergeometric functions associated with root systems. We saw in §8.6 that the hypergeometric system (8.3.9) is a holonomic system of linear partial differential equations of rank $|W|$. For generic values of (λ, k) the system is irreducible. This can be seen for instance from the generic irreducibility of the monodromy representation of the hypergeometric system. For special values of the parameters $\lambda \in \mathfrak{t}^*$ and $k \in \mathcal{K}$ the hypergeometric system (8.3.9) may no longer be irreducible. This happens for example if the system of equations (8.3.9) factorizes via a holonomic system of smaller rank, i.e., if there exists a holonomic system of linear partial differential equations of smaller rank whose solutions are also solutions of (8.3.9). The holonomic systems which appear as factors of the hypergeometric system (8.3.9) are often interesting in their own right.

The hypergeometric function ${}_2F_1$ of matrix argument is a case in point. It was shown by Beerends and Opdam (1993) that the system of differential equations of Theorem 8.7.5 is a factor (in the above sense) of the hypergeometric system (8.3.9) for the root system of type BC_n and a special choice of its parameters, a result that we will explain below. For a good account of this result and of the holonomic system defined by the system of differential equations of Theorem 8.7.5, we refer to Ibukiyama et al. (2012).

It follows in particular that ${}_2F_1$ is an explicit power series expansion at 0 of the BC_n-type hypergeometric function $F(\lambda, k; t)$ for these parameters. We remark that in general such power series expansions are not known. We refer the reader to Beerends and Opdam (1993) and the references therein for a more extensive historical background on this type of hypergeometric series of matrix argument.

Let $e_1, \ldots, e_n$ be the standard basis in $\mathbb{R}^n$. We equip $\mathbb{R}^n$ with the Euclidean inner product $\langle \cdot, \cdot \rangle$ with respect to which the standard basis is an orthonormal basis. The set

$$R_B := \{ \pm e_i, \pm 2e_i, \pm(e_k \pm e_l) \mid i = 1, \ldots, n;\ 1 \le k < l \le n\}$$

forms a root system of type BC_n. The set $S_B := \{e_1 - e_2, e_2 - e_3, \ldots, e_{n-1} - e_n, e_n\}$ is a set of simple roots. The torus is of the form $T = \{(t_1, \ldots, t_n) \mid t_i \in C^\times\}$, where t_i is identified with the character of T corresponding to the root e_i of R_B.

The Weyl group W_B acts naturally on the complex algebraic torus T. It is the hyperoctahedral group $W_B = W_A \ltimes N$, where W_A is the symmetric group of permutations of the coordinates t_i, and $N \approx C_2^n$ is the group of sequences of signs of length n, acting on T by raising t_i to the power given by the ith sign in the sequence. The space $N \backslash T$ of N-orbits in T is isomorphic to the n-dimensional complex affine space. We equip this orbit space with coordinates $x_i := \frac{1}{2} - \frac{1}{4}(t_i + t_i^{-1})$, giving $N \backslash T$ the structure of an n-dimensional complex vector space $V = \mathbb{C}^n$, with linear action by the permutation group W_A. Obviously we have a canonical identification $W_A \backslash V = W_B \backslash T$, and we have

$$\mathbb{C}[V]^{W_A} = \mathbb{C}[x_1, \ldots, x_n]^{W_A} \approx \mathbb{C}[T]^{W_B}. \tag{8.7.3}$$

Given (a, b, c) and α we define multiplicity parameters $k_1 = k_{e_i}$, $k_2 = k_{2e_i}$ and $k_3 = k_{e_i \pm e_j}$ for the root R_B of type BC_n:

$$k_1 = 2c - a - b - 1 - \alpha^{-1}(n-1), \quad k_2 = a + b + \tfrac{1}{2} - c, \quad k_3 = \alpha^{-1}. \tag{8.7.4}$$

In terms of these parameters we have

$$a + b = k_1 + 2k_2 + (n-1)k_3 \quad \text{and} \quad \rho(k) = \tfrac{1}{2}(a+b)\omega_n + \alpha^{-1}\rho_A,$$

where $\omega_n := e_1 + \cdots + e_n$ is the nth fundamental weight with respect to the basis S_B of simple roots, and where $\rho_A := \frac{1}{2}\sum_{i=1}^n (n - 2i + 1)e_i$ is half the sum of the positive-type A-roots $e_i - e_j$ $(i < j)$ of R_B. Observe that ω_n and ρ_A are mutually orthogonal vectors. We define a spectral parameter λ for R_B by

$$\lambda := -a\omega_n + \rho(k). \tag{8.7.5}$$

The main result of Beerends and Opdam (1993) is stated in the following theorem.

Theorem 8.7.6 *Via the identification* (8.7.3) *we consider* ${}_2F_1(a, b; c; x_1, \ldots, x_n; \alpha)$ *as a* W_B-*invariant holomorphic function in an open neighborhood of* $e \in T$. *Then*

$${}_2F_1(a, b; c; x_1, \ldots, x_n; \alpha) = F_B(\lambda, k; t_1, \ldots, t_n), \tag{8.7.6}$$

where F_B *denotes the hypergeometric function for the root system* R_B, *and the parameters* (λ, k) *are defined in terms of* $(a, b, c; \alpha)$ *by* (8.7.4), (8.7.5).

In view of Definition 8.7.4, this result yields a series expansion of the special type of BC_n-Jacobi polynomials which appear on the right-hand side of (8.7.6) in terms of Jack polynomials, with explicit hypergeometric coefficients.

More generally, Macdonald (1989) considered expansions for arbitrary type BC_n-Jacobi polynomials in terms of Jack polynomials. He derived combinatorial expressions for the coefficients as certain tableau sums. Results of this kind can also be derived from the binomial formulae due to Okounkov (1997, 1998) for Koornwinder and Macdonald polynomials in terms of so-called interpolation polynomials; cf. Koornwinder (2015).

One may also express Korányi's second-order operators $\Delta_i(a, b, c; \alpha^{-1})$ directly in terms of the Dunkl–Heckman operators $S_\xi(k)$ (Heckman, 1991) for R_B. These operators are defined by

$$S_\xi(k) := \partial_\xi + \tfrac{1}{2} \sum_{\alpha \in R_{B,+}} k_\alpha \alpha(\xi) \frac{1 + t^{-\alpha}}{1 - t^{-\alpha}} (1 - s_\alpha) = \tfrac{1}{2}(T_\xi(k) + w_0 \circ T_{-\xi}(k) \circ w_0),$$

where $w_0 \colon T \to T$ is given by $w_0(t) := t^{-1}$, the action of the longest Weyl group element of W_B on T. These operators are W_B-equivariant (i.e., for any $w \in W_B$ we have $w \circ S_\xi(k) \circ w^{-1} = S_{w\xi}(k)$), but they do not commute. The W_B-equivariance of the operators $S_\xi(k)$ implies that $S_{e_i}^2(k)$ defines a differential-reflection operator on $\mathbb{C}[T]^N = \mathbb{C}[x_1, \ldots, x_n]$ for every $i = 1, \ldots, n$.

Proposition 8.7.7 *Let $D_i(k)$ be the unique linear partial differential operator on $\mathbb{C}[x_1, \ldots, x_n]$ which coincides with $S_{e_i}^2(k)$ on the subring $\mathbb{C}[x_1, \ldots, x_n]^{W_A}$. For all $i = 1, \ldots, n$ we have*

$$D_i(k) = (\rho(k), e_i)^2 - \Delta_i(a, b, c; \alpha^{-1}).$$

Proof This is a straightforward but tedious direct computation. □

8.7.3 The Missing Euler Integral

The multivariable hypergeometric function associated with a root system generalizes the classical Euler–Gauss $F(a, b; c; z)$ in all its properties, except for one crucial missing insight: the Euler integral representation (8.1.1). For rational parameters a, b, c the integrand $t^{a-1}(1-t)^{c-a-1}(1-tz)^{-b}$ of (8.1.1) is an algebraic function of t, which becomes single valued on a suitable finite cover of the complex plane ramified over the four points $t = 0, 1, \infty, 1/z$. As such it can be viewed as a period of a meromorphic differential on a one-parameter (namely z) family of Riemann surfaces. This is the modern algebraic geometric view on the hypergeometric equation, and has been generalized to the concept of the Gauss–Manin connection (the attribution to Gauss is of course wrong, and should be to Euler, but wrong attributions in mathematics happen quite often).

For multivariable hypergeometric functions of Appell and Lauricella type there are classical integral representations. For integral representations for the *KZ* equation we refer to Chapter 11.

Hence the search for an Euler-type integral representation in the multivariable root system context is urgent, but unfortunately progress has been small. It can be shown that for a "special" spectral parameter $\lambda \in \mathfrak{h}^*$, depending linearly on the coupling parameter $k \in \mathcal{K}$, the hypergeometric system becomes highly reducible and has a subsystem with dimension of the solution space equal to rank(R) + 1. The corresponding monodromy is the reflection representation of the affine Hecke algebra. Let us call this the "special" hypergeometric system associated with R (Couwenberg et al., 2005b).

The natural generalization of the Schwarz map defines a projective structure on the T°/W with T° the complement of the mirrors. For $k \in \mathcal{K}$ positive and sufficiently small it even defines a hyperbolic structure on T°/W with conic singularities along the mirrors. The problem for which of these $k \in \mathcal{K}$ the space T°/W becomes a Heegner divisor complement in a ball quotient can be answered. In the analogous Bessel equation case this has been completely answered in Couwenberg et al. (2005a) and the list is quite substantial. This work generalizes the results of Deligne and Mostow (1986) on the Lauricella F_D hypergeometric function to the root system context. In the toric root system setting, an announcement of similar results was discussed in Couwenberg et al. (2005b) but complete details have not been published yet. The toric setting is interesting because it provides a uniform framework for the period maps of the moduli space of del Pezzo surfaces of degree $d = 3, 2, 1$ to ball quotients of dimension 4, 6, 8 respectively. These period maps were found by Allcock et al. (2002) for $d = 3$ (cubic surfaces), by Kondō (2000) for $d = 2$ (quartic curves) and by Heckman and Looijenga (2002) for $d = 1$ (rational elliptic surfaces).

But despite all this progress on the special hypergeometric system for particular values of $k \in \mathcal{K}$ (satisfying the Schwarz conditions) we do not even have an integral representation for the "special" hypergeometric function for arbitrary $k \in \mathcal{K}$.

References

Allcock, D., Carlson, J. A., and Toledo, D. 2002. The complex hyperbolic geometry of the moduli space of cubic surfaces. *J. Algebraic Geom.*, **11**, 659–724.

van den Ban, E. P., and Schlichtkrull, H. 1997. The most continuous part of the Plancherel decomposition for a reductive symmetric space. *Ann. of Math. (2)*, **145**, 267–364.

Beerends, R. J., and Opdam, E. M. 1993. Certain hypergeometric series related to the root system *BC*. *Trans. Amer. Math. Soc.*, **339**, 581–609.

Bochner, S. 1952. Bessel functions and modular relations of higher type and hyperbolic differential equations. *Comm. Sém. Math. Univ. Lund*, 12–20.

Bourbaki, N. 1968. *Groupes et Algèbres de Lie. Chapitres IV, V et VI*. Hermann, Paris. Also translated in English, Springer, 2002.

Calogero, F. 1971. Solution of the one-dimensional N-body problems with quadratic and/or inversely quadratic pair potentials. *J. Math. Phys.*, **12**, 419–436.

Cherednik, I. 1991. A unification of Knizhnik–Zamolodchikov and Dunkl operators via affine Hecke algebras. *Invent. Math.*, **106**, 411–431.

Constantine, A. G. 1963. Some non-central distribution problems in multivariate analysis. *Ann. Math. Statist.*, **34**, 1270–1285.

Couwenberg, W., Heckman, G., and Looijenga, E. 2005a. Geometric structures on the complement of a projective arrangement. *Publ. Math. Inst. Hautes Études Sci.*, **101**, 69–161.

Couwenberg, W., Heckman, G., and Looijenga, E. 2005b. On the geometry of the Calogero–Moser system. *Indag. Math. (N.S.)*, **16**, 443–459.

Crawley-Boevey, W. 2003. On matrices in prescribed conjugacy classes with no common invariant subspace and sum zero. *Duke Math. J.*, **118**, 339–352.

Deligne, P., and Mostow, G. D. 1986. Monodromy of hypergeometric functions and non-lattice integral monodromy. *Publ. Math. Inst. Hautes Études Sci.*, **63**, 5–89.

Delorme, P. 1999. Espace de Schwartz pour la transformation de Fourier hypergéométrique. *J. Funct. Anal.*, **168**, 239–312. Appendix A by M. Tinfou.

Drinfel'd, V. G. 1986. Degenerate affine Hecke algebras and Yangians. *Functional Anal. Appl.*, **20**, 58–60.

Dunkl, C. F. 1989. Differential-difference operators associated to reflection groups. *Trans. Amer. Math. Soc.*, **311**, 167–183.

Dunkl, C. F., de Jeu, M. F. E., and Opdam, E. M. 1994. Singular polynomials for finite reflection groups. *Trans. Amer. Math. Soc.*, **346**, 237–256.

Flensted-Jensen, M., and Koornwinder, T. 1973. The convolution structure for Jacobi function expansions. *Ark. Mat.*, **11**, 245–262.

Gindikin, S. G., and Karpelevič, F. I. 1962. Plancherel measure for symmetric Riemannian spaces of non-positive curvature (in Russian). *Dokl. Akad. Nauk SSSR*, **145**, 252–255.

Grauert, H., and Remmert, R. 1984. *Coherent Analytic Sheaves*. Grundlehren der Math. Wissenschaften, vol. 265. Springer.

Harish-Chandra. 1958a. Spherical functions on a semisimple Lie group. I. *Amer. J. Math.*, **80**, 241–310.

Harish-Chandra. 1958b. Spherical functions on a semisimple Lie group. II. *Amer. J. Math.*, **80**, 553–613.

Heckman, G. J. 1987. Root systems and hypergeometric functions. II. *Compos. Math.*, **64**, 353–373.

Heckman, G. J. 1991. An elementary approach to the hypergeometric shift operators of Opdam. *Invent. Math.*, **103**, 341–350.

Heckman, G. J. 1997. Dunkl operators. *Astérisque*, No. 245, 223–246. Séminaire Bourbaki, Vol. 1996/97, Exp. No. 828.

Heckman, G., and Looijenga, E. 2002. The moduli space of rational elliptic surfaces. Pages 185–248 of: *Algebraic Geometry 2000, Azumino*. Adv. Stud. Pure Math., vol. 36. Math. Soc. Japan, Tokyo.

Heckman, G. J., and Opdam, E. M. 1987. Root systems and hypergeometric functions. I. *Compos. Math.*, **64**, 329–352.

Heckman, G., and Schlichtkrull, H. 1994. *Harmonic Analysis and Special Functions on Symmetric Spaces*. Academic Press.

Helgason, S. 2000. *Groups and Geometric Analysis*. Amer. Math. Soc. Corrected reprint of the 1984 original.

Helgason, S. 2001. *Differential Geometry, Lie Groups, and Symmetric Spaces*. Amer. Math. Soc. Corrected reprint of the 1978 original.

Herz, C. S. 1955. Bessel functions of matrix argument. *Ann. of Math. (2)*, **61**, 474–523.

Humphreys, J. E. 1990. *Reflection Groups and Coxeter Groups*. Cambridge University Press.

Ibukiyama, T., Kuzumaki, T., and Ochiai, H. 2012. Holonomic systems of Gegenbauer type polynomials of matrix arguments related with Siegel modular forms. *J. Math. Soc. Japan*, **64**, 273–316.

Jack, H. 1970. A class of symmetric polynomials with a parameter. *Proc. Roy. Soc. Edinburgh (Sect. A)*, **69**, 1–18.

James, A. T. 1964. Distributions of matrix variates and latent roots derived from normal samples. *Ann. Math. Statist.*, **35**, 475–501.

Kato, S. 1981. Irreducibility of principal series representations for Hecke algebras of affine type. *J. Fac. Sci. Univ. Tokyo Sect. IA Math.*, **28**, 929–943.

Knop, F., and Sahi, S. 1997. A recursion and a combinatorial formula for Jack polynomials. *Invent. Math.*, **128**, 9–22.

Kondō, S. 2000. A complex hyperbolic structure for the moduli space of curves of genus three. *J. Reine Angew. Math.*, **525**, 219–232.

Koornwinder, T. H. 1974a. Orthogonal polynomials in two variables which are eigenfunctions of two algebraically independent partial differential operators. I, II. *Indag. Math.*, **36**, 48–58, 59–66.

Koornwinder, T. H. 1974b. Orthogonal polynomials in two variables which are eigenfunctions of two algebraically independent partial differential operators. III, IV. *Indag. Math.*, **36**, 357–369, 370–381.

Koornwinder, T. 1975. A new proof of a Paley-Wiener type theorem for the Jacobi transform. *Ark. Mat.*, **13**, 145–159.

Koornwinder, T. H. 2015. Okounkov's BC-type interpolation Macdonald polynomials and their $q = 1$ limit. *Sém. Lothar. Combin.*, **72**, B72a, 27 pp. Corrections in arxiv:1408.5993v5.

Korányi, A. 1991. Hua-type integrals, hypergeometric functions and symmetric polynomials. Pages 169–180 of: *International Symposium in Memory of Hua Loo Keng, Vol. II*. Springer.

Kostant, B. 1963. Lie group representations on polynomial rings. *Amer. J. Math.*, **85**, 327–404.

van der Lek, H. 1983. *The homotopy type of complex hyperplane complements*. Ph.D. thesis, Katholieke Universiteit Nijmegen.

Lelong, P., and Gruman, L. 1986. *Entire Functions of Several Complex Variables*. Springer.

Levin, B. Ya. 1996. *Lectures on Entire Functions*. Translations of Mathematical Monographs, vol. 150. Amer. Math. Soc.

Lusztig, G. 1989. Affine Hecke algebras and their graded version. *J. Amer. Math. Soc.*, **2**, 599–635.

Macdonald, I. G. 1972. The Poincaré series of a Coxeter group. *Math. Ann.*, **199**, 161–174.

Macdonald, I. G. 1982. Some conjectures for root systems. *SIAM J. Math. Anal.*, **13**, 988–1007.

Macdonald, I. G. 1989. *Hypergeometric functions I*. Unpublished manuscript, available at arXiv:1309.4568 (2013).

Macdonald, I. G. 1995. *Symmetric Functions and Hall Polynomials*. Second edn. Oxford University Press.

Matsuo, A. 1992. Integrable connections related to zonal spherical functions. *Invent. Math.*, **110**, 95–121.

Moser, J. 1975. Three integrable Hamiltonian systems connected with isospectral deformations. *Adv. Math.*, **16**, 197–220.

Muirhead, R. J. 1970. Systems of partial differential equations for hypergeometric functions of matrix argument. *Ann. Math. Statist.*, **41**, 991–1001.

Narayanan, E. K., Pasquale, A., and Pusti, S. 2014. Asymptotics of Harish-Chandra expansions, bounded hypergeometric functions associated with root systems, and applications. *Adv. Math.*, **252**, 227–259.

Okounkov, A. 1997. Binomial formula for Macdonald polynomials and applications. *Math. Res. Lett.*, **4**, 533–553.

Okounkov, A. 1998. BC-type interpolation Macdonald polynomials and binomial formula for Koornwinder polynomials. *Transform. Groups*, **3**, 181–207.

Olshanetsky, M. A., and Perelomov, A. M. 1976. Completely integrable Hamiltonian systems connected with semisimple Lie algebras. *Invent. Math.*, **37**, 93–108.

Opdam, E. M. 1988a. Root systems and hypergeometric functions. III. *Compos. Math.*, **67**, 21–49.

Opdam, E. M. 1988b. Root systems and hypergeometric functions. IV. *Compos. Math.*, **67**, 191–209.

Opdam, E. M. 1989. Some applications of hypergeometric shift operators. *Invent. Math.*, **98**, 1–18.

Opdam, E. M. 1993. An analogue of the Gauss summation formula for hypergeometric functions related to root systems. *Math. Z.*, **212**, 313–336.

Opdam, E. M. 1995. Harmonic analysis for certain representations of graded Hecke algebras. *Acta Math.*, **175**, 75–121.

Opdam, E. M. 1999. Cuspidal hypergeometric functions. *Methods Appl. Anal.*, **6**, 67–80.

Opdam, E. M. 2000. *Lecture Notes on Dunkl Operators for Real and Complex Reflection Groups*. MSJ Memoirs, vol. 8. Mathematical Society of Japan.

Oshima, T. 2012. *Fractional Calculus of Weyl Algebra and Fuchsian Differential Equations*. MSJ Memoirs, vol. 28. Mathematical Society of Japan.

Oshima, T., and Shimeno, N. 2010. Heckman–Opdam hypergeometric functions and their specializations. Pages 129–162 of: *New Viewpoints of Representation Theory and Noncommutative Harmonic Analysis*. RIMS Kôkyûroku Bessatsu, B20. Res. Inst. Math. Sci., Kyoto.

Rosenberg, J. 1977. A quick proof of Harish-Chandra's Plancherel theorem for spherical functions on a semisimple Lie group. *Proc. Amer. Math. Soc.*, **63**, 143–149.

Rösler, M. 2010. Positive convolution structure for a class of Heckman–Opdam hypergeometric functions of type *BC*. *J. Funct. Anal.*, **258**, 2779–2800.

Rösler, M., Koornwinder, T., and Voit, M. 2013. Limit transition between hypergeometric functions of type BC and type A. *Compos. Math.*, **149**, 1381–1400.

Schapira, B. 2008. Contributions to the hypergeometric function theory of Heckman and Opdam: sharp estimates, Schwartz space, heat kernel. *Geom. Funct. Anal.*, **18**, 222–250.

Stanley, R. P. 1989. Some combinatorial properties of Jack symmetric functions. *Adv. Math.*, **77**, 76–115.

Stokman, J. V. 2014. The *c*-function expansion of a basic hypergeometric function associated to root systems. *Ann. of Math. (2)*, **179**, 253–299.

Takemura, A. 1984. *Zonal Polynomials*. Institute of Mathematical Statistics, Hayward, CA.

Voit, M. 2015. Product formulas for a two-parameter family of Heckman–Opdam hypergeometric functions of type BC. *J. Lie Theory*, **25**, 9–36.

Vretare, L. 1984. Formulas for elementary spherical functions and generalized Jacobi polynomials. *SIAM J. Math. Anal.*, **15**, 805–833.

Yano, T., and Sekiguchi, J. 1979. The microlocal structure of weighted homogeneous polynomials associated with Coxeter systems. I. *Tokyo J. Math.*, **2**, 193–219.

9

Macdonald–Koornwinder Polynomials

Jasper V. Stokman

9.1 Introduction

In this chapter symmetric and nonsymmetric Macdonald–Koornwinder polynomials are introduced and their basic properties are discussed. These include (bi)orthogonality relations, norm formulas, q-difference(-reflection) equations, duality and evaluation formulas. We develop the theory in such a way that it naturally encompasses all known cases, as well as a new rank-two case. See the first paragraph of §9.3 for the precise meaning of our terminology of Macdonald polynomials, Koornwinder polynomials and Macdonald–Koornwinder polynomials.

Symmetric Macdonald–Koornwinder polynomials are multivariate orthogonal Laurent polynomials. Their rank-one cases are the Askey–Wilson polynomials, the continuous q-Jacobi polynomials and the continuous q-ultraspherical polynomials (Askey and Wilson, 1985), which are three families of classical one-variable q-orthogonal polynomials from the q-Askey scheme (Koekoek et al., 2010). The symmetric Macdonald–Koornwinder polynomials are q-deformations of the symmetric Jacobi polynomials associated with root systems, also known as symmetric Heckman–Opdam polynomials (Chapter 8). They are defined either by orthogonality with respect to an explicit orthogonality measure or as common eigenfunctions of linear, triangular q-difference operators. In general they do not have an explicit expression in terms of products of one-variable basic hypergeometric series, in contrast to the Tratnik-type multivariate q-orthogonal polynomials from, for example, Gasper and Rahman (2005), Rosengren (2001). Instead, the symmetric Macdonald–Koornwinder polynomials associated with classical root systems admit explicit expansion formulas in interpolation polynomials; see Okounkov (1997, 1998a).

The parallels between symmetric Macdonald–Koornwinder polynomials and symmetric Jacobi polynomials associated with root systems are quite strong. From the point of view of applications for instance, symmetric Jacobi polynomials associated with root systems generalize spherical functions on compact symmetric spaces and provide eigenstates for quantum trigonometric Calogero–Moser systems (Chapter 8), while the symmetric Macdonald–Koornwinder polynomials have an interpretation as spherical functions on quantum compact symmetric spaces (Noumi, 1996; Noumi and Sugitani, 1995; Noumi et al., 1997; Letzter, 2004) and give rise to eigenstates for Ruijsenaars' relativistic analogs of quantum trigonometric Calogero–Moser systems (Ruijsenaars, 1987; Cherednik, 1992a, 2005; Kirillov, 1997).

Important special cases of the symmetric Macdonald–Koornwinder polynomials are the GL_n symmetric Macdonald polynomials and the symmetric Koornwinder polynomials. The underlying finite root systems are of type A and type BC, respectively. The symmetric GL_n Macdonald polynomials were introduced by Macdonald (1995, Ch. VI) as a two-parameter family of multivariate orthogonal polynomials in n variables having both the Jack polynomials and the Hall–Littlewood polynomials as limit cases. They have a wealth of applications in combinatorics, algebraic geometry, topology and representation theory (Chapter 10). Macdonald, in an important preprint from 1987 (it appeared in print in 2000; see Macdonald, 2000) introduced root system generalizations of the GL_n Macdonald polynomials. The resulting symmetric Macdonald polynomials are multivariate orthogonal polynomials labeled by so-called admissible pairs of reduced root systems. Recasting the initial data in terms of affine root systems (cf. Cherednik, 2005; Macdonald, 2003) it is natural to speak of an untwisted and a twisted theory of symmetric Macdonald polynomials associated with root systems. An important further extension for nonreduced root systems was constructed in Koornwinder (1992). They are nowadays known as symmetric Koornwinder polynomials. The symmetric Koornwinder polynomials are the only Macdonald–Koornwinder polynomials for which elliptic analogs exist to date; see Rains (2005, 2010) and Chapter 6.

A crucial subsequent development was the definition, due to Macdonald (1996), Cherednik (1995d) and Sahi (1999), of nonsymmetric versions of the Macdonald–Koornwinder polynomials. It was inspired by Heckman's definition of the nonsymmetric variants of the Heckman–Opdam polynomials; see Opdam (1995) and §8.3. A symmetrization procedure turns the nonsymmetric Macdonald–Koornwinder polynomials into the symmetric ones. It is within the nonsymmetric theory that Cherednik's double affine Hecke algebra (Cherednik, 1992a, 1995b) appears as the fundamental algebraic structure underlying the Macdonald–Koornwinder polynomials. The double affine Hecke algebra has been instrumental in obtaining the norm and evaluation formulas for Macdonald–Koornwinder polynomials, which Macdonald had conjectured in the symmetric case in 1987; see Macdonald (2000). Many of these ideas and techniques were developed first for the Jacobi polynomials associated with root systems. See Chapter 8 for a detailed account and references.

The approach in Cherednik (2005) to Macdonald–Koornwinder polynomials using double affine Hecke algebras has been developed for the above-mentioned four different cases (the GL_n case, the untwisted case, the twisted case and the Koornwinder (or $C^\vee C$) case). Cherednik treats the first three cases separately. Macdonald's (2003) exposition covers the last three cases, but various steps still need case-by-case analysis. Haiman (2006) developed a general framework that naturally encompasses the above four cases of the Macdonald–Koornwinder theory. We slightly adjust Haiman's setup and use it to give a uniform treatment of the Macdonald–Koornwinder theory. This theory, besides its four known subclasses mentioned above, includes a new class of rank-two Macdonald–Koornwinder-type polynomials (see §9.3.9).

Before giving a detailed description of the content of the chapter we will first introduce the symmetric GL_n Macdonald polynomials and the symmetric Koornwinder polynomials in the next two subsections, to give the reader a flavor of the type of multivariate orthogonal Laurent polynomials we are dealing with. We end the introductory section by listing various topics on Macdonald–Koornwinder polynomials that we will not be able to treat in this chapter.

9.1.1 Symmetric GL_n Macdonald Polynomials

Macdonald (1988; 1995, Ch. VI) introduced a two-parameter family of symmetric orthogonal polynomials in n variables $t_1, \ldots, t_n$, nowadays often referred to as the Macdonald polynomials. In the general theory of Macdonald–Koornwinder polynomials associated to root data as developed in Macdonald (2000, 2003), Haiman (2006) and in the present chapter, the Macdonald polynomials relate to the symmetric GL_n Macdonald polynomials, which are the symmetric Macdonald–Koornwinder polynomials associated to the GL_n root datum.

The symmetric GL_n Macdonald polynomials P_λ^+ ($\lambda \in \Lambda^+$) form a distinguished two-parameter family of complex linear bases of the algebra $\mathbb{C}[t_1^{\pm 1}, \ldots, t_n^{\pm 1}]^{S_n}$ of S_n-invariant Laurent polynomials in n variables $t_1, \ldots, t_n$. They are labeled by the set Λ^+ of n-tuples $\lambda = (\lambda_1, \ldots, \lambda_n) \in \mathbb{Z}^n$ satisfying $\lambda_1 \geq \lambda_2 \geq \cdots \geq \lambda_n$. The P_λ^+ for $\lambda \in \Lambda^+$ with $\lambda_n \geq 0$ are the Macdonald polynomials from Macdonald (1995, Ch. VI) (in particular, they form a linear basis of the algebra $\mathbb{C}[t_1, \ldots, t_n]^{S_n}$ of symmetric polynomials in $t_1, \ldots, t_n$).

The symmetric GL_n Macdonald polynomials can be defined as common eigenfunctions of q-difference operators or in terms of orthogonality relations, with the inner product either defined analytically or combinatorially. We consider here the analytic approach. See (Macdonald, 1995, Ch. VI, §4, and §10.3) for the combinatorial approach.

Consider the set Λ^+ of n-tuples $\lambda = (\lambda_1, \ldots, \lambda_n) \in \mathbb{Z}^n$ satisfying $\lambda_1 \geq \lambda_2 \geq \cdots \geq \lambda_n$. Write $t = (t_1, \ldots, t_n)$ and $\mathbb{C}[t^{\pm 1}]^{S_n} = \mathbb{C}[t_1^{\pm 1}, \ldots, t_n^{\pm 1}]^{S_n}$. The *symmetric monomials*

$$m_\lambda(t) := \sum_{\mu \in S_n \lambda} t_1^{\mu_1} \cdots t_n^{\mu_n}, \qquad \lambda \in \Lambda^+,$$

with the symmetric group S_n acting on $\mathbb{Z}^n$ by permuting the entries, form a linear basis of $\mathbb{C}[t^{\pm 1}]^{S_n}$. The set Λ^+ is partially ordered by the *dominance order*:

$$\mu \leq \lambda \text{ if } \mu_1 + \cdots + \mu_i \leq \lambda_1 + \cdots + \lambda_i \text{ for } i = 1, \ldots, n-1 \text{ and } \textstyle\sum_{j=1}^n \mu_j = \sum_{j=1}^n \lambda_j.$$

The space $\mathbb{C}[t^{\pm 1}]^{S_n}$ is a dense subspace of the Hilbert space $L^2(T_u, v_+(t)\, d_u t)^{S_n}$ of S_n-invariant L^2-functions on the compact torus $T_u := \{t \in \mathbb{C}^n \mid |t_i| = 1\}$ with $d_u t$ the normalized Haar measure of T_u and with the two-parameter family of weight functions $v_+(t) = v_+(t; \kappa, q)$ ($0 < q, \kappa < 1$) given by

$$v_+(t; \kappa, q) = \prod_{1 \leq i \neq j \leq n} \frac{(t_i/t_j; q)_\infty}{(\kappa^2 t_i/t_j; q)_\infty}$$

(see (9.3.9) for the definition of the q-shifted factorial). Denote by $\langle \cdot, \cdot \rangle$ the associated inner product.

Definition 9.1.1 With Λ^+, $\leq$, m_λ defined as above, the monic *symmetric* GL_n *Macdonald polynomial* $P_\lambda^+(t) = P_\lambda^+(t; \kappa, q)$ of degree $\lambda \in \Lambda^+$ is the unique S_n-invariant Laurent polynomial in the variables $t_1, \ldots, t_n$ satisfying

1. $P_\lambda^+(t) = m_\lambda(t) + \sum_{\mu\in\Lambda^+:\mu<\lambda} d_{\lambda,\mu} m_\mu(t)$ for certain $d_{\lambda,\mu} \in \mathbb{C}$,
2. $\langle P_\lambda^+, m_\mu\rangle = 0$ if $\mu \in \Lambda^+$ and $\mu < \lambda$.

The polynomials P_λ^+ are orthogonal with respect to $\langle\cdot,\cdot\rangle$,

$$\langle P_\lambda^+, P_\mu^+\rangle = 0 \quad \text{if } \lambda \neq \mu.$$

This is clear from the definition only if λ and μ are compatible with respect to the dominance order. The proof in general uses the commuting trigonometric Ruijsenaars–Macdonald q-difference operators (Ruijsenaars, 1987; Macdonald, 1995, Ch. VI, §3):

$$(D_j f)(t) := \sum_{\substack{I\subseteq\{1,\ldots,n\}\\ \#I=j}} \left(\prod_{r\in I, s\notin I} \frac{\kappa^{-1}t_r - \kappa t_s}{t_r - t_s}\right) f(q^{-\sum_{r\in I}\epsilon_r} t), \qquad 1 \leq j \leq n,$$

where $q^{-\sum_{r\in I}\epsilon_r}t$ is the n-vector with entry $q^{-1}t_i$ at $i \in I$ and entry t_j at $j \notin I$. They act on $\mathbb{C}[t^{\pm 1}]^{S_n}$ as triangular linear operators with respect to the partially ordered linear basis $\{m_\lambda(t)\}_{\lambda\in\Lambda^+}$ of symmetric monomials, with the order inherited from the dominance order $\leq$ on the index set Λ^+. The operators D_j are self-adjoint with respect to the inner product $\langle\cdot,\cdot\rangle$. These properties imply that the P_λ^+ is a common eigenfunction of the operators D_j (see §9.3.7 for more details). For the full orthogonality of the polynomials $P_\lambda^+(t)$ it actually suffices to consider only the q-difference operator D_1, since the spectrum of D_1 is simple for generic q. For the symmetric Jacobi polynomials associated with root systems, which are the classical ($q = 1$) analogs of the symmetric Macdonald–Koornwinder polynomials, all the commuting differential operators are needed (see §8.4).

Note that the symmetric GL_n Macdonald polynomials are homogeneous Laurent polynomials. Hence $t^{-\lambda}P_\lambda^+(t)$, with $t^\lambda = t_1^{\lambda_1}\cdots t_n^{\lambda_n}$, depends only on $t_1/t_2, \ldots, t_{n-1}/t_n$. For $n = 2$ the symmetric GL_2 Macdonald polynomial $t^{-\lambda}P_\lambda^+(t)$ is the continuous q-ultraspherical polynomial of degree $\lambda_1 - \lambda_2$ in t_2/t_1. The above results then reduce to the orthogonality relations and the second-order q-difference equation satisfied by the continuous q-ultraspherical polynomials; see §9.3.7 for further details.

9.1.2 Symmetric Koornwinder Polynomials

The symmetric Koornwinder polynomials (Koornwinder, 1992) form a six-parameter family of linear bases of the space $\mathbb{C}[t^{\pm 1}]^{W_0}$ of W_0-invariant Laurent polynomials in n variables $t = (t_1, \ldots, t_n)$, with W_0 the hyperoctahedral group $S_n \ltimes \{\pm 1\}^n$ acting by permutations and inversions of the variables.

In this case $\mathbb{C}[t^{\pm 1}]^{W_0}$ has a linear basis consisting of the W_0-symmetric monomials

$$m_\lambda(t) := \sum_{\mu\in W_0\lambda} t_1^{\mu_1} t_2^{\mu_2}\cdots t_n^{\mu_n}, \qquad \lambda \in \Lambda^+,$$

with $\Lambda^+ := \{\lambda \in \mathbb{Z}^n \mid \lambda_1 \geq \cdots \geq \lambda_n \geq 0\}$ the partitions of length $\leq n$. Here the hyperoctahedral group W_0 acts on $\mathbb{Z}^n$ by permutations and sign changes. We consider Λ^+ as a partially ordered set with respect to the *dominance order*, which in this case is given by

$$\mu \leq \lambda \text{ if } \mu_1 + \cdots + \mu_i \leq \lambda_1 + \cdots + \lambda_i \text{ for } i = 1, \ldots, n.$$

The space $\mathbb{C}[t^{\pm 1}]^{W_0}$ is a dense subspace of the Hilbert space $L^2(T_u, v_+(t)d_ut)^{W_0}$ of W_0-invariant L^2-functions on the compact torus $T_u := \{t \in \mathbb{C}^n \mid |t_i| = 1\}$, with the six-parameter family of weight functions $v_+(t) = v_+(t; a, b, c, d, k, q)$ $(0 < a, b, c, d, q, k < 1)$ given by

$$v_+(t) := \prod_{i=1}^{n} \frac{(t_i^{\pm 2}; q)_\infty}{(at_i^{\pm 1}; q)_\infty (bt_i^{\pm 1}; q)_\infty (ct_i^{\pm 1}; q)_\infty (dt_i^{\pm 1}; q)_\infty} \prod_{1 \leq r < s \leq n} \frac{(t_r t_s^{\pm 1}; q)_\infty (t_r^{-1} t_s^{\pm 1}; q)_\infty}{(kt_r t_s^{\pm 1}; q)_\infty (kt_r^{-1} t_s^{\pm 1}; q)_\infty},$$

where $(uz^{\pm 1}; q)_\infty := (uz; q)_\infty (uz^{-1}; q)_\infty$. Denote by $\langle \cdot, \cdot \rangle$ the associated inner product.

Definition 9.1.2 With Λ^+, $\leq$, m_λ, W_0 defined as above, the monic *symmetric Koornwinder polynomial* $P_\lambda^+(t) = P_\lambda^+(t; a, b, c, d, k, q)$ of degree $\lambda \in \Lambda^+$ is the unique W_0-invariant Laurent polynomial in the variables $t_1, \ldots, t_n$ satisfying
1. $P_\lambda^+(t) = m_\lambda(t) + \sum_{\mu \in \Lambda^+ : \mu < \lambda} d_{\lambda,\mu} m_\mu(t)$ for certain $d_{\lambda,\mu} \in \mathbb{C}$,
2. $\langle P_\lambda^+, m_\mu \rangle = 0$ if $\mu \in \Lambda^+$ and $\mu < \lambda$.

Full orthogonality is again a nontrivial fact. In this case one can establish it by showing that the symmetric Koornwinder polynomials are the eigenfunctions of Koornwinder's multivariable extension of the Askey–Wilson (Askey and Wilson, 1985) second-order q-difference operator (Koornwinder, 1992), given by

$$(Df)(t) := \sum_{i=1}^{n} \sum_{\xi \in \{\pm 1\}} A_i^\xi(t)(f(q^{\xi \epsilon_i} t) - f(t)),$$

$$\text{where} \quad A_i^\xi(t) := \frac{(1 - at_i^\xi)(1 - bt_i^\xi)(1 - ct_i^\xi)(1 - dt_i^\xi)}{(1 - t_i^{2\xi})(1 - qt_i^{2\xi})} \prod_{j \neq i} \frac{(1 - kt_i^\xi t_j)(1 - kt_i^\xi t_j^{-1})}{(1 - t_i^\xi t_j)(1 - t_i^\xi t_j^{-1})}.$$

In fact, D is part of a commutative family $D = D_1, \ldots, D_n$ of algebraically independent linear q-difference operators acting on $\mathbb{C}[t^{\pm 1}]^{W_0}$; see van Diejen (1995) and §9.3.8.

For $n = 1$ the k-dependence drops out and the symmetric Koornwinder polynomials $P_\lambda^+(t)$ reduce to the monic Askey–Wilson polynomials (Askey and Wilson, 1985). See §9.3.8 for further details.

9.1.3 Detailed Description of the Contents

Precise references to the literature are given in the main text.

In §9.2 we give the definition of the affine braid group, affine Weyl group and affine Hecke algebra. We determine an explicit realization of the affine Hecke algebra, which will serve as the starting point for the Cherednik–Macdonald theory on Macdonald–Koornwinder polynomials in the next section. In addition we introduce the initial data. We introduce the space of multiplicity functions associated to the fixed initial data D. We extend the duality on initial data to an isomorphism of the associated spaces of multiplicity functions. We give the basic representation of the extended affine Hecke algebra associated to D, using the explicit realization of the affine Hecke algebra.

In §9.3 we define and study the nonsymmetric and symmetric monic Macdonald–Koornwinder polynomials associated to the initial data D. We first focus on the nonsymmetric polynomials. We characterize them as common eigenfunctions of a family of commuting q-difference reflection operators. These operators are obtained as the images under the basic representation of the elements of the Bernstein–Zelevinsky abelian subalgebra of the extended affine Hecke algebra. We determine the biorthogonality relations of the nonsymmetric monic polynomials and use the finite Hecke symmetrizer to obtain the symmetric monic polynomials. We give the orthogonality relations of the symmetric monic polynomials and show that they are common eigenfunctions of the commuting Macdonald q-difference operators (also known as Ruijsenaars operators in the GL case). We finish this section by describing three cases in detail: the GL case, the $\mathrm{C}^\vee\mathrm{C}$ case and a new nonreduced rank-two case not covered in the treatments by Cherednik (2005) and Macdonald (2003).

In §9.4 we introduce the double affine braid group and the double affine Hecke algebra associated to D and (D, κ) respectively, where κ is a choice of a multiplicity function on R. We lift the duality on initial data to a duality anti-isomorphism on the level of the associated double affine braid groups. We show how it descends to the level of double affine Hecke algebras and how it leads to an explicit evaluation formula for the monic Macdonald–Koornwinder polynomials. We proceed by defining the associated normalized nonsymmetric and symmetric Macdonald–Koornwinder polynomials and deriving their duality and quadratic norms.

In §9.5 we give the norm and evaluation formulas in terms of q-shifted factorials for the GL case, the $\mathrm{C}^\vee\mathrm{C}$ case and for the new nonreduced rank-two case.

In the Appendix we give a short introduction to (the classification of) affine root systems, closely following Macdonald (1972a) but with some adjustments. We close the Appendix with a list of all the affine Dynkin diagrams.

Remark 9.1.3 Four years after the appearance of this chapter as preprint on the arXiv, Ion and Sahi (2015) developed an approach to the theory of double affine braid groups and double affine Hecke algebras based on double affine Coxeter-type data. It leads to a uniform treatment of this theory which is closely related to the treatment in §9.4. The finite root system we use as part of the initial data plays the role of the dual finite root system in Ion and Sahi (2015). What is called untwisted (respectively twisted) in the present chapter, is called twisted (respectively untwisted) in Ion and Sahi (2015).

9.1.4 Further Topics

We list here various important developments involving Macdonald–Koornwinder polynomials that are not discussed in this chapter. For a more extensive discussion of the ramifications of these polynomials we refer to the introductory chapter of Cherednik's book (2005).

1. Shift operators for the Macdonald–Koornwinder polynomials (see, e.g., Cherednik, 1995b; Stokman, 2000a). It leads to explicit evaluation formulas for the constant terms, which are q-analogs of Selberg integrals (Chapter 5).
2. Connections to combinatorics (Chapter 10).
3. Connections to algebraic geometry; see, e.g., Haiman (2001, 2002), Schiffmann and Vasserot (2011).
4. Connections to representation theory; see, e.g., Cherednik (2005, 2003, 2009a), Kasatani (2005a), Ion (2003b, 2006).
5. Applications to harmonic analysis on quantum groups; see, e.g., Noumi (1996), Noumi and Sugitani (1995), Noumi et al. (1997), Letzter (2004), Etingof and Kirillov (1994, 1996, 1998), Oblomkov and Stokman (2005).
6. Connections to quantum integrable systems, such as Ruijsenaars' quantum many body systems (Ruijsenaars, 1987) and integrable one-dimensional spin chains; see, e.g., Cherednik (1992a,b), Kirillov (1997), Sergeev and Veselov (2009), Chalykh (2002), Chalykh and Etingof (2013), Pasquier (2006), Kasatani and Pasquier (2007), Kasatani and Takeyama (2007), Di Francesco and Zinn-Justin (2005), Stokman (2011), Stokman and Vlaar (2015).
7. Integrable probabilistic systems (Macdonald processes); see, e.g., Borodin and Corwin (2014a,b).
8. Applications to torus knot homology (DAHA-Jones polynomials); see, e.g., Cherednik (2013), Cherednik (2016), Cherednik and Danilenko (2016), Elliot and Gukov (2016) and Chapter 10.
9. Limit cases of symmetric Macdonald–Koornwinder polynomials: limits $q \to 1$ to classical multivariable orthogonal polynomials such as the multivariable Jacobi polynomials (Chapter 8); p-adic limits $q \to 0$ to Hall–Littlewood-type polynomials (see, e.g., Macdonald, 1995, 2000; Heckman and Opdam, 1997); Whittaker limits (see, e.g., Cherednik, 2009b); multivariable analogs of limit transitions within the q-Askey scheme (see, e.g., Stokman and Koornwinder, 1997; Stokman, 2000b; Baker and Forrester, 2000).
10. Macdonald–Mehta-type integrals and basic hypergeometric functions associated with root systems, i.e., the q-analogs of the hypergeometric functions associated with root systems discussed in Chapter 8; see, e.g., Cherednik (1997a, 2009b), Stokman (2003, 2014).
11. Interpolation Macdonald–Koornwinder polynomials; see, e.g., Knop (1997); Knop and Sahi (1997), Sahi (1996), Okounkov (1998a,b), Lascoux et al. (2009). Their elliptic versions are discussed in Chapter 6.
12. Special parameter values (e.g., roots of unity); see, e.g., Cherednik (2005, 1995d), van Diejen and Stokman (1998), Kasatani et al. (2006), Kasatani (2005a,b), Feigin et al. (2003), Chalykh (2002), Chalykh and Etingof (2013).
13. Affine and elliptic generalizations; see, e.g., Etingof and Kirillov (1995), Cherednik (1995a), Ruijsenaars (1987) and Komori and Hikami (1997), Rains (2005), Rains (2010), Coskun and Gustafson (2006) respectively. See Chapter 6 for a discussion of the elliptic generalization of the symmetric Koornwinder polynomial.

9.2 The Basic Representation of the Extended Affine Hecke Algebra

We first introduce the affine Hecke algebra and the appropriate initial data for the Cherednik–Macdonald theory on Macdonald–Koornwinder polynomials. Then we introduce the basic representation of the extended affine Hecke algebra, which is fundamental in the development of the theory.

9.2.1 Affine Hecke Algebras

A convenient reference for this subsection is Humphreys (1990). For unexplained notation and terminology regarding affine Weyl groups we refer to the Appendix.

For a generalized Cartan matrix $A = (a_{ij})_{1\le i,j\le r}$ let $M = (m_{ij})_{1\le i,j\le r}$ be the matrix with entries $m_{ii} = 1$ and, for $i \neq j$, $m_{ij} = 2, 3, 4, 6, \infty$ according to whether $a_{ij}a_{ji} = 0, 1, 2, 3, \ge 4$, respectively.

Definition 9.2.1 Let $A = (a_{ij})_{1\le i,j\le r}$ be a generalized Cartan matrix.

1. The *braid group* $\mathcal{B}(A)$ is the group generated by T_i $(1 \le i \le r)$ with defining relations $T_iT_jT_i\cdots = T_jT_iT_j\cdots$ (m_{ij} factors on each side) if $1 \le i \neq j \le r$ (which should be interpreted as no relation if $m_{ij} = \infty$).
2. The *Coxeter group* $\mathcal{W}(A)$ associated to A is the quotient of $\mathcal{B}(A)$ by the normal subgroup generated by T_i^2 $(1 \le i \le r)$.

It is convenient to denote by s_i the element in $\mathcal{W}(A)$ corresponding to T_i for $1 \le i \le r$. They are the *Coxeter generators* of the Coxeter group $\mathcal{W}(A)$.

Let $A = (a_{ij})_{1\le i,j\le r}$ be a generalized Cartan matrix. Suppose that k_i $(1 \le i \le r)$ are nonzero complex numbers such that $k_i = k_j$ if s_i is conjugate to s_j in $\mathcal{W}(A)$. We write k for the collection $\{k_i\}_i$. Let $\mathbb{C}[\mathcal{B}(A)]$ be the complex group algebra of the braid group $\mathcal{B}(A)$.

Definition 9.2.2 The *Hecke algebra* $H(A, k)$ is the complex unital associative algebra given by $\mathbb{C}[\mathcal{B}(A)]/I_k$, where I_k is the two-sided ideal of $\mathbb{C}[\mathcal{B}(A)]$ generated by $(T_i - k_i)(T_i + k_i^{-1})$ for $1 \le i \le r$.

If $k_i = 1$ for all $1 \le i \le r$ then the associated affine Hecke algebra is the complex group algebra of $\mathcal{W}(A)$.

If $w = s_{i_1}s_{i_2}\cdots s_{i_r}$ is a reduced expression in $\mathcal{W}(A)$, i.e., a shortest expression of w as a product of Coxeter generators, then $T_w := T_{i_1}T_{i_2}\cdots T_{i_r} \in H(A, k)$ is well defined (already in the braid group $\mathcal{B}(A)$) and the T_w $(w \in \mathcal{W}(A))$ form a complex linear basis of $H(A, k)$.

Suppose $R_0 \subset V$ is a finite crystallographic root system with ordered basis $\Delta_0 = (\alpha_1, \ldots, \alpha_n)$ and write A_0 for the associated Cartan matrix. Then $W_0 \simeq \mathcal{W}(A_0)$ by mapping the simple reflections $s_{\alpha_i} \in W_0$ to the Coxeter generators s_i of $\mathcal{W}(A_0)$ for $1 \le i \le n$. The associated finite Hecke algebra $H(A_0, k)$ depends only on k and W_0 (as Coxeter group). We will sometimes denote it by $H(W_0, k)$.

Similarly, if R is an irreducible affine root system with ordered basis $\Delta = (a_0, \ldots, a_n)$ and if A is the associated affine Cartan matrix, then $\mathcal{W}(A) \simeq W(R)$ by $s_i \mapsto s_{a_i}$ $(0 \le i \le n)$. Again we write $H(W(R), k)$ for the associated affine Hecke algebra $H(A, k)$.

9.2.2 Realizations of the Affine Hecke Algebra

We use the notations on affine root systems as introduced in the Appendix. The following construction is motivated by Cherednik's polynomial representation (Cherednik, 2005, Thm. 3.2.1) of the affine Hecke algebra and its extension to the nonreduced case by Noumi (1995).

Let $R \subset \widehat{E}$ be an irreducible (possibly nonreduced) affine root system on the affine Euclidean space E of dimension n, with affine Weyl group W. Let $\Delta = (a_0, a_1, \ldots, a_n)$ be an ordered basis of R. Write $A = A(R, \Delta)$ for the associated affine Cartan matrix.

Consider the lattice $\mathbb{Z}R$ in $\widehat{E}$. It is a full W-stable lattice and the simple affine roots form a $\mathbb{Z}$-basis. Denote by F the quotient field $\mathrm{Quot}(\mathbb{C}[\mathbb{Z}R])$ of the complex group algebra $\mathbb{C}[\mathbb{Z}R]$ of $\mathbb{Z}R$. It is convenient to write e^λ $(\lambda \in \mathbb{Z}R)$ for the natural complex linear basis of $\mathbb{C}[\mathbb{Z}R]$. The multiplicative structure of F is determined by $e^0 = 1$, $e^{\lambda+\mu} = e^\lambda e^\mu$ $(\lambda, \mu \in \mathbb{Z}R)$.

The affine Weyl group W canonically acts by field automorphisms on F. On the basis elements e^λ the W-action reads $w(e^\lambda) = e^{w\lambda}$ $(w \in W, \lambda \in \mathbb{Z}R)$. Since W acts by algebra automorphisms on F, we can form the semidirect product algebra $W \ltimes F$.

Let $k\colon a \mapsto k_a\colon R \to \mathbb{C}^* := \mathbb{C} \setminus \{0\}$ be a W-equivariant map, i.e., $k_{wa} = k_a$ for all $w \in W$ and $a \in R$. If $a \in R$ but $2a \notin R$ then we set $k_{2a} := k_a$. Note that $k^{\mathrm{ind}} := k|_{R^{\mathrm{ind}}}$ is a W-equivariant map on R^{ind}, which is determined by its values $k_i := k_{a_i}$ $(0 \le i \le n)$ on the simple affine roots. It satisfies $k_i = k_j$ if s_i is conjugate to s_j in W. Hence we can form the associated affine Hecke algebra $H(W(R), k^{\mathrm{ind}})$.

Theorem 9.2.3 *With the above notation and conventions, there exists a unique algebra monomorphism $\beta = \beta_{R,\Delta,k}\colon H(W(R), k^{\mathrm{ind}}) \hookrightarrow W \ltimes F$ satisfying*

$$\beta(T_i) = k_i s_i + \frac{k_i - k_i^{-1} + (k_{2a_i} - k_{2a_i}^{-1})e^{a_i}}{1 - e^{2a_i}}(1 - s_i), \quad 0 \le i \le n. \tag{9.2.1}$$

The proof uses the Bernstein–Zelevinsky presentation of the affine Hecke algebra, which we present in a slightly more general context in §9.3.1.

Remark 9.2.4 If $2a_i \notin R$ then (9.2.1) simplifies to $\beta(T_i) = k_i s_i + \frac{k_i - k_i^{-1}}{1 - e^{a_i}}(1 - s_i)$.

The notion of similarity of pairs (R, Δ) (see the Appendix) can be extended to triples (R, Δ, k) in the obvious way. The algebra homomorphisms β that are associated to different representatives of the similarity class of (R, Δ, k) are then equivalent in a natural sense. Starting from the next subsection we therefore will focus on the explicit representatives of the similarity classes as described in §A.2.

Recall from the Appendix that the classification of irreducible affine root systems leads to a subdivision of irreducible affine root systems in three types, namely untwisted, twisted and mixed type. It is easy to show that the algebra map $\beta_{R,\Delta,k}$ in case that R is of mixed type (possibly nonreduced) can alternatively be written as $\beta_{(R',\Delta',k')}$ with appropriately chosen triple (R', Δ', k') and with R' of untwisted or of twisted type. In the Cherednik–Macdonald theory, the mixed type can therefore safely be ignored.

9.2.3 Initial Data

It is tempting to believe that the initial data for the Macdonald–Koornwinder polynomials should be similarity classes of irreducible affine root systems R together with a choice of a deformation parameter q and a multiplicity function (playing the role of the free parameters in the theory). In such a parametrization the untwisted and twisted cases should relate to the similarity classes of the irreducible reduced affine root systems of untwisted and twisted type, respectively: the GL case to the irreducible reduced affine root system of type A with a "reductive" extension of the affine Weyl group, and the Koornwinder case with the nonreduced irreducible affine root system of type $C^\vee C$ (we refer here to the classification of affine root systems from Macdonald (1972a); see also the Appendix). It turns out, though, that a more subtle labeling is needed in order to come to a uniform theory capturing all cases and capturing all fundamental properties of the Macdonald–Koornwinder polynomials.

We will take as initial data quintuples $D = (R_0, \Delta_0, \bullet, \Lambda, \Lambda^d)$ with R_0 a finite reduced irreducible root system, Δ_0 an ordered basis of R_0, $\bullet \in \{u, t\}$ (u stands for *untwisted* and t stands for *twisted*) and Λ, Λ^d two lattices satisfying appropriate compatibility conditions with respect to the (co)root lattice of R_0 (see (9.2.2)). We build from D an irreducible affine root system R and an extended affine Weyl group W. The extended affine Weyl group W is simply the semidirect product group $W_0 \ltimes \Lambda^d$ with W_0 the Weyl group of R_0. The affine root system R will be constructed as follows. We associate to R_0 and $\bullet$ the reduced irreducible affine root system $R^\bullet$ of type $\bullet$ with gradient root system R_0. Then R is an irreducible affine root system obtained from $R^\bullet$ by adding $2a$ if $a \in R^\bullet$ has the property that the pairings of the associated coroot $a^\vee$ to elements of Λ take values in $2\mathbb{Z}$ (see (9.2.5)). The Macdonald–Koornwinder polynomials associated to D are the ones naturally related to R in the labeling proposed in the previous paragraph.

Duality will be related to a simple involution on initial data, $D \mapsto D^d = (R_0^d, \Delta_0^d, \bullet, \Lambda^d, \Lambda)$ with R_0^d the coroot system $R_0^\vee$ if $\bullet = u$ and the root system R_0 if $\bullet = t$. This duality is subtle on the level of affine root systems (it can for instance happen that the affine root system R^d associated to D^d is reduced while R is nonreduced). The reason that the present choice of initial data is convenient is the fact that the duality map $D \mapsto D^d$ on initial data naturally lifts to the duality anti-isomorphism of the associated double affine braid group (see Haiman, 2006 and §9.4). This duality anti-isomorphism turns out to be the key tool to prove duality, evaluation formulas and norm formulas for the Macdonald–Koornwinder polynomials.

In this subsection we carefully introduce the initial data and we explain how it relates to affine root systems. As always, we refer for basic notation and facts on affine root systems to the Appendix.

Definition 9.2.5 The set $\mathcal{D}$ of *initial data* consists of quintuples $D = (R_0, \Delta_0, \bullet, \Lambda, \Lambda^d)$ with

1. R_0 a finite set of nonzero vectors in a Euclidean space Z forming a finite, irreducible, reduced crystallographic root system within the real span V of R_0;
2. $\Delta_0 = (\alpha_1, \ldots, \alpha_n)$ an ordered basis of R_0;
3. $\bullet = u$ or $\bullet = t$;
4. Λ and Λ^d full lattices in Z, satisfying

$$\mathbb{Z}R_0 \subseteq \Lambda, \quad (\Lambda, \mathbb{Z}R_0^\vee) \subseteq \mathbb{Z}, \qquad \mathbb{Z}R_0^d \subseteq \Lambda^d, \quad (\Lambda^d, \mathbb{Z}R_0^{d\vee}) \subseteq \mathbb{Z}, \tag{9.2.2}$$

where $R_0^d = \{\alpha^d := \mu_\alpha^\bullet \alpha^\vee\}_{\alpha \in R_0}$ with $\mu_\alpha^u := 1$ $(\alpha \in R_0)$ and $\mu_\alpha^t := |\alpha|^2/2$ $(\alpha \in R_0)$.

Note that $R_0^d = R_0^\vee$ if $\bullet = u$ and $= R_0$ if $\bullet = t$.

We view the vector space $\widehat{V}$ of real-valued affine linear functions on V as the subspace of $\widehat{Z}$ consisting of affine linear functions on Z which are constant on the orthocomplement $V^\perp$ of V in Z. We write c for the constant function 1 on V as well as on Z. In a similar fashion we view the orthogonal group $\mathrm{O}(V)$ as a subgroup of $\mathrm{O}(Z)$ and $\mathrm{O}_c(\widehat{V})$ as a subgroup of $\mathrm{O}_c(\widehat{Z})$, where $\mathrm{O}_c(\widehat{V})$ is the subgroup of linear automorphisms of $\widehat{V}$ preserving c and preserving the natural semi-positive definite form on $\widehat{V}$ (see the Appendix for further details).

Fix $D = (R_0, \Delta_0, \bullet, \Lambda, \Lambda^d) \in \mathcal{D}$. We associate to D a triple (R, Δ, W) of an affine root system $R = R(D)$, an ordered basis $\Delta = \Delta(D)$ of R and an extended affine Weyl group $W = W(D)$ as follows. We first define a reduced irreducible affine root system $R^\bullet \subset \widehat{V}$ to D as

$$R^\bullet := \{m\mu_\alpha^\bullet c + \alpha\}_{m \in \mathbb{Z},\ \alpha \in R_0} \tag{9.2.3}$$

for $\bullet \in \{u, t\}$. In other words, $R^u := \mathcal{S}(R_0)$ and $R^t := \mathcal{S}(R_0^\vee)^\vee$ in the notation of the Appendix (see §A.2). Let $\varphi \in R_0$ (respectively $\theta \in R_0$) be the highest root (respectively highest short root) of R_0 with respect to the ordered basis Δ_0 of R_0. Then the ordered basis $\Delta = \Delta(D)$ of $R^\bullet$ is set to be $\Delta := (a_0, a_1, \ldots, a_n)$ with $a_i := \alpha_i$ for $1 \leq i \leq n$ and

$$a_0 := \begin{cases} c - \varphi & \text{if } \bullet = u, \\ \frac{1}{2}|\theta|^2 c - \theta & \text{if } \bullet = t. \end{cases} \tag{9.2.4}$$

Remark 9.2.6 Suppose that (R', Δ') is a pair consisting of a reduced irreducible affine root system R' and an ordered basis Δ' of R'. If R' is similar to $R^\bullet$ then there exists a similarity transformation realizing $R' \simeq R^\bullet$ and mapping Δ' to $\Delta^\bullet$ as unordered sets.

The affine root system R is the following extension of $R^\bullet$. Define the subset $S = S(D)$ by

$$S := \{i \in \{0, \ldots, n\} \mid (\Lambda, a_i^\vee) = 2\mathbb{Z}\}. \tag{9.2.5}$$

Let $W^\bullet$ be the affine Weyl group of $R^\bullet$. Then we set

$$R = R(D) := R^\bullet \cup \left(\cup_{i \in S} W^\bullet(2a_i)\right), \tag{9.2.6}$$

which is an irreducible affine root system since $\mathbb{Z}R_0 \subseteq \Lambda$. Note that Δ is also an ordered basis of R.

Remark 9.2.7 Note that R is an irreducible affine root system of untwisted or twisted type, but never of mixed type (see the Appendix for the terminology). But the irreducible affine root systems of mixed type are affine root subsystems of the affine root system of type $C^\vee C$, which is the nonreduced extension of the affine root system R^t with R_0 of type B (see §9.3.8 for a detailed description of the affine root system of type $C^\vee C$). Accordingly, special cases of the Koornwinder polynomials are naturally attached to affine root systems of mixed type; see §9.2.2 and Remark 9.3.29.

Remark 9.2.8 The nonreduced extension of the affine root system R^u with R_0 of type B_2 is not an affine root subsystem of the affine root system of type $C^\vee C_2$. It can actually be better viewed as the rank-two case of the family R^u with R_0 of type C_n since, in the corresponding affine Dynkin diagram (see §A.4), the vertex labeled by the affine simple root a_0 is double bonded with the finite Dynkin diagram of R_0. The nonreduced extension of R^u with R_0 of type C_2 was missing in the classification list in Macdonald (1972a). It was added in Macdonald (2003, (1.3.17)) but the associated theory of Macdonald–Koornwinder polynomials was not developed. In the present setup it is a special case of the general theory. We will describe this particular case in detail in §9.3.9.

Finally we define the extended affine Weyl group $W = W(D)$. Write $s_i := s_{a_i}$ $(0 \leq i \leq n)$ for the simple reflections of $W^\bullet$. Note that $s_i = s_{\alpha_i}$ $(1 \leq i \leq n)$ are the simple reflections of the finite Weyl group W_0 of R_0. Furthermore, $s_0 = \tau(\varphi^\vee)s_\varphi$ if $\bullet = u$ and $s_0 = \tau(\theta)s_\theta$ if $\bullet = t$, where $\tau(v)$ stands for the translation by v (see the Appendix). Consequently, $W^\bullet \simeq W_0 \ltimes \tau(\mathbb{Z}R_0^d)$. We omit τ from the notation if no confusion can arise. The *extended affine Weyl group* $W = W(D)$ is now defined as $W := W_0 \ltimes \Lambda^d$. It contains the affine Weyl group $W^\bullet$ of $R^\bullet$ as a normal subgroup, and $W/W^\bullet \simeq \Lambda^d/\mathbb{Z}R_0^d$.

The affine root system $R^\bullet \subset \widehat{Z}$ is W-stable since

$$\tau(\xi)(m\mu_\beta^\bullet c + \beta) = \big(m - (\xi, \beta^{d\vee})\big)\mu_\beta^\bullet c + \beta, \quad m \in \mathbb{Z},\ \beta \in R_0, \tag{9.2.7}$$

and $(\xi, \beta^{d\vee}) \in \mathbb{Z}$ for $\xi \in \Lambda^d$ and $\beta \in R_0$. Moreover, the affine root system R is W-invariant.

We now proceed by giving key examples of initial data. Recall that, for a finite root system $R_0 \subset V$,

$$P(R_0) := \{\lambda \in V \mid (\lambda, \alpha^\vee) \in \mathbb{Z}\ \ \forall \alpha \in R_0\}$$

is the weight lattice of R_0. If $\Delta_0 = (\alpha_1, \ldots, \alpha_n)$ is an ordered basis of R_0 then we write $\varpi_i \in P(R_0)$ $(1 \leq i \leq n)$ for the corresponding fundamental weights, which are characterized by $(\varpi_i, \alpha_j^\vee) = \delta_{i,j}$.

Example 9.2.9

(i) Take an arbitrary finite reduced irreducible root system R_0 in $V = Z$ with ordered basis Δ_0. Choose $\bullet \in \{u, t\}$ and let Λ, Λ^d be lattices in V satisfying

$$\mathbb{Z}R_0 \subseteq \Lambda \subseteq P(R_0), \quad \mathbb{Z}R_0^d \subseteq \Lambda^d \subseteq P(R_0^d).$$

Then $(R_0, \Delta_0, \bullet, \Lambda, \Lambda^d) \in \mathcal{D}$. Note that if $\Lambda = P(R_0)$ then $S = \emptyset$ and $R = R^\bullet$ is reduced. (Cherednik's 2005 theory corresponds to the special case $(\Lambda, \Lambda^d) = (P(R_0), P(R_0^d))$.)

(ii) Take $Z = \mathbb{R}^{n+1}$ with standard orthonormal basis $\{\epsilon_i\}_{i=1}^{n+1}$ and $R_0 = \{\epsilon_i - \epsilon_j\}_{1\le i\neq j\le n+1}$ for the realization of the finite root system of type A_n in Z. Then $V = (\epsilon_1 + \cdots + \epsilon_{n+1})^\perp$. As the ordered basis take $\Delta_0 = (\alpha_1,\ldots,\alpha_n) = (\epsilon_1-\epsilon_2,\ldots,\epsilon_n-\epsilon_{n+1})$. Then $(R_0,\Delta_0,u,\mathbb{Z}^{n+1},\mathbb{Z}^{n+1}) \in \mathcal{D}$. Note that $\theta = \varphi = \epsilon_1 - \epsilon_{n+1}$, hence the simple affine root a_0 of R is $a_0 = c - \epsilon_1 + \epsilon_{n+1}$. This example is naturally related to the GL_{n+1}-type Macdonald polynomials; see §9.3.7.

Given a quintuple $D = (R_0,\Delta_0,\bullet,\Lambda,\Lambda^d)$ we have the dual root system R_0^d with dual ordered basis $\Delta_0^d := (\alpha_1^d,\ldots,\alpha_n^d)$. This extends to an involution $D \mapsto D^d$ on $\mathcal{D}$ with

$$D^d := (R_0^d,\Delta_0^d,\bullet,\Lambda^d,\Lambda) \tag{9.2.8}$$

for $D = (R_0,\Delta,\bullet,\Lambda,\Lambda^d) \in \mathcal{D}$. We call D^d the *initial data dual to D*.

We write $\mu = \mu(D)$ and $\mu^d = \mu(D^d)$ for the function $\mu^\bullet$ on R_0 and R_0^d, respectively. Let $\varpi_i^d \in P(R_0^d)$ $(1 \le i \le n)$ be the fundamental weights with respect to Δ_0^d.

For a given $D = (R_0,\Delta_0,\bullet,\Lambda,\Lambda^d) \in \mathcal{D}$ we thus have a dual quintuple $(R_0^d,\Delta_0^d,\bullet,\Lambda^d,\Lambda) \in \mathcal{D}$, and hence an associated triple (R^d,Δ^d,W^d). Concretely, the highest root φ^d and the highest short root θ^d of R_0^d with respect to Δ_0^d are given by

$$\varphi^d = \begin{cases} \theta^\vee & \text{if } \bullet = u,\\ \varphi & \text{if } \bullet = t,\end{cases} \quad\text{and}\quad \theta^d = \begin{cases} \varphi^\vee & \text{if } \bullet = u,\\ \theta & \text{if } \bullet = t.\end{cases}$$

Hence $\Delta^d = (a_0^d, a_1^d,\ldots,a_n^d)$ with $a_i^d = \alpha_i^d$ $(1 \le i \le n)$ and $a_0^d = \mu_\theta(c - \theta^\vee)$. The dual affine root system $R^d = R(D^d)$ is $R^d = R^{d\bullet} \cup \bigcup_{i\in S^d} W^{d\bullet}(2a_i^d)$ with $S^d = \{i \in \{0,\ldots,n\} \mid (\Lambda^d, a_i^{d\vee}) = 2\mathbb{Z}\}$, with $R^{d\bullet} = \{m\mu_{\alpha^d} + \alpha^d\}_{m\in\mathbb{Z},\alpha\in R_0}$ and with $W^{d\bullet} \simeq W_0 \ltimes \tau(\mathbb{Z}R_0)$ the affine Weyl group of $R^{d\bullet}$. The dual extended affine Weyl group is $W^d = W_0 \ltimes \Lambda$. The simple reflections $s_i^d := s_{a_i^d} \in W^{d\bullet}$ $(0 \le i \le n)$ are $s_i^d = s_i$ for $1 \le i \le n$ and $s_0^d = \tau(\theta)s_\theta$.

Example 9.2.10 The correspondence $R \leftrightarrow R^d$ can turn nonreduced affine root systems into reduced ones. We give here an example of untwisted type. An example of the twisted type will be given in Example 9.3.28.

Take $n \ge 3$ and $R_0 \subseteq V = Z := \mathbb{R}^n$ of type B_n, realized as $R_0 = \{\pm\epsilon_i\} \cup \{\pm\epsilon_i \pm \epsilon_j\}_{i<j}$ (all sign combinations possible), with $\{\epsilon_i\}$ the standard orthonormal basis of V. As the ordered basis of R_0 take $\Delta_0 = (\alpha_1,\ldots,\alpha_{n-1},\alpha_n) = (\epsilon_1-\epsilon_2,\ldots,\epsilon_{n-1}-\epsilon_n,\epsilon_n)$. The highest root is $\varphi = \epsilon_1+\epsilon_2 \in R_0$. We then have $\mathbb{Z}R_0^\vee \subset P(R_0^\vee) = \mathbb{Z}^n = \mathbb{Z}R_0 \subset P(R_0)$ with both sublattices $\mathbb{Z}R_0^\vee \subset \mathbb{Z}^n$ and $\mathbb{Z}^n \subset P(R_0)$ of index two. For $\Lambda = \mathbb{Z}^n = \Lambda^d$ we get the initial data $D = (R_0,\Delta_0,u,\mathbb{Z}^n,\mathbb{Z}^n) \in \mathcal{D}$. Then $S = S(D) = \{n\}$ and $R = R(D)$ is given by

$$R = \{\pm\epsilon_i + mc\}_{1\le i\le n, m\in\mathbb{Z}} \cup \{\pm\epsilon_i \pm \epsilon_j + mc\}_{1\le i<j\le n, m\in\mathbb{Z}} \cup \{\pm 2\epsilon_i + 2mc\}_{1\le i\le n,\, m\in\mathbb{Z}}.$$

It is nonreduced and of untwisted type B_n. We have written it here as the disjoint union of the three $W = W_0 \ltimes \mathbb{Z}^n$-orbits of R. Note that a_0 lies in the orbit $\{\pm\epsilon_i \pm \epsilon_j + mc\}_{1\le i<j\le n, m\in\mathbb{Z}}$.

Dually, $R^d = R(D^d)$ is the reduced affine root system of untwisted type C_n. Concretely,

$$R^d = \{\pm\epsilon_i \pm \epsilon_j + mc\}_{1\le i<j\le n, m\in\mathbb{Z}} \cup \{\pm 2\epsilon_i + 2mc\}_{1\le i\le n, m\in\mathbb{Z}} \cup \{\pm 2\epsilon_i + (2m+1)c\}_{1\le i\le n, m\in\mathbb{Z}},$$

written here as the disjoint union of the three $W^d = W_0 \ltimes \mathbb{Z}^n$-orbits of R^d.

This example shows that basic features of the affine root system can alter under dualization. It turns out though that the number of orbits with respect to the action of the extended affine Weyl group is unaltered. To establish this fact it is convenient to use the concept of a multiplicity function on R.

Definition 9.2.11 Set $\mathcal{M} = \mathcal{M}(D)$ for the complex algebraic group of W-invariant functions $\kappa\colon R \to \mathbb{C}^*$, and $\nu = \nu(D)$ for the complex dimension of $\mathcal{M}$.

Note that $\nu = \nu(D)$ equals the number of W-orbits of R. The value of $\kappa \in \mathcal{M}$ at an affine root $a \in R$ is denoted by κ_a. We call $\kappa \in \mathcal{M}$ a *multiplicity function*. We write $\kappa^\bullet := \kappa|_{R^\bullet}$ for its restriction to $R^\bullet$. For a multiplicity function $\kappa \in \mathcal{M}$ we set $\kappa_{2a} := \kappa_a$ if $a \in R$ is unmultipliable (i.e., $2a \notin R$).

First we need a more precise description of the sets $S = S(D)$ and $S^d = S(D^d)$. It is obtained using the classification of affine root systems (see §A.2 and Macdonald, 1972a).

Lemma 9.2.12 *Let $D = (R_0, \Delta_0, \bullet, \Lambda, \Lambda^d) \in \mathcal{D}$. Set $S_0 = S \cap \{1, \ldots, n\}$.*

(a) *If $\bullet = u$ then $\#S \leq 2$. If $\#S = 2$ then R_0 is of type A_1. If $\#S = 1$ then R_0 is of type B_n $(n \geq 2)$ and $S = S_0 = \{j\}$ with $\alpha_j \in \Delta_0$ the unique simple short root.*

(b) *If $\bullet = t$ then $\#S = 0$ or $\#S = 2$. If $\#S = 2$ then R_0 is either of type A_1 or of type B_n $(n \geq 2)$ and $S = \{0, j\}$ with $\alpha_j \in \Delta_0$ the unique short simple root.*

Note that in both the untwisted and the twisted case, $\alpha_j^d \in W_0(D(a_0^d))$ if $S_0 = \{j\}$.

The following lemma should be compared with Haiman (2006, §5.7).

Lemma 9.2.13 *Let $D \in \mathcal{D}$ and $\kappa \in \mathcal{M}(D)$. Let $\alpha_j \in \Delta_0$ (respectively $\alpha_{j^d}^d \in \Delta_0^d$) be a simple short root. The assignments*

$$\kappa^d_{a_0^d} := \kappa_{2\alpha_j}, \quad \kappa^d_{\alpha_i^d} := \kappa_{\alpha_i}\ (i \in \{1, \ldots, n\}), \quad \kappa^d_{2a_0^d} := \kappa_{2a_0}\ (0 \in S^d), \quad \kappa^d_{2\alpha^d_{j^d}} := \kappa_{a_0}\ (j^d \in S^d)$$

uniquely extend to a multiplicity function $\kappa^d \in \mathcal{M}^d := \mathcal{M}(D^d)$. For fixed $D \in \mathcal{D}$ the map $\kappa \mapsto \kappa^d$ defines an isomorphism $\phi_D\colon \mathcal{M}(D) \xrightarrow{\sim} \mathcal{M}(D^d)$ of complex tori, with inverse ϕ_{D^d}.

Remark 9.2.14 Let $\kappa \in \mathcal{M}(D)$. Recall the convention that $\kappa_{2a} = \kappa_a$ for $a \in R$ such that $2a \notin R$. Then, for all $\alpha \in R_0$, $\kappa_{2\alpha} = \kappa^d_{\mu_{\alpha^d} c + \alpha^d}$.

Corollary 9.2.15 *Let $D \in \mathcal{D}$. The number ν of W-orbits of R is equal to the number ν^d of W^d-orbits of R^d.*

Remark 9.2.16 Returning to Example 9.2.10, note that the correspondence from Lemma 9.2.13 links the orbit $\{\pm\epsilon_i + mc\}$ of R to the orbit $\{2\epsilon_i + 2mc\}$ of R^d, the orbit $\{\pm\epsilon_i \pm \epsilon_j + mc\}$ of R to the orbit $\{\pm\epsilon_i \pm \epsilon_j + mc\}$ of R^d and the orbit $\{\pm 2\epsilon_i + 2mc\}$ of R to the orbit $\{\pm 2\epsilon_i + (2m+1)c\}$ of R^d.

9.2.4 The Basic Representation

We fix throughout this subsection a quintuple $D = (R_0, \Delta_0, \bullet, \Lambda, \Lambda^d) \in \mathcal{D}$ of initial data. Recall that it gives rise to a triple (R, Δ, W) of an irreducible affine root system R containing $R^\bullet$, an ordered basis Δ of R as well as of $R^\bullet$, and an extended affine Weyl group $W = W_0 \ltimes \Lambda^d$. In addition we fix a multiplicity function $\kappa \in \mathcal{M}(D)$ and we write $\kappa^\bullet := \kappa|_{R^\bullet}$. It is a W-equivariant map $R^\bullet \to \mathbb{C}^*$. Write $\kappa_i := \kappa^\bullet_{a_i}$ for $0 \le i \le n$. Note that $\kappa_i = \kappa_j$ if s_i is conjugate to s_j in $W = \Omega \ltimes W^\bullet$.

We write $R^\pm$ and $R^{\bullet\pm}$ for the positive (respectively negative) affine roots of R and $R^\bullet$ with respect to Δ. Since the affine root system $R^\bullet$ is W-stable, we can define the *length* function by

$$l(w) = l_D(w) := \#(R^{\bullet+} \cap w^{-1}R^{\bullet-}), \qquad w \in W.$$

If $w \in W^\bullet = W(R^\bullet)$ then $l(w)$ equals the number of simple reflections s_i $(0 \le i \le n)$ in a reduced expression of w. We have $W = \Omega \ltimes W^\bullet$ with $\Omega = \Omega(D) := \{w \in W \mid l(w) = 0\}$, a subgroup of W. Then $\Omega \simeq \Lambda^d/\mathbb{Z}R_0^d$. The abelian group Ω permutes the simple affine roots a_i $(0 \le i \le n)$, which thus gives rise to an action of Ω on the index set $\{0, \ldots, n\}$. Consequently the action of Ω on $W^\bullet$ by conjugation permutes the set $\{s_i\}_{i=0}^n$ of simple reflections, $ws_iw^{-1} = s_{w(i)}$ for $w \in \Omega$ and $0 \le i \le n$ (cf., e.g., Macdonald, 2003, §2.5). A detailed description of the group Ω in terms of a complete set of representatives of $\Lambda^d/\mathbb{Z}R_0^d$ will be given in §9.3.4.

Extended versions of the affine braid group and of the affine Hecke algebra are defined as follows. Let $A = A(D) := A(R^\bullet, \Delta)$ be the affine Cartan matrix associated to $(R^\bullet, \Delta)$. Recall that the affine braid group $\mathcal{B}^\bullet := \mathcal{B}(A)$ is isomorphic to the abstract group generated by T_w $(w \in W^\bullet)$ with defining relations $T_vT_w = T_{vw}$ for all $v, w \in W^\bullet$ satisfying $l(vw) = l(v) + l(w)$.

Definition 9.2.17

(i) The *extended affine braid group* $\mathcal{B} = \mathcal{B}(D)$ is the group generated by T_w $(w \in W)$ with defining relations $T_vT_w = T_{vw}$ for all $v, w \in W$ satisfying $l(vw) = l(v) + l(w)$.

(ii) The *extended affine Hecke algebra* $H(\kappa^\bullet) = H(D, \kappa^\bullet)$ is the quotient of $\mathbb{C}[\mathcal{B}]$ by the two-sided ideal generated by $(T_i - \kappa_i)(T_i + \kappa_i^{-1})$ $(0 \le i \le n)$.

Similarly to the semidirect product decomposition $W \simeq \Omega \ltimes W^\bullet$ we have $\mathcal{B} \simeq \Omega \ltimes \mathcal{B}^\bullet$ and $H(\kappa^\bullet) \simeq \Omega \ltimes H(W^\bullet, \kappa^\bullet)$, where the action of Ω on $\mathcal{B}$ by group automorphisms (respectively on $H(W^\bullet, \kappa^\bullet)$ by algebra automorphisms) is determined by $w \cdot T_i = T_{w(i)}$ for $w \in \Omega$ and $0 \le i \le n$. For $\omega \in \Omega$ we will denote the element T_ω in the extended affine Hecke algebra $H(\kappa^\bullet)$ simply by ω.

The algebra homomorphism $\beta_{(R,\Delta,\kappa)}\colon H(W^\bullet, \kappa^\bullet) \hookrightarrow W^\bullet \ltimes F \subseteq W \ltimes F$ from §9.2.2 extends to an injective algebra map $\beta_{D,\kappa} : H(\kappa^\bullet) \hookrightarrow W \ltimes F$ by $\beta_{D,\kappa}(T_\omega) = \omega$ $(\omega \in \Omega)$. We will now show that it gives rise to an action of $H(\kappa^\bullet)$ as q-difference reflection operators on a complex torus T_Λ. It is called the *basic representation* of $H(\kappa^\bullet)$. It is fundamental for the development of the Cherednik–Macdonald theory.

The complex torus $T_\Lambda := \mathrm{Hom}_{\mathbb{Z}}(\Lambda, \mathbb{C}^*)$ (of rank $\dim_{\mathbb{R}}(Z)$) is the algebraic group of complex characters of the lattice Λ. The algebra $\mathbb{C}[T_\Lambda]$ of regular functions on T_Λ is isomorphic to the group algebra $\mathbb{C}[\Lambda]$, where the standard basis element e^λ $(\lambda \in \Lambda)$ of $\mathbb{C}[\Lambda]$ is viewed as the regular function $t \mapsto t(\lambda)$ on T_Λ. We write t^λ for the value of e^λ at $t \in T_\Lambda$. Since Λ is W_0-stable, W_0 acts on T_Λ, in turn giving rise to an action of W_0 on $\mathbb{C}[T_\Lambda]$ by algebra automorphisms. Then $w(e^\lambda) = e^{w\lambda}$ $(w \in W_0$ and $\lambda \in \Lambda)$. We now first extend it to an action of the extended affine Weyl group W on $\mathbb{C}[T_\Lambda]$ depending on a fixed parameter $q \in \mathbb{R}_{>0} \setminus \{1\}$.

For $\alpha \in R_0$ set $q_\alpha := q^{\mu_\alpha}$ and define $q^\xi \in T_\Lambda$ ($\xi \in \Lambda^d$) to be the character $\lambda \mapsto q^{(\lambda,\xi)}$ of Λ. The action of W_0 on T_Λ extends to a left W-action $(w,t) \mapsto w_q t$ on T_Λ by

$$\tau(\xi)_q t := q^\xi t, \quad \xi \in \Lambda^d,\ t \in T_\Lambda.$$

Then $(w_q p)(t) := p(w_q^{-1} t)$ ($w \in W$, $p \in \mathbb{C}[T_\Lambda]$) is a W-action by algebra automorphisms on $\mathbb{C}[T_\Lambda]$. In particular,

$$\tau(\xi)_q(e^\lambda) = q^{-(\lambda,\xi)} e^\lambda, \quad \xi \in \Lambda^d,\ \lambda \in \Lambda.$$

It extends to a W-action by field automorphisms on the quotient field $\mathbb{C}(T_\Lambda)$ of $\mathbb{C}[T_\Lambda]$. It is useful to introduce the notation $t_q^{rc+\lambda} := q^r t^\lambda$ ($t \in T_\Lambda$, $r \in \mathbb{R}$, $\lambda \in \Lambda$). Then $(w_q^{-1} t)_q^{rc+\lambda} = t_q^{w(rc+\lambda)}$ for $w \in W$, $t \in T_\Lambda$, $r \in \mathbb{R}$ and $\lambda \in \Lambda$.

We write $W \ltimes_q \mathbb{C}(T_\Lambda)$ for the resulting semidirect product algebra. It canonically acts on $\mathbb{C}(T_\Lambda)$ by q-difference reflection operators. We thus have a sequence of algebra maps

$$H(\kappa^\bullet) \to W \ltimes F \to W \ltimes_q \mathbb{C}(T_\Lambda) \to \mathrm{End}_{\mathbb{C}}(\mathbb{C}(T_\Lambda)),$$

where the first map is $\beta_{D,\kappa}$ and the second map sends $e^{\mu_\alpha mc+\alpha}$ to $q_\alpha^m e^\alpha$ for $m\mu_\alpha c + \alpha \in R^\bullet$. It gives the following result, which is closely related to Haiman (2006, 5.13) in the present generality.

Theorem 9.2.18 *Let $D = (R_0, \Delta_0, \bullet, \Lambda, \Lambda^d) \in \mathcal{D}$ and $\kappa \in \mathcal{M}(D)$. For $a \in R$ we set $\kappa_{2a} := \kappa_a$ if $2a \notin R$. There exists a unique algebra monomorphism*

$$\pi_{\kappa,q} = \pi_{D;\kappa,q} \colon H(D,\kappa^\bullet) \hookrightarrow \mathrm{End}_{\mathbb{C}}(\mathbb{C}[T_\Lambda])$$

satisfying, for $p \in \mathbb{C}[T_\Lambda]$ and $t \in T_\Lambda$,

$$(\pi_{\kappa,q}(T_i)p)(t) = \kappa_{a_i}(s_{i,q}p)(t) + \frac{\kappa_{a_i} - \kappa_{a_i}^{-1} + (\kappa_{2a_i} - \kappa_{2a_i}^{-1})t_q^{a_i}}{1 - t_q^{2a_i}}(p(t) - (s_{i,q}p)(t)), \quad 0 \le i \le n,$$

$$(\pi_{\kappa,q}(\omega)p)(t) = (\omega_q p)(t), \quad \omega \in \Omega.$$

If $2a_i \notin R$ then, by the convention $\kappa_{2a_i} = \kappa_{a_i}$, the first formula reduces to

$$(\pi_{\kappa,q}(T_i)p)(t) = \kappa_{a_i}(s_{i,q}p)(t) + \frac{\kappa_{a_i} - \kappa_{a_i}^{-1}}{1 - t_q^{a_i}}(p(t) - (s_{i,q}p)(t)).$$

The theorem is due to Cherednik (see Cherednik, 2005 and references therein) in the GL_{n+1} case (see Example 9.2.9(ii)) and when $D = (R_0, \Delta_0, \bullet, P(R_0), P(R_0^d))$ with R_0 an arbitrary reduced irreducible root system. The theorem is due to Noumi (1995) for $D = (R_0, \Delta_0, t, \mathbb{Z}R_0, \mathbb{Z}R_0)$ with R_0 of type A_1 or of type B_n ($n \ge 2$). This case is special due to its large degree of freedom ($\nu(D) = 4$ if $n = 1$ and $\nu(D) = 5$ if $n \ge 2$). We will describe this case in detail in §9.3.8.

Remark 9.2.19

(i) From Theorem 9.2.3 one first obtains $\pi_{\kappa,q}$ as an algebra map from $H(\kappa^\bullet)$ to $\mathrm{End}_{\mathbb{C}}(\mathbb{C}(T_\Lambda))$. The image is contained in the subalgebra of endomorphisms preserving $\mathbb{C}[T_\Lambda]$ since, for $\lambda \in \Lambda$, we have $(\lambda, a_i^\vee) \in \mathbb{Z}$ if $2a_i \notin R$ and $(\lambda, a_i^\vee) \in 2\mathbb{Z}$ if $2a_i \in R$.

(ii) Let s be the number of W-orbits of $R \setminus R^{\bullet}$. Extending a W-equivariant map $\kappa^{\bullet}: R^{\bullet} \to \mathbb{C}^*$ to a multiplicity function $\kappa \in \mathcal{M}$ on R amounts to choosing s nonzero complex parameters. Hence, the maps $\pi_{\kappa,q}$ define a family of algebra monomorphisms of the extended affine Hecke algebra $H(\kappa^{\bullet})$ into $\mathrm{End}_{\mathbb{C}}(\mathbb{C}[T_\Lambda])$, parametrized by $s+1$ parameters $\kappa|_{R\setminus R^{\bullet}}$ and q.

9.3 Monic Macdonald–Koornwinder Polynomials

In this section we introduce the monic nonsymmetric and symmetric Macdonald–Koornwinder polynomials. The terminology *Macdonald polynomials* is employed in the literature for the cases that R is reduced (i.e., the cases $D = (R_0, \Delta_0, \bullet, P(R_0), P(R_0^d))$ and the GL_{n+1} case). The *Koornwinder polynomials* correspond to the initial data $D = (R_0, \Delta_0, t, \mathbb{Z}R_0, \mathbb{Z}R_0)$ with R_0 of type A_1 or of type B_n ($n \geq 2$), in which case $R = R(D)$ is nonreduced and of type $\mathrm{C}^\vee\mathrm{C}_n$. To have uniform terminology we will speak of *Macdonald–Koornwinder polynomials* when discussing the theory for arbitrary initial data.

The monic nonsymmetric Macdonald–Koornwinder polynomials will be introduced as the common eigenfunctions in $\mathbb{C}[T_\Lambda]$ of a family of commuting q-difference reflection operators. The operators are obtained as images under the basic representation $\pi_{\kappa,q}$ of elements from a large commutative subalgebra of the extended affine Hecke algebra $H(\kappa^{\bullet})$. A Hecke algebra symmetrizer turns the monic nonsymmetric Macdonald–Koornwinder polynomials into monic symmetric Macdonald–Koornwinder polynomials, which are W_0-invariant regular functions on T_Λ solving a suitable spectral problem of a commuting family of q-difference operators, called Macdonald operators. In addition, we determine in this section the (bi)orthogonality relations of the polynomials.

Throughout this section we fix

1. a quintuple $D = (R_0, \Delta_0, \bullet, \Lambda, \Lambda^d)$ of initial data,
2. a deformation parameter $q \in \mathbb{R}_{>0} \setminus \{1\}$,
3. a multiplicity function $\kappa \in \mathcal{M}(D)$,

and we freely use the associated notation from §9.2.

9.3.1 Bernstein–Zelevinsky Presentation

For a given expression $w = \omega s_{i_1} s_{i_2} \cdots s_{i_r} \in W$ ($\omega \in \Omega$, $0 \leq i_j \leq n$) which is *reduced*, i.e., $r = l(w)$, put $T_w := \omega T_{i_1} T_{i_2} \cdots T_{i_r} \in H(\kappa^{\bullet})$. This is well defined, and $\{T_w\}_{w \in W}$ is a complex linear basis of $H(\kappa^{\bullet})$.

The cones $\Lambda^{d\pm} := \{\xi \in \Lambda^d \mid (\xi, \alpha^{d\vee}) \geq 0 \; \forall \alpha \in R_0^{\pm}\}$ form fundamental domains for the W_0-action on Λ^d. Any $\xi \in \Lambda^d$ can be written as $\xi = \mu - \nu$ with $\mu, \nu \in \Lambda^{d+}$ (and similarly for Λ^{d-}). Furthermore, if $\xi, \xi' \in \Lambda^{d+}$ then $l(\tau(\xi + \xi')) = l(\tau(\xi)) + l(\tau(\xi'))$. It follows that there exists a unique group homomorphism $\xi \mapsto Y^{\xi}: \Lambda^d \to \mathcal{B}$ such that $Y^{\xi} = T_{\tau(\xi)}$ ($\xi \in \Lambda^{d+}$). On the level of the extended affine Hecke algebra it gives rise to an algebra homomorphism

$$p \mapsto p(Y)\colon \mathbb{C}[T_{\Lambda^d}] \to H(\kappa^\bullet), \tag{9.3.1}$$

where $p(Y) = \sum_\xi c_\xi Y^\xi$ if $p(t) = \sum_\xi c_\xi t^\xi$. The image of the map (9.3.1) is denoted by $\mathbb{C}_Y[T_{\Lambda^d}]$.

As in §9.2.4, the lattice Λ^d is W_0-stable, giving rise to a W_0-action on T_{Λ^d}, and hence a W_0-action on $\mathbb{C}[T_{\Lambda^d}]$ by algebra automorphisms.

Denote by H_0 the subalgebra of $H(\kappa^\bullet)$ generated by T_i $(1 \le i \le n)$. We have a natural surjective algebra map $H(W_0, \kappa|_{R_0}) \to H_0$, sending the algebraic generator T_i of $H(W_0, \kappa|_{R_0})$ to $T_i \in H_0$.

The analog of the semidirect product decomposition $W = W_0 \ltimes \Lambda^d$ for the extended affine Hecke algebra $H(\kappa^\bullet)$ is the following *Bernstein–Zelevinsky presentation* of $H(\kappa^\bullet)$ (see Lusztig, 1989).

Theorem 9.3.1

1. *The algebra maps* $\mathbb{C}[T_{\Lambda^d}] \to \mathbb{C}_Y[T_{\Lambda^d}]$ *and* $H(W_0, \kappa|_{R_0}) \to H_0$ *are isomorphisms.*
2. *Multiplication defines a linear isomorphism* $H_0 \otimes \mathbb{C}_Y[T_{\Lambda^d}] \simeq H(\kappa^\bullet)$.
3. *For* $i \in \{1, \ldots, n\}$ *such that* $(\Lambda^d, \alpha_i^{d\vee}) = \mathbb{Z}$, *we have in* $H(\kappa^\bullet)$ *that*

$$p(Y)T_i - T_i(s_ip)(Y) = \left(\frac{\kappa_i - \kappa_i^{-1}}{1 - Y^{-\alpha_i^d}}\right)(p(Y) - (s_ip)(Y)), \quad p \in \mathbb{C}[T_{\Lambda^d}]. \tag{9.3.2}$$

4. *For* $i \in \{1, \ldots, n\}$ *such that* $(\Lambda^d, \alpha_i^{d\vee}) = 2\mathbb{Z}$, *we have in* $H(\kappa^\bullet)$ *that*

$$p(Y)T_i - T_i(s_ip)(Y) = \left(\frac{\kappa_i - \kappa_i^{-1} + (\kappa_0 - \kappa_0^{-1})Y^{-\alpha_i^d}}{1 - Y^{-2\alpha_i^d}}\right)(p(Y) - (s_ip)(Y)), \quad p \in \mathbb{C}[T_{\Lambda^d}]. \tag{9.3.3}$$

These properties characterize $H(\kappa^\bullet)$ *as a unital complex associative algebra.*

With the notion of the dual multiplicity parameter κ^d (see Lemma 9.2.13), the *cross relations* (9.3.2) and (9.3.3) in the affine Hecke algebra $H(\kappa^\bullet)$ can be uniformly written as

$$p(Y)T_i - T_i(s_ip)(Y) = \left(\frac{\kappa^d_{\alpha_i^d} - (\kappa^d_{\alpha_i^d})^{-1} + (\kappa^d_{2\alpha_i^d} - (\kappa^d_{2\alpha_i^d})^{-1})Y^{-\alpha_i^d}}{1 - Y^{-2\alpha_i^d}}\right)(p(Y) - (s_ip)(Y)) \tag{9.3.4}$$

for $p \in \mathbb{C}[T_{\Lambda^d}]$ and $1 \le i \le n$. It follows from the theorem that the center $Z(H(\kappa^\bullet))$ of the extended affine Hecke algebra $H(\kappa^\bullet)$ equals $\mathbb{C}_Y[T_{\Lambda^d}]^{W_0}$.

9.3.2 Monic Nonsymmetric Macdonald–Koornwinder Polynomials

The results in this subsection are from Macdonald (1996), Cherednik (1995d), Sahi (1999) and Haiman (2006). For detailed proofs see, e.g., Macdonald (2003, §§2.8, 4.6, 5.2). We put the following conditions on q and $\kappa \in \mathcal{M}(D)$:

$$0 < q < 1 \text{ and } 0 < \kappa_a < 1 \ (a \in R) \quad \text{or} \quad q > 1 \text{ and } \kappa_a > 1 \ (a \in R). \tag{9.3.5}$$

Set $\eta(x) := 1$ for $x > 0$ and $\eta(x) := -1$ for $x \leq 0$. Define a W_0-equivariant map $\upsilon\colon R_0^d \to \mathbb{R}_{>0}$ (depending on $\kappa^\bullet$) by $\upsilon_{\alpha^d} := \kappa_\alpha^{1/2}\kappa_{\mu_\alpha c+\alpha}^{1/2}$ ($\alpha \in R_0$).

Definition 9.3.2 For $\lambda \in \Lambda$ define $\gamma_{\lambda,q} = \gamma_{\lambda,q}(D;\kappa^\bullet) \in T_{\Lambda^d}$ by

$$\gamma_{\lambda,q} := q^\lambda \prod_{\alpha\in R_0^+} \upsilon_{\alpha^d}^{\eta((\lambda,\alpha^\vee))\alpha^{d\vee}}.$$

In other words, $\gamma_{\lambda,q}^\xi = q^{(\lambda,\xi)} \prod_{\alpha\in R_0^+} \upsilon_{\alpha^d}^{\eta((\lambda,\alpha^\vee))(\xi,\alpha^{d\vee})}$ for all $\xi \in \Lambda^d$.

As a special case we have $\gamma_{\lambda,q} = q^\lambda \prod_{\alpha\in R_0^+} \upsilon_{\alpha^d}^{-\alpha^{d\vee}}$ ($\lambda \in \Lambda^-$), where we use the notation $\Lambda^\pm := \{\lambda \in \Lambda \mid (\lambda,\alpha^\vee) \geq 0 \;\forall\, \alpha \in R_0^\pm\}$.

Write $l^d = l_{D^d}$ for the length function on the dual extended affine Weyl group W^d and $\Omega^d = \Omega(D^d)$ for the subgroup of elements of W^d of length zero with respect to l^d. We have a q-dependent W^d-action on T_{Λ^d} extending the W_0-action by $\tau(\lambda)_q\gamma = q^\lambda\gamma$ for all $\lambda \in \Lambda$ and $\gamma \in T_{\Lambda^d}$. Then $\gamma_{\lambda,q} = \tau(\lambda)_q\gamma_{0,q}$ in T_{Λ^d} if $\lambda \in \Lambda^-$. This generalizes as follows.

Lemma 9.3.3 *We have in T_{Λ^d} that $\gamma_{\lambda,q} = u^d(\lambda)_q\gamma_{0,q}$ ($\lambda \in \Lambda$), where $u^d(\lambda) \in W^d$ is the element of minimal length with respect to l^d in the coset $\tau(\lambda)W_0$.*

The condition (9.3.5) on the parameters, together with Lemma 9.3.3, implies the following lemma.

Lemma 9.3.4 *The map $\lambda \mapsto \gamma_{\lambda,q}\colon \Lambda \to T_{\Lambda^d}$ is injective.*

For later purposes it is convenient to record the following compatibility between the q-dependent W^d-action on $\gamma_{\lambda,q} \in T_{\Lambda^d}$ and the W^d-action $(w\tau(\lambda),\lambda') \mapsto w(\lambda+\lambda')$ on Λ ($w \in W_0$, $\lambda,\lambda' \in \Lambda$).

Proposition 9.3.5 *Let $\lambda \in \Lambda$. Then*

(a) *if $\omega \in \Omega^d$ then $\omega_q\gamma_{\lambda,q} = \gamma_{\omega\lambda,q}$,*
(b) *if $0 \leq i \leq n$ and $s_i^d\lambda \neq \lambda$ then $s_{i,q}^d\gamma_{\lambda,q} = \gamma_{s_i^d\lambda,q}$,*
(c) *if $0 \leq i \leq n$ and $s_i^d\lambda = \lambda$ then $s_{i,q}^d\gamma_{\lambda,q} = \gamma_{\lambda,q}\upsilon_{D(a_i^d)}^{2D(a_i^d)^\vee}$.*

Warning $D(a_i^d) = (Da_i)^d$ holds true for $1 \leq i \leq n$, and for $i = 0$ if $\bullet = t$ (then both sides equal $-\theta$). It is not correct when $i = 0$, $\bullet = u$ and R_0 has two root lengths, since then $(Da_0)^d = -\varphi^\vee$ and $D(a_0^d) = -\theta^\vee$.

For $\lambda,\mu \in \Lambda^+$ we write $\lambda \leq \mu$ if $\mu - \lambda$ can be written as a sum of positive roots $\alpha \in R_0^+$. We also write $\leq$ for the *Bruhat order* of W_0 with respect to the Coxeter generators s_i ($1 \leq i \leq n$); see Humphreys (1990, §5.9). For $\lambda \in \Lambda$ let $\lambda_\pm$ be the unique element in $\Lambda^\pm \cap W_0\lambda$ and write $v(\lambda) \in W_0$ for the element of shortest length such that $v(\lambda)\lambda = \lambda_-$. Then $\tau(\lambda) = u^d(\lambda)v(\lambda)$ in W^d for $\lambda \in \Lambda$.

Definition 9.3.6 Let $\lambda,\mu \in \Lambda$. We write $\lambda \preceq \mu$ if $\lambda_+ < \mu_+$ or if $\lambda_+ = \mu_+$ and $v(\lambda) \geq v(\mu)$.

Note that $\leq$ is a partial order on Λ. Furthermore, if $\lambda \in \Lambda^-$ then $\mu \leq \lambda$ for all $\mu \in W_0\lambda$. For each $\lambda \in \Lambda$ the set of elements $\mu \in \Lambda$ satisfying $\mu \leq \lambda$ thus is contained in the finite set

$$\{\mu \in \Lambda \mid \mu \leq \lambda_-\} = \bigcup_{\mu_+ \in \Lambda^+ : \mu_+ \leq \lambda_+} W_0 \mu_+,$$

which is the smallest saturated subset $\mathrm{Sat}(\lambda_+)$ of Λ containing λ_+ (a subset $X \subseteq \Lambda$ is *saturated* if for each $\alpha \in R_0^+$ and $\lambda \in \Lambda$ we have $\lambda - r\alpha \in X$ for all integers r between zero and $(\lambda, \alpha^\vee)$, including both zero and $(\lambda, \alpha^\vee)$).

We write $p = d_\lambda e^\lambda$ + l.o.t. for an element $p = \sum_{\mu \in \Lambda} d_\mu e^\mu \in \mathbb{C}[T_\Lambda]$ satisfying $d_\mu = 0$ if $\mu \not\leq \lambda$. If in addition $d_\lambda \neq 0$ then we say that p is *of degree* λ.

Proposition 9.3.7 *In* $\mathbb{C}[T_\Lambda]$ *we have* $\pi_{\kappa,q}(r(Y))e^\lambda = r(\gamma_{\lambda,q}^{-1})e^\lambda$ + l.o.t. ($r \in \mathbb{C}[T_{\Lambda^d}]$, $\lambda \in \Lambda$).

Corollary 9.3.8 *For each* $\lambda \in \Lambda$ *there exists a unique* $P_\lambda = P_\lambda(D; \kappa, q) \in \mathbb{C}[T_\Lambda]$ *satisfying* $\pi_{\kappa,q}(r(Y))P_\lambda = r(\gamma_{\lambda,q}^{-1})P_\lambda$ ($r \in \mathbb{C}[T_{\Lambda^d}]$) *and* $P_\lambda = e^\lambda$ + l.o.t.

Definition 9.3.9 $P_\lambda = P_\lambda(D; \kappa, q) \in \mathbb{C}[T_\Lambda]$ is called the monic *nonsymmetric Macdonald–Koornwinder polynomial* of degree $\lambda \in \Lambda$.

For $D = (R_0, \Delta_0, \bullet, P(R_0), P(R_0^d))$ the definition of the nonsymmetric Macdonald–Koornwinder polynomial is due to Macdonald (1996) in the untwisted case ($\bullet = u$) and to Cherednik (1995d) in the general case. For $D = (R_0, \Delta_0, t, \mathbb{Z}R_0, \mathbb{Z}R_0)$ with R_0 of type A_1 or of type B_n ($n \geq 2$) the nonsymmetric Macdonald–Koornwinder polynomials are the nonsymmetric Koornwinder polynomials defined by Sahi (1999). In the present generality (with more flexible choices of lattices Λ and Λ^d) the definition is close to the definition in Haiman (2006, §6). The same references apply for the biorthogonality relations of the nonsymmetric Macdonald–Koornwinder polynomials discussed in the next subsection.

The GL_{n+1} nonsymmetric Macdonald polynomials (corresponding to Example 9.2.9(ii)) are often studied separately; see, e.g., Knop (1997) and Haglund et al. (2008).

9.3.3 Biorthogonality

We assume in this subsection that $\kappa \in \mathcal{M}$ and q satisfy

$$0 < q < 1, \quad 0 < \kappa_a < 1 \ \forall\, a \in R, \quad 0 < \kappa_a \kappa_{2a}^{\pm 1} \leq 1 \ \forall\, a \in R^\bullet, \tag{9.3.6}$$

and for $\lambda \in \Lambda$ we write $P_\lambda := P_\lambda(D, \kappa, q)$, $P_\lambda^\circ := P_\lambda(D, \kappa^{-1}, q^{-1})$, where $\kappa^{-1} \in \mathcal{M}(D)$ is the multiplicity function $a \mapsto \kappa_a^{-1}$. Define

$$c_a(t) = c_a^{\kappa,q}(t; D) := \frac{(1 - \kappa_a \kappa_{2a} t_q^a)(1 + \kappa_a \kappa_{2a}^{-1} t_q^a)}{1 - t_q^{2a}} \in \mathbb{C}(T_\Lambda), \quad a \in R^\bullet. \tag{9.3.7}$$

Then $c_a(w_q^{-1}t) = c_{wa}(t)$ ($w \in W$, $a \in R^\bullet$). In addition, $\pi_{\kappa,q}(T_i) = \kappa_i + \kappa_i^{-1} c_{a_i}(s_{i,q} - 1)$ ($0 \leq i \leq n$).

Since $0 < q < 1$, the infinite product $v := \prod_{a \in R^{\bullet+}} c_a^{-1}$ defines a meromorphic function

$$\begin{aligned} v(t) = &\prod_{\alpha \in R_0^+} \frac{1 - t^{2\alpha}}{(1 - \kappa_\alpha \kappa_{2\alpha} t^\alpha)(1 + \kappa_\alpha \kappa_{2\alpha}^{-1} t^\alpha)} \\ &\times \prod_{\beta \in R_0} \frac{(q_\beta^2 t^{2\beta}; q_\beta^2)_\infty}{(q_\beta^2 \kappa_\beta \kappa_{2\beta} t^\beta, -q_\beta^2 \kappa_\beta \kappa_{2\beta}^{-1} t^\beta, q_\beta \kappa_{\mu_\beta c + \beta} \kappa_{2\mu_\beta c + 2\beta} t^\beta, -q_\beta \kappa_{\mu_\beta c + \beta} \kappa_{2\mu_\beta c + 2\beta}^{-1} t^\beta; q_\beta^2)_\infty} \end{aligned} \tag{9.3.8}$$

on T_Λ. Here we used *q-shifted factorials*

$$(x_1, \ldots, x_m; q)_r := (x_1; q)_r \cdots (x_m; q)_r, \quad (x; q)_r := \prod_{j=0}^{r-1} (1 - q^j x), \ r \in \mathbb{Z}_{\geq 0} \cup \{\infty\}. \tag{9.3.9}$$

Remark 9.3.10 If $\beta \in R_0$ satisfies $(\Lambda^d, \beta^{d\vee}) = \mathbb{Z}$ then $\kappa_{\mu_\beta c + \beta} = \kappa_\beta$ and $\kappa_{2\mu_\beta c + 2\beta} = \kappa_{2\beta}$, and the β-factor in the second line of (9.3.8) simplifies to $\frac{(q_\beta^2 t^{2\beta}; q_\beta^2)_\infty}{(q_\beta \kappa_\beta \kappa_{2\beta} t^\beta, -q_\beta \kappa_\beta \kappa_{2\beta}^{-1} t^\beta; q_\beta)_\infty}$. If in addition $\kappa_{2\beta} = \kappa_\beta$ (for instance, if $2\beta \notin R$) then the β-factor simplifies further to $\frac{(q_\beta t^\beta; q_\beta)_\infty}{(q_\beta \kappa_\beta^2 t^\beta; q_\beta)_\infty}$.

By the conditions (9.3.6) on the parameters, v is a continuous function on the compact torus $T_\Lambda^u = \mathrm{Hom}(\Lambda, S^1) \subset T_\Lambda$, where $S^1 = \{z \in \mathbb{C} \mid |z| = 1\}$. Write $d_u t$ for the normalized Haar measure on T_Λ^u.

Definition 9.3.11 Define a sesquilinear form $\langle \cdot, \cdot \rangle : \mathbb{C}[T_\Lambda] \times \mathbb{C}[T_\Lambda] \to \mathbb{C}$ by

$$\langle p, r \rangle := \int_{T_\Lambda^u} p(t) \overline{r(t)} \, v(t) \, d_u t.$$

Proposition 9.3.12 *Let $p, r \in \mathbb{C}[T_\Lambda]$ and $w \in W$. Then $\langle \pi_{\kappa,q}(T_w) p, r \rangle = \langle p, \pi_{\kappa^{-1}, q^{-1}}(T_w^{-1}) r \rangle$.*

The biorthogonality of the nonsymmetric Macdonald–Koornwinder polynomials readily follows from Proposition 9.3.12.

Theorem 9.3.13 *If $\lambda, \mu \in \Lambda$ and $\lambda \neq \mu$ then $\langle P_\lambda, P_\mu^\circ \rangle = 0$.*

9.3.4 Macdonald Operators

Special cases of what now are known as the Macdonald q-difference operators were explicitly written down by Macdonald (2000) when he introduced the symmetric Macdonald polynomials. Earlier, Ruijsenaars (1987) had introduced these commuting q-difference operators for R_0 of type A as the quantum Hamiltonian of a relativistic version of the quantum trigonometric Calogero–Moser system (see §9.3.7). Koornwinder (1992) introduced a multivariable extension of the second-order Askey–Wilson q-difference operator to define the symmetric Koornwinder polynomials. This case corresponds to $D = (R_0, \Delta_0, t, \mathbb{Z}R_0, \mathbb{Z}R_0)$ with R_0 of type A_1 or of type B_n with $n \geq 2$ (see §9.3.8).

The construction of the whole family of Macdonald q-difference operators using affine Hecke algebras is due to Cherednik (1995b) in the case $D = (R_0, \Delta_0, \bullet, P(R_0), P(R_0^d))$ and due to Noumi (1995) in the case $D = (R_0, \Delta_0, t, \mathbb{Z}R_0, \mathbb{Z}R_0)$ with R_0 of type A_1 or of type B_n

($n \geq 2$). We explain this construction here; see Macdonald (2003, §4.4) for a treatment close to the present one.

In this subsection we assume that $\kappa \in \mathcal{M}$ and q satisfy (9.3.5). Consider the linear map

$$\mathrm{Res}_q\colon \sum_{w \in W_0} D_w w \mapsto \sum_{w \in W_0} D_w\colon \quad W \ltimes_q \mathbb{C}(T_\Lambda) \to \tau(\Lambda^d) \ltimes_q \mathbb{C}(T_\Lambda)$$

where $D_w \in \tau(\Lambda^d) \ltimes_q \mathbb{C}(T_\Lambda)$ ($w \in W_0$). Note that

$$L(p) = (\mathrm{Res}_q(L))(p), \quad p \in \mathbb{C}(T_\Lambda)^{W_0}. \tag{9.3.10}$$

Lemma 9.3.14 *Let $\beta_{\kappa,q}(H_0)'$ be the commutant of the subalgebra $\beta_{\kappa,q}(H_0)$ in $W \ltimes_q \mathbb{C}(T_\Lambda)$. Then Res_q restricts to an algebra homomorphism $\mathrm{Res}_q\colon \beta_{\kappa,q}(H_0)' \to (\tau(\Lambda^d) \ltimes_q \mathbb{C}(T_\Lambda))^{W_0}$, where $(\tau(\Lambda^d) \ltimes_q \mathbb{C}(T_\Lambda))^{W_0}$ is the subalgebra of $W \ltimes_q \mathbb{C}(T_\Lambda)$ consisting of W_0-invariant q-difference operators.*

As $Z(H(\kappa^\bullet)) = \mathbb{C}_Y[T_{\Lambda^d}]^{W_0}$, the lemma implies that the W_0-invariant q-difference operators

$$D_p := \mathrm{Res}_q(\beta_{\kappa,q}(p(Y))) \in (\tau(\Lambda^d) \ltimes_q \mathbb{C}(T_\Lambda))^{W_0}, \quad p \in \mathbb{C}[T_{\Lambda^d}]^{W_0},$$

pairwise commute. The operator D_p is called the *Macdonald q-difference operator* associated to $p \in \mathbb{C}[T_{\Lambda^d}]^{W_0}$.

Define the orbit sums $m_\xi^d(t) := \sum_{\eta \in W_0 \xi} t^\eta \in \mathbb{C}[T_{\Lambda^d}]^{W_0}$ ($\xi \in \Lambda^d$). Then $\{m_\xi^d\}_{\xi \in \Lambda^{d-}}$ is a linear basis of $\mathbb{C}[T_{\Lambda^d}]^{W_0}$. We write D_ξ for $D_{m_\xi^d}$. Set, for $w \in W$,

$$c_w := \prod_{a \in R^{\bullet +} \cap w^{-1} R^{\bullet -}} c_a \in \mathbb{C}(T_\Lambda) \quad \text{and} \quad \kappa_w := \prod_{a \in R^{\bullet +} \cap w^{-1} R^{\bullet -}} \kappa_a.$$

Write $W_{0,\xi}$ for the stabilizer subgroup of ξ in W_0.

Remark 9.3.15 If $\xi \in \Lambda^{d-}$ and $w \in W_{0,\xi}$ then $w(c_{\tau(-\xi)}) = c_{\tau(-\xi)}$ in $\mathbb{C}(T_\Lambda)$.

Proposition 9.3.16 *Let $\xi \in \Lambda^{d-}$. Then*

$$D_\xi = \kappa_{\tau(-\xi)}^{-1} \sum_{w \in W_0/W_{0,\xi}} w(c_{\tau(-\xi)}) \tau(w\xi)_q + \sum_{\eta \in \mathrm{Sat}(\xi_+) \setminus W_0 \xi} g_\eta \tau(\eta)_q$$

for certain $g_\eta \in \mathbb{C}(T_\Lambda)$ satisfying $g_{w\eta} = w(g_\eta)$ for all $w \in W_0$ and $\eta \in \Lambda^d$.

The set of *dominant minuscule weights* in Λ^d is defined by

$$\Lambda^{d+}_{\min} := \{\xi \in \Lambda^d \mid (\xi, \alpha^{d\vee}) \in \{0,1\}\ \forall\, \alpha \in R_0^+\}.$$

Set $\Lambda_0^d := \Lambda^d \cap V^\perp$. We now first give an explicit description of the dominant minuscule weights.

Recall that $Da_0 = -\varphi$ if $\bullet = u$ and $Da_0 = -\theta$ if $\bullet = t$. Hence $-(Da_0)^{d\vee}$ is the highest root of $R_0^{d\vee}$. Consider the expansion $-(Da_0)^{d\vee} = \sum_{i=1}^n m_i \alpha_i^{d\vee}$ of $-(Da_0)^{d\vee} \in R_0^{d\vee}$ with respect to the ordered basis $\Delta_0^{d\vee}$ of $R_0^{d\vee}$. Then $m_i \in \mathbb{Z}_{\geq 1}$ for all i. Set

$$J^+_{\Lambda^d} := \{i \in \{1,\ldots,n\} \mid m_i = 1,\ (\varpi_i^d + V^\perp) \cap \Lambda^d \neq \emptyset\},$$

where $V^\perp$ is the orthocomplement of V in Z.

Proposition 9.3.17

(i) $\Lambda^{d+}_{\min}$ *is a complete set of representatives of* $\Lambda^d/\mathbb{Z}R^d_0$.

(ii) *For* $j\in J^+_{\Lambda^d}$ *choose an element* $\tilde{\varpi}^d_j\in(\varpi^d_j+V^\perp)\cap\Lambda^d$. *Then*

$$\Lambda^{d+}_{\min}=\Lambda^d_0\cup\bigcup_{j\in J^+_{\Lambda^d}}(\tilde{\varpi}^d_j+\Lambda^d_0)\quad\textit{(disjoint union)}.$$

(iii) *For* $\eta\in\Lambda^d$ *let* $u(\eta)\in W$ *be the unique element of minimal length (with respect to* l*) in the coset* $\tau(\eta)W_0$. *Then* $\Omega=\{u(\xi)\mid\xi\in\Lambda^{d+}_{\min}\}$.

Since $-(Da_0)^{d\vee}\in R^{d\vee+}_0$ is the highest root, $-(Da_0)^d\in R^{d+}_0$ is *quasi-minuscule*, meaning that $(-(Da_0)^d,\alpha^{d\vee})\in\{0,1\}$ for all $\alpha^d\in R^{d+}_0\setminus\{-(Da_0)^d\}$.

Corollary 9.3.18 *Let* $w_0\in W_0$ *be the longest Weyl group element.*

(i) *For* $j\in J^+_{\Lambda^d}$ *we have* $D_{w_0\tilde{\varpi}^d_j}=\kappa^{-1}_{\tau(-w_0\tilde{\varpi}^d_j)}\sum_{w\in W_0/W_{0,w_0\tilde{\varpi}^d_j}}w(c_{\tau(-w_0\tilde{\varpi}^d_j)})\tau(ww_0\tilde{\varpi}^d_j)_q$.

(ii) $D_{(Da_0)^d}=\kappa^{-1}_{\tau(-(Da_0)^d)}\sum_{w\in W_0/W_{0,(Da_0)^d}}w(c_{\tau(-(Da_0)^d)})(\tau(w(Da_0)^d)_q-1)+m^d_{(Da_0)^d}(\gamma^{-1}_{0,q})$.

9.3.5 Monic Symmetric Macdonald–Koornwinder Polynomials

In §9.3.2 we introduced the nonsymmetric Macdonald–Koornwinder polynomials, but historically the symmetric Macdonald–Koornwinder polynomials were defined first. The monic symmetric Macdonald polynomials associated to initial data given by quintuples of the form $D=(R_0,\Delta_0,\bullet,P(R_0),P(R^d_0))$ were defined by Macdonald (2000) using the fact that the explicit Macdonald q-difference operator D_ξ for $\xi\in\{w_0\tilde{\varpi}^d_j\}_{j\in J^+_{\Lambda^d}}\cup\{(Da_0)^d\}$ (see Corollary 9.3.18) is a linear operator on $\mathbb{C}[T_\Lambda]^{W_0}$ which is triangular with respect to the suitable partially ordered basis of orbit sums and which has (generically) simple spectrum.

This approach was extended by Koornwinder (1992) to the case corresponding to the initial data $D=(R_0,\Delta,t,\mathbb{Z}R_0,\mathbb{Z}R_0)$ with R_0 of type A_1 or of type B_n $(n\geq 2)$, in which case $D_{-\theta}$ is Koornwinder's multivariable extension of the Askey–Wilson second-order q-difference operator (see §9.3.8). The corresponding symmetric Macdonald–Koornwinder polynomials are the Askey–Wilson polynomials if R_0 is of rank one (Askey and Wilson, 1985), and the symmetric Koornwinder polynomials if R_0 is of higher rank (Koornwinder, 1992).

In this subsection we introduce the monic symmetric Macdonald–Koornwinder polynomials by symmetrizing the nonsymmetric ones; cf. Cherednik (1995b, §4). We assume throughout this subsection that $\kappa\in\mathcal{M}$ and q satisfy (9.3.5). Recall the notation $\kappa_w:=\prod_{a\in R^{\bullet+}\cap w^{-1}R^{\bullet-}}\kappa_a$. It depends only on $\kappa^\bullet=\kappa|_{R^\bullet}$. It satisfies $\kappa_v\kappa_w=\kappa_{vw}$ if $l(vw)=l(v)+l(w)$. Hence there exists a unique linear character $\chi_+\colon H(\kappa^\bullet)\to\mathbb{C}$ satisfying $\chi_+(T_w)=\kappa_w$ for all $w\in W$, the trivial linear character of $H(\kappa^\bullet)$. Define

$$C_+:=\frac{1}{\sum_{w\in W_0}\kappa_w^2}\sum_{w\in W_0}\kappa_wT_w\in H_0(\kappa^\bullet|_{R_0})\subset H(\kappa^\bullet).\tag{9.3.11}$$

The normalization is such that $\chi_+(C_+) = 1$. Then $T_iC_+ = \kappa_iC_+ = C_+T_i$ for $1 \leq i \leq n$, and $C_+^2 = C_+$. The following lemma follows from the explicit expression of $\pi_{\kappa,q}(T_i)$ $(1 \leq i \leq n)$.

Lemma 9.3.19 *The linear endomorphism $\pi_{\kappa,q}(C_+)$ of $\mathbb{C}[T_\Lambda]$ is an idempotent with image $\mathbb{C}[T_\Lambda]^{W_0}$.*

Consider the linear basis $\{m_\lambda\}_{\lambda\in\Lambda^+}$ of $\mathbb{C}[T_\Lambda]^{W_0}$ given by orbit sums, $m_\lambda(t) = \sum_{\mu\in W_0\lambda} t^\mu$. Recall that $P_\lambda = P_\lambda(D;\kappa,q)$ denotes the monic nonsymmetric Macdonald–Koornwinder polynomial of degree $\lambda \in \Lambda$.

Lemma 9.3.20 *If $\lambda \in \Lambda^+$ then $\pi_{\kappa,q}(C_+)P_\lambda = \sum_{\mu\in\Lambda^+:\mu\leq\lambda} c_{\lambda,\mu}m_\mu$ for certain $c_{\lambda,\mu} \in \mathbb{C}$ with $c_{\lambda,\lambda} \neq 0$.*

Definition 9.3.21 The monic *symmetric Macdonald–Koornwinder polynomial* of degree $\lambda \in \Lambda^+$ is defined by $P_\lambda^+ = P_\lambda^+(D;\kappa,q) := c_{\lambda,\lambda}^{-1}\pi_{\kappa,q}(C_+)P_\lambda \in \mathbb{C}[T_\Lambda]^{W_0}$.

Theorem 9.3.22 *P_λ^+, defined above, is the unique element in $\mathbb{C}[T_\Lambda]^{W_0}$ satisfying*
1. *$P_\lambda^+ = \sum_{\mu\in\Lambda^+:\mu\leq\lambda} d_{\lambda,\mu}m_\mu$ with $d_{\lambda,\mu} \in \mathbb{C}$ and $d_{\lambda,\lambda} = 1$,*
2. *$D_pP_\lambda^+ = p(q^{-\lambda}\prod_{\alpha\in R_0^+} \upsilon_{\alpha^d}^{-\alpha^{d\vee}})P_\lambda^+$ for all $p \in \mathbb{C}[T_{\Lambda^d}]^{W_0}$.*

(Note that for $p \in \mathbb{C}[T_{\Lambda^d}]^{W_0}$ and $\lambda \in \Lambda^+$ we have $p(q^{-\lambda}\prod_{\alpha\in R_0^+} \upsilon_{\alpha^d}^{-\alpha^{d\vee}}) = p(\gamma_{w_0\lambda,q}^{-1})$.)

Remark 9.3.23 One may replace condition (2) in Theorem 9.3.22 by the weaker condition

$$D_{(Da_0)^d}P_\lambda^+ = m^d_{(Da_0)^d}\Big(q^{-\lambda}\prod_{\alpha\in R_0^+} \upsilon_{\alpha^d}^{-\alpha^{d\vee}}\Big)P_\lambda^+.$$

Note that the left-hand side of this equation is completely explicit by Corollary 9.3.18.

Explicit expressions of the monic symmetric Macdonald–Koornwinder polynomials when R_0 is of rank one are given in §9.3.7 and §9.3.8.

9.3.6 Orthogonality

In this subsection we assume that $\kappa \in \mathcal{M}$ and q satisfy the conditions (9.3.6). We thus have the monic symmetric Macdonald–Koornwinder polynomials $\{P_\lambda^+\}_{\lambda\in\Lambda^+}$ with respect to the parameters (κ, q), as well as the monic symmetric Macdonald–Koornwinder polynomials with respect to the parameters (κ^{-1}, q^{-1}), in which case we denote them by $\{P_\lambda^{\circ+}\}_{\lambda\in\Lambda^+}$.

Define the W_0-invariant meromorphic function

$$v_+ := \prod_{a\in R^\bullet;a(0)\geq 0} c_a^{-1}$$

on T_Λ. It is related to the weight function v by

$$v = \mathcal{C}v_+, \qquad \mathcal{C} = \mathcal{C}(\cdot;D;\kappa,q) := \prod_{\alpha\in R_0^-} c_\alpha. \tag{9.3.12}$$

One recovers v_+ from v up to a multiplicative constant by symmetrization, in view of the following property of the rational function $\mathcal{C} \in \mathbb{C}(T_\Lambda)$ (cf. Macdonald, 1972b, Theorems (2.8), (2.8 *nr*)).

Lemma 9.3.24 *We have*

$$\sum_{w \in W_0} w\mathcal{C} = \mathcal{C}(\gamma^d_{0,q}) \tag{9.3.13}$$

as identity in $\mathbb{C}(T_\Lambda)$*, where* $\gamma^d_{\xi,q} := \gamma_{\xi,q}(D^d; \kappa^{d\bullet}) \in T_\Lambda$ $(\xi \in \Lambda^d)$*. More generally, if* $\xi \in \Lambda^{d-}$ *and if* (κ, q) *is generic then, as identity in* $\mathbb{C}(T_\Lambda)$*,*

$$\sum_{w \in W_0} w\mathcal{C} = \mathcal{C}(\gamma^d_{\xi,q}) \sum_{\eta \in W_0\xi} \prod_{\alpha \in R_0^+ \cap v(\eta) R_0^-} \frac{c_\alpha(\gamma^d_{\xi,q})}{c_{-\alpha}(\gamma^d_{\xi,q})},$$

where (recall) $v(\eta) \in W_0$ *is the element of shortest length such that* $v(\eta)\eta = \eta_-$*.*

The meromorphic function v_+ on T_Λ reads in terms of q-shifted factorials as follows:

$$v_+(t) = \prod_{\beta \in R_0} \frac{(t^{2\beta}; q_\beta^2)_\infty}{(\kappa_\beta \kappa_{2\beta} t^\beta, -\kappa_\beta \kappa_{2\beta}^{-1} t^\beta, q_\beta \kappa_{\mu_\beta c+\beta} \kappa_{2\mu_\beta c+2\beta} t^\beta, -q_\beta \kappa_{\mu_\beta c+\beta} \kappa_{2\mu_\beta c+2\beta}^{-1} t^\beta; q_\beta^2)_\infty}. \tag{9.3.14}$$

It is a nonnegative real-valued continuous function on T^u_Λ (it is nonnegative since it can be written on T^u_Λ as $v_+(t) = |\delta(t)|^2$ with $\delta(t)$ the expression (9.3.14) with product taken only over the set R_0^+ of positive roots).

Corollary 9.3.25 $\langle \cdot, \cdot \rangle$ *restricts to a positive definite, sesquilinear form on* $\mathbb{C}[T_\Lambda]^{W_0}$*. In fact,*

$$\langle p, r \rangle = \frac{\mathcal{C}(\gamma^d_{0,q})}{\#W_0} \int_{T^u_\Lambda} p(t)\overline{r(t)}\, v_+(t)\, d_u t, \qquad p, r \in \mathbb{C}[T_\Lambda]^{W_0}.$$

Symmetrization of the results on the monic nonsymmetric Macdonald–Koornwinder polynomials by using the idempotent $C_+ \in H_0$ gives the following properties of the monic symmetric Macdonald–Koornwinder polynomials.

Theorem 9.3.26 *Let* $\lambda \in \Lambda^+$*.*

(a) *The symmetric Macdonald–Koornwinder polynomial* $P^+_\lambda \in \mathbb{C}[T_\Lambda]^{W_0}$ *satisfies the following characterizing properties:*
   (i) $P^+_\lambda = \sum_{\mu \in \Lambda^+ : \mu \leq \lambda} d_{\lambda,\mu} m_\mu$ *with* $d_{\lambda,\lambda} = 1$*,*
   (ii) $\langle P^+_\lambda, m_\mu \rangle = 0$ *if* $\mu \in \Lambda^+$ *and* $\mu < \lambda$*.*

(b) $P^{\circ +}_\lambda = P^+_\lambda$*.*

(c) $\langle P^+_\lambda, P^+_\mu \rangle = 0$ *if* $\mu \in \Lambda^+$ *and* $\mu \neq \lambda$*.*

9.3.7 GL_n *Macdonald Polynomials*

Take $n \geq 2$ and $V := (\epsilon_1 + \cdots + \epsilon_n)^\perp \subset \mathbb{R}^n =: Z$ with $\{\epsilon_i\}_{i=1}^n$ the standard orthonormal basis of $\mathbb{R}^n$. Let $R_0 := \{\epsilon_i - \epsilon_j\}_{1 \leq i \neq j \leq n}$ with ordered basis $\Delta_0 := (\alpha_1, \ldots, \alpha_{n-1}) := (\epsilon_1 - \epsilon_2, \ldots, \epsilon_{n-1} - \epsilon_n)$. As lattices take $\Lambda = \mathbb{Z}^n = \Lambda^d$. Then $D := (R_0, \Delta_0, u, \mathbb{Z}^n, \mathbb{Z}^n) \in \mathcal{D}$.

The corresponding irreducible reduced affine root system is $R^u = \{mc + \alpha\}_{m\in\mathbb{Z},\alpha\in R_0}$, and the corresponding additional simple affine root is $a_0 = c - \epsilon_1 + \epsilon_n$. There is no nonreduced extension of R^u involved since $(\Lambda, a_i^\vee) = \mathbb{Z}$ for $i \in \{0, \dots, n\}$. Hence $R = R^u$.

The fundamental weights $\varpi_j = \varpi_j^u$ $(1 \le j \le n-1)$ are given by

$$\varpi_j = -\frac{j}{n}(\epsilon_1 + \cdots + \epsilon_n) + \epsilon_1 + \epsilon_2 + \cdots + \epsilon_j.$$

Define for $1 \le j \le n$ the elements $\tilde{\varpi}_j := \epsilon_1 + \cdots + \epsilon_j$. The orthocomplement $V^\perp$ of V in Z is $\mathbb{R}\tilde{\varpi}_n$ and $\Lambda_0^d := \Lambda^d \cap V^\perp = \mathbb{Z}\tilde{\varpi}_n$. Then $\tilde{\varpi}_j \in (\varpi_j + V^\perp) \cap \Lambda^d$ $(j \in \{1, \dots, n-1\})$. Since $-(Da_0)^{d\vee} = \epsilon_1 - \epsilon_n = \sum_{i=1}^{n-1} \alpha_i$ we conclude that $J^+_{\Lambda^d} = \{1, \dots, n-1\}$. The minuscule dominant weights in Λ^d are thus given by $\Lambda^{d+}_{\min} = \Lambda_0^d \cup \bigcup_{j=1}^{n-1}(\tilde{\varpi}_j + \Lambda_0^d)$. We have $u(\epsilon_1) = \tau(\epsilon_1)s_1 s_2 \cdots s_{n-1}$ in the extended affine Weyl group $W \simeq S_n \ltimes \mathbb{Z}^n$ (where S_n is the symmetric group in n letters). Note also that $u(\epsilon_1)(a_j) = a_{j+1}$ for $0 \le j < n-1$ and $u(\epsilon_1)(a_{n-1}) = a_0$. In addition, we have $u(\epsilon_1)^j = u(\tilde{\varpi}_j)$ for $1 \le j \le n-1$ and $u(\epsilon_1)^n = u(\tilde{\varpi}_n) = \tau(\tilde{\varpi}_n)$ in W. Hence $\mathbb{Z} \simeq \Omega$ by $m \mapsto u(\epsilon_1)^m$.

Observe that $R = R^u$ has one $W(R^u)$-orbit, hence also one W-orbit. The affine Hecke algebra $H(W(R);\kappa)$ and the extended affine Hecke algebra $H(\kappa) = H(D;\kappa)$ thus depend on a single nonzero complex number κ. Also, $\mathbb{Z}$ acts on $H(W(R);\kappa)$ by algebra automorphisms, where $1 \in \mathbb{Z}$ acts on T_i by mapping it to T_{i+1} (reading the subscript modulo n). We write $\mathbb{Z} \ltimes H(W(R);\kappa)$ for the associated semidirect product algebra.

Proposition 9.3.27

(i) *$\mathbb{Z} \ltimes H(W(R);\kappa) \simeq H(\kappa)$ by mapping the generator $1 \in \mathbb{Z}$ to $u(\tilde{\varpi}_1) = u(\epsilon_1)$.*

(ii) *For $1 \le i \le n$ we have $Y^{\epsilon_i} = T_{i-1}^{-1} \cdots T_2^{-1}T_1^{-1}u(\epsilon_1)T_{n-1} \cdots T_{i+1}T_i$ in $H(\kappa)$.*

We now turn to the explicit description of the Macdonald–Ruijsenaars q-difference operators and the orthogonality measure for the GL_n Macdonald polynomials.

The longest Weyl group element $w_0 \in W_0$ maps ϵ_i to ϵ_{n+1-i} for $1 \le i \le n$, hence

$$R^+ \cap \tau(w_0\tilde{\varpi}_j)R^- = \{\epsilon_r - \epsilon_s \mid 1 \le r \le n-j,\ \ n+1-j \le s \le n\}.$$

Write $t_i = t^{\epsilon_i}$ for $t \in T_\Lambda$ and $1 \le i \le n$. Then for $1 \le j \le n-1$ we obtain

$$c_{\tau(-w_0\tilde{\varpi}_j)}(t) = \prod_{\substack{1\le r\le n-j\\ n+1-j\le s\le n}} \frac{1-\kappa^2 t_r t_s^{-1}}{1-t_r t_s^{-1}},$$

and consequently

$$D_{w_0\tilde{\varpi}_j} = \sum_{\substack{I\subset\{1,\dots,n\}:\\ \#I=n-j}} \left(\prod_{r\in I, s\notin I} \frac{1-\kappa^2 t_r t_s^{-1}}{\kappa(1-t_r t_s^{-1})}\right)\tau\Big(\sum_{s\notin I}\epsilon_s\Big)_q = \sum_{\substack{I\subset\{1,\dots,n\}:\\ \#I=j}} \left(\prod_{r\in I, s\notin I} \frac{\kappa^{-1}t_r - \kappa t_s}{t_r - t_s}\right)\tau\Big(\sum_{r\in I}\epsilon_r\Big)_q$$

for $1 \le j \le n-1$ by Corollary 9.3.18(i). These commutative q-difference operators were introduced by Ruijsenaars (1987) as the quantum Hamiltonians of a relativistic version of the trigonometric quantum Calogero–Moser system. These operators also go by the name (trigonometric) *Ruijsenaars* or *Macdonald–Ruijsenaars q-difference operators*.

The corresponding monic *symmetric Macdonald polynomials* $\{P_\lambda\}_{\lambda\in\Lambda^+}$ are parametrized by $\Lambda^+ = \{\lambda \in \mathbb{Z}^n \mid \lambda_1 \geq \lambda_2 \geq \cdots \geq \lambda_n\}$. The orthogonality weight function then becomes

$$v_+(t) = \prod_{1\leq i\neq j\leq n} \frac{(t_i/t_j;q)_\infty}{(\kappa^2 t_i/t_j;q)_\infty}.$$

The GL_2 Macdonald polynomials are essentially the *continuous q-ultraspherical polynomials* (see Cherednik, 2005, Ch. 2 or Macdonald, 2003, §6.3)

$$P_\lambda^+(t) = t_1^{\lambda_1} t_2^{\lambda_2}\, {}_2\phi_1\left(\begin{matrix}\kappa^2, q^{\lambda_2-\lambda_1} \\ q^{1+\lambda_2-\lambda_1}/\kappa^2\end{matrix}; q, \frac{qt_2}{\kappa^2 t_1}\right), \qquad \lambda = (\lambda_1,\lambda_2) \in \mathbb{Z}^2,\ \lambda_1 \geq \lambda_2,$$

where we use standard notation for the basic hypergeometric ${}_r\phi_s$ series (see, e.g., Gasper and Rahman, 2004).

9.3.8 Koornwinder Polynomials

Take $Z = V = \mathbb{R}^n$ with standard orthonormal basis $\{\epsilon_i\}_{i=1}^n$. We realize the root system $R_0 \subset V$ of type B_n as $R_0 = \{\pm\epsilon_i\}_{i=1}^n \cup \{\pm\epsilon_i \pm \epsilon_j\}_{1\leq i<j\leq n}$ (all sign combinations allowed). The W_0-orbits are $\mathcal{O}_l = \{\pm\epsilon_i \pm \epsilon_j\}_{1\leq i<j\leq n}$ and $\mathcal{O}_s = \{\pm\epsilon_i\}_{i=1}^n$. As an ordered basis of R_0 take

$$\Delta_0 := (\alpha_1,\ldots,\alpha_{n-1},\alpha_n) := (\epsilon_1-\epsilon_2,\ldots,\epsilon_{n-1}-\epsilon_n,\epsilon_n).$$

Note that $\mathbb{Z}R_0 = \mathbb{Z}^n$. For $n = 1$ this should be interpreted as $R_0 = \{\pm\epsilon_1\}$, the root system of type A_1, with basis element given by ϵ_1. We consider in this subsection the initial data $D = (R_0, \Delta_0, t, \mathbb{Z}^n, \mathbb{Z}^n) \in \mathcal{D}$.

The associated reduced affine root system is $R^t = \{\frac{1}{2}mc + \alpha\}_{n\in\mathbb{Z},\alpha\in\mathcal{O}_s} \cup \{mc+\beta\}_{m\in\mathbb{Z},\beta\in\mathcal{O}_l}$ (for $n = 1$ it should be read as $R^t = \{\frac{1}{2}mc \pm \epsilon_1\}_{m\in\mathbb{Z}}$, the affine root system R^t with R_0 of type A_1). The associated ordered basis of R^t is $\Delta = (a_0, a_1, \ldots, a_n) = (\frac{1}{2}c - \theta, \alpha_1, \ldots, \alpha_n)$ with $\theta = \epsilon_1 = \sum_{j=1}^n \alpha_j$ the highest short root of R_0 with respect to Δ_0. The affine root system R^t has three W-orbits, $R^t = \widehat{\mathcal{O}}_1 \cup \widehat{\mathcal{O}}_2 \cup \widehat{\mathcal{O}}_3$, where $\widehat{\mathcal{O}}_1 = Wa_0 = (\frac{1}{2}+\mathbb{Z})c + \mathcal{O}_s$, $\widehat{\mathcal{O}}_2 = Wa_i = \mathbb{Z}c + \mathcal{O}_l$ $(1 \leq i < n)$ and $\widehat{\mathcal{O}}_3 = Wa_n = \mathbb{Z}c + \mathcal{O}_s$. If $n = 1$ then $\Delta = (a_0, a_1) = (\frac{1}{2}c - \epsilon_1, \epsilon_1)$ and R^t has two W-orbits $\widehat{\mathcal{O}}_1$ and $\widehat{\mathcal{O}}_3$.

Note that $(\Lambda, a_i^\vee) = 2\mathbb{Z}$ for $i = 0$ and $i = n$, hence $S = \{0, n\} = S^d$ and

$$R = R^t \cup W(2a_0) \cup W(2a_n) = \widehat{\mathcal{O}}_1 \cup \widehat{\mathcal{O}}_2 \cup \widehat{\mathcal{O}}_3 \cup \widehat{\mathcal{O}}_4 \cup \widehat{\mathcal{O}}_5$$

with additional W-orbits $\widehat{\mathcal{O}}_4 = 2\widehat{\mathcal{O}}_1$ and $\widehat{\mathcal{O}}_5 = 2\widehat{\mathcal{O}}_3$. If $n = 1$ then R has four W-orbits $\widehat{\mathcal{O}}_i$ $(i = 1,2,4,5)$. Since $D^d = D$ we have $R^d = R$, $\Delta^d = \Delta$ and $W^d = W$.

Suppose that $\kappa \in \mathcal{M}(D)$ and $q \in \mathbb{C}^*$ satisfy (9.3.5). Fix $1 \leq i < n$. Then $\kappa \in \mathcal{M}(D)$ is determined by five (four in the case $n = 1$) independent numbers $\kappa_{a_0}, \kappa_{a_i}, \kappa_{a_n}, \kappa_{2a_0}$ and κ_{2a_n}. The corresponding Askey–Wilson parameters (Askey and Wilson, 1985) are defined by

$$(a,b,c,d,k) = (\kappa_{a_n}\kappa_{2a_n}, -\kappa_{a_n}\kappa_{2a_n}^{-1}, q^{\frac{1}{2}}\kappa_{a_0}\kappa_{2a_0}, -q^{\frac{1}{2}}\kappa_{a_0}\kappa_{2a_0}^{-1}, \kappa_{a_i}^2) \qquad (9.3.15)$$

(the parameter k drops out in the case $n = 1$). The dual multiplicity function κ^d on $R^d = R$ is then determined by $\kappa^d_{a_0} := \kappa_{2a_n}$, $\kappa^d_{a_i} := \kappa_{a_i}$, $\kappa^d_{a_n} := \kappa_{a_n}$, $\kappa^d_{2a_0} := \kappa_{2a_0}$ and $\kappa^d_{2a_n} := \kappa_{a_0}$. The corresponding Askey–Wilson parameters are

$$(\tilde{a}, \tilde{b}, \tilde{c}, \tilde{d}, \tilde{k}) = (\kappa_{a_n}\kappa_{a_0}, -\kappa_{a_n}\kappa_{a_0}^{-1}, q^{\frac{1}{2}}\kappa_{2a_n}\kappa_{2a_0}, -q^{\frac{1}{2}}\kappa_{2a_n}\kappa_{2a_0}^{-1}, \kappa_{a_i}^2).$$

In terms of the Askey–Wilson parameters this can be expressed as $\tilde{k} = k$ and

$$(\tilde{a}, \tilde{b}, \tilde{c}, \tilde{d}) = \left(\sqrt{q^{-1}abcd}, \frac{ab}{\sqrt{q^{-1}abcd}}, \frac{ac}{\sqrt{q^{-1}abcd}}, \frac{ad}{\sqrt{q^{-1}abcd}}\right).$$

Note that

$$-(Da_0)^{d\vee} = \theta^\vee = 2\alpha_1^\vee + 2\alpha_2^\vee + \cdots + 2\alpha_{n-1}^\vee + \alpha_n^\vee.$$

Furthermore, denoting the fundamental weights of R_0 with respect to the ordered basis Δ_0 by $\{\varpi_i\}_{i=1}^n$, we have $\varpi_n = \frac{1}{2}(\epsilon_1 + \cdots + \epsilon_n) \notin \Lambda^d = \mathbb{Z}^n$. Consequently $J^+_{\Lambda^d} = \emptyset$. Hence the only explicit Macdonald q-difference operator obtainable from Corollary 9.3.18 is $D_{-\theta} = D_{-\epsilon_1}$.

Write

$$\mathcal{D} = \sum_{w \in W_0/W_{0,\epsilon_1}} w(c_{\tau(\epsilon_1)})(\tau(-w\epsilon_1)_q - 1), \tag{9.3.16}$$

so that $D_{-\epsilon_1} = \kappa_{\tau(\epsilon_1)}^{-1}\mathcal{D} + m^d_{-\epsilon_1}(\gamma_{0,q}^{-1})$. Write $t_i = t^{\epsilon_i}$ for $t \in T_{\mathbb{Z}^n}$ and $1 \leq i \leq n$. Since

$$R^{t+} \cap \tau(-\epsilon_1)R^{t-} = \{\epsilon_1, \tfrac{1}{2}c + \epsilon_1\} \cup \{\epsilon_1 \pm \epsilon_j\}_{j=2}^n$$

($= \{\epsilon_1, \frac{1}{2}c + \epsilon_1\}$ if $n = 1$) it follows that $\kappa_{\tau(\epsilon_1)} = \sqrt{q^{-1}abcd}\, k^{n-1}$ and

$$\begin{aligned} c_{\tau(\epsilon_1)}(t) &= c_{\epsilon_1}(t)c_{\frac{1}{2}c+\epsilon_1}(t)\prod_{j=2}^n c_{\epsilon_1-\epsilon_j}(t)c_{\epsilon_1+\epsilon_j}(t) \\ &= \frac{(1-at_1)(1-bt_1)(1-ct_1)(1-dt_1)}{(1-t_1^2)(1-qt_1^2)}\prod_{j=2}^n \frac{(1-kt_1t_j^{-1})(1-kt_1t_j)}{(1-t_1t_j^{-1})(1-t_1t_j)}. \end{aligned} \tag{9.3.17}$$

For $n = 1$ the product over j is not present in (9.3.17). In particular, the operator $\mathcal{D}$ then depends only on q and the four parameters a, b, c, d.

Hence $\mathcal{D}$ is Koornwinder's (1992) second-order q-difference operator

$$(\mathcal{D}f)(t) = \sum_{i=1}^n \Big(A_i(t)((\tau(-\epsilon_i)_q f)(t) - f(t)) + A_i(t^{-1})((\tau(\epsilon_i)_q f)(t) - f(t))\Big),$$

$$A_i(t) = \frac{(1-at_i)(1-bt_i)(1-ct_i)(1-dt_i)}{(1-t_i^2)(1-qt_i^2)}\prod_{j\neq i}\frac{(1-kt_it_j)(1-kt_it_j^{-1})}{(1-t_it_j)(1-t_it_j^{-1})},$$

with the obvious adjustment for $n = 1$, in which case $\mathcal{D}$ is the Askey–Wilson second-order q-difference operator (Askey and Wilson, 1985). This derivation is due to Noumi (1995).

The monic symmetric Macdonald–Koornwinder polynomials associated to (D, κ) are the monic *symmetric Koornwinder polynomials* (Koornwinder, 1992). They can be characterized as follows. Note that $\mathbb{C}[T_{\mathbb{Z}^n}] = \mathbb{C}[z_1^{\pm 1}, \ldots, z_n^{\pm 1}]$ with $z_i = e^{\epsilon_i}$. The finite Weyl group W_0 is the *hyperoctahedral group* $W_0 \simeq S_n \ltimes \{\pm 1\}^n$ acting on $\mathbb{C}[T_{\mathbb{Z}^n}]$ by permutations and inversions of the variables z_i. The symmetric Koornwinder polynomials are then parametrized by

$$\Lambda^+ := \{\lambda \in \mathbb{Z}^n \mid \lambda_1 \geq \lambda_2 \geq \cdots \geq \lambda_n \geq 0\},$$

which we consider a partially ordered set with respect to the *dominance order*: $\lambda \geq \mu$ if $\sum_{j=1}^{i} \lambda_j \geq \sum_{j=1}^{i} \mu_j$ for all i. The symmetric monomials are $m_\lambda = \sum_{\mu \in W_0\lambda} z^\mu \in \mathbb{C}[T_{\mathbb{Z}^n}]^{W_0}$ for $\lambda \in \Lambda^+$, with W_0 acting by permutations and sign changes on $\mathbb{Z}^n$. The monic symmetric Koornwinder polynomial $P_\lambda^+(\cdot) = P_\lambda^+(\cdot; a, b, c, d; q, k)$ of degree $\lambda \in \Lambda^+$ is now uniquely characterized by the eigenvalue equation

$$\mathcal{D}P_\lambda^+ = \left(\sum_{i=1}^{n} \left(q^{-1}abcdk^{2n-i-1}(q^{\lambda_i} - 1) + k^{i-1}(q^{-\lambda_i} - 1)\right)\right)P_\lambda^+$$

and the property that $P_\lambda^+ = m_\lambda + \sum_{\mu \in \Lambda^+; \mu < \lambda} c_{\lambda,\mu} m_\mu$ for certain $c_{\lambda,\mu} \in \mathbb{C}$.

The weight function $v_+(t)$ becomes

$$v_+(t) = \prod_{i=1}^{n} \frac{(t_i^2, t_i^{-2}; q)_\infty}{(at_i, at_i^{-1}, bt_i, bt_i^{-1}, ct_i, ct_i^{-1}, dt_i, dt_i^{-1}; q)_\infty} \prod_{1 \leq r < s \leq n} \frac{(t_r t_s, t_r^{-1} t_s^{-1}, t_r t_s^{-1}, t_r^{-1} t_s; q)_\infty}{(kt_r t_s, kt_r^{-1} t_s^{-1}, kt_r t_s^{-1}, kt_r^{-1} t_s; q)_\infty}. \tag{9.3.18}$$

For $n = 1$ the Koornwinder polynomials are the Askey–Wilson polynomials (Askey and Wilson, 1985). Concretely, with standard notation for basic hypergeometric series (cf. Gasper and Rahman, 2004), we have

$$P_m^+(t) = \frac{(ab, ac, ad; q)_m}{a^m (q^{m-1}abcd; q)_m} \, {}_4\phi_3\left(\begin{matrix} q^{-m}, q^{m-1}abcd, at, at^{-1} \\ ab, ac, ad \end{matrix}; q, q\right), \quad m \in \mathbb{Z}_{\geq 0} = \Lambda^+.$$

The Cherednik–Macdonald theory associated to the Askey–Wilson polynomials was worked out in detail in Noumi and Stokman (2004).

Example 9.3.28 Consider the initial data $D = (R_0, \Delta_0, t, \mathbb{Z}^n, P(R_0)) \in \mathcal{D}$ with the root system $R_0 \subset \mathbb{R}^n$ of type B_n $(n \geq 2)$ and the ordered basis Δ_0 defined in terms of the standard orthonormal basis $\{\epsilon_i\}_{i=1}^n$ of $\mathbb{R}^n$ as above. Compared to the Koornwinder setup just discussed we have thus chosen a different lattice $\Lambda^d = P(R_0)$, which contains $\mathbb{Z}R_0 = \mathbb{Z}^n$ as index-two sublattice. Still $S(D) = \{0, n\}$, hence R is the nonreduced affine root system with five $W_0 \ltimes \mathbb{Z}^n$-orbits $\widehat{\mathcal{O}}_i$ $(1 \leq i \leq 5)$ as introduced above. But the number $\nu = \nu(D)$ of $W_0 \ltimes P(R_0)$-orbits is three: they are given by $\widehat{\mathcal{O}}_2$, $\widehat{\mathcal{O}}_1 \cup \widehat{\mathcal{O}}_3$ and $\widehat{\mathcal{O}}_4 \cup \widehat{\mathcal{O}}_5$.

On the other hand, $R^d = R^t$ since $S(D^d) = \emptyset$, in particular R^d is reduced. It has the three $W^d = W_0 \ltimes \mathbb{Z}^n$-orbits $\widehat{\mathcal{O}}_i$ $(i = 1, 2, 3)$.

Remark 9.3.29 Consider the initial data $D = (R_0, \Delta_0, u, \mathbb{Z}^n, \mathbb{Z}^n)$ with (R_0, Δ_0) of type B_n $(n \geq 3)$ as above. Compared to initial data $D^{\mathrm{C}^\vee\mathrm{C}_n} := (R_0, \Delta_0, t, \mathbb{Z}^n, \mathbb{Z}^n)$ related to Koornwinder polynomials, we thus have only changed the type from twisted to untwisted. Then R^u is the affine root subsystem $\widehat{\mathcal{O}}_2 \cup \widehat{\mathcal{O}}_3$ of the affine root system $R^{\mathrm{C}^\vee\mathrm{C}_n}$ of type $\mathrm{C}^\vee\mathrm{C}_n$ as defined above, and $R = R(D)$ is the nonreduced irreducible affine root subsystem $\widehat{\mathcal{O}}_2 \cup \widehat{\mathcal{O}}_3 \cup \widehat{\mathcal{O}}_5$ of $R(D^{\mathrm{C}^\vee\mathrm{C}_n})$. The dual affine root system $R^d = R(D^d)$ is the reduced irreducible affine root subsystem

$R^d = \widehat{\mathcal{O}}_2 \cup \widehat{\mathcal{O}}_4 \cup \widehat{\mathcal{O}}_5$ (it is of untwisted type, with underlying finite root system $R_0^\vee$ of type C_n). The Macdonald–Koornwinder theory associated to the initial data D thus is the special case of the $C^\vee C_n$ theory when the multiplicity function $\kappa \in \mathcal{M}(D^{C^\vee C_n})$ takes the value 1 at the two orbits $\widehat{\mathcal{O}}_1 = W(a_0)$ and $\widehat{\mathcal{O}}_4 = W(2a_0)$ of $R^{C^\vee C_n} \setminus R$.

9.3.9 A New Class of Nonreduced Rank-Two Macdonald–Koornwinder Polynomials

Consider $R_0 \subset Z = V = \mathbb{R}^2$ the root system of type C_2 given by $R_0 = R_{0,s} \cup R_{0,l}$ with

$$R_{0,s} = \{\pm(\epsilon_1 + \epsilon_2), \pm(\epsilon_1 - \epsilon_2)\}, \quad R_{0,l} = \{\pm 2\epsilon_1, \pm 2\epsilon_2\},$$

where $\{\epsilon_1, \epsilon_2\}$ is the standard orthonormal basis of $\mathbb{R}^2$. As the ordered basis we take $\Delta_0 := (\alpha_1, \alpha_2)$ with $\alpha_1 := \epsilon_1 - \epsilon_2$ and $\alpha_2 := 2\epsilon_2$. Then $\varphi = 2\epsilon_1$ and $\theta = \epsilon_1 + \epsilon_2$. As the quintuple of initial data we take

$$D = (R_0, \Delta_0, \bullet, \Lambda, \Lambda^d) := (R_0, \Delta_0, u, \mathbb{Z}R_0, \mathbb{Z}R_0^\vee).$$

Hence $\Lambda^d = \mathbb{Z}\epsilon_1 \oplus \mathbb{Z}\epsilon_2 \simeq \mathbb{Z}^2$ while $\Lambda = \{\lambda = (\lambda_1, \lambda_2) \in \mathbb{Z}^2 \mid \lambda_1 + \lambda_2 \text{ even}\}$. Hence the simple affine root a_0 of the associated reduced affine root system $R^u = \mathbb{Z}c + R_0$ is $c - 2\epsilon_1$. Then $S = \{1\}$. Hence $R = R^u \cup W(2\alpha_1)$ with $W = W^u = W_0 \ltimes \mathbb{Z}R_0^\vee$. Then R has four W-orbits,

$$W(a_0) = (2\mathbb{Z}+1)c + R_{0,l}, \quad W(\alpha_1) = \mathbb{Z}c + R_{0,s}, \quad W(\alpha_2) = 2\mathbb{Z}c + R_{0,l}, \quad W(2\alpha_1) = 2\mathbb{Z}c + 2R_{0,s}.$$

For the dual initial data $D^d = (R_0^\vee, \Delta_0^\vee, u, \mathbb{Z}R_0^\vee, \mathbb{Z}R_0)$ we have $R^{ud} = \mathbb{Z}c + R_0^\vee$ and furthermore $R^d = R^{ud} \cup W^d(2\alpha_2^\vee)$ with $W^d = W^{ud} = W_0 \ltimes \mathbb{Z}R_0$. The simple affine root a_0^d of R^{ud} is $a_0^d = c - \epsilon_1 - \epsilon_2$. The four W^d-orbits of R^d are

$$W^d(a_0^d) = (2\mathbb{Z}+1)c + R_{0,s}^\vee, \quad W^d(\alpha_1^\vee) = 2\mathbb{Z}c + R_{0,s}^\vee, \quad W^d(\alpha_2^\vee) = \mathbb{Z}c + R_{0,l}^\vee, \quad W^d(2\alpha_2^\vee) = 2\mathbb{Z}c + 2R_{0,l}^\vee.$$

Note that $R \simeq R^d$ but $(R, \Delta) \not\simeq (R^d, \Delta^d)$.

Let $\kappa \in \mathcal{M}(D)$. We write

$$\{a, b, c, d\} := \{\kappa_\theta \kappa_{2\theta}, -\kappa_\theta \kappa_{2\theta}^{-1}, \kappa_\varphi^2, q\kappa_0^2\}, \quad \{\tilde{a}, \tilde{b}, \tilde{c}, \tilde{d}\} = \{\kappa_\varphi \kappa_0, -\kappa_\varphi \kappa_0^{-1}, \kappa_\theta^2, q\kappa_{2\theta}^2\} \qquad (9.3.19)$$

(the dual parameters $(\tilde{a}, \tilde{b}, \tilde{c}, \tilde{d})$ are the parameters (a, b, c, d) with respect to the dual initial data D^d).

We view $\mathbb{C}[T_\Lambda]$ as the subalgebra $\bigoplus_{\lambda \in \Lambda} \mathbb{C}e^\lambda$ of $\mathbb{C}[T_{\mathbb{Z}^2}]$, which induces an embedding of $\mathbb{C}(T_\Lambda)$ as a subfield of $\mathbb{C}(T_{\mathbb{Z}^2})$. More concretely, if $z_i := e^{\epsilon_i}$ for the standard coordinates of $\mathbb{C}[T_{\mathbb{Z}^2}]$, then $\mathbb{C}[T_\Lambda]$ is the subalgebra of $\mathbb{C}[T_{\mathbb{Z}^2}] = \mathbb{C}[z_1^{\pm 1}, z_2^{\pm 1}]$ generated by $z_1 z_2$, z_2^2 and their inverses. We write $t_i = t^{\epsilon_i}$ for $i = 1, 2$ if $t \in T_{\mathbb{Z}^2}$. We sometimes abuse notation by writing $t_1^{\lambda_1} t_2^{\lambda_2}$ for t^λ ($t \in T_\Lambda$, $\lambda \in \Lambda$).

The monic symmetric Macdonald–Koornwinder polynomials associated to (D, κ) can be characterized as eigenfunctions of the q-difference operator $D_{(Da_0)^d} = D_{-\epsilon_1}$ (see Theorem 9.3.22 and Remark 9.3.23). Write

$$\mathcal{D} := \alpha(D_{-\epsilon_1} - m_{-\epsilon_1}^d(\gamma_{0,q}^{-1})), \quad \alpha := \sqrt{q^{-1}a^2b^2cd}. \qquad (9.3.20)$$

Since $R^{u,+} \cap \tau(-\epsilon_1)R^{u,-} = \{\epsilon_1+\epsilon_2, \epsilon_1-\epsilon_2, 2\epsilon_1, c+2\epsilon_1\}$ and $\gamma_{0,q} = \prod_{\alpha\in R_0^+} \upsilon_{\alpha^\vee}^{-\alpha} = (\alpha^{-1}, -ab\alpha^{-1})$, we have by Corollary 9.3.18 that

$$(\mathcal{D}f)(t) = \sum_{i=1}^{2} \Big(A_i(t)((\tau(-\epsilon_i)_q f)(t) - f(t)) + A_i(t^{-1})((\tau(\epsilon_i)_q f)(t) - f(t)) \Big), \quad f \in \mathbb{C}[T_\Lambda]^{W_0},$$

with, for $t \in T_\Lambda$,

$$A_1(t) = \frac{(1-ct_1^2)(1-dt_1^2)}{(1-t_1^2)(1-qt_1^2)} \frac{(1-at_1t_2)(1-at_1t_2^{-1})(1-bt_1t_2)(1-bt_1t_2^{-1})}{(1-t_1^2t_2^2)(1-t_1^2t_2^{-2})},$$
$$A_2(t) = \frac{(1-ct_2^2)(1-dt_2^2)}{(1-t_2^2)(1-qt_2^2)} \frac{(1-at_2t_1)(1-at_2t_1^{-1})(1-bt_2t_1)(1-bt_2t_1^{-1})}{(1-t_2^2t_1^2)(1-t_2^2t_1^{-2})}.$$

The monic symmetric Macdonald–Koornwinder polynomial $P_\lambda^+(\cdot) = P_\lambda^+(\cdot; a,b,c,d;q) \in \mathbb{C}[T_\Lambda]^{W_0}$ associated to (D,κ) ($\lambda \in \Lambda^+$) is the unique eigenfunction of the second-order q-difference operator $\mathcal{D}$ with eigenvalue $\alpha^2(q^{\lambda_1}-1) - \alpha^2(ab)^{-1}(q^{\lambda_2}-1) + (q^{-\lambda_1}-1) - ab(q^{-\lambda_2}-1)$ satisfying $P_\lambda^+ = m_\lambda + \sum_{\mu\in\Lambda^+;\mu<\lambda} c_{\lambda,\mu} m_\mu$ for some $c_{\lambda,\mu} \in \mathbb{C}$; see Remark 9.3.23. Note here that

$$\Lambda^+ = \{\lambda = (\lambda_1,\lambda_2) \in \mathbb{Z}_{\geq 0}^2 \mid \lambda_1 \geq \lambda_2, \ \ \lambda_1+\lambda_2 \text{ even}\}.$$

The P_λ^+ are orthogonal with respect to the pairing

$$\langle p, r\rangle_+ = \int_{T_\Lambda^u} p(t)\overline{r(t)}\, v_+(t)\, d_u t, \quad p, r \in \mathbb{C}[T_\Lambda]^{W_0},$$

with weight function $v_+(t) = \delta(t)\delta(t^{-1})$ given by

$$\delta(t) := \frac{(t_1^2, t_2^2; q)_\infty}{(ct_1^2, ct_2^2, dt_1^2, dt_2^2; q^2)_\infty} \frac{(t_1t_2, t_1t_2^{-1}, -t_1t_2, -t_1t_2^{-1}; q)_\infty}{(at_1t_2, at_1t_2^{-1}, bt_1t_2, bt_1t_2^{-1}; q)_\infty}, \quad t \in T_\Lambda.$$

For $t \in T_{\mathbb{Z}^2}$ it can be rewritten as

$$\delta(t) = \left(\prod_{i=1}^{2} \frac{(t_i^2; q)_\infty}{(\sqrt{c}t_i, -\sqrt{c}t_i, \sqrt{d}t_i, -\sqrt{d}t_i; q)_\infty} \right) \frac{(t_1t_2, t_1t_2^{-1}, -t_1t_2, -t_1t_2^{-1}; q)_\infty}{(at_1t_2, at_1t_2^{-1}, bt_1t_2, bt_1t_2^{-1}; q)_\infty}.$$

As far as we know, $\{P_\lambda^+\}_{\lambda\in\Lambda^+}$ is a four-parameter family of two-variable symmetric Macdonald–Koornwinder polynomials which has not appeared before in the literature.

Subfamilies of $\{P_\lambda^+\}_{\lambda\in\Lambda^+}$ can be related to symmetric Macdonald–Koornwinder polynomials. We give here one example by relating a three-parameter subfamily of $\{P_\lambda\}_{\lambda\in\Lambda^+}$ to the rank-two symmetric Koornwinder polynomial from §9.3.8. To distinguish the Koornwinder case from the present setup we will add a label K; in particular we write $P_\lambda^{K+} = P_\lambda^{K+}(\cdot; a,b,c,d;q,k) \in \mathbb{C}[T_{\mathbb{Z}^2}]^{W_0}$ for the $n=2$ monic symmetric Koornwinder polynomial of degree $\lambda \in \Lambda^{K+}$, where $\Lambda^{K+} = \{(\lambda_1,\lambda_2) \in \mathbb{Z}_{\geq 0}^2 \mid \lambda_1 \geq \lambda_2\}$. By comparison of the characterizations of $P_\lambda \in \mathbb{C}[T_\Lambda]^{W_0}$ and $P_\lambda^{K+} \in \mathbb{C}[T_{\mathbb{Z}^2}]^{W_0}$ for $\lambda \in \Lambda^+$ as an eigenfunction of an explicit second-order q-difference operator $\mathcal{D}$ (9.3.20) and $\mathcal{D}^{\mathrm{K}}$ (9.3.16) respectively, we conclude that for $\lambda \in \Lambda^+$,

$$P_\lambda^+(\cdot; a, -1, \sqrt{c}, -\sqrt{c}, \sqrt{d}, -\sqrt{d}; q) = P_\lambda^{\mathrm{K}+}(\cdot; \sqrt{c}, -\sqrt{c}, \sqrt{d}, -\sqrt{d}; q, a)$$

as elements in $\mathbb{C}[T_\Lambda]^{W_0} \subset \mathbb{C}[z_1^{\pm 1}, z_2^{\pm 1}]^{W_0}$. Possibly there are other ways to relate special cases of P_λ^+ to rank-two symmetric Koornwinder polynomials, for instance through quadratic transformations; cf. Koornwinder (2018, §3.3), Rains and Vazirani (2007).

9.4 Double Affine Hecke Algebras and Normalized Macdonald–Koornwinder Polynomials

Cherednik's double affine braid group and double affine Hecke algebra (Cherednik, 1992a, 1995b) are fundamental for proving properties of the (non)symmetric Macdonald–Koornwinder polynomials such as the evaluation formula, duality and quadratic norms (see Cherednik, 1995c,d). We discuss these results in this section.

The first part closely follows Haiman (2006). We define a double affine braid group and a double affine Hecke algebra depending on q, on the initial data D and, in the case of the double affine Hecke algebra, on the choice of a multiplicity function $\kappa \in \mathcal{M}(D)$. We extend the duality $D \mapsto D^d$ on initial data to an anti-isomorphism of the associated double affine braid groups, and to an algebra anti-isomorphism of the associated double affine Hecke algebras (on the dual side, the double affine Hecke algebra is taken with respect to the dual multiplicity function $\kappa^d \in \mathcal{M}(D^d)$ as defined in Lemma 9.2.13). These anti-isomorphisms are called *duality anti-isomorphisms* and find their origins in the work of Cherednik (1995c,d) (see also Ion, 2003a; Sahi, 1999; Macdonald, 2003; Haiman, 2006 for further results and generalizations).

As a consequence of the duality anti-isomorphism and of the theory of intertwiners we first derive the evaluation formula and duality of the Macdonald–Koornwinder polynomials following closely Cherednik (1995d, 1997b) and Sahi (1999). These results, together with quadratic norm formulas, were first conjectured by Macdonald (see, e.g., Macdonald, 2000, 1991). For R_0 of type A the evaluation formula and duality were proven by Koornwinder (1988) by different methods (see Macdonald, 1995, Ch. VI for a detailed account). Subsequently Cherednik (1995c) established the evaluation formula and duality when $(\Lambda, \Lambda^d) = (P(R_0), P(R_0^d))$. The $\mathrm{C}^\vee\mathrm{C}$ case was established using methods from Koornwinder (1988) by van Diejen (1996) for a suitable subset of multiplicity functions. Cherednik's double affine Hecke algebra methods were extended to the $\mathrm{C}^\vee\mathrm{C}$ case in the work of Noumi (1995) and Sahi (1999), leading to the evaluation formula and duality for all multiplicity functions. Our uniform approach is close that of Haiman (2006). Following Cherednik (1995d, 1997b), we will use the duality anti-isomorphism and intertwiners to establish quadratic norm formulas for the (non)symmetric Macdonald–Koornwinder polynomials.

Throughout this section we fix initial data $D = (R_0, \Delta_0, \bullet, \Lambda, \Lambda^d) \in \mathcal{D}$, a deformation parameter $q \in \mathbb{R}_{>0} \setminus \{1\}$ and a multiplicity function $\kappa \in \mathcal{M}(D)$.

9.4.1 Double affine braid groups, Weyl groups and Hecke algebras

Consider the W-stable additive subgroup $\widehat{\Lambda} := \Lambda + \mathbb{R}c$ of $\widehat{Z}$. In the following definition it is convenient to write $X^{\hat{\lambda}}$ ($\hat{\lambda} \in \widehat{\Lambda}$) for the elements of $\widehat{\Lambda}$. In particular, $X^{\hat{\lambda}}X^{\hat{\mu}} = X^{\hat{\lambda}+\hat{\mu}}$ for $\hat{\lambda}, \hat{\mu} \in \widehat{\Lambda}$ and $X^0 = 1$.

Set $A = A(R, \Delta)$ and $A_0 = A(R_0, \Delta_0)$. The generators of the affine braid group $\mathcal{B}^\bullet = \mathcal{B}(A)$ are denoted by $T_0, T_1, \ldots, T_n$, the generators of $\mathcal{B}_0 = \mathcal{B}(A_0)$ by $T_1, \ldots, T_n$. Recall that elements $T_w \in \mathcal{B}$ ($w \in W^\bullet$) and $T_w \in \mathcal{B}_0$ ($w \in W_0$) can be defined using reduced expressions of w in the Coxeter groups $W^\bullet = W(A^\bullet)$ and $W_0 = W(A_0)$, respectively. Recall furthermore that $\mathcal{B} = \mathcal{B}(D) \simeq \Omega \ltimes \mathcal{B}^\bullet$ denotes the extended affine braid group (cf. Definition 9.2.17).

Definition 9.4.1 The *double affine braid group* $\mathbb{B} = \mathbb{B}(D)$ is the group generated by the groups $\mathcal{B}$ and $\widehat{\Lambda}$ together with the relations

$$T_iX^{\hat{\lambda}} = X^{\hat{\lambda}}T_i \quad \text{if } \hat{\lambda} \in \widehat{\Lambda} \text{ and } 0 \le i \le n \text{ such that } (\hat{\lambda}, a_i^\vee) = 0, \tag{9.4.1}$$

$$T_iX^{\hat{\lambda}}T_i = X^{s_i\hat{\lambda}} \quad \text{if } \hat{\lambda} \in \widehat{\Lambda} \text{ and } 0 \le i \le n \text{ such that } (\hat{\lambda}, a_i^\vee) = 1, \tag{9.4.2}$$

$$\omega X^{\hat{\lambda}} = X^{\omega\hat{\lambda}}\omega \quad \text{if } \hat{\lambda} \in \widehat{\Lambda} \text{ and } \omega \in \Omega. \tag{9.4.3}$$

Note that $X^{\mathbb{R}c}$ is contained in the center of $\mathbb{B}$. It is not necessarily true that any element $g \in \mathbb{B}$ can be written as $g = bX^{\hat{\lambda}}$ (or $g = X^{\hat{\lambda}}b$) with $b \in \mathcal{B}$ and $\hat{\lambda} \in \widehat{\Lambda}$. This possibly fails to be true in the cases where $S(D) \neq \emptyset$ (these are the cases for which $R \neq R^\bullet$, i.e., for which nonreduced extensions of $R^\bullet$ play a role in the theory; cf. §9.2.4). Straightening the elements of $\mathbb{B}$ is always possible as soon as quotients to double affine Weyl groups and double affine Hecke algebras are taken, as we shall see in a moment.

Recall from §9.3.1 the construction of commuting group elements $Y^\xi \in \mathcal{B}$ ($\xi \in \Lambda^d$). It leads to a presentation of $\mathcal{B}$ in terms of the group Λ^d (with its elements denoted by Y^ξ, $\xi \in \Lambda^d$) and the braid group $\mathcal{B}_0 = \mathcal{B}(A_0)$; see e.g., Macdonald (2003, §3.3). On the level of double affine braid groups it implies the following alternative presentation of $\mathbb{B}$ (recall that Da_0 equals $-\varphi$ if $\bullet = u$ and $-\theta$ if $\bullet = t$).

Proposition 9.4.2 *The double affine braid group $\mathbb{B}$ is isomorphic to the group generated by the groups $\mathcal{B}_0$, Λ^d and $\widehat{\Lambda}$, satisfying for $1 \le i \le n$, $\lambda \in \Lambda$ and $\xi \in \Lambda^d$,*

(a) $X^{\mathbb{R}c}$ *is contained in the center,*
(b) 1. $Y^{-\xi}T_i = T_iY^{-\xi}$ *if* $(\xi, \alpha_i^{d\vee}) = 0$,
  2. $T_iX^\lambda = X^\lambda T_i$ *if* $(\lambda, \alpha_i^\vee) = 0$,
(c) 1. $T_iY^{-\xi}T_i = Y^{-s_i\xi}$ *if* $(\xi, \alpha_i^{d\vee}) = 1$,
  2. $T_iX^\lambda T_i = X^{s_i\lambda}$ *if* $(\lambda, \alpha_i^\vee) = 1$,
(d) *if* $(\lambda, a_0^\vee) = 0$ *then* $(Y^{-(Da_0)^d}T_{s_{Da_0}}^{-1})X^\lambda = X^\lambda(Y^{-(Da_0)^d}T_{s_{Da_0}}^{-1})$,
(e) *if* $(\lambda, a_0^\vee) = 1$ *then* $(Y^{-(Da_0)^d}T_{s_{Da_0}}^{-1})X^\lambda(Y^{-(Da_0)^d}T_{s_{Da_0}}^{-1}) = p_{Da_0}^{-1}X^{s_{Da_0}\lambda}$, *where* $p_\alpha := X^{\mu_\alpha c}$ *for* $\alpha \in R_0$.

Recall that $u(\eta) \in W \simeq W_0 \ltimes \Lambda^d$ denotes the element of minimal length in $\tau(\eta)W_0$ for all $\eta \in \Lambda^d$. Then $u(\eta) = \tau(\eta)v(\eta)^{-1}$ with $v(\eta) \in W_0$ the element of minimal length such that $v(\eta)\eta = \eta_-$. Recall furthermore that $\Omega = \{u(\xi) \mid \xi \in \Lambda^{d+}_{\min}\}$.

The identification of the two different sets of generators of $\mathbb{B}$ is as follows: $T_0 = Y^{-(Da_0)^d} T^{-1}_{s_{Da_0}}$ and $u(\xi) = Y^{\xi} T^{-1}_{v(\xi)}$ for $\xi \in \Lambda^{d+}_{\min}$. Conversely, for $\xi = \eta_1 - \eta_2 \in \Lambda^d$ with $\eta_1, \eta_2 \in \Lambda^{d+}$, we have $Y^{\xi} = Y^{\eta_1}(Y^{\eta_2})^{-1}$ and $Y^{\eta_s} = T_{\tau(\eta_s)} = \omega T_{i_1} T_{i_2} \cdots T_{i_r}$ if $\tau(\eta_s) = \omega s_{i_1} s_{i_2} \cdots s_{i_r} \in W$ is a reduced expression ($\omega \in \Omega$ and $0 \le i_j \le n$).

Recall the set $S = S(D)$ given by (9.2.5). Write

$$V_i := X^{-a_i} T_i^{-1} \in \mathbb{B} \quad \forall i \in S. \tag{9.4.4}$$

Recall furthermore that $R = R^\bullet \cup \bigcup_{i \in S} W^\bullet(2a_i)$.

Definition 9.4.3

(i) The *double affine Weyl group* $\mathbb{W} = \mathbb{W}(D)$ is the quotient of $\mathbb{B}$ by the normal subgroup generated by T_j^2 and V_i^2 for $j \in \{0, \ldots, n\}$ and $i \in S$.

(ii) The *double affine Hecke algebra* $\mathbb{H}(\kappa, q) = \mathbb{H}(D; \kappa, q)$ is $\mathbb{C}[\mathbb{B}]/\widehat{I_{\kappa,q}}$, where $\widehat{I_{\kappa,q}}$ is the two-sided ideal of $\mathbb{C}[\mathbb{B}]$ generated by $X^{rc} - q^r$, $(T_j - \kappa_j)(T_j + \kappa_j^{-1})$ and $(V_i - \kappa_{2a_i})(V_i + \kappa_{2a_i}^{-1})$ for $r \in \mathbb{R}$, $j \in \{0, \ldots, n\}$ and $i \in S$.

Recall that W acts on $\widehat{\Lambda}$ by group automorphisms.

Proposition 9.4.4 $\mathbb{W} \simeq W \ltimes \widehat{\Lambda}$.

For the double affine Hecke algebra, note that we have canonical algebra homomorphisms $H(\kappa^\bullet) \to \mathbb{H}(\kappa, q)$ and $\mathbb{C}[\widehat{\Lambda}] \to \mathbb{H}(\kappa, q)$. We write $\tilde{h}$, $\tilde{X}_q^{\hat{\lambda}}$ and $\tilde{X}^{\lambda}$ for the images of $h \in \widehat{H}$, $X^{\hat{\lambda}} \in \mathbb{C}[\widehat{\Lambda}]$ ($\hat{\lambda} \in \widehat{\Lambda}$) and $X^{\lambda} \in \mathbb{C}[\Lambda]$ ($\lambda \in \Lambda$) in $\mathbb{H}(\kappa, q)$, respectively. Define a linear map

$$m\colon H(\kappa^\bullet) \otimes_{\mathbb{C}} \mathbb{C}[\Lambda] \to \mathbb{H}(\kappa, q), \quad h \otimes X^{\lambda} \mapsto \tilde{h}\tilde{X}^{\lambda} \quad (h \in H(\kappa^\bullet),\ \lambda \in \Lambda).$$

If $\hat{\lambda} = \lambda + rc \in \widehat{\Lambda}$ then we interpret $e_q^{\hat{\lambda}}$ as the endomorphism of $\mathbb{C}[T_\Lambda]$ by

$$(e_q^{\hat{\lambda}} p)(t) := t_q^{\hat{\lambda}} p(t) = q^r t^{\lambda} p(t), \quad r \in \mathbb{R},\ \lambda \in \Lambda.$$

Theorem 9.4.5

(i) *We have a unique algebra monomorphism* $\mathbb{H}(\kappa, q) \hookrightarrow \mathrm{End}_{\mathbb{C}}(\mathbb{C}[T_\Lambda])$ *defined by*

$$\tilde{h} \mapsto \pi_{\kappa,q}(h) \quad (h \in H(\kappa^\bullet)), \qquad \tilde{X}_q^{\hat{\lambda}} \mapsto e_q^{\hat{\lambda}} \quad (\hat{\lambda} \in \widehat{\Lambda}).$$

(ii) *The linear map m defines a complex linear isomorphism* $H(\kappa^\bullet) \otimes_{\mathbb{C}} \mathbb{C}[\Lambda] \xrightarrow{\sim} \mathbb{H}(\kappa, q)$.

To simplify notation, we omit the tilde when writing the elements in $\mathbb{H}$. With this convention $X_q^{\hat{\lambda}} = q^r X^{\lambda}$ in $\mathbb{H}(\kappa, q)$ if $\hat{\lambda} = \lambda + rc$.

Together with the Bernstein–Zelevinsky presentation of the extended affine Hecke algebra $H(\kappa^\bullet)$ (see §9.3.1) we conclude that $\mathbb{C}[T_{\Lambda^d}] \otimes_{\mathbb{C}} H_0(\kappa|_{R_0}) \otimes_{\mathbb{C}} \mathbb{C}[T_\Lambda] \simeq \mathbb{H}(\kappa, q)$ as complex vector spaces by mapping $e^{\xi} \otimes h \otimes e^{\lambda}$ to $Y^{\xi} h X^{\lambda}$ ($\xi \in \Lambda^d$, $h \in H_0(\kappa|_{R_0})$ and $\lambda \in \Lambda$). This is the *Poincaré–Birkhoff–Witt property* of the double affine Hecke algebra $\mathbb{H}(\kappa, q)$.

The *dual version of the cross relations* (9.3.2) and (9.3.3), now also including a commutation relation for T_0, is given as follows. Write $p(X) \in \mathbb{H}(\kappa, q)$ for the element corresponding to $p \in \mathbb{C}[T_\Lambda]$. In other words, $p(X) = \sum_{\lambda \in \Lambda} c_\lambda X^{\lambda}$ if $p(t) = \sum_{\lambda \in \Lambda} c_\lambda t^{\lambda}$.

Corollary 9.4.6 *Let $0 \le i \le n$ and $p \in \mathbb{C}[T_\Lambda] \subset \mathbb{H}(\kappa, q)$. Then in $\mathbb{H}(\kappa, q)$ we have*

$$T_i p(X) - (s_{i,q}p)(X)T_i = \left(\frac{\kappa_i - \kappa_i^{-1} + (\kappa_{2a_i} - \kappa_{2a_i}^{-1})X_q^{a_i}}{1 - X_q^{2a_i}}\right)(p(X) - (s_{i,q}p)(X)). \tag{9.4.5}$$

9.4.2 Duality Anti-Isomorphism

Recall from (9.2.8) that we have associated to $D = (R_0, \Delta_0, \bullet, \Lambda, \Lambda^d)$ the dual initial data $D^d = (R_0^d, \Delta_0^d, \bullet, \Lambda^d, \Lambda)$. We define the *dual double affine braid group* by $\mathbb{B}^d := \mathbb{B}(D^d)$. We add a superscript d to elements of $\mathbb{B}^d$ if confusion may arise. So we write ${}^dY^\lambda$ ($\lambda \in \Lambda$), ${}^dX^\xi$ ($\xi \in \Lambda^d$) and T_i^d ($0 \le i \le n$) in $\mathbb{B}^d$. The group generators T_i^d ($1 \le i \le n$) of the homomorphic image of $\mathcal{B}_0$ in $\mathbb{B}^d$ will usually be written without superscripts.

Recall that $Da_0 = -\varphi$ or $-\theta$ if $\bullet = u$ or t, respectively. Then $s_0 = \tau(-Da_0)^d)s_{Da_0} \in W^\bullet$ and $T_0 = Y^{-(Da_0)^d}T_{s_{Da_0}}^{-1} \in \mathbb{B}$. Dually, $D(a_0^d) = -\theta^d$, $s_0^d = \tau(\theta)s_\theta \in W^d$ and $T_0^d = Y^\theta T_{s_\theta}^{-1} \in \mathbb{B}^d$. The following result, in the present generality, is from Haiman (2006, §4) (in that paper the duality isomorphism is constructed, which is related to the duality anti-isomorphism below via an elementary anti-isomorphism). See also Cherednik (1995c,d), Ion (2003a), Sahi (1999) and Macdonald (2003) for special cases.

Theorem 9.4.7 *There exists a unique anti-isomorphism $\delta\colon \mathbb{B} \to \mathbb{B}^d$ satisfying*

$$\begin{aligned} \delta(X^{rc}) &= X^{rc} \quad (r \in \mathbb{R}), & \delta(Y^\xi) &= {}^dX^{-\xi} \quad (\xi \in \Lambda^d),\\ \delta(T_i) &= T_i \quad (i \in \{1, \ldots, n\}), & \delta(X^\lambda) &= {}^dY^{-\lambda} \quad (\lambda \in \Lambda). \end{aligned}$$

Note that $\delta^d = \delta^{-1}$, where δ^d is the duality anti-isomorphism with respect to the dual initial data D^d.

Recall that $v(\lambda)$ ($\lambda \in \Lambda$) is the element in W_0 of smallest length such that $\lambda_- = v(\lambda)\lambda$. Then $\tau(\lambda) = u^d(\lambda)v(\lambda)$ in W^d, where $u^d(\lambda)$ is the shortest element (with respect to the length function l^d on W^d) of the coset $\tau(\lambda)W_0$ in W^d. Then $\Omega = \{u^d(\lambda) \mid \lambda \in \Lambda_{\min}^+\}$, and in $\mathbb{B}^d$ we have ${}^dY^\lambda = u^d(\lambda)T_{v(\lambda)}$. Hence $\delta^d(u^d(\lambda)) = T_{v(\lambda)^{-1}}^{-1}X^{-\lambda}$ in $\mathbb{B}$. On the other hand, $\tau(\theta) = s_0^d s_\theta$ and $l^d(\tau(\theta)) = l(s_\theta) + 1$. Hence $T_0^d = {}^dY^\theta T_{s_\theta}^{-1}$ in $\mathbb{B}^d$ and $\delta^d(T_0^d) = T_{s_\theta}^{-1}X^{-\theta}$ in $\mathbb{B}$.

The next result shows that the duality anti-isomorphism from Theorem 9.4.7 descends to an anti-isomorphism between double affine Hecke algebras. In order to show this, we use the isomorphism $\mathcal{M} \xrightarrow{\sim} \mathcal{M}^d$ from the complex torus $\mathcal{M} = \mathcal{M}(D)$ onto $\mathcal{M}^d = \mathcal{M}(D^d)$ from Lemma 9.2.13.

Theorem 9.4.8 *The anti-isomorphism $\delta\colon \mathbb{B} \to \mathbb{B}^d$ descends to an anti-isomorphism*

$$\delta\colon \mathbb{H}(\kappa, q) \to \mathbb{H}^d(\kappa^d, q) := \mathbb{H}(D^d; \kappa^d, q).$$

For instance, the cross relations (9.4.5) in $\mathbb{H}(\kappa, q)$ for $i \in \{1, \ldots, n\}$ match with the cross relations (9.3.4) in $\mathbb{H}^d(\kappa^d, q)$ through the anti-isomorphism δ. Note that $\delta^d = \delta^{-1}$ also on the level of the double affine Hecke algebra, where δ^d is the duality anti-isomorphism with respect to the dual data (D^d, κ^d).

9.4.3 Evaluation Formulas

We follow closely Cherednik (1995d) (which corresponds to the special case that $(\Lambda, \Lambda^d) = (P(R_0), P(R_0^d)))$. See also Stokman (2000a), Macdonald (2003) and Haiman (2006). We assume in this subsection that q and $\kappa \in \mathcal{M}(D)$ satisfy (9.3.5). We consider the nonsymmetric Macdonald–Koornwinder polynomials $P_\lambda \in \mathbb{C}[T_\Lambda]$ ($\lambda \in \Lambda$) associated to (D, κ, q), as well as the dual nonsymmetric Macdonald–Koornwinder polynomials $P^d_\xi \in \mathbb{C}[T_{\Lambda^d}]$ ($\xi \in \Lambda^d$) associated to (D^d, κ^d, q).

Write $\gamma^d_{\xi,q} = \gamma_{\xi,q}(D^d; \kappa^{d\bullet}) \in T_\Lambda$ ($\xi \in \Lambda^d$) for the spectral points with respect to the dual initial data (D^d, κ^d, q). Concretely, they are given by

$$\gamma^d_{\xi,q} = q^\xi \prod_{\alpha \in R_0^+} {}^d\upsilon_\alpha^{\eta((\xi,\alpha^{d\vee}))\alpha^\vee} \text{ with } {}^d\upsilon_\alpha := (\kappa^d_{\alpha^d})^{\frac{1}{2}} (\kappa^d_{\mu_{\alpha^d} c + \alpha^d})^{\frac{1}{2}} = \kappa_\alpha^{\frac{1}{2}} \kappa_{2\alpha}^{\frac{1}{2}} \text{ for } \alpha \in R_0^+.$$

Cherednik's basic representations

$$\pi_{D;\kappa,q} : H(\kappa^\bullet) \hookrightarrow \mathrm{End}_{\mathbb{C}}(\mathbb{C}[T_\Lambda]), \quad \pi_{D^d;\kappa^d,q} : H(\kappa^{d\bullet}) \hookrightarrow \mathrm{End}_{\mathbb{C}}(\mathbb{C}[T_{\Lambda^d}])$$

extend to algebra maps

$$\hat{\pi} : \mathbb{H}(\kappa, q) \hookrightarrow \mathrm{End}_{\mathbb{C}}(\mathbb{C}[T_\Lambda]), \quad \hat{\pi}^d : \mathbb{H}(\kappa^d, q) \hookrightarrow \mathrm{End}_{\mathbb{C}}(\mathbb{C}[T_{\Lambda^d}])$$

by $\hat{\pi}(X_q^{\hat{\lambda}}) = e_q^{\hat{\lambda}}$ for $\hat{\lambda} \in \widehat{\Lambda}$ and $\hat{\pi}^d({}^dX_q^{\hat{\xi}}) = e_q^{\hat{\xi}}$ for $\hat{\xi} \in \widehat{\Lambda^d}$. Note that

$$\gamma^d_{0,q} = \prod_{\alpha \in R_0^+} {}^d\upsilon_\alpha^{-\alpha^\vee} \in T_\Lambda, \quad \gamma_{0,q} = \prod_{\alpha \in R_0^+} \upsilon_{\alpha^d}^{-\alpha^{d\vee}} \in T_{\Lambda^d}.$$

Definition 9.4.9 Define *evaluation maps* $\mathrm{Ev} : \mathbb{H}(\kappa, q) \to \mathbb{C}$ and $\mathrm{Ev}^d : \mathbb{H}^d(\kappa^d, q) \to \mathbb{C}$ by

$$\mathrm{Ev}(Z) := (\hat{\pi}(Z)1)(\gamma^d_{0,q}) \quad (Z \in \mathbb{H}(\kappa, q)), \qquad \mathrm{Ev}^d(Z) := (\hat{\pi}^d(Z)1)(\gamma_{0,q}) \quad (Z \in \mathbb{H}^d(\kappa^d, q)).$$

The following lemma is crucial for the duality of the (nonsymmetric) Macdonald–Koornwinder polynomials.

Lemma 9.4.10 *For all* $Z \in \mathbb{H}(\kappa, q)$ *we have* $\mathrm{Ev}^d(\delta(Z)) = \mathrm{Ev}(Z)$.

Write $\mathbb{H} = \mathbb{H}(\kappa, q)$ and $\mathbb{H}^d = \mathbb{H}^d(\kappa^d, q)$. Define bilinear forms

$$B : \mathbb{H} \times \mathbb{H}^d \to \mathbb{C}, \quad (Z, \tilde{Z}) \mapsto \mathrm{Ev}(\delta^d(\tilde{Z})Z), \qquad B^d : \mathbb{H}^d \times \mathbb{H} \to \mathbb{C}, \quad (\tilde{Z}, Z) \mapsto \mathrm{Ev}^d(\delta(Z)\tilde{Z}).$$

Corollary 9.4.11 $B(Z, \tilde{Z}) = B^d(\tilde{Z}, Z)$ *for* $Z \in \mathbb{H}$ *and* $\tilde{Z} \in \mathbb{H}^d$.

The following elementary lemma provides convenient tools to derive the evaluation formula for nonsymmetric Macdonald–Koornwinder polynomials.

Lemma 9.4.12 *Let* $p \in \mathbb{C}[T_\Lambda]$, $\tilde{p} \in \mathbb{C}[T_{\Lambda^d}]$, $Z, Z_1, Z_2 \in \mathbb{H}$ *and* $\tilde{Z}, \tilde{Z}_1, \tilde{Z}_2 \in \mathbb{H}^d$. *Then*

(i) $B(Z_1Z_2, \tilde{Z}) = B(Z_2, \delta(Z_1)\tilde{Z})$,
(ii) $B(ZT_i, \tilde{Z}) = \kappa_i B(Z, \tilde{Z})$ *for* $0 \le i \le n$,
(iii) $B((\hat{\pi}(Z)(p))(X), \tilde{Z}) = B(Zp(X), \tilde{Z})$.

Lemma 9.4.12 and Corollary 9.4.11 imply the following proposition.

Proposition 9.4.13

(i) *For* $\xi \in \Lambda^d$ *and* $p \in \mathbb{C}[T_\Lambda]$ *we have* $P^d_\xi(\gamma_{0,q})p(\gamma^d_{\xi,q}) = B(p, P^d_\xi)$.

(ii) *For* $\lambda \in \Lambda$ *and* $\tilde{p} \in \mathbb{C}[T_{\Lambda^d}]$ *we have* $P_\lambda(\gamma^d_{0,q})\tilde{p}(\gamma_{\lambda,q}) = B^d(\tilde{p}, P_\lambda)$.

Corollary 9.4.14 (Duality) *For* $\lambda \in \Lambda$ *and* $\xi \in \Lambda^d$ *we have*

$$P^d_\xi(\gamma_{0,q})P_\lambda(\gamma^d_{\xi,q}) = P_\lambda(\gamma^d_{0,q})P^d_\xi(\gamma_{\lambda,q}).$$

The next aim is to explicitly evaluate $\mathrm{Ev}(P_\lambda) = P_\lambda(\gamma^d_{0,q})$. From the results on the pairing B above and from Proposition 9.3.5 one first derives an important intermediate result which shows how the action of generators of the double affine Hecke algebra on monic nonsymmetric Macdonald–Koornwinder polynomials P_λ can be explicitly expressed in terms of an action on the degree $\lambda \in \Lambda$ of P_λ.

Recall that W^d acts on Λ by $(w\tau(\lambda), \lambda') \mapsto w(\lambda + \lambda')$ ($w \in W_0$ and $\lambda, \lambda' \in \Lambda$).

Proposition 9.4.15

(i) *Let* $i \in \{1, \ldots, n\}$. *If* $s^d_i\lambda = \lambda$ *then* $\hat{\pi}(T_i)P_\lambda = \kappa^d_i P_\lambda$.

(ii) *If* $s^d_0\lambda = \lambda$ *then* $\hat{\pi}(\delta^d(T^d_0))P_\lambda = \kappa^d_0 P_\lambda$.

(iii) *Let* $\lambda \in \Lambda^+_{\min}$, *then* $\hat{\pi}(\delta^d(u^d(\lambda)))1 = \kappa^d_{v(\lambda)}P_{-\lambda_-}$.

(iv) *Suppose* $1 \le i \le n$, $\lambda \in \Lambda$ *such that* $(\lambda, \alpha_i^\vee) > 0$. *Then*

$$\hat{\pi}(T_i)P_\lambda = \frac{\kappa^d_i - (\kappa^d_i)^{-1} + \left(\kappa^d_{2\alpha^d_i} - (\kappa^d_{2\alpha^d_i})^{-1}\right)\gamma^{\alpha^d_i}_{\lambda,q}}{1 - \gamma^{2\alpha^d_i}_{\lambda,q}} P_\lambda + (\kappa^d_i)^{-1}P_{s^d_i\lambda}.$$

(v) *Suppose* $\lambda \in \Lambda$ *such that* $a^d_0(\lambda) > 0$. *Then*

$$\hat{\pi}(\delta^{-1}(T^d_0))P_\lambda = \frac{\kappa^d_0 - (\kappa^d_0)^{-1} + \left(\kappa^d_{2a^d_0} - (\kappa^d_{2a^d_0})^{-1}\right)q_\theta\gamma^{-\theta^d}_{\lambda,q}}{1 - q^2_\theta\gamma^{-2\theta^d}_{\lambda,q}} P_\lambda + \kappa^d_{v(s^d_0\lambda)}(\kappa^d_{v(\lambda)})^{-1}P_{s^d_0\lambda}$$

(*note that* $a^d_0 = \mu_\theta c - \theta^d = \mu_\theta(c - \theta^\vee)$ *and* $s^d_0\lambda = \lambda + (1 - (\lambda, \theta^\vee))\theta$).

The proposition gives the following recursion relations for $\mathrm{Ev}(P_\lambda) = P_\lambda(\gamma^d_{0,q})$.

Corollary 9.4.16

(i) $\mathrm{Ev}(P_\lambda) = (\kappa^d_{v(\lambda)})^{-1}$ $(\lambda \in \Lambda^+_{\min})$.

(ii) $$\mathrm{Ev}(P_{s^d_i\lambda}) = \frac{\left(1 - \kappa^d_i\kappa^d_{2\alpha^d_i}\gamma^{\alpha^d_i}_{\lambda,q}\right)\left(1 + \kappa^d_i(\kappa^d_{2\alpha^d_i})^{-1}\gamma^{\alpha^d_i}_{\lambda,q}\right)}{1 - \gamma^{2\alpha^d_i}_{\lambda,q}}\mathrm{Ev}(P_\lambda) \quad (\lambda \in \Lambda,\ 1 \le i \le n,\ a^d_i(\lambda) > 0).$$

(iii) $$\mathrm{Ev}(P_{s^d_0\lambda}) = \frac{\kappa^d_{v(\lambda)}}{\kappa^d_0\kappa^d_{v(s^d_0\lambda)}} \frac{\left(1 - q_\theta\kappa^d_0\kappa^d_{2a^d_0}\gamma^{-\theta^d}_{\lambda,q}\right)\left(1 + q_\theta\kappa^d_0(\kappa^d_{2a^d_0})^{-1}\gamma^{-\theta^d}_{\lambda,q}\right)}{1 - q^2_\theta\gamma^{-2\theta^d}_{\lambda,q}}\mathrm{Ev}(P_\lambda) \quad (\lambda \in \Lambda,\ a^d_0(\lambda) > 0).$$

Recall the definition $c_a = c_a^{\kappa,q}(\cdot; D) \in \mathbb{C}(T_\Lambda)$ for $a \in R^\bullet$ from (9.3.7). The dual version is denoted by $c_a^d = c_a^{\kappa^d,q}(\cdot; D^d)$ $(a \in R^{d\bullet})$. Concretely, for $a \in R^{d\bullet}$, $c_a^d \in \mathbb{C}(T_{\Lambda^d})$ is given by

$$c_a^d(t) := \frac{(1 - \kappa_a^d \kappa_{2a}^d t_q^a)(1 + \kappa_a^d (\kappa_{2a}^d)^{-1} t_q^a)}{1 - t_q^{2a}}.$$

We also set $c_w^d := \prod_{a \in R^{d\bullet,+} \cap w^{-1}(R^{d\bullet,-})} c_a^d \in \mathbb{C}(T_{\Lambda^d})$ $(w \in W^d)$. An induction argument now gives the explicit evaluation formula for the nonsymmetric Macdonald–Koornwinder polynomials (see Cherednik, 1995d; Stokman, 2000a; Macdonald, 2003).

Theorem 9.4.17 *For $\lambda \in \Lambda$ we have* $\mathrm{Ev}(P_\lambda) = (\kappa_{\tau(\lambda)}^d)^{-1} c_{u^d(\lambda)}^d(\gamma_{0,q})$.

9.4.4 Normalized Nonsymmetric Macdonald–Koornwinder Polynomials and Duality

The treatment in this subsection is close to Cherednik (1995d) and Stokman (2000a), which deal with the case that $(\Lambda, \Lambda^d) = (P(R_0), P(R_0^d))$ and the $\mathrm{C}^\vee\mathrm{C}$ case, respectively. We assume that q and $\kappa \in \mathcal{M}(D)$ satisfy (9.3.5). Then $P_\lambda(\gamma_{0,q}^d) \neq 0$ and $P_\xi^d(\gamma_{0,q}) \neq 0$ for all $\lambda \in \Lambda$ and $\xi \in \Lambda^d$ in view of the evaluation formula (Theorem 9.4.17).

Recall that the Macdonald–Koornwinder polynomials $P_\lambda \in \mathbb{C}[T_\Lambda]$ $(\lambda \in \Lambda)$ and $P_\xi^d \in \mathbb{C}[T_{\Lambda^d}]$ $(\xi \in \Lambda^d)$ satisfy $\hat{\pi}(p(Y))P_\lambda = p(\gamma_{\lambda,q}^{-1})P_\lambda$ and $\hat{\pi}^d(r(^dY))P_\xi^d = r((\gamma_{\xi,q}^d)^{-1})P_\xi^d$ for all $p \in \mathbb{C}[T_{\Lambda^d}]$ and $r \in \mathbb{C}[T_\Lambda]$. This motivates the following notation for the normalized nonsymmetric Macdonald–Koornwinder polynomials.

Definition 9.4.18 (Normalized nonsymmetric Macdonald–Koornwinder polynomials)

$$E(\gamma_{\lambda,q}^{-1}; \cdot) := \frac{P_\lambda}{P_\lambda(\gamma_{0,q}^d)} \in \mathbb{C}[T_\Lambda] \ \ (\lambda \in \Lambda), \qquad E^d((\gamma_{\xi,q}^d)^{-1}; \cdot) := \frac{P_\xi^d}{P_\xi^d(\gamma_{0,q})} \in \mathbb{C}[T_{\Lambda^d}] \ \ (\xi \in \Lambda^d).$$

We denote by $E^\circ(\gamma_{\lambda,q}; \cdot) \in \mathbb{C}[T_\Lambda]$ the normalized Macdonald–Koornwinder polynomial with respect to the inverted parameters (κ^{-1}, q^{-1}) (and similarly for $E^{d\circ}(\gamma_{\xi,q}^d; \cdot)$).

For $\lambda \in \Lambda$ we have $E(\gamma_{\lambda,q}^{-1}; \gamma_{0,q}^d) = 1$. From §9.4.3 we immediately get the following self-duality of the normalized nonsymmetric Macdonald–Koornwinder polynomials (cf. Cherednik, 1995d; Sahi, 1999; Macdonald, 2003; Haiman, 2006).

Corollary 9.4.19 *For all $p \in \mathbb{C}[T_\Lambda]$, $r \in \mathbb{C}[T_{\Lambda^d}]$, $\lambda \in \Lambda$ and $\xi \in \Lambda^d$, we have*

$$B(p, E^d((\gamma_{\xi,q}^d)^{-1}; \cdot)) = p(\gamma_{\xi,q}^d), \qquad B(E(\gamma_{\lambda,q}^{-1}; \cdot), r) = r(\gamma_{\lambda,q}).$$

In particular, $E(\gamma_{\lambda,q}^{-1}; \gamma_{\xi,q}^d) = E^d((\gamma_{\xi,q}^d)^{-1}; \gamma_{\lambda,q})$ *for $\lambda \in \Lambda$ and $\xi \in \Lambda^d$.*

9.4.5 Polynomial Fourier Transform

For the remainder of the chapter we assume that the parameters q and $\kappa \in \mathcal{M}(D)$ satisfy the more restrictive parameter conditions (9.3.6).

In order to compute the norms of the normalized nonsymmetric Macdonald–Koornwinder polynomials it is convenient to formulate the explicit formulas from Proposition 9.4.15 in terms of properties of a Fourier transform whose kernel is given by the normalized nonsymmetric Macdonald–Koornwinder polynomial. For this we first need to consider the adjoint of the double affine Hecke algebra action with respect to the sesquilinear form

$$\langle p_1, p_2\rangle := \int_{T_\Lambda^u} p_1(t)\,\overline{p_2(t)}\,v(t)\,d_ut, \qquad p_1, p_2 \in \mathbb{C}[T_\Lambda],$$

and express it in terms of an explicit antilinear anti-isomorphism of the double affine Hecke algebra.

Lemma 9.4.20 *There exists a unique antilinear antialgebra isomorphism* $^\ddagger\colon \mathbb{H}(\kappa, q) \to \mathbb{H}(\kappa^{-1}, q^{-1})$ *satisfying* $T_w^\ddagger = T_w^{-1}$ $(w \in W)$ *and* $(X^\lambda)^\ddagger = X^{-\lambda}$ $(\lambda \in \Lambda)$. *In addition,*

$$\langle \hat{\pi}_{\kappa,q}(h)p_1, p_2\rangle = \langle p_1, \hat{\pi}_{\kappa^{-1},q^{-1}}(h^\ddagger)p_2\rangle, \qquad h \in \mathbb{H}.$$

Define

$$\mathcal{S} := \{\gamma_{\lambda,q} \mid \lambda \in \Lambda\} \subset T_{\Lambda^d}, \qquad \mathcal{S}^d := \{\gamma^d_{\xi,q} \mid \xi \in \Lambda^d\} \subset T_\Lambda,$$

and write $F(\mathcal{S})$ (respectively $F(\mathcal{S}^d)$) for the space of finitely supported complex-valued functions on $\mathcal{S}$ (respectively $\mathcal{S}^d$). The following lemma follows easily from the results in §9.2.2 and from Proposition 9.3.5.

Lemma 9.4.21 *There exists a unique algebra homomorphism* $\hat{\rho}^d\colon \mathbb{H}^d \to \mathrm{End}_{\mathbb{C}}(F(\mathcal{S}))$ *satisfying*

$$(\hat{\rho}^d(T_i^d)g)(\gamma_{\lambda,q}) = \begin{cases} \kappa_i^d g(\gamma_{\lambda,q}) + (\kappa_i^d)^{-1} c^d_{a_i^d}(\gamma_{\lambda,q})(g(\gamma_{s_i^d\lambda,q}) - g(\gamma_{\lambda,q})) & \text{if } s_i^d\lambda \neq \lambda,\ 0 \le i \le n, \\ \kappa_i^d g(\gamma_{\lambda,q}) & \text{if } s_i^d\lambda = \lambda,\ 0 \le i \le n, \end{cases}$$

$$(\hat{\rho}^d(\omega)g)(\gamma_{\lambda,q}) = g(\gamma_{\omega^{-1}\lambda,q}) \quad \text{if } \omega \in \Omega^d,$$

$$(\hat{\rho}^d({}^dX^\xi)g)(\gamma_{\lambda,q}) = \gamma^\xi_{\lambda,q}\, g(\gamma_{\lambda,q}) \quad \text{if } \xi \in \Lambda^d.$$

Definition 9.4.22 We define the *polynomial Fourier transform* $\mathcal{F}\colon \mathbb{C}[T_\Lambda] \to F(\mathcal{S})$ by

$$(\mathcal{F}p)(\gamma) := \langle p, E^\circ(\gamma;\cdot)\rangle, \qquad p \in \mathbb{C}[T_\Lambda],\ \gamma \in \mathcal{S}.$$

For generators $X \in \mathbb{H}$ and $p \in \mathbb{C}[T_\Lambda]$, we can re-express $\mathcal{F}(\hat{\pi}(X)p)$ as an explicit linear operator acting on $\mathcal{F}p \in F(\mathcal{S})$ by using Lemma 9.4.20, Proposition 9.4.15 and Theorem 9.4.17. It gives the following result (the first part of the theorem follows easily from the duality anti-isomorphism).

Theorem 9.4.23

(i) *The following formulas define an algebra isomorphism* $\Phi\colon \mathbb{H} \to \mathbb{H}^d$:

$$\Phi(T_i) = T_i^d \quad (1 \le i \le n), \qquad \Phi(T_{s_\theta}^{-1}X^{-\theta}) = T_0^d,$$

$$\Phi(T^{-1}_{v(\lambda)^{-1}}X^{-\lambda}) = u^d(\lambda)^{-1} \quad (\lambda \in \Lambda^+_{\min}), \qquad \Phi(Y^\xi) = {}^dX^{-\xi} \quad (\xi \in \Lambda^d).$$

(ii) $\mathcal{F} \circ \hat{\pi}(h) = \hat{\rho}^d(\Phi(h)) \circ \mathcal{F}$ *for all* $h \in \mathbb{H}$.

9.4.6 Intertwiners and Norm Formulas

Define $I_i^d := [T_i^d, {}^dX_q^{a_i^d}] \in \mathbb{H}^d$ $(0 \le i \le n)$ and $I_\omega^d := \omega \in \mathbb{H}^d$ $(\omega \in \Omega^d)$. The following theorem extends results from Cherednik (1997b), Sahi (1999) and Stokman (2000a).

Theorem 9.4.24 *For a reduced expression* $w = \omega s_{i_1}^d s_{i_2}^d \cdots s_{i_r}^d \in W^d$ $(\omega \in \Omega^d,\ 0 \le i_j \le n)$, *the expression*

$$I_w^d := I_\omega^d I_{i_1}^d I_{i_2}^d \cdots I_{i_r}^d \in \mathbb{H}^d$$

is well defined (independent of the choice of reduced expression). In addition,

$$\hat{\pi}^d(I_w^d) = r_w^d \cdot w_q \in \mathrm{End}_{\mathbb{C}}(\mathbb{C}[T_{\Lambda^d}]),$$

where $r_w^d := \prod_{a \in {}^dR^{\bullet +} \cap w({}^dR^{\bullet -})} r_a^d \in \mathbb{C}[T_{\Lambda^d}]$ *with* $r_a^d(t) := (\kappa_a^d)^{-1}(1 - \kappa_a^d \kappa_{2a}^d t_q^a)(1 + \kappa_a^d (\kappa_{2a}^d)^{-1} t_q^a)$ *for* $t \in T_{\Lambda^d}$. *Further, in* $\mathbb{H}^d$,

$$I_w^d I_{w^{-1}}^d = r_w^d({}^dX)(w_q r_{w^{-1}}^d)({}^dX) \quad \text{and} \quad I_w^d p({}^dX) = (w_q p)({}^dX) I_w^d \quad (w \in W^d, p \in \mathbb{C}[T_{\Lambda^d}]).$$

The $I_w^d \in \mathbb{H}^d$ $(w \in W^d)$ are called the *dual intertwiners*. The *intertwiners* are defined by $\mathfrak{I}_w := \delta^d(I_w^d) \in \mathbb{H}$ $(w \in W^d)$. Then in $\mathbb{H}$,

$$\mathfrak{I}_{w^{-1}} \mathfrak{I}_w = r_w^d(Y^{-1})(w_q r_{w^{-1}}^d)(Y^{-1}) \quad \text{and} \quad p(Y^{-1})\mathfrak{I}_w = \mathfrak{I}_w(w_q p)(Y^{-1}) \quad (w \in W^d, p \in \mathbb{C}[T_{\Lambda^d}]).$$

Proposition 9.4.15 and Theorem 9.4.17 give the following result.

Proposition 9.4.25

(i) *If* $\lambda \in \Lambda$ *and* $0 \le i \le n$ *satisfy* $s_i^d \lambda \neq \lambda$ *then* $\hat{\pi}(\mathfrak{I}_i)E(\gamma_{\lambda,q}^{-1};\cdot) = \gamma_{\lambda,q}^{-a_i^d} r_{a_i^d}^d(\gamma_{\lambda,q}) E(\gamma_{s_i^d\lambda,q}^{-1};\cdot)$.

(ii) *If* $\lambda \in \Lambda$ *and* $0 \le i \le n$ *satisfy* $s_i^d \lambda = \lambda$ *then* $\hat{\pi}(\mathfrak{I}_i)E(\gamma_{\lambda,q}^{-1};\cdot) = 0$.

(iii) *If* $\omega \in \Omega^d$ *and* $\lambda \in \Lambda$ *then* $\hat{\pi}(\mathfrak{I}_\omega)E(\gamma_{\lambda,q}^{-1};\cdot) = E(\gamma_{\omega^{-1}\lambda,q}^{-1};\cdot)$.

This proposition shows that intertwiners can be used to create the nonsymmetric Macdonald–Koornwinder polynomial from the constant polynomial $E(\gamma_{0,q}^{-1};\cdot) \equiv 1$ (cf., e.g., Macdonald, 2003, §5.10). We now use this observation to express the norms of the nonsymmetric Macdonald–Koornwinder polynomial in terms of the nonzero constant term

$$\langle 1, 1\rangle = \int_{T_\Lambda^u} v(t)\, d_u t = \frac{\mathcal{C}(\gamma_{0,q}^d)}{\#W_0} \int_{T_\Lambda^u} v_+(t)\, d_u t.$$

The constant term $\langle 1, 1\rangle$ is a q-analog and root system generalization of the Selberg integral. Its explicit evaluation was conjectured by Macdonald (2000) in the case $(\Lambda, \Lambda^d) = (P(R_0), P(R_0^d))$. By various methods it was evaluated in special cases (for references we refer to the detailed discussions in Macdonald (2000), Cherednik (1995b); see Forrester and Warnaar (2008) for a survey on Selberg integrals and their generalizations and applications). A uniform proof in the case $(\Lambda, \Lambda^d) = (P(R_0), P(R_0^d))$ using shift operators was given in Cherednik (1995b, Theorem 0.1) (see Stokman, 2000a for the $C^\vee C$ case). Write

$$N(\lambda) := \frac{\langle E(\gamma_{\lambda,q}^{-1};\cdot), E^\circ(\gamma_{\lambda,q};\cdot)\rangle}{\langle 1, 1\rangle} \quad (\lambda \in \Lambda), \tag{9.4.6}$$

$$c^d_w := \prod_{a\in R^{d\bullet,+}\cap w^{-1}(R^{d\bullet,-})} c^d_a \in \mathbb{C}[T_{\Lambda^d}], \quad c^d_{-w} := \prod_{a\in R^{d\bullet,+}\cap w^{-1}(R^{d\bullet,-})} c^d_{-a} \in \mathbb{C}[T_{\Lambda^d}] \quad (w \in W^d).$$

Warning: $c^d_{-w}(t) = c^d_w(t^{-1})$ is valid only if $w \in W_0$.

Theorem 9.4.26

(i) *For* $\lambda \in \Lambda$, $N(\lambda) = \frac{c^d_{-u^d(\lambda)}(\gamma_{0,q})}{c^d_{u^d(\lambda)}(\gamma_{0,q})}$, *which is nonzero for all* $\lambda \in \Lambda$.

(ii) *The transform* $\mathcal{F}\colon \mathbb{C}[T_\Lambda] \to F(\mathcal{S})$ *is a linear bijection with inverse* $\mathcal{G}\colon F(\mathcal{S}) \to \mathbb{C}[T_\Lambda]$ *given by* $(\mathcal{G}f)(t) := \langle 1,1\rangle^{-1}\sum_{\lambda\in\Lambda} N(\lambda)^{-1} f(\gamma_{\lambda,q})E(\gamma^{-1}_{\lambda,q};t)$ *for* $f \in F(\mathcal{S})$ *and* $t \in T_\Lambda$.

Part (i) of the above theorem should be compared with Cherednik (2005, Prop. 3.4.1) and Macdonald (2003, (5.2.11)) (it originates from Cherednik (1995d) in the case $(\Lambda, \Lambda^d) = (P(R_0), P(R^d_0))$ and Sahi (1999), Stokman (2000a) in the C$^\vee$C case). The first part of the theorem follows from Proposition 9.4.25 and the fact that the intertwiners behave nicely with respect to the anti-involution ‡ (see Lemma 9.4.20): $\mathcal{I}^\ddagger_i = \mathcal{I}_i$ $(0 \le i \le n)$ in $\mathbb{H}(\kappa^{-1}, q^{-1})$ and $\mathcal{I}^\ddagger_\omega = \mathcal{I}_{\omega^{-1}}$ $(\omega \in \Omega^d)$ in $\mathbb{H}(\kappa^{-1}, q^{-1})$. The second part of the theorem is immediate from the first and from the biorthogonality of the nonsymmetric Macdonald–Koornwinder polynomials (see Theorem 9.3.13).

In combination with the evaluation formula (see Theorem 9.4.17) we get the following *norm formula*.

Corollary 9.4.27 *For all* $\lambda \in \Lambda$, $\frac{\langle P_\lambda, P^\circ_\lambda\rangle}{\langle 1,1\rangle} = (\kappa^d_{u^d(\lambda)})^2 c^d_{u^d(\lambda)}(\gamma_{0,q})c^d_{-u^d(\lambda)}(\gamma_{0,q})$.

9.4.7 Normalized Symmetric Macdonald–Koornwinder Polynomials

We still assume that q and $\kappa \in \mathcal{M}(D)$ satisfy (9.3.6). The results in this subsection are from Cherednik (1995d) in the case $(\Lambda, \Lambda^d) = (P(R_0), P(R^d_0))$ and from Sahi (1999), Stokman (2000a) in the C$^\vee$C case.

For $\lambda \in \Lambda^-$ define $E^+(\gamma^{-1}_{\lambda,q};\cdot) \in \mathbb{C}[T_\Lambda]^{W_0}$ by $E^+(\gamma^{-1}_{\lambda,q};\cdot) := \hat{\pi}_{\kappa,q}(C_+)E(\gamma^{-1}_{\lambda,q};\cdot)$, where (recall) $C_+ := \frac{1}{\sum_{w\in W_0}\kappa^2_w}\sum_{w\in W_0}\kappa_w T_w$. We call $E^+(\gamma^{-1}_{\lambda,q};\cdot)$ the *normalized symmetric Macdonald–Koornwinder polynomial* of degree $\lambda \in \Lambda^-$.

Lemma 9.4.28

(i) $E^+(\gamma^{-1}_{\lambda,q};\cdot) = \hat{\pi}(C_+)E(\gamma^{-1}_{w\lambda,q};\cdot)$ *for all* $w \in W_0$ *and* $\lambda \in \Lambda^-$.

(ii) $E^+(\gamma^{-1}_{\lambda,q};\gamma^d_{0,q}) = 1$ *for all* $\lambda \in \Lambda^-$.

(iii) $\{E^+(\gamma^{-1}_{\lambda,q};\cdot)\}_{\lambda\in\Lambda^-}$ *is a basis of* $\mathbb{C}[T_\Lambda]^{W_0}$ *satisfying, for all* $p \in \mathbb{C}[T_{\Lambda^d}]^{W_0}$,

$$D_p(E^+(\gamma^{-1}_{\lambda,q};\cdot)) = p(\gamma^{-1}_{\lambda,q})E^+(\gamma^{-1}_{\lambda,q};\cdot) = p(q^{-\lambda}\gamma^{-1}_{0,q})E^+(\gamma^{-1}_{\lambda,q};\cdot).$$

(iv) $E^+(\gamma^{-1}_{\lambda,q};\cdot) = P^+_{\lambda_+}(\cdot)/P^+_{\lambda_+}(\gamma^d_{0,q})$ *for all* $\lambda \in \Lambda^-$.

(v) $E^+(\gamma^{-1}_{\lambda,q};\gamma^d_{\xi,q}) = E^{+d}((\gamma^d_{\xi,q})^{-1};\gamma_{\lambda,q})$ *for all* $\lambda \in \Lambda^-$ *and* $\xi \in \Lambda^{d-}$.

As before we write superindex $\circ$ to indicate that the parameters (κ, q) are inverted. The nonsymmetric and symmetric Macdonald–Koornwinder polynomials with inverted parameters (κ^{-1}, q^{-1}) can be explicitly expressed in terms of those with parameters (κ, q). The result is as follows (see Cherednik, 2005, §3.3.2 for $(\Lambda, \Lambda^d) = (P(R_0), P(R_0^d))$ and Stokman, 2014, (2.5) in the twisted case).

Proposition 9.4.29

(i) *For all* $\lambda \in \Lambda$, $E^\circ(\gamma_{\lambda,q}; t^{-1}) = \kappa_{w_0}^{-1}(\hat{\pi}_{\kappa,q}(T_{w_0})E(\gamma^{-1}_{-w_0\lambda,q}; \cdot))(t)$ *in* $\mathbb{C}[T_\Lambda]$, *where* $w_0 \in W_0$ *is the longest Weyl group element and* $t \in T_\Lambda$.

(ii) *For all* $\lambda \in \Lambda^-$ *we have in* $\mathbb{C}[T_\Lambda]^{W_0}$ *that*

$$E^{+\circ}(\gamma_{\lambda,q}; t^{-1}) = E^+(\gamma^{-1}_{-w_0\lambda,q}; t), \qquad E^{+\circ}(\gamma_{\lambda,q}; t) = E^+(\gamma^{-1}_{\lambda,q}; t).$$

By means of intertwiners or by Proposition 9.4.15 it is now possible to expand the normalized symmetric Macdonald–Koornwinder polynomials in nonsymmetric ones. This in turn leads to an explicit expression of the quadratic norms of the symmetric Macdonald–Koornwinder polynomials in terms of those of the nonsymmetric ones. Recall the rational function $\mathcal{C}(\cdot) = \mathcal{C}(\cdot; D; \kappa, q) \in \mathbb{C}(T_\Lambda)$ from (9.3.12). Write $\mathcal{C}^d(\cdot) = \mathcal{C}(\cdot; D^d; \kappa^d, q) \in \mathbb{C}(T_{\Lambda^d})$ for its dual version.

Theorem 9.4.30

(i) *For* $\lambda \in \Lambda^-$ *we have* $P^+_{\lambda_+}(t) = \sum_{\mu \in W_0\lambda} \left(\prod_{\alpha \in R_0^+ \cap v(\mu)R_0^-} c^d_\alpha(\gamma_{\lambda,q}) \right) P_\mu(t)$.

(ii) *For* $\lambda \in \Lambda^-$ *we have* $E^+(\gamma^{-1}_{\lambda,q}; t) = \sum_{\mu \in W_0\lambda} \dfrac{\mathcal{C}^d(\gamma_{\mu,q})}{\mathcal{C}^d(\gamma_{0,q})} E(\gamma^{-1}_{\mu,q}; t)$.

(iii) *For* $\lambda, \mu \in \Lambda^-$ *we have* $\dfrac{\langle E^+(\gamma^{-1}_{\lambda,q}; \cdot), E^+(\gamma^{-1}_{\mu,q}; \cdot)\rangle_+}{\langle 1, 1\rangle_+} = \delta_{\lambda,\mu} \dfrac{\mathcal{C}^d(\gamma_{\lambda,q})N(\lambda)}{\mathcal{C}^d(\gamma_{0,q})}$, *where* $\langle p, r\rangle_+ := \int_{T^u_\Lambda} p(t)\overline{r(t)}\, v_+(t)\, d_u t$.

As a consequence we get the following explicit evaluation formulas and quadratic norm formulas for the monic symmetric Macdonald–Koornwinder polynomials. Set

$$N^+(\lambda) := \frac{\langle E^+(\gamma^{-1}_{\lambda,q}; \cdot), E^+(\gamma^{-1}_{\lambda,q}; \cdot)\rangle_+}{\langle 1, 1\rangle_+}, \qquad \lambda \in \Lambda^-.$$

Corollary 9.4.31

(i) *For* $\lambda \in \Lambda^-$ *we have* $P^+_{\lambda_+}(\gamma^d_{0,q}) = \dfrac{\mathcal{C}^d(\gamma_{0,q})c^d_{\tau(\lambda)}(\gamma_{0,q})}{\mathcal{C}^d(\gamma_{\lambda,q})\kappa^d_{\tau(\lambda)}}$.

(ii) *For* $\lambda \in \Lambda^-$ *we have* $N^+(\lambda) = \dfrac{\mathcal{C}^d(\gamma_{\lambda,q})c^d_{-\tau(\lambda)}(\gamma_{0,q})}{\mathcal{C}^d(\gamma_{0,q})c^d_{\tau(\lambda)}(\gamma_{0,q})}$.

Remark 9.4.32 For $\lambda \in \Lambda^-$ we have $\sum_{\mu \in W_0\lambda} N(\mu)^{-1} = N^+(\lambda)^{-1}$.

9.5 Explicit Evaluation and Norm Formulas

We rewrite in this subsection the explicit evaluation formulas and the quadratic norm expressions for the symmetric Macdonald–Koornwinder polynomials in terms of q-shifted factorials (9.3.9). The explicit formulas for the GL_{n+1} symmetric Macdonald polynomials (see §9.3.7) can be immediately obtained as special cases of the explicit formulas below. We keep the conditions (9.3.6) on the parameters (κ, q).

9.5.1 The Twisted Cases

In the case that we have initial data $D = (R_0, \Delta_0, \bullet, \Lambda, \Lambda^d)$ with $\bullet = t$ the evaluation and norm formulas take the following explicit form.

Corollary 9.5.1

(i) *Suppose $\bullet = t$ and $S = \emptyset = S^d$. Then $R = R^\bullet = R^d$, $\kappa = \kappa^d$ and $\kappa_{m\mu_\alpha c+\alpha} = \kappa_\alpha$ for all $m \in \mathbb{Z}$ and $\alpha \in R_0$. Then for all $\lambda \in \Lambda^-$,*

$$P^+_{\lambda_+}(\gamma_{0,q}) = \gamma_{0,q}^{-\lambda} \prod_{\alpha \in R_0^+} \frac{(\kappa_\alpha^2 \gamma_{0,q}^{-\alpha}; q_\alpha)_{-(\lambda,\alpha^\vee)}}{(\gamma_{0,q}^{-\alpha}; q_\alpha)_{-(\lambda,\alpha^\vee)}},$$

$$N^+(\lambda) = \gamma_{0,q}^{2\lambda} \prod_{\alpha \in R_0^+} \left(\frac{1-\gamma_{0,q}^{-\alpha}}{1-q_\alpha^{-(\lambda,\alpha^\vee)}\gamma_{0,q}^{-\alpha}} \right) \frac{(q_\alpha \kappa_\alpha^{-2} \gamma_{0,q}^{-\alpha}; q_\alpha)_{-(\lambda,\alpha^\vee)}}{(\kappa_\alpha^2 \gamma_{0,q}^{-\alpha}; q_\alpha)_{-(\lambda,\alpha^\vee)}}.$$

(ii) *Suppose $\bullet = t$ and $(\Lambda, \Lambda^d) = (\mathbb{Z}R_0, \mathbb{Z}R_0)$ (the $C^\vee C$ case), realized concretely as in §9.3.8. Recall the relabeling of the multiplicity functions κ and κ^d given by $k = \kappa_\varphi^2 = \tilde{k}$ and*

$$\{a, b, c, d\} = \{\kappa_\theta \kappa_{2\theta}, -\kappa_\theta \kappa_{2\theta}^{-1}, q_\theta \kappa_0 \kappa_{2a_0}, -q_\theta \kappa_0 \kappa_{2a_0}^{-1}\},$$
$$\{\tilde{a}, \tilde{b}, \tilde{c}, \tilde{d}\} = \{\kappa_\theta \kappa_0, -\kappa_\theta \kappa_0^{-1}, q_\theta \kappa_{2\theta} \kappa_{2a_0}, -q_\theta \kappa_{2\theta} \kappa_{2a_0}^{-1}\}.$$

Then for $\lambda \in \Lambda^-$ we have

$$P^+_{\lambda_+}(\gamma_{0,q}^d) = (\gamma_{0,q}^d)^{-\lambda} \prod_{\alpha \in R_{0,l}^+} \frac{(\kappa_\varphi^2 \gamma_{0,q}^{-\alpha}; q_\varphi)_{-(\lambda,\alpha^\vee)}}{(\gamma_{0,q}^{-\alpha}; q_\varphi)_{-(\lambda,\alpha^\vee)}} \prod_{\alpha \in R_{0,s}^+} \frac{(\tilde{a}\gamma_{0,q}^{-\alpha}, \tilde{b}\gamma_{0,q}^{-\alpha}, \tilde{c}\gamma_{0,q}^{-\alpha}, \tilde{d}\gamma_{0,q}^{-\alpha}; q_\theta^2)_{-(\lambda,\alpha^\vee)/2}}{(\gamma_{0,q}^{-2\alpha}; q_\theta^2)_{-(\lambda,\alpha^\vee)}},$$

$$N^+(\lambda) = (\gamma_{0,q}^d)^{2\lambda} \prod_{\alpha \in R_{0,l}^+} \frac{1-\gamma_{0,q}^{-\alpha}}{1-q_\varphi^{-(\lambda,\alpha^\vee)}\gamma_{0,q}^{-\alpha}} \frac{(q_\varphi k_\varphi^{-2} \gamma_{0,q}^{-\alpha}; q_\varphi)_{-(\lambda,\alpha^\vee)}}{(k_\varphi^2 \gamma_{0,q}^{-\alpha}; q_\varphi)_{-(\lambda,\alpha^\vee)}}$$
$$\times \prod_{\beta \in R_{0,s}^+} \frac{1-\gamma_{0,q}^{-2\beta}}{1-q_\theta^{-2(\lambda,\beta^\vee)}\gamma_{0,q}^{-2\beta}} \frac{(q_\theta^2 \tilde{a}^{-1}\gamma_{0,q}^{-\beta}, q_\theta^2 \tilde{b}^{-1}\gamma_{0,q}^{-\beta}, q_\theta^2 \tilde{c}^{-1}\gamma_{0,q}^{-\beta}, q_\theta^2 \tilde{d}^{-1}\gamma_{0,q}^{-\beta}; q_\theta^2)_{-(\lambda,\beta^\vee)/2}}{(\tilde{a}\gamma_{0,q}^{-\beta}, \tilde{b}\gamma_{0,q}^{-\beta}, \tilde{c}\gamma_{0,q}^{-\beta}, \tilde{d}\gamma_{0,q}^{-\beta}; q_\theta^2)_{-(\lambda,\beta^\vee)/2}},$$

where $R_{0,s} \subset R_0$ (respectively $R_{0,l} \subset R_0$) are the short (respectively long) roots in R_0. In addition, $q_\varphi = q_\theta^2$.

The formulas in the intermediate case $\bullet = t$ and $S = \emptyset \neq S^d$ (or $S \neq \emptyset = S^d$) are special cases of the $C^\vee C$ case (by choosing an appropriate specialization of the multiplicity function).

9.5.2 The Untwisted Cases

If the initial data is of the form $D = (R_0, \Delta_0, u, \Lambda, \Lambda^d)$ then there are essentially two cases to be considered, namely the case $S = \emptyset = S^d$ and the special rank-two case treated in §9.3.9. Indeed, the cases corresponding to a nonreduced extension of the untwisted affine root system with underlying finite root system R_0 of type B_n ($n \geq 3$) or BC_n ($n \geq 1$) are special cases of the twisted $C^\vee C_n$ case.

Corollary 9.5.2

(i) *Suppose* $\bullet = u$ *and* $S = \emptyset = S^d$*. Then* $R = \mathbb{Z}c + R_0$, $R^d = \mathbb{Z}c + R_0^\vee$ *and* $\kappa_{mc+\alpha} = \kappa_\alpha = \kappa^d_{mc+\alpha^\vee}$ *for all* $m \in \mathbb{Z}$ *and* $\alpha \in R_0$*. Then for all* $\lambda \in \Lambda^-$,

$$P^+_{\lambda_+}(\gamma^d_{0,q}) = \gamma_{0,q}^{-\lambda} \prod_{\alpha \in R_0^+} \frac{(\kappa_\alpha^2 \gamma_{0,q}^{-\alpha^\vee}; q)_{-(\lambda,\alpha^\vee)}}{(\gamma_{0,q}^{-\alpha^\vee}; q)_{-(\lambda,\alpha^\vee)}},$$

$$N^+(\lambda) = \gamma_{0,q}^{2\lambda} \prod_{\alpha \in R_0^+} \frac{1 - \gamma_{0,q}^{-\alpha^\vee}}{1 - q^{-(\lambda,\alpha^\vee)}\gamma_{0,q}^{-\alpha^\vee}} \frac{(q\kappa_\alpha^{-2}\gamma_{0,q}^{-\alpha^\vee}; q)_{-(\lambda,\alpha^\vee)}}{(\kappa_\alpha^2 \gamma_{0,q}^{-\alpha^\vee}; q)_{-(\lambda,\alpha^\vee)}}.$$

(ii) *Suppose* $D = (R_0, \Delta_0, u, \mathbb{Z}R_0, \mathbb{Z}R_0^\vee)$ *with* R_0 *of type* C_2*, realized concretely as in* §9.3.9*. Recall the relabeling* (9.3.19) *of* κ *and* κ^d*. Then for* $\lambda \in \Lambda^-$ *we have*

$$P^+_{\lambda_+}(\gamma^d_{0,q}) = (\gamma^d_{0,q})^{-\lambda} \prod_{\alpha \in R^+_{0,s}} \frac{(\tilde{c}\gamma_{0,q}^{-\alpha^\vee}, \tilde{d}\gamma_{0,q}^{-\alpha^\vee}; q^2)_{-(\lambda,\alpha^\vee)/2}}{(\gamma_{0,q}^{-\alpha^\vee}; q)_{-(\lambda,\alpha^\vee)}} \prod_{\beta \in R^+_{0,l}} \frac{(\tilde{a}\gamma_{0,q}^{-\beta^\vee}, \tilde{b}\gamma_{0,q}^{-\beta^\vee}; q)_{-(\lambda,\beta^\vee)}}{(\gamma_{0,q}^{-2\beta^\vee}; q^2)_{-(\lambda,\beta^\vee)}},$$

$$N^+(\lambda) = (\gamma^d_{0,q})^{2\lambda} \prod_{\alpha \in R^+_{0,s}} \frac{1 - \gamma_{0,q}^{-\alpha^\vee}}{1 - q^{-(\lambda,\alpha^\vee)}\gamma_{0,q}^{-\alpha^\vee}} \frac{(q^2\tilde{c}^{-1}\gamma_{0,q}^{-\alpha^\vee}, q^2\tilde{d}^{-1}\gamma_{0,q}^{-\alpha^\vee}; q^2)_{-(\lambda,\alpha^\vee)/2}}{(\tilde{c}\gamma_{0,q}^{-\alpha^\vee}, \tilde{d}\gamma_{0,q}^{-\alpha^\vee}; q^2)_{-(\lambda,\alpha^\vee)/2}}$$

$$\times \prod_{\beta \in R^+_{0,l}} \frac{1 - \gamma_{0,q}^{-2\beta^\vee}}{1 - q^{-2(\lambda,\beta^\vee)}\gamma_{0,q}^{-2\beta^\vee}} \frac{(q\tilde{a}^{-1}\gamma_{0,q}^{-\beta^\vee}, q\tilde{b}^{-1}\gamma_{0,q}^{-\beta^\vee}; q)_{-(\lambda,\beta^\vee)}}{(\tilde{a}\gamma_{0,q}^{-\beta^\vee}, \tilde{b}\gamma_{0,q}^{-\beta^\vee}; q)_{-(\lambda,\beta^\vee)}},$$

where $R_{0,s} \subset R_0$ (*respectively* $R_{0,l} \subset R_0$) *are the short* (*respectively long*) *roots in* R_0.

A Appendix

A.1 Affine Root Systems

The main reference for this subsection is Macdonald (1972a).

Let E be a real affine space with the associated space of translations V of dimension $n \geq 1$. Fix a real scalar product $(\cdot,\cdot)$ on V and set $|v|^2 := (v, v)$ for all $v \in V$. It turns E into a metric space, called an *affine Euclidean space* (Bruhat and Tits, 1972, (1.3.1)).

A map $\Psi\colon E \to E$ is called an *affine linear endomorphism* of E if there exists a linear endomorphism $d\Psi$ of V such that $\Psi(e + v) = \Psi(e) + d\Psi(v)$ for all $e \in E$ and all $v \in V$. Set $O(E)$ for the group of affine linear isometric automorphisms of E. Let $\tau_E\colon V \to O(E)$ be the group monomorphism defined by $\tau_E(v)(e) := e + v$. Then we have a short exact sequence of groups $\tau_E(V) \hookrightarrow \mathrm{O}(E) \overset{d}{\twoheadrightarrow} \mathrm{O}(V)$. For a subgroup $W \subseteq \mathrm{O}(E)$ let $L_W \subseteq V$ be the additive subgroup such that $\tau_E(L_W) = \mathrm{Ker}(d|_W)$.

A function $a: E \to \mathbb{R}$ is said to be *affine linear* if there exists a linear functional $\alpha: V \to \mathbb{R}$ such that $a(e + v) = a(e) + \alpha(v)$ for $e \in E$ and $v \in V$. Set $\widehat{E}$ for the real $(n+1)$-dimensional vector space of affine linear functions $a: E \to \mathbb{R}$. Let $c \in \widehat{E}$ be the constant function 1. The *gradient* of $a \in \widehat{E}$ is the unique vector $Da \in V$ such that $a(e+v) = a(e) + (Da, v)$ for all $e \in E$ and $v \in V$. The *gradient map* $D: \widehat{E} \to V$ is linear, surjective, with kernel consisting of the constant functions on E.

For $a, b \in \widehat{E}$ set $(a, b) := (Da, Db)$, which defines a semi-positive definite symmetric bilinear form on $\widehat{E}$. The radical consists of the constant functions on E. We write $|a|^2 := (a, a)$ for $a \in \widehat{E}$. Let $\mathrm{O}(\widehat{E})$ be the form-preserving linear automorphisms of $\widehat{E}$ and $\mathrm{O}_c(\widehat{E})$ its subgroup of automorphisms fixing the constant functions.

The contragredient action of $g \in \mathrm{O}(E)$ on $\widehat{E}$, given by $(ga)(e) := a(g^{-1}e)$ $(a \in \widehat{E}, e \in E)$, realizes a group isomorphism $\mathrm{O}(E) \simeq \mathrm{O}_c(\widehat{E})$. Note that $\tau_E(v)a = a - (Da, v)c$ for $a \in \widehat{E}$ and $v \in V$.

A vector $a \in \widehat{E}$ is called *nonisotropic* if $Da \neq 0$. For such a let $s_a: E \to E$ be the orthogonal reflection in the affine hyperplane $a^{-1}(0)$ of E. It is explicitly given by

$$s_a(e) = e - a(e)Da^\vee, \quad e \in E,$$

where $v^\vee := 2v/|v|^2 \in V$ is the covector of $v \in V \setminus \{0\}$. Viewed as an element of $\mathrm{O}_c(\widehat{E})$ it reads

$$s_a(b) = b - (a^\vee, b)a, \quad b \in \widehat{E},$$

where $a^\vee := 2a/|a|^2 \in \widehat{E}$ is the covector of a. For a subset R of nonisotropic vectors in $\widehat{E}$ let $W(R)$ be the subgroup of $\mathrm{O}(E) \simeq \mathrm{O}_c(\widehat{E})$ generated by the orthogonal reflections s_a $(a \in R)$.

Definition A.1 A set R of nonisotropic vectors in $\widehat{E}$ is an *affine root system* on E if
1. R spans $\widehat{E}$,
2. $W(R)$ stabilizes R,
3. $(a^\vee, b) \in \mathbb{Z}$ for all $a, b \in R$,
4. $W(R)$ acts properly on E (i.e., if K_1 and K_2 are two compact subsets of E then $w(K_1) \cap K_2 \neq \emptyset$ for at most finitely many $w \in W(R)$),
5. $L_{W(R)}$ spans V.

The elements $a \in R$ are called *affine roots*. The group $W(R)$ is called the *affine Weyl group* of R. The real dimension n of V is the *rank* of R.

Definition A.2 Let R be an affine root system on E. A nonempty subset $R' \subseteq R$ is called an *affine root subsystem* if $W(R')$ stabilizes R' and if $L_{W(R')}$ generates the real span V' of the set $\{Da\}_{a \in R'}$ of gradients of R' in V.

Remark A.3 With the notation from the previous definitions, let E' be the set of $V'^\perp$-orbits of E, where $V'^\perp$ is the orthocomplement of V' in V. It is an affine Euclidean space with V'

the associated space of translations and with norm induced by the scalar product on V'. Let $F \subseteq \widehat{E}$ be the real span of R'. Then $F \xrightarrow{\sim} \widehat{E'}$ with form-preserving linear isomorphism $a \mapsto a'$ defined by $a'(e + V'^{\perp}) := a(e)$ for all $a \in F$ and $e \in E$. With this identification R' is an affine root system on E'. Furthermore, the corresponding affine Weyl group $W(R')$ is isomorphic to the subgroup of $W(R) \subset \mathrm{O}(E)$ generated by the orthogonal reflections s_a $(a \in R')$.

We call an affine root system R *irreducible* if it cannot be written as a nontrivial orthogonal disjoint union $R' \cup R''$ (orthogonal meaning that $(a, b) = 0$ for all $a \in R'$ and all $b \in R''$). It is called *reducible* otherwise. In that case both R' and R'' are affine root subsystems of R. Each affine root system is an orthogonal disjoint union of irreducible affine root subsystems (cf. Macdonald, 1972a, §3).

Remark A.4 The definition in Macdonald (1972a, §2) of an affine root system is (1)–(4) of Definition A.1. Careful analysis reveals that Macdonald tacitly assumes condition (5), which only follows from the four axioms (1)–(4) if R is irreducible. Through a personal communication I learned that Mark Reeder has independently observed that an extra condition besides (1)–(4) is needed in order to avoid examples of affine root systems given as an orthogonal disjoint union of an affine root system R' and a finite crystallographic root system R'' (compare, on the level of affine Weyl groups, with Bourbaki (1968, Chap. V, §3) and Bruhat and Tits (1972, §1.3)). Mark Reeder proposes adding to (1)–(4) the axiom that for each $\alpha \in D(R)$ there exist at least two affine roots with gradient α. The resulting definition is equivalent to Definition A.1, as well as to the notion of an *échelonnage* from Bruhat and Tits (1972, (1.4.1)) (take their (1.3.2) into account).

An affine root system R is called *reduced* if $\mathbb{R}a \cap R = \{\pm a\}$ for all $a \in R$, and *nonreduced* otherwise. If R is nonreduced then $R = R^{\mathrm{ind}} \cup R^{\mathrm{unm}}$ with R^{ind} (respectively R^{unm}) the reduced affine root subsystem of R consisting of indivisible (respectively unmultipliable) affine roots.

If $R \subset \widehat{E}$ is an affine root system then $R_0 := D(R) \subset V$ is a finite crystallographic root system in V, called the *gradient root system* of R. The associated Weyl group $W_0 = W_0(R_0)$ is the subgroup of $\mathrm{O}(V)$ generated by the orthogonal reflections $s_\alpha \in \mathrm{O}(V)$ in the hyperplanes $\alpha^\perp$ $(\alpha \in R_0)$, which are explicitly given by $s_\alpha(v) = v - (\alpha^\vee, v)\alpha$ for $v \in V$. The Weyl group W_0 coincides with the image of $W(R)$ under the differential d.

We now define an appropriate equivalence relation between affine root systems, called *similarity*. It is a slightly weaker notion of similarity compared to the one used in Macdonald (1972a, §3). This is to render affine root systems similar that differ by a rescaling of the underlying gradient root system; see Remark A.6.

Definition A.5 We call two affine root systems $R \subset \widehat{E}$ and $R' \subset \widehat{E'}$ *similar*, $R \simeq R'$, if there exists a linear isomorphism $T : \widehat{E} \xrightarrow{\sim} \widehat{E'}$ which restricts to a bijection of R onto R' preserving Cartan integers: $((Ta)^\vee, Tb) = (a^\vee, b)$ for $a, b \in R$.

Similarity respects basic notions such as affine root subsystems and irreducibility. If $R \simeq R'$ is realized by the linear isomorphism $T : \widehat{E} \xrightarrow{\sim} \widehat{E'}$ then $Ts_aT^{-1} = s_{Ta}$ for all $a \in R$. In particular, $W(R) \simeq W(R')$. Note that T maps constant functions to constant functions.

Replacing T by $-T$ if necessary, we may assume without loss of generality that $T(c) \in \mathbb{R}_{>0}c'$, where $c \in \widehat{E}$ and $c' \in \widehat{E'}$ denote the constant functions 1 on E and E' respectively. With this additional condition we call T a *similarity transformation* between R and R'.

If R is an affine root system and $\lambda \in \mathbb{R}^* := \mathbb{R} \setminus \{0\}$ then $\lambda R := \{\lambda a\}_{a \in R}$ is an affine root system similar to R (the similarity transformation realizing $R \xrightarrow{\sim} \lambda R$ is scalar multiplication by $|\lambda|$). We call λR a *rescaling* of the affine root system R. If two affine root systems R and R' are similar, then a similarity transformation T between R and a rescaling of R' exists such that $T(c) = c'$. In this case T arises as the contragredient of an affine linear isomorphism from E' onto E. For instance, each affine Weyl group element $w \in W(R)$ is a self-similarity transformation of R in this way.

If the affine root systems $R \subset \widehat{E}$ and $R' \subset \widehat{E'}$ are similar, then so are their gradients $R_0 \subset V$ and $R_0' \subset V'$ (i.e., there exists a linear isomorphism t of V onto V' restricting to a bijection of $R_0 \xrightarrow{\sim} R_0'$ and preserving Cartan integers). Indeed, if T is a similarity transformation between R and R', then the unique linear isomorphism $t\colon V \xrightarrow{\sim} V'$ such that $D \circ T = t \circ D$ realizes the similarity between R_0 and R_0'.

Remark A.6 Let $R \subset \widehat{E}$ be an irreducible affine root system with associated gradient $R_0 \subset V$. Fix an origin $e \in E$. For $\lambda \in \mathbb{R}^* \setminus \{\pm 1\}$ and $a \in R$ define $a_\lambda \in \widehat{E}$ by $a_\lambda(e+v) := a(e) + \lambda(Da, v)$ for all $v \in V$. Then $R_\lambda := \{a_\lambda\}_{a \in R} \subset \widehat{E}$ is an affine root system similar to R, and $R_{\lambda,0} = \lambda R_0$. The affine root systems R and R_λ are not similar if one uses the definition of similarity from Macdonald (1972a, §3).

In the remainder of this Appendix we assume that R is an irreducible affine root system of rank n. Since $W(R)$ acts properly on E, the set of *regular* elements $E_{\mathrm{reg}} := E \setminus \bigcup_{a \in R} a^{-1}(0)$ decomposes as the disjoint union of open n-simplices, called *chambers* of R. For a fixed chamber C there exists a unique $\mathbb{R}$-basis $\Delta = \Delta(C, R)$ of $\widehat{E}$ consisting of indivisible affine roots $a_i = a_{i,C}$ $(0 \leq i \leq n)$ such that $C = \{e \in E \mid a_i(e) > 0 \;\forall\, i \in \{0, \ldots, n\}\}$. The set of roots Δ is called the *basis* of R associated to the chamber C. The affine roots a_i are called *simple affine roots*. Any affine root $a \in R$ can be uniquely written as $a = \sum_{i=0}^n \lambda_i a_i$ with either all $\lambda_i \in \mathbb{Z}_{\geq 0}$ or all $\lambda_i \in \mathbb{Z}_{\leq 0}$. The subset of affine roots of the first type is denoted by R^+ and is called the set of *positive affine roots* with respect to Δ. Then $R = R^+ \cup R^-$ (disjoint union) with $R^- := -R^+$ the subset of *negative affine roots*. The affine Weyl group $W(R)$ is a Coxeter group with the simple reflections s_{a_i} $(0 \leq i \leq n)$ as Coxeter generators.

Definition A.7 A rank-n *affine Cartan matrix* is a rational integral $(n+1) \times (n+1)$ matrix $A = (a_{ij})_{i,j=0}^n$ satisfying the five conditions

1. $a_{ii} = 2$,
2. $a_{ij} \in \mathbb{Z}_{\leq 0}$ if $i \neq j$,
3. $a_{ij} = 0$ implies $a_{ji} = 0$,
4. $\det(A) = 0$ and all the proper principal minors of A are strictly positive,
5. A is indecomposable (i.e., the matrices obtained from A by a simultaneous permutation of its rows and columns are not the direct sum of two nontrivial blocks).

The larger class of rational integral matrices satisfying conditions (1)–(3) are called *generalized Cartan matrices*. They correspond to Kac–Moody Lie algebras; see Kac (1990). The Kac–Moody Lie algebras related to the subclass of affine Cartan matrices are the affine Lie algebras. The affine Cartan matrices have been classified in Kac (1990, Ch. 4).

Fix an ordered basis $\Delta = (a_0, a_1, \ldots, a_n)$ of R. The matrix $A = A(R, \Delta) = (a_{ij})_{0 \leq i,j \leq n}$ defined by $a_{ij} := (a_i^\vee, a_j)$ is an affine Cartan matrix. The coefficients a_{ij} $(0 \leq i, j \leq n)$ are called the *affine Cartan integers* of R.

If $R \simeq R'$ with associated similarity transformation T, then the T-image of an ordered basis Δ of R is an ordered basis Δ' of R'. Let R and R' be irreducible affine root systems with ordered bases Δ and Δ' respectively. We say that (R, Δ) is *similar* to (R', Δ'), $(R, \Delta) \simeq (R', \Delta')$, if the irreducible affine root systems R and R' are similar and if there exists an associated similarity transformation T mapping the ordered basis Δ of R to the ordered basis Δ' of R'. For similar pairs the associated affine Cartan matrices coincide. Moreover, the affine Cartan matrix $A(R, \Delta)$ modulo simultaneous permutations of its row and columns does not depend on the choice of ordered basis Δ. This leads to a map from the set of similarity classes of reduced irreducible affine root systems to the set of affine Cartan matrices up to simultaneous permutations of rows and columns. Using the classification of the affine Cartan matrices from Kac (1990, Ch. 4) and the explicit construction of reduced irreducible affine root systems from Macdonald (1972a) (see also the next subsection), it follows that the map is surjective. It is injective by a straightforward adjustment of the proof for finite crystallographic root systems to the present affine setup (Humphreys, 1972, Prop. 11.1). Hence we obtain the following classification result.

Theorem A.8 *Reduced irreducible affine root systems up to similarity are in bijective correspondence to affine Cartan matrices up to simultaneous permutations of the rows and columns.*

Remark A.9 Affine Cartan matrices also parametrize affine Lie algebras (see Kac, 1990). For a given affine Cartan matrix the set of real roots of the associated affine Lie algebra is the associated irreducible reduced affine root system.

Affine Cartan matrices up to simultaneous permutations of the rows and columns (and hence similarity classes of irreducible reduced affine root systems) can be naturally encoded by affine Dynkin diagrams (Kac, 1990; Macdonald, 1972a). The *affine Dynkin diagram* associated to an affine Cartan matrix $A = (a_{ij})_{i,j=0}^n$ is the graph with $n + 1$ vertices in which we join the ith and jth nodes $(i \neq j)$ by $\max(|a_{ij}|, |a_{ji}|)$ edges. In addition, we put an arrow towards the ith node if $|a_{ij}| > 1$. In §A.4 we list all affine Dynkin diagrams and link this to the classification of Kac (1990).

A.2 Explicit Constructions

The main reference for the results in this subsection is Macdonald (1972a). For an irreducible finite crystallographic root system R_0 of type A, D, E or BC there is exactly one similarity class of reduced irreducible affine root systems whose gradient root system is similar to R_0. For the other types of root systems R_0, there are two such similarity classes of reduced irreducible affine root systems. We now proceed to realize them explicitly.

Let $R_0 \subset V$ be an irreducible finite crystallographic root system (possibly nonreduced). The affine space E is taken to be V with forgotten origin. We will write V for E in the sequel if no confusion is possible.

We identify the space $\widehat{E}$ of affine linear functions on E with $V \oplus \mathbb{R}c$ as a real vector space, with c the constant function identically equal to 1 on V and with $V^* \simeq V$ the linear functionals on V (the identification with V is realized by the scalar product on V). With these identifications,

$$\mathrm{O}_c(\widehat{E}) \simeq \mathrm{O}(E) = \mathrm{O}(V) \ltimes \tau(V),$$

with $\tau(v)(e) = e + v$. Regarding $\tau(v)$ as an element of $\mathrm{O}_c(\widehat{V})$, it is given by

$$\tau(v)a = -(Da, v)c + a, \quad a \in \widehat{V}.$$

Note that the orthogonal reflection $s_a \in \mathrm{O}(E)$ associated to $a = \lambda c + \alpha \in \widehat{E}$ ($\lambda \in \mathbb{R}$, $\alpha \in V \setminus \{0\}$) decomposes as $s_{\lambda c+\alpha} = \tau(-\lambda\alpha^\vee)s_\alpha$.

Consider the subset

$$\mathcal{S}(R_0) := \{mc + \alpha\}_{m\in\mathbb{Z},\alpha\in R_0^{\mathrm{ind}}} \cup \{(2m+1)c + \beta\}_{m\in\mathbb{Z},\beta\in R_0\setminus R_0^{\mathrm{ind}}}$$

of $\widehat{V}$, where $R_0^{\mathrm{ind}} \subseteq R_0$ is the root subsystem of indivisible roots. Then $\mathcal{S}(R_0)$ and $\mathcal{S}(R_0^\vee)^\vee$ are reduced irreducible affine root systems with gradient root system R_0. We call $\mathcal{S}(R_0)$ (respectively $\mathcal{S}(R_0^\vee)^\vee$) the *untwisted* (respectively *twisted*) reduced irreducible affine root system associated to R_0. Note that $\mathcal{S}(R_0) \simeq \mathcal{S}(R_0^\vee)^\vee$ if R_0 is of type A, D, E or BC.

Proposition A.10 *The following reduced irreducible affine root systems form a complete set of representatives of the similarity classes of reduced irreducible affine root systems*:

1. $\mathcal{S}(R_0)$ *with* R_0 *running through the similarity classes of reduced irreducible finite crystallographic root systems (i.e.,* R_0 *of type* A, B, …, G),
2. $\mathcal{S}(R_0^\vee)^\vee$ *with* R_0 *running through the similarity classes of reduced irreducible finite crystallographic root systems having two root lengths (i.e.,* R_0 *of type* B_n $(n \geq 2)$, C_n $(n \geq 3)$, F_4 *and* G_2),
3. $\mathcal{S}(R_0)$ *with* R_0 *a nonreduced irreducible finite crystallographic root system (i.e.,* R_0 *of type* BC_n $(n \geq 1)$).

In view of the above proposition we use the following terminology: a reduced irreducible affine root system R is said to be of *untwisted type* if $R \simeq \mathcal{S}(R_0)$ with R_0 reduced, of *twisted type* if $R \simeq \mathcal{S}(R_0^\vee)^\vee$ with R_0 reduced and of *mixed type* if $R \simeq \mathcal{S}(R_0)$ with R_0 nonreduced. Note that a reduced irreducible affine root system R with gradient root system of type A, D or E is of untwisted and of twisted type.

Suppose that $\Delta_0 = (\alpha_1, \ldots, \alpha_n)$ is an ordered basis of R_0. Let $\varphi \in R_0$ (respectively $\theta \in R_0$) be the associated highest root (respectively the highest short root). Then

$$\Delta := (a_0, a_1, \ldots, a_n) = (c - \varphi, \alpha_1, \ldots, \alpha_n)$$

is an ordered basis of $\mathcal{S}(R_0)$, while

$$\Delta := (a_0, a_1, \ldots, a_n) = \left(\tfrac{1}{2}|\theta|^2 c - \theta, \alpha_1, \ldots, \alpha_n\right)$$

is an ordered basis of $\mathcal{S}(R_0^\vee)^\vee$.

A.3 Nonreduced Irreducible Affine Root Systems

If R is an irreducible affine root system with ordered basis Δ, then Δ is also an ordered basis of the affine root subsystem R^{ind} of indivisible roots. Furthermore, $R \simeq R'$ implies $R^{\mathrm{ind}} \simeq R'^{,\mathrm{ind}}$. To classify nonreduced irreducible affine root systems up to similarity, one thus only needs to understand the possible ways to extend reduced irreducible affine root systems to nonreduced ones.

Let R' be a reduced irreducible affine root system with affine Weyl group $W = W(R')$. Choose an ordered basis $\Delta = (a_0, a_1, \ldots, a_n)$ of R'. Set

$$S := \{a \in \Delta \mid (\mathbb{Z}R', a^\vee) = 2\mathbb{Z}\}. \tag{A.1}$$

Let $S_m \subset S$ with $\#S_m = m \in \{0, \ldots, \#S\}$. Then $R^{(m)} := R' \cup \bigcup_{a \in S_m} W(2a)$ is an irreducible affine root system with $R^{(m),\mathrm{ind}} \simeq R'$.

By consideration of the possible affine Dynkin diagrams associated to (R', Δ) (see §A.4), it follows that the set S (see (A.1)) is of cardinality at most two. It is of cardinality two iff $R' \simeq \mathcal{S}(R_0^\vee)^\vee$ with R_0 of type A_1 or with R_0 of type B_n ($n \geq 2$). It is of cardinality one iff $R' \simeq \mathcal{S}(R_0)$ with R_0 of type B_n ($n \geq 2$) or of type BC_n ($n \geq 1$). Hence the similarity class of $R^{(m)}$ does not depend on the choice of subset $S_m \subseteq S$ of cardinality m, and it does not depend on the choice of ordered basis Δ of R'. The number of W-orbits of $R^{(m)}$ equals the number of W-orbits of R' plus m. The number of similarity classes of irreducible affine root systems R satisfying $R^{\mathrm{ind}} \simeq R'$ is $\#S + 1$.

If R_0 is of type A_1 or of type B_n ($n \geq 2$) we thus have a nonreduced irreducible affine root system in which two W-orbits are added to $\mathcal{S}(R_0^\vee)^\vee$. It is labeled as $\mathrm{C}^\vee\mathrm{C}_n$ by Macdonald (1972a). In the rank-one case it has four W-orbits, otherwise five. A detailed description of this affine root system is given in §9.3.8.

Irreducible affine root subsystems with underlying reduced affine root system $\mathcal{S}(R_0)$ having finite root system R_0 of type BC_n ($n \geq 1$) or of type B_n ($n \geq 3$) can be naturally viewed as affine root subsystems of the affine root system of type $\mathrm{C}^\vee\mathrm{C}_n$. This is not the case for the nonreduced extension of the affine root system $\mathcal{S}(R_0)$ with R_0 of type B_2. It can actually be better viewed as the rank-two case of the family $\mathcal{S}(R_0)$ with R_0 of type C_n since, in the corresponding affine Dynkin diagram, the vertex labeled by the affine simple root a_0 is double bonded with the finite Dynkin diagram of R_0. The nonreduced extension of $\mathcal{S}(R_0)$ with R_0 of type C_2 was missing in the Macdonald (1972a) classification list. It was added in Macdonald (2003, (1.3.17)).

A.4 Affine Dynkin Diagrams

In this subsection we list the connected affine Dynkin diagrams (cf. Macdonald, 1972a, Appendix 1) which, as we have seen, are in one-to-one correspondence to similarity classes of irreducible reduced affine root systems. Each similarity class of irreducible reduced affine root systems has a representative of the form $\mathcal{S}(R_0)$ or $\mathcal{S}(R_0^\vee)^\vee$ for a unique irreducible finite crystallographic root system R_0 up to similarity; see §A.2. Recall that $\mathcal{S}(R_0^\vee)^\vee \simeq \mathcal{S}(R_0)$ if R_0 is of type A, D, E, BC.

We label the connected affine Dynkin diagram by $\widehat{X}$ with X the type of the associated finite root system R_0 if $X \in \{\mathrm{A}, \mathrm{D}, \mathrm{E}, \mathrm{BC}\}$. If the associated finite root system R_0 is of type $X \in \{\mathrm{B}, \mathrm{C}, \mathrm{F}, \mathrm{G}\}$ then we label the connected affine Dynkin diagram by $\widehat{X^u}$ (respectively $\widehat{X^t}$) if the associated irreducible reduced affine root system is $\mathcal{S}(R_0)$ (respectively $\mathcal{S}(R_0^\vee)^\vee$). Since $\mathrm{A}_1 \simeq \mathrm{B}_1 \simeq \mathrm{C}_1$ and $\mathrm{B}_2 \simeq \mathrm{C}_2$, there is some redundancy in the notation. We pick the one which is most convenient to fit into an infinite family of affine Dynkin diagrams. In the terminology of §A.2, the irreducible reduced affine root systems corresponding to affine Dynkin diagrams labeled $\widehat{X}$ with $X \in \{\mathrm{A}, \mathrm{D}, \mathrm{E}\}$ are of untwisted and of twisted type, labeled $\widehat{\mathrm{BC}}$ of mixed type, labeled $\widehat{X^u}$ of untwisted type and labeled $\widehat{X^t}$ of twisted type. In Macdonald (1972a, Appendix 1) the affine Dynkin diagrams labeled $\widehat{\mathrm{B}}_n^t$ and $\widehat{\mathrm{C}}_n^t$ are called type $\mathrm{C}_n^\vee$ and type $\mathrm{B}_n^\vee$ respectively. The remaining relations with the notation and terminology in Macdonald (1972a, Appendix 1) are self-explanatory.

We specify in each affine Dynkin diagram a particular vertex (the gray vertex) which is labeled by the unique affine simple root a_0 in the particular choice of ordered basis Δ of $\mathcal{S}(R_0)$ or $\mathcal{S}(R_0^\vee)^\vee$ as specified in §A.2.

In Kac's notation (see Kac, 1990, §4.8, Tables Aff 1–3) the affine Dynkin diagrams are labeled differently: our label $\widehat{X}$ corresponds to $X^{(1)}$ if $X \in \{\mathrm{A}, \mathrm{D}, \mathrm{E}\}$, and $\widehat{\mathrm{BC}}_n$ corresponds to $\mathrm{A}_{2n}^{(2)}$ $(n \geq 1)$. Our label $\widehat{X^u}$ corresponds to $X^{(1)}$ if $X \in \{\mathrm{B}, \mathrm{C}, \mathrm{F}, \mathrm{G}\}$. Finally, $\widehat{\mathrm{B}}_n^t$ corresponds to $\mathrm{D}_{n+1}^{(2)}$ $(n \geq 2)$, $\widehat{\mathrm{C}}_n^t$ corresponds to $\mathrm{A}_{2n-1}^{(2)}$ $(n \geq 3)$, $\widehat{\mathrm{F}}_4^t$ to $\mathrm{E}_6^{(2)}$ and $\widehat{\mathrm{G}}_2^t$ to $\mathrm{D}_4^{(3)}$.

$\widehat{\mathrm{A}}_1$ and $\widehat{\mathrm{A}}_n$ $(n \geq 2)$

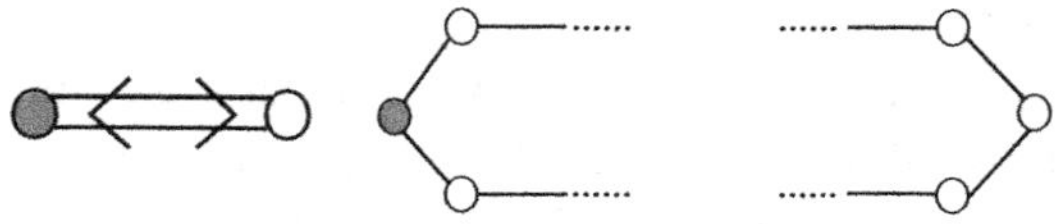

$\widehat{\mathrm{B}}_n^u$ $(n \geq 3)$

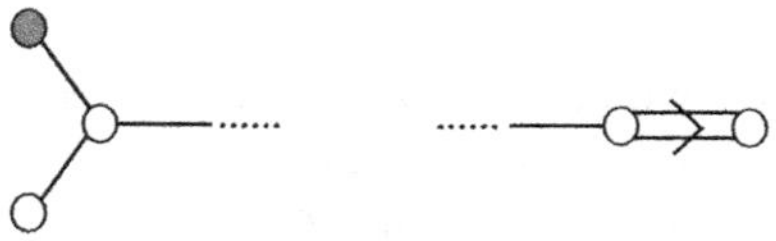

$\widehat{\mathrm{B}}_n^t$ $(n \geq 2)$

$\widehat{\mathrm{BC}}_1$ and $\widehat{\mathrm{BC}}_n$ $(n \geq 2)$

$\widehat{C}^u_n$ $(n \geq 2)$

$\widehat{C}^t_n$ $(n \geq 3)$

$\widehat{D}_n$ $(n \geq 4)$

$\widehat{E}_6$

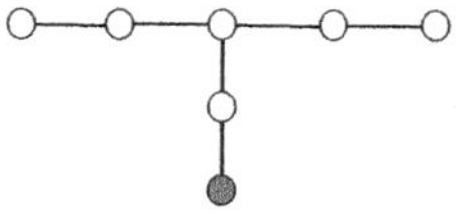

$\widehat{E}_7$ and $\widehat{E}_8$

$\widehat{F}^u_4$ and $\widehat{F}^t_4$

$\widehat{G}^u_2$ and $\widehat{G}^t_2$

Acknowledgements I thank Ian Macdonald, Masatoshi Noumi and Tom Koornwinder for valuable comments on earlier versions of the text.

References

Askey, R., and Wilson, J. 1985. *Some Basic Hypergeometric Orthogonal Polynomials that Generalize Jacobi Polynomials*. Mem. Amer. Math. Soc., vol. 54, no. 319.

Baker, T. H., and Forrester, P. J. 2000. Multivariable Al-Salam & Carlitz polynomials associated with the type *A* q-Dunkl kernel. *Math. Nachr.*, **212**, 5–35.

Borodin, A., and Corwin, I. 2014a. Macdonald processes. *Probab. Theory Related Fields*, **158**, 225–400.

Borodin, A., and Corwin, I. 2014b. Macdonald processes. Pages 292–316 of: *XVIIth International Congress on Mathematical Physics*. World Scientific.

Bourbaki, N. 1968. *Groupes et Algèbres de Lie. Chapitres IV, V et VI*. Hermann, Paris. Also translated in English, Springer, 2002.

Bruhat, F., and Tits, J. 1972. Groupes réductifs sur un corps local. *Inst. Hautes Études Sci. Publ. Math.*, **41**, 5–251.

Chalykh, O. A. 2002. Macdonald polynomials and algebraic integrability. *Adv. Math.*, **166**, 193–259.

Chalykh, O., and Etingof, P. 2013. Orthogonality relations and Cherednik identities for multivariable Baker-Akhiezer functions. *Adv. Math.*, **238**, 246–289.

Cherednik, I. 1992a. Double affine Hecke algebras, Knizhnik–Zamolodchikov equations, and Macdonald's operators. *Int. Math. Res. Not.*, no. 9, 171–180.

Cherednik, I. 1992b. Quantum Knizhnik–Zamolodchikov equations and affine root systems. *Comm. Math. Phys.*, **150**, 109–136.

Cherednik, I. 1995a. Difference-elliptic operators and root systems. *Int. Math. Res. Not.*, 43–58.

Cherednik, I. 1995b. Double affine Hecke algebras and Macdonald's conjectures. *Ann. of Math. (2)*, **141**, 191–216.

Cherednik, I. 1995c. Macdonald's evaluation conjectures and difference Fourier transform. *Invent. Math.*, **122**, 119–145. Erratum in Vol. 125 (1996), p. 391.

Cherednik, I. 1995d. Nonsymmetric Macdonald polynomials. *Int. Math. Res. Not.*, 483–515.

Cherednik, I. 1997a. Difference Macdonald–Mehta conjecture. *Int. Math. Res. Not.*, 449–467.

Cherednik, I. 1997b. Intertwining operators of double affine Hecke algebras. *Selecta Math. (N.S.)*, **3**, 459–495.

Cherednik, I. 2003. Double affine Hecke algebras and difference Fourier transforms. *Invent. Math.*, **152**, 213–303.

Cherednik, I. 2005. *Double Affine Hecke Algebras*. London Mathematical Society Lecture Note Series, vol. 319. Cambridge University Press.

Cherednik, I. 2009a. Nonsemisimple Macdonald polynomials. *Selecta Math. (N.S.)*, **14**, 427–569.

Cherednik, I. 2009b. Whittaker limits of difference spherical functions. *Int. Math. Res. Not.*, 3793–3842.

Cherednik, I. 2013. Jones polynomials of torus knots via DAHA. *Int. Math. Res. Not.*, 5366–5425.

Cherednik, I. 2016. DAHA-Jones polynomials of torus knots. *Selecta Math. (N.S.)*, **22**, 1013–1053.

Cherednik, I., and Danilenko, I. 2016. DAHA and iterated torus knots. *Algebr. Geom. Topol.*, **16**, 843–898.

Coskun, H., and Gustafson, R. A. 2006. Well-poised Macdonald functions W_λ and Jackson coefficients ω_λ on BC_n. Pages 127–155 of: *Jack, Hall–Littlewood and Macdonald Polynomials*. Contemp. Math., vol. 417. Amer. Math. Soc.

Di Francesco, P., and Zinn-Justin, P. 2005. Around the Razumov–Stroganov conjecture: proof of a multi-parameter sum rule. *Electron. J. Combin.*, **12**, Research Paper 6, 27 pp.

van Diejen, J. F. 1995. Commuting difference operators with polynomial eigenfunctions. *Compos. Math.*, **95**, 183–233.

van Diejen, J. F. 1996. Self-dual Koornwinder–Macdonald polynomials. *Invent. Math.*, **126**, 319–339.

van Diejen, J. F., and Stokman, J. V. 1998. Multivariable q-Racah polynomials. *Duke Math. J.*, **91**, 89–136.

Elliot, R., and Gukov, S. 2016. Exceptional knot homology. *J. Knot Theory Ramifications*, **25**, 1640003, 49 pp.

Etingof, P. I., and Kirillov, Jr., A. A. 1994. Macdonald's polynomials and representations of quantum groups. *Math. Res. Lett.*, **1**, 279–296.

Etingof, P. I., and Kirillov, Jr., A. A. 1995. On the affine analogue of Jack and Macdonald polynomials. *Duke Math. J.*, **78**, 229–256.

Etingof, P. I., and Kirillov, Jr., A. A. 1996. Representation-theoretic proof of the inner product and symmetry identities for Macdonald's polynomials. *Compos. Math.*, **102**, 179–202.

Etingof, P., and Kirillov, Jr., A. 1998. On Cherednik–Macdonald–Mehta identities. *Electron. Res. Announc. Amer. Math. Soc.*, **4**, 43–47.

Feigin, B., Jimbo, M., Miwa, T., and Mukhin, E. 2003. Symmetric polynomials vanishing on the shifted diagonals and Macdonald polynomials. *Int. Math. Res. Not.*, 1015–1034.

Forrester, P. J., and Warnaar, S. O. 2008. The importance of the Selberg integral. *Bull. Amer. Math. Soc. (N.S.)*, **45**, 489–534.

Gasper, G., and Rahman, M. 2004. *Basic Hypergeometric Series*. Second edn. Encyclopedia of Mathematics and Its Applications, vol. 96. Cambridge University Press.

Gasper, G., and Rahman, M. 2005. Some systems of multivariable orthogonal Askey–Wilson polynomials. Pages 209–219 of: *Theory and Applications of Special Functions*. Dev. Math., vol. 13. Springer.

Haglund, J., Haiman, M., and Loehr, N. 2008. A combinatorial formula for non-symmetric Macdonald polynomials. *Amer. J. Math.*, **103**, 359–383.

Haiman, M. 2001. Hilbert schemes, polygraphs, and the Macdonald positivity conjecture. *J. Amer. Math. Soc.*, **14**, 941–1006.

Haiman, M. 2002. Vanishing theorems and character formulas for the Hilbert scheme of points in the plane. *Invent. Math.*, **149**, 371–407.

Haiman, M. 2006. Cherednik algebras, Macdonald polynomials and combinatorics. Pages 843–872 of: *International Congress of Mathematicians. Vol. III*. Eur. Math. Soc.

Heckman, G. J., and Opdam, E. M. 1997. Yang's system of particles and Hecke algebras. *Ann. of Math. (2)*, **145**, 139–173. Erratum in Vol. 146 (1997), 749–750.

Humphreys, J. E. 1972. *Introduction to Lie Algebras and Representation Theory*. Graduate Texts in Mathematics, vol. 9. Springer.

Humphreys, J. E. 1990. *Reflection Groups and Coxeter Groups*. Cambridge University Press.

Ion, B. 2003a. Involutions of double affine Hecke algebras. *Compos. Math.*, **139**, 67–84.

Ion, B. 2003b. Nonsymmetric Macdonald polynomials and Demazure characters. *Duke Math. J.*, **116**, 299–318.

Ion, B. 2006. Nonsymmetric Macdonald polynomials and matrix coefficients for unramified principal series. *Adv. Math.*, **201**, 36–62.

Ion, B., and Sahi, S. 2015. *Double affine Hecke algebras and congruence groups*. arXiv:1506.06417.

Kac, V. G. 1990. *Infinite-dimensional Lie Algebras*. Third edn. Cambridge University Press.

Kasatani, M. 2005a. Subrepresentations in the polynomial representation of the double affine Hecke algebra of type GL_n at $t^{k+1}q^{r-1} = 1$. *Int. Math. Res. Not.*, 1717–1742.

Kasatani, M. 2005b. Zeros of symmetric Laurent polynomials of type $(BC)_n$ and Koornwinder–Macdonald polynomials specialized at $t^{k+1}q^{r-1} = 1$. *Compos. Math.*, **141**, 1589–1601.

Kasatani, M., and Pasquier, V. 2007. On polynomials interpolating between the stationary state of a O(n) model and a Q.H.E. ground state. *Comm. Math. Phys.*, **276**, 397–435.

Kasatani, M., and Takeyama, Y. 2007. The quantum Knizhnik–Zamolodchikov equation and non-symmetric Macdonald polynomials. *Funkcial. Ekvac.*, **50**, 491–509.

Kasatani, M., Miwa, T., Sergeev, A. N., and Veselov, A. P. 2006. Coincident root loci and Jack and Macdonald polynomials for special values of the parameters. Pages 207–225 of: *Jack, Hall-Littlewood and Macdonald Polynomials*. Contemp. Math., vol. 417. Amer. Math. Soc.

Kirillov, Jr., A. A. 1997. Lectures on affine Hecke algebras and Macdonald's conjectures. *Bull. Amer. Math. Soc. (N.S.)*, **34**, 251–292.

Knop, F. 1997. Symmetric and non-symmetric quantum Capelli polynomials. *Comment. Math. Helv.*, **72**, 84–100.

Knop, F., and Sahi, S. 1997. A recursion and a combinatorial formula for Jack polynomials. *Invent. Math.*, **128**, 9–22.

Koekoek, R., Lesky, P. A., and Swarttouw, R. F. 2010. *Hypergeometric Orthogonal Polynomials and their q-Analogues*. Springer.

Komori, Y., and Hikami, K. 1997. Quantum integrability of the generalized elliptic Ruijsenaars models. *J. Phys. A*, **30**, 4341–4364.

Koornwinder, T. 1988. *Self-duality for q-ultraspherical polynomials associated with root system A_n*. Unpublished manuscript, availabe at `https://tinyurl.com/y6dmcege`.

Koornwinder, T. H. 1992. Askey–Wilson polynomials for root systems of type *BC*. Pages 189–204 of: *Hypergeometric Functions on Domains of Positivity, Jack Polynomials, and Applications*. Contemp. Math., vol. 138. Amer. Math. Soc.

Koornwinder, T. H. 2018. Quadratic transformations for orthogonal polynomials in one and two variables. Pages 419–447 of: *Representation Theory, Special Functions and Painlevé Equations*. Adv. Stud. Pure Math., vol. 76. Math. Soc. Japan, Tokyo.

Lascoux, A., Rains, E. M., and Warnaar, S. O. 2009. Nonsymmetric interpolation Macdonald polynomials and $\mathfrak{gl}_n$ basic hypergeometric series. *Transform. Groups*, **14**, 613–647.

Letzter, G. 2004. Quantum zonal spherical functions and Macdonald polynomials. *Adv. Math.*, **189**(1), 88–147.

Lusztig, G. 1989. Affine Hecke algebras and their graded version. *J. Amer. Math. Soc.*, **2**, 599–635.

Macdonald, I. G. 1972a. Affine root systems and Dedekind's η-function. *Invent. Math.*, **15**, 91–143.

Macdonald, I. G. 1972b. The Poincaré series of a Coxeter group. *Math. Ann.*, **199**, 161–174.

Macdonald, I. G. 1988. A new class of symmetric functions. *Sém. Lothar. Combin.*, **20**, Paper B20a, 41 pp.

Macdonald, I. G. 1991. *Some conjectures for Koornwinder's orthogonal polynomials*. Unpublished manuscript.

Macdonald, I. G. 1995. *Symmetric Functions and Hall Polynomials*. Second edn. Oxford University Press.

Macdonald, I. G. 1996. Affine Hecke algebras and orthogonal polynomials. *Astérisque*, 189–207. Séminaire Bourbaki, Vol. 1994/95, Exp. No. 797.

Macdonald, I. G. 2000. Orthogonal polynomials associated with root systems. *Sém. Lothar. Combin.*, **45**, B45a, 40 pp. Text of handwritten preprint, 1987.

Macdonald, I. G. 2003. *Affine Hecke Algebras and Orthogonal Polynomials*. Cambridge Tracts in Mathematics, vol. 157. Cambridge University Press.

Noumi, M. 1995. Macdonald–Koornwinder polynomials and affine Hecke rings (in Japanese). *Sūrikaisekikenkyūsho Kōkyūroku*, No. 919, 44–55.

Noumi, M. 1996. Macdonald's symmetric polynomials as zonal spherical functions on some quantum homogeneous spaces. *Adv. Math.*, **123**, 16–77.

Noumi, M., and Stokman, J. V. 2004. Askey–Wilson polynomials: an affine Hecke algebra approach. Pages 111–144 of: *Laredo Lectures on Orthogonal Polynomials and Special Functions*. Nova Sci. Publ., Hauppauge, NY. Also available as arXiv:math/0001033.

Noumi, M., and Sugitani, T. 1995. Quantum symmetric spaces and related q-orthogonal polynomials. Pages 28–40 of: *Group Theoretical Methods in Physics*. World Scientific.

Noumi, M., Dijkhuizen, M. S., and Sugitani, T. 1997. Multivariable Askey-Wilson polynomials and quantum complex Grassmannians. Pages 167–177 of: *Special Functions, q-Series and Related Topics*. Fields Inst. Commun., vol. 14. Amer. Math. Soc.

Oblomkov, A. A., and Stokman, J. V. 2005. Vector valued spherical functions and Macdonald–Koornwinder polynomials. *Compos. Math.*, **141**, 1310–1350.

Okounkov, A. 1997. Binomial formula for Macdonald polynomials and applications. *Math. Res. Lett.*, **4**, 533–553.

Okounkov, A. 1998a. BC-type interpolation Macdonald polynomials and binomial formula for Koornwinder polynomials. *Transform. Groups*, **3**, 181–207.

Okounkov, A. 1998b. (Shifted) Macdonald polynomials: q-integral representation and combinatorial formula. *Compos. Math.*, **112**, 147–182.

Opdam, E. M. 1995. Harmonic analysis for certain representations of graded Hecke algebras. *Acta Math.*, **175**, 75–121.

Pasquier, V. 2006. Quantum incompressibility and Razumov–Stroganov type conjectures. *Ann. Henri Poincaré*, **7**, 397–421.

Rains, E. M. 2005. BC_n-symmetric polynomials. *Transform. Groups*, **10**, 63–132.

Rains, E. M. 2010. Transformations of elliptic hypergeometric integrals. *Ann. of Math. (2)*, **171**, 169–243.

Rains, E. M., and Vazirani, M. 2007. Vanishing integrals of Macdonald and Koornwinder polynomials. *Transform. Groups*, **12**, 725–759.

Rosengren, H. 2001. Multivariable q-Hahn polynomials as coupling coefficients for quantum algebra representations. *Int. J. Math. Math. Sci.*, **28**, 331–358.

Ruijsenaars, S. N. M. 1987. Complete integrability of relativistic Calogero–Moser systems and elliptic function identities. *Comm. Math. Phys.*, **110**, 191–213.

Sahi, S. 1996. Interpolation, integrality, and a generalization of Macdonald's polynomials. *Int. Math. Res. Not.*, 457–471.

Sahi, S. 1999. Nonsymmetric Koornwinder polynomials and duality. *Ann. of Math. (2)*, **150**, 267–282.

Schiffmann, O., and Vasserot, E. 2011. The elliptic Hall algebra, Cherednik–Hecke algebras and Macdonald polynomials. *Compos. Math.*, **147**, 188–234.

Sergeev, A. N., and Veselov, A. P. 2009. Deformed Macdonald–Ruijsenaars operators and super Macdonald polynomials. *Comm. Math. Phys.*, **288**, 653–675.

Stokman, J. V. 2000a. Koornwinder polynomials and affine Hecke algebras. *Int. Math. Res. Not.*, 1005–1042.

Stokman, J. V. 2000b. On *BC* type basic hypergeometric orthogonal polynomials. *Trans. Amer. Math. Soc.*, **352**, 1527–1579.

Stokman, J. V. 2003. Difference Fourier transforms for nonreduced root systems. *Selecta Math. (N.S.)*, **9**, 409–494.

Stokman, J. 2011. Quantum affine Knizhnik–Zamolodchikov equations and quantum spherical functions, I. *Int. Math. Res. Not.*, 1023–1090.

Stokman, J. V. 2014. The c-function expansion of a basic hypergeometric function associated to root systems. *Ann. of Math. (2)*, **179**, 253–299.

Stokman, J. V., and Koornwinder, T. H. 1997. Limit transitions for BC type multivariable orthogonal polynomials. *Canad. J. Math.*, **49**, 373–404.

Stokman, J., and Vlaar, B. 2015. Koornwinder polynomials and the *XXZ* spin chain. *J. Approx. Theory*, **197**, 69–100.

Tratnik, M. V. 1991. Some multivariable orthogonal polynomials of the Askey tableau—continuous families. *J. Math. Phys.*, **32**, 2065–2073.

10

Combinatorial Aspects of Macdonald and Related Polynomials

Jim Haglund

10.1 Introduction

The theory of symmetric functions plays an increasingly important role in modern mathematics, with substantial applications to representation theory, algebraic geometry, special functions, mathematical physics, knot invariants, algebraic combinatorics, statistics and other areas. We give several references to these applications in the following section. Macdonald (1988) introduced a family of symmetric functions with two extra parameters q, t, the Macdonald polynomials, and Macdonald (1995, Ch. VI) described them in great detail. This family contains most of the previously studied families of symmetric functions, such as Schur functions, Hall–Littlewood polynomials, and Jack polynomials, as limiting or special cases.

Macdonald polynomials can be studied from multiple points of view. Haiman (2001) used properties of the Hilbert scheme from algebraic geometry to prove the famous *n! conjecture* of Garsia and Haiman (1993), which says that Macdonald polynomials represent bigraded characters of certain modules of the symmetric group. This implies a Schur-positivity conjecture made by Macdonald (1995, (VI.8.18?)).

Macdonald (in a manuscript from 1987, published in 2000) also gave a construction of orthogonal polynomials associated with any finite root system. These polynomials are symmetric in the sense of being Weyl group invariant and, apart from q, they depend on as many parameters as there are Weyl group orbits. The case of root system A_{n-1} in Macdonald (2000), when extended to the GL_n root system, yields the Macdonald polynomials introduced by Macdonald (1988).

Macdonald (1996, 2003) (the 2003 book gives much more detail) showed that, associated to any affine root system, one can obtain families of nonsymmetric (i.e., not Weyl group invariant) polynomials which yield by symmetrization the polynomials discussed by Macdonald (2000). For the case of root system A one is thus led to systems of nonsymmetric polynomials which are bases for the polynomial ring, and to systems of symmetric polynomials (Macdonald, 1988) which are bases for the ring of symmetric functions. Further improvements in this direction were made by Cherednik (1995). In earlier work Koornwinder (1992) had extended the construction by Macdonald (2000) for the case of root system BC, thus obtaining a very general family of polynomials which contains the Askey–Wilson polynomials, and hence all

Much of this chapter is a condensed version of Chapters 1, 2, 6, 7, and Appendix A of the author's book *The q, t-Catalan numbers and the space of diagonal harmonics: With an appendix on the combinatorics of Macdonald polynomials*,

classical orthogonal polynomials, as special or limit cases. Sahi (1999) treated the corresponding nonsymmetric case. See Chapter 9 for a comprehensive introduction to all these families.

In the setup of Chapter 9, the polynomials introduced by Macdonald in 1988 correspond to a GL_n root system. This chapter focuses on the combinatorics of this case, and also of an associated object, the space of diagonal harmonics DH_n. This space, which has become increasingly important in algebra and representation theory, has a beautiful and remarkably rich combinatorial structure containing two-parameter extensions of Catalan numbers, Schröder numbers, parking functions, and other popular combinatorial objects. Macdonald has commented that the study of root systems and Lie algebras provides an inexhaustible source of wonderful combinatorics. It seems something similar holds for the study of the Hilbert scheme and DH_n.

Section 10.2 contains some background material from the theory of symmetric functions. Section 10.3 contains an overview of the basic analytic and algebraic properties of Macdonald polynomials discovered by Macdonald, Garsia, and Haiman. Section 10.4 covers the combinatorics of the space of diagonal harmonics DH_n, and its connection to Macdonald polynomial theory. In §10.5 we discuss the combinatorial formula of Haglund, Haiman, and Loehr giving the expansion of the Macdonald polynomial into monomials. In §10.6 we discuss some of the many consequences of this formula. A corresponding combinatorial formula for the expansion of the GL_n nonsymmetric Macdonald polynomial into monomials is the subject of §10.7. In §10.8 we discuss how results on the combinatorics of DH_n from §10.4 led to the monomial formula for Macdonald polynomials in §10.5, and §10.9 contains a brief overview of other recent approaches to Macdonald polynomials. Section 10.10 overviews some important recent developments which have not yet been published.

10.2 Basic Theory of Symmetric Functions

This section contains only a brief overview of symmetric function theory; for a more detailed treatment of the subject we refer the reader to Macdonald (1995, Ch. 1) and Stanley (1999, Ch. 7). Given $f(x_1, \ldots, x_n) \in K[x_1, x_2, \ldots, x_n]$ for some field K of characteristic 0, and $\sigma \in S_n$, let $\sigma f := f(x_{\sigma_1}, \ldots, x_{\sigma_n})$. We say that f is a *symmetric polynomial* if $\sigma f = f$ for every σ in the symmetric group S_n. It will be convenient to work with more general functions f depending on countably many indeterminates $x_1, x_2, \ldots$, indicated by $f(x_1, x_2, \ldots)$, in which case we view f as a formal power series in the x_i, and say that it is a *symmetric function* if it is invariant under any permutation of the variables. We let X_n and X stand for the set of variables $\{x_1, \ldots, x_n\}$ and $\{x_1, x_2, \ldots\}$, respectively.

A *partition* λ is a nonincreasing finite sequence $\lambda_1 \geq \lambda_2 \geq \cdots$ of positive integers, and λ_i is called the ith *part* of λ. We let $\ell(\lambda)$ denote the number of parts, $|\lambda| = \sum_i \lambda_i$ the sum of the parts, and we say that λ is a *partition of* $|\lambda|$. For various formulas it will be convenient to assume $\lambda_j = 0$ for $j > \ell(\lambda)$. The *Ferrers graph* (or *diagram*) of λ is an array of unit squares, called *cells*, with λ_i cells in the ith row, with the first cell in each row left-justified, and with the first row at the bottom. We often use λ to refer to its Ferrers graph. We define the *conjugate* partition λ' as the partition whose Ferrers graph is obtained from λ by reflecting across the diagonal $x = y$. See Figure 10.1. By convention $(i, j) \in \lambda$ refers to a cell with (column, row) coordinates (i, j), with the lower left-hand cell of λ having coordinates $(1, 1)$. The notation $x \in \lambda$ means x is a cell in λ. For technical reasons we say that 0 has one partition, the empty set $\emptyset$, with $\ell(\emptyset) = 0 = |\emptyset|$.

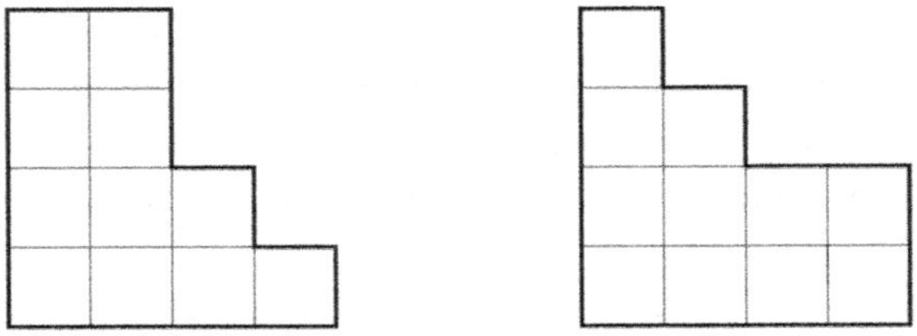

Figure 10.1 On the left, the Ferrers graph of the partition (4, 3, 2, 2), and on the right, that of its conjugate $(4, 3, 2, 2)' = (4, 4, 2, 1)$.

We let Λ^n denote the vector space consisting of symmetric functions in $x_1, x_2, \ldots$ that are homogeneous of degree n. The ring of symmetric functions Λ is the direct sum of the Λ^n. The most basic symmetric functions are the *monomial symmetric functions*, which depend on a partition λ in addition to a set of variables. They are denoted by $m_\lambda(X) = m_\lambda(x_1, x_2, \ldots)$. In a symmetric function it is typical to leave out explicit mention of the variables, with a set of variables being understood from context, so $m_\lambda = m_\lambda(X)$. We illustrate these first by means of examples. We let Par(n) denote the set of partitions of n, and use the notation $\lambda \vdash n$ as an abbreviation for $\lambda \in \text{Par}(n)$. For example,

$$m_{1,1} = \sum_{i<j} x_i x_j, \quad m_{2,1,1}(X_3) = x_1^2 x_2 x_3 + x_1 x_2^2 x_3 + x_1 x_2 x_3^2, \quad m_2(X) = \sum_i x_i^2.$$

In general, $m_\lambda(X)$ is the sum of all distinct monomials in the x_i whose multiset of exponents equals the multiset of parts of λ. Any element of Λ can be expressed uniquely as a linear combination of the m_λ.

We let 1^n stand for the partition consisting of n parts of size 1. The function m_{1^n} is called the nth *elementary symmetric function*, which we denote by e_n. Then $\prod_{i=1}^{\infty}(1 + zx_i) = \sum_{n=0}^{\infty} z^n e_n$, $e_0 = 1$. Another important special case is $m_n = \sum_i x_i^n$, known as the *power-sum symmetric functions*, denoted by p_n. We also define the *complete homogeneous symmetric functions* h_n by $h_n := \sum_{\lambda \vdash n} m_\lambda$, or equivalently $\prod_{i=1}^{\infty}(1 - zx_i)^{-1} = \sum_{n=0}^{\infty} z^n h_n$. For $\lambda \vdash n$, we define $e_\lambda := \prod_i e_{\lambda_i}$, $p_\lambda := \prod_i p_{\lambda_i}$ and $h_\lambda := \prod_i h_{\lambda_i}$. For example,

$$e_{2,1} = \sum_{i<j} x_i x_j \sum_k x_k = m_{2,1} + 3m_{1,1,1}, \quad p_{2,1} = \sum_i x_i^2 \sum_j x_j = m_3 + m_{2,1},$$

$$h_{2,1} = \Big(\sum_i x_i^2 + \sum_{i<j} x_i x_j\Big) \sum_k x_k = m_3 + 2m_{2,1} + 3m_{1,1,1}.$$

It is known that $\{e_\lambda\}_{\lambda \vdash n}$ forms a basis for Λ^n, and so do $\{p_\lambda\}_{\lambda \vdash n}$ and $\{h_\lambda\}_{\lambda \vdash n}$.

Definition 10.2.1 Two simple functions on partitions that will often be used are

$$n(\lambda) := \sum_i (i-1)\lambda_i = \sum_i \binom{\lambda_i'}{2}, \qquad z_\lambda := \prod_i i^{n_i} n_i!\,,$$

where $n_i = n_i(\lambda)$ is the number of parts of λ equal to i.

We let ω denote the ring endomorphism $\omega\colon \Lambda \to \Lambda$ defined by $\omega(p_k) := (-1)^{k-1} p_k$. Thus ω is an involution with $\omega(p_\lambda) = (-1)^{|\lambda|-\ell(\lambda)} p_\lambda$. Also, $\omega(e_n) = h_n$, and more generally $\omega(e_\lambda) = h_\lambda$.

Remark 10.2.2 Identities like $h_{2,1} = m_3 + 2m_{2,1} + 3m_{1,1,1}$ appear at first to depend on a set of variables, but it is customary to view them as polynomial identities in the p_λ. Since the p_k are algebraically independent, we can specialize them to whatever we please, forgetting about the original set of variables X.

We define the *Hall scalar product*, a bilinear form $\Lambda \times \Lambda \to \mathbb{Q}$, by

$$\langle p_\lambda, p_\beta \rangle := z_\lambda \chi(\lambda = \beta), \tag{10.2.1}$$

where, for any logical statement L, we have $\chi(L) := \left\{\begin{smallmatrix} 1 \text{ if } L \text{ is true,} \\ 0 \text{ if } L \text{ is false.}\end{smallmatrix}\right.$ Clearly, $\langle f, g\rangle = \langle g, f\rangle$. Also, $\langle \omega f, \omega g\rangle = \langle f, g\rangle$, which follows from the definition if $f = p_\lambda$, $g = p_\beta$, and by bilinearity for general f, g, since the p_λ form a basis for Λ.

Theorem 10.2.3 (See Macdonald, 1995, §I.4 or Stanley, 1999, Ch. 7) *The h_λ and the m_β are dual with respect to the Hall scalar product, i.e.,*

$$\langle h_\lambda, m_\beta \rangle = \chi(\lambda = \beta). \tag{10.2.2}$$

For any $f \in \Lambda$, and any basis $\{b_\lambda\}_{\lambda \in \mathrm{Par}}$ of Λ, let $f|_{b_\mu}$ denote the coefficient of b_μ when f is expressed in terms of the $\{b_\lambda\}$. Then (10.2.2) implies the following corollary.

Corollary 10.2.4 $\langle f, h_\lambda \rangle = f|_{m_\lambda}$.

10.2.1 Tableaux and Schur Functions

Given $\lambda, \mu \in \mathrm{Par}(n)$, a *semi-standard Young tableau* (or SSYT) of *shape* λ and *weight* μ is a filling of the cells of the Ferrers graph of λ with the elements of the multiset $\{1^{\mu_1} 2^{\mu_2} \cdots\}$ such that the numbers weakly increase across rows and strictly increase up columns. Let $\mathrm{SSYT}(\lambda, \mu)$ denote the set of these fillings, and $K_{\lambda,\mu}$ the cardinality of this set. The $K_{\lambda,\mu}$ are known as the *Kostka numbers*. Our definition also makes sense if our weight is a composition of n, i.e., any finite sequence of nonnegative integers whose sum is n. For example, $K_{(3,2),(2,2,1)} = K_{(3,2),(2,1,2)} = K_{(3,2),(1,2,2)} = 2$ as in Figure 10.2.

If the Ferrers graph of a partition β is contained in the Ferrers graph of λ, denoted $\beta \subseteq \lambda$, let λ/β refer to the subset of cells of λ which are not in β. This is referred to as a *skew shape*. Define an SSYT of shape λ/β and weight ν, where $|\nu| = |\lambda| - |\beta|$, to be a filling of the cells

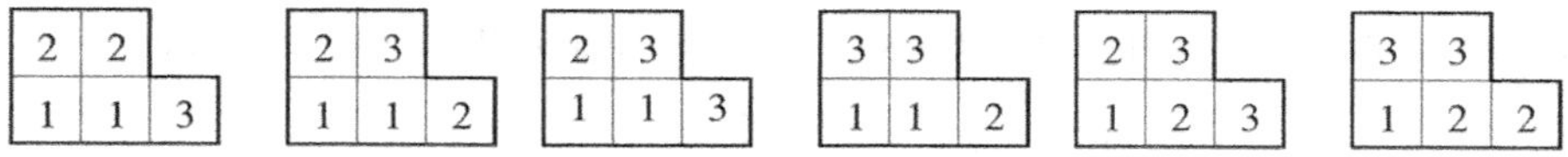

Figure 10.2 Some SSYT of shape $(3, 2)$.

of λ/β with elements of $\{1^{\nu_1}2^{\nu_2}\cdots\}$, again with weak increase across rows and strict increase up columns. We let $\mathrm{SSYT}(\lambda, \mu)$ denote the set of such tableaux, and its cardinality by $K_{\lambda/\beta,\nu}$.

Let $\mathrm{wcomp}(\mu)$ denote the set of all compositions whose multiset of nonzero parts equals the multiset of parts of μ. It follows easily from Figure 10.2 that $K_{(3,2),\alpha} = 2$ for all $\alpha \in \mathrm{wcomp}(2, 2, 1)$. Hence $\sum_{\alpha,T}\prod_i x_i^{\alpha_i} = 2m_{(2,2,1)}$, where the sum is over all tableaux T of shape $(3, 2)$ and weight being some element α of $\mathrm{wcomp}(2, 2, 1)$.

This is a special case of a more general phenomenon. Define $s_\lambda := \sum_{\alpha,T}\prod_i x_i^{\alpha_i}$ ($\lambda \in \mathrm{Par}(n)$), where the sum is over all compositions α of n, and all possible tableaux T of shape λ and weight α. Then

$$s_\lambda = \sum_{\mu \vdash n} K_{\lambda,\mu} m_\mu. \tag{10.2.3}$$

The s_λ, called *Schur functions*, are in Λ and are fundamental to the theory of symmetric functions. Two special cases of (10.2.3) are $s_n = h_n$ (since $K_{n,\mu} = 1$ for all $\mu \in \mathrm{Par}(n)$) and $s_{1^n} = e_n$ (since $K_{1^n,\mu} = \chi(\mu = 1^n)$).

An SSYT of weight 1^n is called *standard*, or an SYT. The set of SYT of shape λ is denoted $\mathrm{SYT}(\lambda)$. Below we list some of the important properties of Schur functions.

Theorem 10.2.5 *Let $\lambda, \mu \in \mathrm{Par}$.*

1. *The Schur functions are orthonormal with respect to the Hall scalar product, i.e., $\langle s_\lambda, s_\mu\rangle = \chi(\lambda = \mu)$. Thus, for any $f \in \Lambda$, $\langle f, s_\lambda\rangle = f|_{s_\lambda}$.*
2. *Action by ω: $\omega(s_\lambda) = s_{\lambda'}$.*
3. (Jacobi–Trudi identity) $s_\lambda = \det(h_{\lambda_i - i + j})_{i,j=1}^{\ell(\lambda)}$, *where $h_0 = 1$ and $h_k = 0$ for $k < 0$.*
4. (Pieri rule) *Let $k \in \mathbb{N}$. Then*

$$s_\lambda h_k = \textstyle\sum_\gamma s_\gamma, \tag{10.2.4}$$

where the sum is over all γ whose Ferrers graph contains λ with $|\gamma/\lambda| = k$ and such that γ/λ is a horizontal strip, *i.e., has no two cells in the same column.*

Note that by applying ω to both sides of (10.2.4) we can get a corresponding expression for $s_\lambda e_k$. For example, $s_{2,1}h_2 = s_{4,1} + s_{3,2} + s_{3,1,1} + s_{2,2,1}$, $s_{2,1}e_2 = s_{2,1,1,1} + s_{2,2,1} + s_{3,1,1} + s_{3,2}$.

10.2.2 Statistics on Tableaux

There is a q-analogue of the Kostka numbers, denoted $K_{\lambda,\mu}(q)$ and also called *Schur coefficients*, which have many applications. They can be defined as the coefficients that arise when expanding Schur functions $s_\lambda(X)$ in terms of Hall–Littlewood polynomials $P_\mu(X;q)$ (Macdonald, 1995, (III.2.6), §III.6):

$$s_\lambda(X) = \sum_\mu K_{\lambda,\mu}(q) P_\mu(X;q). \tag{10.2.5}$$

The $K_{\lambda,\mu}(q)$ are polynomials in q which satisfy $K_{\lambda,\mu}(1) = K_{\lambda,\mu}$. Foulkes (1974) conjectured that $K_{\lambda,\mu} \in \mathbb{N}[q]$. This conjecture was resolved by Lascoux and Schützenberger (1978), who found a statistic *cocharge* to generate these polynomials by (10.2.6). Butler (1994, §2.5) provided a detailed account of their proof, filling in a lot of missing details. A short proof, based on the new combinatorial formula for Macdonald polynomials, is contained in §10.6.

Let $T \in \mathrm{SSYT}(\lambda,\mu)$, and let $\mathrm{charge}(T) := n(\mu) - \mathrm{cocharge}(T)$, with $\mathrm{cocharge}(T)$ to be defined below. The *reading word* $\mathrm{read}(T)$ of T is obtained by reading the entries in T from left to right in the top row of T, then continuing left to right in the second row from the top of T, etc. For example, the tableau in the upper left of Figure 10.2 has reading word 22113. To calculate $\mathrm{cocharge}(T)$, perform the following algorithm on $\mathrm{read}(T)$.

Cocharge Algorithm

1. Start at the end of $\mathrm{read}(T)$ and scan left until you encounter a 1: say this occurs at spot i_1, then $\mathrm{read}(T)_{i_1} = 1$. Start there and scan left until you encounter a 2. If you hit the end of $\mathrm{read}(T)$ before finding a 2, loop around and continue searching left, starting at the end of $\mathrm{read}(T)$. Say the first 2 you find equals $\mathrm{read}(T)_{i_2}$. Now iterate: start at i_2 and search left until you find a 3, etc. Continue in this way until you have found $4, 5, \ldots, \mu'_1$, with μ'_1 occurring at spot $i_{\mu'_1}$. Then the first subword of $\mathrm{read}(T)$ is defined to be the elements of the set $\{\mathrm{read}(T)_{i_1}, \ldots, \mathrm{read}(T)_{i_{\mu'_1}}\}$, listed in the order in which they occur in $\mathrm{read}(T)$ if we start at the beginning of $\mathrm{read}(T)$ and move left to right. For example, if $\mathrm{read}(T) = 21613244153$ is a word of content $\mu = (3, 2, 2, 2, 1, 1)$, then the first subword equals 632415, corresponding to places $3, 5, 6, 8, 9, 10$ of $\mathrm{read}(T)$.

   Next remove the elements of the first subword from $\mathrm{read}(T)$ and find the first subword of what's left. Call this the second subword of $\mathrm{read}(T)$. Remove this and find the first subword in what's left and call this the third subword of $\mathrm{read}(T)$, etc. For the word 21613244153, the subwords are 632415, 2143, 1.
2. The value of $\mathrm{cocharge}(T)$ will be the sum of the values of cocharge on each of the subwords of $\mathrm{read}(T)$. Thus it suffices to assume $\mathrm{read}(T) \in S_m$ for some m, in which case we set

$$\mathrm{cocharge}(\mathrm{read}(T)) = \mathrm{comaj}(\mathrm{read}(T)^{-1}),$$

   where $\mathrm{read}(T)^{-1}$ is the usual inverse in S_m. Here $\mathrm{comaj}(\sigma)$ is equal to the sum of $m - i$ over those i in the *descent set* $\mathrm{Des}(\sigma)$, i.e., over those i for which $\sigma_i > \sigma_{i+1}$.
   For example, if $\sigma = 632415$, then $\sigma^{-1} = 532461$, $\mathrm{cocharge}(\sigma) = 5 + 4 + 1 = 10$, and so $\mathrm{cocharge}(21613244153) = 10 + 4 + 0 = 14$.

Note that to compute charge, we could create subwords in the same manner, and count $m - i$ for each i with $i + 1$ occurring to the right of i instead of to the left. We can now formulate the result of Lascoux and Schützenberger (1978) as

$$\tilde{K}_{\lambda,\mu}(q) := q^{n(\mu)} K_{\lambda,\mu}(1/q) = \sum_{T \in \mathrm{SSYT}(\lambda,\mu)} q^{\mathrm{cocharge}(T)}. \tag{10.2.6}$$

The polynomials $\tilde{K}_{\lambda,\mu}(q)$ have various interpretations in terms of representation theory and geometry (Garsia and Procesi, 1992; Hotta and Springer, 1977; Lusztig, 1981).

In addition to the cocharge statistic, there is a major index statistic on SYT which is often useful. Given an SYT tableau T of shape λ, define a *descent* of T to be a value of i, $1 \le i < |\lambda|$ for which $i+1$ occurs in a row above i in T. Let $\mathrm{maj}(T) := \sum i$ and $\mathrm{comaj}(T) := \sum(|\lambda| - i)$, where the sums are over the descents of T. Then (Stanley, 1999, p. 363)

$$s_\lambda(1, q, q^2, \ldots) = \frac{1}{(q;q)_n} \sum_{T \in \mathrm{SYT}(\lambda)} q^{\mathrm{maj}(T)} = \frac{1}{(q;q)_n} \sum_{T \in \mathrm{SYT}(\lambda)} q^{\mathrm{comaj}(T)},$$

where $(w;q)_n = (1-w)(1-wq)\cdots(1-wq^{n-1})$ is the usual *q-shifted factorial*.

10.2.3 Plethystic Notation

Many of the theorems later in this chapter involving symmetric functions will be expressed in plethystic notation. In this subsection we define this and give several examples in order to acclimate the reader. For more detailed treatments of plethysm see Garsia and Tesler (1996), Garsia et al. (1999) and Macdonald (1995, Ch. I, §8).

Let $E(t_1, t_2, t_3, \ldots)$ be a formal series of rational functions in the parameters $t_1, t_2, \ldots$. We define the *plethystic substitution* of E into p_k, denoted $p_k[E]$, by $p_k[E] := E(t_1^k, t_2^k, \ldots)$. Note the square *plethystic* brackets around E – this is to distinguish $p_k[E]$ from the ordinary kth power sum in a set of variables E, which we have already defined as $p_k(E)$. One thing we need to emphasize is that any minus signs occurring in the definition of E are left as is when replacing the t_i by t_i^k.

Example 10.2.6

1. Inside plethystic brackets, we view a set of variables X as $p_1(X) = x_1 + x_2 + \cdots$. For example, $p_k[X] = p_k(X)$.
2. For z an indeterminate, $p_k[zX] = z^k p_k[X]$.
3. $p_k[X - Y] = \sum_i (x_i^k - y_i^k) = p_k[X] - p_k[Y]$.
4. $p_k\left[\dfrac{X(1-z)}{1-q}\right] = \sum_i \dfrac{x_i^k(1-z^k)}{1-q^k}$.
5. $\prod_i \dfrac{(1-tx_iz)}{(1-x_i)z} = \sum_{n=0}^{\infty} z^n h_n[X(1-t)]$,

   which can be proved by taking $\exp \ln$ of the left-hand side, and expressing everything in terms of the p_k.

Let $Z = (-x_1, -x_2, \ldots)$. Note that $p_k(Z) = \sum_i (-1)^k x_i^k$, which is different from $p_k[-X]$. We need special notation for the case where we wish to replace variables by their negatives inside plethystic brackets. We use the ϵ symbol to denote this, i.e., $p_k[\epsilon X] := \sum_i (-1)^k x_i^k$.

We now extend this definition of plethystic substitution of E into f for an arbitrary $f \in \Lambda$ by first expressing f as a polynomial in the p_k, say $f = \sum_\lambda c_\lambda p_\lambda$ for constants c_λ, then defining $f[E]$ as $f[E] = \sum_\lambda c_\lambda \prod_i p_{\lambda_i}[E]$. We mention that for any $f \in \Lambda$, $\omega(f(X)) = f[-\epsilon X]$.

Some particularly useful plethystic identities are the following *addition formulas*. They can be proved by first expressing them in terms of the p_λ (see Haglund, 2008, Ch. 1). The same identities hold if we replace h_k by e_k throughout.

Theorem 10.2.7 *Let $E = E(t_1, t_2, \ldots)$ and $F = F(w_1, w_2, \ldots)$ be two formal series of rational terms in their indeterminates. Then*

$$h_n[E + F] = \sum_{k=0}^{n} h_k[E]h_{n-k}[F], \quad h_n[E - F] = \sum_{k=0}^{n} h_k[E]h_{n-k}[-F].$$

10.2.4 The Fundamental Basis for the Ring of Quasisymmetric Functions

A multivariate polynomial $f(X)$ is called *quasisymmetric* if the coefficient of $x_{i_1}^{a_1} \cdots x_{i_k}^{a_k}$ in f is equal to the coefficient of $x_{j_1}^{a_1} \cdots x_{j_k}^{a_k}$ in f whenever $1 \le i_1 < i_2 < \cdots < i_k$ and $1 \le j_1 < j_2 < \cdots < j_k$, for all $a_1, a_2, \ldots, a_k \in \mathbb{N}$. For a subset S of $\{1, 2, \ldots, n-1\}$, let

$$F_{n,S}(X) := \sum_{\substack{1 \le a_1 \le a_2 \le \cdots \le a_n \\ a_i = a_{i+1} \implies i \notin S}} x_{a_1} x_{a_2} \cdots x_{a_n}$$

denote Gessel's *fundamental quasisymmetric function*. The $F_{n,S}$ form an important basis for the ring of quasisymmetric functions.

Remark 10.2.8 There is another way to view $F_{n,S}(X)$ which will prove useful later. For any word $b_1 b_2 \cdots b_n$ of positive integers, let the *standardization* $\zeta(b_1 b_2 \cdots b_n)$ denote the permutation in S_n which satisfies $\zeta_i < \zeta_j$ if and only if $b_i \le b_j$, for all $1 \le i < j \le n$. For example, $\zeta(23253) = 13254$. Then for any $\sigma \in S_n$, $F_{n,\mathrm{Des}(\sigma^{-1})}(X)$ is simply the sum of $x_{b_1} x_{b_2} \cdots x_{b_n}$ over all words $b_1 b_2 \cdots b_n$ of positive integers whose standardization $\zeta(b_1 b_2 \cdots b_n)$ equals σ.

10.2.5 Graded Hilbert Series and Characters

Here we assume some basic facts about the representation theory of finite groups which will be familiar to many readers. Good sources for background information on these topics are the books by James and Liebeck (2001) and by Sagan (2001). Let G be a finite group, and V a finite-dimensional complex vector space, with basis $w_1, w_2, \ldots, w_n$. Any linear action of G on V makes V into a complex G-module. A module is called *irreducible* if it has no submodules other than $\{0\}$ and itself. Every finite-dimensional complex G-module V can be expressed as a direct sum of irreducible submodules.

The character of the module (under the given action), which we denote by char(V), is a function on G depending only on the conjugacy class of its argument. The character is called *irreducible* if V is irreducible. The irreducible characters form a basis of the space of conjugation invariant functions on G, so the number of conjugacy classes is equal to the number of irreducible characters. If $V = \bigoplus_{j=1}^{d} V_j$, where each V_j is irreducible, then $\mathrm{char}(V) = \sum_{j=1}^{d} \mathrm{char}(V_j)$.

For the symmetric group S_n the conjugacy class of an element σ is determined by rearranging the lengths of the disjoint cycles of σ into nonincreasing order to form a partition, called the *cycle type* β of σ (see Sagan, 2001). The possible cycle types are precisely the partitions of n. The number of elements in the conjugacy class determined by β is equal to $n!/z_\beta$. For a character χ and an element σ with cycle type β we can write $\chi(\beta)$ instead of $\chi(\sigma)$. Moreover, there is a canonical bijection $\lambda \mapsto \chi^\lambda$ of $\mathrm{Par}(n)$ onto the set of irreducible characters. This bijection can for instance be given by the equivalent identities in Theorem 10.2.9 below.

The dimension of the irreducible S_n-module with character χ^λ is known to be f^λ, the number of SYT of shape λ. For example, in dimension 1 we have two characters, $\chi^{(n)}$ (called the trivial character since $\chi^{(n)}(\mu) = 1$ for all $\mu \vdash n$) and χ^{1^n} (called the sign character since $\chi^{1^n}(\mu) = (-1)^{n-\ell(\mu)}$ which is the sign of any permutation of cycle type μ). One reason that Schur functions are important in the representation theory of S_n is the following.

Theorem 10.2.9 *In the expansion of the p_μ into the s_λ basis, the coefficients are the χ^λ:*

$$p_\mu = \sum_{\lambda \vdash n} \chi^\lambda(\mu) s_\lambda, \qquad s_\lambda = \sum_{\mu \vdash n} z_\mu^{-1} \chi^\lambda(\mu) p_\mu.$$

Let V be a graded subspace of $\mathbb{C}[x_1, \ldots, x_n]$ with respect to the grading of $\mathbb{C}[x_1, \ldots, x_n]$ by degree of homogeneity. Then $V = \bigoplus_{i=0}^{\infty} V^{(i)}$, where $V^{(i)}$ is the finite-dimensional subspace consisting of all elements of V that are homogeneous of degree i in the x_j. We define the *Hilbert series* $\mathrm{Hilb}(V;q)$ of V to be the sum $\mathrm{Hilb}(V;q) = \sum_{i=0}^{\infty} q^i \dim(V^{(i)})$, where dim indicates the dimension as a complex vector space.

Given $f(x_1, \ldots, x_n) \in \mathbb{C}[X_n]$ and $\sigma \in S_n$, set $\sigma f := f(x_{\sigma_1}, \ldots, x_{\sigma_n})$. This defines an action of S_n on $\mathbb{C}[X_n]$. Assume that V is as above and is fixed by the S_n-action, which also respects the grading. We define the *Frobenius series* $\mathrm{Frob}(V;X,q)$ of V to be the symmetric function

$$\mathrm{Frob}(V;X,q) := \sum_{i=0}^{\infty} q^i \sum_{\lambda \in \mathrm{Par}(i)} \mathrm{Mult}(\chi^\lambda, V^{(i)}) s_\lambda(X), \tag{10.2.7}$$

where $\mathrm{Mult}(\chi^\lambda, V^{(i)})$ is the *multiplicity* of the irreducible character χ^λ in the character of $V^{(i)}$ under the action. In other words, if we decompose $V^{(i)}$ into irreducible S_n-submodules, $\mathrm{Mult}(\chi^\lambda, V^{(i)})$ is the number of these submodules whose trace equals χ^λ. We will typically refer to the Frobenius series by the simpler notation $\mathrm{Frob}(V;q)$, leaving out the reference to the implicit set of variables on both sides of (10.2.7).

A polynomial in $\mathbb{C}[X_n]$ is *alternating*, or an *alternate*, if $\sigma f = (-1)^{\mathrm{inv}(\sigma)} f$ ($\sigma \in S_n$), where $\mathrm{inv}(\sigma)$ is the *number of inversions* of σ, i.e., the number of pairs (i, j) with $1 \le i < j \le n$ and $\sigma_i > \sigma_j$. The set of alternates in V forms a subspace denoted by V^ϵ.

Remark 10.2.10 Since the dimension of the representation corresponding to χ^λ equals f^λ, which also equals the coefficient of m_{1^n} in s_λ, by Corollary 10.2.4 and by an easy exercise we successively have that

$$\langle \mathrm{Frob}(V;q), h_{1^n}\rangle = \mathrm{Hilb}(V;q), \quad \langle \mathrm{Frob}(V;q), s_{1^n}\rangle = \mathrm{Hilb}(V^\epsilon;q).$$

Example 10.2.11 Since $\dim(\mathbb{C}[X_n]^{(i)}) = \binom{n+i-1}{i}$, we have $\mathrm{Hilb}(\mathbb{C}[X_n];q) = (1-q)^{-n}$. It is known (take into account the S_n-action and see Bergeron, 2009, §8.5) that

$$\mathrm{Frob}(\mathbb{C}[X_n];q) = \sum_{\lambda\in \mathrm{Par}(n)} s_\lambda \frac{\sum_{T\in \mathrm{SYT}(\lambda)} q^{\mathrm{maj}(T)}}{(q;q)_n}. \tag{10.2.8}$$

10.2.6 The Ring of Coinvariants

The set of symmetric polynomials in $x_1,\dots,x_n$, denoted $\mathbb{C}[X_n]^{S_n}$, which is generated by $1, e_1, \dots, e_n$, is called the *ring of invariants*. Although we will focus on the type-A case, we refer the reader to the excellent book by Humphreys (1990) for general information on how many of these results apply to more general reflection groups. The quotient ring $\mathrm{R}_n := \mathbb{C}[x_1,\dots,x_n]/\langle e_1,e_2,\dots,e_n\rangle$, or equivalently $\mathbb{C}[x_1,\dots,x_n]/\langle p_1,p_2,\dots,p_n\rangle$, obtained by moding out by the ideal generated by all symmetric polynomials of positive degree, is called the *ring of coinvariants*. It is known that $\dim(\mathrm{R}_n) = n!$ as a $\mathbb{C}$-vector space, and moreover that $\mathrm{Hilb}(\mathrm{R}_n;q) = [n]!$, where $[n]! := (q;q)_n/(1-q)^n = (1+q)(1+q+q^2)\cdots(1+q+\cdots+q^{n-1})$. Artin (1944) derived a specific basis for R_n, namely cosets of the elements in the set $\left\{\prod_{1\le i\le n} x_i^{\alpha_i} \mid 0\le \alpha_i \le i-1\right\}$. Also (Stanley, 1979, 2003),

$$\mathrm{Frob}(\mathrm{R}_n;q) = \sum_{\lambda\in \mathrm{Par}(n)} s_\lambda \sum_{T\in \mathrm{SYT}(\lambda)} q^{\mathrm{maj}(T)}. \tag{10.2.9}$$

Let $\Delta := \det\left(x_i^{j-1}\right)_{i,j=1}^n = \prod_{1\le i<j\le n}(x_j - x_i)$ be the *Vandermonde determinant*. The *space of harmonics* H_n can be defined as the $\mathbb{C}$-vector space spanned by Δ and its partial derivatives of all orders. Haiman (1994) provides a detailed proof that H_n is isomorphic to R_n as an S_n-module, and notes that an explicit isomorphism α is obtained by letting $\alpha(h), h\in \mathrm{H}_n$ be the element of R_n represented modulo $\langle e_1,\dots,e_n\rangle$ by h. Thus $\dim(\mathrm{H}_n) = n!$, and moreover the character of H_n under the S_n-action is given by (10.2.9). He also argues that (10.2.9) follows immediately from (10.2.8) and the fact that H_n generates $\mathbb{C}[X_n]$ as a free module over $\mathbb{C}[X_n]^{S_n}$.

10.3 Analytic and Algebraic Properties of Macdonald Polynomials

10.3.1 Macdonald's Original Construction

During the 1980s a number of extensions of Selberg's integral were found (see Chapter 11 and also Forrester and Warnaar (2008) for background on Selberg's integral). Askey (1980) conjectured a q-analogue of the integral, which he proved for $n = 2$ and which was later proved independently by Kadell and Habsieger for general n. Other generalizations involved the insertion of symmetric functions in the x_i into the integrand (see for example Stanton, 1989). One of these extensions, due to Kadell (1988a), involved the insertion of symmetric functions depending on a partition, a set of variables X_n, and another parameter. They are now known as *Jack symmetric functions* since they were first studied by Jack (1970).

In his article Kadell gave evidence that a q-analogue of the Jack symmetric functions existed which featured in a q-analogue of his extension of Selberg's integral. Shortly after, he proved these polynomials existed for $n = 2$ (Kadell, 1988b). The case for general n was solved by Macdonald (1988), and these q-analogues of Jack symmetric functions are now called *Macdonald polynomials*, denoted $P_\lambda(X; q, t)$. A brief discussion of their connection to Kadell's work can also be found in Macdonald (1995, §VI.10, p.387). The $P_\lambda(X; q, t)$ are symmetric functions with coefficients in $\mathbb{Q}(q, t)$. If we let $q = t^\alpha$, divide by $(1 - t)^{|\lambda|}$, and let $t \to 1^-$ in the P_λ we get the Jack symmetric functions with parameter α. Many other important bases of the ring of symmetric functions are also limiting or special cases of the $P_\lambda(X; q, t)$, and their introduction was a major breakthrough in algebraic combinatorics and special functions. In particular, for any q we have $P_\lambda(X; q, q) = s_\lambda(X)$ and $P_\lambda(X; q, 1) = m_\lambda(X)$, we have $P_\lambda(X; 0, t) = P_\lambda(X; t)$ (the Hall–Littlewood polynomial), and we have $P_\lambda(X; 1, t) = \prod_i e_{\lambda'_i}$. Macdonald polynomials have found applications to many areas including algebraic geometry, mathematical physics and representation theory (Carlsson and Mellit, 2018; Gorsky and Neguţ, 2015; Grojnowski and Haiman, 2007; Haglund, 2016; Haiman, 2001, 2002, 2006; Haiman and Woo, 2007).

Here is Macdonald's construction of the $P_\lambda(X; q, t)$. The best reference for their basic properties is Macdonald (1995, §VI.4). The definition involves the following standard partial order on partitions $\lambda, \mu \in \mathrm{Par}(n)$, called *dominance order*: $\lambda \geq \mu \iff \sum_{j=1}^{i} \lambda_j \geq \sum_{j=1}^{i} \mu_j$ for $i \geq 1$.

Theorem 10.3.1 (Macdonald polynomials) *Define a (q, t)-extension of the Hall scalar product* (10.2.1) *by*

$$\langle p_\lambda, p_\mu \rangle_{q,t} := \chi(\lambda = \mu) z_\lambda \prod_{i=1}^{\ell(\lambda)} \frac{1 - q^{\lambda_i}}{1 - t^{\lambda_i}}. \tag{10.3.1}$$

Then the following conditions uniquely define a family $\{P_\lambda(X; q, t)\}_{\lambda \in \mathrm{Par}(n)}$ *of symmetric functions with coefficients in* $\mathbb{Q}(q, t)$:

$$\text{(i)} \quad P_\lambda = \sum_{\mu \leq \lambda} c_{\lambda,\mu} m_\mu, \quad \textit{where } c_{\lambda,\mu} \in \mathbb{Q}(q, t) \textit{ and } c_{\lambda,\lambda} = 1; \tag{10.3.2a}$$

$$\text{(ii)} \quad \langle P_\lambda, P_\mu \rangle_{q,t} = 0 \quad \textit{if } \lambda \neq \mu. \tag{10.3.2b}$$

Remark 10.3.2 Since the (q, t)-extension of the Hall scalar product reduces to the ordinary Hall scalar product when $q = t$, it is clear that $P_\lambda(X; q, q) = s_\lambda(X)$. We also note that since the dominance partial order is not a total order, it is not at all obvious that conditions (10.3.2a) and (10.3.2b) define a *unique* set of polynomials. Indeed, as explained in Macdonald (1995, §VI.4), given any extension of the dominance partial order to a total order, we can apply Gram–Schmidt orthogonalization to obtain a family of symmetric functions satisfying (10.3.2a) and (10.3.2b). His theorem says that we get the same family no matter which extension to a total order we use. Macdonald proves Theorem 10.3.1 by first constructing operators for which the P_λ are simultaneous eigenfunctions with distinct eigenvalues.

Example 10.3.3 For $\mu \vdash n$, $P_\mu(X; 0, t)$ is known as the *Hall–Littlewood polynomial*, denoted $P_\mu(X; t)$. Its *integral form* is defined as $Q_\mu(X; t) := \left(\prod_i (t)_{n_i(\lambda)}\right) P_\mu(X; t)$. If we expand the $Q_\mu(X; t)$ in terms of the Schur basis, the coefficients will be polynomials in $\mathbb{Z}[t]$, but not generally in $\mathbb{N}[t]$, i.e., they will not have positive coefficients. However, if we expand the $Q_\mu(X; t)$ in terms of the $S_\lambda(X; t)$ occurring in Macdonald (1995, (III.4.6)), which are in their turn equal to $s_\lambda[X(1-t)]$ by use of Example 10.2.6(5), then

$$Q_\mu(X; t) = \sum_\lambda K_{\lambda,\mu}(t) S_\lambda(X; t) = \sum_\lambda K_{\lambda,\mu}(t) s_\lambda[X(1-t)] \tag{10.3.3}$$

with $K_{\lambda,\mu}(t)$ as in (10.2.5) and (10.2.6). Indeed, the first equality in (10.3.3) is equivalent to (10.2.5) by the biorthogonalities (Macdonald, 1995, (III.4.9), (III.4.10)).

Given a cell $x \in \lambda$, let the *arm* $a = a(x)$, *leg* $l = l(x)$, *coarm* $a' = a'(x)$ and *coleg* $l' = l'(x)$ (x-dependence usually suppressed) be the number of cells strictly between x and the border of λ in the E, N, W and S directions, respectively, as in Figure 10.3. Also, define

$$B_\mu = B_\mu(q, t) := \sum_{x \in \mu} q^{a'} t^{l'}, \quad \Pi_\mu = \Pi_\mu(q, t) := \prod_{x \in \mu}{}' (1 - q^{a'} t^{l'}), \tag{10.3.4}$$

where a prime symbol $'$ above a product over cells of a partition μ indicates we ignore the corner $(1, 1)$ cell, and $B_\emptyset = 0$, $\Pi_\emptyset = 1$. For example, $B_{(2,2,1)} = 1 + q + t + qt + t^2$ and $\Pi_{(2,2,1)} = (1-q)(1-t)(1-qt)(1-t^2)$. Note that $n(\mu) = \sum_{x \in \mu} l' = \sum_{x \in \mu} l$ (cf. Definition 10.2.1).

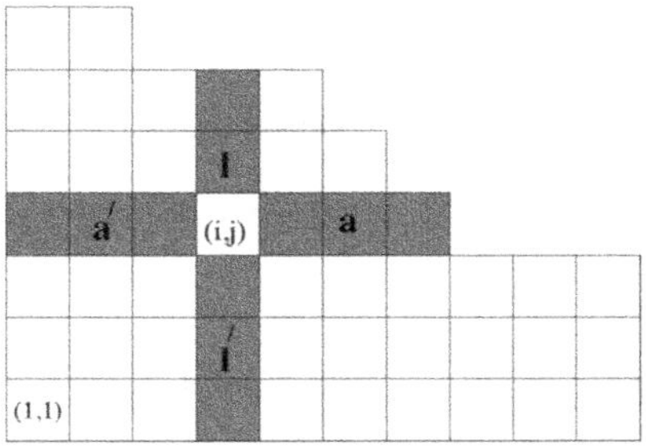

Figure 10.3 The arm a, coarm a', leg l and coleg l' of a cell.

Theorem 10.3.4 below gives some basic results by Macdonald on the P_λ. Its item 2 will soon be particularly useful to us. Recall that for any symmetric function F, square brackets as in $F[Z]$ indicate plethystic substitution, and if $\{t_1, t_2, \ldots\}$ is a set of positive parameters, $F[t_1 + t_2 + \cdots] = F(t_1, t_2, \ldots)$.

Theorem 10.3.4 *Let* $\lambda, \mu \in$ Par.

1. *Let z be an indeterminate. Then* (Macdonald, 1995, (VI.6.17))

$$P_\lambda\left[\frac{1-z}{1-t};q,t\right] = \prod_{x\in\lambda} \frac{t^{l'}-q^{a'}z}{1-q^a t^{l+1}}. \tag{10.3.5}$$

2. (Koornwinder–Macdonald reciprocity) *Assume* $n \geq \max(\ell(\lambda), \ell(\mu))$. *Then* (Macdonald, 1995, (VI.6.6), (VI.6.11))

$$\left(\prod_{x\in\mu} \frac{1-q^a t^{l+1}}{t^{l'}-q^{a'}t^n}\right) P_\mu\left[\sum_{i=1}^n t^{n-i}q^{\lambda_i};q,t\right] \tag{10.3.6}$$

is symmetric in μ, λ, *where as usual we let* $\mu_i = 0$ *for* $i > \ell(\mu)$, $\lambda_i = 0$ *for* $i > \ell(\lambda)$.

3. *For any two sets of variables* X, Y *(see* Garsia and Tesler, 1996, §1),

$$h_n\left[XY\frac{1-t}{1-q}\right] = \sum_{\lambda\vdash n}\left(\prod_{x\in\lambda}\frac{1-q^a t^{l+1}}{1-q^{a+1}t^l}\right)P_\lambda(X;q,t)P_\lambda(Y;q,t), \tag{10.3.7a}$$

$$e_n\,[XY] = \sum_{\lambda\vdash n} P_\lambda(X;q,t)P_{\lambda'}(Y;t,q). \tag{10.3.7b}$$

Identity(10.3.7a) follows from

$$\prod_{i,j}\frac{(tx_iy_j;q)_\infty}{(x_iy_j;q)_\infty} = \sum_\lambda\left(\prod_{x\in\lambda}\frac{1-q^a t^{l+1}}{1-q^{a+1}t^l}\right)P_\lambda(X;q,t)P_\lambda(Y;q,t) \tag{10.3.8}$$

(see Macdonald, 1995, (VI.4.13), (VI.6.19)) by taking the portion of both sides of (10.3.8) of homogeneous degree n in the X- and Y-variables.

Remark 10.3.5 Let $\lambda \vdash n$ and z be an indeterminate. Say that λ has a *hook shape* if $\lambda_2 \leq 1$. Then $s_\lambda[1-z] = 0$ if λ is not a "hook." In fact,

$$s_\lambda[1-z] = \begin{cases}(-z)^r(1-z) & \text{if } \lambda = (n-r, 1^r), 0 \leq r \leq n-1,\\ 0 & \text{otherwise.}\end{cases}$$

This follows by setting $q = t = 0$ in (10.3.5), since $P_\lambda(X;0,0) = s_\lambda$.

10.3.2 The q, t-Kostka Polynomials

Macdonald found that the $P_\lambda(X;q,t)$ have a very mysterious property. Let $J_\mu[X;q,t]$ denote the so-called *Macdonald integral form*, defined as

$$J_\mu(X;q,t) := \left(\prod_{x\in\mu}(1-q^a t^{l+1})\right)P_\mu(X;q,t). \tag{10.3.9}$$

Now expand J_μ in terms of the $s_\lambda[X(1-t)]$:

$$J_\mu(X;q,t) = \sum_{\lambda\vdash|\mu|} K_{\lambda,\mu}(q,t)s_\lambda[X(1-t)] \tag{10.3.10}$$

for some $K_{\lambda,\mu}(q,t) \in \mathbb{Q}(q,t)$. Macdonald (1995, (VI.8.18?)) conjectured that $K_{\lambda,\mu}(q,t) \in \mathbb{N}[q,t]$. This became a famous problem in combinatorics known as *Macdonald's positivity conjecture*.

Part of the fascination for this conjecture is the case $q = 0$, since $J_\mu(X;0,t) = Q_\mu(X;t)$, and so by (10.3.10), (10.3.3) and (10.2.6), we have

$$K_{\lambda,\mu}(0,t) = \sum_{T \in \mathrm{SSYT}(\lambda,\mu)} t^{\mathrm{charge}(T)}. \tag{10.3.11}$$

No two-parameter generalization of charge that generates the $K_{\lambda,\mu}(q,t)$ has ever been found though. Macdonald was also able to show that

$$K_{\lambda,\mu}(1,1) = K_{\lambda,\mu}. \tag{10.3.12}$$

The $K_{\lambda,\mu}(q,t)$ in (10.3.10) are known as the *q,t-Kostka polynomials*.

In the next section we describe a conjecture of Garsia and Haiman which gives a representation-theoretic interpretation for the positivity of the $K_{\lambda,\mu}(q,t)$. This conjecture was proved by Haiman (2001). Thus Macdonald's positivity conjecture was resolved after more than ten years of intensive research.

Macdonald (1995, §VI.8, p. 356) posed a refinement of his positivity conjecture which is still open. Due to (10.3.11) and (10.3.12), one could hope to find statistics $\mathrm{qstat}(T,\mu)$ and $\mathrm{tstat}(T,\mu)$ given by some combinatorial rule such that

$$K_{\lambda,\mu}(q,t) = \sum_{T \in \mathrm{SYT}(\lambda)} q^{\mathrm{qstat}(T,\mu)} t^{\mathrm{tstat}(T,\mu)}. \tag{10.3.13}$$

In Garsia and Haiman's work it is more natural to deal with the polynomials

$$\tilde{K}_{\lambda,\mu}(q,t) := t^{n(\mu)} K_{\lambda,\mu}(q,1/t), \quad \text{so (by (10.2.6))} \tilde{K}_{\lambda,\mu}(0,t) = \sum_{T \in \mathrm{SSYT}(\lambda,\mu)} q^{\mathrm{cocharge}(T)}.$$

Macdonald (1995, §VI.8, Example 2) found a statistical description of the $K_{\lambda,\mu}(q,t)$ whenever $\lambda = (n-k, 1^k)$ is a hook shape. It can be stated as

$$\tilde{K}_{(n-k,1^k),\mu} = e_k[B_\mu - 1]. \tag{10.3.14}$$

For example, $\tilde{K}_{(3,1,1),(2,2,1)}(q,t) = e_2[q + t + qt + t^2] = qt + q^2t + 2qt^2 + t^3 + qt^3$. He also found a statistical description when q is set equal to 1 (Macdonald, 1995, §VI.8, Example 7), and a similar description when $t = 1$. To describe it, say we are given a statistic $\mathrm{stat}(T)$ on skew SYT, an SYT T with n cells and a composition $\alpha = (\alpha_1, \ldots, \alpha_k)$ of n into k parts. Define the α-sectionalization of T to be the set of k skew SYT obtained in the following way. The first element of the set is the portion of T containing the numbers 1 through α_1. The second element is the portion of T containing the numbers $\alpha_1 + 1$ through $\alpha_1 + \alpha_2$, but with α_1 subtracted from each of these numbers, so we end up with a skew SYT of size α_2. In general, the ith element, denoted $T^{(i)}$, is the portion of T containing the numbers $\alpha_1 + \cdots + \alpha_{i-1} + 1$ through $\alpha_1 + \cdots + \alpha_i$, but with $\alpha_1 + \cdots + \alpha_{i-1}$ subtracted from each of these numbers. Then we define the *α-sectionalization* of $\mathrm{stat}(T)$ to be the sum $\mathrm{stat}(T,\alpha) := \sum_{i=1}^{k} \mathrm{stat}(T^{(i)})$.

In the above terminology, Macdonald's formula for the $q = 1$ Kostka numbers can be expressed as $\tilde{K}_{\lambda,\mu}(1,t) = \sum_{T \in \mathrm{SYT}(\lambda)} t^{\mathrm{comaj}(T,\mu')}$. For example, given the tableau T in Figure 10.4 with $\lambda = (4,3,2)$ and (coincidentally) μ also $(4,3,2)$, then $\mu' = (3,3,2,1)$ and the values of $\mathrm{comaj}(T,\mu')$ on $T^{(1)}, \ldots, T^{(4)}$ are $1, 2, 1, 0$, respectively, so $\mathrm{comaj}(T,\mu') = 4$.

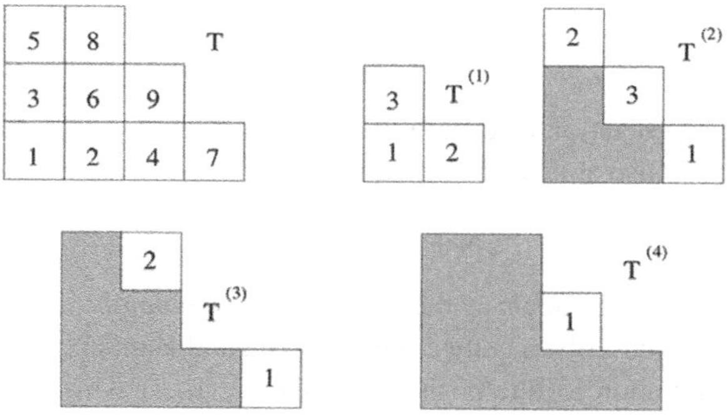

Figure 10.4 The $(3,3,2,1)$-sections of an SYT.

A combinatorial description of the $\tilde{K}_{\lambda,\mu}(q,t)$ when μ is a hook was found by Stembridge (1994). Given a composition α consisting of k parts, define $\mathrm{rev}(\alpha) := (\alpha_k, \alpha_{k-1}, \ldots, \alpha_1)$. Then, if $\mu = (n-k, 1^k)$, Stembridge's result can be expressed as

$$\tilde{K}_{\lambda,\mu}(q,t) = \sum_{T \in \mathrm{SYT}(\lambda)} q^{\mathrm{maj}(T,\mu)} t^{\mathrm{comaj}(T,\mathrm{rev}(\mu'))}. \tag{10.3.15}$$

Macdonald (1995, (VI.8.21), (VI.8.22)) obtained two symmetry relations, which (expressed in terms of the $\tilde{K}_{\lambda,\mu}$) are

$$\tilde{K}_{\lambda,\mu}(q,t) = \tilde{K}_{\lambda,\mu'}(t,q), \tag{10.3.16a}$$

$$\tilde{K}_{\lambda',\mu}(q,t) = t^{n(\mu)} q^{n(\mu')} \tilde{K}_{\lambda,\mu}(1/q, 1/t). \tag{10.3.16b}$$

Fishel (1995) first obtained statistics for the case when μ has two columns. In view of (10.3.16b) this also implies statistics for the case where μ has two rows. Later, Lapointe and Morse (2003c) and Zabrocki (1998) independently found alternate descriptions of this case, but all of these are rather complicated to state. A simpler description of the two-column case, based on the combinatorial formula for Macdonald polynomials in §10.5, is in §10.6.

In 1996 several groups of researchers (Garsia and Remmel, 1998; Garsia and Tesler, 1996; Kirillov and Noumi, 1998; Knop, 1997a,b; Lapointe and Vinet, 1997, 1998; Sahi, 1996) independently proved, using at least three totally different approaches, that $\tilde{K}_{\lambda,\mu}(q,t)$ is a polynomial with integer coefficients, which itself had been a major unsolved problem since 1988.[1]

[1] The first breakthrough on this problem appears to have been work by Lapointe and Vinet (1995, 1996) in 1995, who proved the corresponding integrality result for Jack polynomials. This seemed to have the effect of breaking the ice, since it was shortly after this that the proofs of Macdonald integrality were announced. As in the work by Kadell (1988a,b) on Selberg's integral, this gives another example of how results in Macdonald theory are often preceded by results on Jack polynomials.

We should mention that the Macdonald polynomiality result is immediately implied by the combinatorial formula in §10.5. The paper by Garsia and Remmel (1998) also contains a recursive formula for the $\tilde{K}_{\lambda,\mu}(q,t)$ when λ is an augmented hook, i.e., a hook plus the square $(2,2)$. Their formula immediately implies nonnegativity and by iteration could be used to obtain various combinatorial descriptions for this case. In 1999, G. Tesler (private communication) announced that, by using plethystic methods, he could prove nonnegativity of the case where λ is a *doubly augmented hook*, which is an augmented hook plus either the cell $(2,3)$ or the cell $(3,2)$.

10.3.3 The Garsia–Haiman Modules and the n!-Conjecture

Given any bihomogeneous subspace $W \subseteq \mathbb{C}[X_n, Y_n]$, we define the *bigraded Hilbert series* of W as $\mathrm{Hilb}(W;q,t) := \sum_{i,j\geq 0} t^i q^j \dim(W^{(i,j)})$, where the subspaces $W^{(i,j)}$ consist of those elements of W that are bihomogeneous of degree i in the x-variables and j in the y-variables, so $W = \bigoplus_{i,j\geq 0} W^{(i,j)}$. Also define the *diagonal action* of S_n on W by

$$(\sigma f)(x_1,\ldots,x_n,y_1,\ldots,y_n) := f(x_{\sigma_1},\ldots,x_{\sigma_n},y_{\sigma_1},\ldots,y_{\sigma_n}), \quad \sigma \in S_n,\ f \in W.$$

Clearly the diagonal action fixes the subspaces $W^{(i,j)}$, so we can define the *bigraded Frobenius series* of W as $\mathrm{Frob}(W;q,t) := \sum_{i,j\geq 0} t^i q^j \sum_{\lambda\vdash n} s_\lambda \,\mathrm{Mult}(\chi^\lambda, W^{(i,j)})$. Similarly, let W^ϵ be the subspace of alternating elements in W, and define $\mathrm{Hilb}(W^\epsilon;q,t)k := \sum_{i,j\geq 0} t^i q^j \dim(W^{\epsilon(i,j)})$. As in the case of subspaces of $\mathbb{C}[X_n]$ we have $\mathrm{Hilb}(W^\epsilon;q,t) = \langle \mathrm{Frob}(W^\epsilon;q,t), s_{1^n}\rangle$.

For $\mu \in \mathrm{Par}(n)$, let $(c_1,r_1),\ldots,(c_n,r_n)$ be the $(a'+1, l'+1) = (\text{column}, \text{row})$ coordinates of the cells of μ, taken in some arbitrary order, and let $\Delta_\mu(X_n,Y_n) := \left| y_i^{c_j-1} x_i^{r_j-1} \right|_{i,j=1,n}$. For example,

$$\Delta_{(2,2,1)}(X_5,Y_5) = \begin{vmatrix} 1 & y_1 & x_1 & x_1y_1 & x_1^2 \\ 1 & y_2 & x_2 & x_2y_2 & x_2^2 \\ 1 & y_3 & x_3 & x_3y_3 & x_3^2 \\ 1 & y_4 & x_4 & x_4y_4 & x_4^2 \\ 1 & y_5 & x_5 & x_5y_5 & x_5^2 \end{vmatrix}.$$

Note that, up to sign, $\Delta_{1^n}(X_n,0) = \Delta(X_n)$, the Vandermonde determinant.

For $\mu \vdash n$, let $V(\mu)$ denote the linear span of $\Delta_\mu(X_n,Y_n)$ and its partial derivatives of all orders. Note that, although the sign of Δ_μ may depend on the arbitrary ordering of the cells of μ we started with, $V(\mu)$ is independent of this ordering. Garsia and Haiman (1993) conjectured Theorem 10.3.6 stated below. It was proved by Haiman (2001). It involves the *modified Macdonald polynomial*

$$\tilde{H}_\mu = \tilde{H}_\mu(X;q,t) := \sum_{\lambda\vdash n} \tilde{K}_{\lambda,\mu}(q,t) s_\lambda. \tag{10.3.17}$$

Theorem 10.3.6 *For all $\mu \vdash n$ we have* $\mathrm{Frob}(V(\mu); q, t) = \tilde{H}_{\mu}$.

Note that Theorem 10.3.6 implies $\tilde{K}_{\lambda,\mu}(q,t) \in \mathbb{N}[q,t]$.

Corollary 10.3.7 *For all $\mu \vdash n$,* $\dim V(\mu) = n!$.

Remark 10.3.8 Corollary 10.3.7 was known as the *$n!$ conjecture*. Although Theorem 10.3.6 appears to be much stronger, Haiman proved in 1999 that it is implied by Corollary 10.3.7.

It is clear from the definition of $V(\mu)$ that $\mathrm{Frob}(V(\mu); q, t) = \mathrm{Frob}(V(\mu'); t, q)$. Theorem 10.3.6 thus gives a geometric interpretation to (10.3.16a).

Example 10.3.9 It is known that

$$\tilde{H}_{1^n}(X; q, t) = \mathrm{Frob}(V(1^n); q, t) = (t; t)_n h_n \left[\frac{X}{1-t}\right] \quad \text{and} \quad \tilde{H}_n(X; q, t) = (q; q)_n h_n \left[\frac{X}{1-q}\right].$$

Before Haiman proved the general case using algebraic geometry, Garsia and Haiman (1996b) proved the special case of the $n!$ conjecture when μ is a hook by combinatorial methods. The case where μ is an augmented hook was proved by Reiner (1996).

From (10.3.9) we see that

$$\tilde{H}_{\mu}[X; q, t] = t^{n(\mu)} J_{\mu}\left[\frac{X}{1-1/t}; q, 1/t\right] = t^{-n} P_{\mu}\left[\frac{X}{1-1/t}; q, 1/t\right] \prod_{x \in \mu} (t^{l+1} - q^{a}). \qquad (10.3.18)$$

Macdonald (1995, (VI.6.24)) derived formulas for the coefficients in the expansion of $e_k P_{\mu}(X; q, t)$, and also of $h_k[X\frac{1-t}{1-q}]P_{\mu}(X; q, t)$, in terms of the $P_{\lambda}(X; q, t)$. These expansions reduce to the classical Pieri formulas for Schur functions discussed in Theorem 10.2.5 when $t = q$. When expressed in terms of the J_{μ}, the h_k Pieri rule becomes (Macdonald, 1995, §VI.8, Example 4)

$$h_k\left[X\frac{(1-t)}{1-q}\right] J_{\mu} = \sum_{\substack{\lambda \in \mathrm{Par} \\ \lambda/\mu \text{ is a horizontal } k\text{-strip}}} \frac{\prod_{x \in \mu}(1 - q^{a_{\mu}+\chi(x \in B)} t^{l_{\mu}+\chi(x \notin B)})}{\prod_{x \in \lambda}(1 - q^{a_{\lambda}+\chi(x \in B)} t^{l_{\lambda}+\chi(x \notin B)})} J_{\lambda},$$

where B is the set of columns which contain a cell of λ/μ, where a_{μ}, l_{μ} are the values of a, l when the cell is viewed as part of μ, and where a_{λ}, l_{λ} are the values of a, l when the cell is viewed as part of λ.

10.3.4 The Space of Diagonal Harmonics

Let $p_{h,k}[X_n, Y_n] = \sum_{i=1}^{n} x_i^h y_i^k, h, k \in \mathbb{N}$ denote the *polarized power sum*. It is known that the set $\{p_{h,k}[X_n, Y_n] \mid h + k \geq 0\}$ generates $\mathbb{C}[X_n, Y_n]^{S_n}$, the ring of invariants under the diagonal action. Thus a natural analogue of the quotient ring R_n of coinvariants is the quotient ring DR_n of diagonal coinvariants defined by $\mathrm{DR}_n := \mathbb{C}[X_n, Y_n] / \left\langle \sum_{i=1}^{n} x_i^h y_i^k \right\rangle_{h,k \in \mathbb{Z}; h+k>0}$. By analogy we also define the space of *diagonal harmonics*,

$$\mathrm{DH}_n := \left\{ f \in \mathbb{C}[X_n, Y_n] \,\middle|\, \sum_{i=1}^{n} \frac{\partial^h}{\partial x_i^h} \frac{\partial^k}{\partial y_i^k} f = 0,\ h, k \in \mathbb{Z}, h + k > 0 \right\}.$$

Many of the properties of H_n and R_n carry over to two sets of variables. For example DH_n is a finite-dimensional vector space which is isomorphic to DR_n as an S_n-module (under the diagonal action). The dimension of these spaces turns out to be $(n+1)^{n-1}$, a result which was first conjectured by Haiman (1994) and proved by him in 2002. His proof uses many of the techniques and results from his proof of the $n!$ conjecture. See Stanley (2003) for a nice expository account of the $n!$ theorem and the $(n+1)^{n-1}$ theorem.

Example 10.3.10 An explicit basis for DH_2 is given by $\{1, x_2 - x_1, y_2 - y_1\}$. The elements $x_2 - x_1$ and $y_2 - y_1$ form a basis for DH_2^{ϵ}. Thus $\mathrm{Frob}(DH_2; q, t) = s_2 + (q+t)s_{1^2}$.

The number $(n+1)^{n-1}$ is known to count some interesting combinatorial structures. For example, it counts the number of rooted, labeled trees on $n+1$ vertices with root node labeled 0. It also counts the number of parking functions on n cars. In the next section we discuss a conjecture of Haglund and Loehr which gives a combinatorial description for $\mathrm{Hilb}(DH_n; q, t)$ in terms of statistics on parking functions (Haglund and Loehr, 2005).

We let $M := (1-q)(1-t)$ and $T_\mu := t^{n(\mu)}q^{n(\mu')}$, $w_\mu := \prod_{x\in\mu}(q^a - t^{l+1})(t^l - q^{a+1})$ ($\mu \in \mathrm{Par}$). Haiman derives the $(n+1)^{n-1}$ result as a corollary of the following formula for the Frobenius series of DH_n.

Theorem 10.3.11 (Haiman, 2002) *Let B_μ, Π_μ be as in* (10.3.4). *Then*

$$\mathrm{Frob}(DH_n; q, t) = \sum_{\mu\vdash n} w_\mu^{-1} T_\mu M \tilde{H}_\mu \Pi_\mu B_\mu. \tag{10.3.19}$$

Theorem 10.3.11 was conjectured by Garsia and Haiman (1996a). The conjecture was inspired in part by suggestions from C. Procesi.

From (10.3.14) and the fact that $T_\mu = e_n[B_\mu]$, we have $\langle \tilde{H}_\mu, s_{1^n}\rangle = T_\mu$. Thus, if we define

$$C_n(q,t) := \sum_{\mu\vdash n} w_\mu^{-1} T_\mu^2 M \Pi_\mu B_\mu \tag{10.3.20}$$

then it follows by (10.3.19) that $C_n(q,t) = \langle \mathrm{Frob}(DH_n; q, t), s_{1^n}\rangle = \mathrm{Hilb}(DH_n^{\epsilon}; q, t)$. For instance, from Example 10.3.10 we have $C_2(q,t) = q + t$. The $C_n(q,t)$ are referred to as the *q,t-Catalan sequence*, since Garsia and Haiman (1996a) proved that $C_n(1,1)$ reduces to $C_n = \frac{1}{n+1}\binom{2n}{n}$, the nth Catalan number. The C_n have quite a history and arise very frequently in combinatorics and elsewhere. See Stanley (1999, Solution to Exercise 6.19) and Stanley (2015) for over 210 different objects counted by the Catalan numbers.

10.3.5 The Nabla Operator

We begin this section with a slight generalization of the Koornwinder–Macdonald reciprocity formula, in a form which occurs in a paper by Garsia et al. (1999).

Theorem 10.3.12 *Let $\mu, \lambda \in \mathrm{Par}$, $z \in \mathbb{R}$. Then*

$$\frac{\tilde{H}_\mu[1 + z(MB_\lambda - 1); q, t]}{\prod_{x\in\mu}(1 - zq^{a'}t^{l'})} = \frac{\tilde{H}_\lambda[1 + z(MB_\mu - 1); q, t]}{\prod_{x\in\lambda}(1 - zq^{a'}t^{l'})}. \tag{10.3.21}$$

Proof (Sketch) By cross multiplying, we can rewrite (10.3.21) as a statement saying that two polynomials in z are equal. Letting $z = t^n$ for $n \in \mathbb{N}$, the two polynomials agree by (10.3.6) and two polynomials which agree on infinitely many values must be equal. □

Remark 10.3.13 If $|\mu|, |\lambda| > 0$, we can cancel the factor of $1-z$ in the denominators on both sides of (10.3.21) and then set $z = 1$ to obtain

$$\frac{\tilde{H}_\mu[MB_\lambda; q, t]}{\Pi_\mu} = \frac{\tilde{H}_\lambda[MB_\mu; q, t]}{\Pi_\lambda}. \tag{10.3.22}$$

Another useful special case of (10.3.21) is $\lambda = \emptyset$, which gives $\tilde{H}_\mu[1-z; q, t] = \prod_{x\in\mu}(1-zq^{a'}t^{l'})$.

Let ∇ be the linear operator on symmetric functions which satisfies $\nabla\tilde{H}_\mu = T_\mu\tilde{H}_\mu$. It turns out that many of the results in Macdonald polynomials and diagonal harmonics can be elegantly expressed in terms of ∇. Some of the basic properties of ∇ were first worked out by Bergeron in unpublished notes and more advanced applications followed in a series of papers (Bergeron and Garsia, 1999; Bergeron et al., 1999; Garsia et al., 1999).

Proposition 10.3.14 *If* $n > 0$ *then* $\nabla e_n = \sum_{\mu\vdash n} w_\mu^{-1} T_\mu M \tilde{H}_\mu \Pi_\mu B_\mu$. *Hence Theorem 10.3.11 is equivalent to* $\mathrm{Frob}(\mathrm{DH}_n; q, t) = \nabla e_n$.

Proof (Sketch) We see, by expressing (10.3.7b) in terms of the $\tilde{H}_\mu$ and by using some simple plethystic substitutions, that (10.3.14) is equivalent to $e_n\left[\frac{XY}{M}\right] = \sum_{\mu\vdash n} w_\mu^{-1}\tilde{H}_\mu[X; q, t]\tilde{H}_\mu[Y; q, t]$. Now let $Y = M$, use (10.3.22), and then apply ∇ to both sides. □

10.4 The Combinatorics of the Space of Diagonal Harmonics

10.4.1 The Parking Function Model

A *Dyck path* is a lattice path in the first quadrant of the xy-plane from $(0, 0)$ to (n, n), consisting of unit north N and east E steps, which never goes below the diagonal $x = y$. We let $L^+_{n,n}$ denote the set of all such Dyck paths. A *parking function* σ is a placement of the integers $1, 2, \ldots, n$ (called *cars*) just to the right of the N steps of a Dyck path, in such a way that the numbers are strictly decreasing down columns. The *reading word* of σ is the permutation obtained by reading the cars along diagonals in a SW direction, outside to in, as in Figure 10.5. To a given parking function σ, we associate two statistics area(σ) and dinv(σ). The area *statistic* is defined as the number of squares strictly below the Dyck path and strictly above the diagonal. In Figure 10.5 the number of area cells in a given row is listed on the right of that row. The dinv *statistic* is the number of pairs of cars which form either "primary" or "secondary" inversions. Pairs of cars form a *primary inversion* if they are in the same diagonal, with the larger car in a higher row. Pairs form a *secondary inversion* if they are in successive diagonals, with the larger car in the outer diagonal and in a lower row. For example, for the parking function in Figure 10.5, car 8 forms primary inversions with cars 1 and 5, while car 5 forms a secondary inversion with car 3. The set of inversion pairs for this parking function

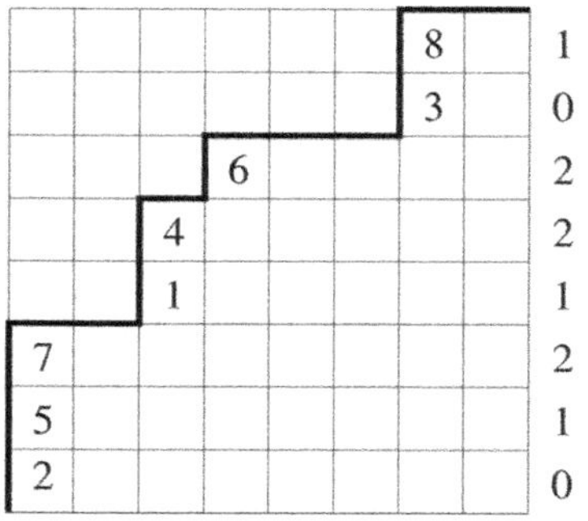

Figure 10.5 A parking function with area = 9, dinv = 6 and reading word 64781532.

is $\{(6,4),(7,1),(8,1),(8,5),(5,3),(3,2)\}$, so dinv = 6 while area = 9. A conjecture by Loehr and the author expresses $\mathrm{Hilb}(\mathrm{DH}_n; q, t)$ as a positive sum of monomials, one for each parking function.

Conjecture 10.4.1 (Haglund and Loehr, 2005; Haglund, 2008, Ch. 5)

$$\mathrm{Hilb}(\mathrm{DH}_n; q, t) = \sum_{\sigma} q^{\mathrm{dinv}(\sigma)} t^{\mathrm{area}(\sigma)}, \tag{10.4.1}$$

where the sum is over all parking functions with n cars.

Remark 10.4.2 Armstrong (2013) has recently introduced a hyperplane arrangement model for $\mathrm{Hilb}(\mathrm{DH}_n; q, t)$ involving a pair of hyperplane arrangements with a statistic associated to each one. See also Armstrong and Rhoades (2012). He gives a bijection with parking functions which sends his pair of hyperplane arrangement statistics to (area′, bounce), another pair of statistics which Haglund and Loehr showed have the same distribution over parking functions as (dinv, area).

Haglund et al. (2005b) introduced a generalization of Conjecture 10.4.1 which gives a combinatorial formula for the coefficient of a monomial symmetric function in the character ∇e_n. Their conjecture, formulated below, involves a *shuffle* of two sequences A and B, which is a permutation where all the elements of A occur in order, and all the elements of B occur in order, but the elements of A and B are intertwined in an arbitrary manner.

Conjecture 10.4.3 (Shuffle conjecture; Haglund et al., 2005b; Haglund, 2008, Ch. 6) *Let β, γ be two compositions with $|\beta|+|\gamma| = n$. Then $\langle \nabla e_n, e_\beta h_\gamma \rangle = \sum_\sigma q^{\mathrm{dinv}(\sigma)} t^{\mathrm{area}(\sigma)}$, where the sum is over all parking functions σ on n cars whose reading word is a shuffle of increasing sequences $(1, 2, \ldots, \gamma_1)$, $(\gamma_1+1, \gamma_1+2, \ldots, \gamma_1+\gamma_2)$, …, and decreasing sequences $(n, n-1, \ldots, n-\beta_1+1)$, $(n-\beta_1, n-\beta_1-1, \ldots, n-\beta_1-\beta_2+1)$, ….*

See §10.10 for a discussion of recent work on this conjecture. If $\beta = (n)$, the shuffle conjecture reduces to a theorem of Garsia and Haglund (2001, 2002) (see also Haglund, 2008, Ch. 3) which gives a combinatorial formula for the sign character of DH_n, or equivalently for the rational function $C_n(q,t)$ defined in (10.3.20). More generally, if $\beta = (n-d)$, $\gamma = (d)$, the shuffle conjecture reduces to the (q, t)-Schröder theorem of Haglund (2004b), which can be described in terms of sums over *Schröder lattice paths* consisting of north, east and diagonal steps. If $\gamma = 1^n$ then the shuffle conjecture reduces to Conjecture 10.4.1.

The shuffle conjecture can also be expressed in the following way.

Conjecture 10.4.4 (Alternate form of the shuffle conjecture)

$$\nabla e_n = \sum_{\pi \in L^+_{n,n}} t^{\mathrm{area}(\pi)} \mathcal{F}_\pi(X;q), \tag{10.4.2}$$

where the sum is over all Dyck paths π *and*

$$\mathcal{F}_\pi(X;q) = \sum_{\sigma \in \mathrm{PF}(\pi)} q^{\mathrm{dinv}(\sigma)} F_{n,\mathrm{Des}(\mathrm{read}(\sigma)^{-1})}(X). \tag{10.4.3}$$

Here PF(π) denotes the set of all parking functions for the Dyck path π. For example, for the parking function of Figure 10.5, the inverse descent set of the reading word is $\{2,3,5\}$, so in (10.4.3) this parking function would be weighted by $q^6 F_{8,\{2,3,5\}}$.

It is not at all obvious that the right-hand side of (10.4.2) is a symmetric function, but in fact each of the $\mathcal{F}_\pi(X;q)$ are symmetric functions. The proof of this relies on the theory of LLT polynomials, which were introduced in Lascoux et al. (1997). Their original construction is described in terms of ribbon tableaux, but we will present an equivalent formulation due to Haiman and his student Bylund, see (Haglund et al., 2005a, Appendix). We start with a grid of dotted lines, all of slope 1, with the vertical distance between successive dotted lines equaling 1. We then embed a tuple $\gamma = (\gamma_1, \gamma_2, \ldots, \gamma_k)$ of skew shapes in this grid such that each square has one of the dotted lines as a diagonal of the square, and we fill each γ_i with an SSYT T_i of that shape. In the example of Figure 10.6, the tuple is (22, 22, 11). If $\mathbf{T} = (T_1, T_2, \ldots, T_k)$ denotes such a tuple of SSYT, we let inv($\mathbf{T}$) denote the number of "inversion pairs" of T. An *inversion pair* is a pair of integers a, b, with $b > a$, and a, b in different skew shapes γ_i, and such that one of the two following conditions holds:

- a, b are on the same diagonal, with b in a column strictly left of a;
- a, b are on successive diagonals, with b strictly *NE* of a, i.e., b is in the diagonal just above that containing a, and in a column strictly to the right of the one containing a.

For example, for the tuple in Figure 10.6, the 5 above the 3 in γ_1 forms inversion pairs of the first type with the 1 and 3 from γ_2, and also with the 2 in γ_3, while the 6 forms inversion pairs with that same 5, 1, and 3, and also the 2 from γ_1.

Theorem 10.4.5 (Lascoux et al., 1997; Haiman–Bylund, see Haglund et al., 2005a) *Given any tuple of skew shapes* $(\gamma_1, \gamma_2, \ldots, \gamma_k)$, *the sum*

$$\sum_{\mathbf{T}=(T_1,\ldots,T_k), T_i \in \mathrm{SSYT}(\gamma_i)} q^{\mathrm{inv}(\mathbf{T})} x^{\mathbf{T}} \tag{10.4.4}$$

is a symmetric function. Here $x^{\mathbf{T}}$ *is the product* $x^{T_1} x^{T_2} \cdots x^{T_k}$ *of the usual x-weights of the SSYT.*

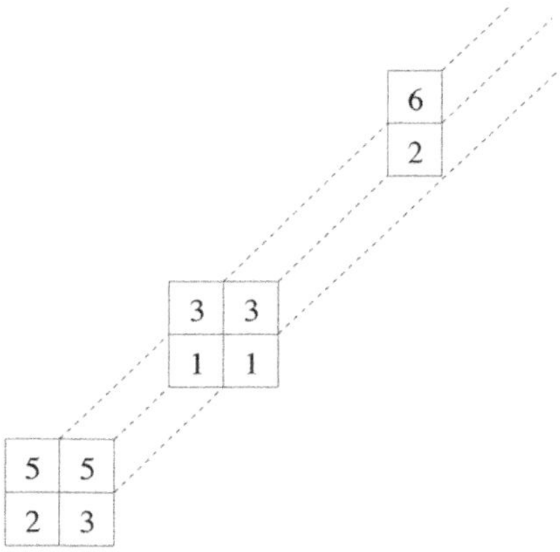

Figure 10.6 A tuple of SSYT occurring in the definition of an LLT product of the shapes (22, 22, 11). This tuple has 13 inversion pairs.

We will refer to the symmetric function (10.4.4) as the *LLT product* of the γ_i. If the multiset of numbers contained in the tuple $\mathbf{T}$ is just the set $\{1, 2, \ldots, n\}$, we say $\mathbf{T} \in \mathrm{SYT}(\gamma)$. Let the reading word read($\mathbf{T}$) of an LLT tuple of SSYT be the word obtained by reading along diagonals, outside to in, and in a NE direction along each diagonal, so for example the reading word of the tuple in Figure 10.6 is 5362513231. Furthermore, let $\zeta(\mathbf{T})$ denote the (unique) element of SYT(γ) whose reading word is the same as the standardization of the reading word of the tuple $\mathbf{T}$. Then clearly $\mathrm{inv}(\mathbf{T}) = \mathrm{inv}(\zeta(\mathbf{T}))$, so by Remark 10.2.8 we have the following corollary.

Corollary 10.4.6 *The LLT product of the γ_i in* (10.4.4) *can be expressed as*

$$\sum_{\mathbf{T}=(T_1,\ldots,T_k)\in \mathrm{SYT}(\gamma)} q^{\mathrm{inv}(\mathbf{T})} F_{n,\mathrm{Des}(\mathrm{read}(\mathbf{T}))^{-1})}(X).$$

Corollary 10.4.7 *The function $\mathcal{F}_\pi(X;q)$ from* (10.4.3) *is a symmetric function.*

Proof For each parking function σ occurring in the definition of F_π, there is a corresponding element $\mathbf{T}(\sigma) \in \mathrm{SYT}(\gamma)$ where $\mathrm{read}(\sigma) = \mathrm{read}(\mathbf{T}(\sigma))$, with the same set of inversion pairs (see Figure 10.7 for an example). Hence, F_π is an LLT product of vertical strips, and is thus symmetric by Corollary 10.4.6. □

Lascoux et al. (1997) conjectured that any LLT product is *Schur positive*, i.e., when expressed in the Schur basis the coefficients are in $\mathbb{N}[q]$. Note that if $q = 1$, the LLT product of the γ_i is just the product of Schur functions $s_{\gamma_1} s_{\gamma_2} \cdots s_{\gamma_k}$, which is Schur positive by the Littlewood–Richardson rule. By results of Leclerc and Thibon (1997) and Kashiwara and Tanisaki (2002), this conjecture is known to be true for the case where each γ_i is a partition (i.e., nonskew) shape, with the lower-left-hand square of each γ_i all on the same diagonal. Haglund et al. (2005b) extended this result somewhat to include the $\mathcal{F}_\pi(X;q)$. Grojnowski

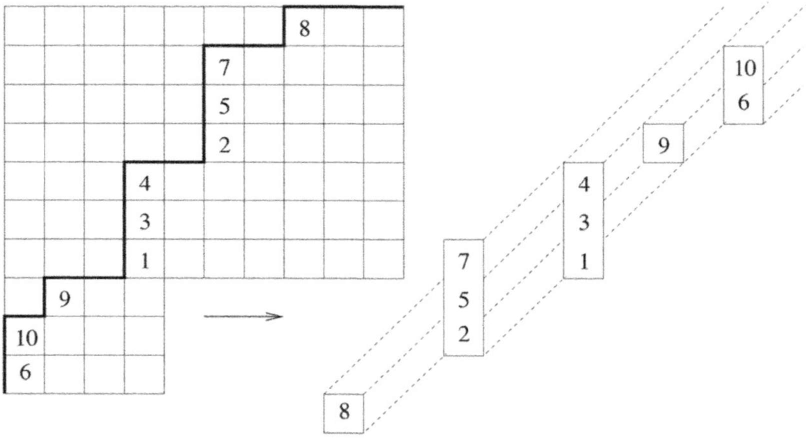

Figure 10.7 An SYT in the LLT product of vertical strips $(1^3/1^2, 1^4/1, 1^3, 1^2/1, 1^2)$ with 5 inversion pairs.

and Haiman (2007) announced a proof of the general case, which uses Kazhdan–Lusztig theory.

Remark 10.4.8 Haglund et al. (2012) have introduced a refinement of the shuffle conjecture known as the *compositional shuffle conjecture*, which says that the portion of the right-hand side of (10.4.2) involving paths π which hit the main diagonal $x = y$ at touch points $(a_1, a_1), (a_1 + a_2, a_1 + a_2), \ldots, (n, n)$ can be expressed as ∇ applied to a *compositional Hall–Littlewood polynomial*. These generalized Hall–Littlewood polynomials are defined using Jing operators; if the composition $(a_1, a_2, \ldots)$ is a partition μ they reduce to $(-1/q)^n \tilde{H}_{\mu'}(X; q, 0)$. Garsia et al. (2012, 2014) have used manipulations of plethystic Macdonald polynomial identities and bijective results of Garsia's student Hicks (2012) to prove many special cases of this conjecture; in particular they obtain a *compositional (q, t)-Schröder theorem*.

Remark 10.4.9 Haiman (1994), in his original work on diagonal harmonics, introduced a more general space $\mathrm{DH}_n^{(k)}$, where k is a positive integer. When $k = 1$ it reduces to DH_n. Haiman (2002) proved that $\mathrm{Frob}(\mathrm{DH}_n^{(k)}) = \nabla^k e_n$, and most of the combinatorial conjectures in this section, have "parameter k" versions, involving lattice paths in an $n \times kn$-rectangle which never go below the diagonal $x = ky$. In particular, there is a parameter k version of the shuffle conjecture (Haglund et al., 2005b), but at the time of this writing even the sign-character case of this remains open.

Recently, a dramatic generalization of the shuffle conjecture, and also the extension of it discussed in Remark 10.4.9, has been introduced. Many different researchers played a role in its formulation. It depends on a pair (m, n) of relatively prime positive integers. The combinatorial side of this conjecture occurs both in work by Hikita (2014) and in unpublished work by D. Armstrong, and it is fully described in Gorsky and Neguţ (2015) and also Garsia et al. (2017).

Let Grid(m, n) be the $n \times m$ grid of labeled squares whose upper-left-hand corner square is labeled with $(n-1)(m-1)-1$, and whose labels decrease by m as you go down columns and by n as you go across rows. For example,

Grid(3, 7) =

11	4	−3
8	1	−6
5	−2	−9
2	−5	−12
−1	−8	−15
−4	−11	−18
−7	−14	−21

To the corners of the squares of Grid(m, n) we associate Cartesian coordinates, where the lower-left-hand corner of the grid has coordinates $(0, 0)$, and the upper-right-hand corner of the grid (m, n). An (m, n)-*Dyck path* is a lattice path of unit N and E steps from $(0, 0)$ to (m, n) which never goes below the line $nx = my$, and we denote the set of such paths by $L^+_{(m,n)}$. One finds that $L^+_{(m,n)}$ is the same as the set of lattice paths π from $(0, 0)$ to (m, n) for which none of the squares with negative labels are above π. For a given π, we let area(π) denote the number of squares in Grid(m, n) with positive labels which are below π. Furthermore, let dinv(π) denote the number of squares in Grid(m, n) which are above π and whose arm and leg lengths satisfy

$$\frac{a}{l+1} < \frac{m}{n} < \frac{a+1}{l}. \tag{10.4.5}$$

Here, by the *arm* of a square s we mean the number of squares in the same row as s, to the right of s and to the left of π. The *leg* of s is the number of squares below s and in its column, and above π. For example, if $(m, n) = (3, 7)$ and $\pi = NNNNNEENNE$, then area$(\pi) = 2$ (corresponding to the squares with labels 2 and 5), and dinv$(\pi) = 2$; the squares with labels 11, 8, 4, 1 have $a = l = 1$; $a = 1, l = 0$; $a = 0, l = 1$; $a = l = 0$, respectively, and so the squares with labels 8 and 11 do not satisfy (10.4.5), while the squares with labels 1 and 4 do.

Let an (m, n)-*parking function* be a path $\pi \in L^+_{(m,n)}$ together with a placement of the integers 1 through n (called *cars*) just to the right of the N steps of π, with strict decrease down columns. For such a pair P, for $1 \le j \le n$, we let rank(j) be the label of the square that contains car j, and we set

$$\text{tdinv}(P) := \left|\{(i, j) \mid 1 \le i < j \le n \text{ and} \operatorname{rank}(i) < \operatorname{rank}(j) < \operatorname{rank}(i) + m\}\right|.$$

Furthermore, define the *reading word* read(P) to be the permutation obtained by listing the cars by decreasing order of their ranks. For example, for the $(3, 7)$-parking function of Figure 10.8, tdinv = 3, with inversion pairs formed by pairs of cars $(6, 7)$, $(4, 6)$ and $(2, 4)$, and the reading word is 7642531.

Let maxtdinv(π) be tdinv of the parking function for π whose reading word is the reverse of the identity, and for any parking function P for π set dinv(P) := dinv(π) + tdinv(P) −

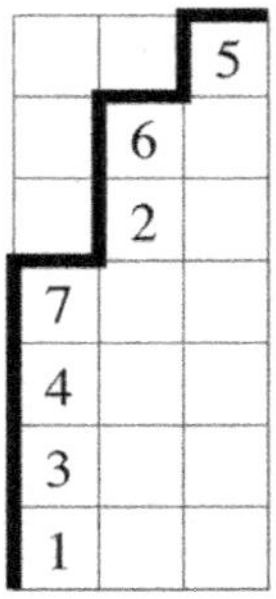

Figure 10.8 A $(3,7)$-parking function.

maxtdinv(π). We remark that, given a $\pi \in L_{n+1,n}$ and $\sigma \in \mathrm{PF}(\pi)$, the above definitions of the dinv and area statistics are the same as the ones given by our original definition, if we simply remove the last E step to form a path $\pi' \in L^+_{n,n}$, and view σ as a parking function for π'. Furthermore, the reading word of σ is also the same in both contexts. Hence, the above construction reduces to the original when $m = n + 1$.

There is an amazing extension of the symmetric function ∇e_n which has emerged from work by Burban and Schiffmann (2012), Schiffmann and Vasserot (2011, 2013), Negut (2014), and others on the elliptic Hall algebra and other related objects in algebraic geometry and string theory. Bergeron et al. (2016a) have given a concrete description of the construction of these symmetric functions by means of a family of plethystic operators on symmetric functions $Q_{(m,n)}$, defined recursively below. They satisfy $Q_{(m+n,n)}(-1)^n = \nabla Q_{(m,n)}(-1)^n$, where the $(-1)^n$ in these relations indicates that they are applied to the constant $(-1)^n$, viewed as an element of Λ. Furthermore, $Q_{(kn+1,n)}(-1)^n = \nabla^k e_n$, so they contain ∇e_n and the symmetric functions from Remark 10.4.9 as special cases.

To construct the $Q_{(m,n)}$, first let D_k be the operator on symmetric functions $F(X)$ defined by

$$D_k F[X] := F\Big[X + \tfrac{M}{z}\Big] \sum_{i\geq 0} (-z)^i e_i[X] \ \Big|_{z^k},$$

where $\ldots|_{z^k}$ means "take the coefficient of z^k in $\ldots$," and again $M = (1-q)(1-t)$. The D_k operators were introduced in Garsia et al. (1999); they form important building blocks in the development of plethystic identities involving Macdonald polynomials. We require the following proposition.

Proposition 10.4.10 (Bergeron et al., 2016b) *For any coprime pair of integers (m,n) with $m, n > 1$, there is a unique pair (a,b) satisfying*

$$(1)\ 1 \leq a \leq m-1, \qquad (2)\ 1 \leq b \leq n-1, \qquad (3)\ mb + 1 = na.$$

Let $c = m - a$ and $d = n - b$. Then both (a,b) and (c,d) are coprime pairs.

For coprime $m, n > 1$ write $\mathrm{Split}(m,n) = (a,b) + (c,d)$, and otherwise set

$$\text{(a)}\ \mathrm{Split}(1,n) = (1, n-1) + (0,1), \qquad \text{(b)}\ \mathrm{Split}(m,1) = (1,0) + (m-1,1).$$

If Split$(m, n) = (a, b) + (c, d)$, recursively set $Q_{(m,n)} := M^{-1}\left[Q_{(c,d)}, Q_{(a,b)}\right]$ with base cases $Q_{(1,0)} = D_0$ and $Q_{(0,1)} = -\underline{e}_1$. Here $\underline{e}_1$ is multiplication by e_1 and $[x, y] = xy - yx$.

Conjecture 10.4.11 (Rational shuffle conjecture; Garsia et al., 2017; Gorsky and Neguţ, 2015) *For any pair of relatively prime positive integers (m, n) and any pair of compositions α, β with $\sum_i \alpha + \sum_j \beta = n$, we have*

$$\langle Q_{(m,n)}(-1)^n, e_\alpha h_\beta \rangle = \sum_{\substack{(m,n) \text{ parking functions } P \\ read(P) \text{ is an } \alpha,\beta \text{ shuffle}}} q^{\mathrm{dinv}(P)} t^{\mathrm{area}(\pi)}, \tag{10.4.6}$$

where the sum is over all (m, n) parking functions P whose reading word is a shuffle of decreasing sequences of lengths $\alpha_1, \alpha_2, \ldots$ and increasing sequences of lengths $\beta_1, \beta_2, \ldots$.

An alternate formulation of Conjecture 10.4.11 is

$$Q_{(m,n)}(-1)^n = \sum_{\pi \in L^+_{m,n}} t^{\mathrm{area}(\pi)} \mathcal{F}_\pi(X; q), \tag{10.4.7}$$

where $\mathcal{F}_\pi(X; q) := \sum_{\sigma \in \mathrm{PF}(\pi)} q^{\mathrm{dinv}(\sigma)} F_{n,\mathrm{Des}(\mathrm{read}(\sigma)^{-1})}(X)$. We leave it as an exercise for the interested reader to show that for any $\pi \in L^+_{(m,n)}$, $\mathcal{F}_\pi(X; q)$ is an LLT product of vertical strips, times a power of q. Hence the right-hand side of (10.4.7) is a Schur positive symmetric function.

Remark 10.4.12 The right-hand side of (10.4.7) first arises in work by Hikita (2014), who proved it is the bigraded Frobenius series of a certain module arising from affine Springer fibers. When $m = n + 1$ though, this module is not obviously isomorphic to DH_n.

Example 10.4.13 When $t = 1/q$, we have (Garsia et al., 2017) $q^{(m-1)(n-1)/2} Q_{(m,n)}(-1)^n = e_n[X[m]_q]/[m]_q$, where $[m]_q := (1 - q^m)/(1 - q)$. As a special case, when $t = 1/q$ we have

$$q^{(m-1)(n-1)/2} \langle Q_{(m,n)}(-1)^n, s_{1^n} \rangle = \begin{bmatrix} n+m-1 \\ n \end{bmatrix}_q, \tag{10.4.8}$$

where $\begin{bmatrix} n+m-1 \\ n \end{bmatrix}_q$ is the q-binomial coefficient. Dennis Stanton has asked for a statistic qstat which would allow us to express the right-hand side of (10.4.8) as a sum of q^{stat} over (m, n)-Dyck paths. In the case $m = n + 1$, MacMahon (1960, p. 214) proved that you can generate this using qstat = maj where, for the computation of maj on a Dyck path π, you write π as a sequence of N and E steps, replace each E by a 1 and each N by a 0, then take the usual maj statistic on words. For general (m, n) there is no known variant of maj which works, but if we assume Conjecture 10.4.6 we can use qstat = dinv $+(m - 1)(n - 1)/2 -$ area.

10.4.2 The Superpolynomial Knot Invariant

Dunfield et al. (2006) hypothesized the existence of a three-parameter knot invariant $\mathcal{P}_K(a, q, t)$, now known as the *superpolynomial knot invariant* of a knot K, which includes the HOMFLY polynomial as a special case. Since then various authors have proposed different possible definitions of the superpolynomial, which are conjecturally all equivalent. These definitions typically involve homology though, and they are difficult to compute.

Let (m,n) be a pair of relatively prime positive integers, and let $T_{(m,n)}$ denote the (m,n) *torus knot*, which is the knot obtained by wrapping a string around the torus at an angle such that by the time you return to the starting point, you have wrapped around m times in one direction and n in the other. An accepted definition of $\mathcal{P}_{T_{(m,n)}}(a,q,t)$ has emerged from work by Aganagic and Shakirov (2012, 2015) (using refined Chern–Simons theory) and Cherednik (2013) (using the double affine Hecke algebra). Gorsky and Neguţ (2015) showed that these two different constructions yield the same three-parameter knot invariant which is now accepted as the definition of the superpolynomial for torus knots. Let

$$\tilde{\mathcal{P}}_{(m,n)}(u,q,t) := q^{(m-1)(n-1)}\mathcal{P}_{T_{(m,n)}}(u,1/q,1/t) \tag{10.4.9}$$

denote the *modified superpolynomial* of $T_{(m,n)}$. It can be described analytically as

$$\sum_{d=0}^{n}(-u)^d\langle Q_{(m,n)}(-1)^n, e_{n-d}h_d\rangle,$$

which is basically the generating function for hook shapes for the symmetric function occurring in Conjecture 10.4.6. Thus, if we assume this conjecture, we have a nice positive expression for the modified superpolynomial for torus knots. See also Gorsky et al. (2014), Haglund (2016) and Oblomkov et al. (2018).

Remark 10.4.14 Garsia et al. (2017) introduced the *compositional rational shuffle conjecture*, which contains both the rational shuffle conjecture and the compositional shuffle conjecture as special cases. The key element in the conjecture is a subtle construction of $Q_{(m,n)}$ operators for (m,n) not relatively prime.

10.4.3 Tesler Matrices and a Polynomial Formula for the Hilbert Series of DH$_n$

Haglund (2011) obtained a new polynomial expression for Hilb(DH$_n$). The expression is in $\mathbb{Z}[q,t]$, and has some negative coefficients, but hopefully further work will lead to a positive expression as in (10.4.1). A *Tesler matrix* of order n is an $n\times n$ upper-triangular matrix of nonnegative integers such that for any j in the range $1\le j\le n$, the sum of all the entries in the jth row of the matrix, minus the sum of all the entries in the jth column strictly above the diagonal, equals 1. Let Q_n denote the set of Tesler matrices of order n. Then the elements of Q_3 are

$$\begin{bmatrix}1&0&0\\0&1&0\\0&0&1\end{bmatrix},\ \begin{bmatrix}1&0&0\\0&0&1\\0&0&2\end{bmatrix},\ \begin{bmatrix}0&1&0\\0&2&0\\0&0&1\end{bmatrix},\ \begin{bmatrix}0&1&0\\0&1&1\\0&0&2\end{bmatrix},\ \begin{bmatrix}0&1&0\\0&0&2\\0&0&3\end{bmatrix},\ \begin{bmatrix}0&0&1\\0&1&0\\0&0&2\end{bmatrix},\ \begin{bmatrix}0&0&1\\0&0&1\\0&0&3\end{bmatrix}.$$

Let $[k]_{q,t} := (t^k-q^k)/(t-q)$ denote the (q,t)-analogue of the integer k, and recall that $M=(1-q)(1-t)$. We associate the weight $\mathrm{wt}(C) := (-M)^{\mathrm{pos}(C)-n}\prod_{c_{ij}>0}[c_{ij}]_{q,t}$ to each Tesler matrix C, where pos(C) is the number of positive entries in C. For example,

$$\mathrm{wt}\begin{bmatrix}0&1&0\\0&1&1\\0&0&2\end{bmatrix} = (t+q)(-M) = -(t+q)(1-q)(1-t).$$

Theorem 10.4.15 (Haglund, 2011)

$$\mathrm{Hilb}(\mathrm{DH}_n) = \sum_{C \in Q_n} \mathrm{wt}(C). \tag{10.4.10}$$

Example 10.4.16 When $n = 3$, (10.4.10) becomes

$$\begin{aligned}\mathrm{Hilb}(\mathrm{DH}_3) = 1 &+ (t+q) + (t+q) - (1-q)(1-t)(t+q) \\ &+ (t+q)(t^2+tq+q^2) + (t+q) + (t^2+tq+q^2).\end{aligned}$$

Note. Formula (10.4.10) is clearly a polynomial. One advantage it has over (10.4.1) is that it is also clearly symmetric in q, t, in fact is a sum of Schur functions in the set of variables $\{q,t\}$. It is known (Bergeron, 2013) that $\mathrm{Hilb}(\mathrm{DH}_n; q, t)$ is a sum of terms of this form, and more generally so is $\langle \mathrm{Frob}(\mathrm{DH}_n; q, t), s_\lambda \rangle$ for any λ, but there is no known combinatorial description of these coefficients. Since $-M = -1 + s_1(q,t) - s_{1,1}(q,t)$, and $[k]_{q,t} = s_{k-1}(q,t)$, (10.4.10) together with the Littlewood–Richardson rule can be used to obtain an expression for $\mathrm{Hilb}(DH_n; q, t)$ as an alternating sum of $s_\lambda(q,t)$. One approach to the problem would be to study how negative terms cancel in this sum.

Remark 10.4.17 Formula (10.4.10) generalizes easily to a formula for $\mathrm{Hilb}(\mathrm{DH}_n^{(k)})$, by summing over matrices whose hook-sums are $1, k, k, \ldots, k$, and using the same weights $\mathrm{wt}(C)$ (Haglund, 2011).

10.4.4 More Recent Work Involving Tesler Matrices

Gorsky and Neguţ (2015) proved the following formula for the modified superpolynomial of the (m,n) torus knot defined in (10.4.9). They derived it from their contour integral identity (Gorsky and Neguţ, 2015, (52)).

Theorem 10.4.18 (Gorsky and Neguţ, 2015) *For any pair of positive, relatively prime integers (m,n),*

$$\tilde{\mathcal{P}}_{(m,n)}(u,q,t) = \sum_{C \in Q_n^{(m)}} \prod_{\substack{1 \le i \le n \\ c_{i,i} > 0}} (1-u) \prod_{1 \le i < n} \left([c_{i,i+1}+1]_{q,t} - [c_{i,i+1}]_{q,t} \right) \prod_{2 \le i+1 < j \le n} (-M)[c_{i,j}]_{q,t}.$$

Garsia and Haglund (2015) derived a Tesler matrix expression for ∇e_n. Mészáros et al. (2016, 2017) introduced the *polytope of Tesler matrices*, whose points with integer coordinates are in bijection with Tesler matrices. Connections between DH_n and polytopes were further developed in Liu et al. (2016). Wilson (2016) obtained Tesler matrix formulas for a broad class of functions of the form $\langle F, h_1^n \rangle$.

10.5 The Expansion of the Macdonald Polynomial into Monomials

In this section we give a combinatorial description of the modified Macdonald polynomial $\tilde{H}_\mu(X; q, t)$ (see (10.3.17)), and discuss its consequences. Let $\mu \vdash n$. We let $\mathrm{dg}(\mu)$ denote the *augmented diagram* of μ, consisting of μ together with a row of squares below μ, referred to as the *basement*, with coordinates $(j, 0)$, $1 \le j \le \mu_1$. Define a *filling* σ of μ to be an assignment of a positive integer to each square of μ. For $s \in \mu$, we let $\sigma(s)$ denote the integer assigned to s, i.e., the integer occupying s. Let the *reading word* $\text{read}(\sigma) = \sigma_1\sigma_2\cdots\sigma_n$ be the word obtained by reading the occupants of μ across rows left to right, starting with the top row and working downwards. Note that the reading word does not include any of the entries in the basement. In this section we assume the basement is occupied by virtual infinity symbols, i.e., $\sigma(j, 0) = \infty$.

For each filling σ of μ, we associate x, q and t weights. The x weight is defined in a similar fashion to SSYT, namely $x^\sigma := \prod_{s\in\mu} x_{\sigma(s)}$. For $s \in \mu$, let North(s) denote the square of μ right above s (if it exists) in the same column, and South(s) the square of dg(μ) directly below s, in the same column. Let the *descent set* of σ, denoted $\text{Des}(\sigma, \mu)$, be the set of squares $s \in \mu$ where $\sigma(s) > \sigma(\text{South}(s))$. (In this section we regard the basement as containing virtual infinity symbols, so no square in the bottom row of σ can be in $\text{Des}(\sigma, \mu)$.) Finally set $\text{maj}(\sigma, \mu) := \sum_{s\in\text{Des}(\sigma,\mu)}(l(s) + 1)$. Note that $\text{maj}(\sigma, 1^n) = \text{maj}(\text{read}(\sigma))$, where maj is the usual *major index statistic*, defined as the sum of those i for which $\sigma_i > \sigma_{i+1}$.

Below, a and l will denote the arm and leg lengths, as in Figure 10.3. We say that a square $u \in \mu$ *attacks* all other squares $v \in \mu$ in its row and strictly to its right, and all other squares $v \in dg(\mu)$ in the row below and strictly to its left. We say that u, v attack each other if u attacks v or v attacks u. An *inversion pair* of σ is a pair of squares u, v where u attacks v and $\sigma(u) > \sigma(v)$. Let $\text{invset}(\sigma, \mu)$ denote the set of inversion pairs of σ, and set

$$\text{inv}(\sigma, \mu) := |\text{invset}(\sigma, \mu)| - \sum_{s\in\text{Des}(\sigma,\mu)} a(s). \tag{10.5.1}$$

For example, if σ is the filling on the left in Figure 10.9 then

$$\begin{aligned}
\text{Des}(\sigma) &= \{(1, 2), (1, 4), (2, 3), (3, 2)\}, \quad \text{maj}(\sigma) = 3 + 1 + 2 + 1 = 7,\\
\text{invset}(\sigma) &= \{\{(1, 4), (2, 4)\}, \{(2, 4), (1, 3)\}, \{(2, 3), (1, 2)\}, \{(1, 2), (3, 2)\},\\
&\qquad \{(2, 2), (3, 2)\}, \{(2, 2), (1, 1)\}, \{(3, 2), (1, 1)\}, \{(2, 1), (3, 1)\}, \{(2, 1), (4, 1)\}\},\\
\text{inv}(\sigma) &= 9 - (2 + 1 + 0 + 0) = 6.
\end{aligned}$$

Note that $\text{inv}(\sigma, (n)) = \text{inv}(\text{read}(\sigma))$, where inv is the usual inversion statistic on words.

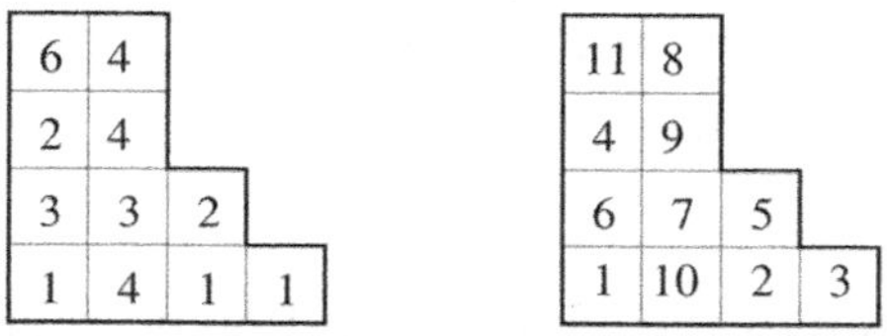

Figure 10.9 On the left, a filling of (4, 3, 2, 2) with reading word 64243321411, and on the right, its standardization.

Definition 10.5.1 For $\mu \vdash n$, let

$$C_\mu(X;q,t) = \sum_{\sigma:\,\mu\to\mathbb{Z}^+} x^\sigma t^{\mathrm{maj}(\sigma,\mu)} q^{\mathrm{inv}(\sigma,\mu)}. \tag{10.5.2}$$

We define the *standardization* of a filling σ, denoted $\zeta(\sigma)$, to be the standard filling whose reading word is the standardization of read(σ). Figure 10.9 gives an example of this. It is immediate from Definition 10.5.1 and Remark 10.2.8 that

$$C_\mu(X;q,t) = \sum_{\tau\in S_n} t^{\mathrm{maj}(\tau,\mu)} q^{\mathrm{inv}(\tau,\mu)} F_{n,\mathrm{Des}(\tau^{-1})}(X), \tag{10.5.3}$$

where we identify a permutation τ with the standard filling whose reading word is τ.

Remark 10.5.2 There is another way to view $\mathrm{inv}(\sigma,\mu)$ which will prove useful to us. Call three squares u, v, w, with $u, v \in \mu$, $w = \mathrm{South}(u)$, and with v in the same row as u and strictly to the right of μ, a *triple*. Given a standard filling σ, we define an orientation on such a triple by starting at the square, either u, v, or w, with the smallest element of σ in it, and going in a circular motion, towards the next largest element, and ending at the largest element. We say the triple is an *inversion triple* or a *coinversion triple* depending on whether this circular motion is counterclockwise or clockwise, respectively. Note that since $\sigma(j,0) = \infty$, if u, v are in the bottom row of σ, they are part of a counterclockwise triple if and only if $\sigma(u) > \sigma(v)$. Extend this definition to (possibly nonstandard) fillings by defining the orientation of a triple to be the orientation of the corresponding triple for the standardized filling $\zeta(\sigma)$. (So for two equal numbers, the one which occurs first in the reading word is regarded as being smaller.) It is an easy exercise to show that $\mathrm{inv}(\sigma,\mu)$ is the number of counterclockwise triples. For example, for the filling in Figure 10.9, the set consisting of the inversion triples is

$$\{\{(1,3),(1,4),(2,4)\},\{(1,2),(1,3),(2,3)\},\{(1,1),(1,2),(3,2)\},$$
$$\{(2,1),(2,2),(3,2)\},\{(2,1),(3,1),(2,0)\},\{(2,1),(4,1),(2,0)\}\}.$$

The following theorem was conjectured by Haglund (2004a) and proved by Haglund et al. (2005a,c). It gives a combinatorial formula for the modified Macdonald polynomial $\tilde{H}_\mu$.

Theorem 10.5.3 $C_\mu(X;q,t) = \tilde{H}_\mu(X;q,t)$ $(\mu \in \mathrm{Par})$.

Remark 10.5.4 Theorem 10.3.6 or (10.3.16a) imply the well-known symmetry relation

$$\tilde{H}_\mu(X;q,t) = \tilde{H}_{\mu'}(X;t,q). \tag{10.5.4}$$

This can be derived fairly easily from the three axioms in Theorem 10.5.5 below. An interesting open question in enumerative combinatorics is to prove this symmetry combinatorially using Theorem 10.5.3. Note that in the case $\mu = (n)$ this question is equivalent to asking for a bijective proof that maj and inv have the same distribution on arbitrary multisets, which is a classical result due to Foata (1968).

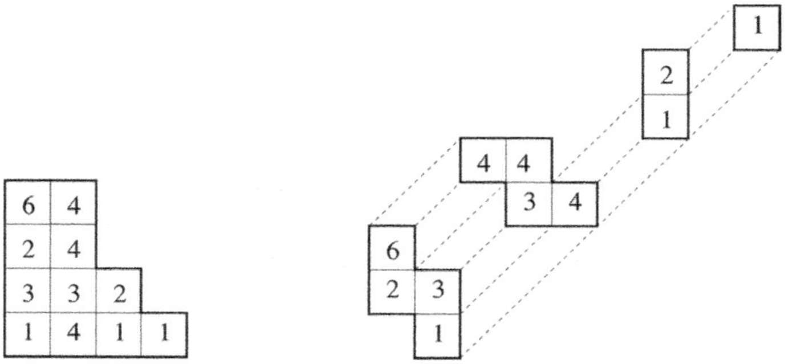

Figure 10.10 On the left, a filling σ, and on the right, the term $\mathbf{T}(\sigma)$ in the corresponding LLT product of ribbons.

10.5.1 Proof of Theorem 10.5.3

Theorem 10.3.1, when translated into a statement about the $\tilde{H}_\mu$ by using (10.3.18), gives (see Haglund et al., 2005a; Haiman, 1999) the following theorem.

Theorem 10.5.5 *The following three conditions uniquely determine a family $\tilde{H}_\mu(X; q, t)$ of symmetric functions:*

$$\tilde{H}_\mu[X(q-1); q, t] = \sum_{\rho \leq \mu'} c_{\rho,\mu}(q, t) m_\rho(X), \tag{10.5.5a}$$

$$\tilde{H}_\mu[X(t-1); q, t] = \sum_{\rho \leq \mu} d_{\rho,\mu}(q, t) m_\rho(X), \tag{10.5.5b}$$

$$\tilde{H}_\mu(X; q, t)|_{x_1^n} = 1. \tag{10.5.5c}$$

Hence, if one can show that $C_\mu(X; q, t)$ is a symmetric function and satisfies the three axioms above, it must be equal to $\tilde{H}_\mu$. The fact that $C_\mu(X; q, t)$ satisfies (10.5.5c) is trivial.

Next we argue that C_μ can be written as a sum of LLT polynomials. Fix a descent set D, and let $G_D(X; q) = \sum_{\sigma:\, \mu \to \mathbb{Z}^+; \mathrm{Des}(\sigma,\mu)=D} q^{\mathrm{inv}(\sigma,\mu)} x^\sigma$. If μ has one column, then G_D is a *ribbon Schur function*, that is, a Schur function of a skew shape containing no 2×2 square blocks of contiguous cells. More generally, $G_D(X; q)$ is an LLT product of ribbons. We illustrate how to transform a filling σ into a term $\mathbf{T}(\sigma)$ in the corresponding LLT product in Figure 10.10. Note that inversion pairs in σ are in direct correspondence with LLT inversion pairs in $\mathbf{T}(\sigma)$. Since the shape of the ribbons in $\mathbf{T}(\sigma)$ depends only on $\mathrm{Des}(\sigma, \mu)$, we have

$$C_\mu(X; q, t) = \sum_D t^L q^{-A} G_D(X; q), \tag{10.5.6}$$

where the sum is over all possible descent sets D of fillings of μ, with

$$L = \sum_{s \in D} (l(s) + 1), \quad A = \sum_{s \in D} a(s).$$

Since LLT polynomials are symmetric functions, C_μ is a symmetric function. Hence, (10.5.3) combined with general results on symmetric functions implies (Haglund et al., 2005a; Haglund, 2008, Ch. 6)

$$\omega^W C_\mu[Z+W;q,t] = \sum_{\beta\in S_n} t^{\mathrm{maj}(\beta,\mu)} q^{\mathrm{inv}(\beta,\mu)} \tilde{F}_{n,\mathrm{Des}(\beta^{-1})}(Z,W), \tag{10.5.7}$$

where ω^W is the involution ω acting on the W-set of variables, leaving the Z-set alone, and

$$\tilde{F}_{n,D}(Z,W) = \sum_{\substack{a_1\le a_2\le\cdots\le a_n \\ a_i=a_{i+1}\in\mathcal{A}_+ \implies i\notin D \\ a_i=a_{i+1}\in\mathcal{A}_- \implies i\in D}} z_{a_1} z_{a_2}\cdots z_{a_n}.$$

Here the indices a_i range over the alphabet $\mathcal{A}_\pm = \{1,\bar{1},2,\bar{2},\ldots\}$, which is the union of $\mathcal{A}_+ = \{1,2,\ldots\}$ and $\mathcal{A}_- = \{\bar{1},\bar{2},\ldots\}$, and by convention $z_{\bar{a}} = w_a$. Formula (10.5.7) is known as the *superization* of C_μ, and $\tilde{F}_{n,D}(Z,W)$ the *super quasisymmetric function*. It is important to note that (10.5.7) holds no matter what ordering we take for the elements of $\mathcal{A}_\pm$; we will be working with the two orderings

$$1<\bar{1}<2<\bar{2}<\cdots<n<\bar{n}, \tag{10.5.8a}$$

$$1<2<\cdots<n<\bar{n}<\cdots<\bar{2}<\bar{1}. \tag{10.5.8b}$$

Assume (10.5.8a) holds. By letting $z_i = qx_i$ and $w_i = -x_i$ in (10.5.7), by an extension of Remark 10.2.8 we get an expression for $C_\mu[X_n(q-1);q,t]$ as a sum over "super fillings" $\tilde{\sigma}$ of $\mathrm{dg}(\mu)$. Haglund et al. (2005a) introduced a sign-reversing involution which pairs super fillings with the same weight but opposite sign, leaving only terms whose monomial weights visibly satisfy the triangularity condition $\rho\le\mu'$ occurring on the right-hand side of (10.5.5a). Next assume (10.5.8b) holds. By letting $z_i = tx_i$ and $w_i = -x_i$ in (10.5.7) we get an expression in terms of super fillings for $C_\mu[X_n(t-1);q,t]$. Haglund et al. (2005a) introduce a different sign-reversing involution for this case, which after cancelation leaves only terms whose monomial weights visibly satisfy the triangularity condition $\rho\le\mu$ occurring on the right-hand side of (10.5.5b). We conclude that $C_\mu(X_n;q,t)$ satisfies all three axioms and hence must equal $\tilde{H}_\mu(X_n;q,t)$. □

Remark 10.5.6 The reader will notice the similarity between the $\mathrm{inv}(\sigma,\mu)$ statistic and the dinv statistic on parking functions from §10.4. In fact, it was elements of the shuffle conjecture combined with known ways of expressing special cases of $\tilde{H}_\mu(X;q,t)$ in terms of LLT polynomials which led Haglund (2004a) to conjecture Theorem 10.5.3. See Haglund (2006) for a detailed description of this story.

10.6 Consequences of Theorem 10.5.3

10.6.1 The Cocharge Formula for Hall–Littlewood Polynomials

In this subsection we show how to derive (10.2.6), Lascoux and Schützenberger's formula for the Schur coefficients of the Hall–Littlewood polynomials, from Theorem 10.5.3. This application was first published by Haglund et al. (2005a,c), although the exposition here is taken mainly from Haglund (2006). We require the following lemma, the proof of which is due to N. Loehr and G. Warrington (private communication, 2003).

Lemma 10.6.1 *Let $\mu \vdash n$. Given multisets M_i, $1 \le i \le \ell(\mu)$ of positive integers with $|M_i| = \mu_i$, there is a unique filling σ with the property that the multiset of elements of σ in the ith row of μ is M_i for $1 \le i \le \ell(\mu)$, and* $\mathrm{inv}(\sigma, \mu) = 0$.

Proof Clearly the elements in the bottom row will generate no inversion triples if and only if they are in monotone nondecreasing order in the reading word. Consider the number to place in square $(1, 2)$, i.e., right above the square $(1, 1)$. Let p be the smallest element of M_2 which is strictly larger than $\sigma(1, 1)$, if it exists, and the smallest element of M_2 otherwise. Then if $\sigma(1, 2) = p$, one sees that $(1, 1)$ and $(1, 2)$ will not form any inversion triples with $(j, 2)$ for any $j > 1$. We can iterate this procedure. In square $(2, 2)$ we place the smallest element of $M_2 - \{p\}$ (the multiset obtained by removing one copy of p from M_2) which is strictly larger than $\sigma(2, 1)$, and so on, until we fill out row 2. Then we let $\sigma(1, 3)$ be the smallest element of M_3 which is strictly larger than $\sigma(1, 2)$, if it exists, and the smallest element of M_3 otherwise, etc. Each square (i, j) cannot be involved in any inversion triples with $(i, j - 1)$ and (k, j) for some $k > i$, so $\mathrm{inv}(\sigma, \mu) = 0$. For example, if $M_1 = \{1, 1, 3, 6, 7\}$, $M_2 = \{1, 2, 4, 4, 5\}$, $M_3 = \{1, 2, 3\}$ and $M_4 = \{2\}$, then the corresponding filling with no inversion triples is given in Figure 10.11. □

Given a filling σ, we construct a word $\mathrm{cword}(\sigma)$ by initializing cword to the empty string, then scanning through $\mathrm{read}(\sigma)$, from the beginning to the end, and each time we encounter a 1, adjoin the number of the row containing this 1 to the beginning of cword. After recording the row numbers of all the 1's in this fashion, we go back to the beginning of $\mathrm{read}(\sigma)$, and adjoin the row numbers of squares containing 2's to the beginning of cword. For example, if σ is the filling in Figure 10.11, then $\mathrm{cword}(\sigma) = 11222132341123$.

2				
3	1	2		
2	4	4	1	5
1	1	3	6	7

Figure 10.11 A filling with no inversion triples.

Assume σ is a filling with no inversion triples. We translate the statistic $\mathrm{maj}(\sigma, \mu)$ into a statistic on $\mathrm{cword}(\sigma)$. Note that $\sigma(1, 1)$ corresponds to the rightmost 1 in $\mathrm{cword}(\sigma)$; denote this 1 by w_{11}. If $\sigma(1, 2) > \sigma(1, 1)$, then $\sigma(1, 2)$ corresponds to the rightmost 2 that is left of w_{11}, otherwise it corresponds to the rightmost 2 (in $\mathrm{cword}(\sigma)$). In any case denote this 2 by w_{12}. More generally, for $i > 1$ the element in $\mathrm{cword}(\sigma)$ corresponding to $\sigma(1, i)$ is the first i encountered when traveling left from $w_{1,i-1}$, looping around, and starting at the right-hand end of $\mathrm{cword}(\sigma)$ if necessary. To find the subword $w_{21}w_{22} \cdots w_{2\mu'_2}$ corresponding to the second column of σ, we do the same algorithm on the word obtained by removing the elements $w_{11}w_{12}\cdots w_{1\mu'_1}$ from cword(σ). After that we remove $w_{21}w_{22}\cdots w_{2\mu'_2}$ and apply the same process to find $w_{31}w_{32}\cdots w_{3\mu'_3}$ etc.

Clearly $\sigma(i,j) \in \mathrm{Des}(\sigma,\mu)$ if and only if w_{ij} occurs to the left of $w_{i,j-1}$ in cword(σ). Thus maj(σ,μ) is transparently equal to the statistic cocharge(cword(σ)) described in the cocharge algorithm in §10.2.2.

We associate a two-line array $A(\sigma)$ to a filling σ with no inversions by letting the upper row $A_1(\sigma)$ be nonincreasing with the same weight as σ, and the lower row $A_2(\sigma)$ be cword(σ). For example, we associate the two-line array $\begin{smallmatrix} 7&6&5&4&4&3&3&2&2&2&1&1&1&1 \\ 1&1&2&2&2&1&3&2&3&4&1&1&2&3 \end{smallmatrix}$ to the filling in Figure 10.11. By construction, below equal entries in the upper row the entries in the lower row are nondecreasing. Since $\tilde{H}_\mu$ is a symmetric function, we can reverse the variables, replacing x_i by x_{n-i+1} for $1 \le i \le n$, without changing the sum. This has the effect of changing $A_1(\sigma)$ into a nondecreasing word, and we end up with an ordered two-line array as in the classic RSK algorithm We can invert this correspondence since from the two-line array we get the multiset of elements in each row of σ, which uniquely determines σ by Lemma 10.6.1. Thus, by Theorem 10.5.3, we get for the *modified Hall–Littlewood polynomials* $\tilde{H}_\mu(X;0,t)$ (see (10.3.17)) that

$$\tilde{H}_\mu(x_1,x_2,\ldots,x_n;0,t) = \sum_{\sigma:\mathrm{inv}(\sigma,\mu)=0} x^{\mathrm{weight}(A_1(\sigma))} t^{\mathrm{cocharge}(A_2(\sigma))} = \sum_{(A_1,A_2)} x^{\mathrm{weight}(A_1)} t^{\mathrm{cocharge}(A_2)}, \tag{10.6.1}$$

where the sum is over ordered two-line arrays satisfying weight$(A_2) = \mu$.

Now it is well known that for any word w of partition weight, we have cocharge(w) = cocharge(read(P_w)), where read(P_w) is the reading word of the insertion tableau P_w under the RSK algorithm (Manivel, 2001, pp. 48–49; Stanley, 1999, p. 417). Hence application of the RSK algorithm to (10.6.1) gives

$$\tilde{H}_\mu(x_1,x_2,\ldots,x_n;0,t) = \sum_{(P,Q)} x^{\mathrm{weight}(Q)} t^{\mathrm{cocharge}(\mathrm{read}(P))},$$

where the sum is over all pairs (P,Q) of SSYT of the same shape with weight$(P) = \mu$. Since the number of different Q tableaux of weight ν matched to a given P tableau of shape λ is equal to the Kostka number $K_{\lambda,\nu}$, we can finish the proof of (10.2.6) by writing

$$\begin{aligned} \tilde{H}_\mu(X;0,t) &= \sum_\nu m_\nu \sum_\lambda \sum_{\substack{P\in \mathrm{SSYT}(\lambda,\mu)\\ Q\in\mathrm{SSYT}(\lambda,\nu)}} t^{\mathrm{cocharge}(\mathrm{read}(P))} \\ &= \sum_\lambda \sum_{P\in\mathrm{SSYT}(\lambda,\mu)} t^{\mathrm{cocharge}(\mathrm{read}(P))} \sum_\nu m_\nu K_{\lambda,\nu} = \sum_\lambda s_\lambda \sum_{P\in\mathrm{SSYT}(\lambda,\mu)} t^{\mathrm{cocharge}(\mathrm{read}(P))}. \end{aligned}$$

10.6.2 Formulas for J_μ

By (10.3.18) we have for the Macdonald integral form J_μ (see (10.3.9)) that

$$J_\mu(Z;q,t) = t^{n(\mu)}\tilde{H}_\mu[Z(1-t);q,1/t] = t^{n(\mu)}\tilde{H}_\mu[Zt(1/t-1);q,1/t]$$
$$= t^{n(\mu)+n}\tilde{H}_{\mu'}[Z(1/t-1);1/t,q],$$

with use of (10.5.4). Given a super filling $\tilde{\sigma}$, let $|\tilde{\sigma}|$ be the filling obtained by replacing each negative letter $\bar{i}$ by the corresponding positive letter i for all i. Say $\tilde{\sigma}$ is *nonattacking* if no two squares containing equal entries in $|\tilde{\sigma}|$ attack each other (in the sense of the paragraph above (10.5.1)). Formula (10.5.7) and the first sign-reversing involution from the proof of Theorem 10.5.3 imply

$$J_\mu(Z;q,t) = \sum_{\text{nonattacking super fillings } \tilde{\sigma} \text{ of } \mu'} z^{|\tilde{\sigma}|} q^{\operatorname{maj}(\tilde{\sigma},\mu')} t^{\operatorname{coinv}(\tilde{\sigma},\mu')} (-t)^{\operatorname{neg}(\tilde{\sigma})}, \tag{10.6.2}$$

where coinv $:= n(\mu) -$ inv is the number of coinversion triples, and we use the ordering $1 < \bar{1} < 2 < \bar{2} < \cdots < n < \bar{n}$.

The following more compact form of (10.6.2) can be obtained by grouping together all the 2^n super fillings $\tilde{\sigma}$ whose absolute value equals a fixed positive filling σ.

Corollary 10.6.2 (Haglund et al., 2005a)

$$J_\mu(Z;q,t) = \sum_{\text{nonattacking fillings } \sigma \text{ of } \mu'} z^{\sigma} q^{\operatorname{maj}(\sigma,\mu')} t^{\operatorname{coinv}(\sigma,\mu')} \times \prod_{\substack{u\in\mu' \\ \sigma(u)=\sigma(\mathrm{South}(u))}} (1-q^{l(u)+1}t^{a(u)+1}) \prod_{\substack{u\in\mu' \\ \sigma(u)\neq\sigma(\mathrm{South}(u))}} (1-t), \tag{10.6.3}$$

where coinv $= n(\mu) -$ inv *is the number of coinversion triples, and each square in the bottom row is included in the last product.*

Example 10.6.3 Let $\mu = (3,3,1)$. Then for the nonattacking filling σ of μ' in Figure 10.12, we have coinversion triples $\{1,2),(2,2),(1,1)\}$, $\{(1,1),(2,1),(1,0)\}$, $\{(1,1),(3,1),(1,0)\}$, so coinv $= 3$. Furthermore, maj $= 3$, squares $(1,1)$, $(1,2)$, $(2,1)$, $(2,3)$ and $(3,1)$ each contribute a $(1-t)$, square $(1,3)$ contributes a $(1-qt^2)$, and $(2,2)$ contributes a $(1-q^2t)$. Thus the term in (10.6.3) corresponding to σ is $x_1x_2^3x_3^2x_4q^3t^3(1-qt^2)(1-q^2t)(1-t)^5$.

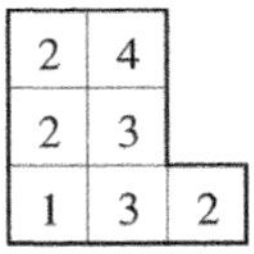

Figure 10.12 A nonattacking filling of $(3,3,1)'$.

The (*integral form*) *Jack polynomials* $J_\mu^{(\alpha)}(Z)$ can be obtained from the Macdonald J_μ by $J_\mu^{(\alpha)}(Z) = \lim_{t\to 1}(1-t)^{-|\mu|} J_\mu(Z; t^\alpha, t)$. If we set $q = t^\alpha$ in (10.6.3) and then divide by $(1-t)^{|\mu|}$ and take the limit as $t \to 1$ we get the following result by Knop and Sahi (1997):

$$J_\mu^{(\alpha)}(Z) = \sum_{\text{nonattacking fillings } \sigma \text{ of } \mu'} z^\sigma \prod_{\substack{u\in\mu' \\ \sigma(u)=\sigma(\text{South}(u))}} (\alpha(l(u)+1) + a(u) + 1). \tag{10.6.4}$$

Remark 10.6.4 There is another formula for J_μ (Haglund, 2008, pp. 132–133) corresponding to the second sign-reversing involution from the proof of Theorem 10.5.3, a formula which gives the expansion of J_μ into fundamental quasisymmetric functions F_α. The terms in the formula are not as elegant as those of (10.6.3) though, and we will not describe it here.

10.6.3 Schur Coefficients

Since by (10.5.6), $\tilde{H}_\mu(X; q, t)$ is a positive sum of LLT polynomials, the result by Grojnowski and Haiman (2007) that LLT polynomials are Schur positive gives a new proof that $\tilde{K}_{\lambda,\mu}(q,t) \in \mathbb{N}[q,t]$. In fact, we also get a natural decomposition of $\tilde{K}_{\lambda,\mu}(q,t)$ into "LLT components." This result is completely geometric though, and it is still hoped that a purely combinatorial formula for the $\tilde{K}_{\lambda,\mu}(q,t)$ of the form (10.3.13) can be found. In this subsection we indicate how such a formula can be obtained when μ has two columns.

By a *final segment* of a word we mean the last k letters of the word, for some k. We say that a filling σ is a *Yamanouchi filling* if, in any final segment of read(σ), there are at least as many i's as $i+1$'s, for all $i \geq 1$. Haglund et al. (2005a) proved the following result.

Theorem 10.6.5 *For any partition μ with $\mu_1 \leq 2$,*

$$\tilde{K}_{\lambda,\mu}(q,t) = \sum_{\sigma \text{ Yamanouchi}} t^{\text{maj}(\sigma,\mu)} q^{\text{inv}(\sigma,\mu)}, \tag{10.6.5}$$

where the sum is over all Yamanouchi fillings of μ.

Other combinatorial formulas for the two-column case are known; see Fishel (1995), Lapointe and Morse (2003c), and Zabrocki (1998, 1999), although (10.6.5) is perhaps the simplest. We also mention that Assaf and Garsia (2009) have found a recursive construction which produces an explicit basis for the Garsia–Haiman modules $V(\mu)$ when μ has at most two columns or is a hook shape. The proof of Theorem 10.6.5 involves a combinatorial construction which groups together fillings which have the same maj and inv statistics. This is carried out with the aid of *crystal graphs*, which occur in the representation theory of Lie algebras. We should mention that in (10.6.5), if we restrict the sum to those fillings with a given descent set, we get the Schur decomposition for the corresponding LLT polynomial.

If, in (10.6.5), we relax the condition that μ has at most two columns, then the equation no longer holds. It is an open problem to find a way of modifying the concept of a Yamanouchi filling such that (10.6.5) is true more generally. A specific conjecture, when μ has three columns, for the $\tilde{K}_{\lambda,\mu}(q,t)$, of the special form (10.3.13), was given by Haglund (2004a), and later proved by Blasiak (2016). In fact, Blasiak's result applies to any LLT product of three skew-shapes, and is the most general result currently known about the combinatorics of LLT Schur coefficients. It builds on joint work by Blasiak and Fomin (2017).

10.7 Nonsymmetric Macdonald Polynomials

As we wrote in §10.1, work by Macdonald (1996) (further developed by Cherednik, 1995; Sahi, 1998; Knop, 1997a; Ion, 2003, among others) and by Sahi (1999) gave a construction of nonsymmetric (i.e., not Weyl group invariant) polynomials, starting from affine root systems and yielding by symmetrization the orthogonal polynomials associated with root systems as introduced by Macdonald (2000) and Koornwinder (1992). See Macdonald (2003) and Chapter 9 for a general treatment of these symmetric and nonsymmetric *Macdonald–Koornwinder polynomials*. In the GL_n case the *nonsymmetric Macdonald polynomials* $E_\alpha(X;q,t)$ form a basis for the polynomial ring $\mathbb{Q}(q,t)[x_1,\dots,x_n]$, and are natural nonsymmetric analogues of the symmetric Macdonald $P_\lambda(X;q,t)$ of Theorem 10.3.1.

In this section we will focus on the combinatorial properties of the GL_n case, where there is a special structure which allows us to assume that α (generally in a weight lattice) is a *composition*, i.e., $\alpha \in \mathbb{N}^n$. Given $\alpha \in \mathbb{N}^n$, let α' denote the *transpose diagram* of α, consisting of the squares $\alpha' := \{(i,j) \mid 1 \le i \le n,\ 1 \le j \le \alpha_i\}$. Furthermore, let $\mathrm{dg}(\alpha')$ denote the augmented transpose diagram obtained by adjoining the basement row of n squares $\{(i,0)\}_{i=1}^n$ below α'. Given $s \in \alpha'$, we let $\mathrm{leg}(s)$ be the number of squares of α' above s and in the same column of s. Define $\mathrm{Arm}(s)$ to be the set of squares of $\mathrm{dg}(\alpha')$ which are either to the right and in the same row as s, and also in a column not taller than the column containing s, or to the left and in the row below the row containing s, and in a column strictly shorter than the column containing s. Then set $\mathrm{arm}(s) := |\mathrm{Arm}(s)|$. For example, for $\alpha = (1,0,3,2,3,0,0,0,0)$, the leg lengths of the squares of $(1,0,3,2,3,0,0,0,0)'$ are listed on the left in Figure 10.13 and the arm lengths on the right. Note that if α is a partition μ, the leg and arm definitions agree with those previously given for μ'.

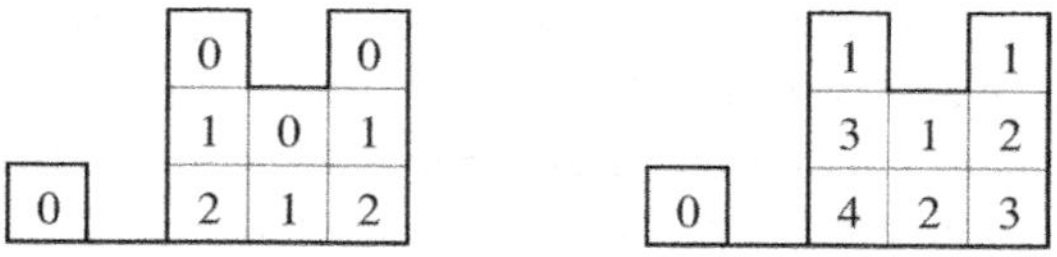

Figure 10.13 The leg lengths (on the left) and the arm lengths (on the right) for $(1,0,3,2,3,0,0,0,0)'$.

Given two polynomials $f(x_1,\dots,x_n;q,t)$ and $g(x_1,\dots,x_n;q,t)$ whose coefficients depend on q, t, define a scalar product

$$\langle f,g\rangle'_{q,t} := \mathrm{CT}\, f(x_1,\dots,x_n;q,t)g(1/x_1,\dots,1/x_n;1/q,1/t)W(x_1,\dots,x_n;q,t),$$

$$W(x_1,\dots,x_n;q,t) := \prod_{1\le i<j\le n} (x_i/x_j;q)_\nu(qx_j/x_i;q)_\nu, \quad t = q^\nu.$$

Here CT means "take the constant term in," and $(z;q)_\nu := (z;q)_\infty/(zq^\nu;q)_\infty$ with $(z;q)_\infty := \prod_{i=0}^{\infty}(1-zq^i)$.

The E_α can be defined in two ways. One way, due to Macdonald (1996), is to introduce a certain partial order "$<$" on compositions, and then show that the E_α are the unique family of polynomials which are triangular in the sense that any monomials x^β that occur in E_α must satisfy $\beta \le \alpha$, and also are orthogonal with respect to $\langle\cdot,\cdot\rangle'_{q,t}$: $\langle E_\alpha, E_\beta\rangle'_{q,t} = 0$ if $\alpha \ne \beta$. (For fixed q, t this can also be viewed as a biorthogonality for the two systems of polynomials $E_\alpha(X;q,t)$ and $E_\beta(X;q^{-1},t^{-1})$.) The P_λ are also orthogonal with respect to $\langle\cdot,\cdot\rangle'_{q,t}$. (The above construction has an extension to general affine root systems, while it is not known whether the orthogonality of the P_λ with respect to the combinatorially defined scalar product in (10.3.1) has a version for other root systems.)

Another way, which is more closely related to the combinatorial formulas we discuss in this section, is to use Cherednik's intertwiner relations, which give recurrence relations which uniquely define the E_α. In the GL_n case, both Knop (1997a) and Sahi (1996) independently found a simplification in one of these two recurrence relations. The Knop–Sahi formula (which is discussed in more detail in Haglund et al., 2008) involves the operators

$$\pi(\alpha_1,\dots,\alpha_n) := (\alpha_n+1,\alpha_1,\dots,\alpha_{n-1}), \qquad \Psi f(x_1,\dots,x_n) := x_1 f(x_2,x_3,\dots,x_n,x_1/q),$$

where f is any polynomial in $\mathbb{Q}(q,t)[x_1,\dots,x_n]$. Also, let $s_i(\alpha)$ be the composition with α_i and α_{i+1} interchanged (if $1 \le i \le n-1$) and α_1 and α_n interchanged (if $i=0$).

Lemma 10.7.1 (Haglund et al., 2008) *The E_α for $\alpha \in \mathbb{N}^n$ are uniquely characterized by the initial value $E_{0^n} = 1$ together with the relations*

$$E_{s_i(\alpha)}(x_1,\dots,x_n;q,t) = \left(T_i + \frac{1-t}{1-q^{l(s)+1}t^{a(s)}}\right)E_\alpha(x_1,\dots,x_n;q,t) \tag{10.7.1a}$$

(where i is such that $\alpha_i > 0$ and $\alpha_{i+1} = 0$) and

$$E_{\pi(\alpha)}(x_1,\dots,x_n;q,t) = q^{\alpha_n}\Psi E_\alpha(x_1,\dots,x_n;q,t). \tag{10.7.1b}$$

In (10.7.1a) *the T_i, $0 \le i \le n-1$ are the usual generators of the affine Hecke algebra, which satisfy the quadratic relation*

$$(T_i - t)(T_i + 1) = 0, \tag{10.7.2}$$

together with the braid relations

$$T_iT_{i+1}T_i = T_{i+1}T_iT_{i+1}, \quad T_iT_j = T_jT_i \quad \text{if} |i-j| > 1,$$

where all indices are modulo n. The T_i act on monomials in the X-variables by $T_ix^\lambda = tx^{s_i(\lambda)} + (t-1)\frac{x^\lambda - x^{s_i(\lambda)}}{1-x^{y_i}}$, where $x^{y_i} = x_i/x_{i+1}$ for $1 \le i \le n-1$ and $x^{y_0} = qx_n/x_1$.

Proof If $\alpha_1 > 0$, then $\alpha = \pi(\beta)$ where $\beta \in \mathbb{N}^N$. By induction on the sum of the parts, we can assume E_β is already determined, and apply (10.7.1a). If $\alpha_1 = 0$ and $\alpha_j > 0$ for some $j > 1$, we can reduce to the case $\alpha_1 > 0$ by repeated application of (10.7.1b). □

For $\alpha \in \mathbb{N}^n$, let $\text{rev}(\alpha) := (\alpha_n, \alpha_{n-1}, \ldots, \alpha_1)$ be the composition obtained by reversing the parts of α, and set

$$\widehat{E}_\alpha(x_1, \ldots, x_n; q, t) := E_{\text{rev}(\alpha)}(x_n, \ldots, x_1; 1/q, 1/t). \tag{10.7.3}$$

This modified version of the E_α is what one gets by specializing the general theory of nonsymmetric Macdonald polynomials (Cherednik, 1995) to GL_n root data (so the $\widehat{E}_\alpha$ is the P_α of Definition 9.3.9). Formula (10.7.3) thus provides the bridge between the two natural conventions on GL_n nonsymmetric Macdonald polynomials.

Marshall (1999), who made a special study of the $\widehat{E}_\alpha$, showed among other things that they satisfy a version of Selberg's integral.

Define the *integral form* nonsymmetric Macdonald polynomials $\widehat{\mathcal{E}}_\alpha$ as

$$\widehat{\mathcal{E}}_\alpha(x_1, \ldots, x_n; q, t) := \widehat{E}_\alpha(x_1, \ldots, x_n; q, t) \prod_{s \in \alpha'} (1 - q^{\text{leg}(s)+1} t^{\text{arm}(s)+1}).$$

Theorem 10.7.2 below describes a combinatorial formula for $\widehat{\mathcal{E}}_\alpha(X; q, t)$ which, in short, is the same as (10.6.3), using the extensions of the definitions of arm, leg, coinv, maj to composition diagrams, and changing the basement to $\sigma(j, 0) = j$.

Given $\alpha \in \mathbb{N}^n$, a *filling* σ of α' is an assignment of integers from the set $\{1, \ldots, n\}$ to the squares of α'. As before, we let the *reading word* $\text{read}(\sigma)$ be the word obtained by reading across rows, left to right, top to bottom. The standardization of a filling σ is the filling whose reading word is the *standardization* of $\text{read}(\sigma)$.

We say that a square $s \in \alpha'$ *attacks* all squares to its right in its row and all squares of $\text{dg}(\alpha')$ to its left in the row below. Call a filling *nonattacking* if there are no pairs of squares (s, u) with $s \in \alpha'$, s attacks u, and $\sigma(s) = \sigma(u)$. Note that, since $\sigma(j, 0) = j$, in any nonattacking filling with s of the form $(k, 1)$ we must have $\sigma(s) \geq k$. Figure 10.14 gives a nonattacking filling of $(1, 0, 3, 2, 3, 0, 0, 0, 0)'$.

As before we let $\text{South}(s)$ denote the square of $\text{dg}(\alpha')$ immediately below s, and let $\text{maj}(\sigma, \alpha')$ denote the sum, over all squares $s \in \alpha'$ satisfying $\sigma(s) > \sigma(\text{South}(s))$, of $\text{leg}(s) + 1$. A triple of α' is three squares u, v, w with $u \in \alpha'$, $v \in \text{Arm}(u)$, and $w = \text{South}(u)$. Note that v, w need not be in α', i.e., they could be in the basement. We determine the orientation of a triple by starting at the smallest and going in a circular motion to the next largest and then to the largest, where if two entries of a triple have equal σ-values then the one that occurs earlier in the reading word is viewed as being smaller. We say that such a triple is a *coinversion triple* if either v is in a column to the right of u, and u, v, w has a clockwise orientation, or v is in a column to the left of u, and u, v, w has a counterclockwise orientation. Let $\text{coinv}(\sigma, \alpha')$ denote the number of coinversion triples of σ. For example, the filling in Figure 10.14 has coinversion triples

$$\begin{aligned} \{&\{(3,2),(3,1),(4,2)\},\{(3,2),(3,1),(5,2)\},\{(3,2),(3,1),(1,1)\},\{(4,2),(4,1),(1,1)\},\\ &\{(5,1),(5,0),(1,0)\},\{(5,1),(5,0),(2,0)\},\{(5,1),(5,0),(4,0)\}\}. \end{aligned} \tag{10.7.4}$$

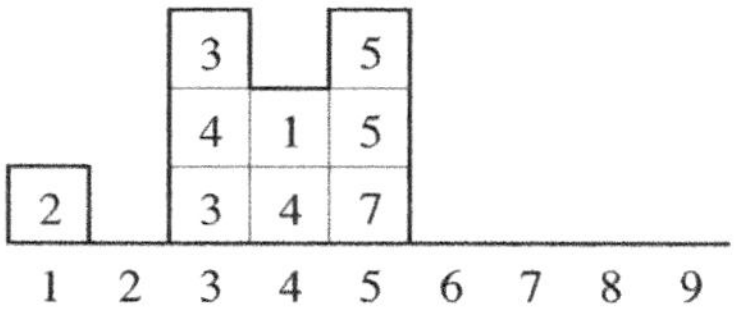

Figure 10.14 A nonattacking filling of $(1, 0, 3, 2, 3, 0, 0, 0, 0)'$.

Theorem 10.7.2 (Haglund et al., 2008) *For $\alpha \in \mathbb{N}^n$, $\sum_i \alpha_i \le n$,*

$$\widehat{\mathcal{E}}_\alpha(x_1, \ldots, x_n; q, t) = \sum_{\substack{\text{nonattacking fillings } \sigma \text{ of } \alpha' \\ \sigma(j,0)=j}} x^\sigma q^{\mathrm{maj}(\sigma,\alpha')} t^{\mathrm{coinv}(\sigma,\alpha')} \\ \times \prod_{\substack{u\in\alpha' \\ \sigma(u)=\sigma(\mathrm{South}(u))}} (1 - q^{\mathrm{leg}(u)+1} t^{\mathrm{arm}(u)+1}) \prod_{\substack{u\in\alpha' \\ \sigma(u)\neq\sigma(\mathrm{South}(u))}} (1 - t), \tag{10.7.5a}$$

where as usual $x^\sigma = \prod_{s\in\alpha'} x_{\sigma(s)}$. Equivalently,

$$\widehat{E}_\alpha(x_1, \ldots, x_n; q, t) = \sum_{\substack{\text{nonattacking fillings } \sigma \text{ of } \alpha' \\ \sigma(j,0)=j}} x^\sigma q^{\mathrm{maj}(\sigma,\alpha')} t^{\mathrm{coinv}(\sigma,\alpha')} \\ \times \prod_{\substack{u\in\alpha' \\ \sigma(u)\neq\sigma(\mathrm{South}(u))}} \frac{1 - t}{1 - q^{\mathrm{leg}(u)+1} t^{\mathrm{arm}(u)+1}}. \tag{10.7.5b}$$

Remark 10.7.3 We can obtain corresponding versions of (10.7.5a) and (10.7.5b) involving the $E_\alpha(X; q, t)$ and its integral form

$$\mathcal{E}_\alpha(x_1, \ldots, x_n; q, t) := E_\alpha(x_1, \ldots, x_n; q, t) \prod_{s\in\mathrm{rev}(\alpha')} (1 - q^{\mathrm{leg}(s)+1} t^{\mathrm{arm}(s)+1})$$

by simply reversing the basement and reversing the parts of α:

$$\mathcal{E}_\alpha(x_1, \ldots, x_n; q, t) = \sum_{\substack{\text{nonattacking fillings } \sigma \text{ of } (\alpha_n, \ldots, \alpha_1)' \\ \sigma(j,0)=n-j+1}} x^\sigma q^{\mathrm{maj}(\sigma,\mathrm{rev}(\alpha)')} t^{\mathrm{coinv}(\sigma,\mathrm{rev}(\alpha)')} \\ \times \prod_{\substack{u\in\mathrm{rev}(\alpha)' \\ \sigma(u)=\sigma(\mathrm{South}(u))}} (1 - q^{\mathrm{leg}(u)+1} t^{\mathrm{arm}(u)+1}) \prod_{\substack{u\in\mathrm{rev}(\alpha)' \\ \sigma(u)\neq\sigma(\mathrm{South}(u))}} (1 - t), \tag{10.7.6a}$$

$$E_\alpha(x_1, \ldots, x_n; q, t) = \sum_{\substack{\text{nonattacking fillings } \sigma \text{ of } (\alpha_n, \ldots, \alpha_1)' \\ \sigma(j,0)=n-j+1}} x^\sigma q^{\mathrm{maj}(\sigma,\mathrm{rev}(\alpha)')} t^{\mathrm{coinv}(\sigma,\mathrm{rev}(\alpha)')} \\ \times \prod_{\substack{u\in\mathrm{rev}(\alpha)' \\ \sigma(u)\neq\sigma(\mathrm{South}(u))}} \frac{(1 - t)}{(1 - q^{\mathrm{leg}(u)+1} t^{\mathrm{arm}(u)+1})}. \tag{10.7.6b}$$

Example 10.7.4 By (10.7.4) the nonattacking filling in Figure 10.14 has coinv $= 7$. There are descents at squares $(1, 1)$, $(3, 2)$ and $(5, 1)$, with maj-values 1, 2 and 3, respectively. The squares (3, 1), (4, 1) and (5, 3) satisfy the condition $\sigma(u) = \sigma(\text{South}(u))$ and contribute factors $(1-q^3t^5)$, $(1-q^2t^3)$ and $(1-qt^2)$, respectively. Hence the total weight associated to this filling in (10.7.5a) is $x_1x_2x_3^2x_4^2x_5^2x_7q^6t^7(1-q^3t^5)(1-q^2t^3)(1-qt^2)(1-t)^6$.

Remark 10.7.5 Let $C_\alpha(x_1,\ldots,x_n;q,t)$ denote the right-hand side of (10.7.6b). The fact that this equals $E_\alpha(x_1,\ldots,x_n;q,t)$ is proved by Haglund et al. (2008) by showing that it satisfies both (10.7.1a) and (10.7.1b). The relation (10.7.1a) actually holds term by term, which is proved by a simple bijective argument. To show C_α also satisfies (10.7.1a) is harder, and utilizes the following result.

Lemma 10.7.6 *For any $G_1, G_2 \in \mathbb{Q}(q,t)X$ and $0 < i < n$, the following conditions are equivalent*:

(i) $G_2 = T_iG_1$;
(ii) $G_1 + G_2$ *and* $tx_{i+1}G_1 + x_iG_2$ *are symmetric in* x_i, x_{i+1}.

Lemma 10.7.6 allows one to reduce (10.7.1a) to a number of technical lemmas involving super fillings and LLT polynomials.

Remark 10.7.7 The polynomials $\widehat{E}_\alpha(x_1,\ldots,x_n;0,0) = \widehat{\mathcal{E}}_\alpha(x_1,\ldots,x_n;0,0)$ are known to equal the "standard bases" of Lascoux and Schützenberger (1990), which arise in the study of Schubert varieties. These polynomials are now referred to as *Demazure atoms* (Haglund et al., 2011b; Mason, 2009) since they decompose Demazure characters. Setting $q = t = 0$ in (10.7.5a) gives a new combinatorial formula for the Demazure atom $\mathcal{A}_\alpha(x_1,\ldots,x_n)$, namely the sum of x^σ over all fillings σ of $\text{dg}(\alpha')$ with no descents and no coinversion triples. Similarly, the special value $E_\alpha(x_1,\ldots,x_n;0,0)$ is known to equal the Demazure character, which by (10.7.6b) can be expressed as the sum of x^σ over all fillings σ of $\text{dg}((\alpha_n,\alpha_{n-1},\ldots,\alpha_1)')$ with no descents, no coinversion triples, and with basement $\sigma(j,0) = n-j+1$.

For any $\alpha \in \mathbb{N}^n$, let α^+ denote the partition obtained by rearranging the parts of α into nonincreasing order. It is well known that the $P_\lambda(x_1,\ldots,x_n;q,t)$ can be expressed as a linear combination of those E_α for which $\alpha^+ = \lambda$. In terms of the $\widehat{E}_\alpha$, this identity takes the form (Marshall, 1999)

$$P_\lambda(X_n;q,t) = \prod_{s\in\lambda'}(1-q^{\text{leg}(s)+1}t^{\text{arm}(s)}) \sum_{\alpha:\alpha^+=\lambda} \frac{\widehat{E}_\alpha(x_1,\ldots,x_n;q,t)}{\prod_{s\in\alpha'}(1-q^{\text{leg}(s)+1}t^{\text{arm}(s)})}. \tag{10.7.7}$$

If we set $q = t = 0$ in (10.7.7), then by Remark 10.3.2 we have the identity $s_\lambda(X_n) = \sum_{\alpha;\alpha^+=\lambda}\mathcal{A}_\alpha(x_1,\ldots,x_n)$. Mason (2008, 2009) has proved this identity bijectively by developing a generalization of the RSK algorithm. Haglund et al. (2011a,b) have used aspects of this generalized RSK algorithm to show that the product of a Schur function and a Demazure atom (character), when expanded in terms of Demazure atoms (characters), has a combinatorial interpretation, which refines the Littlewood–Richardson rule.

Remark 10.7.8 Let $\widehat{E}_\gamma^\sigma(x_1,\ldots,x_n;q,t)$ denote the polynomial obtained by starting with the combinatorial formula (10.7.5a) involving sums over nonattacking fillings, replacing the basement $(1,2,\ldots,n)$ by $(\sigma_1,\sigma_2,\ldots,\sigma_n)$, and keeping other aspects of the formula the same. Then a result first observed by Haiman, studied by Ferreira (2011), and later proved by Alexandersson (2016) says that if $i+1$ occurs to the left of i in the basement $(\sigma_1, \sigma_2, \ldots, \sigma_n)$, then

$$T_i \widehat{E}_\gamma^\sigma(x_1, \ldots, x_n; q, t) = t^A \widehat{E}_\gamma^{\sigma'}(x_1, \ldots, x_n; q, t). \tag{10.7.8}$$

Here A equals 1 if the height of the column of $\widehat{dg}(\gamma)$ above $i+1$ in the basement is greater than or equal to the height of the column above i in the basement, and equals 0 otherwise. Also, σ' is the permutation obtained by interchanging i and $i+1$ in σ. Note the quadratic relation (10.7.2) implies $T_i(T_i + 1 - t) = t$, or $T_i^{-1} = (T_i + 1 - t)/t$. By iterating (10.7.8) one can express $\widehat{E}_\gamma^\sigma$ as a certain simple power of t (depending on γ and σ) times a sequence of T_i applied to $\widehat{E}_\gamma^{n\cdots 21}$, or equivalently as a certain simple power of t times T_σ^{-1} applied to $\widehat{E}_\gamma$, where T_σ^{-1} is the product of the T_i^{-1} occurring in any reduced expression for σ. Now the formula mentioned above for $\widehat{E}_\gamma^{n\cdots 21}$ is the same as the formula in Haglund et al. (2008) for $E_{\gamma_n,\ldots,\gamma_1}$, which shows that one can translate between the E_γ and the $\widehat{E}_\gamma$ using Hecke operators. We should mention that the Hecke algebra and the T_i have played a central role in the subject of nonsymmetric Macdonald polynomials from the outset, as in the work by Macdonald (1996) and Cherednik (1995). Also, the special case $q = t = 0$ of the $\widehat{E}_\gamma^\sigma$ has been studied by Haglund et al. (2013), LoBue and Remmel (2013), and Pun (2016).

Remark 10.7.9 Let μ be a partition, and $\alpha \in \mathbb{N}^n$ with $(\alpha^+)' = \mu$, that is, a diagram obtained by permuting the columns of μ. If we let $\sigma(j, 0) = \infty$, the two involutions from the proof of Theorem 10.5.3 hold for super fillings of α'. It follows that formula (10.5.2) for $\tilde{H}_\mu(X; q, t)$, and formula (10.6.3) for $J_\mu(X; q, t)$ all hold if, instead of summing over fillings of μ, we sum over fillings of α', using the definitions of arm, leg, etc. given earlier in this section.

Remark 10.7.10 Knop and Sahi (1997) obtained a combinatorial formula for the nonsymmetric Jack polynomial, which is a limiting case of (10.7.5a) in the same way that (10.6.4) is a limiting case of (10.6.3). In fact, it was contrasting their formula for the nonsymmetric Jack polynomials with (10.6.3) which led to (10.7.5a).

10.8 The Genesis of the q, t-Catalan Statistics

In this section we outline the empirical steps which led the author to the discovery of the statistics on Dyck paths for the q, t-Catalan $C_n(q, t)$ (Haglund, 2003). Recall that $C_n(q, t)$, introduced by Garsia and Haiman in 1993, was originally defined as the sum of rational functions (10.3.20). Garsia and Haiman proved that $C_n(q, 1)$ equals the sum of $q^{\text{area}(\pi)}$ over all $\pi \in L_{n,n}^+$, and posed the problem of finding a statistic tstat to match with area such that $C_n(q, t) = \sum_{\pi \in L_{n,n}^+} q^{\text{area}(\pi)} t^{\text{tstat}(\pi)}$. By 1998 this problem and the related question of finding statistics to generate $\text{Hilb}(\text{DH}_n; q, t)$ had become fairly well known. At the time, the author was a postdoc at MIT, and both Stanley and Billey suggested he work on the problem. The author tried several different approaches without success, but a year later, while spending a year as a postdoc at UC San Diego, decided to try once more. Since $C_n(q, t) = C_n(t, q)$, tstat had to have the same distribution over Dyck paths as area, but none of the other known statistics on Dyck paths equidistributed with area worked. The method the author was using was to study tables of $C_n(q,t)$, and to try to invent a tstat to pair with area which would match those tables.

After several attempts again met with failure, the author decided to try to find recurrences amongst the tables, in an effort to find rules which would force certain decisions. Note that all paths in $L^+_{n,n}$ which begin with an N and then an E step are clearly in bijection with paths in $L^+_{n-1,n-1}$, and the author noticed that a copy of $t^{n-1}C_{n-1}(q,t)$ seemed to be contained in $C_n(q,t)$ in the sense that $C_n(q,t) - t^{n-1}C_{n-1}(q,t) \in \mathbb{N}[q,t]$.

The author later noticed that copies of $t^{n-k}q^{\binom{k}{2}}C_{n-k}(q,t)$ were contained in $C_n(q,t)$ for all $1 \le k \le n$ and, moreover, $C_n(q,t) - \sum_{k=1}^{n} t^{n-k}q^{\binom{k}{2}}C_{n-k,n-k}(q,t) \in \mathbb{N}[q,t]$. This suggested that for any path which begins with k N steps followed by k E steps, tstat equals $n-k$ plus the value of tstat on the remaining portion of the path, viewed as an element of $L^+_{n-k,n-k}$. In particular, if to any composition α of n into positive parts we associate the "balanced" path $\pi(\alpha)$ consisting of α_1 N steps followed by α_1 E steps, then α_2 N steps followed by α_2 E steps, etc., then $\text{tstat}(\pi(\alpha)) = n - \alpha_1 + n - (\alpha_1 + \alpha_2) + \cdots$. After a month or two of more trial and error, the author finally realized that you can associate a balanced path, called the bounce path, to any Dyck path π, via the algorithm outlined below, and that $\text{tstat}(\pi)$ depends only on the bounce path.

To form the *bounce path* $\text{bounce}(\pi)$, think of shooting a billiard ball straight north from $(0,0)$. Once the billiard ball hits the beginning of an E step of π it ricochets straight east until it hits the main diagonal $x = y$ at a point say (α_1, α_1). It then ricochets straight north and repeats the previous procedure, traveling north until it hits the beginning of an E step, then going east until it hits the diagonal, where it again ricochets north, and so on, until it reaches (n,n). For an example, see Figure 10.15. If the path the billiard ball takes is the balanced path which touches the diagonal at points (α_1,α_1), $(\alpha_1+\alpha_2, \alpha_1+\alpha_2)$, etc., then define the bounce statistic $\text{bounce}(\pi) = (n-\alpha_1)+(n-\alpha_1-\alpha_2)+\cdots$. A short Maple program verified, for $n \le 12$, the conjecture that

$$C_n(q,t) = \sum_{\pi \in L^+_{n,n}} q^{\text{area}(\pi)} t^{\text{bounce}(\pi)}. \tag{10.8.1}$$

While the author and Garsia were trying to prove (10.8.1), Haiman independently found an alternate form of the conjecture involving the statistics area and dinv, as described in §10.4.1. Upon comparing the two conjectures, Haiman and the author quickly realized that they are equivalent, which can be proved bijectively by using what we call the ζ *map* of a Dyck path. Given $\pi \in L^+_{n,n}$, let $R(\pi)$ be the parking function for π whose reading word is the reverse of the identity permutation. Construct another path $\zeta(\pi)$ by first placing the numbers $1, 2, \ldots, n$ along the diagonal of an empty grid of squares, with i in square (i,i). Then let $\zeta(\pi)$ be the unique path with the following property: for each pair (i,j) satisfying $1 \le i < j \le n$, the square of the grid in the column containing i and the row containing j is below the path $\zeta(\pi)$ if and only if, in $R(\pi)$, the rows containing the numbers i, j contribute an inversion to $\text{dinv}(R(\pi))$. See Figure 10.16. We leave it as an exercise for the interested reader to verify that $\text{dinv}(\pi) = \text{area}(\zeta(\pi))$ and $\text{area}(\pi) = \text{bounce}(\zeta(\pi))$. It is still an open problem to prove

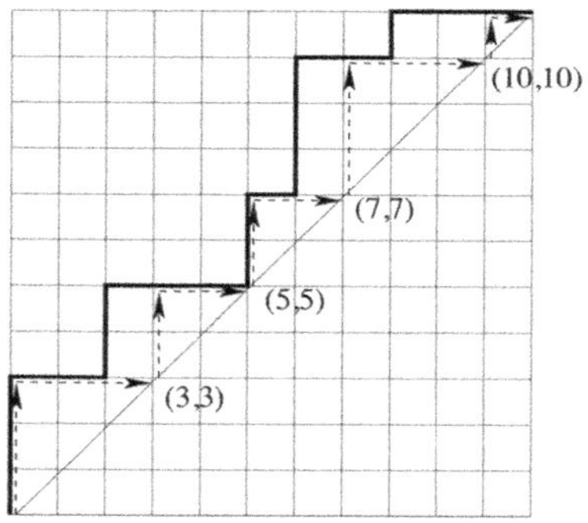

Figure 10.15 The bounce path (dotted line) for a Dyck path.

$C_n(q,t) = C_n(t,q)$ bijectively, perhaps by finding a map on Dyck paths which interchanges the statistics (dinv, area), or interchanges (area, bounce). See Lee et al. (2014) for recent work on this problem.

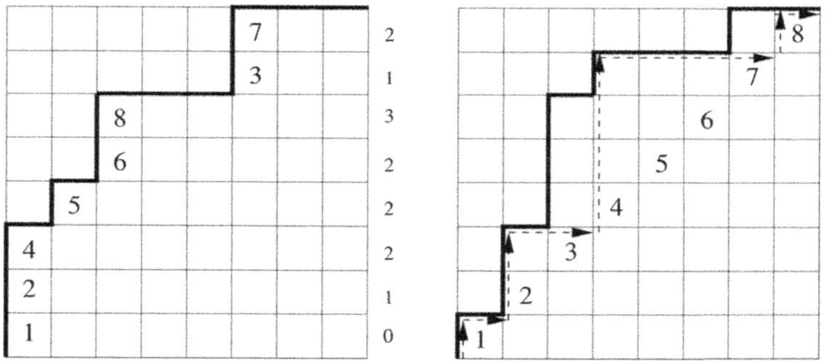

Figure 10.16 On the left, a Dyck path π, and on the right, $\zeta(\pi)$, together with its bounce path.

The bounce path arose independently in work by Andrews et al. (2002), who were calculating the minimal power you needed to raise an ad-nilpotent b-ideal in the Lie algebra $\mathfrak{sl}(n)$ to get 0. They showed that this minimal power equals the number of bounce steps of the bounce path of a certain Dyck path associated to the ideal. In a sequel to this paper, Krattenthaler et al. (2002) (the last three authors of the paper by Andrews et al. (2002)) obtained versions of their results for simple Lie algebras of other types, in particular for type B_n. Attempts by Stump, the author, and others to link the bounce path of type B_n in this paper to a B_n-version of $C_n(q,t)$ defined in unpublished work by Haiman (which uses the Hilbert scheme) have so far been unsuccessful.

10.9 Other Directions

10.9.1 The Formula of Ram and Yip

In view of the combinatorial formulas (10.6.3) and (10.7.6b) for the GL_n (integral form) Macdonald symmetric and nonsymmetric polynomials, a natural question to ask is whether such results exist for versions associated to other affine root systems. Progress in this direction has been made by Ram and Yip (2011), who derive a closed form expression for the E_α for arbitrary affine root systems. Their formula is expressed as a sum over *alcove walks*, which grew out of work by Gaussent and Littelmann (2005) and Littelmann (1994, 1995), and was further developed by Lenart and Postnikov (2007, 2008). They also have a corresponding formula for the symmetric P_μ for arbitrary type. When restricted to the GL_n case, their formula for P_μ in general has more terms than the Haglund–Haiman–Loehr formula (10.6.3), but in this case Lenart (2009) has found a way of grouping terms in the Ram–Yip formula together to get a more compact formula. Rather amazingly, Lenart's more compact formula turns out to be the same as the formula for P_μ obtained as sketched in Remark 10.7.9, with α the reverse of μ. Yip (2012) used the Ram–Yip formula to obtain a *q,t-Littlewood–Richardson rule* (for arbitrary type), which expands a product of a monomial and an E_α in terms of the E_α. The coefficients in this expansion are sums, over alcove walks, of rational functions in q, t. As corollaries she obtains expressions for the product of two E_α, or the product of a symmetric P_λ and an E_α, for arbitrary type.

Recall that Lascoux and Schützenberger's charge statistic arises when expanding the Hall–Littlewood polynomials $P_\mu(X;0,t)$ in terms of the $s_\lambda[X(1-t)]$. It is also known that $P_\mu(X;q,0) = \sum_\lambda K_{\lambda',\mu'}(q)s_\lambda(X)$. Ion (2003) has shown that for general type, the expansion of $P_\mu(X;q,0)$ in terms of Weyl characters has nonnegative coefficients (in type A, a Weyl character is a Schur function). By studying the $t=0$ case of the Ram–Yip formula, Lenart (2010) has developed a version of charge for type C. Lenart and Schilling (2013) have proved that this type-C charge corresponds with the energy function on tensor products of Kirillov–Reshetikhin crystals.

10.9.2 The Probabilistic Interpretation of Diaconis and Ram

Let λ be a partition of k. Diaconis and Ram (2012) introduced a Markov chain on partitions of k whose eigenfunctions are the coefficients of the (GL_n) Macdonald polynomial $P_\lambda(X_n;q,t)$, when expanded in terms of the power-sum basis $\{p_\mu(X_n)\}$. They showed that the stationary distribution of their Markov chain is a new two-parameter family of measures on partitions, which includes the uniform distribution on permutations and the Ewens sampling formula as special cases. By using properties of Macdonald polynomials they obtained a sharp analysis of the rate of convergence of the Markov chain.

10.9.3 k-Schur Functions

For any positive integer k, Lapointe et al. (2003) introduced a family of symmetric functions which depend on a parameter t and reduce to Schur functions when $k=\infty$. These symmetric functions form a basis for a certain subspace of Λ. Several other conjecturally equivalent definitions of this intriguing family have been introduced; they are now commonly called *k-Schur functions* as in Lapointe and Morse (2003a), denoted $s_\lambda^{(k)}(X;t)$ ($\lambda \in \Lambda$, $\lambda_1 \leq k$). An equivalent form of the main conjecture in Lapointe et al. (2003) is that when the modified Macdonald polynomial $\tilde{H}_\mu(X;q,t)$ is expanded into the k-Schur basis with parameter q, i.e., $\{s_\lambda^{(k)}(X;q)\}$, where $k \geq \mu_1'$, the coefficients are in $\mathbf{N}[q,t]$. This and other related conjectures have sparked a large amount of research over the last ten years; see for example Chen (2010), Lam et al. (2010) and Lapointe and Morse (2003b, 2005).

Let the *bandwidth* of an LLT polynomial be the number of dotted diagonal lines which intersect the diagonal of some square in one of the skew shapes in the LLT tuple. For example, for the LLT polynomial in Figure 10.6 the bandwidth is 3, and for the tuple on the right in Figure 10.7 it is 4. It has been suggested that when expanding an LLT polynomial of bandwidth k into the k-Schur function basis $\{s_\lambda^{(k)}(X;q)\}$, the coefficients are in $\mathbf{N}[q]$. By (10.5.6), this refines the conjecture from Lapointe et al. (2003) discussed in the previous paragraph.

k-Schur functions also have other remarkable properties. For example, Lam (2008) proved a conjecture of Shimozono, which says that when $t = 1$ the k-Schur form the Schubert basis for the homology of the loop Grassmannian, a conjecture which was based in part on results by Lapointe and Morse (2008).

10.10 Recent Developments

The Work of Carlsson and Mellit

In August 2015 Carlsson and Mellit posted a preprint on the arXiv which proved the compositional shuffle conjecture of Remark 10.4.8; the work has now appeared, see Carlsson and Mellit (2018). As corollaries they obtain the first proof of Conjecture 10.4.1 (the combinatorial formula for $\mathrm{Hilb}(\mathrm{DH}_n;q,t)$) and more generally the first proof of the shuffle conjecture.

Their method of proof should have other substantial applications. Let DAHA_n denote the positive (i.e., polynomial) part of the GL_n double affine Hecke algebra. This involves two copies of the GL_n affine Hecke algebra: one containing variables $\{y_1,\ldots,y_n\}$ and corresponding T_i-operators (as in (10.7.8)) and another containing variables $\{z_1,\ldots,z_n\}$ and T_i^{-1} operators, together with a number of relations between these elements. Carlsson and Mellit introduced an abstract algebraic structure, the double Dyck path algebra $\mathbb{A}_{q,t}$, which contains a copy of DAHA_n for each $n \geq 1$, together with operators $d_+ = d_+^{(k)}$ and $d_- = d_-^{(k)}$ which map elements of DAHA_k to elements of DAHA_{k+1} and DAHA_{k-1}, respectively.

There is a wonderful action of $\mathbb{A}_{q,t}$ on elements of Λ with coefficients in $\mathbb{Q}(q,t)[y_1,\ldots,y_k]$. The T_i operators act on monomials in the y_i (as in (10.7.8) with x^λ replaced by y^λ), while the action of $d_+^{(k)}$ and $d_-^{(k)}$ is defined using plethysm. The operators d_+ and d_- are constructed such that, if you start at the end of a Dyck path π and create a sequence of operators $L(\pi)$ by tracing the path backward, prepending d_+ to $L(\pi)$ for E steps and d_- to $L(\pi)$ for N steps, then $L(\pi)$ operating on the constant 1 gives a certain LLT product of single cells. Moreover, if for each EN corner of π you replace the corresponding d_-d_+ contribution to $L(\pi)$ by a factor of $(d_-d_+ - d_+d_-)/(q-1)$, then the resulting sequence $M(\pi)$, acting on the constant 1, will yield $\mathcal{F}_{\zeta^{-1}(\pi)}(X;q)$, where $\mathcal{F}$ and ζ are as in (10.4.3) and Figure 10.16.

Say we have a path π which begins with k N steps followed by an E step. Then $M(\pi)$ will begin with k d_- terms. Letting $M'(\pi)$ denote $M(\pi)$ with these k d_- terms removed, then $M'(\pi)$ applied to 1 will be a sum of symmetric functions in X with coefficients in $\mathbb{Q}(q,t)[y_1,\ldots,y_k]$. Carlsson and Mellit show that certain sums of the $M'(\pi)1$, corresponding to elements π for which $\zeta^{-1}(\pi)$ has touch points $(a_1,a_1),(a_1+a_2,a_1+a_2),\ldots,(n,n)$, satisfy a nice recurrence. They also show that the operator ∇ can be expressed using elements of $\mathbb{A}_{q,t}$, and that they can then, by using their commutation relations, prove the compositional shuffle conjecture in two lines.

In a sequel to the Carlsson–Mellit paper, Mellit (2016) claims to have proved the compositional rational shuffle conjecture, which contains the rational shuffle conjecture and compositional shuffle conjecture, and hence all the conjectures from §10.4, as special cases. His proof starts by assuming the properties of the double Dyck path algebra developed by Carlsson and Mellit (2018), then introduces some new ideas. In particular he relates actions of toric braids with parking functions, and exploits the known fact that DAHA$_n$ can be viewed as a quotient of the surface braid group of a torus. One question which the work of Carlsson and Mellit has not yet shed any light on is the problem of finding a combinatorial expression for the Schur coefficients of the $\mathcal{F}_\pi(X;q,t)$ of (10.4.3).

One extension of the shuffle conjecture which is still open is the *Delta conjecture* of Haglund et al. (2018a). For $f \in \Lambda$, let Δ_f be the linear operator defined on the $\tilde{H}_\mu$ basis of modified Macdonald polynomials via $\Delta_f \tilde{H}_\mu(X;q,t) = f[B_\mu(q,t)]\tilde{H}_\mu(X;q,t)$, with $B_\mu(q,t)$ as in (10.3.4). The Delta conjecture gives an elegant combinatorial formula, in terms of parking functions, for $\Delta_{e_k} e_n$, for any $0 \le k \le n$. For $k = n$ it reduces to the shuffle conjecture. To apply the method of Carlsson and Mellit we would have to express $\Delta_{e_k} e_n$ in terms of elements of $\mathbb{A}_{q,t}$, and also find new recurrences. Part of the problem is that there is currently no known way to extend the compositional shuffle conjecture to a compositional Delta conjecture, although Haglund et al. (2018a) does contain such an extension for hook shapes. Recently, Zabrocki (2019) proved this special case. In another recent paper, Haglund et al. (2018b) introduced a quotient ring whose bigraded character equals the symmetric function described by the combinatorial side of the Delta conjecture when $t = 0$. The combinatorics of this $t = 0$ case is controlled by ordered set partitions.

In another direction, Sergel (2016) has proved the square paths conjecture of Loehr and Warrington (2007), which gives a combinatorial interpretation for ∇p_n, and follows from the compositional shuffle conjecture. Her proof uses clever combinatorial manipulations of parking functions. Another extension of the shuffle conjecture due to Loehr and Warrington (2008), which is still open, gives a combinatorial formula, in terms of parking functions for nested Dyck paths, for the expansion of ∇s_λ into monomials, for any λ. There is also a family of conjectures connected to the combinatorics of the character of diagonal harmonics in several sets of variables under the diagonal action of S_n; see Bergeron (2009, 2012, 2013).

Acknowledgements The author would like to thank François Bergeron, Tom Koornwinder, Nick Loehr and Jasper Stokman for many comments and suggestions which led to improvements in this chapter. The author's work on this chapter was supported by NSF grants DMS-0901467, DMS-1200296 and DMS-1600670.

References

Aganagic, M., and Shakirov, S. 2012. Refined Chern–Simons theory and knot homology. Pages 3–31 of: *String-Math 2011*. Proc. Sympos. Pure Math., vol. 85. Amer. Math. Soc.

Aganagic, M., and Shakirov, S. 2015. Knot homology and refined Chern–Simons index. *Comm. Math. Phys.*, **333**, 187–228.

Alexandersson, P. 2016. *Non-symmetric Macdonald polynomials and Demazure-Lusztig operators*. arXiv:1602.05153.

Andrews, G. E., Krattenthaler, C., Orsina, L., and Papi, P. 2002. *ad*-Nilpotent b-ideals in sl(n) having a fixed class of nilpotence: combinatorics and enumeration. *Trans. Amer. Math. Soc.*, **354**, 3835–3853.

Armstrong, D. 2013. Hyperplane arrangements and diagonal harmonics. *J. Comb.*, **4**, 157–190.

Armstrong, D., and Rhoades, B. 2012. The Shi arrangement and the Ish arrangement. *Trans. Amer. Math. Soc.*, **364**, 1509–1528.

Artin, E. 1944. *Galois Theory*. Second edn. Notre Dame Mathematical Lectures, no. 2. University of Notre Dame. Reprinted, 1976.

Askey, R. 1980. Some basic hypergeometric extensions of integrals of Selberg and Andrews. *SIAM J. Math. Anal.*, **11**, 938–951.

Assaf, S., and Garsia, A. 2009. A kicking basis for the two column Garsia–Haiman modules. Pages 103–114 of: *21st International Conference on Formal Power Series and Algebraic Combinatorics (FPSAC 2009)*. DMTCS Proc., Vol. AK. Assoc. DMTCS, Nancy.

Bergeron, F. 2009. *Algebraic Combinatorics and Coinvariant Spaces*. CMS Treatises in Mathematics. Canad. Math. Soc.

Bergeron, F. 2012. *Combinatorics of r-Dyck paths, r-parking functions, and the r-Tamari lattices*. arXiv:1202.6269.

Bergeron, F. 2013. Multivariate diagonal coinvariant spaces for complex reflection groups. *Adv. Math.*, **239**, 97–108.

Bergeron, F., and Garsia, A. M. 1999. Science fiction and Macdonald's polynomials. Pages 1–52 of: *Algebraic Methods and q-Special Functions*. CRM Proc. Lecture Notes, vol. 22. Amer. Math. Soc.

Bergeron, F., Garsia, A.M., Haiman, M., and Tesler, G. 1999. Identities and positivity conjectures for some remarkable operators in the theory of symmetric functions. *Methods Appl. Anal.*, **6**, 363–420.

Bergeron, F., Garsia, A., Sergel Leven, E., and Xin, G. 2016a. Compositional (km, kn)-shuffle conjectures. *Int. Math. Res. Not.*, 4229–4270.

Bergeron, F., Garsia, A., Sergel Leven, E, and Xin, G. 2016b. Some remarkable new plethystic operators in the theory of Macdonald polynomials. *J. Comb.*, **7**, 671–714.

Blasiak, J. 2016. Haglund's conjecture on 3-column Macdonald polynomials. *Math. Z.*, **283**, 601–628.

Blasiak, J., and Fomin, S. 2017. Noncommutative Schur functions, switchboards, and Schur positivity. *Selecta Math. (N.S.)*, **23**, 727–766.

Burban, I., and Schiffmann, O. 2012. On the Hall algebra of an elliptic curve, I. *Duke Math. J.*, **161**, 1171–1231.

Butler, L. M. 1994. *Subgroup Lattices and Symmetric Functions.* Mem. Amer. Math. Soc., vol. 112, no. 539.

Carlsson, E., and Mellit, A. 2018. A proof of the shuffle conjecture. *J. Amer. Math. Soc.*, **31**, 661–697.

Chen, L.-C. 2010. *Skew-linked partitions and a representation-theoretic model for k-Schur.* Ph.D. thesis, Univ. of California at Berkeley.

Cherednik, I. 1995. Nonsymmetric Macdonald polynomials. *Int. Math. Res. Not.*, 483–515.

Cherednik, I. 2013. Jones polynomials of torus knots via DAHA. *Int. Math. Res. Not.*, 5366–5425.

Diaconis, P., and Ram, A. 2012. A probabilistic interpretation of the Macdonald polynomials. *Ann. Prob.*, **40**, 1861–1896.

Dunfield, N. M., Gukov, S., and Rasmussen, J. 2006. The superpolynomial for knot homologies. *Experiment. Math.*, **15**, 129–159.

Ferreira, J. P. 2011. *Row-strict quasisymmetric Schur functions, characterizations of Demazure atoms, and permuted basement nonsymmetric Macdonald polynomials.* Ph.D. thesis, Univ. of California at Davis.

Fishel, S. 1995. Statistics for special q, t-Kostka polynomials. *Proc. Amer. Math. Soc.*, **123**, 2961–2969.

Foata, D. 1968. On the Netto inversion number of a sequence. *Proc. Amer. Math. Soc.*, **19**, 236–240.

Forrester, P. J., and Warnaar, S. O. 2008. The importance of the Selberg integral. *Bull. Amer. Math. Soc. (N.S.)*, **45**, 489–534.

Foulkes, H. O. 1974. A survey of some combinatorial aspects of symmetric functions. Pages 79–92 of: *Permutations.* Gauthier-Villars.

Garsia, A. M., and Haglund, J. 2001. A positivity result in the theory of Macdonald polynomials. *Proc. Nat. Acad. Sci. USA*, **98**, 4313–4316.

Garsia, A. M., and Haglund, J. 2002. A proof of the q, t-Catalan positivity conjecture. *Discrete Math.*, **256**, 677–717.

Garsia, A., and Haglund, J. 2015. A polynomial expression for the character of diagonal harmonics. *Ann. Comb.*, **19**, 693–703.

Garsia, A. M., and Haiman, M. 1993. A graded representation model for Macdonald polynomials. *Proc. Nat. Acad. Sci. USA*, **90**, 3607–3610.

Garsia, A. M., and Haiman, M. 1996a. A remarkable q, t-Catalan sequence and q-Lagrange inversion. *J. Algebraic Combin.*, **5**, 191–244.

Garsia, A. M., and Haiman, M. 1996b. Some natural bigraded S_n-modules and q, t-Kostka coefficients. *Electron. J. Combin.*, **3**, Paper 24, 60 pp.

Garsia, A. M., Haiman, M., and Tesler, G. 1999. Explicit plethystic formulas for Macdonald q, t-Kostka coefficients. *Sém. Lothar. Combin.*, **42**, Paper B42m, 45 pp.

Garsia, A. M., and Procesi, C. 1992. On certain graded S_n-modules and the q-Kostka polynomials. *Adv. Math.*, **94**, 82–138.

Garsia, A. M., and Remmel, J. 1998. Plethystic formulas and positivity for q, t-Kostka coefficients. Pages 245–262 of: *Mathematical Essays in Honor of Gian-Carlo Rota.* Progr. Math., vol. 161. Birkhäuser.

Garsia, A., Sergel Leven, E., Wallach, N., and Xin, G. 2017. A new plethystic symmetric function operator and the rational compositional shuffle conjecture at $t = 1/q$. *J. Combin. Theory Ser. A*, **145**, 57–100.

Garsia, A. M., and Tesler, G. 1996. Plethystic formulas for Macdonald q, t-Kostka coefficients. *Adv. Math.*, **123**, 144–222.

Garsia, A., Xin, G., and Zabrocki, M. 2012. Hall-Littlewood operators in the theory of parking functions and diagonal harmonics. *Int. Math. Res. Not.*, 1264–1299.

Garsia, A. M., Xin, G., and Zabrocki, M. 2014. A three shuffle case of the compositional parking function conjecture. *J. Combin. Theory Ser. A*, **123**, 202–238.

Gaussent, S., and Littelmann, P. 2005. LS galleries, the path model, and MV cycles. *Duke Math. J.*, **127**, 35–88.

Gorsky, E., and Neguţ, A. 2015. Refined knot invariants and Hilbert schemes. *J. Math. Pures Appl. (9)*, **104**, 403–435.

Gorsky, E., Oblomkov, A., Rasmussen, J., and Shende, V. 2014. Torus knots and the rational DAHA. *Duke Math. J.*, **163**, 2709–2794.

Grojnowski, I., and Haiman, M. 2007. *Affine Hecke algebras and positivity of LLT and Macdonald polynomials.* Preprint at `https://math.berkeley.edu/~mhaiman/`.

Haglund, J. 2003. Conjectured statistics for the q,t-Catalan numbers. *Adv. Math.*, **175**, 319–334.

Haglund, J. 2004a. A combinatorial model for the Macdonald polynomials. *Proc. Nat. Acad. Sci. USA*, **101**, 16127–16131.

Haglund, J. 2004b. A proof of the q,t-Schröder conjecture. *Int. Math. Res. Not.*, 525–560.

Haglund, J. 2006. The genesis of the Macdonald polynomial statistics. *Sém. Lothar. Combin.*, **54A**, Paper B54Ao, 16 pp.

Haglund, J. 2008. *The q,t-Catalan Numbers and the Space of Diagonal Harmonics.* University Lecture Series, vol. 41. Amer. Math. Soc.

Haglund, J. 2011. A polynomial expression for the Hilbert series of the quotient ring of diagonal coinvariants. *Adv. Math.*, **227**, 2092–2106.

Haglund, J. 2016. The combinatorics of knot invariants arising from the study of Macdonald polynomials. Pages 579–600 of: *Recent Trends in Combinatorics.* IMA Volumes in Math. and Its Appl., vol. 159. Springer.

Haglund, J., and Loehr, N. 2005. A conjectured combinatorial formula for the Hilbert series for diagonal harmonics. *Discrete Math.*, **298**, 189–204.

Haglund, J., Haiman, M., and Loehr, N. 2005a. A combinatorial formula for Macdonald polynomials. *J. Amer. Math. Soc.*, **18**, 735–761.

Haglund, J., Haiman, M., Loehr, N., Remmel, J. B., and Ulyanov, A. 2005b. A combinatorial formula for the character of the diagonal coinvariants. *Duke Math. J.*, **126**, 195–232.

Haglund, J., Haiman, M., and Loehr, N. 2005c. Combinatorial theory of Macdonald polynomials. I: proof of Haglund's formula. *Proc. Nat. Acad. Sci. USA*, **102**, 2690–2696.

Haglund, J., Haiman, M., and Loehr, N. 2008. A combinatorial formula for non-symmetric Macdonald polynomials. *Amer. J. Math.*, **103**, 359–383.

Haglund, J., Luoto, K., Mason, S., and van Willigenburg, S. 2011a. Quasisymmetric Schur functions. *J. Combin. Theory Ser. A*, **118**, 463–490.

Haglund, J., Luoto, K., Mason, S., and van Willigenburg, S. 2011b. Refinements of the Littlewood–Richardson rule. *Trans. Amer. Math. Soc.*, **363**, 1665–1686.

Haglund, J., Morse, J., and Zabrocki, M. 2012. A compositional shuffle conjecture specifying touch points of the Dyck path. *Canad. J. Math.*, **64**, 822–844.

Haglund, J., Mason, S., and Remmel, J. 2013. Properties of the nonsymmetric Robinson–Schensted–Knuth algorithm. *J. Algebraic Combin.*, **38**, 285–327.

Haglund, J., Remmel, J. B., and Wilson, A. T. 2018a. The Delta conjecture. *Trans. Amer. Math. Soc.*, **370**, 4029–4057.

Haglund, J., Rhoades, B., and Shimozono, M. 2018b. Ordered set partitions, generalized coinvariant algebras, and the Delta conjecture. *Adv. Math.*, **329**, 851–915.

Haiman, M. 1994. Conjectures on the quotient ring by diagonal invariants. *J. Algebraic Combin.*, **3**, 17–76.

Haiman, M. 1999. Macdonald polynomials and geometry. Pages 207–254 of: *New Perspectives in Algebraic Combinatorics*. Math. Sci. Res. Inst. Publ., vol. 38. Cambridge Univ. Press.

Haiman, M. 2001. Hilbert schemes, polygraphs, and the Macdonald positivity conjecture. *J. Amer. Math. Soc.*, **14**, 941–1006.

Haiman, M. 2002. Vanishing theorems and character formulas for the Hilbert scheme of points in the plane. *Invent. Math.*, **149**, 371–407.

Haiman, M. 2006. Cherednik algebras, Macdonald polynomials and combinatorics. Pages 843–872 of: *International Congress of Mathematicians. Vol. III*. Eur. Math. Soc.

Haiman, M., and Woo, A. 2007. Geometry of q and q, t-analogs in combinatorial enumeration. Pages 207–248 of: *Geometric Combinatorics*. IAS/Park City Math. Ser., vol. 13. Amer. Math. Soc.

Hicks, A. S. 2012. Two parking function bijections: a sharpening of the q, t-Catalan and Shröder theorems. *Int. Math. Res. Not.*, 3064–3088.

Hikita, T. 2014. Affine Springer fibers of type A and combinatorics of diagonal coinvariants. *Adv. Math.*, **263**, 88–122.

Hotta, R., and Springer, T. A. 1977. A specialization theorem for certain Weyl group representations and an application to the Green polynomials of unitary groups. *Invent. Math.*, **41**, 113–127.

Humphreys, J. E. 1990. *Reflection Groups and Coxeter Groups*. Cambridge University Press.

Ion, B. 2003. Nonsymmetric Macdonald polynomials and Demazure characters. *Duke Math. J.*, **116**, 299–318.

Jack, H. 1970. A class of symmetric polynomials with a parameter. *Proc. Roy. Soc. Edinburgh (Sect. A)*, **69**, 1–18.

James, G., and Liebeck, M. 2001. *Representations and Characters of Groups*. Second edn. Cambridge University Press.

Kadell, K. 1988a. A proof of some analogues of Selberg's integral for $k = 1$. *SIAM J. Math. Anal.*, **19**, 944–968.

Kadell, K. W. J. 1988b. The q-Selberg polynomials for $n = 2$. *Trans. Amer. Math. Soc.*, **310**, 535–553.

Kashiwara, M., and Tanisaki, T. 2002. Parabolic Kazhdan–Lusztig polynomials and Schubert varieties. *J. Algebra*, **249**, 306–325.

Kirillov, A. N., and Noumi, M. 1998. Affine Hecke algebras and raising operators for Macdonald polynomials. *Duke Math. J.*, **93**, 1–39.

Knop, F. 1997a. Integrality of two variable Kostka functions. *J. Reine Angew. Math.*, **482**, 177–189.

Knop, F. 1997b. Symmetric and non-symmetric quantum Capelli polynomials. *Comment. Math. Helv.*, **72**, 84–100.

Knop, F., and Sahi, S. 1997. A recursion and a combinatorial formula for Jack polynomials. *Invent. Math.*, **128**, 9–22.

Koornwinder, T. H. 1992. Askey–Wilson polynomials for root systems of type BC. Pages 189–204 of: *Hypergeometric Functions on Domains of Positivity, Jack Polynomials, and Applications*. Contemp. Math., vol. 138. Amer. Math. Soc.

Krattenthaler, C., Orsina, L., and Papi, P. 2002. Enumeration of ad-nilpotent b-ideals for simple Lie algebras. *Adv. in Appl. Math.*, **28**, 478–522.

Lam, T. 2008. Schubert polynomials for the affine Grassmannian. *J. Amer. Math. Soc.*, **21**, 259–281.

Lam, T., Lapointe, L., Morse, J., and Shimozono, M. 2010. *Affine Insertion and Pieri Rules for the Affine Grassmannian*. Mem. Amer. Math. Soc., vol. 208, no. 977.

Lapointe, L., and Morse, J. 2003a. Schur function analogs for a filtration of the symmetric function space. *J. Combin. Theory Ser. A*, **101**, 191–224.

Lapointe, L., and Morse, J. 2003b. Schur function identities, their t-analogs, and k-Schur irreducibility. *Adv. Math.*, **180**, 222–247.

Lapointe, L., and Morse, J. 2003c. Tableaux statistics for two part Macdonald polynomials. Pages 61–84 of: *Algebraic Combinatorics and Quantum Groups*. World Scientific.

Lapointe, L., and Morse, J. 2005. Tableaux on $k+1$-cores, reduced words for affine permutations, and k-Schur expansions. *J. Combin. Theory Ser. A*, **112**, 44–81.

Lapointe, L., and Morse, J. 2008. Quantum cohomology and the k-Schur basis. *Trans. Amer. Math. Soc.*, **360**, 2021–2040.

Lapointe, L., and Vinet, L. 1995. A Rodrigues formula for the Jack polynomials and the Macdonald-Stanley conjecture. *Int. Math. Res. Not.*, 419–424.

Lapointe, L., and Vinet, L. 1996. Exact operator solution of the Calogero–Sutherland model. *Comm. Math. Phys.*, **178**, 425–452.

Lapointe, L., and Vinet, L. 1997. Rodrigues formulas for the Macdonald polynomials. *Adv. Math.*, **130**, 261–279.

Lapointe, L., and Vinet, L. 1998. A short proof of the integrality of the Macdonald q,t-Kostka coefficients. *Duke Math. J.*, **91**, 205–214.

Lapointe, L., Lascoux, A., and Morse, J. 2003. Tableau atoms and a new Macdonald positivity conjecture. *Duke Math. J.*, **116**, 103–146.

Lascoux, A., and Schützenberger, M.-P. 1978. Sur une conjecture de H. O. Foulkes. *C. R. Acad. Sci. Paris Sér. A-B*, **286**, A323–A324.

Lascoux, A., and Schützenberger, M.-P. 1990. Keys & standard bases. Pages 125–144 of: *Invariant Theory and Tableaux*. IMA Volumes in Math. and Its Appl., vol. 19. Springer.

Lascoux, A., Leclerc, B., and Thibon, J.-Y. 1997. Ribbon tableaux, Hall-Littlewood functions, quantum affine algebras, and unipotent varieties. *J. Math. Phys.*, **38**, 1041–1068.

Leclerc, B., and Thibon, J.-Y. 1997. Ribbon tableaux, Hall–Littlewood functions, quantum affine algebras, and unipotent varieties. *J. Math. Phys.*, **38**, 1041–1068.

Lee, K., Li, L., and Loehr, N. A. 2014. Combinatorics of certain higher q,t-Catalan polynomials: chains, joint symmetry, and the Garsia–Haiman formula. *J. Algebraic Combin.*, **39**, 749–781.

Lenart, C. 2009. On combinatorial formulas for Macdonald polynomials. *Adv. Math.*, **220**, 324–340.

Lenart, C. 2010. Haglund–Haiman-Loehr type formulas for Hall-Littlewood polynomials of type B and C. *Algebra Number Theory*, **4**, 887–917.

Lenart, C., and Postnikov, A. 2007. Affine Weyl groups in K-theory and representation theory. *Int. Math. Res. Not.*, Paper rnm038, 65 pp.

Lenart, C., and Postnikov, A. 2008. A combinatorial model for crystals of Kac–Moody algebras. *Trans. Amer. Math. Soc.*, **360**, 4349–4381.

Lenart, C., and Schilling, A. 2013. Crystal energy functions via the charge in types A and C. *Math. Z.*, **273**, 401–426.

Littelmann, P. 1994. A Littlewood-Richardson rule for symmetrizable Kac–Moody algebras. *Invent. Math.*, **116**, 329–346.

Littelmann, P. 1995. Paths and root operators in representation theory. *Ann. of Math. (2)*, **142**, 499–525.

Liu, R. I., Mészáros, K., and Morales, A. 2016. *Flow polytopes and the space of diagonal harmonics*. arXiv:1610.08370.

LoBue, J., and Remmel, J. B. 2013. A Murnaghan–Nakayama rule for generalized Demazure atoms. Pages 969–980 of: *25th International Conference on Formal Power Series and Algebraic Combinatorics*. DMTCS Proc., Vol. AS. Assoc. DMTCS, Nancy.

Loehr, N., and Warrington, G. 2007. Square q,t-lattice paths and ∇p_n. *Trans. Amer. Math. Soc.*, **359**, 649–669.

Loehr, N. A., and Warrington, G. S. 2008. Nested quantum Dyck paths and $\nabla(s_\lambda)$. *Int. Math. Res. Not.*, rnm 157, 29 pp.

Lusztig, G. 1981. Green polynomials and singularities of unipotent classes. *Adv. Math.*, **42**, 169–178.

Macdonald, I. G. 1988. A new class of symmetric functions. *Sém. Lothar. Combin.*, **20**, Paper B20a, 41 pp.

Macdonald, I. G. 1995. *Symmetric Functions and Hall Polynomials*. Second edn. Oxford University Press.

Macdonald, I. G. 1996. Affine Hecke algebras and orthogonal polynomials. *Astérisque*, 189–207. Séminaire Bourbaki, Vol. 1994/95, Exp. No. 797.

Macdonald, I. G. 2000. Orthogonal polynomials associated with root systems. *Sém. Lothar. Combin.*, **45**, B45a, 40 pp. Text of handwritten preprint, 1987.

Macdonald, I. G. 2003. *Affine Hecke Algebras and Orthogonal Polynomials*. Cambridge Tracts in Mathematics, vol. 157. Cambridge University Press.

MacMahon, P. A. 1960. *Combinatory Analysis*. Chelsea, New York, NY. Reprint of two volumes (bound as one) first published in 1915, 1916.

Manivel, L. 2001. *Symmetric Functions, Schubert Polynomials and Degeneracy Loci*. SMF/AMS Texts and Monographs, vol. 6. Amer. Math. Soc. Translated from the 1998 French original.

Marshall, D. 1999. Symmetric and nonsymmetric Macdonald polynomials. *Ann. Comb.*, **3**, 385–415.

Mason, S. 2008. A decomposition of Schur functions and an analogue of the Robinson-Schensted-Knuth algorithm. *Sém. Lothar. Combin.*, **57**, Paper B57e, 24 pp.

Mason, S. 2009. An explicit construction of type A Demazure atoms. *J. Algebraic Combin.*, **29**, 295–313.

Mellit, A. 2016. *Toric braids and (m,n)-parking functions*. arXiv:1604.07456v1.

Mészáros, K., Morales, A. H., and Rhoades, B. 2016. The polytope of Tesler matrices. Pages 205–216 of: *Discrete Mathematics and Theoretical Computer Science, Proceedings of FPSAC 2015*. DMTCS.

Mészáros, K., Morales, A. H., and Rhoades, B. 2017. The polytope of Tesler matrices. *Selecta Math. (N.S.)*, **23**, 425–454.

Negut, A. 2014. The shuffle algebra revisited. *Int. Math. Res. Not.*, 6242–6275.

Oblomkov, A., Rasmussen, J., and Shende, V. 2018. The Hilbert scheme of a plane curve singularity and the HOMFLY homology of its link. *Geom. Topol.*, **22**, 645–691. With an appendix by E. Gorsky.

Pun, Y. A. 2016. *On decomposition of the product of Demazure atoms and Demazure characters*. Ph.D. thesis, Univ. of Pennsylvania.

Ram, A., and Yip, M. 2011. A combinatorial formula for Macdonald polynomials. *Adv. Math.*, **226**, 309–331.

Reiner, E. 1996. A proof of the $n!$ conjecture for generalized hooks. *J. Combin. Theory Ser. A*, **75**, 1–22.

Sagan, B. E. 2001. *The Symmetric Group. Representations, Combinatorial Algorithms, and Symmetric Functions*. Second edn. Graduate Texts in Mathematics, vol. 203. Springer.

Sahi, S. 1996. Interpolation, integrality, and a generalization of Macdonald's polynomials. *Int. Math. Res. Not.*, 457–471.

Sahi, S. 1998. The binomial formula for nonsymmetric Macdonald polynomials. *Duke Math. J.*, **94**, 465–477.

Sahi, S. 1999. Nonsymmetric Koornwinder polynomials and duality. *Ann. of Math. (2)*, **150**, 267–282.

Schiffmann, O., and Vasserot, E. 2011. The elliptic Hall algebra, Cherednik–Hecke algebras and Macdonald polynomials. *Compos. Math.*, **147**, 188–234.

Schiffmann, O., and Vasserot, E. 2013. The elliptic Hall algebra and the K-theory of the Hilbert scheme of $\mathbb{A}^2$. *Duke Math. J.*, **162**, 279–366.

Sergel, E. 2016. *A proof of the square paths conjecture*. arXiv:1601.06249.

Stanley, R. P. 1979. Invariants of finite groups and their applications to combinatorics. *Bull. Amer. Math. Soc. (N.S.)*, 475–511.

Stanley, R. P. 1999. *Enumerative Combinatorics, Vol. 2*. Cambridge University Press.

Stanley, R. P. 2003. Recent progress in algebraic combinatorics. *Bull. Amer. Math. Soc.*, **40**, 55–68.

Stanley, R. P. 2015. *Catalan Numbers*. Cambridge University Press.

Stanton, D. (ed). 1989. *q-Series and Partitions*. IMA Volumes in Math. and Its Appl., vol. 18. Springer.

Stembridge, J. R. 1994. Some particular entries of the two-parameter Kostka matrix. *Proc. Amer. Math. Soc.*, **121**, 469–490.

Wilson, A. T. 2016. Generalized Tesler matrices, virtual Hilbert series, and Macdonald polynomial operators. Pages 817–828 of: *Discrete Mathematics and Theoretical Computer Science, Proceedings of FPSAC 2015*. DMTCS.

Yip, M. 2012. A Littlewood–Richardson rule for Macdonald polynomials. *Math. Z.*, **272**, 1259–1290.

Zabrocki, M. 1998. A Macdonald vertex operator and standard tableaux statistics for the two-column (q,t)-Kostka coefficients. *Electron. J. Combin.*, **5**, Paper 45, 46 pp.

Zabrocki, M. 1999. Positivity for special cases of (q,t)-Kostka coefficients and standard tableaux statistics. *Electron. J. Combin.*, **6**, Paper 41, 36 pp.

Zabrocki, M. 2019. A proof of the 4-variable Catalan polynomial of the Delta conjecture. *J. Comb.*, **10**, 599–632.

11

Knizhnik–Zamolodchikov-Type Equations, Selberg Integrals and Related Special Functions

Vitaly Tarasov and Alexander Varchenko

11.1 Introduction

The *Knizhnik–Zamolodchikov* (*KZ*) equations appear in conformal field theory as systems of differential equations for conformal blocks; see Knizhnik and Zamolodchikov (1984). A *KZ equation* is a system of differential equations of the form

$$\kappa \frac{\partial \varphi}{\partial z_i} = K_i(z)\varphi, \quad i = 1, \ldots, n, \tag{11.1.1}$$

where $\varphi(z)$, $z = (z_1, \ldots, z_n)$ is the unknown function with values in a vector space V, $K_i(z) \in \mathrm{End}(V)$ are given linear operators (see for example 11.3.2), and $\kappa \in \mathbb{C}^*$ is a parameter. In the simplest case $\dim V = 2$, the *KZ* equation reduces to the Gauss hypergeometric equation.

For any κ, by consistency of (11.1.1), the *KZ* equation defines a flat connection,

$$\left[\kappa \frac{\partial}{\partial z_i} - K_i(z), \kappa \frac{\partial}{\partial z_j} - K_j(z)\right] = 0, \quad i, j = 1, \ldots, n.$$

This is equivalent to the two sets of equations

$$\frac{\partial K_i}{\partial z_j}(z) = \frac{\partial K_j}{\partial z_i}(z), \quad \left[K_i(z), K_j(z)\right] = 0$$

for all $i, j = 1, \ldots, n$. The general problem is to solve a *KZ* equation.

A *KZ* equation can be identified with equations for flat sections of a Gauss–Manin connection, and the solutions of a *KZ* equation can be given by V-valued multidimensional contour integrals (see §§11.4.2, 11.4.3). A solution of a *KZ* equation has the form

$$\varphi^{(\gamma)}(z) = \int_{\gamma(z)} e^{\Phi(z,t)/\kappa} \omega(z, t)\, dt,$$

where $t = (t_1, \ldots, t_l)$ are integration variables, $\Phi(z, t)$ is a scalar function called the *master function*, $\omega(z, t)$ is a V-valued function called the *weight function*, and $\gamma(z)$ is a "twisted" l-dimensional cycle in $\{z\} \times \mathbb{C}^l$ depending on z horizontally. The functions $\varphi^{(\gamma)}(z, \lambda)$ are called *multidimensional hypergeometric functions*. The simplest of them is the classical Gauss hypergeometric function.

For fixed z, the commuting operators $K_1(z), \ldots, K_n(z)$ on V are called the *Gaudin Hamiltonians*. Their eigenvectors and eigenvalues can be constructed by studying asymptotics of hypergeometric solutions to the *KZ* equation as $\kappa \to 0$. Consider the master function Φ as a function of t depending on parameters z. Given z, for every *critical point* (see §11.4.9) $t \in \mathbb{C}^l$ of Φ, the value $\omega(z,t) \in V$ of the weight function is an eigenvector of the Gaudin Hamiltonians,

$$K_i(z)\omega(z,t) = \frac{\partial\Phi}{\partial z_i}(z,t)\omega(z,t), \quad i = 1,\dots,n.$$

This construction of finding eigenvectors and eigenvalues is the *Bethe ansatz method for the Gaudin model* (Gaudin, 1983; Reshetikhin and Varchenko, 1995).

If a *KZ* equation has a one-dimensional space of solutions, then the corresponding hypergeometric integral is called a *Selberg-type integral*. In all examples a Selberg-type integral can be calculated explicitly. The simplest of them is the *Euler beta integral*

$$\int_0^1 x^{a-1}(1-x)^{b-1}\,dx = \frac{\Gamma(a)\Gamma(b)}{\Gamma(a+b)}, \tag{11.1.2}$$

and more generally, the classical *Selberg integral*

$$\int_{0\leqslant x_k\leqslant\cdots\leqslant x_1\leqslant 1} \prod_{i=1}^{k} x_i^{a-1}(1-x_i)^{b-1} \prod_{1\leqslant i<j\leqslant k} (x_i-x_j)^{2c}\,dx_1\cdots dx_k$$
$$= \prod_{r=0}^{k-1} \frac{\Gamma((r+1)c)}{\Gamma(c)} \frac{\Gamma(a+rc)\Gamma(b+rc)}{\Gamma(a+b+(2k-r-2)c)}. \tag{11.1.3}$$

The integrals (11.1.2), (11.1.3) converge provided

$$\operatorname{Re} a > 0, \quad \operatorname{Re} b > 0, \quad \operatorname{Re} c > -\min\left\{\frac{1}{k}, \frac{\operatorname{Re} a}{k-1}, \frac{\operatorname{Re} b}{k-1}\right\}.$$

The *KZ* equations have a difference analogue called the *quantized KZ* (*qKZ*) equations. A *qKZ equation* is a system of linear difference equations. They were introduced by Frenkel and Reshetikhin (1992) and Smirnov (1992). In the latter reference, it is a system of difference equations describing form-factors in quantum field theory. Frenkel and Reshetikhin (1992) described matrix elements of intertwining operators for a quantum affine algebra. In Idzumi et al. (1993), it is a system of equations for correlation functions of the six-vertex model. The *qKZ* equations have remarkable mathematical properties, as do the *KZ* equations.

The solutions of the *qKZ* equations can be represented by multidimensional q-hypergeometric integrals. In this chapter we consider only the rational *qKZ* equations. The corresponding integrals appear in §11.9.2.

11.2 Representation Theory

The *KZ* equations are associated with arbitrary simple Lie algebras. In this chapter, we restrict ourselves to the case of the Lie algebra $\mathfrak{sl}_2$.

The complex Lie algebra $\mathfrak{sl}_2$ is the three-dimensional Lie algebra spanned by the elements e, f, h satisfying the commutation relations $[h, e] = 2e$, $[e, f] = h$, $[h, f] = 2f$. Denote by $\mathfrak{h} \subset \mathfrak{sl}_2$ the one-dimensional Lie subalgebra generated by h.

Let V be an $\mathfrak{sl}_2$-module. For $m \in \mathbb{C}$, the subspace $V[m] := \{v \in V \mid hv = mv\} \subset V$ is called the *weight subspace* of V of *weight* m. The $\mathfrak{sl}_2$-module V has a *weight decomposition* if $V = \oplus_m V[m]$. A vector $v \in V$ is called *singular* if $ev = 0$.

For $m \in \mathbb{C}$, the *Verma module* M_m is the infinite-dimensional $\mathfrak{sl}_2$-module generated by a vector v_m such that $hv_m = mv_m$ and $ev_m = 0$. The vectors $f^k v_m$ for $k = 0, 1, \dots$ form a basis of M_m. The generators e, f, h act on M_m as follows:

$$e \cdot f^k v_m = k(m - k + 1) f^{k-1} v_m, \quad f \cdot f^k v_m = f^{k+1} v_m, \quad h \cdot f^k v_m = (m - 2k) f^k v_m.$$

If $m \in \mathbb{Z}_{\geqslant 0}$, the subspace of M_m spanned by the vectors $f^k v_m$ for $k > m$ is an $\mathfrak{sl}_2$-submodule isomorphic to M_{-m-2}. The quotient $L_m = M_m / M_{-m-2}$ is an irreducible $\mathfrak{sl}_2$-module with a basis induced by $v_m, f v_m, \dots, f^m v_m$. Any finite-dimensional irreducible $\mathfrak{sl}_2$-module has this form. The subspace of singular vectors of L_m is one-dimensional and generated by v_m.

Any finite-dimensional $\mathfrak{sl}_2$-module has weight decomposition and is isomorphic to the direct sum of irreducible modules, $V = \bigoplus_m L_m \otimes W_m$, where W_m is the multiplicity space of the submodule L_m. The space W_m can be identified with the subspace of singular vectors of weight m: $\operatorname{Sing} V[m] := \{v \in V \mid ev = 0,\ hv = mv\}$.

The tensor product $V = V_1 \otimes \cdots \otimes V_n$ of $\mathfrak{sl}_2$-modules $V_1, \dots, V_n$ is an $\mathfrak{sl}_2$-module. An element $x \in \mathfrak{sl}_2$ acts on V as $x^{(1)} + \cdots + x^{(n)}$, where $x^{(i)} = 1^{\otimes(i-1)} \otimes x \otimes 1^{\otimes(n-i)}$. For $l, m \in \mathbb{Z}_{\geqslant 0}$, $l \geqslant m$, we have $L_l \otimes L_m = L_{l+m} \oplus L_{l+m-2} \oplus \cdots \oplus L_{l-m}$.

Given $m = (m_1, \dots, m_n) \in \mathbb{C}^n$, denote $|m| := m_1 + \cdots + m_n$ and $M^{\otimes m} = M_{m_1} \otimes \cdots \otimes M_{m_n}$. For $J = (j_1, \dots, j_n) \in \mathbb{Z}^n_{\geqslant 0}$, denote $f_J v := f^{j_1} v_{m_1} \otimes \cdots \otimes f^{j_n} v_{m_n} \in M^{\otimes m}$. The vectors $f_J v$ form a basis of $M^{\otimes m}$. We have

$$e \cdot f_J v = \sum_{s=1}^{n} j_s (m_s - j_1 + 1) f_{J-1_s} v, \quad f \cdot f_J v = \sum_{s=1}^{n} f_{J+1_s} v, \quad h \cdot f_J v = (|m| - 2|J|) f_J v,$$

where $J \pm 1_s = (j_1, \dots, j_s \pm 1, \dots, j_n)$.

Let $U(\mathfrak{sl}_2)$ be the universal enveloping algebra of $\mathfrak{sl}_2$. The element

$$\Omega = e \otimes f + f \otimes e + \tfrac{1}{2} h \otimes h \in U(\mathfrak{sl}_2) \otimes U(\mathfrak{sl}_2) \tag{11.2.1}$$

is called the *Casimir element*. For all $x \in \mathfrak{sl}_2$ we have in $U(\mathfrak{sl}_2) \otimes U(\mathfrak{sl}_2)$ that

$$[\Omega, x \otimes 1 + 1 \otimes x] = 0. \tag{11.2.2}$$

11.3 Rational *KZ* Equation and Gaudin Model

Given $\xi \in \mathfrak{h}$ and $z = (z_1, \dots, z_n) \in \mathbb{C}^n$ such that $z_i \neq z_j$ for $i \neq j$, define the *KZ operators* $K_1(z, \xi), \dots, K_n(z, \xi) \in (U(\mathfrak{sl}_2))^{\otimes n}$ by the formula

$$K_i(z, \xi) := \sum_{j \neq i} \frac{\Omega^{(ij)}}{z_i - z_j} + \xi^{(i)}, \tag{11.3.1}$$

where $\Omega^{(ij)} = e^{(i)}f^{(j)} + f^{(i)}e^{(j)} + \frac{1}{2}h^{(i)}h^{(j)}$ is the Casimir element placed at the ith and jth tensor factors. The next proposition follows from (11.2.2).

Proposition 11.3.1 *The KZ operators commute,* $[K_i(z,\xi), K_j(z,\xi)] = 0$ $(i, j = 1,\dots,n)$. *In addition,* $[K_i(z,\xi), h^{(1)} + \cdots + h^{(n)}] = 0$ *and* $[K_i(z,0), x^{(1)} + \cdots + x^{(n)}] = 0$ $(i = 1,\dots,n$ *and* $x \in \mathfrak{sl}_2)$.

Let $V = V_1 \otimes \cdots \otimes V_n$ be a tensor product of $\mathfrak{sl}_2$-modules. By Proposition 11.3.1, for any $\xi \in \mathfrak{h}$, the action of *KZ* operators on V preserves weight subspaces $V[m]$, and for $\xi = 0$, the action of *KZ* operators preserves subspaces $\operatorname{Sing} V[m]$ of singular vectors of given weight.

11.3.1 Rational KZ Equation

Given $\xi \in \mathfrak{h}$, $\kappa \in \mathbb{C}^*$, and a tensor product V of $\mathfrak{sl}_2$-modules, the *KZ differential equation* on a V-valued function $\varphi(z_1,\dots,z_n)$ is the system of differential equations

$$\kappa\frac{\partial\varphi}{\partial z_i} = K_i(z,\xi)\varphi, \quad i = 1,\dots,n \tag{11.3.2}$$

on $U_n = \{z \in \mathbb{C}^n \mid z_i \neq z_j \text{ for } i \neq j\}$. The *KZ* equation defines a flat connection on the trivial bundle $V \times U_n \to U_n$, $[\kappa\frac{\partial}{\partial z_i} - K_i(z,\xi), \kappa\frac{\partial}{\partial z_j} - K_j(z,\xi)] = 0$, called the *KZ connection*. For any $m \in \mathbb{C}$, the subbundle with fiber $V[m] \subset V$ is invariant with respect to the *KZ* connection. If $\xi = 0$, then the subbundle with fiber $\operatorname{Sing} V[m] \subset V$ is also invariant with respect to the *KZ* connection. In this case, for any $x \in \mathfrak{sl}_2$ and any solution $\varphi(z)$ of the *KZ* equation, the function $x\varphi(z)$ is also a solution of the *KZ* equation.

11.3.2 Subbundle of Conformal Blocks

Let $\ell, m_1,\dots,m_n, k$ be nonnegative integers such that

$$\ell > 0, \quad 0 \leqslant m_1,\dots,m_n, m_1 + \cdots + m_n - 2k \leqslant \ell. \tag{11.3.3}$$

Let $p = \ell + 1 + 2k - m_1 - \cdots - m_n$ and $L^{\otimes m} = L_{m_1} \otimes \cdots \otimes L_{m_n}$. For $z \in U_n$, denote

$$W^\ell(z) := \{v \in L^{\otimes m} \mid hv = (|m| - 2k)v,\ ev = 0,\ (z_1e^{(1)} + \cdots + z_ne^{(n)})^p v = 0\}.$$

Clearly, $W^\ell(z) \subset \operatorname{Sing} L^{\otimes m}[|m| - 2k]$. The vector space $W^\ell(z)$ is called the *space of conformal blocks* at level ℓ; see Feigin et al. (1994, 1995).

This definition is not standard. Usually, the space of conformal blocks is defined for given n distinct points on a Riemann surface and n irreducible representations of an affine Lie algebra; see Kazhdan and Lusztig (1993, 1994). If the Riemann surface is the Riemann sphere, then one can describe the space of conformal blocks in terms of finite-dimensional representations of the corresponding finite-dimensional Lie algebra; see Feigin et al. (1994, 1995). We take that description as the definition.

The spaces of conformal blocks $W^\ell(z)$ form a subbundle of $\operatorname{Sing} L^{\otimes m}[|m| - 2k] \times U_n \to U_n$. This subbundle is invariant with respect to the *KZ* connection with parameters $\kappa = \ell + 2$ and $\xi = 0$ due to Knizhnik and Zamolodchikov (1984).

11.3.3 Rational Gaudin Model

For fixed $z_1, \dots, z_n$, the action of the *KZ* operators on a tensor product $V = V_1 \otimes \cdots \otimes V_n$ of $\mathfrak{sl}_2$-modules induces a collection of commuting linear operators on V, called the *Hamiltonians* of the Gaudin model. For any $\xi \in \mathfrak{h}$, they preserve weight subspaces $V[m]$, and for $\xi = 0$, they preserve subspaces $\operatorname{Sing} V[m]$ of singular vectors of given weight.

The main problem of the Gaudin model is to find eigenvectors and eigenvalues of the Hamiltonians.

11.3.4 Dynamical Equation

Introduce a coordinate λ on $\mathfrak{h}$ by $\xi = \lambda h/2$. The *KZ* equation takes the form

$$\kappa \frac{\partial \varphi}{\partial z_i} = \Bigg(\sum_{j \neq i} \frac{\Omega^{(ij)}}{z_i - z_j} + \lambda \frac{h^{(i)}}{2} \Bigg) \varphi, \qquad i = 1, \dots, n.$$

The equation

$$\kappa \frac{\partial \varphi}{\partial \lambda} = \Bigg(\sum_{j=1}^{n} z_j \frac{h^{(j)}}{2} + \frac{fe}{\lambda} \Bigg) \varphi \tag{11.3.4}$$

is called the *dynamical differential equation*. Recall that $x \in \mathfrak{sl}_2$ acts on $V = V_1 \otimes \cdots \otimes V_n$ as $x^{(1)} + \cdots + x^{(n)}$.

Proposition 11.3.2 (Felder et al., 2000) *The dynamical and KZ equations are compatible,*

$$\Bigg[\kappa \frac{\partial}{\partial z_i} - \sum_{j \neq i} \frac{\Omega^{(ij)}}{z_i - z_j} - \lambda \frac{h^{(i)}}{2}, \kappa \frac{\partial}{\partial \lambda} - \sum_{j=1}^{n} z_j \frac{h^{(j)}}{2} - \frac{fe}{\lambda} \Bigg] = 0.$$

The *dynamical Hamiltonian*

$$D(z, \xi) = \sum_{j=1}^{n} z_j \frac{h^{(j)}}{2} + \frac{fe}{\lambda} \in (U(\mathfrak{sl}_2))^{\otimes n} \tag{11.3.5}$$

preserves the weight decomposition of V, $[D(z, \xi), h^{(1)} + \cdots + h^{(n)}] = 0$, and commutes with the Gaudin Hamiltonians, $[D(z, \xi), K_i(z, \xi)] = 0$ for all $i = 1, \dots, n$.

11.3.5 $(\mathfrak{sl}_2, \mathfrak{sl}_2)$ *Duality*

Let $m_1, m_2, \widetilde{m}_1, \widetilde{m}_2$ be nonnegative integers such that $m_1 + m_2 = \widetilde{m}_1 + \widetilde{m}_2$, and write $L^{\otimes m} = L_{m_1} \otimes L_{m_2}$, $L^{\otimes \widetilde{m}} = L_{\widetilde{m}_1} \otimes L_{\widetilde{m}_2}$ for the tensor products of irreducible $\mathfrak{sl}_2$-modules. The weight subspaces $L^{\otimes m}[\widetilde{m}_1 - \widetilde{m}_2]$ and $L^{\otimes \widetilde{m}}[m_1 - m_2]$ are isomorphic, as given by the rule

$$\psi\colon f^{\widetilde{m}_2 - k} v_{m_1} \otimes f^k v_{m_2} \mapsto \frac{(\widetilde{m}_2 - k)!}{(m_2 - k)!} f^{m_2 - k} v_{\widetilde{m}_1} \otimes f^k v_{\widetilde{m}_2}, \quad k = \max(0, \widetilde{m}_2 - m_1), \dots, \min(m_2, \widetilde{m}_2). \tag{11.3.6}$$

Let $z = (z_1, z_2)$, $\tilde{z} = (\lambda/2, -\lambda/2)$, $\xi = \lambda h/2$, $\tilde{\xi} = (z_1 - z_2)h/2$.

Proposition 11.3.3 (Toledano Laredo, 2002; Tarasov and Varchenko, 2002) *The isomorphism ψ intertwines the action of the Gaudin and dynamical Hamiltonians*:

$$D(\tilde{z}, \tilde{\xi})|_{L^{\otimes \tilde{m}}[m_1-m_2]}\psi = \psi\Big(K_1(z,\xi) + \frac{C}{2(z_1-z_2)} - \frac{\lambda}{4}(\widetilde{m}_1 - \widetilde{m}_2)\Big)\Big|_{L^{\otimes m}[\widetilde{m}_1-\widetilde{m}_2]}$$
$$= \psi\Big(-K_2(z,\xi) + \frac{C}{2(z_1-z_2)} + \frac{\lambda}{4}(\widetilde{m}_1 - \widetilde{m}_2)\Big)\Big|_{L^{\otimes m}[\widetilde{m}_1-\widetilde{m}_2]},$$

$$K_1(\tilde{z}, \tilde{\xi})|_{L^{\otimes \tilde{m}}[m_1-m_2]}\psi = \psi\Big(D(z,\xi) - \frac{\widetilde{C}}{2\lambda} + \frac{z_1 m_1 + z_2 m_2 - (z_1+z_2)\widetilde{m}_1}{2}\Big)\Big|_{L^{\otimes m}[\widetilde{m}_1-\widetilde{m}_2]},$$

$$K_2(\tilde{z}, \tilde{\xi})|_{L^{\otimes \tilde{m}}[m_1-m_2]}\psi = \psi\Big(-D(z,\xi) + \frac{\widetilde{C}}{2\lambda} + \frac{z_1 m_1 + z_2 m_2 - (z_1+z_2)\widetilde{m}_2}{2}\Big)\Big|_{L^{\otimes m}[\widetilde{m}_1-\widetilde{m}_2]},$$

where $C = (m_1 + 2)m_2$ and $\widetilde{C} = (\widetilde{m}_1 + 2)\widetilde{m}_2$.

11.4 Hypergeometric Solutions of the Rational *KZ* and Dynamical Equations, and Bethe Ansatz

11.4.1 Master Function and Weight Function

For given $k, n \in \mathbb{Z}_{>0}$, $m = (m_1, \ldots, m_n) \in \mathbb{C}^n$, and $\xi = \lambda h/2$, define the (generally multivalued) *master function*

$$\Phi_{k,n}(t, z, \xi, m) := \lambda\Big(\tfrac{1}{2}\sum_{i=1}^{n} m_i z_i - \sum_{j=1}^{k} t_j\Big) + \tfrac{1}{2}\sum_{i<j} m_i m_j \log(z_i - z_j)$$
$$- \sum_{l=1}^{n}\sum_{i=1}^{k} m_l \log(t_i - z_l) + 2\sum_{i<j}\log(t_i - t_j).$$

For a function $F(t_1, \ldots, t_k)$, denote $\mathrm{Sym}_t[F(t_1, \ldots, t_k)] := \sum_{\sigma\in S_k} F(t_{\sigma(1)}, \ldots, t_{\sigma(k)})$. To any index $J = (j_1, \ldots, j_n)$ of nonnegative integers with $|J| = k$, we assign a rational function

$$A_J(t,z) := \frac{1}{j_1! \cdots j_n!}\mathrm{Sym}_t\Big[\prod_{s=1}^{n}\prod_{i=1}^{j_s}\frac{1}{t_{j_1+\cdots+j_{s-1}+i} - z_s}\Big].$$

The $M^{\otimes m}[|m| - 2k]$-valued function

$$\omega_{k,n}(t,z) = \sum_{J,|J|=k} A_J(t,z) f_J v \tag{11.4.1}$$

is called the *weight function*.

11.4.2 Integral Representations for Solutions for $\xi \neq 0$

Consider the space $\mathbb{C}^k\times\mathbb{C}^n\times\mathfrak{h}$ with coordinates t, z, λ and the projection $\pi\colon C^k\times\mathbb{C}^n\times\mathfrak{h} \to \mathbb{C}^n\times\mathfrak{h}$. Also consider the arrangement in $\mathbb{C}^k\times\mathbb{C}^n\times\mathfrak{h}$ of hyperplanes H_{ij}, $1 \leqslant i < j \leqslant k$; H_i^j, $i = 1, \ldots,$

$k, j = 1, \dots, n$; H^{ij}, $1 \leqslant i < j \leqslant n$; H^0 defined respectively by the equations $t_i = t_j$, $t_i = z_j$, $z_i = z_j$, $\xi = 0$. Denote by $U \subset \mathbb{C}^k \times \mathbb{C}^n \times \mathfrak{h}$ the complement to the arrangement. The master function is a well-defined multivalued function on U.

Fix $\kappa \in \mathbb{C}^*$. The univalued branches of the function $e^{\Phi_{k,n}(t,z,\xi,m)/\kappa}$ over open subsets of U define a one-dimensional local system on U denoted by $\mathcal{L}_\kappa$.

The map $\pi|_U : U \to U_n \times (\mathfrak{h} - 0)$ is a locally trivial bundle with fiber

$$U_{k,n}(z,\xi) := \{t \in \mathbb{C}^k \mid t_i \neq t_j \text{ for } i \neq j \text{ and } t_i \neq z_j \text{ for all } i, j\}.$$

Consider the associated homological vector bundle

$$\amalg_{(z,\xi)\in U_n \times (\mathfrak{h}-0)} H_k(U_{k,n}(z,\xi), \mathcal{L}_\kappa|_{U_{k,n}(z,\xi)}) \to U_n \times (\mathfrak{h} - 0).$$

This bundle has a canonical flat Gauss–Manin connection (the cycles can be deformed from fiber to fiber). We will be interested in (multivalued) horizontal sections on the homological bundle.

Theorem 11.4.1 (Schechtman and Varchenko, 1991a; Felder et al., 2000) *For any horizontal section $\gamma(z,\xi) \in H_k(U_{k,n}(z,\xi), \mathcal{L}_\kappa|_{U_{k,n}(z,\xi)})$ of the homological bundle, the $M^{\otimes m}[|m| - 2k]$-valued function*

$$\varphi^{(\gamma)}(z,\xi) := \int_{\gamma(z,\xi)} e^{\Phi_{k,n}(t,z,\xi,m)/\kappa} \omega_{k,n}(t,z)\, dt_1 \wedge \dots \wedge dt_k \tag{11.4.2}$$

is a solution of the KZ and dynamical equations.

The integrals (11.4.2) (and their coordinates) are called *multidimensional hypergeometric functions*.

11.4.3 Integral Representations for Solutions for $\xi = 0$

For $\xi = 0$, the *KZ* equation takes the form

$$\kappa \frac{\partial \varphi}{\partial z_i} = \sum_{j \neq i} \frac{\Omega^{(ij)}}{z_i - z_j} \varphi, \quad i = 1, \dots, n. \tag{11.4.3}$$

Consider the space $\mathbb{C}^k \times \mathbb{C}^n$ with coordinates t, z and the projection $\pi' : \mathbb{C}^k \times \mathbb{C}^n \to \mathbb{C}^n$. Consider the arrangement in $\mathbb{C}^k \times \mathbb{C}^n$ of hyperplanes H_{ij}, $1 \leqslant i < j \leqslant k$; H_i^j, $j = 1, \dots, n$, $i = 1, \dots, k$; H^{ij}, $1 \leqslant i < j \leqslant n$, defined respectively by the equations $t_i = t_j$, $t_i = z_j$, $z_i = z_j$. Denote by $U' \subset \mathbb{C}^k \times \mathbb{C}^n$ the complement to the arrangement. The master function $\Phi(t, z, 0, m)$ is a well-defined multivalued function on U'.

Fix $\kappa \in \mathbb{C}^*$. The univalued branches of the function $e^{\Phi_{k,n}(t,z,0,m)/\kappa}$ over open subsets of U' define a one-dimensional local system on U' denoted by $\mathcal{L}'_\kappa$.

The map $\pi'|_{U'} : U' \to U_n$ is a locally trivial bundle with fiber

$$U_{k,n}(z) := \{t \in \mathbb{C}^k \mid t_i \neq t_j \text{ for } i \neq j \text{ and } t_i \neq z_j \text{ for all } i, j\}.$$

Consider the associated homological vector bundle

$$\amalg_{z\in U_n} H_k(U_{k,n}(z), \mathcal{L}'_\kappa|_{U_{k,n}(z)}) \to U_n. \tag{11.4.4}$$

This bundle has a canonical flat Gauss–Manin connection.

Theorem 11.4.2 (Date et al., 1990; Schechtman and Varchenko, 1991a) *For any horizontal section* $\gamma(z) \in H_k(U_{k,n}(z), \mathcal{L}'_\kappa|_{U_{k,n}(z)})$ *of the homological bundle* (11.4.4), *the* $M^{\otimes m}[|m| - 2k]$*-valued function*

$$\varphi^{(\gamma)}(z) := \int_{\gamma(z)} e^{\Phi_{k,n}(t,z,0,m)/\kappa}\omega_{k,n}(t,z)\, dt_1 \wedge \cdots \wedge dt_k \tag{11.4.5}$$

takes values in Sing $M^{\otimes m}[|m| - 2k]$ *and is a solution of the KZ equation* (11.4.3).

11.4.4 Solutions with Values in $L^{\otimes m}$

Let $m = (m_1, \ldots, m_n)$ be a vector with nonnegative integer coordinates, $L^{\otimes m}$ the corresponding tensor product of irreducible finite-dimensional $\mathfrak{sl}_2$-modules, and $\pi\colon M^{\otimes m} \to L^{\otimes m}$ the canonical projection.

Proposition 11.4.3 *Let* φ *be a solution of the KZ or dynamical equations with values in* $M^{\otimes m}$. *Then* $\pi(\varphi)$ *is a solution of the KZ or dynamical equations with values in* $L^{\otimes m}$.

11.4.5 Completeness of Hypergeometric Solutions

For generic κ, all solutions of the *KZ* equations are hypergeometric. We formulate one such statement, which is a corollary of a result by Varchenko (1995, Theorem 12.5.5).

Theorem 11.4.4 *Let* $L^{\otimes m}$ *be a tensor product of irreducible finite-dimensional* $\mathfrak{sl}_2$*-modules. Consider the KZ equation with values in* $L^{\otimes m}$ *and parameters* $\kappa \in \mathbb{C}^*$ *and* $\xi = 0$. *Then, for generic* κ, *all solutions of this KZ equation are hypergeometric, that is, given by constructions of §11.4.3 and §11.4.4.*

11.4.6 Remark on Integration Cycles

Let κ and $m_1, \ldots, m_n$ be real, $m_1, \ldots, m_n < 0$, $\kappa > 0$, and $\xi = 0$. Assume that $z \in \mathbb{R}^n$ and denote $U^{\mathbb{R}}_{k,n}(z) := \{t \in \mathbb{R}^k \mid t_i \neq t_j \text{ for } i \neq j \text{ and } t_i \neq z_j \text{ for all } i, j\}$. Let $\gamma(z)$ be a bounded connected component of $U^{\mathbb{R}}_{k,n}(z)$. Fix a univalued branch of the master function $\Phi_{k,n}(t, z, 0, m)$ over $\gamma(z)$. Then the integral (11.4.5) is well defined.

Lemma 11.4.5 *Under the above assumptions, the function* $\varphi^{(\gamma)}(z)$ *given by* (11.4.5) *is a solution of the KZ equation* (11.4.3) *with values in* Sing $M^{\otimes m}[|m| - 2k]$.

A similar statement holds for integral representations in Theorem 11.4.1.

11.4.7 Hypergeometric Solutions and Conformal Blocks

Let ℓ, $m_1, \ldots, m_n$ and k be nonnegative integers satisfying condition (11.3.3). Consider the bundle $\operatorname{Sing} L^{\otimes m}[|m| - 2k] \times U_n \to U_n$ and its subbundle of conformal blocks with fiber $W^\ell(z) \subset \operatorname{Sing} L^{\otimes m}[|m| - 2k]$. The subbundle of conformal blocks is invariant with respect to the *KZ* differential equation with parameters $\kappa = \ell + 2$ and $\xi = 0$.

Theorem 11.4.6 (Feigin et al., 1994, 1995) *Let $\varphi^{(\gamma)}(z)$ be a $\operatorname{Sing} L^{\otimes m}[|m| - 2k]$-valued hypergeometric solution of the KZ equation given by formula* (11.4.5). *Then $\varphi^{(\gamma)}(z) \in W^\ell(z)$.*

11.4.8 Hypergeometric Solutions and $(\mathfrak{sl}_2, \mathfrak{sl}_2)$ Duality

In notation of §11.3.5, Proposition 11.3.3 identifies the *KZ* and dynamical equations on $L_{m_1} \otimes L_{m_2}[\widetilde{m}_1 - \widetilde{m}_2]$ with the dynamical and *KZ* equations on $L_{\widetilde{m}_1} \otimes L_{\widetilde{m}_2}[m_1 - m_2]$, respectively; see Tarasov and Varchenko (2002) for more details. Hypergeometric solutions of the *KZ* and dynamical equations with values in $L_{m_1} \otimes L_{m_2}[\widetilde{m}_1 - \widetilde{m}_2]$ are given by $(\widetilde{m}_1 - \widetilde{m}_2)$-dimensional integrals; see Theorem 11.4.1. Similarly, hypergeometric solutions of the *KZ* and dynamical equations with values in $L_{\widetilde{m}_1} \otimes L_{\widetilde{m}_2}[m_1 - m_2]$ are given by $(m_1 - m_2)$-dimensional integrals. By Proposition 11.3.3, one can identify the $(\widetilde{m}_1 - \widetilde{m}_2)$-dimensional integrals involved and the corresponding $(m_1 - m_2)$-dimensional integrals; see Tarasov and Varchenko (2005c).

Assume that $m_1 + m_2 = \widetilde{m}_1 + \widetilde{m}_2$ tends to infinity so that $\widetilde{m}_1 - \widetilde{m}_2$ is fixed and $m_1 - m_2$ tends to infinity. Then the asymptotics of the hypergeometric integrals in the first integral representation can be calculated by the stationary phase method. That gives the asymptotics with respect to dimension for the hypergeometric integrals in the second integral representation; see the examples given by Tarasov and Varchenko (2005c).

11.4.9 Bethe Ansatz

For fixed z and $\xi = \lambda h/2$, the critical points of the master function are defined by the equations $\partial\Phi_{k,n}/\partial t_i = 0$, $i = 1, \ldots, k$, that is,

$$-\sum_{j=1}^{n} \frac{m_j}{t_i - z_j} + \sum_{j, j \neq i} \frac{2}{t_i - t_j} = \lambda, \quad i = 1, \ldots, k.$$

Theorem 11.4.7 (Gaudin, 1983; Reshetikhin and Varchenko, 1995; Felder et al., 2000) *For any critical point $t^0 = (t_1^0, \ldots, t_k^0)$, the vector $\omega_{k,n}(t^0, z) \in M^{\otimes m}[|m| - 2k]$ is nonzero and is an eigenvector of the Gaudin Hamiltonians,*

$$K_i(z, \xi)\omega_{k,n}(t^0, z) = \frac{\partial\Phi_{k,n}}{\partial z_i}(t^0, z, \xi)\omega_{k,n}(t^0, z), \quad i = 1, \ldots, n.$$

If $\xi \neq 0$, then $\omega_{k,n}(t^0, z)$ is an eigenvector of the dynamical Hamiltonian,

$$D(z, \xi)\omega_{k,n}(t^0, z) = \frac{\partial\Phi_{k,n}}{\partial\lambda}(t^0, z, \xi)\omega_{k,n}(t^0, z).$$

If $\xi = 0$, then $\omega_{k,n}(t^0, z)$ is a singular vector.

11.4.10 Critical Points and Ordinary Differential Operators

Let $m_1, \ldots, m_n$ and k be nonnegative integers such that $|m| - 2k \geqslant 0$. Let $t^0 = (t_1^0, \ldots, t_k^0)$ be a critical point of the master function $\Phi_{k,n}(t, z, 0, m)$. Define the linear second-order differential operator $\mathcal{D}_{t^0}$ (with respect to a new variable u) by the formula

$$\mathcal{D}_{t^0} := \left(\frac{d}{du} - \frac{T'(u)}{T(u)} + \frac{y'(u)}{y(u)}\right)\left(\frac{d}{du} - \frac{y'(u)}{y(u)}\right),$$

where $T(u) = \prod_{j=1}^{n}(u - z_j)^{m_j}$, $y(u) = \prod_{i=1}^{k}(u - t_i^0)$, and $'$ denotes the derivative with respect to u. Then the singular points of $\mathcal{D}_{t^0}$ are $z_1, \ldots, z_n$ and ∞. All singular points are regular. The exponents at infinity are $-k, k - |m| - 1$. The exponents at z_i are $0, m_i + 1$. Clearly, $\mathcal{D}_{t^0} y(u) = 0$.

Theorem 11.4.8 (Scherbak and Varchenko, 2003) *All solutions of the differential equation $\mathcal{D}_{t^0} Y(u) = 0$ are polynomials. More precisely, there is a polynomial $p(u)$ of degree $|m| + 1 - k$ such that $y(u)$, $p(u)$ is a basis of the space of solutions.*

The condition $|m| - 2k \geqslant 0$ means that the tensor product $L_{m_1} \otimes \cdots \otimes L_{m_n}$ may contain the submodule $L_{|m|-2k}$. The correspondence

$$\text{Bethe vector } \omega_{k,n}(t^0, z) \longleftrightarrow \text{critical point } t^0 \longleftrightarrow \text{differential operator } \mathcal{D}_{t^0}$$

is a particular case of the geometric Langlands correspondence; see Mukhin et al. (2009c) and Frenkel (2004).

11.5 Trigonometric *KZ* Equation

11.5.1 Trigonometric KZ Equation and Gaudin Model

Let $\Omega^+ := \frac{1}{4}h \otimes h + e \otimes f$, $\Omega^- := \frac{1}{4}h \otimes h + f \otimes e$, so that $\Omega = \Omega^+ + \Omega^-$. Define the *trigonometric classical r-matrix* by $r(z) = \frac{\Omega^+ z + \Omega^-}{z-1}$.

Given $\xi \in \mathfrak{h}$ and $z = (z_1, \ldots, z_n) \in \mathbb{C}^n$ such that $z_i \neq z_j$ for $i \neq j$, define the trigonometric *KZ* operators $K_1^{\mathrm{tr}}(z, \xi), \ldots, K_n^{\mathrm{tr}}(z, \xi) \in (U(\mathfrak{sl}_2))^{\otimes n}$ by the formula

$$K_i^{\mathrm{tr}}(z, \xi) := \sum_{j \neq i} r^{(ij)}(z_i/z_j) + \xi^{(i)}. \tag{11.5.1}$$

Proposition 11.5.1 *The trigonometric KZ operators commute, $[K_i^{\mathrm{tr}}(z, \xi), K_j^{\mathrm{tr}}(z, \xi)] = 0$ for all $i, j = 1, \ldots, n$, and $[K_i^{\mathrm{tr}}(z, \xi), h^{(1)} + \cdots + h^{(n)}] = 0$ for all $i = 1, \ldots, n$.*

Let $V = V_1 \otimes \cdots \otimes V_n$ be a tensor product of $\mathfrak{sl}_2$-modules. By Proposition 11.5.1, for any $\xi \in \mathfrak{h}$, the action of the trigonometric *KZ* operators on V preserves weight subspaces $V[m]$.

Given $\xi \in \mathfrak{h}$, $\kappa \in \mathbb{C}^*$, and a tensor product of $\mathfrak{sl}_2$-modules, the *trigonometric KZ differential equation* on a V-valued function $\varphi(z_1, \dots, z_n)$ is the system of differential equations

$$\kappa z_i \frac{\partial \varphi}{\partial z_i} = K_i^{\mathrm{tr}}(z, \xi)\varphi, \quad i = 1, \dots, n, \tag{11.5.2}$$

on $U_n^0 := \{z \in \mathbb{C}^n \mid z_i \neq 0 \text{ for all } i;\ z_i \neq z_j \text{ for } i \neq j\}$. The *KZ* equation defines a flat connection on the trivial bundle $V \times U_n^0 \to U_n^0$, $[\kappa z_i \frac{\partial}{\partial z_i} - K_i(z, \xi), \kappa z_j \frac{\partial}{\partial z_j} - K_j(z, \xi)] = 0$, called the *trigonometric KZ connection*. For any $m \in \mathbb{C}$, the subbundle with fiber $V[m] \subset V$ is invariant with respect to this connection.

For a tensor product $V = V_1 \otimes \cdots \otimes V_n$ of $\mathfrak{sl}_2$-modules and fixed $z_1, \dots, z_n$, the action of the trigonometric *KZ* operators on V induces a collection of commuting linear operators, called the *Hamiltonians* of the *trigonometric Gaudin model*.

11.5.2 Difference Dynamical Equation

We say that an $\mathfrak{sl}_2$-module V is *locally finite* if it has weight decomposition with finite-dimensional weight subspaces, and for any $x \in V$ there is a natural number N such that $e^N x = 0$. For example, V is locally finite if it is a tensor product of finite-dimensional or Verma modules.

For $u \in \mathbb{C}$, set $P(u) := \sum_{i=0}^{\infty} f^i e^i \frac{1}{i!} \prod_{j=1}^{i} \frac{1}{u-j-h}$. For a locally finite $\mathfrak{sl}_2$-module V and generic u, the action of $P(u)$ on V is well defined and gives a linear map $V \to V$.

Let $V = V_1 \otimes \cdots \otimes V_n$ be a tensor product of locally finite $\mathfrak{sl}_2$-modules, and $\xi = \lambda h/2$. Consider the linear map $D^{\mathrm{tr}}(z, \xi)\colon V \to V$,

$$D^{\mathrm{tr}}(z, \xi) := (z_1^{h/2})^{(1)} \cdots (z_n^{h/2})^{(n)} P(\lambda). \tag{11.5.3}$$

The *difference dynamical equation* for a V-valued function $\varphi(z, \xi)$ is

$$\varphi(z_1, \dots, z_n, \xi + \kappa h/2) = D^{\mathrm{tr}}(z, \xi)\varphi(z, \xi). \tag{11.5.4}$$

Proposition 11.5.2 (Tarasov and Varchenko, 2000) *The difference dynamical equation is compatible with the trigonometric KZ equation,*

$$\left(\kappa z_i \frac{\partial}{\partial z_i} - K_i^{\mathrm{tr}}(z, \xi + \kappa h/2)\right) D^{\mathrm{tr}}(z, \xi) = D^{\mathrm{tr}}(z, \xi)\left(\kappa z_i \frac{\partial}{\partial z_i} - K_i^{\mathrm{tr}}(z, \xi)\right), \quad i = 1, \dots, n.$$

Hence, if $\varphi(z, \xi)$ is a solution of the trigonometric *KZ* equation with parameter ξ, then $D^{\mathrm{tr}}(z, \xi)\varphi(z, \xi)$ is a solution of the trigonometric *KZ* equation with parameter $\xi + \kappa h/2$.

The element $D^{\mathrm{tr}}(z, \xi)$ is called the *difference dynamical Hamiltonian*. It preserves the weight decomposition of V, $[D^{\mathrm{tr}}(z, \xi), h^{(1)} + \cdots + h^{(n)}] = 0$, and commutes with the trigonometric Gaudin Hamiltonians, $[D^{\mathrm{tr}}(z, \xi), K_i^{\mathrm{tr}}(z, \xi)] = 0$ $(i = 1, \dots, n)$.

11.6 Hypergeometric Solutions of the Trigonometric *KZ* Equation and Bethe Ansatz

11.6.1 Integral Representations for Solutions

For given $k, n \in \mathbb{Z}_{>0}$, $m = (m_1, \ldots, m_n) \in \mathbb{C}^n$, and $\xi = \lambda h/2$, define the *trigonometric master function*

$$\Phi^{\mathrm{tr}}_{k,n}(t, z, \xi, m) := \tfrac{1}{2} \sum_{i=1}^{n} m_i(\lambda + m_i - |m| + 2k) \log z_i - (\lambda - 1 - |m| + 2k) \sum_{i=1}^{k} \log t_i$$

$$+ \tfrac{1}{2} \sum_{i<j} m_i m_j \log(z_i - z_j) - \sum_{l=1}^{n} \sum_{i=1}^{k} m_l \log(t_i - z_l) + 2 \sum_{i<j} \log(t_i - t_j).$$

Consider the space $\mathbb{C}^k \times \mathbb{C}^n$ with coordinates t, z and the projection $\pi\colon \mathbb{C}^k \times \mathbb{C}^n \to \mathbb{C}^n$. Consider the arrangement in $\mathbb{C}^k \times \mathbb{C}^n$ of hyperplanes H_{ij}, $1 \leqslant i < j \leqslant k$; H_i^j, $i = 1, \ldots, k$, $j = 1, \ldots, n$; H_j^0, $j = 1, \ldots, n$; H^{ij}, $1 \leqslant i < j \leqslant n$; H^{0j}, $j = 1, \ldots, n$, defined respectively by the equations $t_i = t_j$, $t_i = z_j$, $t_j = 0$, $z_i = z_j$, $z_j = 0$. Denote by $U^0 \subset \mathbb{C}^k \times \mathbb{C}^n$ the complement to the arrangement. The master function $\Phi^{\mathrm{tr}}(t, z, \xi, m)$ is a well-defined multivalued function on U^0.

Fix $\kappa \in \mathbb{C}^*$. The univalued branches of the function $e^{\Phi_{k,n}(t,z,\xi,m)/\kappa}$ over open subsets of U^0 define a one-dimensional local system on U^0 denoted by $\mathcal{L}^0_\kappa$.

The map $\pi^0|_{U^0}\colon U^0 \to U^0_n$ is a locally trivial bundle with fiber

$$U^0_{k,n}(z) := \{t \in \mathbb{C}^k \mid t_i \neq 0 \text{ for all } i,\ t_i \neq t_j \text{ for } i \neq j,\ t_i \neq z_j \text{ for all } i, j\}.$$

Consider the associated homological vector bundle

$$\amalg_{z \in U^0_n} H_k\big(U^0_{k,n}(z), \mathcal{L}^0_\kappa|_{U^0_{k,n}(z)}\big) \to U^0_n. \tag{11.6.1}$$

This bundle has a canonical flat Gauss–Manin connection.

Theorem 11.6.1 (Markov and Varchenko, 2002) *Fix $\xi \in \mathfrak{h}$. For any horizontal section $\gamma(z) \in H_k(U^0_{k,n}(z), \mathcal{L}^0_\kappa(\xi)|_{U^0_{k,n}(z)})$ of the homological bundle* (11.6.1), *the $M^{\otimes m}[|m| - 2k]$-valued function*

$$\varphi^{(\gamma)}(z) := \int_{\gamma(z)} e^{\Phi^{\mathrm{tr}}_{k,n}(t,z,\xi,m)/\kappa} \omega_{k,n}(t, z)\, dt_1 \wedge \cdots \wedge dt_k, \tag{11.6.2}$$

where $\omega_{k,n}(t, z)$ is the weight function (11.4.1), *is a solution of the trigonometric KZ equation* (11.5.2) *with parameter ξ.*

For any z and ξ we have a canonical isomorphism

$$H_k\big(U^0_{k,n}(z), \mathcal{L}_\kappa(\xi)|_{U^0_{k,n}(z)}\big) \sim H_k\big(U^0_{k,n}(z), \mathcal{L}_\kappa(\xi + \kappa h/2)|_{U^0_{k,n}(z)}\big).$$

Indeed, under the shift $\xi \mapsto \xi + \kappa h/2$, the function $e^{\Phi(t,z,\xi,m)/\kappa}$ is multiplied by a function of the form $c(z) \prod_{i=1}^k t_i^{-1}$, which is a univalued function on $U^0_{k,n}(z)$. Hence, by choosing a horizontal section $\gamma_\xi(z) \in H_k(U^0_{k,n}(z), \mathcal{L}^-_\kappa(\xi)|_{U^0_{k,n}(z)})$ for one particular ξ, we obtain a hypergeometric solution $\varphi^{(\gamma)}(z, \xi + l\kappa h/2)$ of the trigonometric *KZ* equation with parameter $\xi + l\kappa/2$ for any integer l.

Theorem 11.6.2 (Markov and Varchenko, 2002) *Let $\varphi^{(\gamma)}(z, \xi + l\kappa h/2)$, $l \in \mathbb{Z}$ be the sequence of hypergeometric solutions defined as above. Then the functions of the sequence satisfy the difference dynamical equation, $\varphi^{(\gamma)}(z, \xi + (l+1)\kappa h/2) = D^{\mathrm{tr}}(z, \xi + l\kappa h/2)\varphi^{(\gamma)}(z, \xi + l\kappa h/2)$ (for $l \in \mathbb{Z}$).*

11.6.2 Solutions with Values in $L^{\otimes m}$

Let $m = (m_1, \ldots, m_n)$ be a vector with nonnegative integer coordinates, $L^{\otimes m}$ the corresponding tensor product of irreducible finite-dimensional $\mathfrak{sl}_2$-modules, and $\pi\colon M^{\otimes m} \to L^{\otimes m}$ the canonical projection.

Proposition 11.6.3 *Let φ be a solution of the trigonometric KZ or difference dynamical equations with values in $M^{\otimes m}$. Then $\pi(\varphi)$ is a solution of the trigonometric KZ or difference dynamical equations with values in $L^{\otimes m}$.*

11.6.3 Bethe Ansatz

For fixed z and $\xi = \lambda h/2$, the critical points of the trigonometric master function are defined by the equations $\frac{\partial \Phi^{\mathrm{tr}}_{k,n}}{\partial t_i} = 0$, $i = 1, \ldots, k$, that is,

$$\frac{1 + |m| - 2k - \lambda}{t_i} - \sum_{j=1}^{n} \frac{m_j}{t_i - z_j} + \sum_{j, j \neq i} \frac{2}{t_i - t_j} = 0, \quad i = 1, \ldots, k.$$

Theorem 11.6.4 (Markov and Varchenko, 2002) *For any critical point $t^0 = (t^0_1, \ldots, t^0_k)$, the vector $\omega_{k,n}(t^0, z) \in M^{\otimes m}[|m| - 2k]$ is nonzero and is an eigenvector of the trigonometric Gaudin Hamiltonians,*

$$K^{\mathrm{tr}}_i(z, \xi)\omega_{k,n}(t^0, z) = z_i \frac{\partial \Phi^{\mathrm{tr}}_{k,n}}{\partial z_i}(t^0, z, \xi)\omega_{k,n}(t^0, z), \quad i = 1, \ldots, n,$$

and the difference dynamical Hamiltonian,

$$D^{\mathrm{tr}}(z, \xi)\omega_{k,n}(t^0, z) = \exp\Big(\frac{\partial \Phi_{k,n}}{\partial \lambda}(t^0, z, \xi)\Big)\omega_{k,n}(t^0, z).$$

11.7 Knizhnik–Zamolodchikov–Bernard Equation and Elliptic Hypergeometric Functions

11.7.1 Knizhnik–Zamolodchikov–Bernard (KZB) Equation

In the *WZW* model of conformal field theory associated with $\mathfrak{sl}_2$, one defines a holomorphic vector bundle (of conformal blocks) on the moduli space of smooth complex compact curves with marked points labeled by representations of $\mathfrak{sl}_2$. This vector bundle comes with a projectively flat connection, often called the *KZB* (or *KZ*) *connection* (Belavin et al., 1984; Tsuchiya et al., 1989). For curves of genus 0 and 1 this connection is flat. The horizontal sections of the bundle admit integral representations (Schechtman and Varchenko, 1991a; Felder and Varchenko, 1995). For genus 0 curves, the equations for horizontal sections are the rational *KZ* equations with $\xi = 0$ and the integral representations for solutions are given in §11.4.3. In this section we will discuss the genus 1 case. The solutions of the *KZB* equations in this case are called the *multidimensional elliptic hypergeometric functions*.

For genus 1, the *KZB* equation is the following system of equations (Felder and Wieczerkowski, 1996) (with $\mathrm{i} := \sqrt{-1}$):

$$\kappa\frac{\partial u}{\partial z_i} = -h^{(i)}\frac{\partial u}{\partial \lambda} + \sum_{j\neq i} r^{(ij)}(z_i - z_j, \lambda, \tau)u, \quad i = 1, \ldots, n, \tag{11.7.1a}$$

$$2\pi \mathrm{i}\kappa\frac{\partial u}{\partial \tau} = \frac{\partial^2 u}{\partial \lambda^2} + \sum_{i,j} s^{(ij)}(z_i - z_j, \lambda, \tau)u, \tag{11.7.1b}$$

for the unknown function $u(z_1, \ldots, z_n, \lambda, \tau)$ taking values in the zero weight subspace $V[0]$ of a tensor product $V = V_1 \otimes \cdots \otimes V_n$ of $\mathfrak{sl}_2$-modules, $V[0] = \{v \in V \mid hv = 0\}$. The arguments $z_1, \ldots, z_n$, λ, and τ are complex numbers with τ in the upper half-plane H_+, $\operatorname{Im}\tau > 0$, and $r(z, \lambda, \tau)$ and $s(z, \lambda, \tau)$ are some explicitly defined functions with values in $\mathfrak{sl}_2 \otimes \mathfrak{sl}_2$; see the definition of these functions in papers by Felder and Wieczerkowski (1996) and Felder and Varchenko (1995).

The argument $\tau \in H_+$ is the modular parameter of the elliptic curve $\mathbb{C}/(\mathbb{Z} + \tau\mathbb{Z})$. The arguments $z_1, \ldots, z_n$ correspond to points on the elliptic curve. The argument λ corresponds to a point in the Cartan subalgebra $\mathfrak{h} \subset \mathfrak{sl}_2$. Equations (11.7.1a) describe changes of conformal blocks with respect to changes of positions of points. Equation (11.7.1b) describes the deformation of conformal blocks with respect to changes of the modular parameter τ. That equation is called the *KZB heat equation*.

11.7.2 The Case $n = 1$, $V = L_{2p}$

Let $p \in \mathbb{Z}_{\geqslant 0}$ and let $V = L_{2p}$ be the $(2p + 1)$-dimensional irreducible $\mathfrak{sl}_2$-module. Since $V[0] = \mathbb{C} \cdot f^p v_{2p}$, after identification of $V[0]$ with $\mathbb{C}$ the unknown function $u(z_1, \lambda, \tau)$ becomes a scalar. Equation (11.7.1a) says that u does not depend on z_1. The *KZB* heat equation 11.7.1b takes the form

$$2\pi \mathrm{i}\kappa\frac{\partial u}{\partial \tau} = \frac{\partial^2 u}{\partial \lambda^2} + p(p+1)\alpha(\lambda, \tau)u, \tag{11.7.2}$$

where $\alpha(\lambda, \tau)$ is defined in terms of the first *Jacobi theta function* (Whittaker and Watson, 1927),

$$\theta(\lambda,\tau) := -\sum_{j\in\mathbb{Z}} e^{\pi i(j+1/2)^2\tau+2\pi i(j+1/2)(\lambda+1/2)}, \quad \alpha(\lambda,\tau) := \frac{\partial^2}{\partial\lambda^2}(\log\theta(\lambda,\tau)).$$

Note that $\alpha(\lambda,\tau) = -\wp(\lambda,\tau) + c(\tau)$, where $\wp(z,\tau)$ is the *Weierstrass* $\wp$*-function* and $c(\tau)$ is a suitable function of τ. Note also that

$$\frac{\partial^2}{\partial\lambda^2} - p(p+1)\wp(\lambda,\tau) \tag{11.7.3}$$

is the Hamiltonian of the elliptic Calogero–Moser two-body system.

A remarkable fact about all forms of *KZ*-type equations is that their solutions have integral representations. In particular, the integral representation for solutions of (11.7.2) is as follows. Introduce special functions

$$\sigma_\lambda(t,\tau) := \frac{\theta(\lambda-t,\tau)\theta'(0,\tau)}{\theta(\lambda,\tau)\theta(t,\tau)}, \quad E(t,\tau) := \frac{\theta(t,\tau)}{\theta'(0,\tau)}, \quad \theta'(0,\tau) := \frac{\partial\theta}{\partial\lambda}(0,\tau).$$

Define the *elliptic master function* by the formula

$$\Phi_p(t_1,\ldots,t_p,\tau) := -2p\sum_{j=1}^{p}\log E(t_j,\tau) + 2\sum_{1\leqslant i<j\leqslant p}\log E(t_i-t_j,\tau).$$

Let $g(\lambda,\tau)$ be a holomorphic solution of the equation $2\pi i\kappa\frac{\partial g}{\partial\tau} = \frac{\partial^2 g}{\partial\lambda^2}$. For instance, for $\mu\in\mathbb{C}$, $g(\lambda,\tau) = \exp(\lambda\mu + \mu^2\tau/(2\pi i\kappa))$. Let $\Delta_p \subset \mathbb{C}^p$ be the simplex

$$\Delta_p = \{(t_1,t_2,\ldots,t_p)\in\mathbb{R}^p\subset\mathbb{C}^p \mid 0\leqslant t_p\leqslant t_{p-1}\leqslant\cdots\leqslant t_1\leqslant 1\}. \tag{11.7.4}$$

Consider the integral

$$u(\lambda,\tau) = \int_{\Delta_p} e^{\Phi_p(t_1,\ldots,t_p,\tau)/\kappa} g\Big(\lambda + 2\kappa^{-1}\sum_{j=1}^{p} t_j,\tau\Big)\prod_{j=1}^{p}\sigma_\lambda(t_j,\tau)\,dt_1\wedge\cdots\wedge dt_p.$$

The integral is understood in the sense of analytic continuation; see Felder et al. (2003b, §3).

Theorem 11.7.1 (Felder and Varchenko, 1995) *The function* $u(\lambda,\tau)$ *is a solution of the KZB heat equation* (11.7.2).

11.7.3 Bethe Ansatz

Let $\mu\in\mathbb{C}$. Suppose that $t^0 = (t_1^0,\ldots,t_p^0)$ satisfy for $i = 1,\ldots,p$ the critical point equations $\frac{\partial\Phi_p}{\partial t_i}(t^0,\tau) + 2\mu = 0$. Then $\varphi(\lambda,\tau) := e^{\mu\lambda}\prod_{j=1}^{p}\sigma_\lambda(t_j^0,\tau)$ is an eigenfunction of the elliptic Calogero–Moser Hamiltonian (11.7.3):

$$\left(\frac{\partial^2}{\partial\lambda^2} + p(p+1)\alpha(\lambda,\tau)\right)\varphi(\lambda,\tau) = \left(\mu^2 + 2\pi i\frac{\partial\Phi_p}{\partial\tau}(t^0,\tau)\right)\varphi(\lambda,\tau).$$

These Bethe ansatz formulae are equivalent to Hermite's 1872 solution of the Lamé equation; see Whittaker and Watson (1927, §23–71).

11.7.4 Elliptic Hypergeometric Functions Associated with One Marked Point

The *KZB* heat equation (11.7.2) is a linear partial differential equation. Its space of solutions is infinite-dimensional. If $\kappa \in \mathbb{Z}_{>0}$, $p \in \mathbb{Z}_{\geqslant 0}$, and $\kappa \geqslant 2p+2$, then the space of solutions has an important finite-dimensional subspace.

Holomorphic solutions of the *KZB* heat equation (11.7.1b) with the properties

(i) $u(\lambda+2,\tau) = u(\lambda,\tau)$,

(ii) $u(\lambda+2\tau,\tau) = e^{-2\pi i\kappa(\lambda+\tau)}u(\lambda,\tau)$,

(iii) $u(-\lambda,\tau) = (-1)^{p+1}u(\lambda,\tau)$,

(iv) $u(\lambda,\tau) = O((\lambda-m-n\tau)^{p+1})$ as $\lambda \to m+n\tau$ for any $m,n \in \mathbb{Z}$,

are called *conformal blocks* associated with the family of elliptic curves $\mathbb{C}/(\mathbb{Z}+\tau\mathbb{Z})$ with the marked point $z=0$ and the $(2p+1)$-dimensional irreducible $\mathfrak{sl}_2$-module L_{2p}. It is known that the space of conformal blocks has dimension $\kappa-2p-1$. The functions from the space of conformal blocks are called *elliptic hypergeometric functions*.

11.7.5 Integral Representations for Elliptic Hypergeometric Functions

For $\kappa \geqslant 2$ and $n \in \mathbb{Z}$, let $\theta_{\kappa,n}(\lambda,\tau) := \sum_{j\in\mathbb{Z}} \exp\left(2\pi i\kappa(j+n/(2\kappa))^2\tau + 2\pi i\kappa(j+n/(2\kappa))\lambda\right)$ be the theta function of level κ. We have $\theta_{\kappa,n+2\kappa}(\lambda,\tau) = \theta_{\kappa,n}(\lambda,\tau)$.

Let $\Delta_k \subset \mathbb{R}^k \subset \mathbb{C}^k$ be the k-dimensional simplex defined in (11.7.4) and

$$\Delta_{k,p-k} := \{(t_1,t_2,\ldots,t_p) \in \mathbb{C}^p \mid (t_1,\ldots,t_k) \in \Delta_k,\ (t_{k+1}/\tau,\ldots,t_p/\tau) \in \Delta_{p-k}\}.$$

For $0 \leqslant k \leqslant p$, define

$$J^{[k]}_{\kappa,n}(\lambda,\tau) := \int_{\Delta_{k,p-k}} \Phi_\kappa^{1/\kappa}(t_1,\ldots,t_p,\tau)\theta_{\kappa,n}\Big(\lambda+\tfrac{2}{\kappa}\sum_{j=1}^{p} t_j,\tau\Big)\prod_{j=1}^{p}\sigma_\lambda(t_j,\tau)\,dt_j,$$

$$u^{[k]}_{\kappa,n}(\lambda,\tau) := J^{[k]}_{\kappa,n}(\lambda,\tau) + (-1)^{p+1}J^{[k]}_{\kappa,n}(-\lambda,\tau).$$

Theorem 11.7.2 (Felder and Varchenko, 1995) *For $0 \leqslant k \leqslant p$ and any n, the functions $u^{[k]}_{\kappa,n}(\lambda,\tau)$ are solutions of the KZB heat equation* (11.7.2) *and have properties* (i)–(iv).

Theorem 11.7.3 (Felder et al., 2003b) *The set $\{u^{[p]}_{\kappa,n}(\lambda,\tau) \mid p+1 \leqslant n \leqslant \kappa-p-1\}$ is a basis of the space of conformal blocks.*

11.7.6 Elliptic Selberg Integrals

For $\kappa = 2p+2$ the space of conformal blocks is one-dimensional. Furthermore, $u^{[p]}_{\kappa,n} = 0$ unless $n = p+1 \pmod{\kappa}$. The function $u^{[p]}_{2p+2,p+1}(\lambda,\tau)$, given by a p-dimensional integral, can be calculated explicitly. Let $S_p(a,b,c)$ denote the classical p-dimensional Selberg integral (11.1.3) and put

$$C_{p,\kappa,n} = (2\pi)^{p(p+1)/\kappa} e^{-\pi i p(3p-1)/2\kappa} e^{\pi i(p+1)/2} S_p\Big(\frac{n+1}{\kappa}, -\frac{2p}{\kappa}, \frac{1}{\kappa}\Big) \prod_{j=1}^{p} (1 - e^{2\pi i(n+j)/\kappa}).$$

Theorem 11.7.4 (Felder and Varchenko, 1995; Felder et al., 2003a) *We have* $u^{[p]}_{2p+2,p+1}(\lambda,\tau) = C_{p,2p+2,p+1}\theta(\lambda,\tau)^{p+1}$.

To prove the theorem one checks that $\theta(\lambda,\tau)^{p+1}$ is a solution of the *KZB* heat equation (11.7.2) and has properties (i)–(iv). Thus $u^{[p]}_{2p+2,p+1}(\lambda,\tau)$ is proportional to $\theta(\lambda,\tau)^{p+1}$ and the coefficient of proportionality can be determined by taking the limit $\operatorname{Im}\tau \to \infty$.

There are several other elliptic hypergeometric functions which can be calculated explicitly; see Felder et al. (2003a), Rains et al. (2018) and Chapter 6.

11.7.7 Transformations Acting on the Space of Conformal Blocks

The modular group is the group $M = \mathrm{SL}(2,\mathbb{Z})/(\pm I)$, where I is the identity matrix, with generators $T = \left(\begin{smallmatrix}1 & 1\\ 0 & 1\end{smallmatrix}\right)$, $S = \left(\begin{smallmatrix}0 & 1\\ -1 & 0\end{smallmatrix}\right)$ and defining relations $S^2 = 1$, $(ST)^3 = 1$. It acts on the upper half-plane H_+, $T : \tau \mapsto \tau+1$, $S : \tau \mapsto -1/\tau$. Elliptic curves corresponding to parameters $\tau, \tau+1, -1/\tau$ are isomorphic, and the conformal blocks associated to these curves are related.

Introduce four transformations A, B, T, and S acting on functions of λ and τ,

$$\begin{aligned} (Au)(\lambda,\tau) &:= u(\lambda+1,\tau), \qquad & (Bu)(\lambda,\tau) &:= e^{\pi i\kappa(\lambda+\tau/2)} u(\lambda+\tau,\tau), \\ (Tu)(\lambda,\tau) &:= u(\lambda,\tau+1), \qquad & (Su)(\lambda,\tau) &:= e^{-\pi i\kappa\lambda^2/2\tau} \tau^{-1/2-p(p+1)/\kappa} u(\lambda/\tau, -1/\tau), \end{aligned} \tag{11.7.5}$$

where we fix $\arg\tau \in (0,\pi)$.

If $u(\lambda,\tau)$ is a solution of the *KZB* heat equation (11.7.2), then Au, Bu, Tu, and Su are solutions of (11.7.2) too. Moreover, the transformations A, B, T and S preserve the properties (i)–(iv); see Etingof and Kirillov (1995) and Felder et al. (2003a).

The restrictions of the transformations A, B, T and S to the space of conformal blocks satisfy the relations

$$\begin{aligned} &S^2 = (-1)^p i e^{-\pi i p(p+1)/\kappa} I, \quad (ST)^3 = (-1)^p i e^{-\pi i p(p+1)/\kappa} I, \\ &A^2 = I, \quad B^2 = I, \quad AB = (-1)^\kappa BA, \\ &SAS^{-1} = B, \quad SBS^{-1} = A, \quad TAT^{-1} = A, \quad TBT^{-1} = i^\kappa BA, \end{aligned}$$

where I denotes the identity transformation (Felder et al., 2003a). In particular, the transformations S and T define a projective representation of the modular group on the space of conformal blocks.

The transformations A, B, T, and S act on the basis $\{u^{[p]}_{\kappa,n}(\lambda,\tau) \mid p+1 \leqslant n \leqslant \kappa-p-1\}$ of the space of conformal blocks as follows.

Theorem 11.7.5 (Felder et al., 2003b) *We have*

$$Au^{[p]}_{\kappa,n} = (-1)^n u^{[p]}_{\kappa,n}, \quad Bu^{[p]}_{\kappa,n} = -e^{2\pi \mathrm{i} pn/\kappa} u^{[p]}_{\kappa,\kappa-n}, \quad Tu^{[p]}_{\kappa,n} = q^{n^2/2} u^{[p]}_{\kappa,n}, \quad Su^{[p]}_{\kappa,n} = \sum_{m=p+1}^{\kappa-p-1} s_{m,n} u^{[p]}_{\kappa,m},$$

where $q = e^{\pi \mathrm{i}/\kappa}$, $s_{m,n} = e^{-\pi \mathrm{i}/4}(2\kappa)^{-1/2} q^{-(p+1)m}(1-q^{2m})(\prod_{j=1}^{p}(1-q^{2(n-j)}))P^{(p+1)}_{n-p-1}(q^m)$, *and* $P^{(p+1)}_{n-p-1}(x)$ *is the Macdonald polynomial of type* A_1.

Recall that the Macdonald polynomials $P^{(k)}_n(x)$ of type A_1 are Laurent polynomials, depending on two parameters k and n which are nonnegative integers. They are defined by the following conditions:
1. $P^{(k)}_0(x) = 1$;
2. $P^{(k)}_n(x) = x^n + \cdots + x^{-n}$ for $n \geqslant 1$, where the dots are the terms of the intermediate degrees;
3. $P^{(k)}_n(x) = P^{(k)}_n(x^{-1})$;
4. $\langle P^{(k)}_m, P^{(k)}_n\rangle = 0$ for $m \neq n$, where $\langle f, g\rangle$ is the constant term of the Laurent polynomial $f(x)g(x)\prod_{j=0}^{k-1}(1-q^{2j}x^2)(1-q^{2j}x^{-2})$.

The Macdonald polynomials of type A_1 are the continuous q-ultraspherical polynomials; see §9.3.7.

11.8 *qKZ* Equation

The *quantized Knizhnik–Zamolodchikov* (*qKZ*) equations, a.k.a. the *difference KZ equations*, were introduced by Smirnov (1992) and Frenkel and Reshetikhin (1992). They are defined by using the representation theory of Yangians or quantum affine algebras. Here we describe an example associated with the Yangian $Y(\mathfrak{sl}_2)$ in terms of representation theory of the Lie algebra $\mathfrak{sl}_2$.

For every pair M_{m_1}, M_{m_2} of $\mathfrak{sl}_2$ Verma modules and a generic complex number u, there is a unique linear operator $R_{M_{m_1}M_{m_2}}(u) \in \mathrm{End}(M_{m_1} \otimes M_{m_2})$ such that

$$R_{M_{m_1}M_{m_2}}(u)v_{m_1} \otimes v_{m_2} = v_{m_1} \otimes v_{m_2}, \quad [R_{M_{m_1}M_{m_2}}(u), f \otimes 1 + 1 \otimes f] = 0,$$
$$R_{M_{m_1}M_{m_2}}(u)(2uf \otimes 1 + f \otimes h - h \otimes f) = (2uf \otimes 1 - f \otimes h + h \otimes f)R_{M_{m_1}M_{m_2}}(u).$$

The operator $R_{M_{m_1}M_{m_2}}(u)$ preserves the weight decomposition of $M_{m_1} \otimes M_{m_2}$, and the restriction of $R_{M_{m_1}M_{m_2}}(u)$ to any weight subspace defines an operator-valued rational function of u. The operator $R_{M_{m_1}M_{m_2}}(u)$ is called the *rational R-matrix* for the tensor product $M_{m_1} \otimes M_{m_2}$.

The operator $R_{M_{m_1}M_{m_2}}(u)$ commutes with the $\mathfrak{sl}_2$-action on $M_{m_1} \otimes M_{m_2}$:

$$[R_{M_{m_1}M_{m_2}}(u), x \otimes 1 + 1 \otimes x] = 0, \quad x \in \mathfrak{sl}_2, \tag{11.8.1}$$

satisfies the inversion relation $(R_{M_{m_1}M_{m_2}}(u))^{-1} = (R_{M_{m_2}M_{m_1}}(-u))^{(21)}$, and the *Yang–Baxter equation*

$$R^{(12)}_{M_{m_1}M_{m_2}}(u-v)R^{(13)}_{M_{m_1}M_{m_3}}(u)R^{(23)}_{M_{m_2}M_{m_3}}(v) = R^{(23)}_{M_{m_2}M_{m_3}}(v)R^{(13)}_{M_{m_1}M_{m_3}}(u)R^{(12)}_{M_{m_1}M_{m_2}}(u-v). \tag{11.8.2}$$

In addition, $R_{M_{m_1}M_{m_2}}(u) = 1 - \frac{1}{2}m_1 m_2 u^{-1} + \Omega_{M_{m_1}M_{m_2}} u^{-1} + O(u^{-2})$ $(u \to \infty)$, where $\Omega_{M_{m_1}M_{m_2}}$ is the image of the Casimir element (11.2.1) in $\mathrm{End}(M_{m_1} \otimes M_{m_2})$.

The operator $R_{M_{m_1}M_{m_2}}(u)$ is compatible with the canonical projection to the tensor product of irreducible $\mathfrak{sl}_2$-modules, $M_{m_1} \otimes M_{m_2} \to L_{m_1} \otimes L_{m_2}$. We denote by $R_{L_{m_1}L_{m_2}}(u)$ the image of $R_{M_{m_1}M_{m_2}}(u)$ in $\mathrm{End}(L_{m_1} \otimes L_{m_2})$. For example, let $m_1 = m_2 = 1$. Then the $\mathfrak{sl}_2$-module L_1 is spanned by the vectors v_1, fv_1, and the operator $R_{L_1L_1}(u)$ acts as follows:

$$R_{L_1L_1}(u) v_1 \otimes v_1 = v_1 \otimes v_1, \qquad R_{L_1L_1}(u) fv_1 \otimes v_1 = \frac{u}{u+1} fv_1 \otimes v_1 + \frac{1}{u+1} v_1 \otimes fv_1,$$

$$R_{L_1L_1}(u) fv_1 \otimes fv_1 = fv_1 \otimes fv_1, \quad R_{L_1L_1}(u) v_1 \otimes fv_1 = \frac{u}{u+1} v_1 \otimes fv_1 + \frac{1}{u+1} fv_1 \otimes v_1.$$

Given $\xi \in \mathfrak{h}$, $z = (z_1, \dots, z_n) \in \mathbb{C}^n$ such that $z_i \neq z_j$ for $i \neq j$, $m = (m_1, \dots, m_n) \in \mathbb{C}^n$, define the *qKZ operators* $\widehat{K}_1(z, \xi), \dots, \widehat{K}_n(z, \xi) \in \mathrm{End}(M^{\otimes m})$ by the formula

$$\widehat{K}_i(z, \xi) = e^{\xi^{(1)} + \dots + \xi^{(i)}} \overrightarrow{\prod_{1 \leqslant j \leqslant i}} \overleftarrow{\prod_{i < k \leqslant n}} R^{(jk)}_{M_{m_j}M_{m_k}}(z_j - z_k), \tag{11.8.3}$$

where the order of factors in the product is such that j is increasing and k is decreasing from left to right. On replacing $R^{(jk)}_{M_{m_j}M_{m_k}}(z_i - z_k)$ by $R^{(jk)}_{L_{m_j}L_{m_k}}(z_i - z_k)$ in (11.8.3), one defines the *qKZ* operators for the tensor product $L^{\otimes m}$ of irreducible $\mathfrak{sl}_2$-modules.

The next proposition follows from formulae (11.8.1), (11.8.2).

Proposition 11.8.1 (Frenkel and Reshetikhin, 1992; Smirnov, 1992) *The qKZ operators commute,* $[\widehat{K}_i(z, \xi), \widehat{K}_j(z, \xi)] = 0$ *for all* $i, j = 1, \dots, n$. *In addition,* $[\widehat{K}_i(z, \xi), h^{(1)} + \dots + h^{(n)}] = 0$ *and* $[\widehat{K}_i(z, 0), x^{(1)} + \dots + x^{(n)}] = 0$ *for all* $i = 1, \dots, n$ *and all* $x \in \mathfrak{sl}_2$.

Let $V = V_1 \otimes \dots \otimes V_n$ be a tensor product of $\mathfrak{sl}_2$-modules. By Proposition 11.8.1, for any $\xi \in \mathfrak{h}$ the *qKZ* operators preserve weight subspaces $V[k]$, and for $\xi = 0$ the *qKZ* operators preserve subspaces $\mathrm{Sing}\, V[k]$ of singular vectors of given weight.

11.8.1 Rational qKZ Equation

Given $\xi \in \mathfrak{h}$, $\kappa \in \mathbb{C}^*$, and a tensor product V of $\mathfrak{sl}_2$-modules, the *qKZ difference equation* on a V-valued function $\varphi(z_1, \dots, z_n)$ is the system of difference equations

$$\varphi(z_1 + \kappa, \dots, z_i + \kappa, z_{i+1}, \dots, z_n) = \widehat{K}_i(z, \xi)\varphi(z), \quad i = 1, \dots, n. \tag{11.8.4}$$

The *qKZ* equation defines a discrete flat connection on the trivial bundle $V \times \mathbb{C}^n \to \mathbb{C}^n$,

$$[e^{-\kappa(\partial_{z_1} + \dots + \partial_{z_i})} \widehat{K}_i(z, \xi), e^{-\kappa(\partial_{z_1} + \dots + \partial_{z_j})} \widehat{K}_j(z, \xi)] = 0, \quad i, j = 1, \dots, n,$$

called the *qKZ connection*. For any $k \in \mathbb{C}$, the subbundle with fiber $V[k] \subset V$ is invariant with respect to the *qKZ* connection. If $\xi = 0$, then the subbundle with fiber $\mathrm{Sing}\, V[k] \subset V$ is also invariant with respect to the *qKZ* connection. In this case, for any $x \in \mathfrak{sl}_2$ and any solution $\varphi(z)$ of the *qKZ* equation, the function $x\varphi(z)$ is also a solution of the *qKZ* equation.

11.8.2 Dynamical Equation

Let $\xi = \lambda h/2$. The equation

$$\kappa \frac{\partial \varphi}{\partial \lambda} = \left(\sum_{i=1}^{n} \frac{z_i h^{(i)} + f^{(i)} e^{(i)}}{2} + \sum_{1 \leqslant i < j \leqslant n} e^{(i)} f^{(j)} + \frac{fe}{e^\lambda - 1} \right) \varphi \tag{11.8.5}$$

is called the *trigonometric dynamical differential equation*. Recall that $x \in \mathfrak{sl}_2$ acts on $V = V_1 \otimes \cdots \otimes V_n$ as $x^{(1)} + \cdots + x^{(n)}$. The element

$$\widehat{D}(z, \xi) := \sum_{i=1}^{n} \frac{z_i h^{(i)} + f^{(i)} e^{(i)}}{2} + \sum_{1 \leqslant i < j \leqslant n} e^{(i)} f^{(j)} + \frac{fe}{e^\lambda - 1}$$

is called the *trigonometric dynamical Hamiltonian*. It preserves the weight decomposition of V and it commutes with the qKZ operators:

$$[\widehat{D}(z, \xi), h^{(1)} + \cdots + h^{(n)}] = 0, \quad [\widehat{D}(z, \xi), \widehat{K}_i(z, \xi)] = 0 \quad (i = 1, \ldots, n).$$

Proposition 11.8.2 (Tarasov and Varchenko, 2005a) *The trigonometric dynamical and qKZ equations are compatible:*

$$\left[e^{-\kappa(\partial_{z_1} + \cdots + \partial_{z_i})} \widehat{K}_i(z, \xi), \kappa \frac{\partial}{\partial \lambda} - \widehat{D}(z, \xi) \right] = 0, \quad i = 1, \ldots, n.$$

11.8.3 $(\mathfrak{sl}_2, \mathfrak{sl}_2)$ Duality

The qKZ operators and the trigonometric dynamical Hamiltonian correspond respectively to the difference dynamical Hamiltonian (11.5.3) and the trigonometric KZ operators (11.5.1), by the $(\mathfrak{sl}_2, \mathfrak{sl}_2)$ duality described in §11.3.5.

Let $m_1, m_2, \widetilde{m}_1, \widetilde{m}_2$ be nonnegative integers such that $m_1 + m_2 = \widetilde{m}_1 + \widetilde{m}_2$. Consider the tensor products of irreducible $\mathfrak{sl}_2$-modules $L^{\otimes m} = L_{m_1} \otimes L_{m_2}$ and $L^{\otimes \widetilde{m}} = L_{\widetilde{m}_1} \otimes L_{\widetilde{m}_2}$. The isomorphism $\psi \colon L^{\otimes m}[\widetilde{m}_1 - \widetilde{m}_2] \to L^{\otimes \widetilde{m}}[m_1 - m_2]$ of weight subspaces is given by formula (11.3.6). Let $z = (z_1, z_2)$, $\tilde{z} = (e^{\lambda/2}, e^{-\lambda/2})$, $\xi = \lambda h/2$, $\tilde{\xi} = (z_1 - z_2)h/2$.

Proposition 11.8.3 (Tarasov and Varchenko, 2002) *We have*

$$D^{\mathrm{tr}}(\tilde{z}, \tilde{\xi})|_{L^{\otimes \widetilde{m}}[m_1 - m_2]} \psi = e^{-\lambda(\widetilde{m}_1 - \widetilde{m}_2)/4} \left(\prod_{i=0}^{m_1} \frac{z_1 - z_2 + m_2 - i}{z_1 - z_2 - i} \right) \psi \widehat{K}_1(z, \xi)|_{L^{\otimes m}[\widetilde{m}_1 - \widetilde{m}_2]},$$

$$K_1^{\mathrm{tr}}(\tilde{z}, \tilde{\xi})|_{L^{\otimes \widetilde{m}}[m_1 - m_2]} \psi = \psi \left(\widehat{D}(z, \xi) - \frac{(\widetilde{m}_1 + 2)\widetilde{m}_2}{2(e^\lambda - 1)} + \frac{z_1 m_1 + z_2 m_2 - (z_1 + z_2)\widetilde{m}_1}{2} \right) \Bigg|_{L^{\otimes m}[\widetilde{m}_1 - \widetilde{m}_2]},$$

$$K_2^{\mathrm{tr}}(\tilde{z}, \tilde{\xi})|_{L^{\otimes \widetilde{m}}[m_1 - m_2]} \psi = \psi \left(-\widehat{D}(z, \xi) + \frac{(\widetilde{m}_1 + 2)\widetilde{m}_2}{2(e^\lambda - 1)} + \frac{z_1 m_1 + z_2 m_2 - (z_1 + z_2)\widetilde{m}_2}{2} \right) \Bigg|_{L^{\otimes m}[\widetilde{m}_1 - \widetilde{m}_2]}.$$

11.9 Hypergeometric Solutions of the *qKZ* Equations and Bethe Ansatz

11.9.1 Master Function and Weight Function

For $\kappa \in \mathbb{C}^*$, $k, n \in \mathbb{Z}_{>0}$, $m = (m_1, \dots, m_n) \in \mathbb{C}^n$, and $\xi = \lambda h/2$, define the *master function*

$$\widehat{\Phi}_{k,n}(t, z, \xi, m, \kappa) := e^{\lambda(\sum_{l=1}^n m_l z_l/2 - \sum_{i=1}^k t_i)/\kappa} \prod_{l=1}^n \prod_{i=1}^k \frac{\Gamma((t_i - z_l - \frac{1}{2}m_l)/\kappa)}{\Gamma((t_i - z_l + \frac{1}{2}m_l)/\kappa)} \prod_{i<j} \frac{\Gamma((t_i - t_j + 1)/\kappa)}{\Gamma((t_i - t_j - 1)/\kappa)}.$$

To any index $J = (j_1, \dots, j_n)$ of nonnegative integers, $|J| = k$, we assign a rational function

$$\widehat{A}_J(t, z, m) := \frac{1}{j_1! \cdots j_n!} \prod_{i<j} \frac{t_i - t_j}{t_i - t_j - 1}$$

$$\times \operatorname{Sym}_t \left[\prod_{s=1}^n \left(\prod_{i=1}^{j_s} \frac{1}{t_{j_1+\dots+j_{s-1}+i} - z_s + m_s/2} \prod_{r=1}^{s-1} \frac{t_{j_1+\dots+j_{s-1}+i} - z_r - m_r/2}{t_{j_1+\dots+j_{s-1}+i} - z_r + m_r/2} \right) \prod_{i<j} \frac{t_i - t_j - 1}{t_i - t_j} \right].$$

The *weight function* is the $M^{\otimes m}[|m| - 2k]$-valued function $\widehat{\omega}_{k,n}(t, z, m) := \sum_{J, |J|=k} \widehat{A}_J(t, z, m) f_J v$.

11.9.2 Solutions of the qKZ Equation on $M^{\otimes m}$

Set

$$\Theta_{k,n}(t, z, \xi, m, \kappa) := \widehat{\Phi}_{k,n}(t, z, \xi, m, \kappa) e^{2\pi \mathrm{i} \sum_{i=1}^k t_i/\kappa}$$

$$\times \prod_{l=1}^n \prod_{i=1}^k \frac{e^{\pi \mathrm{i}(z_l - t_i)/\kappa}}{\sin(\pi(t_i - z_l + m_l/2)/\kappa)} \prod_{i<j} \frac{\sin(\pi(t_i - t_j)/\kappa)}{\sin(\pi(t_i - t_j - 1)/\kappa)},$$

$$\vartheta_i = e^{2\pi \mathrm{i} t_i/\kappa} (i = 1, \dots, k), \quad \zeta_s = e^{2\pi \mathrm{i} z_s/\kappa} (s = 1, \dots, n),$$

$$\mathbb{I}^k = \{(t_1, \dots, t_k) \in \mathbb{C}^k \mid \operatorname{Re} t_i = 0 \ (i = 1, \dots, k)\}.$$

Denote by $\mathbb{C}[\vartheta, \zeta^{\pm 1}]$ the set of polynomials in $\vartheta_1, \dots, \vartheta_k, \zeta_1^{\pm 1}, \dots, \zeta_n^{\pm 1}$.

Theorem 11.9.1 (Tarasov and Varchenko, 1997a) *Assume that* $0 < \operatorname{Im} \lambda < 2\pi$, $\kappa \in \mathbb{R}_{>0}$, *and* $\operatorname{Re} z_s = 0$, $\operatorname{Re} m_s < 0$ *for all* $s = 1, \dots, n$. *Let* $P \in \mathbb{C}[\vartheta, \zeta^{\pm 1}]$ *have degree at most* $n - 1$ *in each of* $\vartheta_1, \dots, \vartheta_k$. *Then the* $M^{\otimes m}[|m| - 2k]$*-valued function*

$$\varphi^{(P)}(z, \xi) := \int_{\mathbb{I}^k} \Theta_{k,n}(t, z, \xi, m, \kappa) \widehat{\omega}_{k,n}(t, z, m) P(\vartheta, \zeta) \, dt_1 \cdots dt_k \qquad (11.9.1)$$

is a solution of the qKZ and trigonometric dynamical equations.

Theorem 11.9.2 (Tarasov and Varchenko, 1997a) *Assume that* $\xi = 0$, $\kappa \in \mathbb{R}_{<0}$ *and* $\operatorname{Re} z_s = 0$, $\operatorname{Re} m_s < 0$ *for all* $s = 1, \dots, n$. *Let* $P \in \mathbb{C}[\vartheta, \zeta^{\pm 1}]$ *have degree at most* $n - 2$ *in each of* $\vartheta_1, \dots,$ ϑ_k. *Then the* $M^{\otimes m}[|m| - 2k]$*-valued function*

$$\varphi^{(P)}(z) := \int_{\mathbb{I}^k} \Theta_{k,n}(t, z, 0, m, \kappa) \widehat{\omega}_{k,n}(t, z, m) P(\vartheta, \zeta) \, dt_1 \cdots dt_k \qquad (11.9.2)$$

takes values in Sing $M^{\otimes m}[|m| - 2k]$ *and is a solution of the qKZ equation for* $\xi = 0$.

For generic κ, all solutions of the qKZ equation are hypergeometric. This follows from-Tarasov and Varchenko (1997a, Theorems 5.14, 5.15).

It is interesting to compare the hypergeometric solutions (11.4.2), (11.4.5) of the KZ equations and the q-hypergeometric solutions (11.9.1), (11.9.2) of the qKZ equations. The hypergeometric solutions are labeled by elements $\gamma(z,\xi)$ of a homology group while the q-hypergeometric solutions are labeled by polynomials $P(\vartheta,\zeta)$ with certain properties. Thus this space of polynomials is a discrete analogue of the corresponding homology group. On this analogy and discrete analogues of homology and cohomology see Smirnov (1992) and Tarasov and Varchenko (1997a,b).

11.9.3 Solutions of the qKZ Equation on $L^{\otimes m}$

Under the assumptions of Theorems 11.9.1 and 11.9.2, consider the integrals

$$I_J^{(P)}(z,\xi,m) := \int_{\mathbb{I}^k} \Theta_{k,n}(t,z,\xi,m,\kappa)\widehat{A}_J(t,z,m)P(\vartheta,\zeta)\,dt_1\cdots dt_k.$$

As functions of z and m, they can be analytically continued to the complement of the set $\{(z,m) \mid m_i + m_j - z_i + z_j - s \in \kappa\mathbb{Z}_{\geqslant 0}$ for all $i, j = 1,\dots,n,\ s = 0,\dots,k-1\}$.

Proposition 11.9.3 *The statements of Theorems 11.9.1 and 11.9.2 hold true for the $M^{\otimes m}[|m|-2k]$-valued function $\varphi^{(P)}(z,\xi) := \sum_{J,|J|=k} I_J^{(P)}(z,\xi,m)f_J v$.*

If some of the numbers $m_1^0,\dots,m_n^0$ are nonnegative integers, define

$$\mathfrak{J}_{m^0} = \{J = (j_1,\dots,j_n) \mid j_i \leqslant m_i^0 \text{ if } m_i^0 \in \mathbb{Z}_{\geqslant 0},\ i = 1,\dots,n\},$$

and let $\mathcal{P}_{m^0}$ be the set of polynomials $P(\vartheta,\zeta)$ symmetric in $\vartheta_1,\dots,\vartheta_k$ and such that $P(q^{m_i^0}\zeta_i, q^{m_i^0-2}\zeta_i,\dots,q^{-m_i^0}\zeta_i,\vartheta_{m_i^0+2},\dots,\vartheta_k,\zeta_1,\dots,\zeta_n) = 0$, $q = e^{\pi i/\kappa}$, whenever $m_i^0 \in \mathbb{Z}_{\geqslant 0}$.

Theorem 11.9.4 (Mukhin and Varchenko, 2000; Tarasov, 1999) *Let $J \in \mathfrak{J}_{m^0}$ and $P \in \mathcal{P}_{m^0}$ satisfy the assumptions of Theorems 11.9.1 and 11.9.2.*

(i) *The integral $I_J^{(P)}(z,\xi,m)$ can be analytically continued to $m = m^0$ for generic z.*

(ii) *The $L^{\otimes m}[|m|-2k]$-valued function $\varphi^{(P)}(z,\xi) := \sum_{J,|J|=k,J\in\mathfrak{J}_{m^0}} I_J^{(P)}(z,\xi,m)f_J v$ is a solution of the qKZ equation. The function $\varphi^{(P)}(z,\xi)$ satisfies the trigonometric dynamical equation for $\xi \neq 0$ and takes values in* Sing $L^{\otimes m}[|m|-2k]$ *for $\xi = 0$.*

11.9.4 The Limit from the qKZ to the KZ Equation

Suppose that $z_i = \tilde{z}_i/\varepsilon$, $i = 1,\dots,n$, and $\xi = \tilde{\xi}/\varepsilon$. Then for $\varepsilon \to 0$,

$$\widehat{K}_i(z,\xi) = 1 + \varepsilon\left(K_i(\tilde{z},\tilde{\xi}) - \sum_{j,j\neq i}\frac{m_i m_i}{2(\tilde{z}_i - \tilde{z}_j)}\right) + O(\varepsilon^2), \quad i = 1,\dots,n,$$

and $\varepsilon\widehat{D}(z,\xi) = D(\tilde{z},\tilde{\xi}) + O(\varepsilon)$, where $K_1,\ldots,K_n$ are the *KZ* operators (11.3.1), and D is the dynamical Hamiltonian (11.3.5). That is, in the limit $\varepsilon \to 0$, the *qKZ* and trigonometric dynamical equations (11.8.4), (11.8.5) turn into the *KZ* and dynamical equations for a function $\tilde{\varphi}(\tilde{z},\tilde{\xi})$, $\varphi(z,\xi) = \tilde{\varphi}(\tilde{z},\tilde{\xi})\prod_{i<j}(\tilde{z}_i - \tilde{z}_j)^{-m_i m_j/2\kappa}$, in the variables $\tilde{z}$ and $\tilde{\lambda} = \lambda/\varepsilon$.

Let $\tilde{t}_i = t_i/\varepsilon$, $i = 1,\ldots,k$. Then for $\varepsilon \to 0$, we have

$$\kappa \log\widehat{\Phi}(t,z,\xi,m,\kappa) - k(|m| - k + 1)(1 + \log(\varepsilon\kappa)) \to \Phi(\tilde{t},\tilde{z},\tilde{\xi},m) - \sum_{i<j}\frac{m_i m_j}{2}\log(\tilde{z}_i - \tilde{z}_j)$$

and $\widehat{\omega}_{k,n}(t,z,m)\,dt_1\cdots dt_k \to \omega_{k,n}(\tilde{t},\tilde{z})\,d\tilde{t}_1\cdots d\tilde{t}_k$.

Suppose that $\kappa,\lambda,z_1,\ldots,z_n,m_1,\ldots,m_n$ satisfy the assumptions of Theorems 11.9.1, 11.9.2, and $\operatorname{Re}\varepsilon = 0$, $\operatorname{Im}\varepsilon > 0$. Then $\tilde{z}_1,\ldots,\tilde{z}_n$ are real. Assume $\tilde{z}_1 < \cdots < \tilde{z}_n$. Given $J = (j_1,\ldots,j_n)$, $|J| = k$, let

$$P_J(\vartheta,\zeta) = \prod_{s=2}^{n}\left(\prod_{i=1}^{j_s}\vartheta^{s-1}_{j_1+\cdots+j_{s-1}+i}\prod_{r=1}^{s}\zeta_r^{-j_s}\right),$$

and $\gamma_J(\tilde{z}) = \{(\tilde{t}_1,\ldots,\tilde{t}_k)\in\mathbb{R}^k \mid \tilde{z}_s < \tilde{t}_{j_1+\cdots+j_{s-1}+1} < \cdots < \tilde{t}_{j_1+\cdots+j_{s-1}+j_s} < \tilde{z}_{s+1},\ s = 1,\ldots,n\}$.

Theorem 11.9.5 (Tarasov and Varchenko, 1997a) *Under the above-mentioned assumptions,*

$$\varepsilon^{-k(|m|-k+1)/\kappa}\varphi^{(P_J)}(z,\xi) \to C_J\varphi^{(\gamma_J)}(\tilde{z},\tilde{\xi})\prod_{r<s}(\tilde{z}_r - \tilde{z}_s)^{-m_r m_s/2\kappa}, \qquad \varepsilon\to 0,$$

where $\varphi^{(P_J)}$, $\varphi^{(\gamma_J)}$ *are respectively given by* (11.9.1), (11.4.2), *and* C_J *is a suitable constant.*

11.9.5 Hypergeometric Solutions and $(\mathfrak{sl}_2,\mathfrak{sl}_2)$ *Duality*

In the notation of §11.8.3, Proposition 11.8.3 identifies the *qKZ* and trigonometric dynamical equations on $L_{m_1}\otimes L_{m_2}[\widetilde{m}_1-\widetilde{m}_2]$ with the difference dynamical and trigonometric *KZ* equations on $L_{\widetilde{m}_1}\otimes L_{\widetilde{m}_2}[m_1-m_2]$. Solutions of the *qKZ* and trigonometric dynamical equations with values in $L_{m_1}\otimes L_{m_2}[\widetilde{m}_1-\widetilde{m}_2]$ are given by $(\widetilde{m}_1-\widetilde{m}_2)$-dimensional integrals; see Theorem 11.9.4. Similarly, solutions of the trigonometric *KZ* and difference dynamical equations with values in $L_{\widetilde{m}_1}\otimes L_{\widetilde{m}_2}[m_1-m_2]$ are given by (m_1-m_2)-dimensional hypergeometric integrals; see Theorems 11.6.1, 11.6.2. By Proposition 11.8.3, one can identify the $(\widetilde{m}_1-\widetilde{m}_2)$-dimensional integrals and the corresponding (m_1-m_2)-dimensional integrals; see Tarasov and Varchenko (2005b). The simplest such identity is the classical relation for the Gauss hypergeometric function ${}_2F_1(a,b;c;x)$:

$$\begin{aligned}{}_2F_1(a,b;c;x) &= \frac{\Gamma(c)}{\Gamma(a)\Gamma(c-a)}\int_0^1 u^{a-1}(1-u)^{c-a-1}(1-ux)^{-b}\,du\\ &= \frac{1}{2\pi\mathrm{i}}\frac{\Gamma(c)}{\Gamma(a)\Gamma(b)}\int_{-\mathrm{i}\infty-\varepsilon}^{+\mathrm{i}\infty-\varepsilon}(-x)^s\frac{\Gamma(-s)\Gamma(s+a)\Gamma(s+b)}{\Gamma(s+c)}\,ds,\end{aligned} \tag{11.9.3}$$

where $\operatorname{Re}c > \operatorname{Re}a > 0$, $\operatorname{Re}b > 0$, $0 < \varepsilon < \min(\operatorname{Re}a,\operatorname{Re}b)$, $-\pi < \arg(-x) < \pi$.

The above-mentioned identities for hypergeometric integrals can be used to find asymptotics of the hypergeometric integrals with respect to dimension; see an example in Tarasov and Varchenko (2005b).

11.9.6 Bethe Ansatz

For fixed z and $\xi = \lambda h/2$, the difference analogue of the critical point equations for the master function is

$$\lim_{\kappa\to 0} \frac{\widehat{\Phi}_{k,n}(t+\kappa^{(i)}, z, \xi, m, \kappa)}{\widehat{\Phi}_{k,n}(t, z, \xi, m, \kappa)} = 0, \quad i = 1, \dots, k,$$

where $\kappa^{(i)} = (0, \dots, 0, \kappa, 0, \dots, 0)$ with κ located at the ith place. Explicitly, the equations are

$$\prod_{l=1}^{n} \frac{t_i - z_l - m_l/2}{t_i - z_l + m_l/2} = e^{\lambda} \prod_{j, j\neq i} \frac{t_i - t_j - 1}{t_i - t_j + 1}, \quad i = 1, \dots, k. \tag{11.9.4}$$

We call a solution $t^0 = (t_1^0, \dots, t_n^0)$ of equations (11.9.4) *admissible* if $t_i^0 \neq z_l \pm m_l$ for all i, l, and $t_i^0 - t_j^0 \neq 0, 1$ for all $i \neq j$.

Equations (11.9.4) are known in the theory of quantum integrable models as the *Bethe ansatz equations*; see for example Korepin et al. (1993). The next theorem follows from the standard Bethe ansatz calculations for the XXX-type model.

Theorem 11.9.6 *For any admissible solution of equations* (11.9.4)*, the vector* $\widehat{\omega}_{k,n}(t^0, z) \in M^{\otimes m}[|m| - 2k]$ *is nonzero and is an eigenvector of the qKZ operators,*

$$\widehat{K}_i(z, \xi)\widehat{\omega}_{k,n}(t^0, z, m) = e^{\lambda(m_1+\cdots+m_i)} \prod_{l=1}^{i}\prod_{j=1}^{k} \frac{t_j - z_l + m_l/2}{t_j - z_l - m_l/2} \widehat{\omega}_{k,n}(t^0, z, m), \quad i = 1, \dots, n.$$

If $\xi \neq 0$ *then* $\widehat{\omega}_{k,n}(t^0, z, m)$ *is an eigenvector of the trigonometric dynamical Hamiltonian,*

$$\widehat{D}(z, \xi)\widehat{\omega}_{k,n}(t^0, z, m) = \left(\sum_{l=1}^{n} \frac{m_l z_l}{2} - \sum_{j=1}^{k} t_j\right)\widehat{\omega}_{k,n}(t^0, z, m).$$

If $\xi = 0$ *then* $\widehat{\omega}_{k,n}(t^0, z, m)$ *is a singular vector.*

Let $G(x) := x \log x$ and

$$\begin{aligned}\Psi_{k,n}(t, z, \xi, m) := \lambda\left(\sum_{l=1}^{n} m_l z_l/2 - \sum_{i=1}^{k} t_i\right) + \sum_{l=1}^{n}\sum_{i=1}^{k}(G(t_i - z_l - m_l/2) - G(t_i - z_l + m_l/2))\\ + \sum_{i<j}(G(t_i - t_j + 1) - G(t_i - t_j - 1)).\end{aligned}$$

The function $\Psi_{k,n}(t, z, \xi, m)$ describes asymptotics of the master function $\widehat{\Phi}_{k,n}(t, z, \xi, m, \kappa)$ as $\kappa \to 0$. By Stirling's formula,

$$\kappa \log \widehat{\Phi}_{k,n}(t, z, \xi, m, \kappa) - k(|m| - k + 1)(1 + \log \kappa) \to \Psi_{k,n}(t, z, \xi, m),$$

if $\kappa > 0$, $|\arg(t_i - z_l \pm m_l/2)| < \pi$ for all i, l, and $|\arg(t_i - t_j \pm 1)| < \pi$ for all $i < j$.

Equations (11.9.4) can be written as $\exp(\frac{\partial\Psi_{k,n}(t,z,\xi,m)}{\partial t_i}) = 1$ $(i = 1, \dots, k)$; see Yang and Yang (1969) and Varchenko and Tarasov (1995). For the eigenvalues of the qKZ operators and the trigonometric dynamical Hamiltonian, one has

$$\exp\Big(\frac{\partial\Psi_{k,n}(t,z,\xi,m)}{\partial z_l}\Big) = e^{\lambda m_l}\prod_{j=1}^{k}\frac{t_j - z_l + m_l/2}{t_j - z_l - m_l/2}, \qquad \frac{\partial\Psi_{k,n}(t,z,\xi,m)}{\partial\lambda} = \sum_{l=1}^{n}\frac{m_l z_l}{2} - \sum_{j=1}^{k} t_j.$$

11.10 One-integration Examples

11.10.1 Solutions of the KZ Equation for $n = 2$, $\xi \neq 0$

Let $m = (m_1, m_2)$ and $\xi = \lambda h/2$. In this subsection a vector v occurring in the ith tensor factor will mean the vector v_{m_i}. Hypergeometric solutions of the KZ equation with values in the weight subspace $M^{\otimes m}[m_1 + m_2 - 2]$ are

$$\varphi^{(\gamma)}(z,\xi) = e^{\lambda(m_1 z_1 + m_2 z_2)/2\kappa}(z_1 - z_2)^{m_1 m_2/2\kappa} \times \int_{\gamma(z,\xi)} e^{-\lambda t/\kappa}(t - z_1)^{-m_1/\kappa}(t - z_2)^{-m_2/\kappa}\Big(\frac{fv\otimes v}{t - z_1} + \frac{v\otimes fv}{t - z_2}\Big)\,dt; \tag{11.10.1}$$

see (11.4.2). They can be expressed in terms of the *confluent hypergeometric function*

$${}_1F_1(a;b;x) = \frac{\Gamma(b)}{\Gamma(a)\Gamma(b-a)}\int_0^1 e^{ux}u^{a-1}(1-u)^{b-a-1}\,du.$$

The KZ and dynamical differential equations (11.3.2) and (11.3.4) are equivalent to the *confluent hypergeometric equation* $x\frac{d^2F}{dx^2} + (b - x)\frac{dF}{dx} - aF = 0$, satisfied by ${}_1F_1(a;b;x)$.

For real z_1 and z_2 such that $z_1 < z_2$, and $\mathrm{Re}(\lambda/\kappa) > 0$, $\mathrm{Re}(m_1/\kappa) < 0$, $\mathrm{Re}(m_2/\kappa) < 0$, convenient choices for $\gamma(z,\xi)$ in (11.10.1) are the intervals $[z_1, z_2]$ and $[z_2, \infty)$.

11.10.2 Solutions of the KZ Equation for $\xi = 0$

Let $n = 3$ and $m = (m_1, m_2, m_3)$. Hypergeometric solutions of the KZ equation with values in $M^{\otimes m}[m_1 + m_2 + m_3 - 2]$ are

$$\varphi^{(\gamma)}(z) = (z_1 - z_2)^{m_1 m_2/2\kappa}(z_1 - z_3)^{m_1 m_3/2\kappa}(z_2 - z_3)^{m_2 m_3/2\kappa} \times \int_{\gamma(z)} (t - z_1)^{-m_1/\kappa}(t - z_2)^{-m_2/\kappa}(t - z_3)^{-m_3/\kappa}\Big(\frac{fv\otimes v\otimes v}{t - z_1} + \frac{v\otimes fv\otimes v}{t - z_2} + \frac{v\otimes v\otimes fv}{t - z_3}\Big)\,dt; \tag{11.10.2}$$

see (11.4.5). The coordinates of this solution can be expressed in terms of the *Gauss hypergeometric function* ${}_2F_1(a,b;c;x)$; see (11.9.3). Since $\varphi^{(\gamma)}$ takes values in $\mathrm{Sing}\, M^{\otimes m}[|m| - 2]$, there is a linear relation between the three coordinates.

The *KZ* differential equation (11.4.3) is equivalent to the *Gauss hypergeometric equation*

$$x(1-x)\frac{d^2F}{dx^2}+(c-(a+b+1)x)\frac{dF}{dx}-abF=0, \tag{11.10.3}$$

satisfied by ${}_2F_1(a,b;c;x)$.

For real z_1, z_2, z_3 such that $z_1<z_2<z_3$, and $\mathrm{Re}(m_i/\kappa)<0$, $i=1,2,3$, convenient choices for $\gamma(z)$ in (11.10.2) are the intervals $[z_1,z_2]$ and $[z_2,z_3]$.

For general n, hypergeometric solutions of the *KZ* equation (11.4.3) with values in the weight subspace $M^{\otimes m}[|m|-2]$ can be expressed in terms of the *Lauricella hypergeometric function* $F_D^{(n-2)}(a,b_1,\dots,b_{n-2},c;x_1,\dots,x_{n-2})$. On the Lauricella function, see Chapter 3.

11.10.3 Solutions of the Trigonometric KZ Equation

Let $n=2$, $m=(m_1,m_2)$, and $\xi=\lambda h/2$. Hypergeometric solutions of the *qKZ* equation with values in $M^{\otimes m}[m_1+m_2-2]$ are

$$\begin{aligned}\varphi^{(\gamma)}(z)&=z_1^{m_1(\lambda-m_2+2)/2\kappa}z_2^{m_2(\lambda-m_1+2)/2\kappa}(z_1-z_2)^{m_1m_2/2\kappa}\\&\times\int_{\gamma(z)}t^{(m_1+m_2-\lambda-1)/\kappa}(t-z_1)^{-m_1/\kappa}(t-z_2)^{-m_2/\kappa}\left(\frac{fv\otimes v}{t-z_1}+\frac{v\otimes fv}{t-z_2}\right)dt;\end{aligned} \tag{11.10.4}$$

see (11.4.5). They can be expressed in terms of the hypergeometric function ${}_2F_1(a,b;c;x)$. The trigonometric *KZ* equation (11.5.2) is equivalent to the hypergeometric equation (11.10.3). The difference dynamical equation (11.5.4) follows from contiguous relations for the function ${}_2F_1(a,b;c;x)$.

For real z_1 and z_2 such that $0<z_1<z_2$, and for $\mathrm{Re}(m_1/\kappa)<0$, $\mathrm{Re}(m_2/\kappa)<0$, $\mathrm{Re}(m_1+m_2-\lambda)>0$, convenient choices for $\gamma(z,\xi)$ in (11.10.4) are the intervals $[0,z_1]$ and $[z_1,z_2]$.

For general n, hypergeometric solutions of the trigonometric *KZ* equation (11.5.2) with values in the weight subspace $M^{\otimes m}[|m|-2]$ can be expressed in terms of the Lauricella hypergeometric function $F_D^{(n-1)}(a,b_1,\dots,b_{n-1},c;x_1,\dots,x_{n-1})$.

11.10.4 Solutions of the qKZ Equation for n = 2

Let $m=(m_1,m_2)$ and $\xi=\lambda h/2$, and suppose that $0\leqslant\mathrm{Im}\,\lambda<2\pi$. Hypergeometric solutions of the *qKZ* equation with values in $M^{\otimes m}[m_1+m_2-2]$ are

$$\begin{aligned}\varphi^{(P)}(z,\xi)&=e^{\left(\pi\mathrm{i}(z_1+z_2)+\frac12\lambda(m_1z_1+m_2z_2)\right)/\kappa}\int_{-\mathrm{i}\infty}^{+\mathrm{i}\infty}e^{-\lambda t/\kappa}\Gamma\left((t-z_1-\tfrac12m_1)/\kappa\right)\Gamma\left((t-z_2-\tfrac12m_2)/\kappa\right)\\&\times\Gamma\left((z_1-t-\tfrac12m_1)/\kappa\right)\Gamma\left((z_2-t-\tfrac12m_2)/\kappa\right)\left(p_0+p_1e^{2\pi\mathrm{i}t/\kappa}\right)\\&\times\left((t-z_2+m_2)fv\otimes v+(t-z_1-m_1)v\otimes fv\right)dt;\end{aligned} \tag{11.10.5}$$

see (11.9.1). Here p_0, p_1 are complex numbers defining the polynomial $P(\vartheta)=p_0+p_1\vartheta$, and $p_1=0$ if $\mathrm{Im}\,\lambda=0$. The integration contour in (11.10.5) is such that the poles of the first two Γ-functions are on the left and the poles of the last two Γ-functions are on the right.

Solutions (11.10.5) can be expressed in terms of the hypergeometric function ${}_2F_1(a,b;c;x)$. The qKZ equation (11.8.4) follows from contiguous relations for ${}_2F_1(a,b;c;x)$, and the trigonometric dynamical equation (11.8.5) is equivalent to the hypergeometric equation (11.10.3).

For general n, hypergeometric solutions of the qKZ equation (11.8.4) with values in the weight subspace $M^{\otimes m}[|m|-2]$ can be expressed in terms of the generalized hypergeometric function ${}_nF_{n-1}(a_1,\ldots,a_n;b_1,\ldots,b_{n-1};x)$. The qKZ equation (11.8.4) follows from contiguous relations for ${}_nF_{n-1}(a_1,\ldots,a_n;b_1,\ldots,b_{n-1};x)$ and the trigonometric dynamical equation (11.8.5) is equivalent to the nth-order differential equation with respect to x satisfied by ${}_nF_{n-1}(a_1,\ldots,a_n;b_1,\ldots,b_{n-1};x)$.

11.11 Selberg-Type Integrals

11.11.1 Selberg Integral

For $n=1$, the weight subspace $M_{m_1}[m_1-2k]$ is one-dimensional. Thus the KZ and dynamical equations (11.3.2) and (11.3.4) become ordinary first-order linear differential equations, which are elementary to solve. However, the integral (11.4.2) for the hypergeometric solution remains nontrivial. Remarkably, it can be evaluated in a closed form. For real $\kappa>0$ and $\operatorname{Re} m_1<0$, it is tantamount to the following multidimensional generalization of the gamma integral:

$$\int_{0\leqslant x_k\leqslant\cdots\leqslant x_1}\prod_{i=1}^{k}x_i^{a-1}e^{-x_i}\prod_{1\leqslant i<j\leqslant k}(x_i-x_j)^{2c}\,dx_1\cdots dx_k=\prod_{r=0}^{k-1}\frac{\Gamma((r+1)c)}{\Gamma(c)}\Gamma(a+rc). \qquad (11.11.1)$$

For the trigonometric KZ equation (11.5.2), the situation is similar. The multidimensional integral (11.6.2) can be evaluated in a closed form, and for real $\kappa>0$ and $\operatorname{Re} m_1<0$, $\operatorname{Re}(m_1-2k-\lambda)>0$, it reduces to the celebrated *Selberg integral*

$$\begin{aligned}S_k(a,b,c)&=\int_{0\leqslant x_k\leqslant\cdots\leqslant x_1\leqslant 1}\prod_{i=1}^{k}x_i^{a-1}(1-x_i)^{b-1}\prod_{1\leqslant i<j\leqslant k}(x_i-x_j)^{2c}\,dx_1\cdots dx_k\\&=\prod_{r=0}^{k-1}\frac{\Gamma((r+1)c)}{\Gamma(c)}\frac{\Gamma(a+rc)\Gamma(b+rc)}{\Gamma(a+b+(2k-r-2)c)},\end{aligned}$$

evaluated in Selberg (1941, 1944); cf. (11.1.3). Moreover, the difference dynamical equation (11.5.4) turns into the difference equation with respect to a, satisfied by $S_k(a,b,c)$. For $k=1$, the Selberg integral becomes the Euler beta integral; see (11.1.2).

The Selberg integral is one of the most remarkable hypergeometric functions; see for instance Aomoto (1987a,b, 1988), Askey (1980), Dotsenko and Fateev (1984, 1985), and Mehta (2004).

For $n = 2$, the hypergeometric solution $\varphi^{(\gamma)}(z)$, see (11.4.5), of the KZ equation (11.4.3) takes values in the one-dimensional subspace $\operatorname{Sing} M^{\otimes m}[|m| - 2k]$, generated by the vector

$$\sum_{j=0}^{k} \frac{(-1)^j}{j!\,(k-j)!} \prod_{i=0}^{j-1} \frac{1}{m_1 - i} \prod_{i=0}^{k-j-1} \frac{1}{m_2 - i} f^j v_{m_1} \otimes f^{k-j} v_{m_2}.$$

The coordinates of $\varphi^{(\gamma)}(z)$ can be evaluated in a closed form. For real m_1, m_2, and κ such that $m_1 < 0$, $m_2 < 0$, and $\kappa > 0$, the evaluation is equivalent to the equality (see Aomoto, 1987a)

$$\int_{0 \leqslant x_k \leqslant \cdots \leqslant x_1 \leqslant 1} \operatorname{Sym}_x(x_1 \cdots x_\ell) \prod_{i=1}^{k} x_i^{a-1}(1-x_i)^{b-1} \prod_{1 \leqslant i < j \leqslant k} (x_i - x_j)^{2c}\, dx_1 \cdots dx_k$$
$$= S_k(a,b,c) \prod_{r=0}^{\ell-1} \frac{a + (k-r-1)c}{a + b + (2k-r-2)c}, \qquad 0 \leqslant \ell \leqslant k.$$

11.11.2 Mellin–Barnes Integrals

Similarly to hypergeometric solutions of the KZ equations, hypergeometric solutions of the qKZ equations $\varphi^{(P)}(z,\xi)$ for $n = 1$, see (11.9.1), and $\varphi^{(P)}(z)$ for $n = 2$, see (11.9.2), can be evaluated in closed form. In both cases, the polynomial $P(\vartheta, \zeta)$ does not depend on $\vartheta_1, \ldots, \vartheta_k$.

The evaluation of multidimensional integrals (11.9.1) and (11.9.2) produces Mellin–Barnes-type generalizations of the Selberg integral (see Tarasov and Varchenko, 1997a):

$$\int_{\mathbb{I}^k} \prod_{i=1}^{k} \left(u^{2x_i}\Gamma(a+x_i)\Gamma(a-x_i)\right) \prod_{1 \leqslant i < j \leqslant k} \frac{\Gamma(x_i - x_j + c)\Gamma(x_i - x_j + c)}{\Gamma(x_i - x_j)\Gamma(x_i - x_j)}\, dx_1 \cdots dx_k$$
$$= (2\pi \mathrm{i})^k k!\,(u + u^{-1})^{-2ka - k(k-1)c} \prod_{r=0}^{k-1} \frac{\Gamma((r+1)c)}{\Gamma(c)} \Gamma(2a + rc), \qquad \operatorname{Re} a, \operatorname{Re} c, \operatorname{Re} u > 0, \tag{11.11.2a}$$

$$\int_{\mathbb{I}^k} \prod_{i=1}^{k} \left(\Gamma(a_1 + x_i)\Gamma(a_2 + x_i)\Gamma(b_1 - x_i)\Gamma(b_2 - x_i)\right)$$
$$\times \prod_{1 \leqslant i < j \leqslant k} \frac{\Gamma(x_i - x_j + c)\Gamma(x_i - x_j + c)}{\Gamma(x_i - x_j)\Gamma(x_i - x_j)}\, dx_1 \cdots dx_k = (2\pi \mathrm{i})^k k!$$
$$\times \prod_{r=0}^{k-1} \frac{\Gamma((r+1)c)}{\Gamma(c)} \Gamma(a_1 + b_1 + rc)\Gamma(a_1 + b_2 + rc) \frac{\Gamma(a_2 + b_1 + rc)\Gamma(a_2 + b_2 + rc)}{\Gamma(a_1 + a_2 + b_1 + b_2 + (2k - r - 2)c)},$$
$$\operatorname{Re} a_1, \operatorname{Re} a_2, \operatorname{Re} b_1, \operatorname{Re} b_2, \operatorname{Re} c > 0, \tag{11.11.2b}$$

where $\mathbb{I}^k = \{(x_1, \ldots, x_k) \in \mathbb{C}^k \mid \operatorname{Re} x_1 = \cdots = \operatorname{Re} x_k = 0\}$. Integral (11.11.2a) corresponds to integral (11.11.1) after replacing a by $a/2$. Integral (11.11.2b) was evaluated in Gustafson (1990).

11.12 Further Development

There are three kinds of *KZ*-type equations: rational, trigonometric and elliptic, depending on what functions serve as coefficients of the equation. All sorts of *KZ*-type equations (*KZ*, *KZB*, *qKZ*, *qKZB* (difference analogues of *KZB* equations), differential and difference dynamical equations, for the rational and trigonometric cases, as well as elliptic *KZB* and *qKZB* equations) can be associated with any simple complex Lie algebra; see for example Schechtman and Varchenko (1991a), Varchenko (1995), Frenkel and Reshetikhin (1992), Felder and Varchenko (1995), Felder et al. (2000), Etingof and Varchenko (2002), Tarasov and Varchenko (2005a), and Toledano Laredo (2011). Moreover, the rational *KZ* and dynamical differential equations, and the trigonometric *KZ* differential equations, as well as their hypergeometric solutions, can be constructed for any Kac–Moody Lie algebra; see Schechtman and Varchenko (1991a), Varchenko (1995), and Felder et al. (2000). Unlike *KZB* and *qKZB* equations, elliptic *KZ* and *qKZ* equations are known only for the Lie algebra $\mathfrak{gl}_N$; see for example Frenkel and Reshetikhin (1992) and Etingof (1994). How to construct elliptic versions of dynamical equations is an interesting open question.

Hypergeometric solutions for elliptic *KZB* differential equations were obtained by Felder and Varchenko (1995) and for rational difference dynamical equations by Markov and Varchenko (2002). Integral representations for solutions of the *qKZ* equations associated with $\mathfrak{gl}_N$ were constructed by Varchenko and Tarasov (1995). Hypergeometric solutions of the rational and trigonometric *qKZ* equations associated with $\mathfrak{sl}_2$ were studied in more detail by Tarasov and Varchenko (1997a,b) and Tarasov (1999). Integral representations for solutions of the trigonometric *qKZ* equations involve q-Gamma functions rather than the usual Gamma functions.

The elliptic *qKZB* equations associated with $\mathfrak{sl}_2$ and their hypergeometric solutions were studied by Felder et al. (1997, 1999) and Felder and Varchenko (1999, 2001, 2002, 2004). In particular, the elliptic Macdonald polynomials of type A_1 were introduced by Felder and Varchenko (2004), in terms of theta-hypergeometric integrals. Etingof and Kirillov (1995) introduced affine Macdonald polynomials as traces of intertwiners of quantum affine algebras and stated rough analogues of Macdonald's conjectures. Sun (2016) established the precise connection between the elliptic Macdonald polynomials of type A_1 and affine Macdonald polynomials for $U_q(\widehat{\mathfrak{sl}_2})$. Rains et al. (2018) refined the statement of the denominator and evaluation conjectures for affine Macdonald polynomials proposed by Etingof and Kirillov (1995) and they proved the first nontrivial cases of these conjectures. These were related to evaluations of certain theta hypergeometric integrals defined by Felder and Varchenko (2004), and the evaluations were performed by well-chosen applications of the elliptic beta integral introduced by Spiridonov (2001).

The monodromy of rational and trigonometric *KZ* equations can be described in terms of the representation theory of quantum groups; see Kohno (1987, 1988), Drinfel'd (1990), and Kazhdan and Lusztig (1993, 1994). The integral representations for solutions of the *KZ* equations provide a topological way to establish this connection; see Schechtman and Varchenko

(1991b), Bezrukavnikov et al. (1998), and Varchenko (1995). A similar description of the monodromy of the rational differential dynamical equations for simple Lie algebras was done by Toledano Laredo (2002) and Millson and Toledano Laredo (2005).

For the *qKZ* equations the monodromy group is replaced by the transition matrices between asymptotic solutions. The transition matrices for the *qKZ* equation considered in §11.8 can be described in terms of the trigonometric R-matrices; see Tarasov and Varchenko (1997a). The simplest example is the formula for the infinite product of 2×2 matrices, obtained by Reshetikhin and Faddeev (1983):

$$\lim_{n\to\infty}(H^n A(u+n)A(u+n-1)\cdots A(u-n)H^n) = \begin{pmatrix} 1 & \dfrac{b\sin(\pi a)}{a\sin(\pi u)} \\ \dfrac{c\sin(\pi a)}{a\sin(\pi u)} & -1 \end{pmatrix},$$

where $H = \left(\begin{smallmatrix} 1 & 0 \\ 0 & -1 \end{smallmatrix}\right)$, $A(u) = \left(\begin{smallmatrix} 1 & b/u \\ c/u & -1 \end{smallmatrix}\right)$, $a^2 = -bc$. The transition matrices for the trigonometric *qKZ* equation associated with the Lie algebra $\mathfrak{sl}_2$ were described by Tarasov and Varchenko (1997b) in terms of elliptic R-matrices.

For Selberg-type integrals associated with more general *KZ* and *qKZ* equations, see Tarasov and Varchenko (2003) and Warnaar (2008, 2009, 2010).

Interrelations of the Bethe ansatz, critical points of the master function, differential and difference equations with polynomial solutions only, and Schubert calculus are discussed by Mukhin and Varchenko (2003, 2004) and Mukhin et al. (2008, 2009a,b,c, 2011, 2012).

The trigonometric dynamical differential equations and the *qKZ* equations appear in the theory of equivariant quantum cohomology of Nakajima varieties; see Braverman et al. (2011), Maulik and Okounkov (2012), and Gorbounov et al. (2013). In particular, the quantum differential equation for the cotangent bundle of a partial flag variety can be identified with the trigonometric dynamical differential equations for the tensor product of vector representations of the Lie algebra $\mathfrak{sl}_N$ and, hence, solved in terms of multidimensional hypergeometric and q-hypergeometric integrals (Tarasov and Varchenko, 2014). The fact that the quantum differential equation can be solved in terms of integrals depending on parameters gives an example of mirror symmetry in Givental's spirit.

The trigonometric *qKZ* difference equations and the corresponding trigonometric dynamical difference equations appear in the study of the equivariant K-theory of Nakajima varieties; see Okounkov and Smirnov (2016) and Okounkov (2017).

References

Aomoto, K. 1987a. Jacobi polynomials associated with Selberg integrals. *SIAM J. Math. Anal.*, **18**, 545–549.

Aomoto, K. 1987b. On the complex Selberg integral. *Quart. J. Math. Oxford Ser. (2)*, **38**, 385–399.

Aomoto, K. 1988. Correlation functions of the Selberg integral. Pages 591–605 of: *Ramanujan Revisited*. Academic Press.

Askey, R. 1980. Some basic hypergeometric extensions of integrals of Selberg and Andrews. *SIAM J. Math. Anal.*, **11**, 938–951.

Belavin, A. A., Polyakov, A. M., and Zamolodchikov, A. B. 1984. Infinite conformal symmetry in two-dimensional quantum field theory. *Nuclear Phys. B*, **241**, 333–380.

Bezrukavnikov, R., Finkelberg, M., and Schechtman, V. 1998. *Factorizable Sheaves and Quantum Groups*. Lecture Notes in Math., vol. 1691. Springer.

Braverman, A., Maulik, D., and Okounkov, A. 2011. Quantum cohomology of the Springer resolution. *Adv. Math.*, **227**, 421–458.

Date, E., Jimbo, M., Matsuo, A., and Miwa, T. 1990. Hypergeometric-type integrals and the $\mathfrak{sl}(2,\mathbf{C})$ Knizhnik–Zamolodchikov equation. *Internat. J. Modern Phys. B*, **4**, 1049–1057.

Dotsenko, V. S., and Fateev, V. A. 1984. Conformal algebra and multipoint correlation functions in 2D statistical models. *Nuclear Phys. B*, **240**, 312–348.

Dotsenko, V. S., and Fateev, V. A. 1985. Four-point correlation functions and the operator algebra in 2D conformal invariant theories with central charge $C \leqslant 1$. *Nuclear Phys. B*, **251**, 691–734.

Drinfel'd, V. G. 1990. Quasi-Hopf algebras. *Leningrad Math. J.*, **1**, 1419–1457. English translation of the Russian original.

Etingof, P. I. 1994. Representations of affine Lie algebras, elliptic r-matrix systems, and special functions. *Comm. Math. Phys.*, **159**, 471–502.

Etingof, P. I., and Kirillov, Jr., A. A. 1995. On the affine analogue of Jack and Macdonald polynomials. *Duke Math. J.*, **78**, 229–256.

Etingof, P., and Varchenko, A. 2002. Dynamical Weyl groups and applications. *Adv. Math.*, **167**, 74–127.

Feigin, B., Schechtman, V., and Varchenko, A. 1994. On algebraic equations satisfied by hypergeometric correlators in WZW models. I. *Comm. Math. Phys.*, **163**, 173–184.

Feigin, B., Schechtman, V., and Varchenko, A. 1995. On algebraic equations satisfied by hypergeometric correlators in WZW models. II. *Comm. Math. Phys.*, **170**, 219–247.

Felder, G., and Varchenko, A. 1995. Integral representation of solutions of the elliptic Knizhnik–Zamolodchikov–Bernard equations. *Int. Math. Res. Not.*, no. 5, 221–233.

Felder, G., and Varchenko, A. 1999. Resonance relations for solutions of the elliptic QKZB equations, fusion rules, and eigenvectors of transfer matrices of restricted interaction-round-a-face models. *Commun. Contemp. Math.*, **1**, 335–403.

Felder, G., and Varchenko, A. 2001. The q-deformed Knizhnik–Zamolodchikov–Bernard heat equation. *Comm. Math. Phys.*, **221**, 549–571.

Felder, G., and Varchenko, A. 2002. q-deformed KZB heat equation: completeness, modular properties and $\mathrm{SL}(3,\mathbb{Z})$. *Adv. Math.*, **171**, 228–275.

Felder, G., and Varchenko, A. 2004. Hypergeometric theta functions and elliptic Macdonald polynomials. *Int. Math. Res. Not.*, 1037–1055.

Felder, G., and Wieczerkowski, C. 1996. Conformal blocks on elliptic curves and the Knizhnik–Zamolodchikov–Bernard equations. *Comm. Math. Phys.*, **176**, 133–161.

Felder, G., Tarasov, V., and Varchenko, A. 1997. Solutions of the elliptic qKZB equations and Bethe ansatz. I. Pages 45–75 of: *Topics in Singularity Theory*. Amer. Math. Soc. Transl. Ser. 2, vol. 180. Amer. Math. Soc. Also available as arXiv:q-alg/9606005 (1996).

Felder, G., Tarasov, V., and Varchenko, A. 1999. Monodromy of solutions of the elliptic quantum Knizhnik–Zamolodchikov–Bernard difference equations. *Internat. J. Math.*, **10**, 943–975.

Felder, G., Markov, Y., Tarasov, V., and Varchenko, A. 2000. Differential equations compatible with KZ equations. *Math. Phys. Anal. Geom.*, **3**, 139–177.

Felder, G., Stevens, L., and Varchenko, A. 2003a. Elliptic Selberg integrals and conformal blocks. *Math. Res. Lett.*, **10**, 671–684.

Felder, G., Stevens, L., and Varchenko, A. 2003b. Modular transformations of the elliptic hypergeometric functions, Macdonald polynomials, and the shift operator. *Mosc. Math. J.*, **3**, 457–473, 743.

Frenkel, E. 2004. Opers on the projective line, flag manifolds and Bethe ansatz. *Mosc. Math. J.*, **4**, 655–705, 783.

Frenkel, I. B., and Reshetikhin, N. Yu. 1992. Quantum affine algebras and holonomic difference equations. *Comm. Math. Phys.*, **146**, 1–60.

Gaudin, M. 1983. *La Fonction d'Onde de Bethe*. Masson, Paris.

Gorbounov, V., Rimányi, R., Tarasov, V., and Varchenko, A. 2013. Quantum cohomology of the cotangent bundle of a flag variety as a Yangian Bethe algebra. *J. Geom. Phys.*, **74**, 56–86.

Gustafson, R. A. 1990. A generalization of Selberg's beta integral. *Bull. Amer. Math. Soc. (N.S.)*, **22**, 97–105.

Idzumi, M., Tokihiro, T., Iohara, K., Jimbo, M., Miwa, T., and Nakashima, T. 1993. Quantum affine symmetry in vertex models. *Internat. J. Modern Phys. A*, **8**, 1479–1511.

Kazhdan, D., and Lusztig, G. 1993. Tensor structures arising from affine Lie algebras. I, II. *J. Amer. Math. Soc.*, **6**, 905–947, 949–1011.

Kazhdan, D., and Lusztig, G. 1994. Tensor structures arising from affine Lie algebras. III, IV. *J. Amer. Math. Soc.*, **7**, 335–381, 383–453.

Knizhnik, V. G., and Zamolodchikov, A. B. 1984. Current algebra and Wess–Zumino model in two dimensions. *Nuclear Phys. B*, **247**, 83–103.

Kohno, T. 1987. Monodromy representations of braid groups and Yang–Baxter equations. *Ann. Inst. Fourier (Grenoble)*, **37**, 139–160.

Kohno, T. 1988. Linear representations of braid groups and classical Yang–Baxter equations. Pages 339–363 of: *Braids*. Contemp. Math., vol. 78. Amer. Math. Soc.

Korepin, V. E., Bogoliubov, N. M., and Izergin, A. G. 1993. *Quantum Inverse Scattering Method and Correlation Functions*. Cambridge University Press.

Markov, Y., and Varchenko, A. 2002. Hypergeometric solutions of trigonometric KZ equations satisfy dynamical difference equations. *Adv. Math.*, **166**, 100–147.

Maulik, D., and Okounkov, A. 2019. Quantum groups and quantum cohomology. *Astérisque*, **408**, 209 pp.

Mehta, M. L. 2004. *Random Matrices*. Third edn. Elsevier/Academic Press.

Millson, J. J., and Toledano Laredo, V. 2005. Casimir operators and monodromy representations of generalised braid groups. *Transform. Groups*, **10**, 217–254.

Mukhin, E., and Varchenko, A. 2000. The quantized Knizhnik–Zamolodchikov equation in tensor products of irreducible sl_2-modules. Pages 347–384 of: *Calogero–Moser–Sutherland Models*. CRM Ser. Math. Phys. Springer.

Mukhin, E., and Varchenko, A. 2003. Solutions to the *XXX* type Bethe ansatz equations and flag varieties. *Cent. Eur. J. Math.*, **1**, 238–271.

Mukhin, E., and Varchenko, A. 2004. Critical points of master functions and flag varieties. *Commun. Contemp. Math.*, **6**, 111–163.

Mukhin, E., Tarasov, V., and Varchenko, A. 2008. Spaces of quasi-exponentials and representations of $\mathfrak{gl}_N$. *J. Phys. A*, **41**, 194017, 28 pp.

Mukhin, E., Tarasov, V., and Varchenko, A. 2009a. The B. and M. Shapiro conjecture in real algebraic geometry and the Bethe ansatz. *Ann. of Math. (2)*, **170**, 863–881.

Mukhin, E., Tarasov, V., and Varchenko, A. 2009b. Bethe algebra of homogeneous *XXX* Heisenberg model has simple spectrum. *Comm. Math. Phys.*, **288**, 1–42.

Mukhin, E., Tarasov, V., and Varchenko, A. 2009c. Schubert calculus and representations of the general linear group. *J. Amer. Math. Soc.*, **22**, 909–940.

Mukhin, E., Tarasov, V., and Varchenko, A. 2011. Bethe algebra of the $\mathfrak{gl}_{N+1}$ Gaudin model and algebra of functions on the critical set of the master function. Pages 307–324 of: *New Trends in Quantum Integrable Systems*. World Scientific.

Mukhin, E., Tarasov, V., and Varchenko, A. 2012. Three sides of the geometric Langlands correspondence for $\mathfrak{gl}_N$ Gaudin model and Bethe vector averaging maps. Pages 475–511 of: *Arrangements of Hyperplanes—Sapporo 2009*. Adv. Stud. Pure Math., vol. 62. Math. Soc. Japan, Tokyo.

Okounkov, A. 2017. Lectures on K-theoretic computations in enumerative geometry. Pages 251–380 of: *Geometry of Moduli Spaces and Representation Theory*. IAS/Park City Math. Ser., vol. 24. Amer. Math. Soc.

Okounkov, A., and Smirnov, A. 2016. *Quantum difference equation for Nakajima varieties*. arXiv:1602.09007.

Rains, E. M., Sun, Y., and Varchenko, A. 2018. Affine Macdonald conjectures and special values of Felder-Varchenko functions. *Selecta Math. (N.S.)*, **24**, 1549–1591.

Reshetikhin, N. Yu., and Faddeev, L. D. 1983. Hamiltonian structures for integrable field theory models. *Theoret. Math. Phys.*, **56**, 847–862. Translation of Russian original.

Reshetikhin, N., and Varchenko, A. 1995. Quasiclassical asymptotics of solutions to the KZ equations. Pages 293–322 of: *Geometry, Topology, & Physics*. Int. Press, Cambridge, MA.

Schechtman, V. V., and Varchenko, A. N. 1991a. Arrangements of hyperplanes and Lie algebra homology. *Invent. Math.*, **106**, 139–194.

Schechtman, V. V., and Varchenko, A. N. 1991b. Quantum groups and homology of local systems. Pages 182–197 of: *Algebraic Geometry and Analytic Geometry*. Springer.

Scherbak, I., and Varchenko, A. 2003. Critical points of functions, $\mathfrak{sl}_2$ representations, and Fuchsian differential equations with only univalued solutions. *Mosc. Math. J.*, **3**, 621–645, 745.

Selberg, A. 1941. Über einen Satz von A. Gelfond. *Arch. Math. Naturvid.*, **44**, 159–170.

Selberg, A. 1944. Bemerkinger om et multipelt integral (in Norwegian). *Norsk Mat. Tidsskr.*, **26**, 71–78.

Smirnov, F. A. 1992. *Form Factors in Completely Integrable Models of Quantum Field Theory*. World Scientific.

Spiridonov, V. P. 2001. On the elliptic beta function. *Russian Math. Surveys*, **56**, 185–186.

Sun, Y. 2016. Traces of intertwiners for quantum affine $\mathfrak{sl}_2$ and Felder–Varchenko functions. *Comm. Math. Phys.*, **347**, 573–653.

Tarasov, V. 1999. Bilinear identity for q-hypergeometric integrals. *Osaka J. Math.*, **36**, 409–436.

Tarasov, V., and Varchenko, A. 1997a. Geometry of q-hypergeometric functions as a bridge between Yangians and quantum affine algebras. *Invent. Math.*, **128**, 501–588.

Tarasov, V., and Varchenko, A. 1997b. Geometry of q-hypergeometric functions, quantum affine algebras and elliptic quantum groups. *Astérisque*, **246**, 135 pp.

Tarasov, V., and Varchenko, A. 2000. Difference equations compatible with trigonometric KZ differential equations. *Int. Math. Res. Not.*, 801–829.

Tarasov, V., and Varchenko, A. 2002. Duality for Knizhnik–Zamolodchikov and dynamical equations. *Acta Appl. Math.*, **73**, 141–154.

Tarasov, V., and Varchenko, A. 2003. Selberg-type integrals associated with $\mathfrak{sl}_3$. *Lett. Math. Phys.*, **65**, 173–185.

Tarasov, V., and Varchenko, A. 2005a. Dynamical differential equations compatible with rational qKZ equations. *Lett. Math. Phys.*, **71**, 101–108.

Tarasov, V., and Varchenko, A. 2005b. Identities between q-hypergeometric and hypergeometric integrals of different dimensions. *Adv. Math.*, **191**, 29–45.

Tarasov, V., and Varchenko, A. 2005c. Identities for hypergeometric integrals of different dimensions. *Lett. Math. Phys.*, **71**, 89–99.

Tarasov, V., and Varchenko, A. 2014. Hypergeometric solutions of the quantum differential equation of the cotangent bundle of a partial flag variety. *Cent. Eur. J. Math.*, **12**, 694–710.

Toledano Laredo, V. 2002. A Kohno–Drinfeld theorem for quantum Weyl groups. *Duke Math. J.*, **112**, 421–451.

Toledano Laredo, V. 2011. The trigonometric Casimir connection of a simple Lie algebra. *J. Algebra*, **329**, 286–327.

Tsuchiya, A., Ueno, K., and Yamada, Y. 1989. Conformal field theory on universal family of stable curves with gauge symmetries. Pages 459–566 of: *Integrable Systems in Quantum Field Theory and Statistical Mechanics*. Academic Press.

Varchenko, A. 1995. *Multidimensional Hypergeometric Functions and Representation Theory of Lie Algebras and Quantum Groups*. World Scientific.

Varchenko, A. N., and Tarasov, V. O. 1995. Jackson integral representations for solutions of the Knizhnik–Zamolodchikov quantum equation. *St. Petersburg Math. J.*, **6**, 275–313. Translation from the Russian original.

Warnaar, S. O. 2008. Bisymmetric functions, Macdonald polynomials and $\mathfrak{sl}_3$ basic hypergeometric series. *Compos. Math.*, **144**, 271–303.

Warnaar, S. O. 2009. A Selberg integral for the Lie algebra A_n. *Acta Math.*, **203**, 269–304.

Warnaar, S. O. 2010. The $\mathfrak{sl}_3$ Selberg integral. *Adv. Math.*, **224**, 499–524.

Whittaker, E. T., and Watson, G. N. 1927. *A Course of Modern Analysis*. Fourth edn. Cambridge University Press.

Yang, C. N., and Yang, C. P. 1969. Thermodynamics of a one-dimensional system of bosons with repulsive delta-function interaction. *J. Math. Phys.*, **10**, 1115–1122.

12

$9j$-Coefficients and Higher

Joris Van der Jeugt

12.1 Introduction

$3j$-Coefficients (or $3j$-symbols), $6j$-coefficients, $9j$-coefficients and higher (referred to as $3nj$-coefficients) play a crucial role in various physical applications dealing with the quantization of angular momentum. This is because the quantum operators of angular momentum satisfy the $\mathfrak{su}(2)$ commutation relations. So the $3nj$-coefficients in this chapter are $3nj$-coefficients of the Lie algebra $\mathfrak{su}(2)$. For these coefficients, we shall emphasize their hypergeometric expressions and their relations to discrete orthogonal polynomials. Note that $3nj$-coefficients can also be considered for other Lie algebras. For positive discrete series representations of $\mathfrak{su}(1,1)$, the $3nj$-coefficients carry different labels but have the same structure as those of $\mathfrak{su}(2)$ (Rasmussen, 1975; Van der Jeugt, 2003), and the related orthogonal polynomials are the same. For other Lie algebras, the definition of $3j$-coefficients (i.e., coupling coefficients or Clebsch–Gordan coefficients related to the decomposition of tensor products of irreducible representations) is more involved, since in general multiplicities appear in the decomposition of tensor products (de Swart, 1963; Biedenharn et al., 1967). Note that there is also a vast literature on the q-analogues of $3nj$-coefficients in the context of quantum groups or quantized enveloping algebras: for the quantum universal enveloping algebra $U_q(\mathfrak{su}(2))$, the $3j$- and $6j$-coefficients are straightforward q-analogues of those of $\mathfrak{su}(2)$, and the related discrete orthogonal polynomials are the corresponding q-orthogonal polynomials in terms of basic hypergeometric series (Kirillov and Reshetikhin, 1989; Koelink and Koornwinder, 1989;123 Ališauskas, 2000, 2001).

Here, we shall be dealing only with $3nj$-coefficients for $\mathfrak{su}(2)$. First, we give a short summary of the relevant class of representations of the Lie algebra $\mathfrak{su}(2)$. An important notion is the tensor product of such representations. In the tensor product decomposition, the important Clebsch–Gordan coefficients appear. $3j$-Coefficients are proportional to these Clebsch–Gordan coefficients. We give some useful expressions (as hypergeometric series) and their relation to Hahn polynomials. Next, the tensor product of three representations is considered, and the relevant Racah coefficients (or $6j$-coefficients) are defined. The explicit expression of a Racah coefficient as a hypergeometric series of ${}_4F_3$ type and the connection with Racah polynomials and their orthogonality is given. $9j$-Coefficients are defined in the context of the tensor product of four representations. They are related to a discrete orthogonal polynomial in two variables (but no expression as a hypergeometric double sum is known). Finally, we consider the general tensor product of $(n + 1)$ representations and "generalized recoupling coefficients" or $3nj$-coefficients.

There are several standard books on quantum theory of angular momentum and $3nj$-coefficients. A classical reference is the book by Edmonds (1957), and an interesting set of historical papers on the subject was collected by Biedenharn and Van Dam (1965). The books by Biedenharn and Louck (1981a,b) treat the subject thoroughly. An excellent collection of formulas is found in the book by Varshalovich et al. (1988). Srinivasa Rao and Rajeswari (1993) emphasize the connection with hypergeometric series and some special topics such as zeros and the numerical computation. Lecture notes by Van der Jeugt (2003) give a self-contained mathematical introduction for $3nj$-coefficients of both the Lie algebras $\mathfrak{su}(2)$ and $\mathfrak{su}(1,1)$.

12.2 Representations of the Lie Algebra $\mathfrak{su}(2)$

An introduction to Lie algebras and their representations is not given here; it can be found e.g. in the book by Humphreys (1972). We shall just recall some basic notions related to $\mathfrak{sl}(2,\mathbb{C})$ and $\mathfrak{su}(2)$. As a vector space, the Lie algebra $\mathfrak{sl}(2,\mathbb{C})$ consists of all traceless complex (2×2) matrices. In this matrix form, the Lie algebra bracket $[x, y]$ is the commutator $xy - yx$. We consider the following standard basis of $\mathfrak{sl}(2,\mathbb{C})$:

$$J_0 := \begin{pmatrix} 1/2 & 0 \\ 0 & -1/2 \end{pmatrix}, \quad J_+ := \begin{pmatrix} 0 & 1 \\ 0 & 0 \end{pmatrix}, \quad J_- := \begin{pmatrix} 0 & 0 \\ 1 & 0 \end{pmatrix}.$$

The basic commutation relations then read $[J_0, J_\pm] = \pm J_\pm$, $[J_+, J_-] = 2J_0$. In the universal enveloping algebra of $\mathfrak{sl}(2,\mathbb{C})$, the following element (called the *Casimir operator*) is *central* (i.e., it commutes with every element):

$$C := J_+J_- + J_0^2 - J_0 = J_-J_+ + J_0^2 + J_0.$$

A $*$-*operation* on a complex Lie algebra is a conjugate-linear anti-automorphic involution. With a $*$-operation there is associated a real subalgebra (real form) consisting of all elements x in the complex Lie algebra for which $x^* = -x$. For $\mathfrak{sl}(2,\mathbb{C})$, there exist two nonequivalent $*$-operations, one corresponding to $\mathfrak{su}(2)$ and one to $\mathfrak{su}(1,1)$. For the real form $\mathfrak{su}(2)$, this is explicitly given by $J_0^* = J_0$, $J_\pm^* = J_\mp$. It consists of the matrices $\left(\begin{smallmatrix} ia & b \\ -\bar{b} & -ia \end{smallmatrix}\right)$ $(a \in \mathbb{R},\, b \in \mathbb{C})$.

A *representation* of the Lie algebra $\mathfrak{g}$ in a finite-dimensional complex vector space V is a linear map $\phi\colon \mathfrak{g} \to \mathrm{End}(V)$ such that $\phi([x, y]) = \phi(x)\phi(y) - \phi(y)\phi(x)$ for all elements $x, y \in \mathfrak{g}$. Then V is called the *representation space*. It is convenient to use the language of modules, and thus to refer to V as a $\mathfrak{g}$-module and to the action $\phi(x)(v)$ as $x \cdot v$ $(v \in V)$. The representation ϕ or the representation space V is *irreducible* if V has no nontrivial invariant subspaces under the action of $\mathfrak{g}$. It is *completely reducible* if V is a direct sum of irreducible representation subspaces. A representation ϕ of a real Lie algebra $\mathfrak{g}$ is *unitary* if V is a vector space with hermitian inner product $\langle\cdot,\cdot\rangle$ and $\langle x \cdot v, w\rangle = -\langle v, x \cdot w\rangle$ for all $x \in \mathfrak{g}$ and all $v, w \in V$. In this case, one also refers to V as a unitary representation. Unitary representations are completely reducible.

The following lists all irreducible unitary representations of $\mathfrak{su}(2)$.

Theorem 12.2.1 *For every $j \in \frac{1}{2}\mathbb{N} = \{0, \frac{1}{2}, 1, \frac{3}{2}, 2, \dots\}$, there is a unique (up to equivalence) irreducible unitary representation of $\mathfrak{su}(2)$ of dimension $2j+1$. An orthonormal basis for the corresponding representation space D_j is denoted by $\{e_m^{(j)} \mid m = -j, -j+1, \dots, j\}$. The action of $J_0, J_\pm$ is given by*

$$J_0 e_m^{(j)} = m e_m^{(j)}, \quad J_+ e_m^{(j)} = \sqrt{(j-m)(j+m+1)}\, e_{m+1}^{(j)}, \quad J_- e_m^{(j)} = \sqrt{(j+m)(j-m+1)}\, e_{m-1}^{(j)}. \tag{12.2.1}$$

For the Casimir operator, one has $C^* = C$ and $Ce_m^{(j)} = j(j+1)e_m^{(j)}$.

In the following, D_j will denote both the representation space and the representation on that vector space.

12.3 Clebsch–Gordan Coefficients and 3j-Coefficients

An important notion to introduce is the concept of the *tensor product* of two unitary representations of a real Lie algebra $\mathfrak{g}$. Let V and W be $\mathfrak{g}$-modules, and let $V \otimes W$ be the tensor product of the underlying vector spaces. Recall that if V and W have respective bases $v_1, v_2, \dots$ and $w_1, w_2, \dots$, then $V \otimes W$ has a basis consisting of the vectors $v_i \otimes w_j$. Now $V \otimes W$ has the structure of a $\mathfrak{g}$-module by defining

$$x \cdot (v \otimes w) := x \cdot v \otimes w + v \otimes x \cdot w. \tag{12.3.1}$$

This tensor product space is naturally equipped with an inner product by $\langle v \otimes w, v' \otimes w' \rangle := \langle v, v' \rangle \langle w, w' \rangle$. If V and W are unitary representations, then the tensor product representation is also unitary with respect to this inner product.

Turning to the case of $\mathfrak{su}(2)$, let $j_1, j_2 \in \frac{1}{2}\mathbb{N}$, and consider the tensor product $D_{j_1} \otimes D_{j_2}$, sometimes denoted by $(j_1) \otimes (j_2)$. A set of basis vectors of $D_{j_1} \otimes D_{j_2}$ is given by

$$e_{m_1}^{(j_1)} \otimes e_{m_2}^{(j_2)}, \quad m_1 = -j_1, -j_1+1, \dots, j_1; \; m_2 = -j_2, -j_2+1, \dots, j_2.$$

This basis is often referred to as the *uncoupled basis*. The total dimension of $D_{j_1} \otimes D_{j_2}$ is $(2j_1+1)(2j_2+1)$. The action of the $\mathfrak{su}(2)$ basis elements on these vectors is determined by (12.3.1).

In general, the module $D_{j_1} \otimes D_{j_2}$ is not irreducible, but it is completely reducible. Its irreducible components are again representations D_j of the form given in Theorem 12.2.1: $D_{j_1} \otimes D_{j_2} = \oplus_j D_j$. Herein, j takes the values $|j_1 - j_2|, |j_1 - j_2| + 1, \dots, j_1 + j_2$ (each value once, i.e., there is no multiplicity in the decomposition). So it must be possible to write a basis of $D_{j_1} \otimes D_{j_2}$ in terms of the standard basis vectors of the representations D_j appearing in the decomposition. This basis is referred to as the *coupled basis*. The coefficients expressing the coupled basis vectors in terms of the uncoupled basis vectors are known as the *Clebsch–Gordan coefficients* of $\mathfrak{su}(2)$. They first appeared in the work of Clebsch and Gordan on invariant theory of algebraic forms. But it was Wigner (von Neumann and Wigner, 1928;

Wigner, 1931, 1965) who studied these coefficients systematically and who introduced the related $3j$-coefficients.

Theorem 12.3.1 *The tensor product* $D_{j_1} \otimes D_{j_2}$ *decomposes into irreducible unitary representations* D_j *of* $\mathfrak{su}(2)$, $D_{j_1} \otimes D_{j_2} = \bigoplus_{j=|j_1-j_2|}^{j_1+j_2} D_j$. *An orthonormal basis of* $D_{j_1} \otimes D_{j_2}$ *is given by the vectors*

$$e_m^{(j_1 j_2)j} = \sum_{m_1} C^{j_1,j_2,j}_{m_1,m-m_1,m} e^{(j_1)}_{m_1} \otimes e^{(j_2)}_{m-m_1} \qquad (|j_1 - j_2| \le j \le j_1 + j_2, \; -j \le m \le j), \tag{12.3.2}$$

where the coefficients $C^{j_1,j_2,j}_{m_1,m-m_1,m}$ *are* Clebsch–Gordan coefficients *of* $\mathfrak{su}(2)$, *for which an expression is given below by* (12.3.4). *The action of* J_0, $J_\pm$ *on the basis vectors* $e_m^{(j_1 j_2)j}$ *is the standard action* (12.2.1) *of the representation* D_j.

In (12.3.2), the summation index m_1 runs from $-j_1$ to j_1 in steps of 1, and such that $-j_2 \le m - m_1 \le j_2$: thus $\max(-j_1, m - j_2) \le m_1 \le \min(j_1, m + j_2)$. The Clebsch–Gordan coefficient $C^{j_1,j_2,j}_{m_1,m_2,m}$ can be considered as a real function of six arguments from $\frac{1}{2}\mathbb{N}$. Following Theorem 12.3.1, these arguments satisfy the following conditions:

(c1) (j_1, j_2, j) forms a *triad*, i.e., $-j_1 + j_2 + j$, $j_1 - j_2 + j$ and $j_1 + j_2 - j$ are nonnegative integers;

(c2) m_1 is a *projection* of j_1, i.e., $m_1 \in \{-j_1, -j_1 + 1, \ldots, j_1\}$ (and similarly, m_2 is a projection of j_2 and m is a projection of j);

(c3) $m = m_1 + m_2$.

Usually, one extends the definition by saying that $C^{j_1,j_2,j}_{m_1,m_2,m} = 0$ if one of the conditions (c1), (c2), (c3) is not satisfied. Then one can write

$$e_m^{(j_1 j_2)j} = \sum_{m_1,m_2} C^{j_1,j_2,j}_{m_1,m_2,m} e^{(j_1)}_{m_1} \otimes e^{(j_2)}_{m_2}. \tag{12.3.3}$$

An explicit formula for $C^{j_1,j_2,j}_{m_1,m_2,m}$ can, for instance, be obtained by using a differential operator realization of $\mathfrak{su}(2)$ and a polynomial realization of the basis vectors of the representations. One finds

$$\begin{aligned} C^{j_1,j_2,j}_{m_1,m_2,m} &= \sqrt{(2j+1)}\,\Delta(j_1, j_2, j)\delta(j_1, m_1, j_2, m_2, j, m) \\ &\times \sum_k \frac{(-1)^k}{k!\,(j_1-m_1-k)!\,(j_1+j_2-j-k)!\,(j_2+m_2-k)!\,(j-j_2+m_1+k)!\,(j-j_1-m_2+k)!}, \end{aligned} \tag{12.3.4}$$

where

$$\begin{aligned} \Delta(j_1, j_2, j) &= \sqrt{\frac{(-j_1 + j_2 + j)!\,(j_1 - j_2 + j)!\,(j_1 + j_2 - j)!}{(j_1 + j_2 + j + 1)!}}, \\ \delta(j_1, m_1, j_2, m_2, j, m) &= \sqrt{(j_1 - m_1)!\,(j_1 + m_1)!\,(j_2 - m_2)!\,(j_2 + m_2)!\,(j - m)!\,(j + m)!}\,. \end{aligned} \tag{12.3.5}$$

This rather symmetrical form is due to van der Waerden (1932) and Racah (1942). The expression is generally valid (that is, for all arguments satisfying (c1)–(c3)). The summation is over all integer k-values such that the factorials in the denominator of (12.3.4) are nonnegative.

It is clear that the summation in (12.3.4) can be rewritten in terms of a terminating ${}_3F_2$ series of unit argument; indeed, assume that $j - j_2 + m_1 \geq 0$ and $j - j_1 - m_2 \geq 0$; then this sum equals

$$((j_1 - m_1)!\,(j_1 + j_2 - j)!\,(j_2 + m_2)!\,(j - j_2 + m_1)!\,(j - j_1 - m_2)!)^{-1} \tag{12.3.6}$$

$$\times\, {}_3F_2\left(\begin{matrix} -j_1 + m_1, -j_1 - j_2 + j, -j_2 - m_2 \\ j - j_2 + m_1 + 1, j - j_1 - m_2 + 1 \end{matrix}; 1\right). \tag{12.3.7}$$

Once the Clebsch–Gordan coefficient is rewritten as a terminating ${}_3F_2(1)$, one can use a transformation (Andrews et al., 1999, Corollary 3.3.4) (known as *Sheppard's transformation*, and sometimes referred to as *Thomae's transformation*) to find yet other formulas. Application of Sheppard's transformation to (12.3.7), while keeping $-j_1 + m_1$ as negative numerator parameter, yields

$$C^{j_1,j_2,j}_{m_1,m_2,m} = C'\,{}_3F_2\left(\begin{matrix} -j_1 + m_1, -j_1 - j_2 + j, -j_1 - j_2 - j - 1 \\ -2j_1, -j_1 - j_2 + m \end{matrix}; 1\right); \tag{12.3.8}$$

herein C' is some constant determined in (12.3.10). This expression is generally valid under conditions (c1)–(c3).

Various symmetries can be deduced for Clebsch–Gordan coefficients. Some of these symmetries follow by replacing the summation index in (12.3.4). Others can be deduced by performing permutations of the numerator and/or denominator parameters in a ${}_3F_2(1)$ expression as in (12.3.8). In order to express these symmetries, one often introduces the so-called $3j$-*coefficient* (due to Wigner)

$$\begin{pmatrix} j_1 & j_2 & j \\ m_1 & m_2 & -m \end{pmatrix} := \frac{(-1)^{j_1 - j_2 + m}}{\sqrt{2j+1}} C^{j_1,j_2,j}_{m_1,m_2,m}.$$

From (12.3.4) one finds the classical expression for the $3j$-coefficient:

Proposition 12.3.2 *Let $j_1, j_2, j_3, m_1, m_2, m_3 \in \frac{1}{2}\mathbb{N}$. If (j_1, j_2, j_3) forms a triad, m_i is a projection of j_i $(i = 1, 2, 3)$ and $m_1 + m_2 + m_3 = 0$, then the $3j$-coefficient is determined by*

$$\begin{pmatrix} j_1 & j_2 & j_3 \\ m_1 & m_2 & m_3 \end{pmatrix} = (-1)^{j_1 - j_2 - m_3} \Delta(j_1, j_2, j_3)\delta(j_1, m_1, j_2, m_2, j_3, m)$$
$$\times \sum_k \frac{(-1)^k}{k!\,(j_1 - m_1 - k)!\,(j_1 + j_2 - j_3 - k)!\,(j_2 + m_2 - k)!\,(j_3 - j_2 + m_1 + k)!\,(j_3 - j_1 - m_2 + k)!}. \tag{12.3.9}$$

In all other cases, the $3j$-coefficient is zero. In (12.3.9), the summation is over all integers such that the arguments in the factorials are nonnegative. Alternatively, one can write

$$\begin{pmatrix} j_1 & j_2 & j_3 \\ m_1 & m_2 & m_3 \end{pmatrix} = (-1)^{j_1 - j_2 - m_3} \frac{(2j_1)!\,(j_1 + j_2 + m_3)!\,(j_3 - m_3)!}{(j_1 - j_2 + j_3)!\,(j_1 + j_2 - j_3)!}$$
$$\times \frac{\Delta(j_1, j_2, j_3)}{\delta(j_1, m_1, j_2, m_2, j_3, m_3)}\,{}_3F_2\left(\begin{matrix} m_1 - j_1, j_3 - j_1 - j_2, -j_1 - j_2 - j_3 - 1 \\ -2j_1, -j_1 - j_2 - m_3 \end{matrix}; 1\right). \tag{12.3.10}$$

The symmetries of the $3j$-coefficient can be described through the corresponding *Regge array* (Regge, 1958):

$$\begin{pmatrix} j_1 & j_2 & j_3 \\ m_1 & m_2 & m_3 \end{pmatrix} = R_{3j}\begin{bmatrix} -j_1+j_2+j_3 & j_1-j_2+j_3 & j_1+j_2-j_3 \\ j_1-m_1 & j_2-m_2 & j_3-m_3 \\ j_1+m_1 & j_2+m_2 & j_3+m_3 \end{bmatrix}.$$

In this 3×3 array, all entries are nonnegative integers such that for each row and each column the sum of the entries equals $J = j_1 + j_2 + j_3$; conversely, every 3×3 array with nonnegative integers such that all row and column sums are the same, corresponds to a Regge array or a $3j$-coefficient. The symmetries are easy to describe in terms of the Regge array. They generate a group of 72 symmetries, described as follows:

- The Regge array is invariant under transposition.
- Under permutation of the rows (resp. columns), the Regge array remains invariant up to sign. For cyclic permutations this sign is $+1$; for noncyclic permutations this sign is $(-1)^J$.

Observe that for certain special values of the arguments, the single sum expression in equation (12.3.9) reduces to a single term. This is the case, e.g., when $m_1 = j_1$. Many such *closed form expressions* (i.e., without a summation expression) are listed in Varshalovich et al. (1988).

The coupled and uncoupled basis vectors in (12.3.3) are orthonormal bases for $D_{j_1}\otimes D_{j_2}$. So the matrix relating these two bases in (12.3.3) is orthogonal. This implies that the Clebsch–Gordan coefficients satisfy the following orthogonality relations:

$$\sum_{m_1,m_2} C^{j_1,j_2,j}_{m_1,m_2,m} C^{j_1,j_2,j'}_{m_1,m_2,m'} = \delta_{j,j'}\delta_{m,m'}, \tag{12.3.11a}$$

$$\sum_{j,m} C^{j_1,j_2,j}_{m_1,m_2,m} C^{j_1,j_2,j}_{m_1',m_2',m} = \delta_{m_1,m_1'}\delta_{m_2,m_2'}. \tag{12.3.11b}$$

These relations can also be expressed by means of $3j$-coefficients, e.g.

$$\sum_{m_1,m_2}(2j_3+1)\begin{pmatrix} j_1 & j_2 & j_3 \\ m_1 & m_2 & m_3 \end{pmatrix}\begin{pmatrix} j_1 & j_2 & j_3' \\ m_1 & m_2 & m_3' \end{pmatrix} = \delta_{j_3,j_3'}\delta_{m_3,m_3'}.$$

The orthogonality relations for $\mathfrak{su}(2)$ Clebsch–Gordan coefficients or $3j$-coefficients are actually related to the (discrete) orthogonality of Hahn polynomials. Consider the expression (12.3.8), and let us write

$$N = 2j_1,\quad x = j_1 - m_1,\quad n = j_1 + j_2 - j,\quad \alpha = m - j_1 - j_2 - 1,\quad \beta = -j_1 - j_2 - m - 1.$$

Suppose j_1, j_2 and m are fixed numbers, with $j_2 - j_1 \geq |m|$. Then m_1 can vary between $-j_1$ and j_1, and j can vary between $j_2 - j_1$ and $j_2 + j_1$. In terms of the new variables, this means that N is a fixed nonnegative integer, α and β are fixed (with $\alpha,\beta \leq -N-1$); the quantities x and n are nonnegative integers with $0 \leq x \leq N$ and $0 \leq n \leq N$. The ${}_3F_2$ series appearing in (12.3.8) is then of the form

$$Q_n(x;\alpha,\beta,N) = {}_3F_2\left(\begin{matrix} -x, -n, n+\alpha+\beta+1 \\ -N, \alpha+1 \end{matrix}; 1\right).$$

Herein, $Q_n(x;\alpha,\beta,N)$ is the *Hahn polynomial* of degree n in the variable x; see Koekoek et al. (2010, §9.5). Interestingly, the orthogonality (12.3.11a) of $\mathfrak{su}(2)$ Clebsch–Gordan coefficients (or $3j$-coefficients) is equivalent to the orthogonality (Koekoek et al., 2010, (9.5.2)) of Hahn polynomials. In a similar way, one can verify that the orthogonality relation (12.3.11b) is equivalent to the orthogonality relation (Koekoek et al., 2010, (9.6.2)) of dual Hahn polynomials. The relation between Hahn polynomials and $\mathfrak{su}(2)$ $3j$-coefficients was known to some people, but appeared explicitly only in 1981 (Koornwinder, 1981). There, the relationship is established in the context of representations of the Lie group SU(2) rather than in terms of representations of the Lie algebra $\mathfrak{su}(2)$, as here.

To conclude this section, let us mention that the action of J_+ or J_- on (12.3.3) yields certain recurrence relations for Clebsch–Gordan coefficients or $3j$-coefficients. Appropriately combined recurrence relations then lead to the classical 3-term recurrence relation of Hahn or dual Hahn polynomials (Koekoek et al., 2010, §§9.5, 9.6).

12.4 Racah Coefficients and $6j$-Coefficients

Consider the tensor product of three irreducible unitary representations of $\mathfrak{su}(2)$,

$$D_{j_1}\otimes D_{j_2}\otimes D_{j_3} = (D_{j_1}\otimes D_{j_2})\otimes D_{j_3} = D_{j_1}\otimes (D_{j_2}\otimes D_{j_3}). \tag{12.4.1}$$

Clearly, a basis for this tensor product is given by $e^{(j_1)}_{m_1}\otimes e^{(j_2)}_{m_2}\otimes e^{(j_3)}_{m_3}$, where m_i is a projection of j_i. This is the *uncoupled basis*. In order to decompose the actual tensor product $D_{j_1}\otimes D_{j_2}\otimes D_{j_3}$ into irreducible $\mathfrak{su}(2)$ representations, one can proceed in two ways. First, decompose $D_{j_1}\otimes D_{j_2}$ into irreducibles, say $\oplus D_{j_{12}}$, and then decompose each tensor product $D_{j_{12}}\otimes D_{j_3}$ into irreducibles D_j. Secondly, decompose $D_{j_2}\otimes D_{j_3}$ into irreducibles, say $\oplus D_{j_{23}}$, and then decompose each tensor product $D_{j_1}\otimes D_{j_{23}}$ into irreducibles D_j. So one can immediately define two sets of orthonormal basis vectors for the irreducible components of (12.4.1), corresponding to these two coupling schemes:

$$\begin{aligned} e^{((j_1j_2)j_{12}j_3)j}_m &= \sum_{m_{12},m_3} C^{j_{12},j_3,j}_{m_{12},m_3,m} e^{(j_1j_2)j_{12}}_{m_{12}}\otimes e^{(j_3)}_{m_3} \\ &= \sum_{\substack{m_1,m_2,m_3\\ m_1+m_2+m_3=m}} C^{j_1,j_2,j_{12}}_{m_1,m_2,m_{12}} C^{j_{12},j_3,j}_{m_{12},m_3,m} e^{(j_1)}_{m_1}\otimes e^{(j_2)}_{m_2}\otimes e^{(j_3)}_{m_3} \end{aligned} \tag{12.4.2a}$$

and

$$\begin{aligned} e^{(j_1(j_2j_3)j_{23})j}_m &= \sum_{m_1,m_{23}} C^{j_1,j_{23},j}_{m_1,m_{23},m} e^{(j_1)}_{m_1}\otimes e^{(j_2j_3)j_{23}}_{m_{23}} \\ &= \sum_{\substack{m_1,m_2,m_3\\ m_1+m_2+m_3=m}} C^{j_2,j_3,j_{23}}_{m_2,m_3,m_{23}} C^{j_1,j_{23},j}_{m_1,m_{23},m} e^{(j_1)}_{m_1}\otimes e^{(j_2)}_{m_2}\otimes e^{(j_3)}_{m_3}. \end{aligned} \tag{12.4.2b}$$

Let us denote the matrix transforming the basis (12.4.2a) into (12.4.2b) by U. Its matrix elements are given by $\langle e^{(j_1(j_2j_3)j_{23})j}_m, e^{((j_1j_2)j_{12}j_3)j'}_{m'}\rangle$. From the action of the $\mathfrak{su}(2)$ Casimir operator C and of the $\mathfrak{su}(2)$ diagonal operator J_0, it is easy to see that this element is zero if $j'\neq j$ and if

$m' \neq m$. Furthermore, by the action of J_+ one verifies that this element is independent of m. So one can write

$$\langle e_m^{(j_1(j_2j_3)j_{23})j}, e_{m'}^{((j_1j_2)j_{12}j_3)j'} \rangle = \delta_{j,j'}\delta_{m,m'} U_{j_3,j,j_{23}}^{j_1,j_2,j_{12}}. \tag{12.4.3}$$

The coefficients $U_{j_3,j,j_{23}}^{j_1,j_2,j_{12}}$ are called the *Racah coefficients*. So one can write

$$e_m^{((j_1j_2)j_{12}j_3)j} = \sum_{j_{23}} U_{j_3,j,j_{23}}^{j_1,j_2,j_{12}} e_m^{(j_1(j_2j_3)j_{23})j}, \tag{12.4.4a}$$

and vice versa, since U is an orthogonal matrix,

$$e_m^{(j_1(j_2j_3)j_{23})j} = \sum_{j_{12}} U_{j_3,j,j_{23}}^{j_1,j_2,j_{12}} e_m^{((j_1j_2)j_{12}j_3)j}. \tag{12.4.4b}$$

The orthogonality of the matrix is also expressed by

$$\sum_{j_{12}} U_{j_3,j,j_{23}}^{j_1,j_2,j_{12}} U_{j_3,j,j'_{23}}^{j_1,j_2,j_{12}} = \delta_{j_{23},j'_{23}}, \qquad \sum_{j_{23}} U_{j_3,j,j_{23}}^{j_1,j_2,j_{12}} U_{j_3,j,j_{23}}^{j_1,j_2,j'_{12}} = \delta_{j_{12},j'_{12}}.$$

An expression for the Racah coefficient follows from (12.4.3), (12.4.2a) and (12.4.2b):

$$U_{j_3,j,j_{23}}^{j_1,j_2,j_{12}} = \sum_{\substack{m_1,m_2,m_3 \\ m_1+m_2+m_3=m}} C_{m_1,m_2,m_{12}}^{j_1,j_2,j_{12}} C_{m_{12},m_3,m}^{j_{12},j_3,j} C_{m_2,m_3,m_{23}}^{j_2,j_3,j_{23}} C_{m_1,m_{23},m}^{j_1,j_{23},j}. \tag{12.4.5}$$

Herein, m is an arbitrary but fixed projection of j. The sum is over m_1, m_2 and m_3 such that $m_1 + m_2 + m_3 = m$; m_{12} stands for $m_1 + m_2$ and m_{23} for $m_2 + m_3$. So this is a double sum over the product of four Clebsch–Gordan coefficients. This is clearly a rather complicated object; Racah was the first to simplify this expression by various summation manipulations and to rewrite it as a single sum (Racah, 1942).

The Racah coefficient has a number of symmetries that can be deduced from symmetry properties of Clebsch–Gordan coefficients. In this context, it is appropriate to introduce the so-called $6j$*-coefficient*. Wigner (1965) was the first to introduce the $6j$-coefficient in his Princeton Lectures (1940), published much later in 1965. They take the form

$$\begin{Bmatrix} a & b & c \\ d & e & f \end{Bmatrix} := (-1)^{a+b+d+e} \frac{U_{d,e,f}^{a,b,c}}{\sqrt{(2c+1)(2f+1)}},$$

where (a, b, c), (d, e, c), (d, b, f) and (a, e, f) are triads. Then the $6j$-coefficient is invariant under any permutation of its columns, or under the interchange of the upper and lower arguments in each of any two columns.

Even more, one can also use the Regge symmetries of the Clebsch–Gordan coefficients, and obtain similar symmetries for the $6j$-coefficient. In order to describe these, let the *Regge array* for the $6j$-coefficient be defined as the 3×4 array (Regge, 1959)

$$\begin{Bmatrix} a & b & c \\ d & e & f \end{Bmatrix} = R_{6j} \begin{bmatrix} d+e-c & b+d-f & a+e-f & a+b-c \\ d+f-b & c+d-e & a+c-b & a+f-e \\ e+f-a & b+c-a & c+e-d & b+f-d \end{bmatrix}.$$

Then the value of the Regge array is invariant under any permutation of its rows or columns. Note that the arguments of the Regge array are such that all entries are nonnegative integers, and the differences between corresponding elements of rows (resp. columns) are constant. Conversely, every 3×4 array of nonnegative integers with this property corresponds to a Regge array, or a $6j$-coefficient.

As mentioned, Racah managed to obtain a single sum expression for the Racah coefficient (or $6j$-coefficient). A simple method to obtain this single sum expression from (12.4.5) is outlined by Vilenkin and Klimyk (1991) or by Van der Jeugt (2003). The final result is the following proposition.

Proposition 12.4.1 *Let $a, b, c, d, e, f \in \frac{1}{2}\mathbb{N}$, where (a, b, c), (d, e, c), (d, b, f) and (a, e, f) are triads. Then the $6j$-coefficient is given by*

$$\begin{Bmatrix} a & b & c \\ d & e & f \end{Bmatrix} = \Delta(a,b,c)\Delta(c,d,e)\Delta(a,e,f)\Delta(b,d,f)$$
$$\times \sum_k \frac{(-1)^k (1+k)!}{(k-t_1)!\,(k-t_2)!\,(k-t_3)!\,(k-t_4)!\,(c_1-k)!\,(c_2-k)!\,(c_3-k)!}. \qquad (12.4.6)$$

Herein, the t_i correspond to the four triad sums ($t_1 = a+b+c$, $t_2 = d+e+c$, $t_3 = d+b+f$, $t_4 = a+e+f$) and the c_i correspond to the sums of two columns ($c_1 = a+d+b+e$, $c_2 = a+d+c+f$, $c_3 = b+e+c+f$). The sum is over all integer k-values such that all factorials assume nonnegative arguments, and Δ is defined by (12.3.5).

By replacement of the summation variable in (12.4.6), one can (under certain assumptions) rewrite the sum in terms of a ${}_4F_3$ series:

$$\begin{Bmatrix} a & b & c \\ d & e & f \end{Bmatrix} = \frac{(-1)^{b+c+e+f}\Delta(a,b,c)\Delta(c,d,e)\Delta(a,e,f)\Delta(b,d,f)}{(b+c-a)!\,(c-d+e)!\,(e+f-a)!\,(b-d+f)!}$$
$$\times \frac{(1+b+c+e+f)!}{(a+d-b-e)!\,(a+d-c-f)!}$$
$$\times\, {}_4F_3\left({a-b-c, d-b-f, a-e-f, d-c-e \atop -b-e-c-f-1, a+d-b-e+1, a+d-c-f+1} ; 1 \right). \qquad (12.4.7)$$

This expression is valid for $a+d \geq b+e$ and $a+d \geq c+f$ (which can always be assumed after applying a symmetry corresponding to a permutation of columns of the $6j$-coefficient). The ${}_4F_3$ in (12.4.7) is a terminating balanced ${}_4F_3$ of unit argument; for such series there exist transformation formulas due to Whipple (1926) and Andrews et al. (1999, Theorem 3.3.3). This allows the relation between $6j$-coefficients and ${}_4F_3$ series to be written in various forms; see Varshalovich et al. (1988). One of these forms is

$$\begin{Bmatrix} a & b & c \\ d & e & f \end{Bmatrix} = (-1)^{b+c+e+f} \frac{(2b)!\,(b+c-e+f)!\,(b+c+e+f+1)!}{\nabla(b,a,c)\nabla(c,d,e)\nabla(f,a,e)\nabla(b,d,f)}$$
$$\times\, {}_4F_3\left({a-b-c, d-b-f, -a-b-c-1, -b-d-f-1 \atop -2b, -b-c+e-f, -b-c-e-f-1} ; 1 \right), \qquad (12.4.8)$$

where

$$\nabla(a,b,c) = \sqrt{(a+b-c)!\,(a-b+c)!\,(a+b+c+1)!\,/\,(-a+b+c)!}\,. \qquad (12.4.9)$$

Expression (12.4.8) is valid for all possible arguments of the $6j$-coefficient, provided of course that (a, b, c), (d, e, c), (d, b, f) and (a, e, f) are triads.

Let us rewrite:

$$n = -a + b + c, \quad x = b - d + f, \quad \alpha \equiv -N - 1 = -1 - b - c + e - f,$$
$$\beta = -1 + f - b - c - e, \quad \gamma = -2b - 1, \quad \delta = -2f - 1.$$

Let b, c, e and f be fixed numbers (parameters), with $e - f \geq |b - c|$ and $e - c \geq |b - f|$. Then a and d can be thought of as variables, with a varying between $e - f$ and $b + c$, and d running from $e - c$ to $b + f$. In terms of the new parameters/variables, this means that N is a fixed nonnegative integer parameter, and x and n are nonnegative integer variables with $0 \leq x \leq N$ and $0 \leq n \leq N$. The ${}_4F_3$-series of (12.4.8) is of the following form:

$$R_n(\lambda(x); \alpha, \beta, \gamma, \delta) := {}_4F_3\left({-n, n + \alpha + \beta + 1, -x, x + \gamma + \delta + 1 \atop \alpha + 1, \beta + \delta + 1, \gamma + 1} ; 1 \right), \quad \lambda(x) := x(x + \gamma + \delta + 1).$$

This is the *Racah polynomial* $R_n(\lambda(x)) \equiv R_n(\lambda(x); \alpha, \beta, \gamma, \delta)$; see Koekoek et al. (2010, (9.2.1)). The orthogonality of the Racah coefficients – appropriately rewritten – is equivalent to the orthogonality relation (Koekoek et al., 2010, (9.2.2)) of Racah polynomials. The orthogonality of Racah coefficients is of historical importance: it motivated J. A. Wilson to introduce Racah polynomials in his Ph.D. thesis (Wilson, 1978). Soon afterwards, this led Askey and Wilson to q-Racah polynomials and Askey–Wilson polynomials (Askey and Wilson, 1979, 1985).

The coupled vectors (12.4.2a), (12.4.2b) in $D_{j_1} \otimes D_{j_2} \otimes D_{j_3}$, related by Racah coefficients, are examples of *binary coupling schemes* (Biedenharn and Louck, 1981b; Van der Jeugt, 2003). In obvious notation, this reads

$e_m^{((j_1 j_2) j_{12} j_3) j} =$ 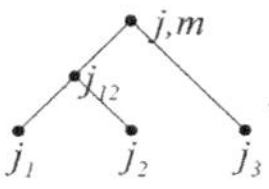

, $e_m^{(j_1 (j_2 j_3) j_{23}) j} =$ 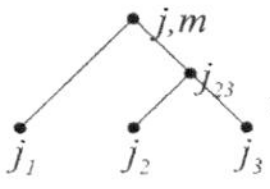

, (12.4.10)

and to express one type in terms of the other, one uses Racah coefficients; see (12.4.4a) or (12.4.4b).

Consider now the tensor product of four $\mathfrak{su}(2)$ representations $D_{j_1} \otimes D_{j_2} \otimes D_{j_3} \otimes D_{j_4}$. Using (12.4.4a) twice, according to the order

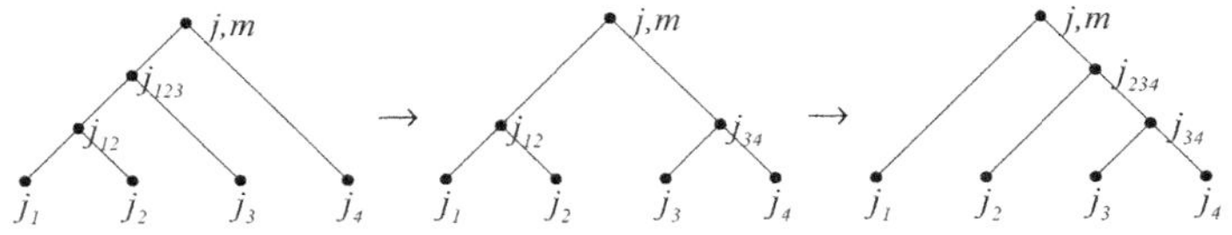

one has

$$e_m^{(((j_1 j_2) j_{12} j_3) j_{123} j_4) j} = \sum_{j_{34}, j_{234}} U^{j_{12}, j_3, j_{123}}_{j_4, j, j_{34}} \, U^{j_1, j_2, j_{12}}_{j_{34}, j, j_{234}} \, e_m^{(j_1 (j_2 (j_3 j_4) j_{34}) j_{234}) j}. \quad (12.4.11)$$

Alternatively, one can use (12.4.4a) three times according to a different order:

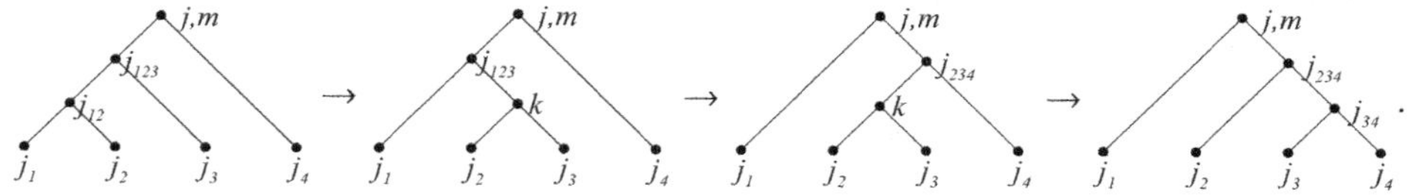

Then

$$e_m^{(((j_1j_2)j_{12}j_3)j_{123}j_4)j} = \sum_{k,j_{234},j_{34}} U_{j_3,j_{123},k}^{j_1,j_2,j_{12}} U_{j_4,j,j_{234}}^{j_1,k,j_{123}} U_{j_4,j_{234},j_{34}}^{j_2,j_3,k} e_m^{(j_1(j_2(j_3j_4)j_{34})j_{234})j}. \tag{12.4.12}$$

Comparison of (12.4.11) with (12.4.12) yields (Biedenharn, 1953; Elliott, 1953) the following theorem.

Theorem 12.4.2 *The Racah coefficients of* $\mathfrak{su}(2)$ *satisfy the following identity, known as the* Biedenharn–Elliott identity:

$$U_{j_{34},j,j_{234}}^{j_1,j_2,j_{12}} U_{j_4,j,j_{34}}^{j_{12},j_3,j_{123}} = \sum_k U_{j_3,j_{123},k}^{j_1,j_2,j_{12}} U_{j_4,j,j_{234}}^{j_1,k,j_{123}} U_{j_4,j_{234},j_{34}}^{j_2,j_3,k}.$$

In terms of $6j$*-coefficients, this can be rewritten as*

$$\sum_x (-1)^{J+x}(2x+1)\begin{Bmatrix} a & b & x \\ c & d & p \end{Bmatrix}\begin{Bmatrix} c & d & x \\ e & f & q \end{Bmatrix}\begin{Bmatrix} e & f & x \\ b & a & r \end{Bmatrix} = \begin{Bmatrix} p & q & r \\ e & a & d \end{Bmatrix}\begin{Bmatrix} p & q & r \\ f & b & c \end{Bmatrix},$$

where $J = a+b+c+d+e+f+p+q+r$*, and all labels are representation labels (elements of* $\frac{1}{2}\mathbb{N}$*). The sum is over all* x *(in steps of 1) with* $\max(|a-b|, |c-d|, |e-f|) \le x \le \min(a+b, c+d, e+f)$*.*

The Biedenharn–Elliott identity can be considered as a master identity for special functions, since it gives rise to many known identities for orthogonal polynomials as limit cases (sometimes after analytic continuation); see e.g. Van der Jeugt (2003, §5.5).

12.5 The 9*j*-Coefficient

Consider again the tensor product of four $\mathfrak{su}(2)$ representations $D_{j_1} \otimes D_{j_2} \otimes D_{j_3} \otimes D_{j_4}$, and the vectors corresponding to the following couplings:

$$v = e_m^{((j_1j_2)j_{12}(j_3j_4)j_{34})j} = \tag{12.5.1a}$$

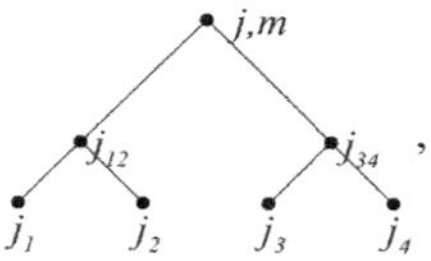

$$v' = e_m^{((j_1j_3)j_{13}(j_2j_4)j_{24})j} = \tag{12.5.1b}$$

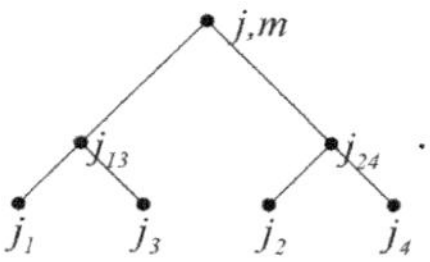

In terms of Clebsch–Gordan coefficients, these read

$$v = \sum_{\substack{m_1,m_2,m_3,m_4 \\ j_{12},j_{34}}} C^{j_1,j_2,j_{12}}_{m_1,m_2,m_{12}} C^{j_3,j_4,j_{34}}_{m_3,m_4,m_{34}} C^{j_{12},j_{34},j}_{m_{12},m_{34},m} e^{j_1}_{m_1} \otimes e^{j_2}_{m_2} \otimes e^{j_3}_{m_3} \otimes e^{j_4}_{m_4}, \qquad (12.5.2a)$$

$$v' = \sum_{\substack{m_1,m_2,m_3,m_4 \\ j_{13},j_{24}}} C^{j_1,j_3,j_{13}}_{m_1,m_3,m_{13}} C^{j_2,j_4,j_{24}}_{m_2,m_4,m_{24}} C^{j_{13},j_{24},j}_{m_{13},m_{24},m} e^{j_1}_{m_1} \otimes e^{j_2}_{m_2} \otimes e^{j_3}_{m_3} \otimes e^{j_4}_{m_4}. \qquad (12.5.2b)$$

The matrix relating the basis (12.5.1a) and (12.5.1b) consists of the $9j$*-coefficients*. More precisely, one defines (Wigner, 1965)

$$\begin{Bmatrix} j_1 & j_2 & j_{12} \\ j_3 & j_4 & j_{34} \\ j_{13} & j_{24} & j \end{Bmatrix} := \frac{\langle e_m^{((j_1 j_2) j_{12} (j_3 j_4) j_{34}) j}, e_m^{((j_1 j_3) j_{13} (j_2 j_4) j_{24}) j} \rangle}{\sqrt{(2j_{12}+1)(2j_{34}+1)(2j_{13}+1)(2j_{24}+1)}}. \qquad (12.5.3)$$

Since the two bases (12.5.2a), (12.5.2b) are orthonormal, the corresponding transformation matrix is orthogonal, and this leads to orthogonality relations of $9j$-coefficients:

$$\sum_{j_{12},j_{34}} (2j_{12}+1)(2j_{34}+1) \begin{Bmatrix} j_1 & j_2 & j_{12} \\ j_3 & j_4 & j_{34} \\ j_{13} & j_{24} & j \end{Bmatrix} \begin{Bmatrix} j_1 & j_2 & j_{12} \\ j_3 & j_4 & j_{34} \\ j'_{13} & j'_{24} & j \end{Bmatrix} = \frac{\delta_{j_{13},j'_{13}} \delta_{j_{24},j'_{24}}}{(2j_{13}+1)(2j_{24}+1)}. \qquad (12.5.4)$$

From (12.5.2a), (12.5.2b) it is clear that the $9j$-coefficient can be written as a multiple sum over the product of six Clebsch–Gordan coefficients. Rewriting this in terms of $3j$-coefficients yields, in appropriate notation,

$$\begin{Bmatrix} a & b & c \\ d & e & f \\ g & h & j \end{Bmatrix} = \sum_{\text{all } m} \begin{pmatrix} a & b & c \\ m_a & m_b & m_c \end{pmatrix} \begin{pmatrix} d & e & f \\ m_d & m_e & m_f \end{pmatrix} \begin{pmatrix} g & h & j \\ m_g & m_h & m_j \end{pmatrix} \begin{pmatrix} a & d & g \\ m_a & m_d & m_g \end{pmatrix} \begin{pmatrix} b & e & h \\ m_b & m_e & m_h \end{pmatrix} \begin{pmatrix} c & f & j \\ m_c & m_f & m_j \end{pmatrix}. \qquad (12.5.5)$$

Alternatively, just as in (12.4.12), the two vectors (12.5.1a) and (12.5.1b) can be related through the product of Racah coefficients, according to

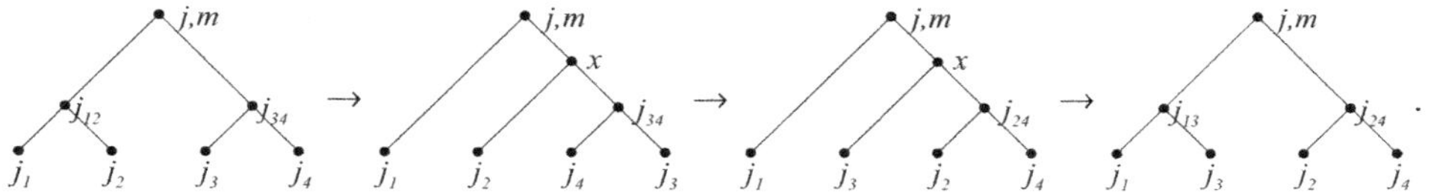

This leads to a single sum over the product of three Racah coefficients for the $9j$-coefficient. When rewritten in terms of $6j$-coefficients, this gives

$$\begin{Bmatrix} a & b & c \\ d & e & f \\ g & h & j \end{Bmatrix} = \sum_x (-1)^{2x}(2x+1) \begin{Bmatrix} a & b & c \\ f & j & x \end{Bmatrix} \begin{Bmatrix} d & e & f \\ b & x & h \end{Bmatrix} \begin{Bmatrix} g & h & j \\ x & a & d \end{Bmatrix}. \qquad (12.5.6)$$

From symmetry properties of $3j$- and $6j$-coefficients and by (12.5.5) and (12.5.6), one obtains the symmetries of $9j$-coefficients. The $9j$-coefficient is, up to sign, invariant with respect to permutations of its columns, permutations of its rows and under transposition. Even permutations (and transposition) leave the $9j$-coefficient unchanged, whereas odd permutations introduce a factor $(-1)^J$, where J is the sum of all nine arguments of the $9j$-coefficient.

Just as the orthogonality of $3j$- and $6j$-coefficients can be related to the orthogonality of a discrete polynomial, this can be done for the $9j$-coefficient. For this purpose, write the arguments as

$$\begin{Bmatrix} a & b & a+b-x \\ c & d & c+d-y \\ a+c-m & b+d-n & a+b+c+d-N \end{Bmatrix},$$

where m, n, x and y take nonnegative integer values with $m+n \le N$ and $x+y \le N$. This expression can be written as a factor times $R_{m,n}(x,y) \equiv R_{m,n}(x,y;\alpha,\beta,\gamma,\delta,N)$, where $R_{m,n}(x,y)$ is a polynomial of degree $N-n$ in $\lambda(x) = x(x+\alpha+\beta+1)$ and of degree $N-m$ in $\mu(y) = y(y+\gamma+\delta+1)$, and where $(\alpha,\beta,\gamma,\delta)$ is equal to $(-2a-1,-2b-1,-2c-1,-2d-1)$ (Van der Jeugt (2000); see also Rahman (2009) for a different approach). The orthogonality relation (12.5.4) then reads

$$\sum_{x=0}^{N}\sum_{y=0}^{N-x} w(x,y)R_{m,n}(x,y)R_{m',n'}(x,y) = \delta_{m,m'}\delta_{n,n'}h_{m,n}, \tag{12.5.7}$$

where $w(x,y)$ is some expression in x, y and the five parameters α, β, γ, δ, N, and $h_{m,n}$ is an expression involving m, n and the same five parameters. So the $R_{m,n}(x,y)$ are a discrete version of orthogonal polynomials on the triangle (see §2.3.3 for the continuous case). In terms of the common convention for Pochhammer symbols, the weight function is

$$\begin{aligned} w(x,y) &= \frac{(-1)^{x+y}(-N)_{x+y}}{x!\,y!\,(\alpha+\beta+\gamma+\delta+N+3)_{x+y}}\,\frac{(\alpha+1)_x(\delta+1)_y}{(\beta+1)_x(\gamma+1)_y} \\ &\times \frac{(\alpha+\beta+1)_x(\alpha+\beta+2)_{2x}}{(\alpha+\beta+1)_{2x}(\alpha+\beta+N+2)_{x-y}}\,\frac{(\gamma+\delta+1)_y(\gamma+\delta+2)_{2y}}{(\gamma+\delta+1)_{2y}(\gamma+\delta+N+2)_{y-x}}; \end{aligned}$$

the expression for $h_{m,n}$ is more complicated (Van der Jeugt, 2000).

Although (12.5.7) looks like a neat two-variable extension of the orthogonality of Hahn and Racah polynomials, the setback is that the known forms of the expression $R_{m,n}(x,y)$ are complicated. Although the transposition symmetry for the $9j$-coefficient implies a duality between (m,n) and (x,y), none of the known expressions for $R_{m,n}(x,y)$ (Van der Jeugt, 2000) displays this duality explicitly (see also Genest and Vinet, 2014). Furthermore, no difference operators are known for which $R_{m,n}(x,y)$ are eigenfunctions. Ideally, one would expect a double sum expression of hypergeometric type for $R_{m,n}(x,y)$. So far, such an expression is not available. This is because all known expressions of $9j$-coefficients are rather involved. One such form is obtained as follows: one starts from a formula similar to (12.5.5) but expressing the product of one $3j$-coefficient with the $9j$-coefficient as an essentially double sum over the product of five $3j$-coefficients. Then, making appropriate choices for the projection numbers appearing in these $3j$-coefficients (i.e., choices that reduce some $3j$'s to closed forms), this gives for the $9j$-coefficient an expression as a double sum over the product of three $3j$-coefficients (times factors). So using (12.3.9), this reduces to a fivefold summation expression. It is not too difficult to see that one of these summations can be performed due to Vandermonde's theorem (Andrews et al., 1999, Corollary 2.2.3), leaving a complicated fourfold summation expression for the $9j$-coefficient. This fourfold expression can be found in some books, e.g. by Varshalovich et al. (1988) and by Jucys and Bandzaĭtis (1977).

Ališauskas and Jucys (1971) went on manipulating this fourfold sum expression, changing summation variables in several ways, and by this tour de force they managed to perform one further sum (again using Vandermonde's theorem) and finally ended up with a *triple sum series* for the $9j$-coefficient (Ališauskas and Jucys, 1971). Their method was later reproduced in the book by Jucys and Bandzaĭtis (1977). Much later, Rosengren (1999) deduced Ališauskas's triple sum series in a simpler way: starting from the single sum over the product of three $6j$-coefficients (12.5.6), expressing these coefficients as single sums through (12.4.6) or an alternative single sum, then manipulating the summation variables such that one summation can be performed using a summation formula for a very-well-poised ${}_4F_3(-1)$ series (Andrews et al., 1999, Corollary 3.5.3) following from a limit of Dougall's formula (Andrews et al., 1999, Theorem 3.5.1), and thus ending up with a triple sum series.

Ališauskas (2000, 2001) later derived several triple sum series, also for the q-case. We present one form here:

$$\begin{Bmatrix} a & b & c \\ d & e & f \\ g & h & j \end{Bmatrix} = (-1)^{c+f-j} \frac{\nabla(d,a,g)\nabla(b,e,h)\nabla(j,g,h)}{\nabla(d,e,f)\nabla(b,a,c)\nabla(j,c,f)} \times \sum_{x,y,z} X_x Y_y Z_z \frac{(a+d-h+j-y-z)!}{(-b+d-f+h+x+y)!\,(-a+b-f+j+x+z)!} \tag{12.5.8}$$

with

$$X_x = (-1)^x \frac{(2f-x)!\,(d+e-f+x)!\,(c-f+j+x)!}{x!\,(e+f-d-x)!\,(c+f-j-x)!},$$
$$Y_y = (-1)^y \frac{(-b+e+h+y)!\,g+h-j+y)!}{y!\,(2h+1+y)!\,(b+e-h-y)!\,(g-h+j-y)!},$$
$$Z_z = (-1)^z \frac{(2a-z)!\,(-a+b+c+z)!}{z!\,(a+d+g+1-z)!\,(a+d-g-z)!\,(a-b+c-z)!}.$$

Herein $\nabla(a,b,c)$ has been defined in (12.4.9) and the sum in (12.5.8) is over all integer values of x, y and z such that all factorials in the summation are nonnegative. Srinivasa Rao and Rajeswari (1993) rewrote this expression as a triple hypergeometric series (generalizations of Appell's series; see Chapter 3), and deduced some identities from this for so-called stretched $9j$-coefficients (Srinivasa Rao and Van der Jeugt, 1994).

12.6 Beyond $9j$: Graphical Methods

More generally, one can consider the tensor product of $n+1$ irreducible unitary representations of $\mathfrak{su}(2)$ and their related $3nj$-coefficients. In the tensor product $V = D_{j_1} \otimes D_{j_2} \otimes \cdots \otimes D_{j_{n+1}}$, a basis (the uncoupled basis) is given by $e_{m_1}^{(j_1)} \otimes e_{m_2}^{(j_2)} \otimes \cdots \otimes e_{m_{n+1}}^{(j_{n+1})}$, where m_i is a projection of j_i. Just as in (12.5.1a), coupled basis vectors can be defined by means of *binary coupling schemes*. The idea is a simple extension of the two ways in which the tensor product of three representations can be "coupled"; see (12.4.1). Of course, as n increases, the number of ways that representations can be coupled also increases. A *$3nj$-coefficient* is then, as in (12.5.3), proportional to the inner product of two vectors corresponding to different couplings. For example, when $n = 4$ there are essentially two distinct $12j$-coefficients (that do not reduce to products of $9j$- and/or $6j$-coefficients). The $12j$-coefficient of the first type corresponds to the inner product of the vectors described by

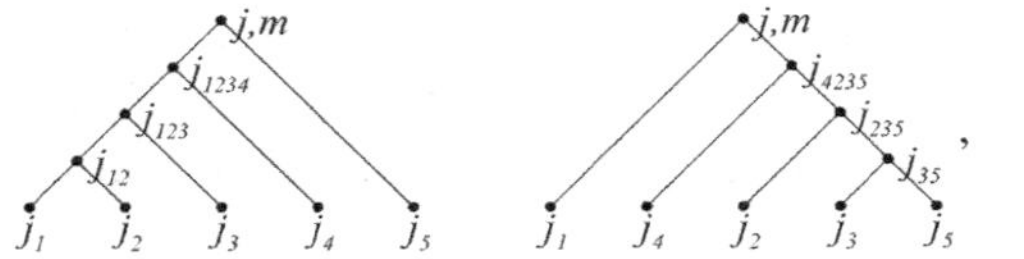

and the $12j$-coefficient of the second type to

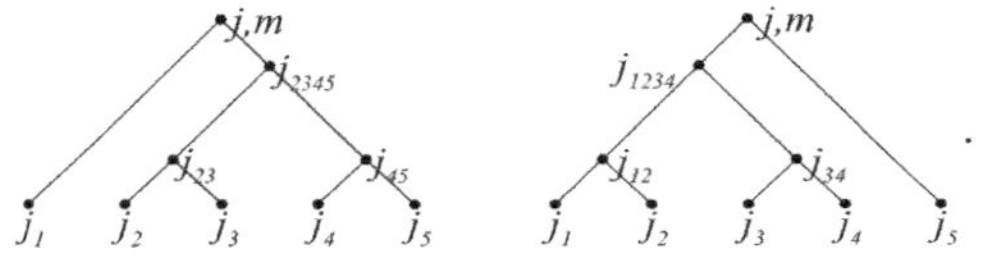

Since $3nj$-coefficients relate two sets of orthonormal basis vectors, they satisfy an orthogonality relation like (12.5.4). For $12j$-coefficients, this orthogonality relation involves a triple sum; see e.g. Varshalovich et al. (1988, §10.13). As far as we know, orthogonality relations for $12j$-coefficients or higher have not explicitly been related to the orthogonality of discrete multivariable polynomials. But there exist other interesting interpretations: for example, $3nj$-coefficients have been identified as connection coefficients between orthogonal polynomials in n variables (Lievens and Van der Jeugt, 2002).

From the above example of $12j$-coefficients, it is clear that in general a $3nj$-coefficient is determined by two binary coupling schemes T_1 and T_2 on $n+1$ elements. Just as in the case of the $9j$-coefficient (see the sequence of binary coupling schemes preceding (12.5.6)), one can find a sequence of binary coupling schemes starting with T_1 and ending with T_2, such that two consecutive elements in the sequence are related through an elementary transformation (i.e., a transformation turning the left-hand side of (12.4.10) into its right-hand side). Consequently, as in (12.5.6), this yields an expression of the $3nj$-coefficient as a (multiple) sum over products of $6j$-coefficients. This method is usually referred to as the *method of trees*; see Biedenharn and Louck (1981b, Topic 12). It involves combinatorial problems (enumeration of all binary coupling schemes), and interesting graph-theoretical problems (e.g. finding the shortest sequence to go from T_1 to T_2 by means of elementary transformations); see Fack et al. (1999).

The method of trees, as described here briefly, is still quite general: e.g. it could also be applied to $3nj$-coefficients of $\mathfrak{su}(1,1)$ (Van der Jeugt, 2003). For $3nj$-coefficients of $\mathfrak{su}(2)$, there exists however a more powerful method, namely that of Jucys graphs. The *Jucys graph* of a $3nj$-coefficient is obtained by "gluing" the $n+1$ leaves of the binary coupling trees T_1 and T_2 together (thereby deleting these leaves as vertices of the resulting graph) and connecting their two roots by an extra edge, thus yielding a cubic graph. On such a cubic graph, various transformations or rules can be applied, which reduce the cubic graph (four basic reduction rules are sufficient; see Fack et al., 1997). Each reduction rule yields a certain contribution to a formula, and in this way new expressions can be obtained for $3nj$-coefficients. This graphical method of Jucys has become an art of its own, leading to magnificent formulas relating sums over products of $3nj$-coefficients; see Jucys et al. (1962), Varshalovich et al. (1988) and Brink and Satchler (1993).

References

Ališauskas, S. 2000. The triple sum formulas for $9j$ coefficients of SU(2) and $u_q(2)$. *J. Math. Phys.*, **41**, 7589–7610.

Ališauskas, S. 2001. $3nj$ coefficients of $u_q(2)$ and multiple basic hypergeometric series. *Phys. Atomic Nuclei*, **64**, 2164–2169.

Ališauskas, S. J., and Jucys, A. P. 1971. Weight lowering operators and the multiplicity-free isoscalar factors for the group R_5. *J. Math. Phys.*, **12**, 594–605.

Andrews, G. E., Askey, R., and Roy, R. 1999. *Special Functions*. Encyclopedia of Mathematics and Its Applications, vol. 71. Cambridge University Press.

Askey, R., and Wilson, J. 1979. A set of orthogonal polynomials that generalize the Racah coefficients or 6-j symbols. *SIAM J. Math. Anal.*, **10**, 1008–1016.

Askey, R., and Wilson, J. 1985. *Some Basic Hypergeometric Orthogonal Polynomials that Generalize Jacobi Polynomials*. Mem. Amer. Math. Soc., vol. 54, no. 319.

Biedenharn, L. C. 1953. An identity by the Racah coefficients. *J. Math. Phys.*, **31**, 287–293.

Biedenharn, L. C., and Louck, J. D. 1981a. *Angular Momentum in Quantum Physics*. Encyclopedia of Mathematics and Its Applications, vol. 8. Addison-Wesley.

Biedenharn, L. C., and Louck, J. D. 1981b. *The Racah–Wigner Algebra in Quantum Theory*. Encyclopedia of Mathematics and Its Applications, vol. 9. Addison-Wesley.

Biedenharn, L. C., and Van Dam, H. (eds). 1965. *Quantum Theory of Angular Momentum. A Collection of Reprints and Original Papers*. Academic Press.

Biedenharn, L. C., Giovannini, A., and Louck, J. D. 1967. Canonical definition of Wigner coefficients in U_n. *J. Math. Phys.*, **8**, 691–700.

Brink, D. M., and Satchler, G. R. 1993. *Angular Momentum*. Third edn. Oxford University Press.

Edmonds, A. R. 1957. *Angular Momentum in Quantum Mechanics*. Investigations in Physics, vol. 4. Princeton University Press.

Elliott, J. P. 1953. Theoretical studies in nuclear structure, V: the matrix elements of non-central forces with an application to the 2p-shell. *Proc. Roy. Soc. A*, **281**, 345–370.

Fack, V., Pitre, S. N., and Van der Jeugt, J. 1997. Calculation of general recoupling coefficients using graphical methods. *Comput. Phys. Comm.*, **101**, 155–170.

Fack, V., Lievens, S., and Van der Jeugt, J. 1999. On rotation distance between binary coupling trees and applications for $3nj$-coefficients. *Comput. Phys. Comm.*, **119**, 99–114.

Genest, V. X., and Vinet, L. 2014. The generic superintegrable system on the 3-sphere and the $9j$ symbols of $\mathfrak{su}(1, 1)$. *SIGMA*, **10**, Paper 108, 28 pp.

Humphreys, J. E. 1972. *Introduction to Lie Algebras and Representation Theory*. Graduate Texts in Mathematics, vol. 9. Springer.

Jucys, A. P., and Bandzaĭtis, A. A. 1977. *Theory of Angular Momentum in Quantum Mechanics* (in Russian). Izdat. "Mokslas", Vilnius.

Jucys, A. P., Levinson, I. B., and Vanagas, V. V. 1962. *Mathematical Apparatus of the Theory of Angular Momentum*. Israel Program for Scientific Translations, Jerusalem. Translated from the 1960 Russian original.

Kirillov, A. N., and Reshetikhin, N. Yu. 1989. Representations of the algebra $U_q(\mathrm{sl}(2))$, q-orthogonal polynomials and invariants of links. Pages 285–339 of: *Infinite-dimensional Lie Algebras and Groups*. Adv. Ser. Math. Phys., vol. 7. World Scientific.

Koekoek, R., Lesky, P. A., and Swarttouw, R. F. 2010. *Hypergeometric Orthogonal Polynomials and their q-Analogues*. Springer.

Koelink, H. T., and Koornwinder, T. H. 1989. The Clebsch–Gordan coefficients for the quantum group $\mathrm{S}_\mu\mathrm{U}(2)$ and q-Hahn polynomials. *Indag. Math.*, **51**, 443–456.

Koornwinder, T. H. 1981. Clebsch–Gordan coefficients for SU(2) and Hahn polynomials. *Nieuw Arch. Wisk. (3)*, **29**, 140–155.

Lievens, S., and Van der Jeugt, J. 2002. $3nj$-coefficients of su(1, 1) as connection coefficients between orthogonal polynomials in n variables. *J. Math. Phys.*, **43**, 3824–3849.

von Neumann, J., and Wigner, E. 1928. Zur Erklärung einiger Eigenschaften der Spektren aus der Quantenmechanik des Drehelektrons, III. *Z. Physik*, **51**, 844–858.

Racah, G. 1942. Theory of complex spectra. II. *Phys. Rev.*, **62**, 438–462.

Rahman, M. 2009. A q-analogue of the 9-j symbols and their orthogonality. *J. Approx. Theory*, **161**, 239–258.

Rasmussen, W. 1975. Identity of the SU(1, 1) and SU(2) Clebsch-Gordan coefficients coupling unitary discrete representations. *J. Phys. A*, **8**, 1038–1047.

Regge, T. 1958. Symmetry properties of Clebsch–Gordan's coefficients. *Nuovo Cimento*, **10**, 544–545.

Regge, T. 1959. Symmetry properties of Racah's coefficients. *Nuovo Cimento*, **11**, 116–117.

Rosengren, H. 1999. Another proof of the triple sum formula for Wigner $9j$-symbols. *J. Math. Phys.*, **40**, 6689–6691.

Srinivasa Rao, K., and Rajeswari, V. 1993. *Quantum Theory of Angular Momentum* (Selected topics). Springer.

Srinivasa Rao, K., and Van der Jeugt, J. 1994. Stretched 9-j coefficients and summation theorems. *J. Phys. A*, **27**, 3083–3090.

de Swart, J. J. 1963. The octet model and its Clebsch-Gordan coefficients. *Rev. Mod. Phys.*, **35**, 916–939.

Van der Jeugt, J. 2000. Hypergeometric series related to the 9-j coefficient of $\mathfrak{su}(1, 1)$. *J. Comput. Appl. Math.*, **118**, 337–351.

Van der Jeugt, J. 2003. $3nj$-coefficients and orthogonal polynomials of hypergeometric type. Pages 25–92 of: *Orthogonal Polynomials and Special Functions*. Lecture Notes in Math., vol. 1817. Springer.

Varshalovich, D. A., Moskalev, A. N., and Khersonskiĭ, V. K. 1988. *Quantum Theory of Angular Momentum*. World Scientific.

Vilenkin, N. Ja., and Klimyk, A. U. 1991. *Representation of Lie Groups and Special Functions. Vol. 1*. Kluwer.

van der Waerden, B. L. 1932. *Die Gruppentheoretische Methode in der Quantenmechanik.* Springer.

Whipple, F. J. W. 1926. Well-poised series and other generalized hypergeometric series. *Proc. London Math. Soc. (2)*, **25**, 525–544.

Wigner, E. 1931. *Gruppentheorie und ihre Anwendung auf die Quantenmechanik der Atomspektren.* Vieweg, Braunschweig.

Wigner, E. 1965. On the matrices which reduce the Kronecker products of representations of S.R. groups. Pages 87–133 of: *Quantum Theory of Angular Momentum.* Academic Press.

Wilson, J. A. 1978. *Hypergeometric series recurrence relations and some new orthogonal functions.* Ph.D. thesis, University of Wisconsin–Madison.

Index